QUANTUM FIELD THEO
NON-EQUILIBRIUM ST

This text introduces the real-time approach to non-equilibrium statistical mechanics and the quantum field theory of non-equilibrium states in general. After a lucid introduction to quantum field theory and Green's functions, Schwinger's closed time path technique is developed, followed eventually by the real-time formulation and its Feynman diagram technique. The formalism is employed to derive quantum kinetic equations by using the quasi-classical Green's function technique, and is applied to study renormalization effects, non-equilibrium superconductivity, and quantum effects in disordered conductors.

The book offers two ways of learning how to study non-equilibrium states of many-body systems: the mathematical, canonical way, and an intuitive way using Feynman diagrams. The latter provides an easy introduction to the powerful functional methods of field theory. The usefulness of Feynman diagrams, even in a classical context, is shown by studies of classical stochastic dynamics such as vortex dynamics in disordered superconductors. The book demonstrates that quantum fields and Feynman diagrams are the universal language for studying fluctuations, be they of quantum or thermal origin, or even purely statistical.

Complete with numerous exercises to aid self-study, this textbook is suitable for graduate students in statistical mechanics, condensed matter physics, and quantum field theory in general.

JØRGEN RAMMER is a professor in the Department of Physics at Umeå University, Sweden. He has also worked in Denmark, Germany, Norway, Canada and the USA. His past research interests are partly reflected in the topics of this book; his main current interests are in decoherence and charge transport in nanostructures.

QUANTUM FIELD THEORY OF NON-EQUILIBRIUM STATES

JØRGEN RAMMER
Umeå University, Sweden

CAMBRIDGE
UNIVERSITY PRESS

CAMBRIDGE UNIVERSITY PRESS
Cambridge, New York, Melbourne, Madrid, Cape Town,
Singapore, São Paulo, Delhi, Tokyo, Mexico City

Cambridge University Press
The Edinburgh Building, Cambridge CB2 8RU, UK

Published in the United States of America by Cambridge University Press, New York

www.cambridge.org
Information on this title: www.cambridge.org/9780521188005

First published 2007
First paperback edition 2011

A catalogue record for this publication is available from the British Library

ISBN 978-0-521-87499-1 Hardback
ISBN 978-0-521-18800-5 Paperback

Contents

Preface   xi

1 Quantum fields   1
   1.1 Quantum mechanics   2
   1.2 *N*-particle system   5
        1.2.1 Identical particles   6
        1.2.2 Kinematics of fermions   9
        1.2.3 Kinematics of bosons   11
        1.2.4 Dynamics and probability current and density   13
   1.3 Fermi field   14
   1.4 Bose field   23
        1.4.1 Phonons   25
        1.4.2 Quantizing a classical field theory   26
   1.5 Occupation number representation   29
   1.6 Summary   31

2 Operators on the multi-particle state space   33
   2.1 Physical observables   33
   2.2 Probability density and number operators   37
   2.3 Probability current density operator   40
   2.4 Interactions   42
        2.4.1 Two-particle interaction   42
        2.4.2 Fermion–boson interaction   45
        2.4.3 Electron–phonon interaction   45
   2.5 The statistical operator   48
   2.6 Summary   52

3 Quantum dynamics and Green's functions   53
   3.1 Quantum dynamics   53
        3.1.1 The Schrödinger picture   54
        3.1.2 The Heisenberg picture   56
   3.2 Second quantization   60
   3.3 Green's functions   62
        3.3.1 Physical properties and Green's functions   62
        3.3.2 Stable of one-particle Green's functions   64

9.2   Generating functional 270
    9.2.1   Functional differentiation 272
    9.2.2   From diagrammatics to differential equations 274
9.3   Connection to operator formalism 281
9.4   Fermions and Grassmann variables 282
9.5   Generator of connected amplitudes 284
    9.5.1   Source derivative proof 284
    9.5.2   Combinatorial proof 290
    9.5.3   Functional equation for the generator 294
9.6   One-particle irreducible vertices 296
    9.6.1   Symmetry broken states 301
    9.6.2   Green's functions and one-particle irreducible vertices . . 302
9.7   Diagrammatics and action 306
9.8   Effective action and skeleton diagrams 307
9.9   Summary 312

10 Effective action   **313**
10.1   Functional integration 313
    10.1.1   Functional Fourier transformation 314
    10.1.2   Gaussian integrals 315
    10.1.3   Fermionic path integrals 319
10.2   Generators as functional integrals 320
    10.2.1   Euclid versus Minkowski 323
    10.2.2   Wick's theorem and functionals 324
10.3   Generators and 1PI vacuum diagrams 330
10.4   1PI loop expansion of the effective action 333
10.5   Two-particle irreducible effective action 339
    10.5.1   The 2PI loop expansion of the effective action 346
10.6   Effective action approach to Bose gases 351
    10.6.1   Dilute Bose gases 351
    10.6.2   Effective action formalism for bosons 352
    10.6.3   Homogeneous Bose gas 356
    10.6.4   Renormalization of the interaction 359
    10.6.5   Inhomogeneous Bose gas 363
    10.6.6   Loop expansion for a trapped Bose gas 365
10.7   Summary 372

11 Disordered conductors   **373**
11.1   Localization 373
    11.1.1   Scaling theory of localization 374
    11.1.2   Coherent backscattering 377
11.2   Weak localization 388
    11.2.1   Quantum correction to conductivity 388
    11.2.2   Cooperon equation 392
    11.2.3   Quantum interference and the Cooperon 398
    11.2.4   Quantum interference in a magnetic field 402

11.2.5 Quantum interference in a time-dependent field 404
11.3 Phase breaking in weak localization 408
11.3.1 Electron–phonon interaction 410
11.3.2 Electron–electron interaction 416
11.4 Anomalous magneto-resistance 423
11.4.1 Magneto-resistance in thin films 424
11.5 Coulomb interaction in a disordered conductor 428
11.6 Mesoscopic fluctuations 437
11.7 Summary 448

12 Classical statistical dynamics   **449**
12.1 Field theory of stochastic dynamics 450
12.1.1 Langevin dynamics 450
12.1.2 Fluctuating linear oscillator 451
12.1.3 Quenched disorder 454
12.1.4 Dynamical index notation 455
12.1.5 Quenched disorder and diagrammatics 457
12.1.6 Over-damped dynamics and the Jacobian 459
12.2 Magnetic properties of type-II superconductors 460
12.2.1 Abrikosov vortex state 460
12.2.2 Vortex lattice dynamics 462
12.3 Field theory of pinning 464
12.3.1 Effective action 467
12.4 Self-consistent theory of vortex dynamics 469
12.4.1 Hartree approximation 470
12.5 Single vortex 472
12.5.1 Perturbation theory 473
12.5.2 Self-consistent theory 474
12.5.3 Simulations 476
12.5.4 Numerical results 476
12.5.5 Hall force 482
12.6 Vortex lattice 487
12.6.1 High-velocity limit 488
12.6.2 Numerical results 489
12.6.3 Hall force 492
12.7 Dynamic melting 493
12.8 Summary 500

Appendices   **501**

A Path integrals   **503**

B Path integrals and symmetries   **511**

C Retarded and advanced Green's functions   **513**

D Analytic properties of Green's functions   **517**

Bibliography   523
Index   531

Preface

The purpose of this book is to provide an introduction to the applications of quantum field theoretic methods to systems out of equilibrium. The reason for adding a book on the subject of quantum field theory is two-fold: the presentation is, to my knowledge, the first to extensively present and apply to non-equilibrium phenomena the real-time approach originally developed by Schwinger, and subsequently applied by Keldysh and others to derive transport equations. Secondly, the aim is to show the universality of the method by applying it to a broad range of phenomena. The book should thus not just be of interest to condensed matter physicists, but to physicists in general as the method is of general interest with applications ranging the whole scale from high-energy to soft condensed matter physics. The universality of the method, as testified by the range of topics covered, reveals that the language of quantum fields is the universal description of fluctuations, be they of quantum nature, thermal or classical stochastic. The book is thus intended as a contribution to unifying the languages used in separate fields of physics, providing a universal tool for describing non-equilibrium states.

Chapter 1 introduces the basic notions of quantum field theory, the bose and fermi quantum fields operating on the multi-particle state spaces. In Chapter 2, operators on the multi-particle space representing physical quantities of a many-body system are constructed. The detailed exposition in these two chapters is intended to ensure the book is self-contained. In Chapter 3, the quantum dynamics of a many-body system is described in terms of its quantum fields and their correlation functions, the Green's functions. In Chapter 4, the key formal tool to describe non-equilibrium states is introduced: Schwinger's closed time path formulation of non-equilibrium quantum field theory, quantum statistical mechanics. Perturbation theory for non-equilibrium states is constructed starting from the canonical operator formalism presented in the previous chapters. In Chapter 5 we develop the real-time formalism necessary to deal with non-equilibrium states; first in terms of matrices and eventually in terms of two different types of Green's functions. The diagram representation of non-equilibrium perturbation theory is constructed in a way that the different aspects of spectral and quantum kinetic properties appear in a physically transparent and important fashion for non-equilibrium states. The equivalence of the real-time and imaginary-time formalisms are discussed in detail. In Chapter 6 we consider the coexistence regime between equilibrium and non-equilibrium states, the linear response regime. In Chapter 7 we develop and apply the quantum kinetic equation approach to the normal state, and in particular consider electrons in metals and semiconductors. As applications we consider the Boltzmann limit, and then phenomena beyond the Boltzmann theory, such as renormalization of transport coefficients due to interactions. In Chapter 8 we consider non-equilibrium superconductivity. In particular we introduce the quasi-classical Green's function technique so efficient for the description of superfluids. We derive the quantum kinetic equation describing elastic and inelastic scattering in superconductors. The time-dependent Ginzburg–Landau equation is obtained for a dirty superconductor. As an application of the quasi-classical theory, we consider the phenomena of conversion of normal currents to supercurrents and the corresponding charge imbalance.

Unlike Schwinger, not stooping to the paganism of using diagrams, we shall, like the boys in the basement, take heavy advantage of using Feynman diagrams. By introducing Feynman diagrams, the most developed of our senses can become functional in the pursuit of understanding quantum dynamics, an addition that shall make its pursuit easier also for non-equilibrium situations. Though the picture of reality that the representation of perturbation theory in terms of Feynman diagrams inspires might be a figment of the imagination, its usefulness for developing physical intuition has amply proved its value, as witnessed first in elementary particle physics. We develop the diagrammatics for non-equilibrium states, and show that the additional rules for the universal vertex display the two important features of quantum statistics and spectral properties of the interacting particles in an explicit fashion. In Chapter 9 we shall take the stand of formulating the laws of physics in terms of propagators and vertices and their Feynman diagrams representing probability amplitudes as dictated by the superposition principle. In fact, we take the Shakespearian approach and construct quantum dynamics in terms of Feynman diagrams by invoking the only two options for a particle: *to act or not to interact.* From this diagrammatic starting point, and employing the intuitive appeal of diagrammatic arguments, we then construct the formalism of non-equilibrium quantum field theory in terms of the powerful functional methods; first in terms of the generating functional and functional differentiation technique. In Chapter 10 we then introduce the final tool in the functional arsenal: functional integration, and arrive at the effective action description of general non-equilibrium states. As an application of the effective action approach we consider the dilute Bose gas, and the case of a trapped Bose–Einstein condensate. In Chapter 11 we consider quantum transport properties of disordered conductors, weak localization and interaction effects. In particular we show how the quasi-classical Green's function technique used in describing non-equilibrium properties of a dirty superconductor can be utilized to describe the destruction of phase coherence in the normal state due to non-equilibrium effects and interactions. Finally, in Chapter 12, we consider the classical limit of the developed general non-equilibrium quantum field theory. We consider classical stochastic dynamics and show that field theoretic methods and diagrammatics are useful tools even in the classical context. As an example we consider the flux flow properties of the Abrikosov lattice in a type-II superconductor. We thus demonstrate the fact that quantum field theory, through its diagrammatics and functional formulations, is the universal language for describing fluctuations whatever their nature.

Readers' guide. Firstly, readers bothered by the old-fashioned habit of footnotes can simply skip them; they are either quick reminders or serve the purpose of providing a general perspective. The book can be read chronologically but, like any fox hole, it has two entrances. For the reader whose interest is the general structure of quantum field theories, the book offers the possibility to jump directly to Chapter 9 where a quantum field theory is defined in terms of its propagators and vertices and their resulting Feynman diagrams as dictated by the superposition principle. The powerful methods of generating functionals are then constructed from the diagrammatics. However, the reader acquainted with Chapter 4 will then have at hand the general quantum field theory applicable to non-equilibrium states.

The scope of the book is not so much to dwell on a detailed application of the non-equilibrium theory to a single topic, but rather to show the versatility and universality of the method by applying it to a broad range of core topics of physics. One purpose of the book is to demonstrate the utility of Feynman diagrams in non-equilibrium quantum statistical mechanics using an approach appealing to physical intuition. The real-time description of non-equilibrium quantum statistical mechanics is therefore adopted, and the diagrammatic technique for systems out of equilibrium is developed systematically, and a representation most appealing to physical intuition applied. Though most examples are taken from condensed matter physics, the book is intended to contribute to the cross-fertilization between all the fields of physics studying the influence of fluctuations, be they quantum or thermal or purely statistical, and to establish that the convenient technique to use is in fact that of non-equilibrium quantum field theory. The book should therefore be of interest to a wide audience of physicists; in particular the book is intended to be self-contained so that students of physics and physicists in general can benefit from its detailed expositions. It is even contended that the method is of importance for other fields such as chemistry, and of course useful for electrical engineers.

A complete allocation of the credit for the progress in developing and applying the real-time description of non-equilibrium states has not been attempted. However, the references, in particular the cited review articles, should make it possible for the interested reader to trace this information.

The book is intended to be sufficiently broad to serve as a text for a one- or two-semester graduate course on non-equilibrium statistical mechanics or condensed matter theory. It is also hoped that the book can serve as a useful reference book for courses on quantum field theory, physics of disordered systems, and quantum transport in general. It is hoped that this attempt to make the exposition as lucid as possible will be successful to the point that the book can be read by students with only elementary knowledge of quantum and statistical mechanics, and read with benefit on its own. Exercises have been provided in order to aid self-instruction.

I am grateful to Dr. Joachim Wabnig for providing figures.

Jørgen Rammer

1

Quantum fields

Quantum field theory is a necessary tool for the quantum mechanical description of processes that allow for transitions between states which differ in their particle content. Quantum field theory is thus quantum mechanics of an arbitrary number of particles. It is therefore mandatory for relativistic quantum theory since relativistic kinematics allows for creation and annihilation of particles in accordance with the formula for equivalence of energy and mass. Relativistic quantum theory is thus inherently dealing with many-body systems. One may, however, wonder why quantum field theoretic methods are so prevalent in condensed matter theory, which considers non-relativistic many-body systems. The reason is that, though not mandatory, it provides an efficient way of respecting the quantum statistics of the particles, i.e. the states of identical fermions or bosons must be antisymmetric and symmetric, respectively, under the interchange of pairs of identical particles. Furthermore, the treatment of spontaneously symmetry broken states, such as superfluids, is facilitated; not to mention critical phenomena in connection with phase transitions. Furthermore, the powerful functional methods of field theory, and methods such as the renormalization group, can by use of the non-equilibrium field theory technique be extended to treat non-equilibrium states and thereby transport phenomena.

It is useful to delve once into the underlying mathematical structure of quantum field theory, but the upshot of this chapter will be very simple: just as in quantum mechanics, where the transition operators, $|\phi\rangle\langle\psi|$, contain the whole content of quantum kinematics, and the *bra* and *ket* annihilate and create states in accordance with

$$(|\phi\rangle\langle\psi|)\,|\chi\rangle \;=\; \langle\psi|\chi\rangle\,|\phi\rangle \tag{1.1}$$

we shall find that in quantum field theory two types of operators do the same job. One of these operators, the creation operator, $a^\dagger$, is similar in nature to the *ket* in the transition operator, and the other, the annihilation operator, a, is similar to the action of the *bra* in Eq. (1.1), annihilating the state it operates on. Then the otherwise messy obedience of the quantum statistics of particles becomes a trivial matter expressed through the anti-commutation or commutation relations of the creation and annihilation operators.

1.1   Quantum mechanics

A short discussion of quantum mechanics is first given, setting the scene for the notation. In quantum mechanics, the state of a physical system is described by a vector, $|\psi\rangle$, providing a complete description of the system. The description is unique modulo a phase factor, i.e. the state of a physical system is properly represented by a ray, the equivalence class of vectors $e^{i\varphi}|\psi\rangle$, differing only by an overall phase factor of modulo one.

We consider first a single particle. Of particular intuitive importance are the states where the particle is definitely at a given spatial position, say $\mathbf{x}$, the corresponding state vector being denoted by $|\mathbf{x}\rangle$. The projection of an arbitrary state onto such a position state, the scalar product between the states,

$$\psi(\mathbf{x}) \;=\; \langle\mathbf{x}|\psi\rangle\,, \tag{1.2}$$

specifies the probability amplitude, the so-called wave function, whose absolute square is the probability for the event that the particle is located at the position in question.[1] The states of definite spatial positions are delta normalized

$$\langle\mathbf{x}|\mathbf{x}'\rangle \;=\; \delta(\mathbf{x}-\mathbf{x}')\,. \tag{1.3}$$

Of equal importance is the complementary representation in terms of the states of definite momentum, the corresponding state vectors denoted by $|\mathbf{p}\rangle$. Analogous to the position states they form a complete set or, equivalently, they provide a resolution of the identity operator, $\hat{I}$, in terms of the momentum state projection operators

$$\int d\mathbf{p}\,|\mathbf{p}\rangle\langle\mathbf{p}| \;=\; \hat{I}\,. \tag{1.4}$$

The appearance of an integral in Eq. (1.4) assumes space to be infinite, and the (conditional) probability amplitude for the event of the particle to be at position $\mathbf{x}$ *given* it has momentum $\mathbf{p}$ is specified by the plane wave function

$$\langle\mathbf{x}|\mathbf{p}\rangle \;=\; \frac{1}{(2\pi\hbar)^{3/2}}\,e^{\frac{i}{\hbar}\mathbf{p}\cdot\mathbf{x}}\,, \tag{1.5}$$

the transformation between the complementary representations being Fourier transformation. The states of definite momentum are therefore also delta normalized[2]

$$\langle\mathbf{p}|\mathbf{p}'\rangle \;=\; \delta(\mathbf{p}-\mathbf{p}')\,. \tag{1.6}$$

The possible physical momentum values are represented as eigenvalues, $\hat{\mathbf{p}}|\mathbf{p}\rangle = \mathbf{p}|\mathbf{p}\rangle$, of the operator

$$\hat{\mathbf{p}} \;=\; \int d\mathbf{p}\,\mathbf{p}\,|\mathbf{p}\rangle\langle\mathbf{p}| \tag{1.7}$$

---

[1] Treating space as a continuum, the relevant quantity is of course the probability for the particle being in a small volume around the position in question, $P(\mathbf{x})\Delta\mathbf{x} = |\psi(\mathbf{x})|^2\Delta\mathbf{x}$, the absolute square of the wave function denoting a probability *density*.

[2] If the particle is confined in space, say confined in a box as often assumed, the momentum states are Kronecker normalized, $\langle\mathbf{p}|\mathbf{p}'\rangle = \delta_{\mathbf{p},\mathbf{p}'}$.

representing the physical quantity *momentum*. Similarly for the position of a particle. The average value of a physical quantity is thus specified by the matrix element of its corresponding operator, say the average position in state $|\psi\rangle$ is given by the three real numbers composing the vector $\langle\psi|\hat{\mathbf{x}}|\psi\rangle$. In physics it is customary to interpret a scalar product as the value of the *bra*, a linear functional on the state vector space, on the vector, *ket*, in question.[3]

The complementarity of the position and momentum descriptions is also expressed by the commutator, $[\hat{\mathbf{x}}, \hat{\mathbf{p}}] \equiv \hat{\mathbf{x}}\,\hat{\mathbf{p}} - \hat{\mathbf{p}}\,\hat{\mathbf{x}}$, of the operators representing the two physical quantities, being the c-number specified by the quantum of action

$$[\hat{\mathbf{x}}, \hat{\mathbf{p}}] = i\hbar \, . \tag{1.8}$$

The fundamental position and momentum representations refer only to the kinematical structure of quantum mechanics. The dynamics of a system is determined by the Hamiltonian $\hat{H} = H(\hat{\mathbf{p}}, \hat{\mathbf{x}})$, the operator specified according to the correspondence principle by Hamilton's function $H(\hat{\mathbf{p}}, \hat{\mathbf{x}})$, i.e. for a non-relativistic particle of mass m in a potential $V(\mathbf{x})$ the Hamiltonian, the energy operator, is

$$\hat{H} = \frac{\hat{\mathbf{p}}^2}{2m} + V(\hat{\mathbf{x}}) \, . \tag{1.9}$$

It can often be convenient to employ the eigenstates of the Hamiltonian

$$\hat{H}\,|\epsilon_\lambda\rangle = \epsilon_\lambda\,|\epsilon_\lambda\rangle \, . \tag{1.10}$$

The completeness of the states of definite energy, $|\epsilon_\lambda\rangle$, is specified by *their* resolution of the identity

$$\sum_\lambda |\epsilon_\lambda\rangle\langle\epsilon_\lambda| = \hat{I} \tag{1.11}$$

here using a notation corresponding to the case of a discrete spectrum.

At each instant of time a complete description is provided by a state vector, $|\psi(t)\rangle$, thereby defining an operator, the time-evolution operator connecting state vectors at different times

$$|\psi(t)\rangle = \hat{U}(t, t')\,|\psi(t')\rangle \, . \tag{1.12}$$

Conservation of probability, conservation of the length of a state vector, or its normalized scalar product $\langle\psi(t)|\psi(t)\rangle = 1$, under time evolution, determines the evolution operator to be unitary, $U^{-1}(t, t') = U^\dagger(t, t')$. The dynamics is given by the Schrödinger equation

$$i\hbar\,\frac{d|\psi(t)\rangle}{dt} = \hat{H}\,|\psi(t)\rangle \tag{1.13}$$

and for an isolated system the evolution operator is thus the unitary operator

$$\hat{U}(t, t') = e^{-\frac{i}{\hbar}\hat{H}(t-t')} \, . \tag{1.14}$$

Here we have presented the operator calculus approach to quantum dynamics, the equivalent path integral approach is presented in Appendix A.

---

[3]For a detailed introduction to quantum mechanics we direct the reader to chapter 1 in reference [1].

In order to describe a physical problem we need to specify particulars, typically in the form of an initial condition. Such general initial condition problems can be solved through the introduction of the Green's function. The Green's function $G(\mathbf{x}, t; \mathbf{x}', t')$ represents the solution to the Schrödinger equation for the particular initial condition where the particle is definitely at position $\mathbf{x}'$ at time t'

$$\lim_{t \searrow t'} \psi(\mathbf{x}, t) = \delta(\mathbf{x} - \mathbf{x}') = \langle \mathbf{x}, t' | \mathbf{x}', t' \rangle . \tag{1.15}$$

The solution of the Schrödinger equation corresponding to this initial condition therefore depends parametrically on $\mathbf{x}'$ (and t'), and is by definition the conditional probability density amplitude for the dynamics in question[4]

$$\psi_{\mathbf{x}', t'}(\mathbf{x}, t) = \langle \mathbf{x}, t | \mathbf{x}', t' \rangle = \langle \mathbf{x} | \hat{U}(t, t') | \mathbf{x}' \rangle \equiv G(\mathbf{x}, t; \mathbf{x}', t') . \tag{1.16}$$

The Green's function, defined to be a solution of the Schrödinger equation, satisfies

$$\left(i\hbar \frac{\partial}{\partial t} - H(-i\hbar \nabla_{\mathbf{x}}, \mathbf{x}) \right) G(\mathbf{x}, t; \mathbf{x}', t') = 0 \tag{1.17}$$

where, according to Eq. (1.3), the Hamiltonian in the position representation, H, is specified by the position matrix elements of the Hamiltonian

$$\langle \mathbf{x} | \hat{H} | \mathbf{x}' \rangle = H(-i\hbar \nabla_{\mathbf{x}}, \mathbf{x}) \, \delta(\mathbf{x} - \mathbf{x}') . \tag{1.18}$$

The Green's function, G, is the kernel of the Schrödinger equation on integral form (being a first order differential equation in time)

$$\psi(\mathbf{x}, t) = \int d\mathbf{x}' \, G(\mathbf{x}, t; \mathbf{x}', t') \, \psi(\mathbf{x}', t') \tag{1.19}$$

as identified in terms of the matrix elements of the evolution operator by using the resolution of the identity in terms of the position basis states

$$\langle \mathbf{x} | \psi(t) \rangle = \int d\mathbf{x}' \langle \mathbf{x} | \hat{U}(t, t') | \mathbf{x}' \rangle \langle \mathbf{x}' | \psi(t') \rangle . \tag{1.20}$$

The Green's function propagates the wave function, and we shall therefore also refer to the Green's function as the propagator. It completely specifies the quantum dynamics of the particle.

We note that the partition function of thermodynamics and the trace of the evolution operator are related by analytical continuation:

$$\begin{aligned} Z &= \operatorname{Tr} e^{-\hat{H}/kT} = \int d\mathbf{x} \, \langle \mathbf{x} | e^{-\hat{H}/kT} | \mathbf{x} \rangle = \operatorname{Tr} \hat{U}(-i\hbar/kT, 0) \\ &= \int d\mathbf{x} \, G(\mathbf{x}, -i\hbar/kT; \mathbf{x}, 0) \end{aligned} \tag{1.21}$$

---

[4]In the continuum limit the Green's function is not a normalizable solution of the Schrödinger equation, as is clear from Eq. (1.15).

showing that the partition function is obtained from the propagator at the imaginary time $\tau = -i\hbar/kT$. The formalisms of thermodynamics, i.e. equilibrium statistical mechanics, and quantum mechanics are thus equivalent, a fact we shall take advantage of throughout. The physical significance is the formal equivalence of quantum and thermal fluctuations.

Quantum mechanics can be formulated without the use of operators, viz. using Feynman's path integral formulation. In Appendix A, the path integral expressions for the propagator and partition function for a single particle are obtained. Various types of Green's functions and their properties for the case of a single particle are discussed in Appendix C, and their analytical properties are considered in Appendix D.

1.2   N-particle system

Next we consider a physical system consisting of N particles. If the particles in an assembly are distinguishable, i.e. different species of particles, an orthonormal basis in the N-particle state space $H^{(N)} = H_1 \otimes H_2 \otimes \cdots \otimes H_N$ is the (tensor) product states, for example specified in terms of the momentum quantum numbers of the particles

$$|\mathbf{p}_1, \mathbf{p}_2, \cdots, \mathbf{p}_N\rangle \equiv |\mathbf{p}_1\rangle \otimes |\mathbf{p}_2\rangle \otimes \cdots \otimes |\mathbf{p}_N\rangle \equiv |\mathbf{p}_1\rangle|\mathbf{p}_2\rangle \cdots |\mathbf{p}_N\rangle . \quad (1.22)$$

We follow the custom of suppressing the tensorial notation.

Formally everything in the following, where an N-particle system is considered, is equivalent no matter which complete set of single-particle states are used. In practice the choice follows from the context, and to be specific we shall mainly explicitly employ the momentum states, the choice convenient in practice for a spatially translational invariant system.[5] These states are eigenstates of the momentum operators

$$\hat{\mathbf{p}}_i |\mathbf{p}_1, \mathbf{p}_2, \cdots, \mathbf{p}_N\rangle = \mathbf{p}_i |\mathbf{p}_1, \mathbf{p}_2, \cdots, \mathbf{p}_N\rangle , \quad (1.23)$$

where tensorial notation for operators are suppressed, i.e.

$$\hat{\mathbf{p}}_i = \hat{I}_1 \otimes \cdots \hat{I}_{i-1} \otimes \hat{\mathbf{p}}_i \otimes \hat{I}_{i+1} \otimes \cdots \hat{I}_N , \quad (1.24)$$

each operating in the one-particle subspace dictated by its index. In particular the N-particle momentum states are eigenstates of the total momentum operator

$$\hat{\mathbf{P}}_N = \sum_{i=1}^{N} \hat{\mathbf{p}}_i \quad (1.25)$$

---

[5] In the next sections we shall mainly use the momentum basis, and refer in the following to the quantum numbers labeling the one-particle states as *momentum*, although any complete set of quantum numbers could equally well be used. The N-tuple $(\mathbf{p}_1, \mathbf{p}_2, \cdots, \mathbf{p}_N)$ is a complete description of the N-particle system if the particles do not posses internal degrees of freedom. In the following, where we for example have electrons in mind, we suppress for simplicity of notation the spin labeling and simply assume it is absorbed in the momentum labeling. If the particles have additional internal degrees of freedom, such as color and flavor, these are included in a similar fashion. If more than one type of species is to be considered simultaneously the species type, say quark and gluon, must also be indicated.

corresponding to the total momentum eigenvalue

$$\mathbf{P} = \sum_{i=1}^{N} \mathbf{p}_i \, . \tag{1.26}$$

The position representation of the momentum states is specified by the plane wave functions, Eq. (1.5), the scalar product of the momentum states and the analogous N-particle states of definite positions being

$$\psi_{\mathbf{p}_1,\dots,\mathbf{p}_N}(\mathbf{x}_1,\dots,\mathbf{x}_N) = \langle \mathbf{x}_1, \mathbf{x}_2, \dots, \mathbf{x}_N | \mathbf{p}_1, \mathbf{p}_2, \dots, \mathbf{p}_N \rangle = \prod_{i=1}^{N} \langle \mathbf{x}_i | \mathbf{p}_i \rangle$$

$$= \left(\frac{1}{(2\pi\hbar)^{3/2}} \right)^N e^{\frac{i}{\hbar} \mathbf{p}_1 \cdot \mathbf{x}_1} e^{\frac{i}{\hbar} \mathbf{p}_2 \cdot \mathbf{x}_2} \dots e^{\frac{i}{\hbar} \mathbf{p}_N \cdot \mathbf{x}_N} \, . \tag{1.27}$$

1.2.1 Identical particles

For an assembly of identical particles a profound change in the above description is needed. In quantum mechanics true identity between objects are realized, viz. elementary particle species, say electrons, are profoundly identical, i.e. there exists nothing in Nature which can distinguish any two electrons. Identical particles are indistinguishable. States which differ only by two identical particles being interchanged are thus described by the same ray.[6] As a consequence of their indistinguishability, assemblies of identical particles are described by states which with respect to interchange of pairs of identical particles are either antisymmetric or symmetric

$$|\mathbf{p}_1, \mathbf{p}_2, \dots, \mathbf{p}_N \rangle = \pm |\mathbf{p}_2, \mathbf{p}_1, \dots, \mathbf{p}_N \rangle \, , \tag{1.28}$$

this leaving the probability for a set of momenta of the particles, $P(\mathbf{p}_1, \mathbf{p}_2, \dots, \mathbf{p}_N)$, a function symmetric with respect to interchange of any pair of the identical particles.

A word on notation: the particle we call the *first* particle is in the momentum state specified by the first argument, and the particle we call the Nth particle is in the momentum state specified by the Nth argument. Particles whose states are symmetric with respect to interchange are called bosons , and for the antisymmetric case called fermions.[7]

---

[6]The quantum state with all of the electrons in the Universe interchanged will thus be the same as the present one. A radical invariance property of systems of identical particles!

[7]Quantum statistics and the spin degree of freedom of a particle are intimately connected as relativistic quantum field theory demands that bosons have integer spin, whereas particles with half-integer spin are fermions. This so-called spin-statistics connection seems in the present non-relativistic quantum theory quite mysterious, i.e. unintelligible. It only gets its explanation in the relativistic quantum theory, which we usually connect with high energy phenomena, where for any particle relativity, through Lorentz invariance, requires the existence of an anti-particle of the same mass and opposite charge (some neutral particles, such as the photon, are their own anti-particle). Then, in fermion anti-fermion pair production the particles must be antisymmetric with already existing particles as unitarity, i.e. conservation of probability, requires such a minus sign [2]. Historically, the exclusion principle, which is a direct consequence of Fermi statistics, was discovered by Pauli before the advent of relativistic quantum theory as a vehicle to explain the periodic properties of the elements. Pauli was also the first to show that the spin-statistics relation is a consequence of Lorentz invariance, causality and energy and norm positivity.

Any N-particle state $|\mathbf{p}_1, \mathbf{p}_2, \ldots, \mathbf{p}_N\rangle$ can be mapped into a state which is either symmetric or antisymmetric with respect to interchange of any two particles. To obtain the symmetric state we simply apply the symmetrization operator $\hat{S}$ which symmetrizes an N-particle state according to

$$\hat{S}\,|\mathbf{p}_1, \mathbf{p}_2, \ldots, \mathbf{p}_N\rangle \;=\; \frac{1}{N!}\sum_P |\mathbf{p}_{P_1}\rangle|\mathbf{p}_{P_2}\rangle \cdots |\mathbf{p}_{P_N}\rangle \tag{1.29}$$

and the antisymmetrization operator $\hat{A}$ antisymmetrizes according to

$$\hat{A}\,|\mathbf{p}_1, \mathbf{p}_2, \ldots, \mathbf{p}_N\rangle \;=\; \frac{1}{N!}\sum_P (-1)^{\zeta_P}|\mathbf{p}_{P_1}\rangle|\mathbf{p}_{P_2}\rangle \cdots |\mathbf{p}_{P_N}\rangle\,. \tag{1.30}$$

The summations are over all permutations P of the particles. Permutations form a group, and any permutation can be build by successive transpositions which only permute a pair. In the case of antisymmetrization, each term appears with the sign of the permutation in question

$$\text{sign}(P) \;=\; \prod_{1\le i<j\le N} \frac{j-i}{P_j-P_i}\,. \tag{1.31}$$

We have written this in terms of the number ζ_P which counts the number of transpositions needed to build the permutation P, since $\text{sign}(P) = (-1)^{\zeta_P}$.

If the single-particle state labels in the N-particle state to be symmetrized on the left in Eq. (1.29) are permuted, the same symmetrized state results, since if P' can be any of the $N!$ permutations, then $P'P$ for fixed permutation P will run through them all, $\hat{S}\,|\mathbf{p}_{P_1}, \mathbf{p}_{P_2}, \cdots \mathbf{p}_{P_N}\rangle = \hat{S}\,|\mathbf{p}_1, \mathbf{p}_2, \ldots, \mathbf{p}_N\rangle$.

We note that the sign of a product of permutations, $Q = P'P$, equals the product of the signs of the two permutations, $\text{sign}(Q) = \text{sign}(P')\cdot\text{sign}(P)$, and a permutation and its inverse have the same sign (owing to their equal number of transpositions), $\zeta_{P^{-1}} = \zeta_P$. Antisymmetrization of a permuted state gives the same antisymmetric state multiplied by the sign of the permutation permuting the original N-particle state since

$$\hat{A}\,|\mathbf{p}_{P_1}, \mathbf{p}_{P_2}, \cdots \mathbf{p}_{P_N}\rangle \;=\; \frac{1}{N!}\sum_{P'} (-1)^{\zeta_{P'}}|\mathbf{p}_{Q_1}\rangle|\mathbf{p}_{Q_2}\rangle \cdots |\mathbf{p}_{Q_N}\rangle \tag{1.32}$$

and as P' runs through all the permutations so does $Q = P'P$, and therefore

$$\hat{A}\,|\mathbf{p}_{P_1}, \mathbf{p}_{P_2}, \cdots \mathbf{p}_{P_N}\rangle \;=\; (-1)^{\zeta_P}\frac{1}{N!}\sum_Q (-1)^{\zeta_Q}|\mathbf{p}_{Q_1}\rangle|\mathbf{p}_{Q_2}\rangle \cdots |\mathbf{p}_{Q_N}\rangle$$

$$=\; (-1)^{\zeta_P}\,\hat{A}\,|\mathbf{p}_1, \mathbf{p}_2, \ldots, \mathbf{p}_N\rangle\,. \tag{1.33}$$

Therefore, if any two single-particle states are identical, the antisymmetrized state vector equals the zero vector, since the two states obtained by permuting the two identical labels are identical and yet upon antisymmatrization they differ by a minus sign, i.e. Pauli's exclusion principle for fermions: no two fermions can occupy the same state.

Further, according to Eq. (1.33), applying the antisymmetrization operator twice

$$\hat{A}^2 |\mathbf{p}_1, \mathbf{p}_2, \ldots, \mathbf{p}_N\rangle = \hat{A}\frac{1}{N!}\sum_P (-1)^{\zeta_P} |\mathbf{p}_{P_1}\rangle|\mathbf{p}_{P_2}\rangle \cdots |\mathbf{p}_{P_N}\rangle$$

$$= \frac{1}{N!}\sum_P (-1)^{\zeta_P}(-1)^{\zeta_P} \hat{A} |\mathbf{p}_1, \mathbf{p}_2, \ldots, \mathbf{p}_N\rangle$$

$$= \hat{A} |\mathbf{p}_1, \mathbf{p}_2, \ldots, \mathbf{p}_N\rangle \tag{1.34}$$

gives the same state as applying it only once, i.e. the symmetrization operators are projectors, $\hat{A}^2 = \hat{A}$, $\hat{S}^2 = \hat{S}$. The presence of the factor $1/N!$ in the definitions, Eq. (1.29) and Eq. (1.30), is thus there to ensure the operators are normalized projectors. Representing mutually exclusive symmetry properties, they are orthogonal projectors, their product is the operator that maps any vector onto the zero vector

$$\hat{A}\,\hat{S} = \hat{0} = \hat{S}\,\hat{A} \tag{1.35}$$

since symmetrizing an antisymmetric state, or vice versa, gives the zero vector.

The symmetrization operators are hermitian, $\hat{A}^\dagger = \hat{A}$, $\hat{S}^\dagger = \hat{S}$, as verified for example for $\hat{A}$ by first noting that according to the definition of the adjoint operator

$$\langle\mathbf{p}_1, \ldots, \mathbf{p}_N|\hat{A}^\dagger|\mathbf{p}_1', \mathbf{p}_2', \ldots, \mathbf{p}_N'\rangle = \langle\mathbf{p}_1', \ldots, \mathbf{p}_N'|\hat{A}|\mathbf{p}_1, \mathbf{p}_2, \ldots, \mathbf{p}_N\rangle^*$$

$$= \frac{1}{N!}\sum_P (-1)^{\zeta_P} \langle\mathbf{p}_1'|\mathbf{p}_{P_1}\rangle^* \cdots \langle\mathbf{p}_N'|\mathbf{p}_{P_N}\rangle^*$$

$$= \frac{(-1)^{\zeta_S}}{N!}\langle\mathbf{p}_{S_1}|\mathbf{p}_1'\rangle \cdots \langle\mathbf{p}_{S_N}|\mathbf{p}_N'\rangle \tag{1.36}$$

the matrix element being nonzero only if the set $\{\mathbf{p}_i'\}_{i=1,\ldots,N}$ is a permutation of the set $\{\mathbf{p}_i\}_{i=1,\ldots,N}$, S being the permutation that brings the set $\{\mathbf{p}_i\}_{i=1,\ldots,N}$ into the set $\{\mathbf{p}_i'\}_{i=1,\ldots,N}$, $\mathbf{p}_{S_i} = \mathbf{p}_i'$. Permuting both sets of indices by the inverse permutation S^{-1} of S, and using that $\zeta_{S^{-1}} = \zeta_S$, we get

$$\langle\mathbf{p}_1, \ldots, \mathbf{p}_N|\hat{A}^\dagger|\mathbf{p}_1', \mathbf{p}_2', \ldots, \mathbf{p}_N'\rangle = \frac{1}{N!}(-1)^{\zeta_{S^{-1}}}\langle\mathbf{p}_1|\mathbf{p}_{S_1^{-1}}'\rangle \cdots \langle\mathbf{p}_N|\mathbf{p}_{S_N^{-1}}'\rangle$$

$$= \frac{1}{N!}\sum_P (-1)^{\zeta_P}\langle\mathbf{p}_1, \ldots, \mathbf{p}_N|\mathbf{p}_{P_1}', \ldots, \mathbf{p}_{P_N}'\rangle$$

$$= \langle\mathbf{p}_1, \ldots, \mathbf{p}_N|\hat{A}|\mathbf{p}_1', \ldots, \mathbf{p}_N'\rangle . \tag{1.37}$$

Exercise 1.1. Show that the adjoint of a product of linear operators A and B is the product of their adjoint operators in opposite sequence

$$(A\,B)^\dagger = B^\dagger A^\dagger \tag{1.38}$$

and generalize to the case of an arbitrary number of operators.

Exercise 1.2. The vector space of state vectors, the *kets*, and the dual space of linear functionals on the state space, the *bras*, are isomorphic vector spaces, which we express by the adjoint operation, $|\psi\rangle^\dagger = \langle\psi|$ and $\langle\psi|^\dagger = |\psi\rangle$. This mapping is anti-linear and isomorphic, and we use the same symbol as for the adjoint of an operator.

Show that for arbitrary state vectors and operators on the state space the relationship $(\hat{X}|\psi\rangle)^\dagger = \langle\psi|\hat{X}^\dagger$. An operator being its own adjoint, $\hat{X}^\dagger = \hat{X}$, is said to be a hermitian operator and its eigenvalues are real, such operators being of primary importance in quantum mechanics.

Exercise 1.3. Show that the symmetrization operator, $\hat{S}$, is hermitian.

The linear operators $\hat{S}$ and $\hat{A}$ project any state onto either of the two orthogonal subspaces of symmetric or antisymmetric states.[8] The state space for a physical system consisting of N identical particles is thus not $H^{(N)}$, the N-fold product of the one-particle state space, but either the symmetric subspace, $\mathcal{B}^{(N)}$, for bosons, or antisymmetric subspace, $\mathcal{F}^{(N)}$, for fermions, obtained by projecting the states of $H^{(N)}$ by either type of symmetrization operator depending on the statistics of the particles in question.

1.2.2 Kinematics of fermions

Let us introduce the orthogonal, normalized up to a phase factor, antisymmetric basis states in the antisymmetric N-particle state space $\mathcal{F}^{(N)}$

$$
\begin{aligned}
|\mathbf{p}_1 \wedge \mathbf{p}_2 \wedge \cdots \wedge \mathbf{p}_N\rangle &\equiv \sqrt{N!}\,\hat{A}\,|\mathbf{p}_1, \mathbf{p}_2, \ldots, \mathbf{p}_N\rangle \\
&= \frac{1}{\sqrt{N!}} \sum_P (-1)^{\zeta_P} |\mathbf{p}_{P_1}\rangle \otimes |\mathbf{p}_{P_2}\rangle \otimes \cdots \otimes |\mathbf{p}_{P_N}\rangle \\
&= \frac{1}{\sqrt{N!}} \sum_P (-1)^{\zeta_P} |\mathbf{p}_{P_1}\rangle |\mathbf{p}_{P_2}\rangle \cdots |\mathbf{p}_{P_N}\rangle \\
&= \frac{1}{\sqrt{N!}} \sum_P (-1)^{\zeta_P} |\mathbf{p}_{P_1}, \mathbf{p}_{P_2}, \ldots, \mathbf{p}_{P_N}\rangle .
\end{aligned}
\tag{1.39}
$$

We demonstrate that they are orthogonal by using the properties of the antisymmetrization operator (we first for simplicity of the Kronecker function assume box normalization, i.e, the momentum values are discrete)

$$
\begin{aligned}
\langle \mathbf{p}_1 \wedge \cdots \wedge \mathbf{p}_N | \mathbf{p}'_1 \wedge \cdots \wedge \mathbf{p}'_N \rangle &= N! \langle \mathbf{p}_1, \ldots, \mathbf{p}_N | \hat{A}^\dagger \hat{A} | \mathbf{p}'_1, \ldots, \mathbf{p}'_N \rangle \\
&= N! \langle \mathbf{p}_1, \ldots, \mathbf{p}_N | \hat{A} | \mathbf{p}'_1, \ldots, \mathbf{p}'_N \rangle
\end{aligned}
$$

---

[8]Only for the case of *two* particles do the two subspaces of symmetric and antisymmetric states span the original state space, $H^{(2)} = H \otimes H$. In general, the other subspaces for the case of more than two particles do not seem to be state spaces for systems of identical particles.

$$= \langle \mathbf{p}_1, \cdots , \mathbf{p}_N | \sum_P (-1)^{\zeta_P} | \mathbf{p}'_{P_1}, \cdots , \mathbf{p}'_{P_N} \rangle$$

$$= \begin{cases} (-1)^{\zeta_S} & \{\mathbf{p}'\}_i \equiv \{\mathbf{p}\}_i \\ \\ 0 & \text{otherwise} \end{cases} \qquad (1.40)$$

where $\{\mathbf{p}'_i\}_{i=1,\ldots,N} \equiv \{\mathbf{p}_i\}_{i=1,\ldots,N}$ is short for *the labels* $\{\mathbf{p}'_i\}_{i=1,\ldots,N}$ *being a permutation of the labels* $\{\mathbf{p}_i\}_{i=1,\ldots,N}$, and S the permutation that takes the set $\{\mathbf{p}_i\}_{i=1,\ldots,N}$ into $\{\mathbf{p}'_i\}_{i=1,\ldots,N}$, $\mathbf{p}_{S_i} = \mathbf{p}'_i$. Or simply in words, only if the primed set of momenta is a permutation of the unprimed set is the scalar product of the states nonzero (we have of course assumed that all momenta are different since otherwise for fermions the vector is the zero-vector).

Incidentally we have

$$\langle \mathbf{p}_1 \wedge \mathbf{p}_2, \wedge \cdots \wedge \mathbf{p}_N | \, \mathbf{p}'_1, \mathbf{p}'_2, .., \mathbf{p}'_N \rangle = \begin{cases} \frac{1}{\sqrt{N!}} (-1)^{\zeta_S} & \{\mathbf{p}'\}_i \equiv \{\mathbf{p}\}_i \\ \\ 0 & \text{otherwise} \end{cases} \qquad (1.41)$$

expressing that additional permutations are redundant, for example an additional antisymmetrization is redundant as expressed by the second equality sign in Eq. (1.40), or equivalently that the symmetrization operators are hermitian projectors.

The scalar product of antisymmetric states is the determinant of the $N \times N$ matrix with entries $\langle \mathbf{p}_i | \mathbf{p}'_j \rangle$

$$\langle \mathbf{p}_1 \wedge \cdots \wedge \mathbf{p}_N | \mathbf{p}'_1 \wedge \cdots \wedge \mathbf{p}'_N \rangle = \det(\langle \mathbf{p}_i | \mathbf{p}'_j \rangle)$$

$$= \sum_P (-1)^{\zeta_P} \langle \mathbf{p}_1 | \mathbf{p}'_{P_1} \rangle \cdots \langle \mathbf{p}_N | \mathbf{p}'_{P_N} \rangle , \qquad (1.42)$$

the Slater determinant.

In the operator calculus perturbation theory for a single particle, the resolution of the identity plays a crucial efficient role. For an assembly of identical particles this role will be taken over by the commutation rules for the quantum fields we shall shortly introduce. The resolutions of the identity on the symmetrized subspaces reflect the redundancy of antisymmetrized or symmetrized states. Though not of much practical use, we include them for completeness (the resolution of the identity makes a short appearance in Section 3.1.1). The resolution of the identity on the antisymmetric state space can be written in terms of the N-state identity operator since the identity operator commutes with any operator

$$1 = \hat{A} \, \hat{I}^{(N)} \, \hat{A} = \hat{A} \left(\hat{I}_1 \otimes \hat{I}_2 \otimes \hat{I}_3 \otimes \cdots \otimes \hat{I}_N \right) \hat{A}^\dagger$$

$$= \hat{A} \sum_{\mathbf{p}_1,\ldots,\mathbf{p}_N} |\mathbf{p}_1\rangle\langle\mathbf{p}_1| \otimes |\mathbf{p}_2\rangle\langle\mathbf{p}_2| \otimes \cdots \otimes |\mathbf{p}_N\rangle\langle\mathbf{p}_N| \, \hat{A}^\dagger$$

$$= \hat{A} \sum_{\mathbf{p}_1,\ldots,\mathbf{p}_N} |\mathbf{p}_1, \mathbf{p}_2, \ldots, \mathbf{p}_N\rangle\langle\mathbf{p}_1, \mathbf{p}_2, \ldots, \mathbf{p}_N| \, \hat{A}^\dagger$$

$$= \frac{1}{N!} \sum_{\mathbf{p}_1,\ldots,\mathbf{p}_N} |\mathbf{p}_1 \wedge \mathbf{p}_2 \wedge \cdots \wedge \mathbf{p}_N\rangle\langle\mathbf{p}_1 \wedge \mathbf{p}_2 \wedge \cdots \wedge \mathbf{p}_N|$$

$$= \sum_{|\mathbf{p}_1|<|\mathbf{p}_2|<\cdots<|\mathbf{p}_N|} |\mathbf{p}_1 \wedge \mathbf{p}_2 \wedge \cdots \wedge \mathbf{p}_N\rangle\langle\mathbf{p}_1 \wedge \mathbf{p}_2 \wedge \cdots \wedge \mathbf{p}_N| \, . \quad (1.43)$$

In obtaining the last equality we have used the fact that if the momenta of the particles are interchanged in the N-particle state to be antisymmetrized, the same antisymmetric state vector is obtained modulo a phase factor ±1, for example

$$\hat{A}\,|\mathbf{p}_1,\mathbf{p}_2,\ldots,\mathbf{p}_N\rangle \;=\; -\,\hat{A}\,|\mathbf{p}_2,\mathbf{p}_1,\ldots,\mathbf{p}_N\rangle \, . \quad (1.44)$$

In the sum in the second last expression in Eq. (1.43), there are thus $N!$ identical terms.

The symmetrization phase factor in Eq. (1.40) can always be chosen to equal 1 by considering proper orderings in the definition of the basis states, thereby removing the redundancy in the general definition, Eq. (1.39), of the basis states. For example, if we choose to use only basis vectors where the momenta appear ordered according to the ordering $|\mathbf{p}_1| < |\mathbf{p}_2| < \cdots < |\mathbf{p}_N|$, this restriction on defining the set of basis states $|\mathbf{p}_1 \wedge \mathbf{p}_2 \wedge \cdots \wedge \mathbf{p}_N\rangle$ results in them forming an ortho*normal* basis in the antisymmetric state space $\mathcal{F}_N$, as also expressed by the last equality in Eq. (1.43).

1.2.3 Kinematics of bosons

We now turn to a discussion of the state space relevant for N identical bosons. In the symmetric state space, $\mathcal{B}^{(N)}$, we introduce the symmetric orthogonal basis states

$$|\mathbf{p}_1 \vee \mathbf{p}_2 \vee \cdots \vee \mathbf{p}_N\rangle \;\equiv\; \sqrt{N!}\; \hat{S}\,|\mathbf{p}_1,\mathbf{p}_2,\ldots,\mathbf{p}_N\rangle$$

$$= \frac{1}{\sqrt{N!}} \sum_P |\mathbf{p}_{P_1}\rangle \otimes |\mathbf{p}_{P_2}\rangle \otimes \cdots \otimes |\mathbf{p}_{P_N}\rangle$$

$$= \frac{1}{\sqrt{N!}} \sum_P |\mathbf{p}_{P_1}\rangle|\mathbf{p}_{P_2}\rangle \cdots |\mathbf{p}_{P_N}\rangle$$

$$= \frac{1}{\sqrt{N!}} \sum_P |\mathbf{p}_{P_1},\mathbf{p}_{P_2},\ldots,\mathbf{p}_{P_N}\rangle \, . \quad (1.45)$$

All derivations of formulas to be obtained for symmetric basis states runs equivalent to those for antisymmetric basis states. For example,

$$\langle\mathbf{p}_1 \vee \cdots \vee \mathbf{p}_N|\mathbf{p}'_1 \vee \cdots \vee \mathbf{p}'_N\rangle \;=\; \sum_P \langle\mathbf{p}_1|\mathbf{p}'_{P_1}\rangle \cdots \langle\mathbf{p}_N|\mathbf{p}'_{P_N}\rangle$$

$$= \mathrm{per}(\langle\mathbf{p}_i|\mathbf{p}_{j'}\rangle) \, , \quad (1.46)$$

where the last equality defines the *permanent* of the $N \times N$ matrix which has the entries $\langle\mathbf{p}_i|\mathbf{p}'_j\rangle$.

In fact, the bose and fermi cases, i.e. the symmetric and antisymmetric basis states, can be treated simultaneously if we introduce the factor $(\epsilon)^{\zeta_P}$ inside the summation sign

$$|\mathbf{p}_1 \Diamond \cdots \Diamond \mathbf{p}_N\rangle \equiv \frac{1}{\sqrt{N!}} \sum_P (\epsilon)^{\zeta_P} |\mathbf{p}_{P_1}, \mathbf{p}_{P_2}, \ldots, \mathbf{p}_{P_N}\rangle \qquad (1.47)$$

since then the fermi case corresponds to $\epsilon = -1$ and the bose case to $\epsilon = +1$, and $\Diamond$ stands for $\vee$ or $\wedge$ for the bose and fermi cases, respectively.

The states introduced in Eq. (1.45) provide a resolution of the identity in the symmetric state space, $\mathcal{B}^{(N)}$, specified by

$$\begin{aligned}
1 &= \hat{S}\, \hat{I}^{(N)}\, \hat{S} = \hat{S}\left(\hat{I}_1 \otimes \hat{I}_2 \otimes \hat{I}_3 \otimes \cdots \otimes \hat{I}_N\right) \hat{S}^\dagger \\[2mm]
&= \hat{S} \sum_{\mathbf{p}_1,\ldots,\mathbf{p}_N} |\mathbf{p}_1\rangle\langle\mathbf{p}_1| \otimes |\mathbf{p}_2\rangle\langle\mathbf{p}_2| \otimes \cdots \otimes |\mathbf{p}_N\rangle\langle\mathbf{p}_N|\, \hat{S}^\dagger \\[2mm]
&= \hat{S} \sum_{\mathbf{p}_1,\ldots,\mathbf{p}_N} |\mathbf{p}_1, \mathbf{p}_2, \ldots, \mathbf{p}_N\rangle\langle\mathbf{p}_1, \mathbf{p}_2, \ldots, \mathbf{p}_N|\, \hat{S}^\dagger \\[2mm]
&= \frac{1}{N!} \sum_{\mathbf{p}_1,\ldots,\mathbf{p}_N} |\mathbf{p}_1 \vee \mathbf{p}_2 \vee \cdots \vee \mathbf{p}_N\rangle\langle\mathbf{p}_1 \vee \mathbf{p}_2 \vee \cdots \vee \mathbf{p}_N| . \qquad (1.48)
\end{aligned}$$

The symmetric states introduced in Eq. (1.45) are not normalized in general, since for bosons the momenta need not differ. Of course, if the momentum values are all different, the state $|\mathbf{p}_1 \vee \mathbf{p}_2 \vee \cdots \vee \mathbf{p}_N\rangle$ is a sum of $N!$ normalized N-particle states which are all orthogonal to each other, and the state is therefore normalized in view of the overall prefactor. However, if say n_1 of the momentum values equals $\mathbf{p}_1$ and all the rest are different, the state $|\mathbf{p}_1 \vee \mathbf{p}_2 \vee \cdots \vee \mathbf{p}_N\rangle$ will be a sum of $N!/n_1!$ N-particle states each orthogonal to each other but now appearing with the prefactor $n_1!$, since permutations among the identical labels produce the same N-particle state. In general, if n_i is the number of times $\mathbf{p}_i$ occurs among the vectors $\mathbf{p}_1, \mathbf{p}_2, \ldots, \mathbf{p}_N$, n_j being equal to 0 if the momentum value $\mathbf{p}_j$ does not appear, then the set of ordered vectors, choosing for example the ordering according to $|\mathbf{p}_1| \leq |\mathbf{p}_2| \leq \cdots \leq |\mathbf{p}_N|$,

$$\frac{1}{\sqrt{n_1! n_2! \cdots n_N!}}\, |\mathbf{p}_1 \vee \mathbf{p}_2 \vee \cdots \vee \mathbf{p}_N\rangle = \sqrt{\frac{N!}{n_1! n_2! \cdots n_N!}}\, \hat{S}\, |\mathbf{p}_1, \mathbf{p}_2, \ldots, \mathbf{p}_N\rangle \quad (1.49)$$

constitute an orthonormal basis for the symmetric state space.

Equivalently we can state for the scalar product in Eq. (1.46)

$$\langle \mathbf{p}_1 \vee \cdots \vee \mathbf{p}_N | \mathbf{p}'_1 \vee \cdots \vee \mathbf{p}'_N \rangle = \begin{cases} n_1! n_2! \cdots & \{\mathbf{p}'\}_i \equiv \{\mathbf{p}\}_i \\[2mm] 0 & \text{otherwise.} \end{cases} \qquad (1.50)$$

The resolution of the identity in the symmetric N-particle state space can therefore also be expressed in terms of orthonormal states according to

$$\hat{I}_S^{(N)} = \frac{1}{n_1! n_2! n_3! \cdots} \sum_{|\mathbf{p}_1| \leq |\mathbf{p}_2| \leq \cdots \leq |\mathbf{p}_N|} |\mathbf{p}_1 \vee \mathbf{p}_2 \vee \cdots \vee \mathbf{p}_N\rangle\langle\mathbf{p}_1 \vee \mathbf{p}_2 \vee \cdots \vee \mathbf{p}_N| . \quad (1.51)$$

1.2.4 Dynamics and probability current and density

The quantum dynamics of an N-particle system of identical particles is given by the Schrödinger equation

$$i\hbar \frac{\partial \psi(\mathbf{x}_1, \mathbf{x}_2, \ldots; t)}{\partial t} = H \psi(\mathbf{x}_1, \mathbf{x}_2, \ldots; t), \quad (1.52)$$

where H is the Hamiltonian in the position representation for the N-particle system. For example, for the case of N non-relativistic electrons interacting through instantaneous two-particle interaction the Hamiltonian is

$$H = \sum_{i=1}^{N} \frac{1}{2m} \left(\frac{\hbar}{i} \frac{\partial}{\partial \mathbf{x}_i} \right)^2 + \frac{1}{2} \sum_{i \neq j} V(\mathbf{x}_i - \mathbf{x}_j). \quad (1.53)$$

In non-relativistic quantum mechanics the even or odd character of a wave function is preserved in time as any Hamiltonian for identical particles is symmetric in the degrees of freedom, here in the momenta and positions, but as well as other degrees of freedom in general (this is the meaning of identity of particles, no interaction can distinguish them). So if even- or oddness of a wave function is the state of affairs at one moment in time it will stay this way for all times.[9]

All physical properties are expressible in terms of the wave function, for example the average density of the particles, or rather the probability for the event that one of the particles is at position $\mathbf{x}$, is

$$n(\mathbf{x}, t) = \sum_{i=1}^{N} \frac{1}{N} \int \prod_{j=1}^{N} d\mathbf{x}_j \, \delta(\mathbf{x}_i - \mathbf{x}) |\psi(\mathbf{x}_1, \mathbf{x}_2, \ldots; t)|^2 = \int \prod_{i \neq 1} d\mathbf{x}_i \, |\psi(\mathbf{x}, \mathbf{x}_2, \ldots; t)|^2$$

$$(1.54)$$

where the last equality follows from the symmetry of the wave function for identical particles.

Taking the time derivative of the probability density, and using the Schrödinger equation, gives the continuity equation

$$\frac{\partial n(\mathbf{x}, t)}{\partial t} + \nabla_{\mathbf{x}} \cdot \mathbf{j}(\mathbf{x}, t) = 0, \quad (1.55)$$

where the probability current density is[10]

$$\mathbf{j}(\mathbf{x}, t) = \frac{e\hbar}{2mi} \sum_{i=1}^{N} \frac{1}{N} \int \prod_{j=1}^{N} d\mathbf{x}_j \, \delta(\mathbf{x}_i - \mathbf{x}) \left(\nabla_{\mathbf{x}_i} - \nabla_{\mathbf{x}_i'} \right) \psi(\mathbf{x}_1, \ldots; t) \, \psi^*(\mathbf{x}_1', \ldots; t) \Big|_{\mathbf{x}_i' = \mathbf{x}_i}$$

$$= \frac{e\hbar}{2mi} \int \prod_{i \neq 1} d\mathbf{x}_i \, \left(\nabla_{\mathbf{x}} - \nabla_{\mathbf{x}'} \right) \psi(\mathbf{x}, \mathbf{x}_2, \ldots; t) \, \psi^*(\mathbf{x}', \mathbf{x}_2', \ldots; t) \Big|_{\mathbf{x}' = \mathbf{x}}, \quad (1.56)$$

---

[9] For the cases of more than two identical particles there are other time invariant subspaces than the symmetric and antisymmetric ones. They do not seem to be of physical relevance.

[10] In the presence of a vector potential the formula must be amended with the diamagnetic term, see Exercise 1.4 and Section 2.3.

again the last equality follows from the symmetry of the wave function for identical particles. The Schrödinger equation guarantees conservation of probability, i.e. the continuity equation, Eq. (1.55), as a consequence of the Hamiltonian being hermitian.

Exercise 1.4. The Hamiltonian for a charged spinless particle coupled to a vector potential, $\mathbf{A}$, is[11]

$$\hat{H} = \frac{1}{2m} \left(\frac{\hbar}{i} \frac{\partial}{\partial \mathbf{x}} - e\mathbf{A}(\mathbf{x}, t) \right)^2 . \tag{1.57}$$

Show that the probability current density for the particle in state ψ is

$$\mathbf{j}(\mathbf{x}, t) = \frac{1}{2m} \left(\frac{\hbar}{i} \nabla_{\mathbf{x}} - \frac{\hbar}{i} \nabla_{\mathbf{x}'} - 2e\mathbf{A}(\mathbf{x}, t) \right) \psi(\mathbf{x}, t) \, \psi^*(\mathbf{x}', t) \Big|_{\mathbf{x}'=\mathbf{x}} . \tag{1.58}$$

Rarely can the dynamics of an N-particle system of identical particles be solved exactly. When it comes to performing actual approximate calculations, the quantum statistics of the particles will even in the non-relativistic quantum theory of an interacting N-particle system cause havoc, and a more flexible vehicle for respecting the quantum statistics of identical particles is convenient. We now turn to introduce these, the quantum fields. In relativistic quantum theory and conveniently for many-body systems, the quantum fields instead of the wave function become the carriers of the dynamics, as we will discuss in Chapter 3.

1.3   Fermi field

We introduce the fermion creation operator, $a_{\mathbf{p}}^{\dagger}$, corresponding to momentum value $\mathbf{p}$, as the linear mapping of $\mathcal{F}^{(N)}$ into $\mathcal{F}^{(N+1)}$ defined for an arbitrary (not necessarily ordered) basis vector by[12]

$$a_{\mathbf{p}}^{\dagger} |\mathbf{p}_1 \wedge \mathbf{p}_2 \wedge \cdots \wedge \mathbf{p}_N\rangle \equiv |\mathbf{p} \wedge \mathbf{p}_1 \wedge \mathbf{p}_2 \wedge \cdots \wedge \mathbf{p}_N\rangle , \tag{1.59}$$

i.e. it maps an antisymmetrized N-particle state into the antisymmetrized $(N+1)$-particle state where an additional fermion has momentum $\mathbf{p}$. The choice of placing $\mathbf{p}$ at the front is, of course, arbitrary. The other popular choice is to place it at the end. This reflects that a creation operator, like a state vector, is defined only modulo a phase factor.

If in the N-fermion state the momentum state $\mathbf{p}$ is already occupied, i.e. *exactly* one of the $\mathbf{p}_i$s equals $\mathbf{p}$, then owing to the antisymmetric nature of the state

$$a_{\mathbf{p}}^{\dagger} |\mathbf{p}_1 \wedge \mathbf{p}_2 \wedge \cdots \wedge \mathbf{p}_N\rangle = 0_{N+1} , \tag{1.60}$$

---

[11]The form of the Hamiltonian follows from gauge invariance; i.e. the gauge transformation of the electromagnetic field, $\mathbf{A}(\mathbf{x}, t) \to \mathbf{A}(\mathbf{x}, t) + \nabla\Lambda(\mathbf{x}, t)$, $\phi(\mathbf{x}, t) \to \phi(\mathbf{x}, t) - \dot{\Lambda}(\mathbf{x}, t)$, and the transformation of the wave function $\psi(\mathbf{x}, t) \to \psi(\mathbf{x}, t) e^{\frac{ie}{\hbar}\Lambda(\mathbf{x}, t)}$, leaves all physical quantities invariant. The gauge invariance of quantum mechanics is a consequence of the wave function obtained by the above phase transformation equally well represents the probability distribution of the particle.

[12]As emphasized, the label on the creation operator could refer to any state; usually though, it refers to a complete set of single-particle states.

the zero vector of state space $H^{(N+1)}$. This is the expedience with which the fermion creation operators respect Pauli's exclusion principle.

We introduce the *sum* of state spaces $\mathcal{F}^{(N)}$ and $\mathcal{F}^{(N+1)}$. For example, $\mathcal{F}^{(1)} + \mathcal{F}^{(2)}$ consists of states spanned by 2-tuple states, $(|\mathbf{p}\rangle, |\mathbf{p}_1 \wedge \mathbf{p}_2\rangle)$, and is equipped with the scalar product, which is the sum of the scalar products in the subspaces, i.e. for the above vector and $(c_1|\mathbf{p}'\rangle, c_2|\mathbf{p}_1' \wedge \mathbf{p}_2'\rangle)$ the scalar product is

$$(\langle\mathbf{p}|, \langle\mathbf{p}_1 \wedge \mathbf{p}_2|)(c_1|\mathbf{p}'\rangle, c_2|\mathbf{p}_1' \wedge \mathbf{p}_2'\rangle) = c_1\langle\mathbf{p}|\mathbf{p}'\rangle + c_2\langle\mathbf{p}_1 \wedge \mathbf{p}_2|\mathbf{p}_1' \wedge \mathbf{p}_2'\rangle .$$
(1.61)

In order for an operator to represent an observable physical quantity it must map a state space onto itself. In order to facilitate this experience for the fermion creation operators,[13] the multi-particle space or Fock space $\mathcal{F}$ (named after the Soviet physicist Vladimir Fock), is introduced as the *sum* of the state spaces[14]

$$\mathcal{F} = \sum_{N=0}^{\infty} \mathcal{F}^{(N)} ,$$
(1.62)

where by definition $\mathcal{F}^{(0)}$ is the set of complex numbers.

The inclusion of $\mathcal{F}^{(0)}$ is demanded in relativistic quantum theory since relativistic kinematics predicts the creation and annihilation of particles. The zero vector in Fock space can not represent the state of absence of any particle.[15] Particle species not present in the initial and final states must, before and after a reaction, be in a state in their respective Fock spaces so that their scalar products equal one, thereby not influencing the probabilities for the various possible reactions. Since none of these particle states is initially and finally occupied, though they may appear virtually in intermediate states to facilitate the reaction, and since the zero vector does not respect the above property, the state where particles of a given species are absent, the vacuum state for these particles is represented by (choosing the simplest phase choice)

$$|0\rangle \equiv (1, 0_1, 0_2, \ldots) .$$
(1.63)

Even for a non-relativistic system, the vacuum state is a convenient vehicle for generating all states of the multi-particle space as we will see shortly.

The set of basis states of the Fock space consists of the vacuum state and all the basis vectors of each N-particle subspace. In the Fock space, states of the type $(0, |\mathbf{p}_1\rangle, 0_2, |\mathbf{p}_1' \wedge \mathbf{p}_2' \wedge \mathbf{p}_3'\rangle, \ldots, |\mathbf{p}_1 \wedge \mathbf{p}_2, \wedge \cdots \wedge \mathbf{p}_N\rangle, \ldots)$ are thus encountered, superposition of states with different number of particles. In accordance with the definition of the scalar product of states in the multi-particle space, it can only

---

[13]Whether a fermi field is an observable, i.e. a measurable quantity, is doubtful. For example, it does not have a classical limit as states can at most be singly occupied. A bose field (introduced in Section 1.4) on the other hand is an observable, since any number of bosons can occupy a single state and the average value of a bose field can thus be nonzero, an example being the classical state of light, the coherent state, created by a laser.

[14]In mathematical terms, the state space is a Hilbert space, and the Fock space is a Hilbert sum of Hilbert spaces, and itself a Hilbert space.

[15]The zero vector in the Fock space is of course $(0, 0_1, 0_2, \ldots) \equiv 0$, for which the obvious short notation has been introduced.

be nonzero if the states have components with the same number of particles. An N-particle basis state in the multi-particle space is usually shortened according to $(0, 0_1, 0_2, \ldots, |\mathbf{p}_1 \wedge \mathbf{p}_2 \wedge \cdots \wedge \mathbf{p}_N\rangle, 0_{N+1}, 0_{N+2}, \ldots) \rightarrow |\mathbf{p}_1 \wedge \mathbf{p}_2 \wedge \cdots \wedge \mathbf{p}_N\rangle$.

For the vacuum state, the creation operator operates also in accordance with its general prescription of adding a particle

$$a_\mathbf{p}^\dagger |0\rangle = |0, |\mathbf{p}\rangle, 0_2, 0_3, \ldots\rangle. \tag{1.64}$$

The state vector (using the abbreviated notation introduced above)

$$a_{\mathbf{p}_1}^\dagger \cdots a_{\mathbf{p}_N}^\dagger |0\rangle = |\mathbf{p}_1 \wedge \mathbf{p}_2 \wedge \cdots \wedge \mathbf{p}_N\rangle \tag{1.65}$$

is an antisymmetric N-particle basis state in the multi-particle state space, provided that all the momenta are different of course, otherwise it is the zero-vector.[16] The bracket notation appears a little clumsy in this context, and the notation

$$\Phi_{\mathbf{p}_1, \ldots, \mathbf{p}_N} = a_{\mathbf{p}_1}^\dagger \cdots a_{\mathbf{p}_N}^\dagger \Phi_0 \tag{1.66}$$

is often used, where Φ_0 denotes the vacuum. For a state which is a superposition of states with different number of particles we can also express it in terms of the vacuum state, for example $(0, |\mathbf{p}_1\rangle, |\mathbf{p}_1' \wedge \mathbf{p}_2'\rangle, 0_3, \ldots) = (a_{\mathbf{p}_1}^\dagger + a_{\mathbf{p}_1'}^\dagger a_{\mathbf{p}_2'}^\dagger)|0\rangle$.

By construction, any two fermion creation operators, $a_\mathbf{p}^\dagger$ and $a_{\mathbf{p}'}^\dagger$, anti-commute, i.e.

$$\{a_\mathbf{p}^\dagger, a_{\mathbf{p}'}^\dagger\} \equiv a_\mathbf{p}^\dagger a_{\mathbf{p}'}^\dagger + a_{\mathbf{p}'}^\dagger a_\mathbf{p}^\dagger = 0, \tag{1.67}$$

meaning that operating with the anti-commutator $\{a_\mathbf{p}^\dagger, a_{\mathbf{p}'}^\dagger\}$ on any vector in Fock space produces the zero vector in Fock space, $0 \equiv (0, 0_1, 0_2, \ldots)$, just as multiplying any vector in Fock space by the number 0 does. This follows immediately from the fact that operating with the anti-commutator on any basis vector, say $(0, 0_1, 0_2, \ldots, |\mathbf{p}_1 \wedge \mathbf{p}_2 \wedge \cdots \wedge \mathbf{p}_N\rangle, 0_{N+1}, 0_{N+2}, \ldots)$, gives the sum of two vectors which differ only by a minus sign (or if the momentum labels in the anti-commutator are equal, the sum of two zero vectors). For the case $\mathbf{p}' = \mathbf{p}$, the anti-commutation relation Eq. (1.67) becomes $a_\mathbf{p}^\dagger a_\mathbf{p}^\dagger = -a_\mathbf{p}^\dagger a_\mathbf{p}^\dagger$ and therefore by itself $a_\mathbf{p}^\dagger a_\mathbf{p}^\dagger = 0$. This is Pauli's exclusion principle expressed in terms of the creation operator: two fermions can not be accommodated in the same state.

We then introduce the fermion annihilation operator, $a_\mathbf{p}$, as the adjoint of the fermion creation operator $a_\mathbf{p}^\dagger$. Since the creation operator maps an N-particle state into an $(N+1)$-particle state, the annihilation operator, being the adjoint, will map an N-particle state into an $(N-1)$-particle state. To understand its properties we can restrict attention to the basis vectors of the subspaces $\mathcal{F}^{(N)}$ and $\mathcal{F}^{(N-1)}$ of the Fock space, and we have

$$\langle \mathbf{p}_2' \wedge \cdots \wedge \mathbf{p}_N' | a_\mathbf{p} | \mathbf{p}_1 \wedge \cdots \wedge \mathbf{p}_N \rangle^* = \langle \mathbf{p}_1 \wedge \cdots \wedge \mathbf{p}_N | a_\mathbf{p}^\dagger | \mathbf{p}_2' \wedge \cdots \wedge \mathbf{p}_N' \rangle$$

$$= \langle \mathbf{p}_1 \wedge \cdots \wedge \mathbf{p}_N | \mathbf{p} \wedge \mathbf{p}_2' \wedge \cdots \wedge \mathbf{p}_{N-1}' \rangle$$

$$= \det(\langle \mathbf{p}_i | \mathbf{p}_j' \rangle), \tag{1.68}$$

---

[16] With the chosen ordering convention of the previous section it is the ground state for N non-interacting fermions.

where, in the last equality, we have introduced the notation $\mathbf{p}'_1 = \mathbf{p}$, and used Eq. (1.42). Expanding the determinant in terms of its first column we get

$$
\langle \mathbf{p}'_2 \wedge \cdots \wedge \mathbf{p}'_N | a_\mathbf{p} | \mathbf{p}_1 \wedge \cdots \wedge \mathbf{p}_N \rangle^* = \sum_{n=1}^{N} (-1)^{n-1} \langle \mathbf{p}_n | \mathbf{p} \rangle \det(\langle \mathbf{p}_i | \mathbf{p}'_j \rangle^{(n)})
$$

$$
= \sum_{n=1}^{N} (-1)^{n-1} \langle \mathbf{p}_n | \mathbf{p} \rangle \det(\langle \mathbf{p}_i | \mathbf{p}'_j \rangle^{(n)}) ,
$$

$$(1.69)$$

where the sub-determinant, $\det(\langle \mathbf{p}_i | \mathbf{p}'_j \rangle^{(n)})$, is the determinant of the matrix Eq. (1.68), with row n and the first column removed. Using $\langle \mathbf{p} | \mathbf{p}' \rangle^* = \langle \mathbf{p}' | \mathbf{p} \rangle$ we get

$$
\langle \mathbf{p}'_2 \wedge \cdots \wedge \mathbf{p}'_N | a_\mathbf{p} | \mathbf{p}_1 \wedge \cdots \wedge \mathbf{p}_N \rangle = \sum_{n=1}^{N} (-1)^{n-1} \langle \mathbf{p} | \mathbf{p}_n \rangle \det(\langle \mathbf{p}'_j | \mathbf{p}_i \rangle^{(n)})
$$

$$(1.70)$$

and using Eq. (1.42) for the $(N-1)$-particle case, the right-hand side can be rewritten as

$$
\sum_{n=1}^{N} (-1)^{n-1} \langle \mathbf{p} | \mathbf{p}_n \rangle \langle \mathbf{p}'_2 \wedge \cdots \wedge \mathbf{p}'_N | \mathbf{p}_1 \wedge \cdots (\text{no } \mathbf{p}_n) \,.. \wedge \mathbf{p}_N \rangle \qquad (1.71)
$$

and we have

$$
a_\mathbf{p} | \mathbf{p}_1 \wedge \cdots \wedge \mathbf{p}_N \rangle = \sum_{n=1}^{N} (-1)^{n-1} \langle \mathbf{p} | \mathbf{p}_n \rangle | \mathbf{p}_1 \wedge \cdots (\text{no } \mathbf{p}_n) \cdots \wedge \mathbf{p}_N \rangle .
$$

$$(1.72)$$

Thus operating with the fermion annihilation operator labeled by $\mathbf{p}$ on an N-particle basis state produces the zero vector unless exactly one of the momentum values equals $\mathbf{p}$, and in that case it equals the $(N-1)$-particle state where none of the fermions occupies the originally occupied momentum state $\mathbf{p}$. The annihilation operator $a_\mathbf{p}$ thus annihilates the particle in state $\mathbf{p}$. In the simplest of situations we have

$$
a_\mathbf{p} | \mathbf{p} \rangle = | 0 \rangle . \qquad (1.73)
$$

Annihilating the single-particle state turns it into the vacuum state.

In particular it follows from Eq. (1.68) that operating with any fermion annihilation operator on the vacuum state produces the zero vector

$$
a_\mathbf{p} | 0 \rangle = 0 . \qquad (1.74)
$$

According to Eq. (1.67), the fermion annihilation operators anti-commute

$$\{a_{\mathbf{p}}, a_{\mathbf{p}'}\} = 0 . \tag{1.75}$$

For the case $\mathbf{p}' = \mathbf{p}$, the anti-commutation relation Eq. (1.75) has the consequence $a_{\mathbf{p}} a_{\mathbf{p}} = 0$, expressing the exclusion principle: no two identical fermions can occupy the same momentum state.

Next we inquire into the relations obtained by subsequent operations with fermion creation and annihilation operators, and calculate, according to Eq. (1.72),

$$
\begin{aligned}
a_{\mathbf{p}'} a_{\mathbf{p}}^{\dagger} |\mathbf{p}_1 \wedge \cdots \wedge \mathbf{p}_N\rangle &= a_{\mathbf{p}'} |\mathbf{p} \wedge \mathbf{p}_1 \wedge \cdots \wedge \mathbf{p}_N\rangle \\
&= \langle \mathbf{p}'|\mathbf{p}\rangle |\mathbf{p}_1 \wedge \cdots \wedge \mathbf{p}_N\rangle \\
&+ \sum_{n=1}^{N} (-1)^n \langle \mathbf{p}'|\mathbf{p}_n\rangle |\mathbf{p} \wedge \mathbf{p}_1 \wedge \cdots (\text{ no } \mathbf{p}_n) \cdots \wedge \mathbf{p}_N\rangle
\end{aligned}
$$

$$\tag{1.76}$$

and similarly

$$
\begin{aligned}
a_{\mathbf{p}}^{\dagger} a_{\mathbf{p}'} |\mathbf{p}_1 \wedge \cdots \wedge \mathbf{p}_N\rangle &= a_{\mathbf{p}}^{\dagger} \sum_{n=1}^{N} (-1)^{n-1} \langle \mathbf{p}'|\mathbf{p}_n\rangle |\mathbf{p}_1 \wedge \cdots (\text{ no } \mathbf{p}_n) \cdots \wedge \mathbf{p}_N\rangle \\
&= \sum_{n=1}^{N} (-1)^{n-1} \langle \mathbf{p}'|\mathbf{p}_n\rangle |\mathbf{p} \wedge \mathbf{p}_1 \wedge \cdots (\text{ no } \mathbf{p}_n) \cdots \wedge \mathbf{p}_N\rangle ,
\end{aligned}
$$

$$\tag{1.77}$$

and by adding the two equations we realize the relation

$$\{a_{\mathbf{p}'}, a_{\mathbf{p}}^{\dagger}\} = \langle \mathbf{p}'|\mathbf{p}\rangle . \tag{1.78}$$

The anti-commutator of fermion creation and annihilation operators is not an operator but a *c*-number, i.e. proportional to the identity operator. This is the fundamental relation obeyed by the fermion creation and annihilation operators, and its virtue is that it makes respecting the quantum statistics a trivial matter. When doing calculations for fermion processes, we can in fact, as we show later, forget all the previous index-nightmare Fock state vector formalism, and we need only remember the fundamental anti-commutation relation.

We note that, according to Eq. (1.77) and Eq. (1.72),

$$
a_{\mathbf{p}}^{\dagger} a_{\mathbf{p}} |\mathbf{p}_1 \wedge \cdots \wedge \mathbf{p}_N\rangle = \begin{cases} |\mathbf{p}_1 \wedge \cdots \wedge \mathbf{p}_N\rangle & \text{if exactly one of the } \mathbf{p}_i's \text{ equals } \mathbf{p} \\ 0_{N-1} & \text{otherwise} , \end{cases}
$$

$$\tag{1.79}$$

i.e. the operator $a_{\mathbf{p}}^{\dagger} a_{\mathbf{p}}$ counts the number of particles in state $\mathbf{p}$, i.e. the eigenvalue of the operator is either 1 or 0, depending on the state in question being occupied or not.

The operator $n_{\mathbf{p}} = a_{\mathbf{p}}^\dagger a_{\mathbf{p}}$ is therefore referred to as the number operator for state or mode $\mathbf{p}$. The number of particles counted in the vacuum state is correctly zero. One readily verifies (see Exercise 1.6 below), that all the mode number operators commute and each number operator has only two eigenvalues, 0 or 1. The total set of momentum state number operators, $\{n_{\mathbf{p}}\}_{\mathbf{p}}$, thus constitutes a complete set of commuting operators as specifying the eigenvalues for each number operator uniquely specifies a basis vector. They can therefore be used to define a representation, as discussed in Section 1.5.

Had we used any other complete set of single particle states, say labeled by index λ, we would analogously have obtained for the commutation relations for the operators creating and annihilating particles in states λ_1 and λ_2

$$\{a_{\lambda_1}, a_{\lambda_2}^\dagger\} = \langle \lambda_1 | \lambda_2 \rangle = \delta_{\lambda_1, \lambda_2}, \tag{1.80}$$

where the set of chosen single-particle states here is assumed orthonormal and discrete unless we use compact notation to include a continuum as well, Kronecker including delta. An example could be that of the energy eigenstates. In the case of momentum states, we encounter in Eq. (1.78) either a Kronecker function or a delta function depending on whether the particles are confined or not.

Since the creation and annihilation operators are defined in terms of operations on state vectors, they inherit their invariance with respect to a global phase transformation

$$a_\lambda \to e^{i\phi} a_\lambda \quad, \quad a_\lambda^\dagger \to e^{-i\phi} a_\lambda^\dagger. \tag{1.81}$$

Note that indeed all the anti-commutation relations remain invariant under the phase transformation.

Exercise 1.5. Show for arbitrary operators A, B and C the relations

$$[A, BC] = B[A, C] + [A, B]C = [A, B]C - B[C, A] \tag{1.82}$$

and analogously for $[AB, C]$, and in terms of anti-commutators

$$[A, BC] = \{A, B\}C - B\{C, A\}. \tag{1.83}$$

Exercise 1.6. Let us familiarize ourselves with the consequences of the algebra of creation and annihilation operators

$$\{a_{\lambda_1}^\dagger, a_{\lambda_2}^\dagger\} = 0 = \{a_{\lambda_1}, a_{\lambda_2}\} \tag{1.84}$$

and, according to Eq. (1.80) for different state labels, creation and annihilation operators also anti-commute as

$$\{a_{\lambda_1}, a_{\lambda_2}^\dagger\} = 0. \tag{1.85}$$

It therefore suffices to consider a single pair of creation and annihilation operators, denoted $a = a_\lambda$, and we have $\{a, a^\dagger\} = 1$. As a consequence of the anti-commutation relations, Eq. (1.84),

$$a^2 = 0 = (a^\dagger)^2 \tag{1.86}$$

and verify therefore

$$(a^\dagger a)^2 = a^\dagger a. \tag{1.87}$$

Show that for any c-number ϵ

$$e^{\epsilon a^\dagger a} = aa^\dagger + e^\epsilon a^\dagger a. \tag{1.88}$$

Show that for the number operator, $n = a^\dagger a$, we have its characteristic equation

$$n(n-1) = 0 \tag{1.89}$$

demonstrating that its eigenvalues can be either zero or one.

Show that for different state labels, the number operators commute as

$$[n_\lambda, n_{\lambda'}] = 0 \tag{1.90}$$

even though the creation and annihilation operators all anti-commute, and the number operators behave as bose operators. Or, in general, polynomials containing an even number of anti-commuting operators behave algebraically as numbers.

Exercise 1.7. Show that the infinite product state

$$|\text{BCS}\rangle = \prod_{\mathbf{p}} \left(u_{\mathbf{p}} + v_{\mathbf{p}} \, a^\dagger_{\mathbf{p}\uparrow} a^\dagger_{-\mathbf{p}\downarrow} \right) |0\rangle \tag{1.91}$$

is normalized provided that $|u_{\mathbf{p}}|^2 + |v_{\mathbf{p}}|^2 = 1$ for all $\mathbf{p}$.

If, instead of momentum states, position states had been used we would analogously have encountered creation operators which create fermions at definite positions, for example

$$a^\dagger_{\mathbf{x}_1} a^\dagger_{\mathbf{x}_2} \cdots a^\dagger_{\mathbf{x}_N} |0\rangle = |\mathbf{x}_1 \wedge \mathbf{x}_2 \wedge \cdots \wedge \mathbf{x}_N\rangle. \tag{1.92}$$

Creation operators of different representations are related through their transformation functions. For position and momentum we have, according to Eq. (1.5),

$$|\mathbf{x}\rangle = \int d\mathbf{p} \, |\mathbf{p}\rangle \langle \mathbf{p}|\mathbf{x}\rangle = \int \frac{d\mathbf{p}}{(2\pi\hbar)^{3/2}} \, e^{-\frac{i}{\hbar}\mathbf{P}\cdot\mathbf{x}} |\mathbf{p}\rangle \tag{1.93}$$

and therefore the relationship

$$a^\dagger_{\mathbf{x}} = \int \frac{d\mathbf{p}}{(2\pi\hbar)^{3/2}} \, e^{-\frac{i}{\hbar}\mathbf{P}\cdot\mathbf{x}} \, a^\dagger_{\mathbf{p}}. \tag{1.94}$$

In accordance with tradition, instead of using the notation $a^\dagger_{\mathbf{x}}$ we shall introduce

$$\psi^\dagger(\mathbf{x}) = a^\dagger_{\mathbf{x}}. \tag{1.95}$$

To obtain the inverse relation we use

$$|\mathbf{p}\rangle = \int d\mathbf{x} \, |\mathbf{x}\rangle \langle \mathbf{x}|\mathbf{p}\rangle = \int \frac{d\mathbf{x}}{(2\pi\hbar)^{3/2}} \, e^{\frac{i}{\hbar}\mathbf{x}\cdot\mathbf{P}} |\mathbf{x}\rangle \tag{1.96}$$

and we have the relationships between the operators creating particles with definite position and momentum

$$\psi^\dagger(\mathbf{x}) = \int \frac{d\mathbf{p}}{(2\pi\hbar)^{3/2}} e^{-\frac{i}{\hbar}\mathbf{P}\cdot\mathbf{x}} a_{\mathbf{p}}^\dagger \quad , \quad a_{\mathbf{p}}^\dagger = \int \frac{d\mathbf{x}}{(2\pi\hbar)^{3/2}} e^{\frac{i}{\hbar}\mathbf{P}\cdot\mathbf{x}} \psi^\dagger(\mathbf{x}) . \qquad (1.97)$$

Taking the adjoint we obtain analogously for the annihilation operators or quantum fields

$$\psi(\mathbf{x}) = \int \frac{d\mathbf{p}}{(2\pi\hbar)^{3/2}} e^{\frac{i}{\hbar}\mathbf{P}\cdot\mathbf{x}} a_{\mathbf{p}} \quad , \quad a_{\mathbf{p}} = \int \frac{d\mathbf{x}}{(2\pi\hbar)^{3/2}} e^{-\frac{i}{\hbar}\mathbf{P}\cdot\mathbf{x}} \psi(\mathbf{x}) . \qquad (1.98)$$

How one prefers to keep track of factors of $2\pi\hbar$ in the above *Fourier transformations*, is, of course, a matter of taste. With the above convention determined by the fundamental choice of Eq. (1.5), no such factors appear in the fundamental anti-commutation relation, Eq. (1.78). If the fermions are confined to a box of volume V we shall use (guided by our preference for fields to have the same dimensions as wave functions)

$$\psi(\mathbf{x}) = \frac{1}{\sqrt{V}} \sum_{\mathbf{P}} e^{\frac{i}{\hbar}\mathbf{P}\cdot\mathbf{x}} a_{\mathbf{p}} \quad , \quad a_{\mathbf{p}} = \frac{1}{\sqrt{V}} \int_V d\mathbf{x}\, e^{-\frac{i}{\hbar}\mathbf{P}\cdot\mathbf{x}} \psi(\mathbf{x}) , \qquad (1.99)$$

leaving the fundamental anti-commutation relation Eq. (1.78)

$$\{a_{\mathbf{p}'}, a_{\mathbf{p}}^\dagger\} = \delta_{\mathbf{P},\mathbf{P}'} , \qquad (1.100)$$

where the discrete allowed momentum values are specified by the boundary condition for the states, say periodic boundary conditions.

One readily verifies, as a consequence of the analogous Eq. (1.78), or by using Eq. (1.97) and Eq. (1.98), the fundamental anti-commutation relations for Fermi fields in the position representation[17]

$$\{\psi(\mathbf{x}), \psi^\dagger(\mathbf{x}')\} = \delta(\mathbf{x} - \mathbf{x}') \qquad (1.101)$$

and

$$\{\psi(\mathbf{x}), \psi(\mathbf{x}')\} = 0 = \{\psi^\dagger(\mathbf{x}), \psi^\dagger(\mathbf{x}')\} . \qquad (1.102)$$

For equal position the latter two equations have, as a consequence, $\psi(\mathbf{x})\psi(\mathbf{x}) = 0$ and $\psi^\dagger(\mathbf{x})\psi^\dagger(\mathbf{x}) = 0$, expressing the exclusion principle: no two identical fermions can occupy the same position.

For the N-particle basis state with particles at the indicated locations we shall also use the notation

$$\Phi_{\mathbf{x}_1 \mathbf{x}_2 \ldots \mathbf{x}_N} = \psi_{\mathbf{x}_1}^\dagger \psi_{\mathbf{x}_2}^\dagger \cdots \psi_{\mathbf{x}_N}^\dagger |0\rangle \qquad (1.103)$$

---

[17]If the particles represented by the fields have internal degrees of freedom, say spin, we have

$$\{\psi(\mathbf{x},\sigma_z), \psi^\dagger(\mathbf{x}',\sigma_z')\} = \delta_{\sigma_z,\sigma_z'}\, \delta(\mathbf{x} - \mathbf{x}') .$$

Often the notation $\psi_{\sigma_z}(\mathbf{x}) = \psi(\mathbf{x}, \sigma_z)$ is used.

where $|0\rangle$ denotes the vacuum state for the fermions.

As stressed, any complete set of single-particle states, not just the position or momentum states, could have been employed. For example in the absence of translation invariance and using the single-particle energy eigenstates we have analogously for the quantum fields

$$\psi(\mathbf{x}) = \sum_{\lambda} \langle \mathbf{x}|\lambda\rangle \, a_{\lambda} = \sum_{\lambda} \psi_{\lambda}(\mathbf{x}) \, a_{\lambda} \ , \quad a_{\lambda} = \int d\mathbf{x} \ \psi_{\lambda}^*(\mathbf{x}) \, \psi(\mathbf{x}) \ , \qquad (1.104)$$

where $\psi_{\lambda}(\mathbf{x})$ are the orthonormal eigenstates of a single-particle Hamiltonian.

Instead of characterizing the quantum statistics of a collection of fermions in terms of the antisymmetry of their state vectors, which as we have seen is a bit messy or at least requires a substantial amount of indices-writing, it is now taken care of by the simple *anti-commutation* relations for the creation and annihilation operators. The price paid for this enormous simplification is of course that the operators now are operators on a super-space, the multi-particle space. As shown in the next section, the implementation for bosons is identical to the above except that the quantum statistics is taken care of by the *commutation* relations of the creation and annihilation operators.

Exercise 1.8. For N non-interacting spin one-half fermions, an ideal Fermi gas, the ground state is obtained from the vacuum state according to

$$|G_0\rangle = \left(\prod_{\sigma, |\mathbf{p}| < p_F} a_{\mathbf{p},\sigma}^{\dagger} \right) |0\rangle \ , \qquad (1.105)$$

i.e all the states below the Fermi energy, $\epsilon_F = p_F^2/2m$, are occupied in accordance with Pauli's exclusion principle, and all states above are empty for the case of the ground state. Pictorially, the ground state is that of a filled sphere in momentum space, the Fermi sea, with the Fermi surface separating occupied and unoccupied states.

Show that the one-particle Green's function or density matrix becomes

$$G_{\sigma}(\mathbf{x} - \mathbf{x}') \equiv \langle G_0 | \psi_{\sigma}^{\dagger}(\mathbf{x}) \psi_{\sigma}(\mathbf{x}') | G_0 \rangle$$

$$= \frac{3n}{2} \frac{\sin k_F |\mathbf{x} - \mathbf{x}'| - k_F |\mathbf{x} - \mathbf{x}'| \cos k_F |\mathbf{x} - \mathbf{x}'|}{(k_F |\mathbf{x} - \mathbf{x}'|)^3} \ , \qquad (1.106)$$

where n is the density of the fermions, and $k_F = p_F/\hbar$, and in the considered three dimensions $k_F^3 = 3\pi^2 n$. The considered amplitude specifies the overlap between the state where a particle with spin σ at position $\mathbf{x}'$ has been removed from the ground state and the state where a particle with spin σ at position $\mathbf{x}$ has been removed from the ground state. Or equivalently, it specifies the amplitude for transition to the ground state of the state where a particle with spin σ at position $\mathbf{x}'$ has been removed from the ground state and subsequently a particle with spin σ has been added at position $\mathbf{x}$.

At small spatial separation

$$\langle G_0|\psi_\sigma^\dagger(\mathbf{x})\psi_\sigma^\dagger(\mathbf{x}')|G_0\rangle \simeq \frac{n}{2}\left(1 - \frac{(k_F|\mathbf{x}-\mathbf{x}'|)^2}{10}\right) \tag{1.107}$$

and at $\mathbf{x} = \mathbf{x}'$ it counts the density of fermions per spin at the position in question.

Show that the pair correlation function is related to the one-particle density matrix according to

$$\langle G_0|\psi_\sigma^\dagger(\mathbf{x})\psi_{\sigma'}^\dagger(\mathbf{x}')\psi_{\sigma'}(\mathbf{x}')\psi_\sigma(\mathbf{x})|G_0\rangle = \begin{cases} \left(\frac{n}{2}\right)^2 & \sigma' \neq \sigma \\ \left(\frac{n}{2}\right)^2 - G_\sigma^2(\mathbf{x}-\mathbf{x}') & \sigma' = \sigma. \end{cases} \tag{1.108}$$

Interpret the result and note in particular the anti-bunching of non-interacting fermions: the avoidance of identical fermions to be at the same position in space, a *repulsion* solely due to the exchange symmetry, the exclusion principle at work in real space.

So far the creation operators are just a kinematic gadget giving an equivalent way of describing the N-particle state space for arbitrary N, since for example

$$a_{\mathbf{p}_1}^\dagger a_{\mathbf{p}_2}^\dagger \cdots a_{\mathbf{p}_N}^\dagger|0\rangle = |\mathbf{p}_1 \wedge \mathbf{p}_2 \wedge \cdots \wedge \mathbf{p}_N\rangle \tag{1.109}$$

specifies the basis states in terms of the creation operators and the vacuum state. In Chapter 2, we shall show how operators representing physical quantities can be expressed in terms of the creation and annihilation operators, and thereby realize in Chapter 3 their usefulness in describing quantum dynamics in the most general case where the number of particles is not conserved. But first we consider the kinematics for the case where the identical particles are bosons.

1.4 Bose field

The bose particle creation operator, $a_{\mathbf{p}}^\dagger$, is introduced according to its action on the basis states of Eq. (1.45)

$$a_{\mathbf{p}}^\dagger|\mathbf{p}_1 \vee \mathbf{p}_2 \vee \cdots \vee \mathbf{p}_N\rangle \equiv |\mathbf{p} \vee \mathbf{p}_1 \vee \mathbf{p}_2 \vee \cdots \vee \mathbf{p}_N\rangle \tag{1.110}$$

and the adjoint operates according to

$$a_{\mathbf{p}}|\mathbf{p}_1 \wedge \cdots \wedge \mathbf{p}_N\rangle = \sum_{n=1}^{N} \langle \mathbf{p}|\mathbf{p}_n\rangle|\mathbf{p}_1 \wedge \cdots (\text{ no } \mathbf{p}_n) \cdots \wedge \mathbf{p}_N\rangle, \tag{1.111}$$

i.e. it annihilates a particle in state $\mathbf{p}$, and is referred to as the bose annihilation operator. As previously noted, the derivation is equivalent to the antisymmetric case.

Since no minus signs ever occur, the bose creation and annihilation operators satisfy the commutation relations (the analogous equations to Eq. (1.76) and Eq. (1.77) are now *subtracted* to give the following result)

$$[a_{\mathbf{p}'}, a_{\mathbf{p}}^\dagger] = \langle \mathbf{p}'|\mathbf{p}\rangle \tag{1.112}$$

and

$$[a_{\mathbf{p}'}^{\dagger}, a_{\mathbf{p}}^{\dagger}] = 0 = [a_{\mathbf{p}'}, a_{\mathbf{p}}] . \tag{1.113}$$

We note that, according to the equation for bosons analogous to Eq. (1.77), the operator $a_{\mathbf{p}}^{\dagger} a_{\mathbf{p}}$ counts the number of particles in state $\mathbf{p}$

$$a_{\mathbf{p}}^{\dagger} a_{\mathbf{p}} |\mathbf{p}_1 \vee \cdots \vee \mathbf{p}_N\rangle = n_{\mathbf{p}} |\mathbf{p}_1 \vee \cdots \vee \mathbf{p}_N\rangle \tag{1.114}$$

where $n_{\mathbf{p}}$ denotes the number of particles in momentum state $\mathbf{p}$ in the basis state $|\mathbf{p}_1 \vee \cdots \vee \mathbf{p}_N\rangle$, i.e. the number of $\mathbf{p}_i$s which are equal to $\mathbf{p}$, and the operator $n_{\mathbf{p}} = a_{\mathbf{p}}^{\dagger} a_{\mathbf{p}}$ is referred to as the number operator for state or mode $\mathbf{p}$. In contrast to the case of fermions, the boson number operators have besides the eigenvalue 0 all natural numbers as eigenvalues. As in the case of fermions, the total set of momentum state number operators, $\{n_{\mathbf{p}}\}_{\mathbf{p}}$, thus constitute a complete set of commuting operators giving rise to a representation as discussed in Section 1.5.

Quite analogous to the case of fermions, creation and annihilation operators with respect to position can be introduced. For the N-particle basis state with particles at the indicated locations we have

$$\Phi_{\mathbf{x}_1 \mathbf{x}_2 \ldots \mathbf{x}_N} = \psi_{\mathbf{x}_1}^{\dagger} \psi_{\mathbf{x}_2}^{\dagger} \ldots \psi_{\mathbf{x}_N}^{\dagger} |0\rangle , \tag{1.115}$$

where $|0\rangle$ denotes the vacuum state for the bosons.

Kinematically, independent boson fields are assumed to commute, and bose fields commute with fermi fields (at equal times).

Though already stated, the expression for the resolution of the identity is not of much practical use; the job has been taken over by the creation and annihilation operators, we include it for completeness. The resolution of the identity in the multi-particle space takes the form (and identically for fermions by using the antisymmetric states)

$$\begin{aligned}
1 &= \frac{1}{N!} \sum_{N=0}^{\infty} \sum_{\mathbf{p}_1, \mathbf{p}_2, \ldots, \mathbf{p}_N} |\mathbf{p}_1 \vee \mathbf{p}_2 \vee \cdots \vee \mathbf{p}_N\rangle\langle \mathbf{p}_1 \vee \mathbf{p}_2 \vee \cdots \vee \mathbf{p}_N| \\
&= \sum_{N=0}^{\infty} \frac{1}{n_1! \cdots n_N!} \sum_{\mathbf{p}_1 \leq \mathbf{p}_2 \leq \cdots} |\mathbf{p}_1 \vee \mathbf{p}_2 \vee \cdots \vee \mathbf{p}_N\rangle\langle \mathbf{p}_1 \vee \mathbf{p}_2 \vee \cdots \vee \mathbf{p}_N| ,
\end{aligned} \tag{1.116}$$

where the term $N = 0$ denotes the projection operator onto the vacuum, $|0\rangle\langle 0|$.

Exercise 1.9. Compact notation encompassing both bosons and fermions can sometimes be convenient. Writing an anti-commutator $\{A, B\} \equiv [A, B]_+$, the double valued variable, $s = \pm$, comprises both anti-commutators and commutators, $[A, B]_s$, and distinguishes the two types of quantum statistics. Show that

$$[n_{\mathbf{p}}, a_{\mathbf{p}}^{\dagger}]_s = a_{\mathbf{p}}^{\dagger} , \quad [n_{\mathbf{p}}, a_{\mathbf{p}}]_s = s\, a_{\mathbf{p}} . \tag{1.117}$$

1.4.1 Phonons

The bose field does not occur only in connection with the elementary bosonic particles of the standard model, but can be useful in describing collective phenomena such as the long wave length oscillations of the ions in say a metal or a semiconductor, and we turn to see how this comes about. The Hamiltonian describing the ions of mass M and density n_i in a crystal lattice is given by the kinetic energy term for the ions and an effective ion–ion interaction determined by the screened Coulomb interaction. Expanding the effective ion–ion interaction potential to lowest, quadratic, order, neglecting anharmonic effects and thus only accounting for small oscillations of the ions, the Hamiltonian can be diagonalized by an orthogonal transformation rendering it equivalent to that of a set of independent harmonic oscillators. In this long wave length description, the background dynamics can be described by a continuum limit quantum field, the quantum displacement field, $\mathbf{u}(\mathbf{x})$, a coarse-grained description of the ionic displacements at position $\mathbf{x}$. For longer than interatomic distance, the screened Coulomb interaction is effectively a delta function, $V_{\text{eff}}(\mathbf{x} - \mathbf{x}') = Z^2/2N_0\,\delta(\mathbf{x} - \mathbf{x}')$, and together with the kinetic energy of the background ions, the background Hamiltonian functional valid for small displacements then becomes[18]

$$H_{\text{b}} = \int d\mathbf{x} \left[\frac{1}{2Mn_i} \left(\Pi(\mathbf{x})\right)^2 + \frac{Mn_i c^2}{2} \left(\nabla_{\mathbf{x}} \cdot \mathbf{u}(\mathbf{x})\right)^2 \right], \qquad (1.118)$$

where the components of the momentum density and the displacement field inherit the canonical commutation relations of the ions

$$[\Pi_\alpha(\mathbf{x}),\, \mathbf{u}_\beta(\mathbf{x})] = \frac{\hbar}{i}\,\delta_{\alpha\beta}\,\delta(\mathbf{x} - \mathbf{x}') \qquad (1.119)$$

and the sound velocity is given by

$$c^2 = \frac{Zn}{2N_0 M} = \frac{Z}{3}\frac{m}{M}\,v_{\text{F}}^2 \qquad (1.120)$$

where $n = Zn_i$ is the equilibrium electron density and m the electron mass. We note that the longitudinal sound velocity is typically smaller by a factor of 100 than the Fermi velocity, v_{F}. The continuum description of the oscillations of the in fact discretely located ions appeared because the ions were assumed to exhibit only small oscillations.

The Hamiltonian describing the dynamics of the background is in fact just a set of harmonic oscillators, as obtained by diagonalizing the Hamiltonian. Introducing the normal mode operators

$$a_{\mathbf{k}} = \left(\frac{2Mn_i\omega_{\mathbf{k}}}{\hbar V}\right)^{1/2} \frac{\mathbf{k}\cdot\mathbf{c}_{\mathbf{k}}}{k}, \quad \text{where} \quad \mathbf{c}_{\mathbf{k}} + \mathbf{c}_{-\mathbf{k}}^\dagger = \int_V d\mathbf{x}\, e^{-i\mathbf{x}\cdot\mathbf{k}}\,\mathbf{u}(\mathbf{x}) \qquad (1.121)$$

---

[18] For details of these arguments, starting from the quantum mechanics of the individual ions and then taking the continuum limit, we refer the reader to, for example, chapter 10 of reference [1].

or

$$\mathbf{u}(\mathbf{x}) = \frac{1}{V} \sum_{\mathbf{k} \neq 0, |\mathbf{k}| < k_D} \left(\mathbf{c}_{\mathbf{k}} \, e^{i\mathbf{k} \cdot \mathbf{x}} + \mathbf{c}^{\dagger}_{-\mathbf{k}} \, e^{-i\mathbf{k} \cdot \mathbf{x}} \right) \tag{1.122}$$

the background Hamiltonian becomes the free longitudinal phonon Hamiltonian

$$H_{\mathrm{ph}} = H_{\mathrm{b}} = \sum_{|\mathbf{k}| \leq k_D} \hbar \omega_{\mathbf{k}} \left(a^{\dagger}_{\mathbf{k}} a_{\mathbf{k}} + \frac{1}{2} \right) \tag{1.123}$$

with linear dispersion $\omega_{\mathbf{k}} = c \, |\mathbf{k}|$, and the operators satisfy the harmonic oscillator normal mode commutation relations

$$[a_{\mathbf{k}}, a^{\dagger}_{\mathbf{k}'}] = \delta_{\mathbf{k}, \mathbf{k}'} \quad , \quad [a^{\dagger}_{\mathbf{k}}, a^{\dagger}_{\mathbf{k}'}] = 0 \quad , \quad [a_{\mathbf{k}}, a_{\mathbf{k}'}] = 0 \tag{1.124}$$

inherited from the canonical commutation relations for the position and momentum operators of the individual ions. A quantum of an oscillator, a quantized sound mode, is referred to as a phonon. In the Debye model, the lattice vibrations are assumed to have linear dispersion all the way to the cut-off wave vector k_D.

However, instead of the above quantum mechanical argument, we can also here take the opportunity to discuss the classical field theory of oscillations in an isotropic elastic medium, and then obtain the corresponding quantum field theory by quantizing the dynamics of the normal modes. This trick can then be elevated to give us the quantum theory of the electromagnetic field.

1.4.2   Quantizing a classical field theory

As an example of quantizing a classical field theory we consider an elastic isotropic medium of volume V specified by its longitudinal sound velocity c and mass density ρ. In terms of the displacement field, $\mathbf{u}(\mathbf{x}, t)$, describing the displacement of the background matter at position $\mathbf{x}$ at time t, we have for small displacements

$$\frac{\delta n_{\mathrm{b}}(\mathbf{x}, t)}{n_{\mathrm{i}}} = - \nabla \cdot \mathbf{u}(\mathbf{x}, t) \, , \tag{1.125}$$

where $\delta n_{\mathrm{b}}(\mathbf{x}, t)$ is the deviation of the medium density from the average density n_{i}. Newton's equation and the continuity equation leads for small $\delta n_{\mathrm{b}}(\mathbf{x}, t)$ to this density disturbance satisfying the wave equation

$$\left(\Delta_{\mathbf{x}} - \frac{1}{c^2} \frac{\partial^2}{\partial t^2} \right) \delta n_{\mathrm{b}}(\mathbf{x}, t) = 0 \tag{1.126}$$

or the dynamics of the elastic medium is equivalently, through the principle of least action, described by the Lagrange functional

$$L[\mathbf{u}, \dot{\mathbf{u}}] = \frac{\rho}{2} \int d\mathbf{x} \left[\left(\frac{\partial \mathbf{u}(\mathbf{x}, t)}{\partial t} \right)^2 - c^2 \left(\nabla \cdot \mathbf{u}(\mathbf{x}, t) \right)^2 \right] \, . \tag{1.127}$$

In accordance with the assumed isotropy of the elastic medium, it exhibits no shear or vorticity, sustaining only longitudinal waves

$$\nabla_{\mathbf{x}} \times \mathbf{u}(\mathbf{x}, t) = \mathbf{0} \quad , \quad \mathbf{k} \times \mathbf{u}_{\mathbf{k}}(t) = \mathbf{0} \quad , \quad \mathbf{k} \parallel \mathbf{u}_{\mathbf{k}}(t) \, . \tag{1.128}$$

The classical equations of motion for the displacement field of the medium, the Lagrange field equations following from Hamilton's principle of least action, can be expressed in terms of the displacement field

$$\Box\, \mathbf{u}(\mathbf{x},t) = 0 \qquad , \qquad \ddot{\mathbf{u}}_{\mathbf{k}}(t) + c^2 k^2 \mathbf{u}_{\mathbf{k}}(t) = 0 \tag{1.129}$$

specified by the d'Alembertian

$$\Box = \left(\triangle_{\mathbf{x}} - \frac{1}{c^2}\frac{\partial^2}{\partial t^2}\right). \tag{1.130}$$

The solution is, for example, the running normal mode expansion with periodic boundary conditions

$$\mathbf{u}(\mathbf{x},t) = \frac{1}{V}\sum_{\mathbf{k}\neq 0}[\mathbf{c}_{\mathbf{k}}(t)\,e^{i\mathbf{k}\cdot\mathbf{x}} + \mathbf{c}_{\mathbf{k}}^*(t)\,e^{-i\mathbf{k}\cdot\mathbf{x}}] \quad , \quad \mathbf{c}_{\mathbf{k}}(t) = \mathbf{c}_{\mathbf{k}}\,e^{-i\omega_{\mathbf{k}}t} \quad , \quad \omega_{\mathbf{k}} = c|\mathbf{k}| \tag{1.131}$$

or equivalently for the Fourier components

$$\mathbf{u}_{\mathbf{k}}(t) = \mathbf{c}_{\mathbf{k}}(t) + \mathbf{c}_{-\mathbf{k}}^*(t) , \tag{1.132}$$

as the vector field $\mathbf{u}(\mathbf{x},t)$ is real.

Introducing the momentum density of the medium[19]

$$\Pi(\mathbf{x},t) \equiv \rho\,\frac{\partial \mathbf{u}(\mathbf{x},t)}{\partial t} = \frac{\delta L[\mathbf{u},\dot{\mathbf{u}}]}{\delta \dot{\mathbf{u}}(\mathbf{x},t)} \tag{1.133}$$

and recalling that the Hamilton and Lagrange functions are related through a Legendre transformation (see, for example, Eq. (3.46) or Eq. (A.10)), we have the Hamilton functional for the dynamics of the elastic medium

$$H_{\mathrm{b}} = \int d\mathbf{x}\left[\frac{1}{\rho}\left(\Pi(\mathbf{x},t)\right)^2 + \frac{\rho c^2}{2}\left(\nabla\cdot\mathbf{u}(\mathbf{x},t)\right)^2\right]. \tag{1.134}$$

Introducing

$$a_{\mathbf{k}}(t) = \left(\frac{\rho\,\omega_{\mathbf{k}}}{\hbar V}\right)^{1/2}\frac{\mathbf{k}\cdot\mathbf{c}_{\mathbf{k}}(t)}{k} \tag{1.135}$$

we obtain

$$H_{\mathrm{b}} = \frac{1}{2}\sum_{\mathbf{k}}\hbar\omega_{\mathbf{k}}(a_{\mathbf{k}}(t)\,a_{\mathbf{k}}^*(t) + a_{\mathbf{k}}^*(t)\,a_{\mathbf{k}}(t)). \tag{1.136}$$

The classical theory is now quantized by letting the normal mode expansion coefficients become operators, $a_{\mathbf{k}}^*(t) \to a_{\mathbf{k}}^\dagger(t)$, which satisfy the harmonic oscillator creation and annihilation equal time commutation relations[20]

$$[a_{\mathbf{k}}(t), a_{\mathbf{k}'}^\dagger(t)] = \delta_{\mathbf{k},\mathbf{k}'} \quad , \quad [a_{\mathbf{k}}^\dagger(t), a_{\mathbf{k}'}^\dagger(t)] = 0 \quad , \quad [a_{\mathbf{k}}(t), a_{\mathbf{k}'}(t)] = 0 . \tag{1.137}$$

---

[19]Functional differentiation is discussed in section 9.2.1.
[20]This so-called second quantization procedure is discussed further in Section 3.2.

There is for the present purpose nothing conspicuous about this quantization procedure, as it gives the same Hamiltonian as the one derived quantum mechanically in the previous section, where these commutation relations are directly inherited from the canonical commutation relations for the position and momentum operators of the individual ions of the material. The continuum description was appropriate since only long wave length oscillations, long compared with the inter-ionic distance, were of relevance.

In classical physics, kinematics and dynamics of physical quantities are expressed in terms of the same quantities. Kinematics, i.e. the description of the physical state of an object is, in classical physics, intuitive, described in terms of the position and velocity of an object, $(\mathbf{x}(t), \mathbf{v}(t))$: we can point to the position of an object and from its motion construct its velocity. The classical dynamics is expressed in terms of the time dependence of the positions and velocities (or momenta) of the concerned objects, say in terms of Hamilton's equations. In quantum mechanics, dynamics and kinematics can be separated, as is the case in the Schrödinger picture, where the dynamics is carried by a state vector and the physical properties of a system by operators. When quantizing a classical theory, the Hamiltonian is thus obtained in the so-called Heisenberg picture, where the operators representing physical quantities are time dependent and also carrying the dynamics of the system. The quantized elastic medium Hamiltonian is therefore expressed in the Heisenberg picture, and the Hamiltonian in the Schrödinger picture is here obtained simply by removing the time variable, recalling $a_{\mathbf{k}}(t) = a_{\mathbf{k}} e^{-i\omega_{\mathbf{k}}t}$, i.e. we implement the commutation relations for the $a_{\mathbf{k}}$ quantities, and we recover the expression in Eq. (1.123) for the phonon Hamiltonian. The Schrödinger and Heisenberg pictures are discussed in detail in Section 3.1.2.

A similar prescription in fact works for quantizing the free Maxwell equations of classical electrodynamics, producing the quantum theory of electromagnetism, quantum electrodynamics or QED, as discussed in Exercise 1.10, where the quanta of the field are Einstein's photons. In the case of phonons, the quanta describe the quantum states of small oscillations of an assembly of atoms as described by the Schrödinger equation. However, for the case of electromagnetism, the photons do not refer to any dynamics of a medium. The non-relativistic field theory of interacting electrons, described by the Hamiltonian for Coulomb interaction, Eq. (1.53), is only a limiting case of QED, but the one relevant for the dynamics of, say, electrons in solids. In the next chapter we shall therefore take the approach to non-relativistic quantum field theory which starts from the known interactions of an N-particle system and then construct their forms on the multi-particle state spaces. However, when we eventually consider the dynamics of a quantum field theory in terms of its Feynman diagrammatics in Chapter 9, all theories appear on an equal footing, particulars are just embedded in the various indices possibilities for propagators and vertices.

Exercise 1.10. Maxwell's equations, the classical equations of motion for the electromagnetic field, can for vacuum, the free theory, be obtained from Hamilton's principle of least action with the Lagrange density (SI units are employed)

$$\mathcal{L} = \frac{4\pi}{2}\left(\epsilon_0\,\mathbf{E}^2 + \mu_0\,\mathbf{B}^2\right). \tag{1.138}$$

Representing the electric field solely in terms of a vector potential, $\varphi = 0$, and choosing the Coulomb or radiation gauge, $\nabla \cdot \mathbf{A} = 0$, show that the Lagrange density becomes

$$\mathcal{L} = \frac{4\pi}{2} \left(\epsilon_0 \dot{\mathbf{A}}^2 + \mu_0 (\nabla \times \mathbf{A})^2 \right) \tag{1.139}$$

and the Euler–Lagrange equation becomes

$$\left(\nabla^2 - \frac{1}{c^2} \frac{\partial^2}{\partial t^2} \right) \mathbf{A}(\mathbf{x}, t) = 0 \,, \tag{1.140}$$

where $c = 1/\sqrt{\mu_0 \epsilon_0}$ denotes the velocity of light. Note that manifest Lorentz and gauge invariance have been sacrificed in the Coulomb gauge. Expressing the solution in terms of running normal modes, obtain that the Hamiltonian for free photons has the form

$$H_{\text{ph}} = \int \frac{d\mathbf{k}}{(2\pi)^3} \sum_{p=1,2} \hbar c |\mathbf{k}| \, a_{\mathbf{k}p}^\dagger \, a_{\mathbf{k}p} \,, \tag{1.141}$$

where since we are in the transverse gauge two perpendicular polarizations occur, and the creation and annihilation operators for photons with wave vectors $\mathbf{k}$ and polarizations p satisfy the commutation relations

$$[a_{\mathbf{k}p}, a_{\mathbf{k}'p'}^\dagger] = (2\pi)^3 \, \delta(\mathbf{k} - \mathbf{k}') \, \delta_{pp'} \,. \tag{1.142}$$

1.5   Occupation number representation

In this section we make a side remark which is not necessary for understanding any of the further undertakings; we just include it for its historical relevance, since this is how quantum field theory traditionally was presented, originating in the treatment of the electromagnetic field and emulated for fermions in many textbooks.

The operator

$$N_\lambda = a_\lambda^\dagger a_\lambda \tag{1.143}$$

counts, as noted in the previous sections, the number of particles in state λ in any N-particle basis state expressed in terms of these states

$$\Phi_{\lambda_1 \lambda_2 \dots \lambda_N} = a_{\lambda_1}^\dagger \, a_{\lambda_2}^\dagger \cdots a_{\lambda_N}^\dagger \, |0\rangle \,. \tag{1.144}$$

The set of numbers, $\{n_{\lambda_i}\}_i$, counted in the basis states by the set of number operators $\{N_{\lambda_i}\}_i$ therefore uniquely characterizes the basis states, and the set of these operators therefore forms a complete set of commuting operators in the corresponding multi-particle space, symmetric or antisymmetric. They therefore give rise to a representation, the occupation number representation. As a basis set in the multi-particle space, we can therefore equally well use the occupation number representation, where the orthonormal basis states are defined by this complete set of commuting operators, and simply are labeled by stating how many particles are present in any of the single particle states λ, $|n_{\lambda_1}, n_{\lambda_2}, n_{\lambda_3}, \ldots\rangle$. These states are related to our previous basis states according to

$$|n_{\lambda_1}, n_{\lambda_2}, n_{\lambda_3}, \ldots\rangle \equiv \frac{1}{\sqrt{n_1! n_2! n_3! \cdots}} |\lambda_1 \diamond \cdots \diamond \lambda_1 \diamond \lambda_2 \diamond \cdots \diamond \lambda_2 \diamond \lambda_3 \diamond \cdots \diamond \lambda_3 \diamond \cdots\rangle,$$

(1.145)

where n_i is the number of times state λ_i occurs, and $\diamond$ stands for $\vee$ or $\wedge$ for the bose or fermi case, respectively. In the fermion case, each λ_i can of course at most occur once, i.e. $n_i = 0$ or $n_i = 1$.

We note that if, as in the following, the λ-label refers to the single-particle energy states, the sum of single-particle energies

$$E_0(\{n_\lambda\}_\lambda) = \sum_\lambda \epsilon_\lambda n_\lambda$$

(1.146)

of an assembly of identical particle is the energy eigenvalue of the free Hamiltonian

$$H_0 = \sum_\lambda \epsilon_\lambda N_\lambda$$

(1.147)

in state $|n_{\lambda_1}, n_{\lambda_2}, n_{\lambda_3}, \ldots\rangle$, i.e. the single-particle or free Hamiltonian can be expressed in terms of the number operators.

The occupation number representation is not necessary, since the introduction of the workings of the creation and annihilation operators as done in the previous sections is easier. However, one notices that, for the case of bosons, the creation operator operates on an occupation number eigenstate according to[21]

$$a^\dagger_{\lambda_i} |n_{\lambda_1}, n_{\lambda_2}, n_{\lambda_3}, \ldots\rangle = \sqrt{n_i + 1} \, |n_{\lambda_1}, n_{\lambda_2}, n_{\lambda_3}, \ldots, n_{\lambda_i} + 1, \ldots\rangle \quad (1.148)$$

and the annihilation operator according to

$$a_{\lambda_i} |n_{\lambda_1}, n_{\lambda_2}, n_{\lambda_3}, \ldots\rangle = \sqrt{n_i} \, |n_{\lambda_1}, n_{\lambda_2}, n_{\lambda_3}, \ldots, n_{\lambda_i} - 1, \ldots\rangle \quad (1.149)$$

and we realize that, in the bose case, creation and annihilation operators act analogously to creation and annihilation operators for a harmonic oscillator. The quanta in a harmonic oscillator thus have an equivalent interpretation in terms of particles occupying the energy states of a harmonic oscillator. Here emerges the reason for the success of Einstein's revolutionary interpretation of Planck's lumps of energy in the electromagnetic field as particles. This is how the quanta of the electromagnetic field oscillators are interpreted as particles, viz. photons. Vice versa, the collective small oscillations of lattice ions performed by the atoms or ions in a solid can be represented in terms of harmonic oscillators, the so-called phonons as described in Section 1.4.1, or equivalently has identical properties to particles obeying bose statistics. For bosons such as photons, i.e. in quantum optics, the occupation, or just number representation, is of course of fundamental relevance.

---

[21] For fermions additional sign factors appear as discussed in Exercise 1.11.

Exercise 1.11. Consider the case of fermions and define the basis states in the number representation in terms of the vacuum state economically according to

$$|n_1, n_2, n_3, \ldots, n_\infty\rangle = (a_1^\dagger)^{n_1} (a_2^\dagger)^{n_2} (a_3^\dagger)^{n_3} \ldots (a_\infty^\dagger)^{n_\infty} |0\rangle \qquad (1.150)$$

where the n_is can take on the values 0 or 1.

Show that, for $n_s = 1$,

$$a_s |n_1, n_2, n_3, \ldots, n_\infty\rangle = (-1)^{S_s} (a_1^\dagger)^{n_1} \ldots (a_s a_s^\dagger) \ldots (a_\infty^\dagger)^{n_\infty} |0\rangle, \qquad (1.151)$$

where $S_s = n_1 + n_2 + \cdots + n_{s-1}$ counts how many anti-commutations it takes to move a_s to its place displayed on the right-hand side. If $n_s = 0$, the annihilation operator a_s could be moved all the way to act on the vacuum, producing the zero-vector.

Use the above observations to show that

$$a_s |\ldots, n_s, \ldots\rangle = (-1)^{S_s} \sqrt{n_s} \, |\ldots, n_s - 1, \ldots\rangle \qquad (1.152)$$

and

$$a_s^\dagger |\ldots, n_s, \ldots\rangle = (-1)^{S_s} \sqrt{n_s + 1} \, |\ldots, n_s + 1, \ldots\rangle \qquad (1.153)$$

and thereby

$$N_s |\ldots, n_s, \ldots\rangle = n_s |\ldots, n_s, \ldots\rangle. \qquad (1.154)$$

Here we have used modulo one-notation in the n_s-state labeling: $1 + 1 = 0$ and $0 - 1 = 0$. These relations are therefore similar to those for bosons except that obnoxious sign factors occur owing to the Fermi statistics. The occupation number representation for fermions is therefore not attractive as the *wedge* is not explicit.

1.6 Summary

In this chapter we have considered the quantum mechanical description of systems which can be in superposition of states with an arbitrary content of particles. To deal with such situations, endemic to relativistic quantum theory, quantum fields were introduced, describing the creation and annihilation of particles. The states in the multi-particle state space could be simply expressed by operating with the creation field on the vacuum state, the state corresponding to absence of particles. The whole kinematics of a many-body system is thus expressed in terms of just these two operators. Our first encounter with a quantum field theory was the case of quantized lattice vibrations, phonons, and equivalent to the quantum mechanics of a set of harmonic oscillators, and the archetype resulting from the scheme of quantizing a classical field theory. The scheme was then exploited to quantize the electromagnetic field where the quanta of the field, the photons, were *particles* with two internal spin or polarization or helicity states. In the case of phonons, the continuum quantum field description was only an appropriate long wave length description, whereas in the case of photons the quantum field theory is truly a description of a system with an infinite number of degrees of freedom. In the next chapter we shall consider non-relativistic many-body systems, and the task is therefore not to assess the form of the Hamiltonian, but the more mundane task of elevating a known N-particle Hamiltonian to its form on the multi-particle state space.

As we develop the various topics of the book the following conclusion will emerge: quantum fields are the universal vehicle for describing fluctuations whatever their nature, being quantum or thermal or purely classical stochastic.

2

Operators on the multi-particle state space

A physical property A is characterized by the total set of possible values $\{a\}_a$ it can exhibit. In quantum mechanics, the same information is expressed by the operator representing the physical quantity in question, expressed by the weighted sum of projection operators

$$\hat{A} = \sum_a a \, |a\rangle\langle a| \tag{2.1}$$

weighted by the eigenvalues of the operator in question.[1] We now want to find the expression for the operator on the multi-particle space whose restriction to any N-particle subspace reduces to the operator in question for the system consisting of N identical bosons or fermions. We show that all operators for an N-particle system are lifted very simply to the multi-particle space through an expression in terms of the creation and annihilation operators in a way analogous to the *bra* and *ket* expression in Eq. (2.1).

2.1 Physical observables

In quantum mechanics, physical properties are represented by operators, say momentum by an operator denoted $\hat{\mathbf{p}}$, and for an N-particle system their total momentum is represented by the operator in Eq. (1.25), denoted $\hat{\mathbf{P}}_N$. We now want to find the expression for the operator on the multi-particle space whose restriction to any N-particle subspace reduces to the total momentum operator for the N identical particles. In the following we consider the case of fermions; as usual for kinematics the case of bosons is a trivial corollary. We have for the operation of the total momentum

---

[1] For details on the construction of operators from values of physical outcomes, we refer to chapter 1 of reference [1].

operator on a general antisymmetric N-particle basis state

$$\hat{\mathbf{P}}_N \, |\lambda_1 \wedge \lambda_2 \wedge \cdots \wedge \lambda_N\rangle \;=\; \left(\sum_{i=1}^{N} \hat{\mathbf{p}}_i\right) \frac{1}{\sqrt{N!}} \sum_P (-1)^{\zeta_P} |\lambda_{P_1}\rangle |\lambda_{P_2}\rangle \cdots |\lambda_{P_N}\rangle$$

$$=\; \sum_P (-1)^{\zeta_P} \Big((\hat{\mathbf{p}}|\lambda_{P_1}\rangle) |\lambda_{P_2}\rangle \cdots |\lambda_{P_N}\rangle$$

$$+\; |\lambda_{P_1}\rangle (\hat{\mathbf{p}}|\lambda_{P_2}\rangle) \cdots |\lambda_{P_N}\rangle + \cdots$$

$$+\; |\lambda_{P_1}\rangle |\lambda_{P_2}\rangle \cdots (\hat{\mathbf{p}}|\lambda_{P_N}\rangle) \Big) . \tag{2.2}$$

Presently we are discussing the one-body momentum operator

$$\hat{\mathbf{p}} \;=\; \int d\mathbf{p} \, \mathbf{p} \, |\mathbf{p}\rangle\langle\mathbf{p}| \tag{2.3}$$

but it is in fact appropriate first to access how the general one-body transition operator $|\mathbf{p}'\rangle \langle \mathbf{p}|$ is implemented, and thereby the whole operator algebra.[2]

For a general one-body operator, $\hat{f}^{(1)}$, the corresponding operator for the N-particle system

$$\hat{F}_N^{(1)} \;=\; \sum_{i=1}^{N} \hat{f}_i^{(1)} \tag{2.4}$$

operates according

$$\hat{F}_N^{(1)} |\lambda_1 \wedge \lambda_2 \wedge \cdots \wedge \lambda_N\rangle \;=\; |f^{(1)} \lambda_1 \wedge \lambda_2 \wedge \cdots \wedge \lambda_N\rangle$$

$$+\; |\lambda_1 \wedge f^{(1)} \lambda_2 \wedge \cdots \wedge \lambda_N\rangle + \cdots$$

$$+\; |\lambda_1 \wedge \lambda_2 \wedge \cdots \wedge f^{(1)} \lambda_N\rangle , \tag{2.5}$$

where $f^{(1)} \lambda$ labels the state which $\hat{f}^{(1)}$ maps the state labeled by λ into

$$|f^{(1)} \lambda\rangle \;=\; \hat{f}^{(1)} |\lambda\rangle . \tag{2.6}$$

The operator, $\hat{F}_N^{(\mathbf{pp}')}$, on the N-particle space corresponding to the one-body operator $\hat{f}^{(1)} = |\mathbf{p}'\rangle\langle\mathbf{p}|$ thus operates according to

$$\hat{F}_N^{(\mathbf{pp}')} |\lambda_1 \wedge \lambda_2 \wedge \cdots \wedge \lambda_N\rangle \;=\; \langle\mathbf{p}|\lambda_1\rangle \, |\mathbf{p}' \wedge \lambda_2 \wedge \cdots \wedge \lambda_N\rangle$$

$$+\; \langle\mathbf{p}|\lambda_2\rangle \, |\lambda_1 \wedge \mathbf{p}' \wedge \lambda_3 \wedge \cdots \wedge \lambda_N\rangle + \cdots$$

$$+\; \langle\mathbf{p}|\lambda_N\rangle \, |\lambda_1 \wedge \lambda_2 \wedge \cdots \wedge \lambda_{N-1} \wedge \mathbf{p}'\rangle. \tag{2.7}$$

---

[2]We are here guided by the knowledge that a *bra* has the feature of an annihilation operator and a *ket* has the feature of a creation operator, and the transition operators constitute the basis of the measurement algebra of a quantum system, i.e. completeness of a basis in the state space is expressed, in the λ-representation, by the identity $\sum_\lambda |\lambda\rangle\langle\lambda| = 1$, and the set $\{|\lambda\rangle\langle\lambda'|\}_{\lambda,\lambda'}$ is a basis in the dual space, the space of linear operators on the state space. For details see chapter 1 in reference [1].

Since by antisymmetrization

$$(-1)^{n-1} |\mathbf{p}' \wedge \lambda_1 \wedge \cdots \wedge (\text{ no } \lambda_n) \wedge \cdots \wedge \lambda_N\rangle$$

$$= |\lambda_1 \wedge \cdots \wedge \lambda_{n-1} \wedge \mathbf{p}' \wedge \lambda_{n+1} \wedge \cdots \wedge \lambda_N\rangle \qquad (2.8)$$

we have

$$\hat{F}_N^{(\mathbf{p}\mathbf{p}')}|\lambda_1 \wedge \cdots \wedge \lambda_N\rangle = \sum_{n=1}^{N} \langle \mathbf{p}|\lambda_n\rangle(-1)^{n-1}|\lambda_1 \wedge \cdots (\text{no } \lambda_n \text{ instead } \mathbf{p}') \cdots \wedge \lambda_N\rangle,$$

$$(2.9)$$

but according to Eq. (1.77) this is the same state which is obtained when operating with the operator $a_{\mathbf{p}'}^\dagger a_{\mathbf{p}}$ so that

$$\hat{F}_N^{(\mathbf{p}\mathbf{p}')}|\lambda_1 \wedge \cdots \wedge \lambda_N\rangle = a_{\mathbf{p}'}^\dagger a_{\mathbf{p}}|\lambda_1 \wedge \cdots \wedge \lambda_N\rangle. \qquad (2.10)$$

We have thus established how to implement a one-body operator onto the multi-particle space so that its restriction to any N-particle subspace is the corresponding N-particle operator. The implementation for bosons is identical to the above, as usual the derivation is completely analogous, in fact simpler since no minus signs occur.

There is of course nothing special about momentum labels; the formal machinery, i.e. the combinatorics, works for any set of one-particle states, say labeled by μ, so that corresponding to the one-particle operator $|\mu_2\rangle\langle\mu_1|$ corresponds the operator $F^{(1)}$ in the multi-particle space

$$F^{(1)} = a_{\mu_2}^\dagger a_{\mu_1}. \qquad (2.11)$$

An arbitrary one-particle operator has, in an arbitrary basis, the form

$$\hat{f}^{(1)} = \sum_{\lambda,\lambda'} |\lambda\rangle\langle\lambda|\hat{f}^{(1)}|\lambda'\rangle\langle\lambda'| \qquad (2.12)$$

and by linearity the corresponding operator $F^{(1)}$ in the multi-particle space is thus

$$F^{(1)} = \sum_{\lambda,\lambda'} \langle\lambda|\hat{f}^{(1)}|\lambda'\rangle a_\lambda^\dagger a_{\lambda'}. \qquad (2.13)$$

We note that if $\hat{f}^{(1)}$ is hermitian in the one-particle state space, as is $F^{(1)}$ in the multi-particle state space.

The total momentum operator $\mathbf{P}$ in the multi-particle space is thus

$$\mathbf{P} = \int d\mathbf{p}\, \mathbf{p}\, a_{\mathbf{p}}^\dagger a_{\mathbf{p}} = \int d\mathbf{x}\, \psi^\dagger(\mathbf{x}) \frac{\hbar}{i}\nabla_{\mathbf{x}}\, \psi(\mathbf{x}) \qquad (2.14)$$

expressed in either the momentum or position representation of the field.

Exercise 2.1. Show that the commutator of the total momentum operator and the field is

$$[\psi(\mathbf{x}), \mathbf{P}] = \frac{\hbar}{i} \nabla_\mathbf{x} \psi(\mathbf{x}) \tag{2.15}$$

or equivalently

$$\psi(\mathbf{x}) = e^{-\frac{i}{\hbar} \mathbf{x} \cdot \mathbf{P}} \psi(\mathbf{0}) e^{\frac{i}{\hbar} \mathbf{x} \cdot \mathbf{P}} . \tag{2.16}$$

We now have the prescription for mapping any one-particle operator into the corresponding operator on the multi-particle space. For a non-relativistic particle of mass m in a potential V the Hamiltonian is

$$\hat{H} = \frac{\hat{\mathbf{p}}^2}{2m} + V(\hat{\mathbf{x}}, t) = \frac{\hat{\mathbf{p}}^2}{2m} + \int d\mathbf{x} \, \hat{n}(\mathbf{x}) V(\mathbf{x}, t) , \tag{2.17}$$

where in the second equality we have introduced the probability density operator for a particle, $\hat{n}(\mathbf{x}) = \delta(\hat{\mathbf{x}} - \mathbf{x})$ (recall Section 1.2.4). In the position representation the Hamiltonian has the matrix elements

$$\langle \mathbf{x} | \left(\frac{\hat{\mathbf{p}}^2}{2m} + V(\hat{\mathbf{x}}, t) \right) | \mathbf{x}' \rangle = \left(\frac{1}{2m} \left(\frac{\hbar}{i} \right)^2 \frac{\partial^2}{\partial \mathbf{x}^2} + V(\mathbf{x}, t) \right) \delta(\mathbf{x} - \mathbf{x}') \tag{2.18}$$

and according to Eq. (2.13) the corresponding Hamiltonian on the multi-particle space becomes

$$H = \int d\mathbf{x} \, \psi^\dagger(\mathbf{x}) \left(\frac{1}{2m} \left(\frac{\hbar}{i} \right)^2 \frac{\partial^2}{\partial \mathbf{x}^2} + V(\mathbf{x}, t) \right) \psi(\mathbf{x}) . \tag{2.19}$$

We note that the single particle properties can be expressed in terms of the occupation number operators. For the case of energy, the energy eigenstates should be used, recall Eq. (1.147) and see Exercise 2.2, and of course for the case of momentum, Eq. (2.14), the reference states should be the momentum states.

Exercise 2.2. Show that the kinetic energy operator for an assembly of non-relativistic free identical particles of mass m

$$H = \int d\mathbf{x} \, \psi^\dagger(\mathbf{x}) \left(\frac{1}{2m} \left(\frac{\hbar}{i} \right)^2 \frac{\partial^2}{\partial \mathbf{x}^2} \right) \psi(\mathbf{x}) \tag{2.20}$$

in the momentum and energy representation has the form

$$H = \sum_\mathbf{P} \epsilon_\mathbf{p} a_\mathbf{p}^\dagger a_\mathbf{p} = \sum_\mathbf{P} \epsilon_\mathbf{p} n_\mathbf{p} , \tag{2.21}$$

where $\epsilon_\mathbf{p} = \mathbf{p}^2/2m$ is the kinetic energy of the free particle with momentum $\mathbf{p}$. The *sum* over momenta occurs, one momentum state per momentum volume $\Delta \mathbf{p} = (2\pi\hbar)^3/V$ in three dimensions, as the particles are assumed enclosed in a box of volume V.

Exercise 2.3. Show that the average value of the kinetic energy operator for an electron gas consisting of N electrons in the ground state, i.e. the energy of N free electrons in the ground state, is

$$\sum_{\mathbf{p}} \epsilon_{\mathbf{p}} \langle a_{\mathbf{p}}^{\dagger} a_{\mathbf{p}} \rangle = \frac{3}{5} \frac{p_{\mathrm{F}}^2}{2m} N, \qquad (2.22)$$

where p_{F} is the Fermi momentum, the radius of the sphere of occupied momentum states (in three dimensions $p_{\mathrm{F}} = \hbar(3\pi^2 n)^{1/3}$, where $n = N/V$ is the density of the electrons).

Exercise 2.4. Show that the vacuum state is non-degenerate and uniquely characterized by all the eigenvalues of the state number operators $n_{\mathbf{p}}$ being zero.

Exercise 2.5. For the quantities discussed so far, a possible (say) spin degree of freedom of the particles did not have its two spin states discriminated, and its presence was left implicit in the notation. To consider a situation where spin states needs to be specified explicitly, consider (say) electrons interacting with the magnetic moments of impurities. The interaction of an electron interacting with the magnetic moments of impurities is

$$V_{\alpha,\alpha'}^{(\mathrm{sf})}(\mathbf{x}) = \sum_{a} u(\mathbf{x} - \mathbf{x}_a) \, \mathbf{S}_a \cdot \boldsymbol{\sigma}_{\alpha,\alpha'} \qquad (2.23)$$

where $\mathbf{x}_a$ is the location of a magnetic impurity with spin $\mathbf{S}_a$ and $\boldsymbol{\sigma}$ represents the electron spin. In the multi-particle space, the interaction of the impurity spins and the electrons thus becomes

$$V_{\mathrm{sf}} = \sum_{a} \int d\mathbf{x} \, u(\mathbf{x} - \mathbf{x}_a) \, \psi_{\alpha}^{\dagger}(\mathbf{x}) \, \mathbf{S}_a \cdot \boldsymbol{\sigma}_{\alpha,\alpha'} \, \psi_{\alpha'}(\mathbf{x}) . \qquad (2.24)$$

Show it can be rewritten in the form

$$V_{\mathrm{sf}} = \sum_{a} \int d\mathbf{x} \, u(\mathbf{x} - \mathbf{x}_a) \left(S^{-} \psi_{\uparrow}^{\dagger}(\mathbf{x}) \psi_{\downarrow}(\mathbf{x}) + S^{+} \psi_{\downarrow}^{\dagger}(\mathbf{x}) \psi_{\uparrow}(\mathbf{x}) \right.$$

$$\left. + \; S^{z} \left(\psi_{\uparrow}^{\dagger}(\mathbf{x}) \psi_{\uparrow}(\mathbf{x}) - \psi_{\downarrow}^{\dagger}(\mathbf{x}) \psi_{\downarrow}(\mathbf{x}) \right) \right), \qquad (2.25)$$

where $S^{\pm} = S^x \pm i S^y$.

Density and current density operators are important as well as their coupling to external fields, and we now turn to their construction in the multi-particle state space.

2.2 Probability density and number operators

The one-particle probability density operator (recall Section 1.2.4)

$$\hat{n}(\mathbf{x}) = \delta(\hat{\mathbf{x}} - \mathbf{x}) = |\mathbf{x}\rangle\langle\mathbf{x}| \qquad (2.26)$$

maps according to the general prescription, Eq. (2.13), to the operator on the multi-particle space

$$n(\mathbf{x}) = \int d\mathbf{x}_1 \int d\mathbf{x}_2 \, \langle \mathbf{x}_1 | \hat{n}(\mathbf{x}) | \mathbf{x}_2 \rangle \, \psi^\dagger(\mathbf{x}_1) \, \psi(\mathbf{x}_2) \tag{2.27}$$

and therefore the probability density operator in the multi-particle space is

$$n(\mathbf{x}) = \psi^\dagger(\mathbf{x}) \, \psi(\mathbf{x}) \ . \tag{2.28}$$

By construction this operator reduces in each N-particle subspace to the N-particle density operator

$$\hat{n}(\mathbf{x}) = \sum_{i=1}^{N} \delta(\hat{\mathbf{x}}_i - \mathbf{x}) \ . \tag{2.29}$$

The identity operator in the one-particle state space[3]

$$\hat{I} = \int d\mathbf{x} \, |\mathbf{x}\rangle\langle\mathbf{x}| = \int d\mathbf{x} \, \hat{n}(\mathbf{x}) \tag{2.30}$$

becomes in the multi-particle space according to the general prescription, Eq. (2.13), the operator

$$N = \int d\mathbf{x} \, \psi^\dagger(\mathbf{x}) \, \psi(\mathbf{x}) = \int d\mathbf{p} \, a_\mathbf{p}^\dagger \, a_\mathbf{p} \ , \tag{2.31}$$

which is the operator that counts the total number of particles in each N-particle state, the total number operator. For example, according to the equation analogous to Eq. (1.77) but in the position representation

$$\begin{aligned}
N \, |\mathbf{x}_1 \wedge \mathbf{x}_2\rangle &= \int d\mathbf{x} \, \psi^\dagger(\mathbf{x}) \, \psi(\mathbf{x}) \, |\mathbf{x}_1 \wedge \mathbf{x}_2\rangle \\[2mm]
&= \int d\mathbf{x} \, (\langle\mathbf{x}|\mathbf{x}_1\rangle \, |\mathbf{x} \wedge \mathbf{x}_2\rangle - \langle\mathbf{x}|\mathbf{x}_2\rangle \, |\mathbf{x} \wedge \mathbf{x}_1\rangle) \\[2mm]
&= 2 \, |\mathbf{x}_1 \wedge \mathbf{x}_2\rangle \ .
\end{aligned} \tag{2.32}$$

Or more efficiently by just using the basic anti-commutation or commutation relation for the fields, we obtain by consecutively anti-commuting or commuting, depending on the particles being fermions or bosons, the $\psi(\mathbf{x})$-operator in the number operator to the right and eventually killing the vacuum, that for the basis state Eq. (1.115)

$$N \, \Phi_{\mathbf{x}_1 \, \mathbf{x}_2 \, \dots \, \mathbf{x}_n} = n \, \Phi_{\mathbf{x}_1 \, \mathbf{x}_2 \, \dots \, \mathbf{x}_n} \ . \tag{2.33}$$

For the case of the vacuum state the eigenvalue is zero, there are no particles in the vacuum. In non-relativistic quantum mechanics, the total number operator for each set of species is always conserved; however, this is of course not the case in relativistic quantum theory. We note that the vacuum state has zero energy and momentum (and of course, as noted, zero number of particles).

---

[3]The physical interpretation of the identity operator in the one-particle state space is the number operator, counting the particle number in any one-particle state, $\hat{I} |\psi\rangle = 1 |\psi\rangle$.

Exercise 2.6. Show that if $|\psi_n\rangle$ represents a state with n particles, $N|\psi_n\rangle = n|\psi_n\rangle$, then

$$N\,\psi^\dagger(\mathbf{x})\,|\psi_n\rangle = (n+1)\,\psi^\dagger(\mathbf{x})\,|\psi_n\rangle \qquad (2.34)$$

i.e. $\psi^\dagger(\mathbf{x})\,|\psi_n\rangle$ is a state with $(n+1)$ particles.

Since $\psi(\mathbf{x})$ removes a particle from any state, the relationship

$$\psi(\mathbf{x})\,f(N) = f(N+1)\,\psi(\mathbf{x}) \qquad (2.35)$$

is valid for an arbitrary function, f, of the number operator. In particular we have

$$e^{-\alpha N}\,\psi(\mathbf{x})\,e^{\alpha N} = e^{\alpha}\,\psi(\mathbf{x})\,. \qquad (2.36)$$

Exercise 2.7. Show that the number operator for electrons in the momentum representation takes the form

$$N = \sum_\sigma \int d\mathbf{p}\, a^\dagger_{\mathbf{p}\sigma}\, a_{\mathbf{p}\sigma}\,. \qquad (2.37)$$

Exercise 2.8. The state considered in Exercise 1.7 on page 20 is the famous BCS-state, which describes remarkably well the ground state properties of many s-wave superconductors as realized by J. Bardeen, L. N. Cooper and J. R. Schrieffer (in 1957). Note its total disrespect for the sacred conservation law of non-relativistic Fermi systems, the conservation of the number of particles or, equivalently, we can say that the state corresponds to a situation with broken global gauge invariance.

For the reader interested in BCS-ology (which is further investigated in Section 8.1), verify for the average of the number operator

$$\langle N\rangle \equiv \langle \text{BCS}|N|\text{BCS}\rangle = 2\sum_{\mathbf{p}} |v_{\mathbf{p}}|^2 \qquad (2.38)$$

and for the variance

$$\langle (N - \langle N\rangle)^2\rangle = \langle N^2\rangle - \langle N\rangle^2 = 4\sum_{\mathbf{p}} u_{\mathbf{p}}^2 v_{\mathbf{p}}^2\,. \qquad (2.39)$$

Show that $a^\dagger_{\mathbf{p}\uparrow}|\text{BCS}\rangle$ and $a_{-\mathbf{p}\downarrow}|\text{BCS}\rangle$ represent the same state and that they are orthogonal to the state $|\text{BCS}\rangle$. The BCS-pairing state consists of linear superpositions of particle and hole states. Show that, as a consequence, anomalous moments are non-vanishing in the state $|\text{BCS}\rangle$, for example

$$\langle \text{BCS}|a_{\mathbf{p}\uparrow}\, a_{-\mathbf{p}\downarrow}|\text{BCS}\rangle = -u_{\mathbf{p}}^* v_{\mathbf{p}}\,. \qquad (2.40)$$

2.3 Probability current density operator

For a single particle, the probability current density operator is, according to Eq. (1.58),

$$\hat{\mathbf{j}}(\mathbf{x}) = \frac{1}{m}\{\hat{\mathbf{p}}, \hat{n}(\mathbf{x})\} . \tag{2.41}$$

For particles carrying electric charge, e, the electric current density operator or charge current density operator in the presence of a vector potential, $\mathbf{A}(\mathbf{x}, t)$, is specified in terms of the kinematic momentum operator ($\hat{\mathbf{p}}_{\text{can}}$ being the operator satisfying the canonical commutation relation, Eq. (1.8))

$$\hat{\mathbf{p}}_t^{\text{kin}} = \hat{\mathbf{p}}_{\text{can}} - e\mathbf{A}(\hat{\mathbf{x}}, t) \tag{2.42}$$

and the charge current density operator is

$$\hat{\mathbf{j}}_t(\mathbf{x}) = \frac{e}{2m}\{\hat{\mathbf{p}}_t^{\text{kin}}, \hat{n}(\mathbf{x})\} . \tag{2.43}$$

The current density operator then has two distinct parts

$$\hat{\mathbf{j}}_t(\mathbf{x}) = \hat{\mathbf{j}}_{\text{p}}(\mathbf{x}) + \hat{\mathbf{j}}_t^{\text{d}}(\mathbf{x}) \tag{2.44}$$

consisting of the so-called paramagnetic current density operator (or simply the current density operator in the absence of a vector potential)

$$\hat{\mathbf{j}}_{\text{p}}(\mathbf{x}) = \frac{e}{2m}\{\hat{\mathbf{p}}_{\text{can}}, \hat{n}(\mathbf{x})\} \tag{2.45}$$

and in the present case a time-dependent so-called diamagnetic current density operator

$$\hat{\mathbf{j}}_t^{\text{d}}(\mathbf{x}) = -\frac{e^2}{2m}\{\mathbf{A}(\hat{\mathbf{x}}, t), \hat{n}(\mathbf{x})\} = -\frac{e^2}{m}\,\hat{n}(\mathbf{x})\,\mathbf{A}(\hat{\mathbf{x}}, t) , \tag{2.46}$$

the last equality sign following from the fact that the two operators commute.

For particles carrying electric charge e, the electric current density operator on the multi-particle space is therefore, according to Eq. (2.12),

$$\hat{\mathbf{j}}_{A(t)}(\mathbf{x}, t) = \psi^\dagger(\mathbf{x})\,\hat{\mathbf{j}}_{A(t)}(\mathbf{x}, t)\,\psi(\mathbf{x}) , \tag{2.47}$$

where

$$\hat{\mathbf{j}}_{A(t)}(\mathbf{x}, t) = \hat{\mathbf{j}}^{(1)}(\mathbf{x}, t) - \frac{e^2}{m}\,\mathbf{A}(\mathbf{x}, t) \tag{2.48}$$

is the one-particle current density operator in the position representation in the presence of an external vector potential $\mathbf{A}(\mathbf{x}, t)$ and

$$\hat{\mathbf{j}}^{(1)}(\mathbf{x}, t) = \frac{e\hbar}{2mi}\left(\overrightarrow{\frac{\partial}{\partial\mathbf{x}}} - \overleftarrow{\frac{\partial}{\partial\mathbf{x}}}\right) , \tag{2.49}$$

the arrows indicating on which field, to the left or right, the differential operator operates.

The interaction between an electron and an electromagnetic field represented by a vector potential, $\mathbf{A}(\mathbf{x}, t)$, can be written in terms of the current density operator and the density operator

$$\hat{H}_{A(t)} = -\int d\mathbf{x}\, \hat{\mathbf{j}}_t(\mathbf{x}) \cdot \mathbf{A}(\mathbf{x}, t) - \frac{e^2}{2m}\int d\mathbf{x}\, \hat{n}(\mathbf{x})\, \mathbf{A}^2(\mathbf{x}, t)$$

$$= -\int d\mathbf{x}\, \hat{\mathbf{j}}_p(\mathbf{x}) \cdot \mathbf{A}(\mathbf{x}, t) + \frac{e^2}{2m}\int d\mathbf{x}\, \hat{n}(\mathbf{x})\, \mathbf{A}^2(\mathbf{x}, t) , \qquad (2.50)$$

which becomes the operator on the multi-particle space

$$H_{A(t)} = -\int d\mathbf{x}\, \mathbf{j}_{A(t)}(\mathbf{x}, t) \cdot \mathbf{A}(\mathbf{x}, t) - \frac{e^2}{2m}\int d\mathbf{x}\, n(\mathbf{x})\, \mathbf{A}^2(\mathbf{x}, t)$$

$$= -\int d\mathbf{x}\, \mathbf{j}_p(\mathbf{x}) \cdot \mathbf{A}(\mathbf{x}, t) + \frac{e^2}{2m}\int d\mathbf{x}\, n(\mathbf{x})\, \mathbf{A}^2(\mathbf{x}, t) , \qquad (2.51)$$

where

$$\mathbf{j}_p(\mathbf{x}) = \psi^\dagger(\mathbf{x})\, \hat{\mathbf{j}}^{(1)}(\mathbf{x}, t)\, \psi(\mathbf{x}) . \qquad (2.52)$$

The total Hamiltonian on the multi-particle space for an assembly of charged identical particles interacting with a classical electromagnetic field is thus[4]

$$H_{A\phi} = \int d\mathbf{x}\, \psi^\dagger(\mathbf{x}) \left(\frac{1}{2m}\left(\frac{\hbar}{i}\frac{\partial}{\partial \mathbf{x}} - e\mathbf{A}(\mathbf{x}, t) \right)^2 + e\phi(\mathbf{x}, t) \right) \psi(\mathbf{x}) . \qquad (2.53)$$

Physical observables as well as their couplings to classical fields are thus represented on the multi-particle space by operators quadratic in the fields.

Exercise 2.9. Show that the current density operator for electrons in the momentum representation takes the form

$$\mathbf{j}_p(\mathbf{x}) = e\sum_\sigma \int \frac{d\mathbf{p}'}{(2\pi\hbar)^{3/2}} \int \frac{d\mathbf{p}}{(2\pi\hbar)^{3/2}} \frac{\mathbf{p}' + \mathbf{p}}{2m} e^{-\frac{i}{\hbar}(\mathbf{p}' - \mathbf{p})\cdot\mathbf{x}} a^\dagger_{\mathbf{p}'\sigma} a_{\mathbf{p}\sigma} . \qquad (2.54)$$

Having established how to implement one-particle operators on the multi-particle space, we now turn to implement more complicated operators, viz. those describing interactions.

---

[4]Here expressed in the so-called Schrödinger picture, where the dynamics is carried by state vectors and the quantum field operators are time independent. The opposite scenario, the Heisenberg picture, will be discussed in Chapter 3.

2.4   Interactions

In this section we shall consider interactions between particles. In relativistic quantum theory, the forms of interaction are determined by Lorentz invariance and expressed in terms of polynomials of the field operators describing the creation and annihilation of particles. Our main interest shall be the typical interactions that are relevant in condensed matter physics, and there the task is to obtain their form in the multi-particle space knowing their form on any N-particle system. We therefore start by considering the case of two-body interaction, and consider fermions as the case of bosons follows as a simple corollary.

2.4.1   Two-particle interaction

If the identical particles, say fermions, interact through an instantaneous two-body potential, $V^{(2)}(\mathbf{x}_i, \mathbf{x}_j)$, the interaction between two fermions is represented in the antisymmetric two-particle state space by the operator

$$\hat{V}^{(2)} = \frac{1}{2} \int d\mathbf{x}_1 \int d\mathbf{x}_2 \, |\mathbf{x}_1 \wedge \mathbf{x}_2\rangle V^{(2)}(\mathbf{x}_1, \mathbf{x}_2)\langle \mathbf{x}_1 \wedge \mathbf{x}_2| \tag{2.55}$$

since

$$\hat{V}^{(2)}|\mathbf{x}_1 \wedge \mathbf{x}_2\rangle = \frac{1}{2}V^{(2)}(\mathbf{x}_1, \mathbf{x}_2)\,|\mathbf{x}_1 \wedge \mathbf{x}_2\rangle - \frac{1}{2}V^{(2)}(\mathbf{x}_2, \mathbf{x}_1)\,|\mathbf{x}_2 \wedge \mathbf{x}_1\rangle$$

$$= V^{(2)}(\mathbf{x}_1, \mathbf{x}_2)\,|\mathbf{x}_1 \wedge \mathbf{x}_2\rangle \tag{2.56}$$

and thereby

$$\langle \mathbf{x}_1 \wedge \mathbf{x}_2|\hat{V}^{(2)}|\psi\rangle = V^{(2)}(\mathbf{x}_1, \mathbf{x}_2)\,\psi(\mathbf{x}_1, \mathbf{x}_2)\,, \tag{2.57}$$

where $\psi(\mathbf{x}_1, \mathbf{x}_2)$ is the wave function describing the state of the two fermions, by construction it is an antisymmetric function of its arguments.

We now show that the two-body interaction which operates on an N-particle basis state according to

$$\hat{V}_N^{(2)}\,|\mathbf{x}_1 \wedge \mathbf{x}_2 \wedge \cdots \wedge \mathbf{x}_N\rangle = \sum_{i<j} V^{(2)}(\mathbf{x}_i, \mathbf{x}_j)\,|\mathbf{x}_1 \wedge \mathbf{x}_2 \wedge \cdots \wedge \mathbf{x}_N\rangle$$

$$= \frac{1}{2}\sum_{i \neq j} V^{(2)}(\mathbf{x}_i, \mathbf{x}_j)\,|\mathbf{x}_1 \wedge \mathbf{x}_2 \wedge \cdots \wedge \mathbf{x}_N\rangle \tag{2.58}$$

in the Fock space is represented by the operator

$$V = \frac{1}{2} \int d\mathbf{x} \int d\mathbf{x}'\, \psi^\dagger(\mathbf{x})\, \psi^\dagger(\mathbf{x}')\, V^{(2)}(\mathbf{x}, \mathbf{x}')\, \psi(\mathbf{x}')\, \psi(\mathbf{x})\,. \tag{2.59}$$

First we note that by twice applying the equation analogous to Eq. (1.72) for the annihilation operator in the position representation we get

$$\psi(\mathbf{x}')\psi(\mathbf{x})|\mathbf{x}_1 \wedge \cdots \wedge \mathbf{x}_N\rangle = \psi(\mathbf{x}')\sum_{n=1}^{N}(-1)^{n-1}\delta(\mathbf{x}-\mathbf{x}_n)|\mathbf{x}_1 \wedge \cdots (\text{ no } \mathbf{x}_n)\cdots \wedge \mathbf{x}_N\rangle$$

$$= \sum_{n=1}^{N}(-1)^{n-1}\delta(\mathbf{x}-\mathbf{x}_n)\sum_{m=1,(m\neq n)}^{N}(-1)^{m-\theta(n-m)}\delta(\mathbf{x}'-\mathbf{x}_m)$$

$$\times \quad |\mathbf{x}_1 \wedge \mathbf{x}_2 \wedge \cdots (\text{ no } \mathbf{x}_n \text{ and } \mathbf{x}_m)\cdots \wedge \mathbf{x}_N\rangle , \qquad (2.60)$$

where θ denotes the step function. The second statistics exponent factor is m if $m > n$ and of the usual form $m - 1$ if $m < n$, simply adjusting to when operating with the second annihilation operator the labeling of the state vector differs from the one used in the definition Eq. (1.72). Then by operating with creation operators we get

$$\psi^\dagger(\mathbf{x})\psi^\dagger(\mathbf{x}')\psi(\mathbf{x}')\psi(\mathbf{x})|\mathbf{x}_1 \wedge \cdots \wedge \mathbf{x}_N\rangle = \sum_{m(\neq n)}\delta(\mathbf{x}-\mathbf{x}_n)\delta(\mathbf{x}'-\mathbf{x}_m)$$

$$\times \quad (-1)^{n-1}(-1)^{m-\theta(n-m)}|\mathbf{x} \wedge \mathbf{x}' \wedge \mathbf{x}_1 \wedge \mathbf{x}_2 \wedge \cdots (\text{ no } \mathbf{x}_n \text{ and } \mathbf{x}_m)\cdots \wedge \mathbf{x}_N\rangle$$

$$= \sum_{m(\neq n)}\delta(\mathbf{x}-\mathbf{x}_n)\delta(\mathbf{x}'-\mathbf{x}_m)|\mathbf{x}_1 \wedge \mathbf{x}_2 \wedge \mathbf{x}_3 \wedge \cdots \wedge \mathbf{x}_N\rangle \qquad (2.61)$$

and multiplying with $V^{(2)}(\mathbf{x},\mathbf{x}')$ and integrating over $\mathbf{x}$ and $\mathbf{x}'$ in Eq. (2.59) therefore reproduces Eq. (2.58). Clearly, the operator V on the multi-particle space is hermitian since the function $V^{(2)}(\mathbf{x},\mathbf{x}')$ is real.

We note that the perhaps more intuitive guess for the two-body interaction in terms of the density operator

$$V' = \frac{1}{2}\int d\mathbf{x} \int d\mathbf{x}'\, n(\mathbf{x})\,V^{(2)}(\mathbf{x},\mathbf{x}')\,n(\mathbf{x}') \qquad (2.62)$$

differs from the correct expression, Eq. (2.59), by a self-energy term

$$V' = V + \frac{1}{2}\int d\mathbf{x}\, V^{(2)}(\mathbf{x},\mathbf{x})\,n(\mathbf{x}) , \qquad (2.63)$$

which, for example, for the case of Coulomb interaction would be infinite unless no particles are present, in which case it becomes the other extreme, viz. zero.

The two-particle interaction part of the Hamiltonian, Eq. (2.59), is so-called normal-ordered, i.e. all annihilation operators appear to the right of any creation operator. We recall that the one-body part of the Hamiltonian is also normal-ordered, as are those representing physical observables. We note that, as a consequence, the vacuum state has zero energy and momentum.

The derived expression, Eq. (2.59), for two-body interaction of fermions is of course the same for two-particle interaction of bosons, the derivation being identical, in fact simpler since no minus sign is involved in the interchange of two bosons.

The Hamiltonian for non-relativistic identical particles interacting through an instantaneous two-body interaction is thus

$$H = \int d\mathbf{x}\, \psi^\dagger(\mathbf{x}) \left(\frac{1}{2m} \left(\frac{\hbar}{i} \right)^2 \frac{\partial^2}{\partial \mathbf{x}^2} \right) \psi(\mathbf{x})$$

$$+ \frac{1}{2} \int d\mathbf{x} \int d\mathbf{x}'\, \psi^\dagger(\mathbf{x}) \psi^\dagger(\mathbf{x}') V^{(2)}(\mathbf{x}, \mathbf{x}') \psi(\mathbf{x}') \psi(\mathbf{x}) . \tag{2.64}$$

Exercise 2.10. Show that the Hamiltonian for an assembly of particles interacting through two-particle interaction commutes with the number operator.

Exercise 2.11. Show that if the two-particle potential is translational invariant

$$V^{(2)}(\mathbf{x}, \mathbf{x}') = V(\mathbf{x} - \mathbf{x}') , \tag{2.65}$$

we have in the momentum representation for the operator on multi-particle space

$$V = \frac{1}{2} \int \frac{d\mathbf{q}}{(2\pi\hbar)^3} \int d\mathbf{p} \int d\mathbf{p}'\, V(-\mathbf{q})\, a^\dagger_{\mathbf{p}-\mathbf{q}} a^\dagger_{\mathbf{p}'+\mathbf{q}} a_{\mathbf{p}'} a_{\mathbf{p}} , \tag{2.66}$$

where $V(\mathbf{q})$ is the Fourier transform of the real potential $V(\mathbf{x})$

$$V(\mathbf{q}) = \int d\mathbf{x}\, e^{-\frac{i}{\hbar}\mathbf{x}\cdot\mathbf{q}}\, V(\mathbf{x}) . \tag{2.67}$$

If the potential furthermore is inversion symmetric, $V(-\mathbf{x}) = V(\mathbf{x})$, we obtain

$$V = \frac{1}{2} \int \frac{d\mathbf{q}}{(2\pi\hbar)^3} \int d\mathbf{p} \int d\mathbf{p}'\, V(\mathbf{q})\, a^\dagger_{\mathbf{p}+\mathbf{q}} a^\dagger_{\mathbf{p}'-\mathbf{q}} a_{\mathbf{p}'} a_{\mathbf{p}} . \tag{2.68}$$

If the particles possess spin and their two-body interaction is spin dependent, the interaction in the multi-particle space becomes

$$V = \frac{1}{2} \sum_{\alpha\alpha',\beta\beta'} \int d\mathbf{x} \int d\mathbf{x}'\, \psi^\dagger_\alpha(\mathbf{x}) \psi^\dagger_\beta(\mathbf{x}') V_{\alpha\alpha',\beta\beta'}(\mathbf{x}, \mathbf{x}') \psi_{\beta'}(\mathbf{x}') \psi_{\alpha'}(\mathbf{x}) , \tag{2.69}$$

where, in accordance with custom, the spin degree of freedom appears as an index.

Exercise 2.12. Consider a piece of metal of volume $\mathcal{V}$ and describe it in the Sommerfeld model where the ionic charge is assumed smeared out to form a fixed uniform neutralizing background charge density.

Show that, in the momentum representation, the operator on the multi-particle space representing the interacting electrons is

$$V = \frac{1}{2\mathcal{V}} \sum_{\mathbf{q} \neq 0, \mathbf{p}, \mathbf{p}', \sigma\sigma'} \frac{e^2}{\epsilon_0 q^2} a^\dagger_{\mathbf{p}+\mathbf{q},\sigma} a^\dagger_{\mathbf{p}'-\mathbf{q},\sigma'} a_{\mathbf{p}',\sigma'} a_{\mathbf{p},\sigma} , \tag{2.70}$$

i.e. the interaction with the background charge eliminates the $(\mathbf{q} = 0)$-term in the Coulomb interaction.

2.4.2 Fermion–boson interaction

In relativistic quantum theory the creation and annihilation operators, the quantum fields, are necessary to describe dynamics, since particle can be created and annihilated. Relativistic quantum theory is thus inherently dealing with many-body systems. In a non-relativistic quantum theory the introduction of the multi-particle space is never mandatory, but is of convenience since it allows for an automatic way of respecting the quantum statistics of the particles even when interactions are present. It is also quite handy, but again not mandatory, when it comes to the description of symmetry broken states such as the cases of condensed states of fermions in superconductors and superfluid ^{3}He, and for describing Bose–Einstein condensates of bosons.

The generic interaction between fermions and bosons is of the form

$$H_{\mathrm{b-f}} = g \int d\mathbf{x}\, \psi^\dagger(\mathbf{x})\, \phi(\mathbf{x})\, \psi(\mathbf{x}) , \tag{2.71}$$

where $\psi(\mathbf{x})$ is the fermi field and $\phi(\mathbf{x})$ is the real (hermitian) bose field, and the interaction is characterized by some coupling constant g, and possibly dressed up in some indices characteristic for the fields in question, such as Minkowski and spinor in the case of QED.[5] The fermi and bose fields commute since they operate on their respective multi-particle spaces making up the total product multi-particle state space.[6] For the fermion–boson interaction which shall be of interest in the following, viz. the electron–phonon interaction, Eq. (2.71) is also a relevant form.

2.4.3 Electron–phonon interaction

Of importance later is the interaction between electrons and the quantized lattice vibrations in, say, a metal or semiconductor, the electron–phonon interaction. For illustration it suffices to consider the jellium model where the electrons couple only to longitudinal compressional charge configurations of the lattice ions, the longitudinal phonons. A deformation of the ionic charge distribution in a piece of matter, will create an effective potential felt by an electron at point $\mathbf{x}_e$, which in the jellium model

---

[5] Even the standard model has only fermionic interactions of this form. The fully indexed theory will be addressed in Chapter 9.

[6] If a theory contains two or more kinematically independent fermion species their corresponding fields are taken to anti-commute.

is given by the deformation potential[7]

$$V(\mathbf{x}_e) = \frac{n}{2N_0} \, \nabla_{\mathbf{x}_e} \cdot \mathbf{u}(\mathbf{x}_e) \, , \tag{2.72}$$

where $\mathbf{u}$ is the displacement field of the background ionic charge, N_0 is the density of electron states at the Fermi energy *per* spin (in three dimensions $N_0 = mp_F/2\pi^2\hbar^3$), and n is the electron density. The quantized lattice dynamics leads to the electron–phonon interaction in the jellium model becoming (recall Eq. (1.131))

$$\hat{V}_{e-ph}(\mathbf{x}_e) = \frac{n}{2N_0} \, \nabla_{\mathbf{x}_e} \cdot \hat{\mathbf{u}}(\mathbf{x}_e) = \frac{i}{2}\sqrt{\frac{\hbar}{N_0 V}} \sum_{|\mathbf{k}|\leq k_D} \sqrt{\omega_\mathbf{k}} \, [\hat{a}_\mathbf{k} \, e^{i\mathbf{k}\cdot\mathbf{x}_e} - \hat{a}_\mathbf{k}^\dagger \, e^{-i\mathbf{k}\cdot\mathbf{x}_e}] \tag{2.73}$$

where the harmonic oscillator creation and annihilation operators satisfy the commutation relations, $[\hat{a}_\mathbf{k}, \hat{a}_{\mathbf{k}'}^\dagger] = \delta_{\mathbf{k},\mathbf{k}'}$, and describe the weakly perturbed collective ionic oscillations (recall Sections 1.4.1 and 1.4.2). We assume a finite lattice of volume V. The set of harmonic oscillators is in its multi-particle description thus specified by the phonon field operator

$$\phi(\mathbf{x}) \equiv c\sqrt{Mn_i} \, \nabla_\mathbf{x} \cdot \mathbf{u}(\mathbf{x}) = i\sqrt{\frac{\hbar}{2V}} \sum_{|\mathbf{k}|\leq k_D} \sqrt{\omega_\mathbf{k}} \, [a_\mathbf{k} \, e^{i\mathbf{k}\cdot\mathbf{x}} - a_\mathbf{k}^\dagger \, e^{-i\mathbf{k}\cdot\mathbf{x}}] \, , \tag{2.74}$$

which is a real scalar bose field whose quanta, the phonons, are equivalent to bose particles, the bose field in the multi-particle space of longitudinal phonons. The interaction between the lattice of ions and an electron is thus transmitted in discrete units, the quanta we called phonons. In accordance with custom we leave out hats on operators on a multi-particle space; the phonon creation and annihilation operators of course satisfy the above stated canonical commutation relations as well as those of Eq. (1.113).[8] The (longitudinal) phonon field, Eq. (2.74), is a real or hermitian field, $\phi^\dagger(\mathbf{x}) = \phi(\mathbf{x})$, and contains a sum of creation and annihilation operators. Except for the explicit upper (ultraviolet) cut off, imposed by the finite lattice constant, it is thus analogous to the field describing a spin zero particle.

The electron–phonon interaction in the product of multi-particle spaces for electrons and phonons is according to Eq. (2.72) given in terms of the phonon field and the electron density reflecting that the electrons couple to the (screened) ionic charge deformations (or equivalently, Eq. (2.72) is a one-body operator for the electrons since it is a potential-coupling)[9]

$$V_{e-ph} = g\int d\mathbf{x} \, n_e(\mathbf{x}) \, \phi(\mathbf{x}) = g\int d\mathbf{x} \, \psi^\dagger(\mathbf{x}) \, \phi(\mathbf{x}) \, \psi(\mathbf{x}) \, , \tag{2.75}$$

---

[7]The electron–phonon interaction is an effective collective description of the underlying screened electron–ion Coulomb interaction. For the argument leading to the expression of the deformation potential see, for example, chapter 10 of reference [1].

[8]Phonons refer to collective oscillations of the ions and their screening cloud of electrons, similarly as the effective Coulomb electron–electron interaction describes the interaction between electrons and their screening clouds. Such objects are referred to as quasi-particles.

[9]That the electron–phonon interaction takes this form is the reason for introducing the phonon field, Eq. (2.74), instead of using the displacement field.

where the electron–phonon interaction coupling constant, g, in the jellium model is given by

$$g^2 = \frac{1}{2N_0} = \frac{4}{9} \frac{\epsilon_F^2}{Mn_i c^2} \qquad (2.76)$$

and for the last rewriting in Eq. (2.75) we have used the fact that fermi and bose fields commute since they are operators on different parts of the product space consisting of the (tensorial) product of the multi-particle space for fermions and bosons, respectively. The electron field operates on its Fock space and the bose field operates on its multi-particle space.

We note that in the jellium model, the electron–phonon interaction is local just as in relativistic interactions,[10] but here in the context of solid state physics it is only an approximation to an in general non-local interaction between the electrons and the ionic charge deformations. Furthermore, in general the phonon field is not a scalar field as a real crystal supports besides longitudinal also transverse vibrations. The general form of the electron–phonon interaction is

$$V_{e-ph} = \sum_{\mathbf{k},\mathbf{k'},\mathbf{q},b,\sigma} g_{\mathbf{k}\mathbf{k'}\mathbf{q}b} c^\dagger_{\mathbf{k'}\sigma} c_{\mathbf{k}\sigma} \left(a_{\mathbf{q}b} + a^\dagger_{-\mathbf{q}b} \right), \qquad (2.77)$$

where c and a are the electron and phonon fields, respectively, and in addition to the two transverse phonon branches, optical branches can in general be present if the unit cell of the crystal contains several atoms. Owing to the presence of the periodic crystal lattice, the momentum is no longer a good quantum number, and instead states are labeled by the Bloch or so-called crystal wave vector as defined by the translations respecting the crystal symmetry. The coupling function, $g_{\mathbf{k}\mathbf{k'}\mathbf{q}b}$, vanishes unless the crystal wave vector is conserved modulo a reciprocal lattice vector, $\mathbf{k'} = \mathbf{k} + \mathbf{q} + \mathbf{K}$. The new type of interaction processes, corresponding to $\mathbf{K} \neq \mathbf{0}$, so-called Umklapp-processes, are the signature of the periodic crystal structure.

The phonons and electrons have dynamics of their own as described by the Hamiltonians of Eq. (1.123) and Eq. (2.64), and we have thus arrived at the Hamiltonian describing electrons and phonons.

Exercise 2.13. Interaction between photons and electrons is obtained by minimal coupling, $\mathbf{P} \to \mathbf{P} - e\mathbf{A}$, where the photon field in the Schrödinger picture is specified by (recall Exercise 1.10 on page 28)

$$\mathbf{A}(\mathbf{x}) = \int \frac{d\mathbf{k}}{(2\pi)^3} \sqrt{\frac{\hbar}{2c|\mathbf{k}|}} \sum_{p=1,2} \left(a_{\mathbf{k}p} e^{i\mathbf{k}\cdot\mathbf{x}} + a^\dagger_{\mathbf{k}p} e^{-i\mathbf{k}\cdot\mathbf{x}} \right) \mathbf{e}_p(\mathbf{k}), \qquad (2.78)$$

where in the transverse gauge the two perpendicular unit polarization vectors, $\mathbf{e}_p(\mathbf{k})$, are also perpendicular to the wave vector, $\mathbf{k}$, of the photon.

---

[10]In relativistic quantum theory the *form* of the interactions can be inferred from the symmetry properties of the system. In condensed matter physics the interactions typically originate in the Coulomb interaction; this is the case for the electron–phonon interaction, which originates in the Coulomb interaction between the electrons and nuclei constituting a piece of material such as a metal.

The total electron–photon Hamiltonian, for the case of non-relativistic electrons, then becomes

$$H = H_{\mathrm{ph}} + H_{\mathrm{el}} + H_{\mathrm{el-ph}} \tag{2.79}$$

where

$$H_{\mathrm{el}} + H_{\mathrm{el-ph}} = \frac{1}{2m}\left(\mathbf{P} - e\mathbf{A}(\mathbf{x})\right)^2 \tag{2.80}$$

and $\mathbf{P}$ is the total momentum operator for the electrons, Eq. (2.14).

Show that the electron–photon interaction can be written in the form

$$H_{\mathrm{el-ph}} = -\int d\mathbf{x}\, \mathbf{j}_{\mathrm{p}}(\mathbf{x}) \cdot \mathbf{A}(\mathbf{x}) + \frac{e^2}{2m}\int d\mathbf{x}\, n(\mathbf{x})\, \mathbf{A}^2(\mathbf{x}) \tag{2.81}$$

where the current and density operators for the electrons are specified in Sections 2.3 and 2.2.

2.5   The statistical operator

Up until now, we have described the physical states of a system in terms of state vectors in the multi-particle state space. A general state vector, $|\Psi\rangle$, can be expanded on the basis vectors (using for once the resolution of the identity on the multi-particle state space)

$$|\Psi\rangle = \sum_{N=0}^{\infty} \frac{1}{N!} \sum_{\mathbf{p}_1, \ldots, \mathbf{p}_N} \langle \mathbf{p}_1 \vee \mathbf{p}_2 \vee \cdots \vee \mathbf{p}_N | \Psi \rangle\, |\mathbf{p}_1 \vee \mathbf{p}_2 \vee \cdots \vee \mathbf{p}_N\rangle \tag{2.82}$$

or expressed in terms of the vacuum state with the help of our new, so far only kinematic gadget, the field operator

$$|\Psi\rangle = \sum_{N=0}^{\infty} \sum_{\mathbf{p}_1, \ldots, \mathbf{p}_N} c(\mathbf{p}_1, \ldots, \mathbf{p}_N)\, a_{\mathbf{p}_1}^{\dagger} \cdots a_{\mathbf{p}_N}^{\dagger} |0\rangle\,, \tag{2.83}$$

where the cs are the complex amplitude coefficients specifying the state. Equivalently the state vector could be expressed in terms of the field operator in the position representation.

The description of a system in terms of its wave function is not the generic one as systems often can not be considered isolated, and instead will be in a mixture of states described by a statistical operator or density matrix (at a certain moment in time)

$$\rho \equiv \sum_k p_k |\psi_k\rangle\langle\psi_k| \tag{2.84}$$

allowing only the statements for the events of the system that it occurs with probability p_k in the quantum states $|\psi_k\rangle$ in the multi-particle state.[11] Certainly this is

---

[11]In general a statistical operator is specified by a set of (normalized) state vectors, $|\psi_1\rangle$, $|\psi_2\rangle$, ..., $|\psi_n\rangle$, not necessarily orthogonal, and a set of non-negative numbers adding up to one, $\sum_{i=1}^{n} p_i = 1$, according to $\rho = \sum_{i=1}^{n} p_i |\psi_i\rangle\langle\psi_i|$. Since the statistical operator is hermitian and non-negative, it is always possible to find an orthonormal set of states $|\phi_1\rangle$, $|\phi_2\rangle$, ..., $|\phi_N\rangle$, so that $\rho = \sum_{n=1}^{N} \pi_n |\phi_n\rangle\langle\phi_n|$, where $\pi_n \geq 0$ and $\sum_{n=1}^{N} \pi_n = 1$.

the generic situation in condensed matter physics and statistical physics in general. A diagonal element of the statistical operator, $\langle \psi | \rho | \psi \rangle$, thus gives the probability for the occurrence of the arbitrary state $|\psi\rangle$.

For the evaluation of the average value of a physical quantity represented by the operator A, a mixture adds an additional purely statistical averaging, as the quantum average value in a state is weighted by the probability of occurrence of the state

$$\langle A \rangle \equiv \sum_k p_k \langle \psi_k | A | \psi_k \rangle . \tag{2.85}$$

In view of Eq. (2.84), the average value for a mixture can be expressed in terms of the statistical operator according to

$$\langle A \rangle = \text{Tr}(\rho \, A) \tag{2.86}$$

where Tr denotes the trace in the multi-particle state, i.e. the sum of all diagonal elements evaluated in an arbitrary basis, generalizing the matrix element formula for the average value in a pure state $|\psi\rangle$, $\langle A \rangle = \text{Tr}(|\psi\rangle\langle\psi| \, A)$.

The statistical operator is seen to be hermitian and positive, $\langle \psi | \rho | \psi \rangle \geq 0$, and has unit trace for a normalized probability distribution. The statistical operator is only idempotent, i.e. a projector, for the case of a pure state, $\rho = |\psi\rangle\langle\psi|$. For a mixture we have $\rho^2 \neq \rho$, and $\text{Tr}\,\rho^2 < 1$.

In practice the most important mixture of states will be the one corresponding to a system in thermal equilibrium at a temperature T (including as a limiting case the zero temperature situation where the system definitely is in its ground state). In that case, applying the zeroth law of thermodynamics (that two systems in equilibrium at temperature T will upon being brought in thermal contact be in equilibrium at the same temperature) gives, for the thermal equilibrium statistical operator (Boltzmann's constant, the converter between energy and absolute temperature scale, is denoted k)

$$\rho_T = \frac{1}{Z} \, e^{-\mathcal{H}/kT} , \tag{2.87}$$

where the normalization factor

$$Z(T, V, N_\text{s}) = \text{Tr}\, e^{-\mathcal{H}/kT} = e^{-F(T,V,N_\text{s})/kT} \tag{2.88}$$

expressing the normalization of the probability distribution of Boltzmann weights, is the partition function. Here Tr denotes the trace in the multi-particle state space of the physical system of interest, but since the number of particles is fixed there is only the contributions from the corresponding N-particle subspace

$$e^{-\mathcal{H}/kT} = \sum_n e^{-E_n/kT} \tag{2.89}$$

as we are describing the system in the canonical ensemble. As the temperature approaches absolute zero, the term with minimum energy, the non-degenerate ground state energy, dominates the sum, and at zero temperature, the average value of a physical quantity becomes the N-particle ground state average value. The partition function or equivalently the free energy, F, are functions of the macroscopic parameters, the temperature, T, and the volume, V, and the number of particles, N_s, of different species in the system, and it contains all thermodynamic information.

When a system consists of a huge number of particles, it is more convenient to perform calculations in the *grand* canonical ensemble, where instead of the inconvenient constrain of a fixed number of particles, their chemical potential and average number of particles are specified. In the grand canonical ensemble, the system is thus described in the multi-particle state space. The system is thus imagined coupled to particle reservoirs, or a subsystem is considered. The system can exchange particles with the reservoirs which are described by their chemical potentials (assuming in general several particle species present). This feature is simply included by introducing Lagrange multipliers, i.e. tacitly understanding that single-particle energies are measured relative to their chemical potential, $H \rightarrow H - \sum_s \mu_s N_s$. In this case, the partition function instead of being a function of the number of particles, is a function of the chemical potentials for the species in question

$$Z_{\mathrm{gr}}(T, \mu_s) = \mathrm{Tr}\, e^{-(\mathcal{H} - \mu_s N_s)/kT} = e^{-\Omega(T, \mu_s)/kT} \tag{2.90}$$

specified by the average number of particles according to

$$\langle N_s \rangle = -\left(\frac{\partial \Omega}{\partial \mu_s}\right)_{T,V}. \tag{2.91}$$

For systems of particles where the total number of *particles* is not conserved, i.e. where the Hamiltonian and the total number operator do not commute, such as for phonons and photons, the chemical potential vanishes, and the grand canonical ensemble is employed. For degenerate fermions, such as electrons in a metal, the chemical potential is a huge energy compared to relevant temperatures, viz. tied to the Fermi energy as discussed in Exercise 2.15.

The average value in the grand canonical ensemble of a physical quantity represented by the operator A is thus

$$\langle A \rangle = \mathrm{Tr}(\rho A) = \sum_{N,n} e^{(\Omega - E_n + \mu N)/kT} \langle N, E_n | A | E_n, N \rangle. \tag{2.92}$$

As the following exercise shows, if alternatively attempted in the canonical ensemble, calculations run smoothly in the grand canonical ensemble free of the constraint of a fixed number of particles. In the thermodynamic limit, using either of course gives the same results as the fluctuations in the particle number in the grand canonical ensemble around the mean value then is negligible.

Exercise 2.14. Show that the grand canonical partition function for non-interacting non-relativistic fermions or bosons of mass m contained in a volume V is given by

$$\ln Z_{\mathrm{gr}}^{(0)}(T, V, \mu) = \mp \sum_{\mathbf{p}} \ln\left(1 \mp e^{-(\epsilon_{\mathbf{p}} - \mu)/kT}\right) = e^{-\Omega_0/kT} \quad, \quad \epsilon_{\mathbf{p}} = \frac{\mathbf{p}^2}{2m} \tag{2.93}$$

and the average number of particles are specified by the Bose–Einstein or Fermi–Dirac distributions, respectively,

$$\langle N \rangle = -\left(\frac{\partial \Omega_0}{\partial \mu}\right)_{T,V} = \sum_{\mathbf{p}} \frac{1}{e^{(\epsilon_\mathbf{p} - \mu)/kT} \mp 1} \qquad (2.94)$$

from which one readily verifies that the thermodynamic potential is specified by the pressure, P, and volume of the system according to

$$\Omega_0(T, \mu, V) = -PV . \qquad (2.95)$$

Exercise 2.15. Show that for a system of non-interacting degenerate fermions, i.e. at temperatures where $kT \ll \mu$, the chemical potential is pinned to the Fermi energy, $\epsilon_F = \hbar^2 (3\pi^2 n)^{2/3}/2m$, as

$$\mu = \epsilon_F \left(1 - a \left(\frac{T}{T_F}\right)^2\right) , \qquad (2.96)$$

where the fermions of mass m are assumed residing in three spatial dimensions (in which case the constant of order one is $a = \pi^2/12$) with a density n, and $T_F = \epsilon_F/k$ is the Fermi temperature which for a metal, in view of the large density of conduction electrons, is seen to be huge, typically of the order 10^4 K.

Exercise 2.16. At zero temperature, a system of fermions such as a metal contains high-speed electrons, all states below the Fermi energy are fully occupied, a reservoir for injecting electrons into other conductors. For bosons the opposite, coming to rest, can happen. First we observe that for non-interacting bosons the chemical potential can not be positive, $\mu \leq 0$, as dictated by the Bose–Einstein distribution function for occupation of energy levels. As the temperature decreases, the chemical potential increases and becomes vanishingly small at and below the temperature T_0 determined by the density, n, and mass, m, of the bosons (say, spinless and enclosed in a volume V) according to

$$n = \frac{N}{V} = \frac{(m)^{3/2}}{\sqrt{2}\,\pi^2 \hbar^3} \int_0^\infty d\epsilon \, \frac{\epsilon^{1/2}}{e^{\epsilon/kT_0} - 1} . \qquad (2.97)$$

At this temperature, the lowest energy level, $\epsilon_{\mathbf{p}=0} = 0$, starts to be macroscopically occupied.

Show that at temperatures below T_0, the number of bosons in the lowest level is (the population of the other levels are governed by the Bose–Einstein distribution)

$$N_0 = N \left(1 - \left(\frac{T}{T_0}\right)^{3/2}\right) . \qquad (2.98)$$

In the degenerate region at temperatures below T_0, the bosons comes to rest, the phenomenon of Bose–Einstein condensation (1925), the bosons become ordered in momentum space. The total condensation at zero temperature is of course a trivial feature of the quantum statistics of *non-interacting* bosons. Using the model of non-interacting bosons to estimate T_0 for the case of ^{4}He gives $T_0 \simeq 3.2$ K, quite close to the temperature 2.2 K of the λ-transition where liquid Helium becomes a superfluid (discovered 1928 and proposed a Bose–Einstein condensate by Fritz London, 1938).

2.6   Summary

In this chapter we have constructed the operators of relevance on the multi-particle space, and shown how they are expressed in terms of the quantum fields, the creation and annihilation fields. The kinematics of a many-body system, its possible quantum states and the operators representing its physical quantities, is thus expressed in terms of these two objects. The Hamiltonians on the multi-particle state space were constructed for the generic types of many-body interactions. We now turn to consider the dynamics of many-body systems described by their quantum fields on a multi-particle state space. In particular the quantum dynamics of a quantum field theory describing systems out of equilibrium. This will lead us to the study of correlation functions for quantum fields, the Green's functions for non-equilibrium states.

3

Quantum dynamics and Green's functions

In the previous chapter we studied the kinematics of many-body systems, and the form of operators representing the physical properties of a system, all of which were embodied by the quantum field. In this chapter we shall study the quantum dynamics of many-body systems, which can also be embodied by the quantum fields. We shall employ the fact that the quantum dynamics of a system, instead of being described in terms of the dynamics of the states or the evolution operator, i.e. as previously done through the Schrödinger equation, can instead be carried by the quantum fields. The quantum dynamics is therefore expressed in terms of the correlation functions or Green's functions of the quantum fields evaluated with respect to some state of the system. In particular we shall consider the general case of quantum dynamics for arbitrary non-equilibrium states. After introducing various types of Green's functions and relating them to measurable quantities, we will discuss the simplifications reigning for the special case of equilibrium states.

3.1 Quantum dynamics

Quantum dynamics can be described in different ways since quantum mechanics is a linear theory and the dynamics described by a unitary transformation of states.[1] This will come in handy in the next chapter when we study a quantum theory in terms of its perturbative expansion using the so-called interaction picture. Here we first discuss the Schrödinger and Heisenberg pictures.

---

[1]This should be contrasted with classical mechanics where dynamics is specified in terms of the physical quantities themselves, the generic case being intractable nonlinear partial differential equations. We shall study such a classical situation in Chapter 12 with the help of methods borrowed from quantum field theory, and where in addition the classical system interacts with an environment as described by a stochastic force.

3.1.1   The Schrödinger picture

Having the Hamiltonian on the multi-particle space at hand we can consider the dynamics described in the multi-particle space. An arbitrary state in the multi-particle space has at any time in question the expansion on, say the position basis states

$$|\Psi(t)\rangle \;=\; \sum_{N=0}^{\infty} \int dx_1 dx_2 \ldots dx_N \; \psi_N(x_1, x_2, \ldots, x_N, t)\, |x_1 \diamondsuit x_2 \cdots \diamondsuit x_N\rangle \,, \qquad (3.1)$$

where $\diamondsuit$ stands for $\vee$ or $\wedge$ for the bose or fermi cases respectively. The coefficients

$$\psi_N(x_1, x_2, \ldots, x_N, t) \;=\; \langle x_1 \diamondsuit x_2 \cdots \diamondsuit x_N | \Psi(t)\rangle \qquad (3.2)$$

are the wave functions describing each N-particle system, and they are symmetric or antisymmetric due to the symmetry properties of the basis states $|x_1 \diamondsuit x_2 \cdots \diamondsuit x_N\rangle$, i.e. no new state is produced by using *non-symmetric* coefficients $\psi_N(x_1, x_2, \ldots, x_N, t)$.

The dynamics of a multi-particle particle state is specified by the Schrödinger equation in the multi-particle space

$$i\hbar \frac{d|\Psi(t)\rangle}{dt} \;=\; \mathcal{H}(t)\, |\Psi(t)\rangle \,, \qquad (3.3)$$

where $\mathcal{H}(t)$ is the Hamiltonian on the multi-particle space, which can be explicitly time dependent due to external forces as our interest will be to consider non-equilibrium states. In the multi-particle space, the dynamics of all N-particle systems are described simultaneously since the above equation contains the infinite set of equations, $N = 0, 1, 2, \ldots$, which in the position representation are

$$i\hbar \frac{\partial \psi_N(x_1, x_2, \ldots, x_N, t)}{\partial t} \;=\; \sum_{M=0}^{\infty} \int dx_1' dx_2' \ldots dx_M' \; \psi_M(x_1', x_2', \ldots, x_M', t)$$

$$\times\; \langle x_1 \diamondsuit x_2 \cdots \diamondsuit x_N | \mathcal{H}(t) | x_1' \diamondsuit x_2' \cdots \diamondsuit x_M'\rangle \,. \qquad (3.4)$$

The even or odd character of a wave function is preserved in time as any Hamiltonian for identical particles is symmetric in the $\hat{p}_i$s and $\hat{x}_i$s as well as other degrees of freedom (this is the meaning of identity of particles, no interaction can distinguish them), so if even- or oddness of a wave function is the state of affairs at one moment in time it will stay this way for all times.[2]

For the case of two-particle interaction, Eq. (2.59), the Hamiltonian has only nonzero matrix elements between configurations with the same number of particles, the total number of particles is a conserved quantity, and the infinite set of equations, Eq. (3.4), splits into independent equations describing systems with the definite number of particles $N = 0, 1, 2, \ldots$.[3] For $N = 0$, the vacuum state, we have for the

---

[2]Time-invariant subspaces other than the symmetric and anti-symmetric ones do not seem to be physically relevant.

[3]For the case of an N-particle system, the multi-particle space is then not needed, we could discuss it solely in terms of its N-particle state space.

c-number representing its wave function

$$i\hbar \frac{d\psi_0(t)}{dt} = \langle 0|\mathcal{H}(t)|0\rangle \, \psi_0(t) = 0 \,, \tag{3.5}$$

where the last equality sign follows from the fact that since the Hamiltonian for two-particle interaction, Eq. (2.59), operates first with the annihilation operator on the vacuum state, it annihilates it. Since Hamiltonians are normal-ordered, this result is quite general: the vacuum state is without dynamics.

In the case of electron–phonon interaction, the Hamiltonian has off-diagonal elements with respect to the phonon multi-particle subspace. The number of phonons is owing to interaction with the electrons not conserved; an electron can emit or absorb phonons just like an electron in an excited state of an atom can emit a photon in the decay to a lower energy state. The chemical potential of phonons thus vanishes.

Instead of describing the dynamics in terms of the state vector we can introduce the time development or time evolution operator[4]

$$|\psi(t)\rangle = U(t,t') \, |\psi(t')\rangle \tag{3.6}$$

connecting the state vectors at the different times in question where they provide a complete description of the system. Solving the Schrödinger equation, Eq. (3.3), by iteration gives

$$U(t,t') = Te^{-\frac{i}{\hbar}\int_{t'}^{t} d\bar{t}\,\mathcal{H}(\bar{t})} \,, \tag{3.7}$$

where T denotes time-ordering.[5] The time-ordering operation orders a product of time-dependent operators into its time-descending sequence (displayed here for the case of three operators)

$$T(A(t_1)\,B(t_2)\,C(t_3)) = \begin{cases} A(t_1)\,B(t_2)\,C(t_3) & \text{for} \quad t_1 > t_2 > t_3 \\ B(t_2)\,A(t_1)\,C(t_3) & \text{for} \quad t_2 > t_1 > t_3 \\ \text{etc.} & \text{etc.} \end{cases} \tag{3.8}$$

In case of fermi fields, the time-ordering (and anti-time-ordering which we shortly encounter) shall by definition include a product of minus signs, one for each interchange of fermi fields. Since the Hamiltonian contains an even number of fermi fields, no statistical factors are thus involved in interchanging Hamiltonians referring to different moments in time under the time-ordering symbol.

The Schrödinger equation, Eq. (3.3), then gives the equation of motion for the time evolution operator[6]

$$i\hbar \frac{\partial U(t,t')}{\partial t} = \mathcal{H}(t)\,U(t,t') \,. \tag{3.9}$$

---

[4]It is amazing how compactly quantum dynamics can be captured, encapsulated in the single object, the evolution operator.

[5]For details see, for example, chapter 2 of reference [1].

[6]From a mathematical point of view, convergence properties of limiting processes for operator sequences A_t are inherited from the topology of the vector space; i.e. convergence is defined by convergence of an arbitrary vector $A_t|\psi\rangle$. The dual space of linear operator on the multi-particle state space can also be equipped with its own topology by introducing the scalar product for operators A and B, $\text{Tr}(B^\dagger A)$. But the result of differentiating is gleaned immediately from simple algebraic properties.

We note the semi-group property of the evolution operator

$$U(t, t'') \, U(t'', t') = U(t, t') \tag{3.10}$$

and the unitarity of the evolution operator, $U^\dagger(t, t') = U^{-1}(t, t')$, as a state vector has the scalar product with itself of modulus one enforced by the probability interpretation of the state vector. As a consequence, $U^\dagger(t, t') = U(t', t)$.

Exercise 3.1. Show that

$$U^\dagger(t, t') \;\equiv\; [U(t, t')]^\dagger = \tilde{T} e^{\frac{i}{\hbar} \int_{t'}^{t} d\bar{t}\, \mathcal{H}(\bar{t})} , \tag{3.11}$$

where the anti-time-ordering symbol, $\tilde{T}$, orders the time sequence oppositely as compared with the time-ordering symbol, T, as the adjoint inverts the order of a sequence of operators. Use this (or the unitarity of the evolution operator, $I = U^\dagger(t, t') \, U(t, t')$) to verify

$$-i\hbar \frac{\partial U^\dagger(t, t')}{\partial t} = U^\dagger(t, t') \, \mathcal{H}(t) \quad , \quad -i\hbar \frac{\partial U(t, t')}{\partial t'} = U(t, t') \, \mathcal{H}(t') . \tag{3.12}$$

The dynamics of a mixture is described by the time dependence of the statistical operator which, according to Eq. (3.6), is

$$\rho(t) = U(t, t') \, \rho(t') \, U^\dagger(t, t') \tag{3.13}$$

and the statistical operator satisfies the von Neumann equation

$$i\hbar \frac{d\rho(t)}{dt} = [\mathcal{H}(t), \rho(t)] . \tag{3.14}$$

A diagonal element of the statistical operator, $\langle \psi | \rho(t) | \psi \rangle$, gives the probability for the occurrence of the arbitrary state $|\psi\rangle$ at time t (explaining the use of the word density matrix).

An important set of mixtures in practice for an isolated system (i.e. the Hamiltonian is time independent) is the stationary states in which all physical properties are time independent. The statistical operator is thus for stationary states a function of the Hamiltonian of the system, $\rho = \rho(\mathcal{H})$.

For an isolated system, the evolution operator takes the simple form

$$U(t, t') = e^{-\frac{i}{\hbar}(t - t')\mathcal{H}} . \tag{3.15}$$

The generator of time displacements is the only operator in the Heisenberg picture which, in general, is time independent, and the quantity it represents we call the energy. For an isolated system, the Hamiltonian thus represents the energy.

3.1.2 The Heisenberg picture

Instead of having the dynamics described by an equation of motion for a state vector or realisticly by a statistical operator, the Schrödinger picture discussed above, it is convenient to transfer the dynamics to the physical quantities, resembling in this feature the dynamics of classical physics. In this so-called Heisenberg picture, the state of the system

$$|\psi_{\mathcal{H}}\rangle \equiv U^\dagger(t, t_{\mathrm{r}}) |\psi(t)\rangle = |\psi(t_{\mathrm{r}})\rangle \tag{3.16}$$

is time independent, according to Eq. (3.3) and Eq. (3.12), whereas the operators representing physical quantities are time dependent

$$A_{\mathcal{H}}(t) \equiv U^\dagger(t, t_{\mathrm{r}}) A U(t, t_{\mathrm{r}}) . \tag{3.17}$$

At the arbitrary reference time, t_{r}, the two pictures coincide, the evolution operator satisfying $U(t, t) = 1$.

We note that if $\{|a\rangle\}_a$ is the set of eigenstates of the operator A then the Heisenberg operator has the same spectrum but different eigenstates

$$A_{\mathcal{H}}(t) |a, t\rangle = a |a, t\rangle \quad , \quad |a, t\rangle \equiv U^\dagger(t, t_{\mathrm{r}}) |a\rangle . \tag{3.18}$$

The operator representing a physical quantity in the Heisenberg picture satisfies the equation of motion[7]

$$i\hbar \frac{\partial A_{\mathcal{H}}(t)}{\partial t} = [A_{\mathcal{H}}(t), \mathcal{H}_{\mathcal{H}}(t)] , \tag{3.19}$$

where

$$\mathcal{H}_{\mathcal{H}}(t) \equiv U^\dagger(t, t_{\mathrm{r}}) \mathcal{H}(t) U(t, t_{\mathrm{r}}) . \tag{3.20}$$

Introducing the field in the Heisenberg picture

$$\psi_{\mathcal{H}}(\mathbf{x}, t) = U^\dagger(t, t_{\mathrm{r}}) \psi(\mathbf{x}) U(t, t_{\mathrm{r}}) \tag{3.21}$$

we obtain its equation of motion

$$i\hbar \frac{\partial \psi_{\mathcal{H}}(\mathbf{x}, t)}{\partial t} = [\psi_{\mathcal{H}}(\mathbf{x}, t), \mathcal{H}(t)] . \tag{3.22}$$

Often context allows us to leave out the subscript, writing $\psi(\mathbf{x}, t) = \psi_{\mathcal{H}}(\mathbf{x}, t)$.

In the Heisenberg picture, only the equal time anti-commutator or commutator, for fermions or bosons, respectively, of the fields is in general a simple quantity, of course the c-number function:

$$[\psi(\mathbf{x}, t), \psi^\dagger(\mathbf{x}', t)]_s = \delta(\mathbf{x} - \mathbf{x}') . \tag{3.23}$$

At unequal times, the anti-commutator or commutator of the fields are, owing to interactions, complicated operators whose unravelling will be done in terms of the correlation functions of the fields, the Green's functions we introduce in the next section.

---

[7] If the Schrödinger operator is time dependent, such as can be the case for the current operator in the presence of a time-dependent vector potential representing a classical field, of course an additional term appears.

Exercise 3.2. Show that the probability density for a particle to be at position $\mathbf{x}$ at time t

$$n(\mathbf{x}, t) = \mathrm{Tr}(\rho(t)\, \psi^\dagger(\mathbf{x})\, \psi(\mathbf{x})) \tag{3.24}$$

can be rewritten in terms of the fields in the Heisenberg picture

$$n(\mathbf{x}, t) = \mathrm{Tr}(\rho\, \psi^\dagger(\mathbf{x}, t)\, \psi(\mathbf{x}, t)) , \tag{3.25}$$

where ρ is an arbitrary statistical operator at the reference time where the two pictures coincide.

For an isolated system, where the Hamiltonian is time independent, the quantum field (or any other) operator in the Heisenberg picture is related to the operator in the Schrödinger picture in accordance with Eq. (3.17), which in that case becomes (the coincidence with the Schrödinger picture is chosen to be at time $t = 0$)

$$\psi(\mathbf{x}, t) = e^{\frac{i}{\hbar}Ht}\, \psi(\mathbf{x})\, e^{-\frac{i}{\hbar}Ht} . \tag{3.26}$$

Exercise 3.3. Show that the time evolution of a free field in the Heisenberg picture specified by the free or kinetic energy Hamiltonian in Eq. (2.21), is given by

$$a_{\mathbf{p}}(t) = a_{\mathbf{p}}\, e^{-\frac{i}{\hbar}\epsilon_{\mathbf{p}}t} , \tag{3.27}$$

where $\epsilon_{\mathbf{p}} = \mathbf{p}^2/2m$ is the kinetic energy of the free particle with momentum $\mathbf{p}$, and the coincidence with the Schrödinger picture is chosen to be at time $t = 0$.

Show the commutation relations for the free fields is

$$[\psi_0(\mathbf{x}, t), \psi_0^\dagger(\mathbf{x}', t')]_s = \frac{1}{V} \sum_{\mathbf{p}} e^{\frac{i}{\hbar}\mathbf{p}\cdot(\mathbf{x}-\mathbf{x}') - \frac{i}{\hbar}\epsilon_{\mathbf{p}}(t-t')} . \tag{3.28}$$

For the case of an isolated system of particles interacting through an instantaneous two-particle interaction, Eq. (2.59), the Hamiltonian transformed according to Eq. (3.17) can be expressed in terms of the fields in the Heisenberg picture

$$\mathcal{H}_{\mathcal{H}}(t) = \int d\mathbf{x}\, \psi_{\mathcal{H}}^\dagger(\mathbf{x}, t) \left(\frac{1}{2m} \left(\frac{\hbar}{i} \frac{\partial}{\partial \mathbf{x}} \right)^2 \right) \psi_{\mathcal{H}}(\mathbf{x}, t)$$

$$+ \frac{1}{2} \int d\mathbf{x} \int d\mathbf{x}'\, \psi_{\mathcal{H}}^\dagger(\mathbf{x}, t)\, \psi_{\mathcal{H}}^\dagger(\mathbf{x}', t)\, V^{(2)}(\mathbf{x}, \mathbf{x}')\, \psi_{\mathcal{H}}(\mathbf{x}', t)\, \psi_{\mathcal{H}}(\mathbf{x}, t) \tag{3.29}$$

and according to Eq. (3.19), $\mathcal{H}(t) = \mathcal{H}$, i.e. the Hamiltonian in the Heisenberg picture *is* the Hamiltonian, representing the energy of the system.

Our interest shall be the case of non-equilibrium situations where a system is coupled to external classical fields, for example the coupling of current and density of charged particles to electromagnetic fields as represented by the Hamiltonian

$$H_{A,\phi}(t) = \int d\mathbf{x}\, \psi_{\mathcal{H}}^\dagger(\mathbf{x}, t) \left(\frac{1}{2m} \left(\frac{\hbar}{i} \frac{\partial}{\partial \mathbf{x}} - e\mathbf{A}(\mathbf{x}, t) \right)^2 + e\phi(\mathbf{x}, t) \right) \psi_{\mathcal{H}}(\mathbf{x}, t) , \tag{3.30}$$

where the quantum fields are in the Heisenberg picture.

Considering the case of two-particle interaction, and using the operator identities

$$[A, BC] = [A, B]C - B[C, A] = \{A, B\}C - B\{C, A\} \qquad (3.31)$$

for bose or fermi fields, respectively, and their commutation relations, the equation of motion for the field in the Heisenberg picture becomes

$$i\hbar \frac{\partial \psi(\mathbf{x}, t)}{\partial t} = h(t)\,\psi(\mathbf{x}, t) + \int d\mathbf{x}'\, V^{(2)}(\mathbf{x}, \mathbf{x}')\, \psi^\dagger(\mathbf{x}', t)\psi(\mathbf{x}', t)\,\psi(\mathbf{x}, t)\,, \qquad (3.32)$$

where $h = h(-i\nabla_\mathbf{x}, \mathbf{x}, t)$ is the free single-particle Hamiltonian, which can be time-dependent due to external classical fields. For example, for the case of a charged particle coupled to an electromagnetic field

$$h(-i\hbar\nabla_\mathbf{x}, \mathbf{x}, t) = \frac{1}{2m}\left(\frac{\hbar}{i}\frac{\partial}{\partial \mathbf{x}} - e\mathbf{A}(\mathbf{x}, t)\right)^2 + e\varphi(\mathbf{x}, t)\,. \qquad (3.33)$$

The dynamics of a system, specified by the time dependence of the quantum field in the Heisenberg picture, is thus described in terms of higher-order expressions in the field operators.

Exercise 3.4. Multiply Eq. (3.32) from the left by $\psi^\dagger(\mathbf{x}, t)$, and obtain also the adjoint of this construction. Obtain the continuity or charge conservation equation in the multi-particle space

$$\frac{\partial n(\mathbf{x}, t)}{\partial t} + \nabla_\mathbf{x} \cdot \mathbf{j}(\mathbf{x}, t) = 0\,, \qquad (3.34)$$

where

$$n(\mathbf{x}, t) = \psi^\dagger(\mathbf{x}, t)\,\psi(\mathbf{x}, t) \qquad (3.35)$$

and

$$\mathbf{j}(\mathbf{x}, t) = \frac{\hbar}{2mi}\left(\psi^\dagger(\mathbf{x}, t)\,\nabla_\mathbf{x}\psi(\mathbf{x}, t) - (\nabla_\mathbf{x}\psi^\dagger(\mathbf{x}, t))\,\psi(\mathbf{x}, t)\right) \qquad (3.36)$$

are the probability current and density operators on the multi-particle space in the Heisenberg picture.

Exercise 3.5. Show that the commutation relation for the displacement field operator in the Heisenberg picture at equal times is

$$\left[u_\alpha\mathbf{x}, t), n_\mathrm{i} M \frac{\partial u_\beta(\mathbf{x}', t)}{\partial t}\right]_- = i\hbar\delta_{\alpha\beta}\,\delta(\mathbf{x} - \mathbf{x}') \qquad (3.37)$$

reflecting the canonical commutation relations of non-relativistic quantum mechanics for the position and momentum operators of the ions in a lattice.

Exercise 3.6. Show that the phonon field in the Heisenberg picture satisfies the equal-time commutation relation (neglecting the ultraviolet or Debye cut-off, $\omega_\mathrm{D} \to \infty$)

$$\left[\phi(\mathbf{x}, t), \frac{\partial \phi(\mathbf{x}', t)}{\partial t}\right]_- = -i\frac{\hbar c^2}{2}\Delta_\mathbf{x}\,\delta(\mathbf{x} - \mathbf{x}')\,. \qquad (3.38)$$

Exercise 3.7. Show that, for an isolated system of identical particles interacting through an instantaneous two-body interaction, $V(\mathbf{x} - \mathbf{x}')$, the field operator in the Heisenberg picture, say in the momentum representation, satisfies the equation of motion (recall Exercise 2.10 on page 44)

$$i\hbar \frac{da_\mathbf{p}(t)}{dt} = \epsilon_\mathbf{p}\, a_\mathbf{p}(t) + \int \frac{d\mathbf{q}}{(2\pi\hbar)^3} \int d\mathbf{p}'\, V(-\mathbf{q})\, a^\dagger_{\mathbf{p}'+\mathbf{q}}(t)\, a_{\mathbf{p}'}(t)\, a_{\mathbf{p}+\mathbf{q}}(t) . \tag{3.39}$$

Show that the Hamiltonian in the Heisenberg picture can be expressed in the form

$$\mathcal{H}(t) = \frac{1}{2} \sum_\mathbf{p} \left(\epsilon_\mathbf{p}\, a^\dagger_\mathbf{p}(t)\, a_\mathbf{p}(t) + i\hbar a^\dagger_\mathbf{p}(t) \frac{da_\mathbf{p}(t)}{dt} \right) . \tag{3.40}$$

Exercise 3.8. Obtain the equation of motion for the electron and phonon fields in the Heisenberg picture for the case of longitudinal electron–phonon interaction.

Any property of a physical system is expressed in terms of a correlation function of field operators taken with respect to the state in question. In Section 3.3, we turn to introduce these, the Green's functions. But first, we will take a short historical detour.

3.2   Second quantization

Quantum field theory, as presented in the previous chapters, is simply the quantum mechanics of an arbitrary number of particles. For the non-relativistic case the practical task was to lift the N-particle description to the multi-particle state space. Quantum fields are often referred to as *second quantization*, which in view of our general introduction of quantum field theory for many-body systems is of course a most unfortunate choice of language. The misnomer has its origin in the following analogy.

Consider the Schrödinger equation for a single particle, say in a potential

$$i\hbar \frac{\partial \psi(\mathbf{x}, t)}{\partial t} = \left(\frac{1}{2m} \left(\frac{\hbar}{i} \frac{\partial}{\partial \mathbf{x}} \right)^2 + V(\mathbf{x}, t) \right) \psi(\mathbf{x}, t) . \tag{3.41}$$

Next, interpret the equation as a classical field equation *à la* Maxwell's equations. A difference is, of course, that the field is *complex*, and in the case of the electromagnetic field there are additional field components. The Schrödinger equation, Eq. (3.41), can be derived from the variational principle

$$\delta \int dt \int d\mathbf{x}\, \mathcal{L} = 0 , \tag{3.42}$$

where the *Lagrange* density is

$$\mathcal{L} = \psi^*(\mathbf{x}, t)\, i\hbar \frac{\partial \psi(\mathbf{x}, t)}{\partial t} - \frac{\hbar^2}{2m} \nabla_\mathbf{x} \psi^*(\mathbf{x}, t) \cdot \nabla_\mathbf{x} \psi(\mathbf{x}, t) - \psi^*(\mathbf{x}, t)\, V(\mathbf{x}, t)\, \psi(\mathbf{x}, t).$$
$$\tag{3.43}$$

3.2. Second quantization

The conjugate field variable is then

$$\Pi(\mathbf{x}, t) = \frac{\partial \mathcal{L}}{\partial \frac{\partial \psi(\mathbf{x}, t)}{\partial t}} = i\hbar \psi^*(\mathbf{x}, t) \qquad (3.44)$$

in analogy with the canonical momenta in classical mechanics

$$\mathbf{p} = \frac{\partial L}{\partial \dot{\mathbf{x}}} \qquad (3.45)$$

and the variables of the field, $\mathbf{x}$, is in the analogy equivalent to the labeling, i, of the mechanical degrees of freedom, and $\Pi(\mathbf{x})$ corresponds to p_i.

Analogous to Hamilton's function

$$H = \sum_i p_i \dot{x}_i - L \qquad (3.46)$$

enters the Hamilton *function* for the *classical* Schrödinger field

$$\mathcal{H} = \int d\mathbf{x} \int dt \left(\Pi(\mathbf{x}, t) \frac{\partial \psi(\mathbf{x}, t)}{\partial t} - \mathcal{L} \right). \qquad (3.47)$$

In analogy with the canonical commutation relations

$$[p_i, x_j] = \frac{\hbar}{i} \delta_{ij} \quad , \quad [x_i, x_j] = 0 \quad , \quad [p_i, p_j] = 0 \qquad (3.48)$$

the quantum field theory of the corresponding species is then obtained from the *classical* Schrödinger field by imposing the quantization relations for the quantum fields (not distinguishing them in notation from their *classical* counterparts)

$$[\Pi(\mathbf{x}, t), \psi(\mathbf{x}', t)] = \frac{\hbar}{i} \delta(\mathbf{x} - \mathbf{x}') \qquad (3.49)$$

and

$$[\psi(\mathbf{x}, t), \psi(\mathbf{x}', t)] = 0 \quad , \quad [\Pi(\mathbf{x}, t), \Pi(\mathbf{x}', t)] = 0. \qquad (3.50)$$

Since according to Eq. (3.45), $\Pi(\mathbf{x}, t) = i\hbar \psi^\dagger(\mathbf{x}, t)$, these are the commutation relations for a bose field, Eq. (1.101) and Eq. (1.102). The Hamiltonian, Eq. (3.47), is seen to be identical to the Hamiltonian operator on the multi-particle space, Eq. (2.19).

In this presentation the bose particles emerge as quanta of the field in analogy to the quanta of light in the analogous second quantization of the electromagnetic field (recall Section 1.4.2). The quantum field theory of fermions is similarly constructed as quanta of a field, but this time anti-commutation relations are assumed for the field.[8]

---

[8] A practicing quantum field theorist need thus not carry much baggage, short-cutting the road by *second quantization*.

3.3   Green's functions

An exact solution of a quantum field theory amounts, according to Eq. (3.32), to knowing all the correlation functions of the field variables; needless to say a mission impossible in general. We shall refer to these correlation functions generally as Green's functions.[9] We shall also use the word propagator interchangeably for the various types of Green's functions.

To get an intuitive feeling for the simplest kind of Green's function, the single-particle propagator, consider adding at time $t_{1'}$ a particle in state $\mathbf{p}_{1'}$ to the arbitrary state $|\Psi(t_{1'})\rangle$, i.e. we obtain the state $a_{\mathbf{p}_{1'}}^\dagger |\Psi(t_{1'})\rangle$, which at time t has evolved to the state

$$|\Psi_{\mathbf{p}_{1'},t_{1'}}(t)\rangle \;=\; e^{-iH(t-t_{1'})} \, a_{\mathbf{p}_{1'}}^\dagger \, |\Psi(t_{1'})\rangle \;=\; e^{-iHt} a_{\mathbf{p}_{1'}}^\dagger(t_{1'}) |\Psi_H\rangle \,, \qquad (3.51)$$

where in the last equality we have introduced the creation operator and state vector in the Heisenberg picture (choosing the time of coincidence with the Schrödinger picture at time $t_r = 0$). Similarly, we could consider the state where a particle at time t_1 is added in state $\mathbf{p}_1$. The amplitude for the event that the first constructed state is revealed in the other state at the arbitrary (and irrelevant) moment in time t is then

$$\langle \Psi_{\mathbf{p}_1 t_1}(t) | \Psi_{\mathbf{p}_{1'},t_{1'}}(t) \rangle \;=\; \langle \Psi_H | a_{\mathbf{p}_1}(t_1) \, a_{\mathbf{p}_{1'}}^\dagger(t_{1'}) | \Psi_H \rangle \qquad (3.52)$$

and the single-particle Green's function is a measure of the persistence, in time span $|t_1 - t_{1'}|$, of the single-particle character of the excitation consisting of adding a particle to the system (or determining the persistence of a hole state when removing a particle upon the interchange $a \leftrightarrow a^\dagger$).

3.3.1   Physical properties and Green's functions

Physical quantities for a many-body system such as the average (probability) density of the particles or the average particle (probability) current density are specified in terms of one-particle Green's functions. For a system in an arbitrary state described by the statistical operator ρ, the average density at a space-time point for a particle species of interest is (recall the result of Exercise 3.2 on page 58, which amounted to employing the cyclical property of the trace)

$$n(\mathbf{x}, t) \;=\; \sum_{\sigma_z} \text{Tr}(\rho \, \psi_{\mathcal{H}}^\dagger(\mathbf{x}, \sigma_z, t) \, \psi_{\mathcal{H}}(\mathbf{x}, \sigma_z, t)) \,, \qquad (3.53)$$

where the quantum field describing the particles $\psi_{\mathcal{H}}(\mathbf{x}, \sigma_z, t)$ is in the Heisenberg picture with respect to the arbitrary Hamiltonian $\mathcal{H}(t)$, and Tr denotes the trace in the multi-particle state space of the physical system in question. The reference time where the Schrödinger and Heisenberg pictures coincide is chosen as the moment where the state is specified, i.e. when the arbitrary statistical operator, ρ, representing the state of the system is specified. Here σ_z describes an internal degree of

---

[9]Thus using the notion in a broader sense than in mathematics, where it denotes the fundamental solution of a linear partial differential equation as discussed in Appendix C.

freedom of the identical particles in question. For example, in the case of electrons this is the spin degree of freedom, and the density is the sum of the density of electrons with spin up and down, respectively.[10] The average density is expressed in terms of the diagonal element of the so-called *G-lesser* Green's function

$$n(\mathbf{x}, t) = \pm i \sum_{\sigma_z} G^<(\mathbf{x}, \sigma_z, t, \mathbf{x}, \sigma_z, t) , \tag{3.54}$$

where[11]

$$\begin{aligned} G^<(\mathbf{x}, \sigma_z, t, \mathbf{x}'\sigma_z', t') &= \mp i \, \mathrm{Tr}(\rho \, \psi_{\mathcal{H}}^\dagger(\mathbf{x}', \sigma_z', t') \, \psi_{\mathcal{H}}(\mathbf{x}, \sigma_z, t)) \\ &\equiv \mp i \, \langle \psi_{\mathcal{H}}^\dagger(\mathbf{x}', \sigma_z', t') \, \psi_{\mathcal{H}}(\mathbf{x}, \sigma_z, t) \rangle , \end{aligned} \tag{3.55}$$

where upper (lower) sign corresponds to bosons (fermions), respectively, and we have introduced the notation that the bracket means trace of the operators in question weighted with respect to the state of the system, all quantities in the Heisenberg picture,

$$\langle \ldots \rangle \equiv \mathrm{Tr}(\rho \ldots) . \tag{3.56}$$

For the case of a pure state, $\rho = |\Psi\rangle\langle\Psi|$, we see that $G^<(\mathbf{x}, t, \mathbf{x}', t')$ is the amplitude for the transition at time t' to the state $\psi_{\mathcal{H}}(\mathbf{x}', \sigma_z', t')|\Psi\rangle$, where a particle with spin σ_z' is removed at position $\mathbf{x}'$ from state $|\Psi\rangle$, *given* the system at time t is in the state $\psi_{\mathcal{H}}(\mathbf{x}, \sigma_z, t)|\Psi\rangle$ where a particle with spin σ_z is removed at position $\mathbf{x}$ (assuming $t < t'$, otherwise we are dealing with the complex conjugate of the opposite transition). Equivalently, it is the amplitude to remain in the state $|\Psi\rangle$ after removing at time t a particle with spin σ_z at position $\mathbf{x}$ and restoring at time t' a particle with spin σ_z' at position $\mathbf{x}'$. For the case of a mixture, an additional statistical averaging over the distribution of initial states takes place. Average quantities, such as the probability density, can thus be expressed in terms of the one-particle Green's function.

The average electric current density for an assembly of identical fermions having charge e in an electric field represented by the vector potential $\mathbf{A}$ is, according to Eq. (2.47),

$$\begin{aligned} \mathbf{j}(\mathbf{x}, t) &= \frac{e\hbar}{2m} \sum_{\sigma_z} \left(\frac{\partial}{\partial \mathbf{x}} - \frac{\partial}{\partial \mathbf{x}'} \right) G^<(\mathbf{x}, \sigma_z, t, \mathbf{x}', \sigma_z, t) \bigg|_{\mathbf{x}'=\mathbf{x}} \\ &+ i \frac{e^2}{m} \mathbf{A}(\mathbf{x}, t) \sum_{\sigma_z} G^<(\mathbf{x}, \sigma_z, t, \mathbf{x}, \sigma_z, t) , \end{aligned} \tag{3.57}$$

the particles assumed to have an internal degree of freedom, say spin as is the case for electrons.

---

[10] One can, of course, also encounter situations where interest is in the density of electrons of a given spin, in which case one studies $n(\mathbf{x}, \sigma_z, t) = \mathrm{Tr}(\rho \psi_{\mathcal{H}}^\dagger(\mathbf{x}, \sigma_z, t) \, \psi_{\mathcal{H}}(\mathbf{x}, \sigma_z, t))$.

[11] The annoying presence of the imaginary unit is for later convenience with respect to the Feynman diagram rules. However, one is entitled to the choice of favorite for defining Green's functions, and the corresponding adjustment of the list of Feynman rules.

From the equation of motion for the field operator, Eq. (3.32), the equation of motion for the Green's function G-*lesser* becomes, for the case of two-particle interaction (assuming spin independent interaction so the spin degree of freedom is suppressed or using inclusive notation),

$$\left(i\hbar \frac{\partial}{\partial t} - h(t) \right) G^<(\mathbf{x}, t, \mathbf{x}', t') = \int d\mathbf{x}''\, V^{(2)}(\mathbf{x}, \mathbf{x}'')\, G_{(2)}(\mathbf{x}, t, \mathbf{x}'', t, \mathbf{x}'', t, \mathbf{x}', t') ,$$

$$(3.58)$$

where

$$G_{(2)}(\mathbf{x}, t, \mathbf{x}'', t, \mathbf{x}'', t, \mathbf{x}', t') = \pm i \langle \psi(\mathbf{x}, t)\, \psi(\mathbf{x}'', t)\, \psi^\dagger(\mathbf{x}'', t)\, \psi^\dagger(\mathbf{x}', t') \rangle \qquad (3.59)$$

is a so-called two-particle Green's function since it involves the propagation of two particles. The dynamics of a system, specified by the time dependence of the one-particle Green's function, is thus described in terms of higher-order correlation functions in the field operators. The equation of motion for the one-particle Green's function thus leads to an infinite hierarchy of equations for correlation functions containing ever increasing numbers of field operators, describing the correlations set up in the system by the interactions.[12] Since there is no closed set of equations for reduced quantities such as Green's functions, approximations are, in practice, needed in order to obtain information about the system. On some occasions the system provides a small parameter that allows controlled approximations; a case to be studied later is that of electron–phonon interaction in metals. In less controllable situations one in despair appeals to the tendency of higher-order correlations to average out for a many-particle system, when it comes to such average properties as densities and currents, so that the hierarchy of correlations can be broken off self-consistently at low order. We shall discuss such situations in Section 10.6 and in Chapter 12 in the context of applying the effective action approach to such differing situations as a trapped Bose–Einstein condensate and classical statistical dynamics, respectively.

3.3.2   Stable of one-particle Green's functions

The correlation function G-*lesser* appeared in the previous section most directly as related to average properties such as densities and currents. However, we shall encounter various types of quantum field correlation functions, i.e. various kinds of Green's functions that appear for reasons of their own. For definiteness we collect them all here, though they are not needed until later. The rest of this chapter can thus be skipped on a first reading if one shares the view that things should not be called upon before needed.

We shall also encounter the so-called G-*greater* Green's function

$$G^>(\mathbf{x}, t, \mathbf{x}', t') = -i\, \langle \psi_{\mathcal{H}}(\mathbf{x}, t)\, \psi_{\mathcal{H}}^\dagger(\mathbf{x}', t') \rangle = -i \mathrm{Tr}(\rho\, \psi_{\mathcal{H}}(\mathbf{x}, t)\, \psi_{\mathcal{H}}^\dagger(\mathbf{x}', t')) , \quad (3.60)$$

the amplitude for the process of an added particle at position $\mathbf{x}$ at time t *given* a particle is added at position $\mathbf{x}'$ at time t', the one-particle propagator in the presence of interaction with all the other particles.

---

[12]Analogous to the BBGKY-hierachy in classical kinetics or for any description of a system in terms of a reduced, i.e. partially traced out, quantity.

We shall later, for reasons of calculation in perturbation theory, encounter the time-ordered Green's function

$$G(\mathbf{x}, t, \mathbf{x}', t') = -i \langle T(\psi_{\mathcal{H}}(\mathbf{x}, t)\, \psi_{\mathcal{H}}^{\dagger}(\mathbf{x}', t')) \rangle \qquad (3.61)$$

and we note (valid for both bosons and fermions, recalling the minus sign convention when two fermi fields are interchanged)

$$G(\mathbf{x}, t, \mathbf{x}', t') = \begin{cases} G^{<}(\mathbf{x}, t, \mathbf{x}', t') & t' > t \\ G^{>}(\mathbf{x}, t, \mathbf{x}', t') & t > t' . \end{cases} \qquad (3.62)$$

In perturbation theory, the time-ordered Green's function appears because of the crucial role of time-ordering in the evolution operator, Eq. (3.7). Quantum dynamics is ruled by operators, non-commuting objects. However, as shown in Chapter 5, the necessity of the time-ordered Green's function is only in one version of perturbation theory, and then an additional analytic continuation needs to be invoked. Or, if one is interested only in ground state properties, then perturbation theory can be formulated in closed form involving only the time-ordered Green's function. The general real-time perturbation theory valid for non-equilibrium situations will be formulated in Chapter 5 in terms of essentially two Green's functions, and in a way which displays physical information of systems most transparently.

Finally, in this set-up we shall also later encounter the anti-time-ordered Green's function

$$\tilde{G}(\mathbf{x}, t, \mathbf{x}', t') = -i \langle \tilde{T}(\psi_{\mathcal{H}}(\mathbf{x}, t)\, \psi_{\mathcal{H}}^{\dagger}(\mathbf{x}', t')) \rangle , \qquad (3.63)$$

where $\tilde{T}$ anti-time orders, i.e. orders oppositely to that of T. We note that the time-ordered and anti-time-ordered Green's functions can be expressed in terms of G-greater and G-lesser, for example

$$\tilde{G}(\mathbf{x}, t, \mathbf{x}', t') = \theta(t - t')\, G^{<}(\mathbf{x}, t, \mathbf{x}', t') + \theta(t' - t)\, G^{>}(\mathbf{x}, t, \mathbf{x}', t') , \qquad (3.64)$$

where θ is the step or Heaviside function.

Recalling Eq. (3.58), we note for the free Green's functions the relations

$$G_0^{-1}(\mathbf{x}, t)\, G_0^{<}(\mathbf{x}, t, \mathbf{x}', t') = 0 \quad , \quad G_0^{-1}(\mathbf{x}, t)\, G_0^{>}(\mathbf{x}, t, \mathbf{x}', t') = 0 \qquad (3.65)$$

and for the time-ordered

$$G_0^{-1}(\mathbf{x}, t)\, G_0(\mathbf{x}, t, \mathbf{x}', t') = \hbar\, \delta(\mathbf{x} - \mathbf{x}')\, \delta(t - t') \qquad (3.66)$$

and anti-time-ordered

$$G_0^{-1}(\mathbf{x}, t)\, \tilde{G}_0(\mathbf{x}, t, \mathbf{x}', t') = -\hbar\, \delta(\mathbf{x} - \mathbf{x}')\, \delta(t - t') , \qquad (3.67)$$

where

$$G_0^{-1}(\mathbf{x}, t) = \left(i\hbar \frac{\partial}{\partial t} - h\left(\frac{\hbar}{i} \nabla_{\mathbf{x}}, \mathbf{x}, t \right) \right) , \qquad (3.68)$$

which for the case of a charged particle coupled to an electromagnetic field is

$$G_0^{-1}(\mathbf{x}, t) = \left(i\hbar \frac{\partial}{\partial t} - \frac{1}{2m} \left(\frac{\hbar}{i} \frac{\partial}{\partial \mathbf{x}} - e\mathbf{A}(\mathbf{x}, t) \right)^2 - e\varphi(\mathbf{x}, t) \right) . \tag{3.69}$$

Introducing

$$G_0^{-1}(\mathbf{x}, t, \mathbf{x}', t') = G_0^{-1}(\mathbf{x}, t)\, \delta(\mathbf{x} - \mathbf{x}')\, \delta(t - t') \tag{3.70}$$

we obtain a quantity on equal footing with the Green's function, the inverse free Green's function (here in the position representation) as

$$(\hat{G}_0^{-1} \otimes \hat{G}_0)(\mathbf{x}, t, \mathbf{x}', t') = \hbar\, \delta(\mathbf{x} - \mathbf{x}')\, \delta(t - t') , \tag{3.71}$$

where $\otimes$ signifies matrix multiplication in the spatial and time variables, i.e. internal integrations over space and for the latter internal integration from minus to plus infinity of times.

Exercise 3.9. The equation of motion for the free phonon field is (recall Section 2.4.3)

$$\Box \phi(\mathbf{x}, t) = 0 . \tag{3.72}$$

Show that the time-ordered free phonon Green's function

$$D_0(\mathbf{x}, t, \mathbf{x}', t') = -i\langle T(\phi(\mathbf{x}, t)\, \phi(\mathbf{x}', t')) \rangle \tag{3.73}$$

therefore satisfies the equation of motion

$$\Box D_0(\mathbf{x}, t, \mathbf{x}', t') = \frac{\hbar}{i} \triangle_{\mathbf{x}}\, \delta(\mathbf{x} - \mathbf{x}')\, \delta(t - t') . \tag{3.74}$$

Exercise 3.10. From the equation of motion for the field operator, show that the equation of motion for the time-ordered Green's function is

$$\left(i\hbar \frac{\partial}{\partial t} - h_0(t) \right) G(\mathbf{x}, t, \mathbf{x}', t') = \hbar\, \delta(\mathbf{x} - \mathbf{x}')\delta(t - t')$$

$$- i\langle T([\psi(\mathbf{x}, t), H_i(t)]\, \psi^\dagger(\mathbf{x}', t')) \rangle , \tag{3.75}$$

where $H_i(t)$ is the interaction part of the Hamiltonian in the Heisenberg picture.

Other combinations of field correlations will be of importance in Chapter 5 when the real-time perturbation theory of general non-equilibrium states are considered, viz. the retarded Green's function

$$G^{\mathrm{R}}(\mathbf{x}, t, \mathbf{x}', t') = -i\theta(t - t')\langle [\psi(\mathbf{x}, t), \psi^\dagger(\mathbf{x}', t')]_{\mp} \rangle$$

$$= \theta(t - t')\left(G^{>}(\mathbf{x}, t, \mathbf{x}', t') - G^{<}(\mathbf{x}, t, \mathbf{x}', t') \right) \tag{3.76}$$

and advanced Green's functions

$$G^A(\mathbf{x}, t, \mathbf{x}', t') = i\theta(t' - t)\langle[\psi(\mathbf{x}, t), \psi^\dagger(\mathbf{x}', t')]_\mp\rangle$$

$$= -\theta(t' - t)\left(G^>(\mathbf{x}, t, \mathbf{x}', t') - G^<(\mathbf{x}, t, \mathbf{x}', t')\right) \quad (3.77)$$

and the Keldysh or kinetic Green's function

$$G^K(\mathbf{x}, t, \mathbf{x}', t') = -i\langle[\psi(\mathbf{x}, t), \psi^\dagger(\mathbf{x}', t')]_\pm\rangle$$

$$= G^>(\mathbf{x}, t, \mathbf{x}', t') + G^<(\mathbf{x}, t, \mathbf{x}', t'), \quad (3.78)$$

where upper and lower signs, as usual, are for bose and fermi fields, respectively. Introducing the notation $\bar{s} = -s$, the two kinds of statistics can be combined leaving the Green's functions in the forms

$$G^R(\mathbf{x}, t, \mathbf{x}', t') = -i\theta(t - t')\langle[\psi(\mathbf{x}, t), \psi^\dagger(\mathbf{x}', t')]_{\bar{s}}\rangle \quad (3.79)$$

and

$$G^A(\mathbf{x}, t, \mathbf{x}', t') = i\theta(t' - t)\langle[\psi(\mathbf{x}, t), \psi^\dagger(\mathbf{x}', t')]_{\bar{s}}\rangle \quad (3.80)$$

and

$$G^K(\mathbf{x}, t, \mathbf{x}', t') = -i\langle[\psi(\mathbf{x}, t), \psi^\dagger(\mathbf{x}', t')]_s\rangle = G^S(\mathbf{x}, t, \mathbf{x}', t') \quad (3.81)$$

where the superscript on the last Green's function also could remind us of it being symmetric with respect to the quantum statistics.

Exercise 3.11. Show that the density, up to a state independent constant, can be expressed in terms of the kinetic Green's function according to

$$n(\mathbf{x}, t) = \pm i \sum_{\sigma_z} G^K(\mathbf{x}, \sigma_z, t, \mathbf{x}, \sigma_z, t) . \quad (3.82)$$

Exercise 3.12. Show that the current density can be expressed in terms of the kinetic Green's function according to (in the absence of a vector potential)

$$\mathbf{j}(\mathbf{x}, t) = \frac{e\hbar}{2m}\left(\frac{\partial}{\partial\mathbf{x}} - \frac{\partial}{\partial\mathbf{x}'}\right)G^K(\mathbf{x}, t, \mathbf{x}', t)\bigg|_{\mathbf{x}'=\mathbf{x}} . \quad (3.83)$$

The presence of a vector potential just adds the diamagnetic term (recall Eq. (3.57)) in accordance with gauge invariance, $-i\hbar\nabla \to -i\hbar\nabla - e\mathbf{A}$.

We note the relationship

$$G^R(\mathbf{x}, t, \mathbf{x}', t') - G^A(\mathbf{x}, t, \mathbf{x}', t') = G^>(\mathbf{x}, t, \mathbf{x}', t') - G^<(\mathbf{x}, t, \mathbf{x}', t') \quad (3.84)$$

irrespective of the quantum statistics of the particles. The above combination is of such importance that we introduce the additional notation for the spectral weight function

$$A(\mathbf{x},t,\mathbf{x}',t') = i(G^{\mathrm{R}}(\mathbf{x},t,\mathbf{x}',t') - G^{\mathrm{A}}(\mathbf{x},t,\mathbf{x}',t')) = \langle[\psi(\mathbf{x},t),\psi^{\dagger}(\mathbf{x}',t')]_{\mp}\rangle$$

$$= i\left(G^{>}(\mathbf{x},t,\mathbf{x}',t') - G^{<}(\mathbf{x},t,\mathbf{x}',t')\right) . \tag{3.85}$$

We note as a consequence of the equal time anti-commutation or commutation relations of the field operators, that the spectral function at equal times satisfies

$$A(\mathbf{x},t,\mathbf{x}',t) = \delta(\mathbf{x}-\mathbf{x}') \tag{3.86}$$

irrespective of the state of the system.

Exercise 3.13. Introduce the mixed or Wigner coordinates[13]

$$\mathbf{R} = \frac{\mathbf{x}+\mathbf{x}'}{2} \quad , \quad \mathbf{r} = \mathbf{x}-\mathbf{x}' \tag{3.87}$$

and

$$T = \frac{t+t'}{2} \quad , \quad t = t-t' . \tag{3.88}$$

Show that the spectral weight function expressed in these variables satisfies the sum-rule

$$\int\limits_{-\infty}^{\infty} \frac{dE}{2\pi} A(E,\mathbf{p},\mathbf{R},T) = 1 . \tag{3.89}$$

Exercise 3.14. Verify the relations, valid for both bosons and fermions,

$$G^{\mathrm{A}}(\mathbf{x},t,\mathbf{x}',t') = \left(G^{\mathrm{R}}(\mathbf{x}',t',\mathbf{x},t)\right)^{*} \tag{3.90}$$

and

$$G^{\mathrm{K}}(\mathbf{x},t,\mathbf{x}',t') = -\left(G^{\mathrm{K}}(\mathbf{x}',t',\mathbf{x},t)\right)^{*} \tag{3.91}$$

and

$$A(\mathbf{x},t,\mathbf{x}',t') = \left(A(\mathbf{x}',t',\mathbf{x},t)\right)^{*} \tag{3.92}$$

and

$$G^{<}(\mathbf{x},t,\mathbf{x}',t') = -\left(G^{<}(\mathbf{x}',t',\mathbf{x},t)\right)^{*} \tag{3.93}$$

and

$$G^{>}(\mathbf{x},t,\mathbf{x}',t') = -\left(G^{>}(\mathbf{x}',t',\mathbf{x},t)\right)^{*} . \tag{3.94}$$

Note the relations are valid for arbitrary states.

---

[13]There will be more about Wigner coordinates in Section 7.2.

For the case of a hermitian bose field, such as the phonon field, additional useful relations exist

$$D^R(\mathbf{x}, t, \mathbf{x}', t') = D^A(\mathbf{x}', t', \mathbf{x}, t) \tag{3.95}$$

and

$$D^K(\mathbf{x}, t, \mathbf{x}', t') = D^K(\mathbf{x}', t', \mathbf{x}, t) \tag{3.96}$$

and

$$D^>(\mathbf{x}, t, \mathbf{x}', t') = D^<(\mathbf{x}', t', \mathbf{x}, t) . \tag{3.97}$$

We thus have that $D^{R(A)}(\mathbf{x}, t, \mathbf{x}', t')$ are real functions, whereas $D^K(\mathbf{x}, t, \mathbf{x}', t')$ is purely imaginary.

Above, the Green's function are displayed in terms of the fields in the position representation. Equally, we can introduce the Green's function displayed in the momentum representation, related to the above by Fourier transformation, or for that matter in any representation specified by a complete set of states, say the energy representation specified in terms of the eigenstates of the Hamiltonian.

Correlation functions of the quantum fields can be obtained by differentiation of a generating functional. For example, to generate time-ordered Green's functions we introduce

$$Z[\eta, \eta^*] = \left\langle T e^{i \int d\mathbf{x} \int_{-\infty}^{\infty} dt \, (\psi(\mathbf{x},t) \eta(\mathbf{x},t) + \psi^\dagger(\mathbf{x},t) \eta^*(\mathbf{x},t))} \right\rangle \tag{3.98}$$

generating for example the time-ordered Green's function for bosons, Eq. (3.61), by differentiating twice with respect to the complex c-number source function η,[14] to give

$$G(\mathbf{x}, t; \mathbf{x}', t') = i \frac{\delta^2 Z[\eta, \eta^*]}{\delta \eta^*(\mathbf{x}', t') \delta \eta(\mathbf{x}, t)} \bigg|_{\eta=0=\eta^*} = -i\langle T(\psi(\mathbf{x}, t) \psi^\dagger(\mathbf{x}', t')) \rangle . \tag{3.99}$$

The generating functional is a device we shall consider in detail in Chapter 9.

The Green's functions introduced in this section are the correlation functions for the case of an arbitrary state. Before we embark on the construction of the general non-equilibrium perturbation theory and its diagrammatic representation starting from the canonical formalism as presented here and in the first two chapters, we consider briefly equilibrium theory, in particular the general property characterizing equilibrium.[15]

---

[14] For the case of fermions, the sources must be anti-commuting c-numbers, so-called Grassmann variables. We elaborate on this point in Chapter 9.

[15] In Chapter 9 we proceed the other way around, and the reader inclined to take diagrammatics as a starting point of a physical theory can thus start from there.

3.4   Equilibrium Green's functions

In this section we shall consider a system in thermal equilibrium. In that case the state of the system is specified by the Boltzmann statistical operator, Eq. (2.87), characterized by its macroscopic parameter, the temperature T.

In thermal equilibrium, the correlation functions of a system are subdued to a boundary condition in imaginary time as specified by the fluctuation–dissipation theorem.[16] In the canonical ensemble, for example the relation

$$\langle \psi_H(\mathbf{x}, t)\, \psi_H^\dagger(\mathbf{x}', t') \rangle \;=\; \langle \psi_H^\dagger(\mathbf{x}', t')\, \psi_H(\mathbf{x}, t + i\beta) \rangle \tag{3.100}$$

is valid, where $\beta = \hbar/kT$, as a consequence of the cyclic invariance of the trace as the bracket denotes the average

$$\langle \ldots \rangle \;\equiv\; \mathrm{Tr}\left(\frac{e^{-H/kT}}{\mathrm{Tr}(e^{-H/kT})} \cdots \right). \tag{3.101}$$

The relationship in Eq. (3.100) can, for example, be stated in terms of the Green's functions as

$$G^<(\mathbf{x}, t + i\beta, \mathbf{x}', t') \;=\; \pm G^>(\mathbf{x}, t, \mathbf{x}', t') \tag{3.102}$$

valid for arbitrary interactions among the particles in the system.

Exercise 3.15. Show that in the grand canonical ensemble, for example the following relation, is valid

$$\langle \psi_H(\mathbf{x}, t - i\beta)\, \psi_H^\dagger(\mathbf{x}', t') \rangle \;=\; e^{\frac{\beta\mu}{\hbar}} \langle \psi_H^\dagger(\mathbf{x}', t')\, \psi_H(\mathbf{x}, t) \rangle \tag{3.103}$$

in which case the average is

$$\langle \ldots \rangle \;\equiv\; \mathrm{Tr}\left(\frac{e^{-(H-\mu N)/kT}}{\mathrm{Tr}(e^{-(H-\mu N)/kT}} \cdots \right) \tag{3.104}$$

and the Hamiltonian and the total number operator commute if the chemical potential is nonzero (recall Eq. (2.36)). Stated in terms of Green's functions we have

$$G^<(\mathbf{x}, t, \mathbf{x}', t') \;=\; \pm e^{\frac{\beta\mu}{\hbar}} G^>(\mathbf{x}, t - i\beta, \mathbf{x}', t'), \tag{3.105}$$

where in the grand canonical ensemble for example

$$
\begin{aligned}
G^<(\mathbf{x}, t, \mathbf{x}', t') \;&=\; \mp i \langle \psi_H^\dagger(\mathbf{x}', t')\, \psi_H(\mathbf{x}, t) \rangle \\[2mm]
&=\; \frac{\mp i}{\mathrm{Tr}(e^{-(H-\mu N)/kT})} \, \mathrm{Tr}(e^{-(H-\mu N)/kT} \psi_H^\dagger(\mathbf{x}', t')\, \psi_H(\mathbf{x}, t)).
\end{aligned}
\tag{3.106}
$$

---

[16] Additional discussion of the fluctuation–dissipation theorem and its importance in linear response theory are continued in Chapter 6. That the operators in Eq. (3.100) are the field operators is immaterial; the relationship is valid for arbitrary operators.

The importance of the canonical ensembles should be stressed for the validity of these fluctuation–dissipation relations or so-called Kubo–Martin–Schwinger boundary conditions. They state that the Green's functions are anti-periodic or periodic in imaginary time depending on the particles being fermions or bosons, the interval of periodicity being set by the inverse temperature. This is the crucial observation for the Euclidean or imaginary-time formulation of quantum statistical mechanics, as further discussed in Section 5.7.

In thermal equilibrium, correlation functions only depend on the difference between the times, $t - t'$, i.e. they are invariant with respect to displacements in time. If in addition the equilibrium state is translationally invariant, then all Green's functions are specified according to

$$G(\mathbf{x}, t, \mathbf{x}', t') = \int \frac{d\mathbf{p}}{(2\pi\hbar)^3} \int_{-\infty}^{\infty} \frac{dE}{2\pi} e^{\frac{i}{\hbar}(\mathbf{p} \cdot (\mathbf{x} - \mathbf{x}') - E(t - t'))} G(\mathbf{p}, E) \tag{3.107}$$

or equivalently

$$G(\mathbf{p}, E, \mathbf{p}', E') = 2\pi (2\pi\hbar)^3 \delta(\mathbf{p} - \mathbf{p}') \delta(E - E') G(\mathbf{p}, E). \tag{3.108}$$

For example,

$$G^K(\mathbf{p}, E) = -i \int_{-\infty}^{\infty} d(t - t') e^{\frac{i}{\hbar}(t - t')E} \langle [a_\mathbf{p}(t), a_\mathbf{p}^\dagger(t')]_\pm \rangle. \tag{3.109}$$

The relationship in Eq. (3.105) then takes the form of the detailed balancing condition

$$G^<(\mathbf{p}, E) = \pm e^{-\frac{E - \mu}{kT}} G^>(\mathbf{p}, E). \tag{3.110}$$

Exercise 3.16. Show that, for free bosons or fermions specified by the Hamiltonian in Eq. (2.21), we have

$$G_0^R(\mathbf{p}, E) - G_0^A(\mathbf{p}, E) = -2\pi i \, \delta(E - \epsilon_\mathbf{p}) \quad , \quad \epsilon_\mathbf{p} = \frac{p^2}{2m}. \tag{3.111}$$

Exercise 3.17. Show that, for free longitudinal phonons,, specified by the Hamiltonian in Eq. (1.123), we have

$$D_0^R(\mathbf{k}, \omega) - D_0^A(\mathbf{k}, \omega) = -2\pi i \, \omega_\mathbf{k}^2 \, \mathrm{sign}(\omega) \, \delta(\omega^2 - \omega_\mathbf{k}^2) \, \theta(\omega_D - |\omega|) \tag{3.112}$$

where the sign-function, $\mathrm{sign}(x) = \theta(x) - \theta(-x) = x/|x|$, is plus or minus one, depending on the sign of the argument.

Instead of defining the unitary transformation to the Heisenberg picture according to Eq. (3.26), we can let it be governed by the grand canonical Hamiltonian, $H - \mu N$, and we have, according to Eq. (2.36),[17]

$$\psi_\mu(\mathbf{x}, t) = e^{-\frac{i}{\hbar}\mu N t}\, \psi_H(\mathbf{x}, t)\, e^{\frac{i}{\hbar}\mu N t} = e^{\frac{i}{\hbar}\mu t}\, \psi_H(\mathbf{x}, t) \,. \tag{3.113}$$

Defining the grand canonical Green's functions in terms of these fields, we observe that they are related to those defined according to Eq. (3.106), or those in the canonical ensemble in the thermodynamic limit, according to

$$G_\mu(\mathbf{x}, t, \mathbf{x}', t') = G(\mathbf{x}, t, \mathbf{x}', t')\, e^{\frac{i}{\hbar}\mu(t-t')} \,. \tag{3.114}$$

Since average densities and currents are expressed in terms of the equal time Green's function, formulas have the same appearance in both ensembles.

For the Fourier transformed Green's function with respect to time, the transition to the grand canonical ensemble thus corresponds to the substitution $E \to E + \mu$, as energies will appear measured from the chemical potential. The detailed balancing condition, Eq. (3.110), can therefore, for a translationally invariant state, equivalently be stated in the form

$$G^<(E, \mathbf{p}) = \pm e^{-E/kT}\, G^>(E, \mathbf{p}) \tag{3.115}$$

and we have dropped the chemical potential index as these are the Green's functions we shall use in the following. The absence of the chemical potential in the exponential shows that the relationships are specified in the grand canonical ensemble, where energies are measured relative to the chemical potential (upper and lower signs refer as usual to bosons and fermions, respectively).

In thermal equilibrium, the kinetic Green's function and the retarded and advanced Green's functions, or rather the spectral weight function, are thus related for the case of fermions according to

$$G^K(E, \mathbf{p}) = (G^R(E, \mathbf{p}) - G^A(E, \mathbf{p}))\tanh\frac{E}{2kT} \,. \tag{3.116}$$

In thermal equilibrium, all Green's functions can thus be specified once a single of them is known, say the retarded Green's functions, and the quantum statistics of the particles is then reflected in relations governed by the fluctuation–dissipation type relationship such as in Eq. (3.116). Occasionally we keep Boltzmann's constant, k, explicitly, the non-essential converter between energy and temperature scales.

Exercise 3.18. Show that, for bosons in equilibrium at temperature T, the fluctuation–dissipation theorem reads

$$G^K(E, \mathbf{p}) = (G^R(E, \mathbf{p}) - G^A(E, \mathbf{p}))\coth\frac{E}{2kT} \,. \tag{3.117}$$

---

[17]The number operator is assumed to commute with the Hamiltonian. If the number operator does not commute with the Hamiltonian, such as for phonons, the description is of course in the grand canonical ensemble and the chemical potential vanishes.

Exercise 3.19. Show that

$$iG^{>}(E,\mathbf{p}) = (1 \pm f_{\mp}(E)) A(E,\mathbf{p}) \tag{3.118}$$

and

$$\pm iG^{<}(E,\mathbf{p}) = f_{\mp}(E) A(E,\mathbf{p}), \tag{3.119}$$

where the functions

$$f_{\mp}(E) = \frac{1}{e^{E/kT} \mp 1} \tag{3.120}$$

denote either the Bose–Einstein distribution or Fermi–Dirac distribution, for bosons and fermions respectively.

Exercise 3.20. Show that the average energy in the thermal equilibrium state for the case of two-body interaction between fermions (recall Exercise 3.7 on page 60), for example, can be expressed as

$$\langle H \rangle = \frac{1}{2} \sum_{\mathbf{p}} \left\langle \left. \left(\epsilon_{\mathbf{p}} a_{\mathbf{p}}^{\dagger}(t) a_{\mathbf{p}}(t) + i\hbar a_{\mathbf{p}}^{\dagger}(t) \frac{da_{\mathbf{p}}(t)}{dt} \right) \right|_{t=0} \right\rangle$$

$$= -i\frac{1}{2} \sum_{\mathbf{p}} \left. \left(i\hbar \frac{d}{dt} + \epsilon_{\mathbf{p}} \right) G^{<}(\mathbf{p},t) \right|_{t=0}$$

$$= -i\frac{1}{2} \sum_{\mathbf{p}} \int_{-\infty}^{\infty} \frac{dE}{2\pi} (E + \epsilon_{\mathbf{p}}) G^{<}(\mathbf{p},E)$$

$$= \frac{1}{2} \sum_{\mathbf{p}} \int_{-\infty}^{\infty} \frac{dE}{2\pi} (E + \epsilon_{\mathbf{p}}) A(\mathbf{p},E) f(E), \tag{3.121}$$

where f is the Fermi function, and thereby for the energy density

$$\frac{\langle H \rangle}{V} = \frac{1}{2} \int \frac{d\mathbf{p}}{(2\pi\hbar)^3} \int_{-\infty}^{\infty} \frac{dE}{2\pi} (E + \epsilon_{\mathbf{p}}) A(\mathbf{p},E) f(E). \tag{3.122}$$

For a system in thermal equilibrium, the correlation function

$$\langle \psi_{\mu}(\mathbf{x},t) \psi_{\mu}^{\dagger}(\mathbf{x}',t') \rangle = \text{Tr}(e^{(\Omega-(H-\mu N))/kT} \psi_{\mu}(\mathbf{x},t) \psi_{\mu}^{\dagger}(\mathbf{x}',t')) \tag{3.123}$$

can be spectrally decomposed by inserting a complete set of energy states in the multi-particle space

$$(H - \mu N)|E_n, N\rangle = (E_n - \mu N)|E_n, N\rangle \tag{3.124}$$

giving

$$\langle \psi_\mu(\mathbf{x}, t)\, \psi_\mu^\dagger(\mathbf{x}', t') \rangle = \sum_{N,n,m} e^{(\Omega - E_n + \mu N)/kT} e^{i(t-t')(E_n - E_m + \mu)} \times$$

$$\langle N, E_n | \psi(\mathbf{x}, t = 0)) | E_m, N+1 \rangle \times$$

$$\langle N+1, E_m | \psi^\dagger(\mathbf{x}', t' = 0) | E_n, N \rangle. \qquad (3.125)$$

From this expression we observe that the *G-greater* Green's function $G^>(\mathbf{x}, t, \mathbf{x}', t')$ considered as a function of imaginary times is an analytic function in the region, $-1/kT < \Im m(t - t') < 0$, if the exponential $\exp\{-E_n(1/kT + i(t - t'))\}$ dominates the convergence of the sum.

Exercise 3.21. Show similarly that $G^<(\mathbf{x}, t, \mathbf{x}', t')$ is an analytic function in the region of imaginary times, $0 < \Im m(t - t') < 1/kT$.

Assuming a translational invariant system and using Eq. (2.16) we have

$$\langle \psi_\mu(\mathbf{x}, t)\, \psi_\mu^\dagger(\mathbf{x}', t') \rangle = \sum_{N,n,m} e^{(\Omega - E_n + \mu N)/kT} e^{i(t-t')(E_n - E_m + \mu)}$$

$$\times\ e^{-i(\mathbf{x} - \mathbf{x}')\mathbf{p}_{nm}} \langle N, E_n | \psi(\mathbf{0}, 0) | E_m, N+1 \rangle$$

$$\times\ \langle N+1, E_m | \psi^\dagger(\mathbf{0}, 0) | E_n, N \rangle, \qquad (3.126)$$

where $\mathbf{p}_{nm} = \mathbf{P}_n - \mathbf{P}_m$ is the difference between the total momentum eigenvalues for the two states in question. For the Fourier transform we then have

$$\langle a_\mathbf{p}(t)\, a_\mathbf{p}^\dagger(t') \rangle = (2\pi)^3 \sum_{N,n,m} e^{(\Omega - E_n + \mu N)/kT} e^{i(t-t')(E_n - E_m + \mu)}$$

$$\times \delta(\mathbf{p} - \mathbf{p}_{mn}) |\langle N, E_n | \psi(\mathbf{0}, 0)) | E_m, N+1 \rangle|^2. \qquad (3.127)$$

Noting the analyticity in the upper-half ω-plane of the following function, we have for real values of ω

$$\int_{-\infty}^{\infty} dt\, \theta(t)\, e^{i\omega t} = \frac{1}{\omega + i\delta}, \qquad (3.128)$$

where $\delta = 0^+$, or equivalently

$$\theta(t) = \int_{-\infty}^{\infty} \frac{d\omega}{-2\pi i} \frac{e^{-i\omega t}}{\omega + i\delta}. \qquad (3.129)$$

Therefore for the retarded and advanced Green's functions we have the spectral representations

$$G^{R(A)}(\mathbf{p}, E) = (2\pi)^3 e^{\Omega/kT} \sum_{N_n, n, m} e^{-(E_n - \mu N)/kT} \times$$

$$\left(\delta(\mathbf{p} - \mathbf{p}_{mn}) \frac{|\langle N_n, E_n | \psi(\mathbf{0}, 0) | E_m, N_m \rangle|^2}{E + E_{mn} \, (\overset{+}{\scriptscriptstyle -}) \, i\delta} \mp (n \leftrightarrow m) \right), \quad (3.130)$$

where $E_{mn} = E_m - E_n + \mu(N_m - N_n)$, and we recall that $N_m = N_n \pm 1$. The retarded (advanced) Green's function is thus analytic in the upper (lower) half-plane for the energy variable E. The simple poles for the retarded (advanced) Green's function are thus spread densely just below (above) the real axis, and in the thermodynamic limit this spectrum of simple poles coalesces into a continuum creating a branch cut for the functions along the real axis.

In equilibrium all propagators are thus specified in terms of a single Green's function, say the retarded or equivalently the spectral function, as by analyticity the retarded and advanced Green's functions satisfy the causality or Kramers–Kronig relations for their real and imaginary parts, or compactly

$$G^{R(A)}(E, \mathbf{p}) = \int_{-\infty}^{\infty} \frac{dE'}{-2\pi i} \frac{G^R(E', \mathbf{p}) - G^A(E', \mathbf{p})}{E - E' \, (\overset{+}{\scriptscriptstyle -}) \, i0}$$

$$= \int_{-\infty}^{\infty} \frac{dE'}{2\pi} \frac{A(E', \mathbf{p})}{E - E' \, (\overset{+}{\scriptscriptstyle -}) \, i0}. \quad (3.131)$$

The spectral weight function, has according to Eq. (3.130), the spectral decomposition

$$A(\mathbf{p}, E) = -(2\pi)^4 \sum_{N_n, n, m} e^{(\Omega - (E_n - \mu N_n))/kT}$$

$$\times \left(\delta(\mathbf{p} - \mathbf{p}_{mn}) \, \delta(E - E_{mn}) |\langle N_n, E_n | \psi(\mathbf{0}, 0) | E_m, N_m \rangle|^2 \right.$$

$$\left. \mp (n \leftrightarrow m) \right) \quad (3.132)$$

or equivalently

$$A(\mathbf{p}, E) = (2\pi)^4 \sum_{N_n, n, m} e^{(\Omega - (E_n - \mu N_n))/kT} \left(1 \mp e^{-\frac{E_{mn}}{kT}} \right)$$

$$\times \delta(\mathbf{p} - \mathbf{p}_{mn}) \, \delta(E - E_{mn}) |\langle N_n, E_n | \psi(\mathbf{0}, 0) | E_m, N_m \rangle|^2, \quad (3.133)$$

where the upper and lower sign is for bosons and fermions, respectively.

The analytic properties of the retarded (or advanced) Green's function determines the analytic properties of all the other introduced Green's functions, and are further studied in Section 5.6.

The three Green's functions, $G^{R,A,K}$, thus carry different information about the many-body system: $G^{R,A}$ the spectral properties and G^K in addition the quantum statistics of the concerned particles. In Chapter 5 we will construct the diagrammatic perturbation theory that, even for non-equilibrium states, keeps these important features explicit.

Exercise 3.22. Show that for large energy variable, $E \rightarrow \infty$, the retarded and advanced Green's functions always have the asymptotic behavior

$$G^{R(A)}(E, \mathbf{p}) \simeq \frac{1}{E}. \tag{3.134}$$

In the absence of interactions, i.e. for free bosons or fermions specified by the Hamiltonian in Eq. (2.21), one readily obtains for the spectral weight function, Eq. (3.85),

$$A_0(\mathbf{p}, E) = 2\pi\,\delta(E - \xi_\mathbf{p}) \quad, \quad \xi_\mathbf{p} = \epsilon_\mathbf{p} - \mu = \frac{\mathbf{p}^2}{2m} - \mu, \tag{3.135}$$

and according to the fluctuation–dissipation theorem, all one-particle Green's function are then immediately obtained.

In the presence of interactions, the delta-spike in the spectral weight function will be broadened and a tail appears, however, subject to the general sum-rule of Eq. (3.89) which for the equilibrium state reads

$$\int\limits_{-\infty}^{\infty} \frac{dE}{2\pi}\, A(E, \mathbf{p}) = 1. \tag{3.136}$$

Exercise 3.23. The quantum statistics of particles have, according to the above, a profound influence on the form of the Green's function. Show that, for the case of non-interacting fermions at zero temperature, the Fermi surface is manifest in the time-ordered Green's function, Eq. (3.61), according to (say) in the canonical ensemble,

$$G_0(E, \mathbf{p}) = \frac{1}{E - \epsilon_\mathbf{p} + i\delta\,\text{sign}(|\mathbf{p}| - p_F)}, \tag{3.137}$$

where $\delta = 0^+$, and the sign-function, $\text{sign}(x) = \theta(x) - \theta(-x) = x/|x|$, is plus or minus one depending on the sign of the argument. The grand canonical case corresponds to the substitution $\epsilon_\mathbf{p} \rightarrow \xi_\mathbf{p} = \epsilon_\mathbf{p} - \mu$.

Exercise 3.24. For N non-interacting bosons in a volume V at zero temperature, they all occupy the lowest energy level corresponding to the label $\mathbf{p} = \mathbf{0}$. In the field operator, $\psi(\mathbf{x}) = \xi_0 + \psi'(\mathbf{x})$, the creation operator for the lowest energy level is singled out, $\xi_0 = a_0/\sqrt{V}$, and ξ_0 and $\xi_0^\dagger$ can, for a non-interacting system in the thermodynamic limit, be regarded as c-numbers, $[\xi_0, \xi_0^\dagger] = 1/V$.

Show that the time ordered Green's function for non-interacting bosons in the ground state, $|\Phi_N\rangle = (N!)^{-1/2}(a_0^\dagger)^N|0\rangle$, is given by

$$G_0(\mathbf{x}, t, \mathbf{x}', t') = G^{(0)}(t - t') + G_0'(\mathbf{x}, t, \mathbf{x}', t'), \tag{3.138}$$

where

$$G^{(0)}(t - t') = -i\langle\Phi_N|T(\xi_0(t)\,\xi_0^\dagger(t'))|\Phi_N\rangle \tag{3.139}$$

is specified at $t' = t + 0$ by $iG^{(0)}(0^-) = n$, where $n = N/V$ is the density of the bosons, and $G_0'(\mathbf{x}, t, \mathbf{x}', t')$ is specified by its Fourier transform

$$G_0'(E, \mathbf{p}) = \frac{1}{E - \epsilon_\mathbf{p} + i\delta} \tag{3.140}$$

corresponding to *G-lesser* vanishing for the field ψ', or equivalently, the density of the bosons is solely provided by the occupation of the lowest energy level. The presence of a Bose–Einstein condensate at low temperatures thus leads to such modifications for boson Green's function expressions.[18]

As mentioned, perturbation theory and diagrammatic summation schemes are the main tools in unraveling the effects of interactions on the equilibrium properties of a system. This has been dealt with in textbooks mainly using the imaginary-time formalism (which we will discuss in Section 5.7.1), and unfortunately most numerously in the so-called Matsubara technique. This technique, which is based on a purely mathematical feature, lacks physical transparency. A main purpose of this book is to show that the real-time technique, which is based on the basic feature of quantum dynamics, has superior properties in terms of physical insight. Furthermore, there is no need to delve into equilibrium theory Feynman diagrammatics since it will be a simple corollary of the general real-time non-equilibrium theory we now turn to develop.

3.5 Summary

In this chapter we have shown that by transforming to the Heisenberg picture, the quantum dynamics of a many-body system can be described by the time development of the field operator in the Heisenberg picture. The measurable physical quantities of a system were thus expressed in terms of strings of Heisenberg operators weighted

---

[18]If the ground state of a system of interacting bosons has no condensed phase, standard zero-temperature perturbation theory can not be applied. In the opposite case, the zero-momentum fields can be treated as external fields. This leads to additional vertices in the Feynman diagrammatics as encountered in Section 10.6.

with respect to a state, generally a mixture described by a statistical operator. The dynamics of such systems are therefore described in terms of the correlation functions of field operators, the Green's functions of the theory.

In thermal equilibrium, the fluctuation–dissipation relation leads to simplification as all the one-particle or two-point Green's functions can be expressed in terms of the spectral weight function. Different schemes pertaining to equilibrium can be devised for calculating equilibrium Green's functions but we shall not entertain them here as they will be simple corollaries of the general non-equilibrium theory presented in the next chapter.

The equations of motion for Green's functions of interacting quantum fields involve ever increasing higher-order correlation functions. The rest of the book is devoted to the study and calculation of Green's functions for non-equilibrium states using diagrammatic and functional methods. We therefore turn to develop the formalism necessary for obtaining information about the properties of systems when they are out of equilibrium.

4

Non-equilibrium theory

In this chapter we will develop the general formalism necessary for dealing with non-equilibrium situations. We first formulate the non-equilibrium problem, and discuss why the standard method applicable for the study of ground state properties fails for arbitrary states. We then introduce the closed time path formulation, and construct the perturbation theory for the closed time path or contour ordered Green's function valid for non-equilibrium states. The diagrammatic perturbation theory in the closed time path formulation is then formulated, and generic types of interaction are considered.[1]

4.1 The non-equilibrium problem

Let us consider an arbitrary physical system described by the Hamiltonian H. Since we consider non-equilibrium quantum field theory, the Hamiltonian acts on the multi-particle state space introduced in the first chapter, consisting of products of multi-particle spaces for the species involved. The generic non-equilibrium problem can be construed as follows: far in the past, prior to time t_0, the system can be thought of as having been brought to the equilibrium state characterized by temperature T. The state of the system is thus at time t_0 described by the statistical operator[2]

$$\rho(H) \;\; = \;\; \frac{e^{-H/kT}}{\text{Tr}(e^{-H/kT})}, \tag{4.1}$$

where Tr denotes the trace in the multi-particle state space of the physical system in question. At times larger than $t = t_0$, a possible time-dependent mechanical perturbation, described by the Hamiltonian $H'(t)$, is applied to the system. The

---

[1] In this chapter we follow the exposition given in reference [3].

[2] We can also imagine and treat the case where the particles in the system are coupled to particle reservoirs described by their chemical potentials as this is simply included by tacitly understanding that single-particle energies are measured relative to their chemical potentials, $H \to H - \sum_s \mu_s N_s$, i.e. shifting from the canonical to the grand canonical ensemble. In fact, in actual calculations it is the more convenient choice, as discussed in Sections 2.5 and 3.4.

total Hamiltonian is thus

$$\mathcal{H}(t) \;=\; H \,+\, H'(t)\,. \tag{4.2}$$

The simplest non-equilibrium problem is concerned with the calculation of some average value of a physical quantity A at times $t > t_0$. The state of the system is evolved to

$$\rho(t) \;=\; U(t,t_0)\,\rho(H)\,U^{\dagger}(t,t_0)\,, \tag{4.3}$$

where (recall Eq. (3.7))

$$U(t,t') \;=\; T e^{-\frac{i}{\hbar}\int_{t'}^{t} d\bar{t}\,\mathcal{H}(\bar{t})} \tag{4.4}$$

is the evolution operator, and the average value of the quantity of interest is thus

$$\langle A(t)\rangle \;=\; \mathrm{Tr}(\rho(t)\,A)\,, \tag{4.5}$$

where A is the operator representing the physical quantity in question in the Schrödinger picture. Transforming to the Heisenberg picture

$$\langle A(t)\rangle \;=\; \mathrm{Tr}(\rho(H)\,A_{\mathcal{H}}(t)) \;=\; \langle A_{\mathcal{H}}(t)\rangle\,, \tag{4.6}$$

where, as discussed in Section 3.1.2, $A_{\mathcal{H}}(t)$ denotes the operator representing the physical quantity in question in the Heisenberg picture with respect to $\mathcal{H}(t)$, and we have chosen the reference time in Eq. (3.17) to be t_0. The average value is typically a type of quantity of interest for a macroscopic system, i.e. a system consisting of a huge number of particles. For example, for the average (probability) density of the particle species described by the quantum field ψ, we have according to Eq. (2.28),[3]

$$n(\mathbf{x},t) \;=\; \langle \psi_{\mathcal{H}}^{\dagger}(\mathbf{x},t)\,\psi_{\mathcal{H}}(\mathbf{x},t)\rangle\,. \tag{4.7}$$

The average density is seen to be equal to the equal-time and equal-space value of the *G-lesser* Green's function, $G^{<}$, introduced in Section 3.3

$$n(\mathbf{x},t) \;=\; \mp iG^{<}(\mathbf{x},t,\mathbf{x},t)\,, \tag{4.8}$$

where upper (lower) sign is for bosons and fermions, respectively.

If fluctuations are of interest or importance we encounter higher order correlation functions, generically according to Section 2.1, then appears the trace over products of pairs of Heisenberg field operators for particle species weighted by the initial state. If one is interested in the probability that a certain sequence of properties are measured at different times, one encounters arbitrary long products of Heisenberg operators.[4] Since physical quantities are expressed in terms of the quantum fields of the particles and interactions in terms of their higher-order correlations, of interest are the correlation functions, the so-called non-equilibrium Green's functions.

Owing to interactions, memory of the initial state of a subsystem is usually rapidly lost. We shall not in practice be interested in transient properties but rather steady

---

[3]Possible spin degrees of freedom are suppressed, or imagined absorbed in the spatial variable.

[4]See chapter 1 of reference [1] for a discussion of such probability connections or histories with a modern term.

states, where the dependence on the initial state is lost, and the time dependence is governed by external forces. Initial correlations can be of interest in their own right, even in many-body systems.[5] In fact, for all of the following, the statistical operator in the previous formulae, say Eq. (4.6), could have been chosen as arbitrary. This would lead to additional features which we point out as we go along, and in practice each case then has to be dealt with on an individual basis.

The equation of motion for the one-particle Green's function leads to an infinite hierarchy of equations for correlation functions containing ever increasing numbers of field operators, describing the correlations between the particles set up in the system by the interactions and external forces. In order to calculate the effects of interactions, we now embark on the construction of perturbation theory and the diagrammatic representation of non-equilibrium theory starting from the canonical formalism presented in the first chapter. But first we describe why the zero temperature, i.e. ground state, formalism is not capable of dealing with general non-equilibrium situations, before embarking on finding the necessary remedy, and eventually construct non-equilibrium perturbation theory and its corresponding diagrammatic representation.

4.2 Ground state formalism

To see the need for the closed time path description consider the problem of perturbation theory. The Hamiltonian of a system

$$H = H_0 + H^{(i)} \qquad (4.9)$$

consists of a term quadratic in the fields, H_0, describing the free particles, and a complicated term, $H^{(i)}$, describing interactions.

Constructing perturbation theory for zero temperature quantum field theory, i.e. the system is in its ground state $|G\rangle$, only the time-ordered Green's function

$$G(\mathbf{x}, t, \mathbf{x}', t') = -i\langle T(\psi_H(\mathbf{x}, t)\,\psi_H^\dagger(\mathbf{x}', t'))\rangle = -i\langle G|T(\psi_H(\mathbf{x}, t)\,\psi_H^\dagger(\mathbf{x}', t'))|G\rangle \qquad (4.10)$$

needs to be considered. Here $\psi_H(\mathbf{x}, t)$ is the field operator in the Heisenberg picture with respect to H for one of the species of particles described by the Hamiltonian.[6] The time-ordered Green's function contains more information than seems necessary for calculating mean or average values, since for times $t < t'$ it becomes the *G-lesser* Green's function

$$G(\mathbf{x}, t, \mathbf{x}', t') = G^<(\mathbf{x}, t, \mathbf{x}', t') \qquad (4.11)$$

---

[5] All transient effects for the above chosen initial condition are of course included. Whether this choice is appropriate for the study of transient effects depends on the given physical situation.

[6] For a reader not familiar with zero temperature quantum field theory, no such thing is required. It will be a simple corollary of the more powerful formalism presented in Section 4.3.2, and developed to its final real-time formalism presented in Chapter 5. The reason for the usefulness of the time ordering operation is to be expected remembering the crucial appearance of time-ordering in the evolution operator. Also, under the governing of the time-ordering symbol, operators can be commuted without paying a price except for the possible quantum statistical minus signs in the case of fermions.

and thereby are all average values of physical quantities specified once the time-ordered Green's function is known for $t < t'$. However, a perturbation theory involving *only* the *G-lesser* Green's function can not be constructed.

The time-ordered Green's function can, instead of being expressed in terms of the field operator $\psi_H(\mathbf{x}, t)$, i.e. in the Heisenberg picture with respect to H, be expressed in terms of the field operators $\psi_{H_0}(\mathbf{x}, t)$, the Heisenberg picture with respect to H_0 or the so-called interaction picture,

$$\psi_{H_0}(\mathbf{x}, t) = e^{\frac{i}{\hbar} H_0(t - t_r)}\, \psi(\mathbf{x})\, e^{-\frac{i}{\hbar} H_0(t - t_r)} \tag{4.12}$$

as they are related according to the unitary transformation

$$\psi_H(\mathbf{x}, t) = U^\dagger(t, t_r)\, \psi_{H_0}(\mathbf{x}, t)\, U(t, t_r)\,, \tag{4.13}$$

where

$$U(t, t_r) = T\, e^{-i \int_{t_r}^{t} d\bar{t}\, H_{H_0}^{(i)}(\bar{t})} \tag{4.14}$$

is the evolution operator in the interaction picture (leaving out for brevity an index to distinguish it from the full evolution operator $\exp\{-iH(t - t_r)\}$) and

$$H_{H_0}^{(i)}(t) = e^{iH_0(t - t_r)}\, H^{(i)}\, e^{-iH_0(t - t_r)}\,. \tag{4.15}$$

This is readily seen by noting that the expression on the right-hand side in Eq. (4.13) satisfies the first-order in time differential equation

$$i\hbar\, \frac{\partial \psi(\mathbf{x}, t)}{\partial t} = [\psi(\mathbf{x}, t), H]\,, \tag{4.16}$$

the same equation satisfied by the field $\psi_H(\mathbf{x}, t)$, and at the reference time t_r, the two operators are seen to coincide (coinciding with the field in the Schrödinger picture, $\psi(\mathbf{x})$).

Transforming to the interaction picture, and using the semi-group property of the evolution operator, $U(t, t'')\, U(t'', t') = U(t, t')$,[7] and the relation $U^\dagger(t, t') = U(t', t)$, the time ordered Green's function can be expressed in the form

$$G(\mathbf{x}, t, \mathbf{x}', t') = -i \left\langle U^\dagger(t, t_r)\psi_{H_0}(\mathbf{x}, t) U(t, t')\psi_{H_0}^\dagger(\mathbf{x}', t') U(t', t_r) \right\rangle \theta(t - t')$$

$$\pm\; i \left\langle U^\dagger(t', t_r)\psi_{H_0}^\dagger(\mathbf{x}', t') U(t', t)\psi_{H_0}(\mathbf{x}, t) U(t, t_r) \right\rangle \theta(t' - t) \tag{4.17}$$

which can also be expressed on the form (t_m denotes $\max\{t, t'\}$)

$$G(\mathbf{x}, t, \mathbf{x}', t') = -i \left\langle U^\dagger(t_m, t_r) T \left(\psi_{H_0}(\mathbf{x}, t)\psi_{H_0}^\dagger(\mathbf{x}', t') U(t_m, t_r) \right) \right\rangle \tag{4.18}$$

since the time-ordering symbol places the operators in the original order.[8]

---

[7] For a detailed discussion of the evolution operator and the Heisenberg and interaction pictures we refer to chapter 2 of reference [1].

[8] In fact the operator identity

$$T(\psi_H(\mathbf{x}, t)\, \psi_H^\dagger(\mathbf{x}', t')) = U^\dagger(t_m, t_r) T \left(\psi_{H_0}(\mathbf{x}, t)\, \psi_{H_0}^\dagger(\mathbf{x}', t')\, U(t_m, t_r) \right)$$

is valid since only transformation of operators was involved, and nowhere is advantage taken of the averaging with respect to the state in question.

Usually, say in a scattering experiment realized in a particle accelerator, only transitions from an initial state in the far past are of interest so that the reference time is chosen in the far past, $t_r = -\infty$, and inserting $1 = U(t_m, \infty)U(\infty, t_m)$ after $U^\dagger$ gives[9]

$$G(\mathbf{x}, t, \mathbf{x}', t') = -i\langle U^\dagger(\infty, -\infty) T(\psi_{H_0}(\mathbf{x}, t)\,\psi_{H_0}^\dagger(\mathbf{x}', t')\,U(\infty, -\infty))\rangle . \quad (4.19)$$

If the average is with respect to the ground state of the system, one can make use of the trick of adiabatic switching, i.e. the interaction is assumed turned on and off adiabatically, say by the substitution $H_{H_0}^{(i)}(t) \to e^{-\epsilon|t|}H_{H_0}^{(i)}(t)$. The non-interacting (non-degenerate) ground state $|G_0\rangle$, $H_0|G_0\rangle = E_0|G_0\rangle$, is evolved by the full adiabatic evolution operator U_ϵ into the normal ground state of the interacting system at time $t = 0$, $|G\rangle_\epsilon = U_\epsilon(0, -\infty)|G_0\rangle$. The ϵ on the evolution operator indicates that the interaction is turned on and off adiabatically. In perturbation theory it can then be shown, that in the limit of $\epsilon \to 0$, the true interacting ground state at time $t = 0$ is obtained modulo a phase factor that is obtained from the limiting expression of turning the interaction on and off adiabatically, (the Gell-Mann–Low theorem [4]),[10]

$$U_\epsilon(\infty, -\infty)|G_0\rangle = e^{i\phi}|G_0\rangle \quad , \quad e^{i\phi} = \langle G_0|U_\epsilon(\infty, -\infty)|G_0\rangle . \quad (4.20)$$

As a consequence, the time-ordered Green's function, Eq. (4.10), can be expressed in terms of the non-interacting ground state and the fields in the interaction picture according to

$$G(\mathbf{x}, t, \mathbf{x}', t') = -i\,\frac{\langle G_0|T(\psi_{H_0}(\mathbf{x}, t)\,\psi_{H_0}^\dagger(\mathbf{x}', t')\,U(\infty, -\infty))|G_0\rangle}{\langle G_0|U(\infty, -\infty)|G_0\rangle} . \quad (4.21)$$

In the next section we will show that the artifice of turning the interaction on and off adiabatically is not needed when using the closed time path formulation and generalizing time-ordering to contour-ordering, and it can also be avoided by using functional methods as in Chapter 9, and plays no role in the non-equilibrium formalism. In describing a scattering experiment, adiabatic switching is of course an innocent initial and final boundary condition as the particles are then free.[11]

Since the Gell-Mann–Low theorem fails for states other than the ground state, and thus even for an equilibrium state at finite temperature, we are in general stuck

---

[9]In fact as an operator identity

$$T(\psi_H(\mathbf{x}, t)\,\psi_H^\dagger(\mathbf{x}', t')) = U^\dagger(\infty, -\infty) T(\psi_{H_0}(\mathbf{x}, t)\,\psi_{H_0}^\dagger(\mathbf{x}', t')\,U(\infty, -\infty)).$$

[10]Clearly, it is important that no dissipation or irreversible effects takes place. Contrarily, in statistical physics, reduced dynamics is the main interest, i.e. certain degrees of freedom are left unobserved and emission and absorption takes place, technically partial traces occurs.

[11]As will become clear from the following sections, the denominator in Eq. (4.21) is diagrammatically the sum of all the vacuum diagrams that therefore cancel all the disconnected diagrams in the numerator, and one obtains the standard connected Feynman diagrammatics for the time-ordered Green's function for a system at zero temperature such as is relevant in, say, QED. In QED one works with the so-called scattering matrix or S-matrix, $S(\infty, -\infty)$, defined in terms of the full evolution operator, $S(t, t') = e^{iH_0 t}U(t, t')e^{-iH_0 t'}$, so that the matrix elements of the S-matrix are expressed in terms of the free-particle states.

with the operator $U^\dagger(\infty, -\infty)$ inside the averaging in Eq. (4.19) and Eq. (4.18). At finite temperatures and *a fortiori* for non-equilibrium states, a perturbation theory involving only one kind of a real-time Green's functions can not be obtained. In order to construct a single object which contains all the dynamical information we shall follow Schwinger and introduce the closed time path formulation [5].

4.3   Closed time path formalism

Let us return to the non-equilibrium situation of Section 4.1 where the dynamics is determined by a time dependent Hamiltonian $\mathcal{H}(t) = H + H'(t)$, where H is the Hamiltonian for the isolated system of interest and $H'(t)$ is a time-dependent perturbation acting on it. The unitary transformation relating operators in the Heisenberg pictures governed by the Hamiltonians $\mathcal{H}(t)$ and H, respectively, is specified by the unitary transformation

$$O_{\mathcal{H}}(t) = V^\dagger(t, t_0)\, O_H(t)\, V(t, t_0) \quad , \quad V(t, t_0) = T e^{-\frac{i}{\hbar}\int_{t_0}^t d\bar{t}\, H'_H(\bar{t})} \tag{4.22}$$

and

$$H'_H(t) = U_H^\dagger(t, t_0)\, H'(t)\, U_H(t, t_0) \quad , \quad U_H(t, t_0) = e^{-\frac{i}{\hbar}H(t-t_0)} \tag{4.23}$$

and we have chosen t_0 as reference time where the two pictures coincide. This relation between the two pictures is obtained by first comparing both pictures to the Schrödinger picture obtaining

$$O_{\mathcal{H}}(t) = U_{\mathcal{H}}^\dagger(t, t_0)\, U_H(t, t_0)\, O_H(t)\, U_H^\dagger(t, t_0)\, U_{\mathcal{H}}(t, t_0) \,, \tag{4.24}$$

where

$$U_{\mathcal{H}}(t, t_0) = T e^{-\frac{i}{\hbar}\int_{t_0}^t d\bar{t}\, \mathcal{H}(\bar{t})} \tag{4.25}$$

is the evolution operator corresponding to the Hamiltonian $\mathcal{H}(t)$. Then one notes that $V(t, t_0)$ and $U_H^\dagger(t, t_0)\, U_{\mathcal{H}}(t, t_0)$ satisfy the same first-order in time differential equation and the same initial condition. We have thus obtained Dyson's formula

$$V(t, t_0) = U_H^\dagger(t, t_0)\, U_{\mathcal{H}}(t, t_0) \tag{4.26}$$

or explicitly

$$T e^{-\frac{i}{\hbar}\int_{t'}^t d\bar{t}\, H'_H(\bar{t})} = e^{\frac{i}{\hbar}H(t-t_0)}\, T e^{-\frac{i}{\hbar}\int_{t'}^t d\bar{t}\, \mathcal{H}(\bar{t})} . \tag{4.27}$$

Here Dyson's formula appeared owing to unitary transformations between Heisenberg and interaction pictures, but once conjectured it can of course immediately be established by direct differentiation. Dyson's formula is useful in many contexts, be the time variable real or imaginary, and also for equilibrium states such as when phase transitions are studied in, for instance, a renormalization group treatment. We shall in fact apply Dyson's formula for imaginary times in Section 4.3.2.

We now introduce the contour, the closed time path, which starts at t_0 and proceeds along the real time axis to time t and then back again to t_0, the closed contour c_t as depicted in Figure 4.1.

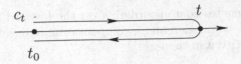

Figure 4.1 The closed time path contour c_t.

We then show that the transformation between the two Heisenberg pictures, Eq. (4.24), can be expressed on closed contour form as (units are chosen to set $\hbar$ equal to one at our convenience)

$$O_{\mathcal{H}}(t) = T_{c_t}\left(e^{-i \int_{c_t} d\tau\, H'_H(\tau)} O_H(t) \right) , \tag{4.28}$$

where τ denotes the contour variable proceeding from t_0 along the real-time axis to t and then back again to t_0, i.e. the variable on c_t. The contour ordering symbol T_{c_t} orders products of operators according to the position of their contour time argument on the closed contour, earlier contour time places an operator to the right.

The crucial equivalence of Eq. (4.24) and Eq. (4.28), which form a convenient basis for formulating perturbation theory in the closed time path formalism, is based on the algebra of operators under the contour ordering being equivalent to the algebra of numbers.[12] Expanding the exponential in Eq. (4.28) gives

$$O_{\mathcal{H}}(t) = \sum_{n=0}^{\infty} \frac{(-i)^n}{n!} \int_{c_t} d\tau_1 \ldots \int_{c_t} d\tau_n\, T_{c_t}\left(H'_H(\tau_1) \ldots H'_H(\tau_n) O_H(t) \right). \tag{4.29}$$

Let us consider the nth order term. In order to verify Eq. (4.28), we note that the contour can be split into forward and backward parts

$$c_t = \vec{c} + \overleftarrow{c} . \tag{4.30}$$

Splitting the contour into forward and backward contours gives 2^n terms. Out of these there are $n!/(m!(n-m)!)$ terms $(m = 0,1,2,\ldots,n)$, which contain m integrations over the forward contour, and the rest of the factors, $n-m$, have integratons over the backward contour. Since they differ only by a different dummy integration labeling they all give the same contribution and

$$\int_{c_t} d\tau_1 \ldots \int_{c_t} d\tau_n\, T_{c_t}\left(H'_H(\tau_1) \ldots H'_H(\tau_n) O_H(t) \right) = \sum_{m=0}^{n} \frac{n!}{m!(n-m)!}$$

$$\times \int_{\overleftarrow{c}} d\tau_{m+1} \ldots \int_{\overleftarrow{c}} d\tau_n\, T_{\overleftarrow{c}}\left(H'_H(\tau_{m+1}) \ldots H'_H(\tau_n) \right) O_H(t)$$

$$\times \int_{\vec{c}} d\tau_1 \ldots \int_{\vec{c}} d\tau_m\, T_{\vec{c}}\left(H'_H(\tau_1) \ldots H'_H(\tau_m) \right) , \tag{4.31}$$

---

[12]Even though the Hamiltonian for fermions contains non-commuting objects, the fermi fields, they appear in pairs and quantum statistical minus signs do not occur.

where $T_{\vec{c}}$ and $T_{\overleftarrow{c}}$ denotes contour ordering on the forward and backward parts, respectively. Adding a summation and a compensating Kronecker function the nth-order term can be rewritten in the form[13]

$$
\int_{c_t} d\tau_1 \ldots \int_{c_t} d\tau_n \, T_{c_t} \left(H'_H(\tau_1) \ldots H'_H(\tau_n) O_H(t) \right) = \sum_{k=0}^{\infty} \sum_{m=0}^{\infty} \frac{n!}{m! \, k!} \delta_{n, k+m}
$$

$$
\times \int_{\overleftarrow{c}} d\tau_1 \ldots \int_{\overleftarrow{c}} d\tau_k \, T_{\overleftarrow{c}} \left(H'_H(\tau_1) \ldots H'_H(\tau_k) \right) O_H(t)
$$

$$
\times \int_{\vec{c}} d\tau'_1 \ldots \int_{\vec{c}} d\tau'_m \, T_{\vec{c}} \left(H'_H(\tau'_1) \ldots H'_H(\tau'_m) \right) .
$$
(4.32)

The summation over n in Eq. (4.29) is now trivial, giving

$$
T_{c_t} \left(e^{-i \int_{c_t} d\tau H'_H(\tau)} O_H(t) \right) = \sum_{k=0}^{\infty} \sum_{m=0}^{\infty} \frac{(-i)^k (-i)^m}{m! \, k!}
$$

$$
\times \int_{\overleftarrow{c}} d\tau_1 \ldots \int_{\overleftarrow{c}} d\tau_k \, T_{\overleftarrow{c}} \left(H'_H(\tau_1) \ldots H'_H(\tau_k) \right) O_H(t)
$$

$$
\times \int_{\vec{c}} d\tau'_1 \ldots \int_{\vec{c}} d\tau'_m \, T_{\vec{c}} \left(H'_H(\tau'_1) \ldots H'_H(\tau'_m) \right)
$$
(4.33)

and thereby

$$
T_{c_t} \left(e^{-i \int_{c_t} d\tau H'_H(\tau)} O_H(t) \right) = T_{\overleftarrow{c}} \left(e^{-i \int_{\overleftarrow{c}} d\tau H'_H(\tau)} \right) O_H(t) \, T_{\vec{c}} \left(e^{-i \int_{\vec{c}} d\tau H'_H(\tau)} \right).
$$
(4.34)

Parameterizing the forward and backward contours according to

$$
\tau(t') = t' \qquad t' \epsilon \, [t_0, t] ,
$$
(4.35)

we get

$$
T_{\vec{c}} \left(e^{-i \int_{\vec{c}} d\tau H'_H(\tau)} \right) = T \, e^{-i \int_{t_0}^{t} dt' \, H'_H(t')} = V(t, t_0)
$$
(4.36)

and

$$
T_{\overleftarrow{c}} \left(e^{-i \int_{\overleftarrow{c}} d\tau H'_H(\tau)} \right) = \tilde{T} \, e^{i \int_{t_0}^{t} dt' \, H'_H(t')} = V^\dagger(t, t_0)
$$
(4.37)

i.e. contour ordering along the forward contour is identical to ordinary time ordering, $T_{\vec{c}} = T$, whereas contour ordering along the backward contour corresponds to anti-time ordering, $T_{\overleftarrow{c}} = \tilde{T}$. The equivalence of Eq. (4.24) and Eq. (4.28) has thus been established. We have shown that the times in $V^\dagger(t, t_0)$ corresponds to contour times

---

[13]Under the ordering operation, the algebra of non-commuting objects reigning the operators is not important, and the consideration is essentially the algebra of showing $\exp(a+b) = \exp(a)\exp(b)$.

lying on the backward part, and the times in $V(t, t_0)$ corresponds to contour times lying on the forward part.

We shall now use Eq. (4.28) to introduce the contour variable instead of the time variable. We hereby embark on Schwinger's closed time path formulation of non-equilibrium quantum statistical mechanics originally introduced in reference [5].[14] We shall thereby develop the diagrammatic perturbative structure of the closed time path or contour ordered Green's function.

4.3.1 Closed time path Green's function

A generalization offers itself, which will lead to a single object in terms of which non-equilibrium perturbation theory can be formulated. The trick will be to *democratize* the status of all times appearing in the time-ordered Green's function, Eq. (4.18), i.e. the original real times t and t' will be perceived to reside on the closed time path or contour. The one-particle Green's function in Eq. (4.18) contains two times; let us now denote them t_1 and $t_{1'}$. We introduce the contour, which starts at t_0 and proceeds along the real-time axis through t_1 and $t_{1'}$ and then back again to t_0, the closed contour c as depicted in Figure 4.2, $c = \vec{c} + \overleftarrow{c}$.[15] We have hereby freed the time variables, which hitherto were tied to the real axis, to lie on either the forward or return part of the contour, and we introduce the contour variable τ to signify this two-valued choice of the time variable, examples of which are given in Figure 4.2.[16]

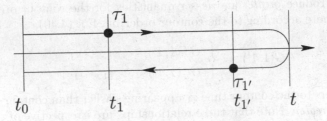

Figure 4.2 Examples of real times being elevated to contour times.

We are thus led to study the closed time path Green's function or the contour-ordered Green's function

$$G(\mathbf{x}_1, s_{z_1}, \tau_1, \mathbf{x}_{1'}, s_{z_{1'}}, \tau_{1'}) = -i \frac{\mathrm{Tr}(e^{-H/kT} T_c(\psi_{\mathcal{H}}(\mathbf{x}_1, s_{z_1}, \tau_1)\, \psi_{\mathcal{H}}^\dagger(\mathbf{x}_{1'}, s_{z_{1'}}, \tau_{1'})))}{\mathrm{Tr}(e^{-H/kT})}$$

$$(4.38)$$

---

[14] Reviews of the closed time path formalism stressing various applications are, for example, those of references [6], [7] and [8].

[15] If we discussed a correlation function involving more than two fields, the contour should stretch all the way to the maximum time value, or in fact we can let the contour extend from t_0 to $t = \infty$ and back again to t_0, since, as we soon realize, beyond $\max(t_1, t_{1'}, \ldots)$ the forward and backward evolutions take each other out, producing simply the identity operator.

[16] For mathematical rigor, i.e. proper convergence, both the forward and backward contours should be conceived of as being located infinitesimally below the real axis. This will be witnessed by the analytical continuation procedure discussed in Section 5.7, but in practice this consideration will not be necessary.

where τ_1 and $\tau_{1'}$ can lie on either the forward or backward parts of the closed contour. We have had a particle with spin in mind, say the electron, but introducing the condensed notation $1 = (\mathbf{x}_1, s_{z_1}, \tau_1)$ we have[17]

$$G(1, 1') = -i \langle T_c(\psi_{\mathcal{H}}(1) \psi_{\mathcal{H}}^\dagger(1')) \rangle = -i Z^{-1} \mathrm{Tr}(e^{-H/kT} T_c(\psi_{\mathcal{H}}(1) \psi_{\mathcal{H}}^\dagger(1'))) \quad (4.39)$$

at which stage any particle could be under discussion as the only relevant thing in the rest of the section is the contour variable. A contour ordering symbol T_c has been introduced, which orders operators according to the position of their contour-time argument on the closed contour, for example for the case of two contour times

$$T_c(\psi(\mathbf{x}_1, \tau_1) \psi^\dagger(\mathbf{x}_{1'}, \tau_{1'})) = \begin{cases} \psi(\mathbf{x}_1, \tau_1) \psi^\dagger(\mathbf{x}_{1'}, \tau_{1'}) & \tau_1 \overset{c}{>} \tau_{1'} \\ \mp \psi^\dagger(\mathbf{x}_{1'}, \tau_{1'}) \psi(\mathbf{x}_1, \tau_1) & \tau_{1'} \overset{c}{>} \tau_1 \end{cases} \quad (4.40)$$

where the upper (lower) sign is for fermions (bosons) respectively. An obvious notation for ordering along the contour has been introduced, viz. $\tau_1 \overset{c}{>} \tau_{1'}$ means that τ_1 is further along the contour c than $\tau_{1'}$ irrespective of their corresponding numerical values on the real axis. The contour ordering thus orders an operator sequence according to the contour position; operators with earliest contour times are put to the right. The algebra of bose fields under the contour ordering is thus like the algebra of (complex) numbers, whereas the algebra of fermi fields under the contour ordering is like the Grassmann algebra of anti-commuting numbers.[18]

We also introduce *greater* and *lesser* quantities for the contour ordered Green's function, and note according to the contour ordering, Eq. (4.40),

$$G(1, 1') = \begin{cases} G^<(1, 1') & \tau_{1'} \overset{c_t}{>} \tau_1 \\ G^>(1, 1') & \tau_1 \overset{c_t}{>} \tau_{1'} . \end{cases} \quad (4.41)$$

Here *lesser* refers to the contour time τ_1 appearing earlier than contour time $\tau_{1'}$, and vice versa for *greater*. Note that these relationships are irrespective of the numerical relationship of their corresponding real time values: if the contour times in $G^<(1, 1')$ and $G^>(1, 1')$ are identified with their corresponding real times we recover their corresponding real-time Green's functions discussed in Section 3.3.

Transforming from the Heisenberg picture with respect to the Hamiltonian $\mathcal{H}(t)$ to the Heisenberg picture with respect to the Hamiltonian H, gives, according to Eq. (4.28),

$$G^>(1, 1') = -i \langle \psi_{\mathcal{H}}(1) \psi_{\mathcal{H}}^\dagger(1') \rangle$$

$$= -i \left\langle T_{c_{t_1}} \left(e^{-i \int_{c_{t_1}} d\tau\, H_H'(\tau)} \psi_H(1) \right) \left(T_{c_{t_{1'}}} e^{-i \int_{c_{t_{1'}}} d\tau\, H_H'(\tau)} \psi_H^\dagger(1') \right) \right\rangle$$

---

[17]In the following we shall consider the fields as entering the Green's function, however, for the following it could be any type of operators and any number of products, $G(1, 2, 3, \ldots) = \langle T_c(A_{\mathcal{H}}(1) B_{\mathcal{H}}^\dagger(2) C_{\mathcal{H}}^\dagger(3) \ldots) \rangle$. Note that if the operators represent physical quantities, they are specified in terms of the fields, and we are back to strings of field operators modulo the operations specific to the quantities in question.

[18]In Chapter 10 we shall in fact show that in view of this, quantum field theory can, instead of being formulated in terms of quantum field *operators*, be formulated in terms of scalar or Grassmann *numbers* by the use of path integrals.

$$= -i \left\langle T_{c_{t_1}+c_{t_{1'}}} \left(e^{-i \int_{c_{t_1}+c_{t_{1'}}} d\tau\, H'_H(\tau)} \psi_H(1)\, \psi_H^\dagger(1') \right) \right\rangle, \qquad (4.42)$$

where the contours c_{t_1} $(c_{t_{1'}})$ starts at t_0 and passes through t_1 $(t_{1'})$, respectively, and returns to t_0. In the last equality the combined contour, $c_{t_1} + c_{t_{1'}}$, depicted in Figure 4.3, has been introduced. It stretches from t_0 to $\min\{t_1, t_{1'}\}$ and back to t_0 and then forward to $\max\{t_1, t_{1'}\}$ before finally returning back again to t_0. The contributions from the hatched parts depicted in Figure 4.3 cancel since for this part the field operators at times t_1 and $t_{1'}$ are not involved and a closed contour appears which gives the unit operator, or equivalently $U^\dagger(t_1, t_0)\, U(t_1, t_0) = 1$, and the last equality in Eq. (4.42) is established. By the same argument, the contour could be extended from $\max\{t_1, t_{1'}\}$ all the way to plus infinity before returning to t_0, and we encounter the general real-time contour.

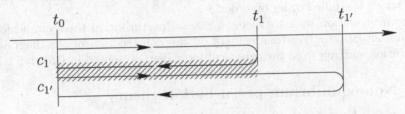

Figure 4.3 Parts of contour evolution operators canceling in Eq. (4.42).

We have an analogous situation for $G^<(1, 1')$, and we have shown that

$$G(1, 1') = -i\langle T_c(\psi_{\mathcal{H}}(1)\, \psi_{\mathcal{H}}^\dagger(1'))\rangle$$

$$= -i \left\langle T_c \left(e^{-i \int_c d\tau\, H'_H(\tau)} \psi_H(1)\, \psi_H^\dagger(1') \right) \right\rangle, \qquad (4.43)$$

where the contour c starts at t_0 and stretches through $\max(t_1, t_{1'})$ (or all the way to plus infinity) and back again to t_0. By introducing the closed contour and contour ordering we have managed to bring all operators under the ordering operation, which will prove very useful when it comes to deriving the perturbation theory for the contour-ordered Green's function.

Exercise 4.1. From the equation of motion for the field operator, show that the equation of motion for the contour-ordered Green's function is

$$\left(i\hbar \frac{\partial}{\partial \tau} - h_0(\tau) + \mu \right) G(\mathbf{x}, \tau, \mathbf{x}', \tau') = \hbar\, \delta(\mathbf{x} - \mathbf{x}')\, \delta_c(\tau - \tau')$$

$$- i \langle T_c([\psi^\dagger(\mathbf{x}, \tau), H_i(\tau)]\psi^\dagger(\mathbf{x}', \tau'))\rangle, \quad (4.44)$$

where h_0 denotes the single-particle Hamiltonian, and we have introduced the contour delta function

$$\delta_c(\tau - \tau') = \begin{cases} \delta(\tau - \tau') & \text{for } \tau \text{ and } \tau' \text{ on forward branch} \\ -\delta(\tau - \tau') & \text{for } \tau \text{ and } \tau' \text{ on return branch} \\ 0 & \text{for } \tau \text{ and } \tau' \text{ on different branches} \end{cases} \qquad (4.45)$$

and $H_i(\tau)$ is the interaction part of the Hamiltonian in the Heisenberg picture (recall Exercise 3.10 on page 66).

The equation of motion for the Green's function leads, as noted in Section 3.3, to an infinite hierarchy of equations for correlation functions containing an ever increasing number of field operators describing the correlations between the particles set up in the system by the interactions and external forces. Needless to say, an exact solution of a quantum field theory is a mission impossible in general. At present, the only general method available for gaining knowledge from the fundamental principles about the dynamics of a system is the perturbative study. This goes for non-equilibrium states *a fortiori*, and we shall now construct the perturbation theory valid for non-equilibrium states. This consists of dividing the Hamiltonian into one part representing a simpler well-understood problem and a nontrivial part, the effect of which is studied order by order.

In the next section we construct the general perturbation theory valid for non-equilibrium situations. We thus embark on the construction of the diagrammatic representation starting from the canonical formalism presented in Chapter 1.

4.3.2   Non-equilibrium perturbation theory

We now proceed to obtain the perturbation theory expressions for the contour-ordered Green's functions. The Hamiltonian of the system, Eq. (4.9) consists of a term quadratic in the fields, H_0, describing the free particles, and a complicated term, $H^{(i)}$, describing interactions. To get an expression ready-made for a perturbative expansion of the contour-ordered Green's function, the Hamiltonian in the weighting factor needs to be quadratic in the fields, i.e. we need to transform the operators in Eq. (4.42) to the interaction picture with respect to H_0. Quite analogous to the manipulations in the previous section we have

$$O_{\mathcal{H}}(t) = T_{c_t}\left(e^{-i\int_{c_t} d\tau\,(H^{(i)}_{H_0}(\tau)+H'_{H_0}(\tau))}\,O_{H_0}(t)\right), \tag{4.46}$$

where we have further, or directly, transformed from the Heisenberg picture with respect to the Hamiltonian H to the Heisenberg picture with respect to the free Hamiltonian H_0, the relation being equivalent to that in Eq. (4.28). The operator $H'_{H_0}(\tau)$ is thus the mechanical external perturbation in the Heisenberg picture with respect to H_0.[19] We have thus analogous to the derivation of the expression Eq. (4.42) for the contour-ordered Green's function, Eq. (4.39), that the contour-ordered Green's function in the interaction picture is

$$G(1,1') = -i\,\frac{\mathrm{Tr}\left(e^{-\beta H}\,T_c\left(e^{-i\int_c d\tau(H^{(i)}_{H_0}(\tau)+H'_{H_0}(\tau))}\,\psi_{H_0}(1)\,\psi^\dagger_{H_0}(1')\right)\right)}{\mathrm{Tr}\left(e^{-\beta H}\right)}. \tag{4.47}$$

We have introduced the notation $\beta = 1/kT$ for the inverse temperature.

---

[19] We shall later take advantage of the artifice of employing different dynamics on the forward and backward paths, making the closed time path formulation a powerful functional tool.

We can now employ Dyson's formula, Eq. (4.27), for the case of a time-independent Hamiltonian, H, and imaginary times, to express the Boltzmann weighting factor in terms of the weighting factor for the free theory

$$e^{-\beta H} = e^{-\beta H_0} \, T_{c_a} \, e^{-i \int_{t_0}^{t_0 - i\beta} d\tau \, H_{H_0}^{(i)}(\tau)} \tag{4.48}$$

where T_{c_a} contour orders along the contour stretching down into the lower complex time plane from t_0 to $t_0 - i\beta$, the appendix contour c_a as depicted in Figure 4.4. We then get the expression

$$iG(1,1') = \frac{\mathrm{Tr}\left(e^{-\beta H_0}\left(T_{c_a} e^{-i \int_{t_0}^{t_0 - i\beta} d\tau H_{H_0}^{(i)}(\tau)}\right) T_c \left(e^{-i \int_c d\tau (H_{H_0}^{(i)}(\tau) + H'_{H_0}(\tau))} \psi_{H_0}(1) \psi_{H_0}^\dagger(1')\right)\right)}{\mathrm{Tr}\left(e^{-\beta H_0} \, T_{c_a} e^{-i \int_{t_0}^{t_0 - i\beta} d\tau H_{H_0}^{(i)}(\tau)}\right)} \tag{4.49}$$

ready-made for a perturbative expansion of the contour-ordered Green's function valid for the non-equilibrium case. The term involving imaginary times stretching down into the lower complex time plane from t_0 to $t_0 - i\beta$ can be brought under *one* contour ordering by adding the appendix contour c_a to the contour c giving in total the contour c_i as depicted in Figure 4.4, and we thus have

$$G(1,1') = -i \frac{\mathrm{Tr}\left(e^{-\beta H_0} T_{c_i}\left(e^{-i \int_{c_i} d\tau H_{H_0}^{(i)}(\tau)} e^{-i \int_c d\tau H'_{H_0}(\tau)} \psi_{H_0}(1)\,\psi_{H_0}^\dagger(1')\right)\right)}{\mathrm{Tr}\left(e^{-\beta H_0} T_{c_i}\left(e^{-i \int_{c_i} d\tau H_{H_0}^{(i)}(\tau)} e^{-i \int_c d\tau H'_{H_0}(\tau)}\right)\right)}. \tag{4.50}$$

The contour c_i stretches from t_0 to $\max\{t_1, t_{1'}\}$ (or infinity) and back again to t_0 and has in addition to the contour c the additional appendix c_a, i.e. stretches further down into the lower complex time plane from t_0 to $t_0 - i\beta$, as depicted in Figure 4.4.

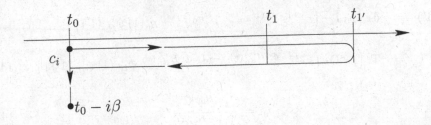

Figure 4.4 The contour c_i.

In the numerator we have used the fact that, under contour ordering, operators can be commuted, leaving operator algebra identical to that of numbers, so that for example

$$T_c\left(e^{-i \int_c d\tau \, (H_{H_0}^{(i)}(\tau) + H'_{H_0}(\tau))}\right) = T_c\left(e^{-i \int_c d\tau \, H_{H_0}^{(i)}(\tau)} e^{-i \int_c d\tau \, H'_{H_0}(\tau)}\right). \tag{4.51}$$

The expression in Eq. (4.50) is of a form for which we can use Wick's theorem to obtain the perturbative expansion of the contour-ordered Green's function and the associated Feynman diagrammatics. Before we show Wick's theorem in the next section, some general remarks are in order.

In the denominator in Eq. (4.50), we introduced a closed contour contribution, that of contour c, stretching from t_0 to $\max\{t_1, t_{1'}\}$ (or infinity) and back again to t_0, which since no operators interrupts at intermediate times is just the identity operator

$$T_c \left(e^{-i \int_c d\tau \, (H_{H_0}^{(i)}(\tau) + H'_{H_0}(\tau))} \right) = 1 . \tag{4.52}$$

This was done in order for the expression in Eq. (4.50) to be written on the form where the usual combinatorial arguments applies to show that unlinked or disconnected diagrams originating in the numerator are canceled by the vacuum diagrams from the denominator. However, for the non-equilibrium states of interest here, such features are actually artificial relics of the formalisms used in standard zero and finite time formalisms. A reader not familiar with these combinatorial arguments need not bother about these remarks since we shall now specialize to the situation where this feature is absent.[20]

We note that only interactions are alive on the appendix contour part, c_a, whereas the external perturbation vanishes on this part of the contour. If we are not interested in transient phenomena in a system or physics on short time scales of the order of the collision time scale due to the interactions, we can let t_0 approach minus infinity, $t_0 \to -\infty$, and the contribution from the imaginary part of the contour c_i vanishes. The physical argument is that a propagator with one of its arguments on the imaginary time appendix is damped on the time scale of the scattering time of the system. Thus as the initial time, t_0, where the system is perturbed by the external field, retrudes back into the past beyond the microscopic scattering times of the system, then effectively $t_0 \to -\infty$, and contributions due to the imaginary appendix part c_a of the contour vanish.[21] The denominator in Eq. (4.49) thus reduces to the partition function for the non-interacting system and we finally have for the contour-ordered or closed time path Green's function

$$G(1,1') = \text{Tr} \left(\rho_0 T_C \left(e^{-i \int_C d\tau \, (H_{H_0}^{(i)}(\tau) + H'_{H_0}(\tau))} \, \psi_{H_0}(1) \, \psi_{H_0}^\dagger(1') \right) \right)$$

$$= \text{Tr} \left(\rho_0 T_C \left(e^{-i \int_C d\tau \, H_{H_0}^{(i)}(\tau)} \, e^{-i \int_C d\tau \, H'_{H_0}(\tau)} \, \psi_{H_0}(1) \, \psi_{H_0}^\dagger(1') \right) \right) , \tag{4.53}$$

where

$$\rho_0 = \frac{e^{-H_0/kT}}{\text{Tr} \, e^{-H_0/kT}} \tag{4.54}$$

---

[20] In Section 9.5, where we start studying physics from scratch in terms of diagrammatics, the cancellation of the vacuum diagrams is discussed in detail. There, both a diagrammatic proof as well as the combinatorial proof relevant for the present discussion are given for the cancellation of the numerator by the separated off vacuum diagrams of the numerator.

[21] If the interactions are turned on adiabatically, then as the arbitrary initial time is retruding back into the past, $t_0 \to -\infty$, the interaction vanishes in the past, and therefore vanishes on the imaginary appendix part of the contour. However, there is no need to appeal to adiabatic coupling since interaction always has the physical effect of intrinsic damping. We note that at ever increasing temperatures, the appendix contour contribution disappears, since thermal fluctuations then immediately wipe out initial correlations.

is the statistical operator for the equilibrium state of the non-interacting system at the temperature T. The last equality sign follows since the algebra of Hamiltonians under contour ordering is equivalent to that of numbers. The contour C appearing in Eq. (4.53) is Schwinger's closed time path [5], the Schwinger–Keldysh or real-time contour, which starts at time $t = -\infty$ and proceeds to time $t = \infty$ and then back again to time $t = -\infty$, as depicted in Figure 4.5.

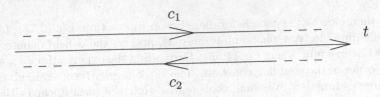

Figure 4.5 The Schwinger–Keldysh closed time path or real-time closed contour.

We note that non-equilibrium perturbation theory in fact has a simpler structure than the standard equilibrium theory as there is no need for canceling of unlinked or disconnected diagrams. The contour evolution operator for a closed loop is one: in the perturbative expansion for the denominator in Eq. (4.50)

$$D = \text{Tr}\left(e^{-\beta H_0} T_c \left(e^{-i\int_c d\tau\, H_{H_0}^{(i)}(\tau)}\, e^{-i\int_c d\tau\, H'_{H_0}(\tau)}\right)\right) \tag{4.55}$$

only the identity term corresponding to no evolution survives, all other terms comes in two, one with a minus sign, and the sum cancels. We shall take advantage of the absence of this so-called *denominator*-problem in Chapter 12, and this aspect of the presented non-equilibrium theory is a very important aspect in the many applications of the closed time path formalism: from the dynamical approach to perform quenched disorder average to the field theory of classical statistical dynamics.[22]

Before turning to obtain the full diagrammatics of non-equilibrium perturbation theory, let us acquaint ourselves with lowest order terms. The simplest kind of coupling is that of particles to an external classical field $V(\mathbf{x}, t)$. In that case the contour ordered Green's function has the form ready for a perturbative expansion

$$G_C(1,1') = \text{Tr}\left(\rho_0\, T_C\left(e^{-i\int_C d\tau \int d\mathbf{x}\, V(\mathbf{x},\tau)\, \psi_{H_0}^\dagger(\mathbf{x},\tau)\, \psi_{H_0}(\mathbf{x},\tau)}\, \psi_{H_0}(1)\, \psi_{H_0}^\dagger(1')\right)\right). \tag{4.56}$$

Expanding the exponential we get strings of, say, fermi field operators traced and weighted with respect to the free statistical operator. The zeroth-order term just gives the free contour Green's function

$$G_C^{(0)}(1,1') = -i\text{Tr}\left(\rho_0\, T_C\left(\psi_{H_0}(1)\, \psi_{H_0}^\dagger(1')\right)\right). \tag{4.57}$$

---

[22]This is an appealing alternative in the quantum field theoretic treatment of quenched disorder, more physically appealing than the obscure Replica trick or supersymmetry methods, the latter being limited to systems without interactions.

For the first-order term we have the expression

$$G_C^{(1)}(1,1') = (-i)^2 \int d\mathbf{x}_2 \int_C d\tau_2 \, V(2) \mathrm{Tr} \left(\rho_0 \, T_C \left(\psi_{H_0}^\dagger(2) \, \psi_{H_0}(2) \, \psi_{H_0}(1) \, \psi_{H_0}^\dagger(1') \right) \right).$$

$$(4.58)$$

Wick's theorem will provide us with the recipe for decomposing such averages over strings of fermi or bose field operators into products involving the free contour Green's function, Eq. (4.57).

Let us then look at the generic fermion–boson interaction, Eq. (2.71), (or equivalently, the jellium electron–phonon interaction), and let the ψ-field denote the fermi field in the Green's function in Eq. (4.53). Expanding the exponential we get terms in increasing order of the coupling constant. The zeroth-order term again gives the free contour Green's function. We shall soon realize that the term linear in the phonon or boson field vanishes, and therefore consider the second order term[23]

$$G^{(2)}(1,1') = -i\mathrm{Tr} \left(e^{-\beta H_0} T_{c_i} \left(\frac{(-i)^2}{2!} \int_{c_i} d\tau_3 \, H_{H_0}^{(i)}(\tau_3) \int_{c_i} d\tau_2 \, H_{H_0}^{(i)}(\tau_2) \, \psi_{H_0}(1) \, \psi_{H_0}^\dagger(1') \right) \right)$$

$$= \frac{i}{2!} g^2 \int_{c_i} d\tau_3 d\tau_2 \mathrm{Tr} \left(e^{-\beta H_0} T_{c_i} \left(\psi^\dagger(\mathbf{x}, \tau_3) \, \psi_{H_0}(\mathbf{x}, \tau_3) \, \phi_{H_0}(\mathbf{x}, \tau_3) \right.\right.$$

$$\times \quad \psi_{H_0}^\dagger(\mathbf{x}, \tau_2) \, \psi_{H_0}(\mathbf{x}, \tau_2) \, \phi_{H_0}(\mathbf{x}, \tau_2) \, \psi_{H_0}(1) \, \psi_{H_0}^\dagger(1') \Big) \Big) . \qquad (4.59)$$

This has the form of a trace over a string of fermi and bose operators under the contour ordering symbol weighted by the Hamiltonian for the free fields which is Gaussian. The trace over these independent degrees of freedom splits into a product of two separate traces containing only fermi or bose fields weighted by their respective free field Hamiltonians, $H_0 = H_{\mathrm{f}}^{(0)} + H_{\mathrm{b}}^{(0)}$. Higher-order terms in the expansion have the same form, they just contain strings with a larger number of fields. In perturbation theory the task is to evaluate such terms. Owing to the Gaussian nature of the average there is a simple prescription for the form of an arbitrary such term. We now turn to show this.

4.3.3   Wick's theorem

In the previous section we realized that the quantities appearing in perturbation theory for the contour-ordered Green's function are strings of field operators weighted by the statistical operator for the free theory. The way to decompose such Gaussian averages of strings we refer to as Wick's theorem, honoring its precursor in QED.[24] The perturbative expansion of the contour-ordered Green's function now proceeds

---

[23] An exception to this is the case of Bose–Einstein condensation. In that case it can be convenient to work with states which are superpositions of states with different number of particles, allowing nonzero field averages due to spontaneously broken gauge symmetry. This case is discussed further in Section 10.6.

[24] We are here dealing with Wick's theorem at the factory floor, which is good for a start: constructing the decomposition. In Chapters 9 and 10 we will proceed from the top down, having explicit expressions for the construction.

by expanding the exponentials. When we expand the exponential in Eq. (4.50) or Eq. (4.53), products of interaction Hamiltonians appear under contour ordering. The generic case for the perturbative expansion to nth order of the contour-ordered Green's function is the trace of products, or strings, of the field operators of the theory in the interaction picture weighted by the free part of the Hamiltonian, a quadratic form in these fields. For example, in the case of electron–phonon interaction a string of n phonon fields and $2n$ fermi fields occurs; see Eq. (4.126). The weighted trace over the fermi and bose fields separates into the two traces over these independent degrees of freedom. To be explicit, let us first consider the trace over the bose degrees of freedom, and of interest is therefore the calculation of the weighted trace of a string of contour-ordered bose field operators, ordered along a contour C.[25] We introduce the representation of the bose field in terms of its creation and annihilation operators as in Eq. (2.74) and encounter strings of creation and annihilation operators[26]

$$S = \text{tr}(\rho_T\, T_C(c(\tau_n)\, c(\tau_{n-1}) \, \cdots \, c(\tau_2)\, c(\tau_1)))$$

$$\equiv \langle T_C(c(\tau_n)\, c(\tau_{n-1}) \, \cdots \, c(\tau_2)\, c(\tau_1))\rangle\,, \qquad (4.60)$$

where the cs denote either a creation or annihilation operator, a or $a^\dagger$, and

$$\rho_T = \frac{e^{-H_b^{(0)}/kT}}{\text{Tr}\, e^{-H_b^{(0)}/kT}} \qquad (4.61)$$

is the statistical operator for the equilibrium state of the non-interacting bosons or phonons at temperature T, and

$$H_b^{(0)} = \sum_q h_q = \sum_q \epsilon_q\, a_q^\dagger\, a_q \qquad (4.62)$$

or for the grand canonical ensemble, substituting in Eq. (4.62) $H_b^{(0)} \to H_b^{(0)} - \mu_b N_b$ (i.e. we measure energies from the chemical potential, $\omega_q = \epsilon_q - \mu$) and we have introduced the notation

$$\langle \ldots \rangle = \text{tr}(\rho_T \cdots) \qquad (4.63)$$

where tr denotes trace with respect to the bose species under consideration. As in Eq. (4.60) we suppress whenever possible reference to the, for argument's sake, irrelevant state labels, here momentum or wave vectors (and possibly spin and longitudinal and transverse phonon labels).

The contour ordering symbol, T_C, orders the operators according to their position on the contour C (earlier contour positions orders operators to the right) so that, for example, for two bose operators indexed by contour times τ and τ'

$$T_C(c(\tau)\, c(\tau')) = \begin{cases} c(\tau)\, c(\tau') & \text{for} \quad \tau >_C \tau' \\[2mm] c(\tau')\, c(\tau) & \text{for} \quad \tau' >_C \tau \end{cases} \qquad (4.64)$$

---

[25] In the following the contour C can be the real-time contour depicted in Figure 4.5 or the contour depicted in Figure 4.4, allowing us to include the general case of transient phenomena.
[26] Although the Hamiltonian contains fields at equal times, we can in the course of the argument assume them infinitesimally split, and all the contour time variables can thus be considered different.

where the upper identity is for contour time τ being further along the contour than τ' and the lower identity being the ordering for the opposite case (for the fermi case, we should remember the additional minus sign for interchange of fermi operators).

Such an ordered expression of bose field operators as in Eq. (4.60) can now be decomposed according to Wick's theorem, which relies only on the simple property

$$[c_q, \rho_T] = \rho_T c_q \left[\exp\{\lambda_c \omega_q / k_B T\} - 1\right] \tag{4.65}$$

valid for a Hamiltonian quadratic in the bose field ($\lambda_c = \pm 1$, depending upon whether c_q is a creation or an annihilation operator for state q). We now turn to prove Wick's theorem, which is the statement that the quadratically weighted trace of a contour-ordered string of creation and annihilation operators can be decomposed into a sum over all possible pairwise products

$$\langle T_C(c(\tau_n) c(\tau_{n-1}) \cdots c(\tau_2) c(\tau_1))) \rangle = \sum_{\text{a.p.p.}} \prod_{q,q'} \langle T_{c_t}(c_q(\tau) c_{q'}(\tau')) \rangle \tag{4.66}$$

where the sum is over all possible ways of picking pairs (a.p.p.) among the n operators, not distinguishing ordering within pairs. Equivalently, Wick's theorem states that the trace of a contour-ordered string of creation and annihilation operators weighted with a quadratic Hamiltonian has the Gaussian property. The expressions on the right are free thermal equilibrium contour-ordered Green's functions, quantities for which we have explicit expressions.

Before proving Wick's theorem and the relation Eq. (4.65), we first observe some preliminary results. Different q-labels describe different momentum degrees of freedom, so operators for different qs commute, and algebraic manipulations with commuting operators are just as for usual numbers giving for example

$$\rho_T = \prod_q \rho_q^T, \tag{4.67}$$

where we have introduced the thermal statistical operator for each mode

$$\rho_q^T = z_q^{-1} e^{-h_q / kT} \tag{4.68}$$

and the partition function for the single mode

$$z_q = \frac{1}{1 - e^{-\omega_q / kT}}. \tag{4.69}$$

The independence of each mode degree of freedom, as expressed by the commutation of operators corresponding to different degrees of freedom, gives

$$c_q \rho_T = \left(\prod_{q'(\neq q)} \rho_{q'}^T\right) c_q \rho_q^T. \tag{4.70}$$

Now, using the commutation relations for the creation and annihilation operators we have

$$c_q h_q = (h_q - \lambda_c \omega_q) c_q, \tag{4.71}$$

where

$$\lambda_c = \begin{cases} +1 & \text{for} \quad c_q = a_q^\dagger \\ -1 & \text{for} \quad c_q = a_q \end{cases} . \tag{4.72}$$

Using Eq. (4.71) repeatedly gives

$$c_q h_q^n = (h_q - \lambda_c \omega_q)^n c_q \tag{4.73}$$

and upon expanding the exponential function and re-exponentiating we can commute through to get

$$c_q \rho_T = e^{\lambda_c \omega_q / kT} \rho_T c_q \tag{4.74}$$

so that for the commutator of interest we have the property stated in Eq. (4.65).

We then prove for an arbitrary operator A that in the bose case

$$\langle [c_q, A] \rangle = (1 - e^{\lambda_c \omega_q / kT}) \langle c_q A \rangle \tag{4.75}$$

as we first note, by using the cyclic invariance property of the trace, that

$$\langle [c_q, A] \rangle = -tr([c_q, \rho_T] A) \tag{4.76}$$

and then by using Eq. (4.65) we get Eq. (4.75).

Exercise 4.2. Show that for the case of fermions

$$\langle \{c_q, A\} \rangle = (1 + e^{\lambda_c \omega_q / kT}) \langle c_q A \rangle . \tag{4.77}$$

Employing Eq. (4.75) with $A = 1, 1, a, a^\dagger$, respectively, we observe that all the following averages vanish

$$0 = \langle a(t) \rangle = \langle a^\dagger(t) \rangle = \langle a(t) a(t') \rangle = \langle a^\dagger(t) a^\dagger(t') \rangle \tag{4.78}$$

and as a consequence the average value of the interaction energy vanishes, $\langle H_i(t) \rangle = 0$, for the case of fermion–boson interaction (and electron–phonon interaction). These equalities are valid for any state diagonal in the total number of particles, i.e. a state with a definite number of particles.

Repeating the algebraic manipulations leading to Eq. (4.74), or by analytical continuation of the result, we have

$$c_q(t) = c_q e^{-itH_b^{(0)}} = e^{i\lambda_c \omega_q t} e^{-itH_b^{(0)}} c_q \tag{4.79}$$

from which we get that the creation and annihilation operators in the interaction picture have a simple time dependence in terms of a phase factor

$$c_q(t) = e^{itH_b^{(0)}} c_q e^{-itH_b^{(0)}} = c_q e^{i\lambda_c \omega_q t} . \tag{4.80}$$

The commutators formed by creation and annihilation operators in the interaction picture are thus c-numbers, the only non-vanishing one being specified by

$$[a_q(t), a_{q'}^\dagger(t')] = \delta_{q,q'} e^{-i\omega_q (t-t')} . \tag{4.81}$$

According to Eq. (4.75) we thereby have

$$\langle a_q(t)\, a_{q'}^\dagger(t')\rangle = (1 - e^{-\omega_q/kT})^{-1}\langle [a_q(t), a_{q'}^\dagger(t')]\rangle$$

$$= \delta_{q,q'}\,(1 - e^{-\omega_q/kT})^{-1}e^{-i\omega_q(t-t')}$$

$$= \delta_{q,q'}\,(n(\omega_q) + 1)e^{-i\omega_q(t-t')}$$

$$\equiv i\,D^{>}_{qq'}(t,t')\,, \tag{4.82}$$

where the Bose–Einstein distribution appears as specified by the Bose function

$$n(\omega_q) = \frac{1}{e^{\omega_q/kT} - 1} = \frac{1}{e^{(\epsilon_q - \mu_b)/kT} - 1}\,. \tag{4.83}$$

Exercise 4.3. Show that, for the opposite ordering of the creation and annihilation operators, the correlation function is

$$i\,D^{<}_{qq'}(t,t') \equiv \langle a_{q'}^\dagger(t')\, a_q(t)\rangle = \mathrm{tr}(\rho_T\, a_{q'}^\dagger(t')\, a_q(t))$$

$$= n(\omega_q)\,\delta_{q,q'}\,e^{-i\omega_q(t-t')}\,. \tag{4.84}$$

Exercise 4.4. Show that, for the case of fermi operators, the correlation functions are

$$G^{<}_{qq'}(t,t') \equiv i\langle a_{q'}^\dagger(t')\, a_q(t)\rangle = i\,\mathrm{tr}(\rho_T\, a_{q'}^\dagger(t')\, a_q(t))$$

$$= if(\epsilon_q)\,\delta_{q,q'}\,e^{-i\epsilon_q(t-t')} \tag{4.85}$$

and

$$G^{>}_{qq'}(t,t') \equiv -i\langle a_q(t)\, a_{q'}^\dagger(t')\rangle = -i\,\mathrm{tr}(\rho_T\, a_q(t)\, a_{q'}^\dagger(t'))$$

$$= -i(1 - f(\epsilon_q))\,\delta_{q,q'}\,e^{-i\epsilon_q(t-t')}\,, \tag{4.86}$$

where $f(\epsilon_q)$ is the Fermi function

$$f(\epsilon_q) = \frac{1}{e^{(\epsilon_q - \mu)/kT} + 1}\,. \tag{4.87}$$

Exercise 4.5. Show that, for the case of fermi operators,

$$\langle \{a_q(t), a_{q'}^\dagger(t')\}\rangle = \delta_{q,q'}\,e^{-i\epsilon_q(t-t')}\,. \tag{4.88}$$

If the string S, Eq. (4.60), contains an odd number of operators, the expression equals zero since the expectation value is with respect to the thermal equilibrium state.[27] For an odd number of operators we namely encounter a matrix element between states with different number of particles or quanta; for example,

$$\langle a_q a_q^\dagger a_q \rangle = Z^{-1} \sum_{\{n_q\}_q} e^{-E(\{n_q\}_q)/kT} (\sqrt{n_q})^3 \langle n_q | n_q - 1 \rangle = 0 , \qquad (4.89)$$

which is zero by orthogonality of the different energy eigenstates.

As an example of using Wick's theorem we write down the term we encounter at fourth order in the coupling to the bosons (we suppress, for the present consideration, the immaterial q labels)

$$\mathrm{tr}(\rho_T T_{c_t}(a(\tau_1) a^\dagger(\tau_2) a(\tau_3) a^\dagger(\tau_4))) = \langle T_{c_t}(a(\tau_1) a^\dagger(\tau_2)) \rangle \langle T_{c_t}(a(\tau_3) a^\dagger(\tau_4)) \rangle$$

$$+ \langle T_{c_t}(a(\tau_1) a^\dagger(\tau_4)) \rangle \langle T_{c_t}(a(\tau_3) a^\dagger(\tau_2)) \rangle .$$

$$(4.90)$$

Here we have deleted terms that do not pair creation and annihilation operators, because such terms, just as above, lead to matrix elements between orthogonal states:

$$\langle T_{c_t}(a(\tau) a(\tau')) \rangle = 0 = \langle T_{c_t}(a^\dagger(\tau) a^\dagger(\tau')) \rangle . \qquad (4.91)$$

At the fourth-order level the ordered Gaussian decomposition can of course be obtained by noting that only by pairing equal numbers of creation and annihilation operators can the number of quanta stay conserved and the matrix element be nonzero as we have the expression

$$\mathrm{tr}(\rho_T T_{c_t}(a(\tau_1) a^\dagger(\tau_2) a(\tau_3) a^\dagger(\tau_4)))$$

$$= \sum_{\{n_q\}_q} e^{-E(\{n_q\}_q)/kT} \langle \{n_q\}_q | T_{c_t}(a(\tau_1) a^\dagger(\tau_2) a(\tau_3) a^\dagger(\tau_4)) | \{n_q\}_q \rangle . \qquad (4.92)$$

Wick's theorem is the generalization of this simple observation.

We now turn to the general proof of Wick's theorem for the considered case of bosons. Wick's theorem is trivially true for $N = 1$ (and for $N = 2$ according to the above consideration), and we now turn to prove Wick's theorem by induction. Let us therefore consider an N-string with $2N$ operators

$$S_N = \langle T_C(c(\tau_{2N}) c(\tau_{2N-1}) \cdots c(\tau_2) c(\tau_1)) \rangle . \qquad (4.93)$$

We can assume that the contour-time labeling already corresponds to the contour-ordered one, since the bose operators can be moved freely around under the contour

---

[27]This would not be the case for, say, photons in a coherent state in which case the substitution $c \to c - \langle c \rangle$ is needed. Also in describing a Bose–Einstein condensate it is convenient to work with a superposition of states containing a different number of particles so that $\langle c \rangle$ is non-vanishing, a situation we shall deal with in due time. For the case of electron–phonon interaction we thus assume no linear term in the phonon Hamiltonian, which would correspond to a displaced oscillator, or that such a term is effectively removed by redefining the equilibrium position of the oscillator.

ordering, or otherwise we just relabel the indices, and we have[28]

$$S_N = \left\langle \prod_{n=1}^{2N} c(\tau_n) \right\rangle = \left\langle c(\tau_{2N}) \prod_{n=1}^{2N-1} c(\tau_n) \right\rangle. \tag{4.94}$$

We then use the above proved relation, Eq. (4.75), to rewrite

$$S_N = \left(1 - e^{\lambda_c \omega_q / kT}\right)^{-1} \left\langle [c(\tau_{2N}), \prod_{n=1}^{2N-1} c(\tau_n)] \right\rangle. \tag{4.95}$$

In the first term in the commutator we commute $c(\tau_{2N})$ to the right

$$\left[c(\tau_{2N}), \prod_{n=1}^{2N-1} c(\tau_n) \right] = c(\tau_{2N-1}) c(\tau_{2N}) \prod_{n=1}^{2N-2} c(\tau_n) + [c(\tau_{2N}), c(\tau_{2N-1})] \prod_{n=1}^{2N-2} c(\tau_n)$$

$$- \left(\prod_{n=1}^{2N-1} c(\tau_n) \right) c(\tau_{2N}). \tag{4.96}$$

We now keep commuting $c(\tau_{2N})$ through in the first term repeatedly, each time generating a commutator, and eventually ending up with canceling the last term in Eq. (4.96), so that

$$\left[c(\tau_{2N}), \prod_{n=1}^{2N-1} c(\tau_n) \right] = \sum_{n=1}^{2N-1} [c(\tau_{2N}), c(\tau_n)] \prod_{\substack{m=1 \\ m(\neq n)}}^{2N-1} c(\tau_m). \tag{4.97}$$

Then we use that the commutator is a c-number, which according to Eq. (4.75) we can rewrite as

$$[c_q(\tau_{2N}), c_{q'}(\tau_n)] = \delta_{q,q'} \left(1 - e^{\lambda_{c_q} \omega_q / kT}\right) \langle c_q(\tau_{2N}) c_{q'}(\tau_n) \rangle \tag{4.98}$$

and being a c-number it can be taken outside the thermal average in Eq. (4.95), and we obtain

$$S_N = \sum_{n=1}^{2N-1} \langle c(\tau_{2N}) c(\tau_n) \rangle \left\langle \prod_{\substack{m=1 \\ m(\neq n)}}^{2N-1} c(\tau_m) \right\rangle$$

$$= \sum_{n=1}^{2N-1} \langle T_{c_t}(c(\tau_{2N}) c(\tau_n)) \rangle \left\langle T_{c_t} \left(\prod_{\substack{m=1 \\ m(\neq n)}}^{2N-1} c(\tau_m) \right) \right\rangle, \tag{4.99}$$

where we reintroduce the contour ordering. By assumption the second factor can be written as a sum over all possible pairs (on a.p.p.-form), and by induction the N

---

[28] For fermions interchange of fields involves a minus sign, and an overall sign factor occurs, $(-1)^{\zeta_P}$, where ζ_P is the sign of the permutation P bringing the string of fields to a contour time-ordered form.

case is then precisely seen to be of that form too. We note, that to prove Wick's theorem we have only exploited that the weight was a quadratic form, leaving the commutator a c-number.[29]

The contour label uniquely specifies from which term in the spatial representation of the bose field it originates, and since Eq. (4.99) is valid for both creation and annihilation operators, and therefore for any linear combinations of such, we have therefore shown[30]

$$\langle T_C(\phi(\mathbf{x}_{2n}, \tau_{2n}) \, \phi(\mathbf{x}_{2n-1}, \tau_{2n-1}) \cdots \phi(\mathbf{x}_2, \tau_2) \, \phi(\mathbf{x}_1, \tau_1)))\rangle$$

$$= \sum_{\text{a.p.p.}} \prod_{i \neq j} \langle T_C(\phi(\mathbf{x}_i, \tau_i)\phi(\mathbf{x}_j, \tau_j)))\rangle \equiv \sum_{\text{a.p.p.}} \prod_{i \neq j} i^N \, D_0(\mathbf{x}_i, \tau_i; \mathbf{x}_j, \tau_j) . \quad (4.100)$$

The index on the contour-ordered Green's functions indicates they are the free ones. Performing the trace over a string of bose operators weighted by a quadratic form therefore corresponds to pairing the operators together pairwise in all possible ways.[31]

For the case of fermi operators, the proof of Wick's theorem runs analogous to the above, in fact the bose and fermi cases can be handled in unison if we unite Eq. (4.75) and Eq. (4.77) by introducing the notation

$$\langle [c_q, A]_s \rangle = \left(1 + s \, e^{\lambda_c \omega_q / kT}\right) \langle c_q \, A \rangle , \quad (4.101)$$

where $s = \mp$ signifies the case of bose and fermi statistics, respectively. The arguments relating Eq. (4.94) to Eq. (4.106) run identical with commutators replaced by anti-commutators and a minus sign, or for treating the two cases simultaneous s is introduced. For the combined case we have

$$S_N = \left\langle \prod_{n=1}^{2N} c(\tau_n) \right\rangle = \left\langle c(\tau_{2N}) \prod_{n=1}^{2N-1} c(\tau_n) \right\rangle$$

$$= \left(1 + s \, e^{\lambda_c(\tau_{2N}) \omega_q / kT}\right)^{-1} \left\langle [c(\tau_{2N}), \prod_{n=1}^{2N-1} c(\tau_n)]_s \right\rangle \quad (4.102)$$

and

$$\left\langle [c(\tau_{2N}), \prod_{n=1}^{2N-1} c(\tau_n)]_s \right\rangle = \left\langle -s(c(\tau_{2N-1}) \, c(\tau_{2N}) - s[c(\tau_{2N}), c(\tau_{2N-1})]_s) \prod_{n=1}^{2N-2} c(\tau_n) \right\rangle$$

$$+ s \left\langle \left(\prod_{n=1}^{2N-1} c(\tau_n) \right) c(\tau_{2N}) \right\rangle$$

---

[29] If the weight was not quadratic, we would have encountered correlations that must be handled additionally.

[30] A reader familiar with the standard $T = 0$ or finite temperature imaginary-time Wick's theorem, will recognize that their validity just represents special cases of the above proof.

[31] The presented general version of Wick's theorem is capable of dealing with many-body systems of bosons, irrespective of the absence or presence of a Bose–Einstein condensate, if one employs the grand canonical ensemble.

$$= (-s) \left\langle c(\tau_{2N-1}) \, c(\tau_{2N}) \prod_{n=1}^{2N-2} c(\tau_n) \right\rangle$$

$$+ \, [c(\tau_{2N}), c(\tau_{2N-1})]_s \left\langle \prod_{n=1}^{2N-2} c(\tau_n) \right\rangle$$

$$+ \, s \left\langle \left(\prod_{n=1}^{2N-1} c(\tau_n) \right) c(\tau_{2N}) \right\rangle , \tag{4.103}$$

where the (anti- or) commutator, being a c-number, can be taken outside the operator averaging. We now keep (anti- or) commuting $c(\tau_{2N})$ through in the first term repeatedly, each time generating a (anti- or) commutator and a factor (-s), and eventually ending up with canceling the last term, so that

$$\left[c(\tau_{2N}), \prod_{n=1}^{2N-1} c(\tau_n) \right]_s = \sum_{n=1}^{2N-1} (-s)^{n-1} [c(\tau_{2N}), c(\tau_n)]_s \prod_{\substack{m=1 \\ m\,(\neq n)}}^{2N-1} c(\tau_m) . \tag{4.104}$$

Then we use the fact that the (anti- or) commutator is a c-number, which we can rewrite

$$[c_q(\tau_{2N}), c_{q'}(\tau_n)]_s = \delta_{q,q'} \left(1 + s \, e^{\lambda_{c_q} \omega_q / kT} \right) \langle c_q(\tau_{2N}) \, c_{q'}(\tau_n) \rangle \tag{4.105}$$

and taking it outside the thermal average we obtain

$$S_N = \sum_{n=1}^{2N-1} \langle c(\tau_{2N}) \, c(\tau_n) \rangle \left\langle \prod_{\substack{m=1 \\ m\,(\neq n)}}^{2N-1} c(\tau_m) \right\rangle$$

$$= \sum_{n=1}^{2N-1} \langle T_{c_t}(c(\tau_{2N}) \, c(\tau_n)) \rangle \left\langle T_{c_t} \left(\prod_{\substack{m=1 \\ m\,(\neq n)}}^{2N-1} c(\tau_m) \right) \right\rangle . \tag{4.106}$$

For the case of a fermi field we thus obtain the analogous result to Eq. (4.100)

$$\langle T_C(\psi(\mathbf{x}_{2n}, \tau_{2n}) \, \psi(\mathbf{x}_{2n-1}, \tau_{2n-1}) \ldots \psi(\mathbf{x}_2, \tau_2) \, \psi(\mathbf{x}_1, \tau_1)) \rangle$$

$$= \sum_{\text{a.p.p.}} \prod_{i \neq j} (-1)^{\zeta_P} \langle T_C(\psi(\mathbf{x}_i, \tau_i) \, \psi(\mathbf{x}_j, \tau_j)) \rangle$$

$$\equiv \sum_{\text{a.p.p.}} \prod_{i \neq j} (-1)^{\zeta_P} i^N G_0(\mathbf{x}_i, \tau_i; \mathbf{x}_j, \tau_j) , \tag{4.107}$$

where the quantum statistical factor $(-1)^{\zeta_P}$ counts the number of transpositions relating the orderings on the two sides. For the case of a state with a definite number of particles, only if fermi creation and annihilation fields are paired do we get a non-vanishing contribution.[32] In the last equality, the free contour ordered Green's function is introduced.[33]

In the perturbative expansion of the Green's functions, the quantum fields, and their associated multi-particle spaces, have left the stage, absorbed in the expressions for the free propagators.

The perturbative expansion lends itself to suggestive diagrammatics, the Feynman diagrammatics for non-equilibrium systems, which we now turn to introduce.

Exercise 4.6. Consider a harmonic oscillator, where $\hat{x}(t)$ is the position operator in the Heisenberg picture, and show that, for the generating functional we have

$$Z[f_t] \equiv \left\langle T e^{i \int_{-\infty}^{\infty} dt \, f_t \, \hat{x}(t)} \right\rangle = \text{tr}\left(\rho_T \, T e^{i \sqrt{\frac{\hbar}{2M\omega_q}} \int_{-\infty}^{\infty} dt \, f_t (\hat{a}(t) + \hat{a}^\dagger(t))} \right)$$

$$= e^{-\frac{1}{2} \int_{-\infty}^{\infty} dt \int_{-\infty}^{\infty} dt' \, f_t \, \langle T(\hat{x}(t)\,\hat{x}(t')) \rangle f_{t'}} . \tag{4.108}$$

In Chapter 9 we will consider the generating functional technique for quantum field theory. Quantum mechanics is then the case of the zero dimensional field theory.

4.4 Non-equilibrium diagrammatics

Empowered by Wick's theorem, we can envisage the whole perturbative expansion of the contour ordered Green's function. Writing down the nth-order contribution from the expansion of the exponential in Eq. (4.53) containing the interaction, and employing Wick's theorem, we get expressions involving products of propagators and vertices. However, the expressions resulting from perturbation theory quickly become unwieldy. A convenient method of representing perturbative expressions by diagrams was invented by Feynman. Besides the appealing aspect of representing perturbative expressions by drawings, the diagrammatic method can also be used directly for reasoning and problem solving. The easily recognizable topology of diagrams makes the diagrammatic method a powerful tool for constructing approximation schemes as well as exact equations that may hold true beyond perturbation theory. Furthermore, by elevating the diagrams to be a representation of possible alternative physical processes, the diagrammatic representation becomes a suggestive tool providing physical intuition into quantum dynamics. In this section we construct the general diagrammatic perturbation theory valid for non-equilibrium situations. We shall illustrate the diagrammatics by considering the generic cases.

---

[32]The use of states with a non-definite number of fermions, as useful in the theory of superconductivity, would lead to the appearance of so-called anomalous Green's functions, as we discuss in Chapter 8.

[33]Minus the imaginary unit provided N-fold times from the expansion of the exponential function containing the interaction, explains why the imaginary unit was introduced in the definition of the contour-ordered Green's function in the first place. However, one is of course entitled to keep track of factors at one's taste.

4.4.1 Particles coupled to a classical field

The simplest kind of coupling is that of an assembly of identical particle species coupled to an external classical field, $V(\mathbf{x}, t)$. In that case the contour-ordered Green's function, written in the form ready for a perturbative expansion, Eq. (4.53) or Eq. (4.50), has the form

$$
G_C(1, 1') = -i \mathrm{Tr} \left(\rho_0 \, T_c \left(e^{-i \int_c d\tau \int d\mathbf{x} \, V(\mathbf{x}, \tau) \psi_{H_0}^\dagger (\mathbf{x}, \tau) \psi_{H_0}(\mathbf{x}, \tau)} \; \psi_{H_0}(1) \, \psi_{H_0}^\dagger(1') \right) \right),
$$

(4.109)

where c is the contour that starts at t_0 and stretches through $\max(t_1, t_{1'})$ and back again to t_0, as depicted in Figure 4.4. If t_0 is taken to be in the far past, $t_0 \to -\infty$, we obtain the real-time contour of Figure 4.5. Expanding the exponential we get strings of, say, fermion operators subdued to the contour-ordering operation and thermally weighted by the Hamiltonian for the free field, which is Gaussian as ρ_0 is given by Eq. (4.54). Higher-order terms in the expansion have the same form, they just contain strings with a larger number of fields. In perturbation theory the task is to evaluate such terms, or rather first break them down into Gaussian products as accomplished by Wick's theorem, i.e. decomposed into a product of free thermal equilibrium contour-ordered Green's functions.

For the first-order term, Eq. (4.58), we have according to Wick's theorem the expression

$$
G_C^{(1)}(1, 1') = \int d\mathbf{x}_2 \int_c d\tau_2 \, G_C^{(0)}(1, 2) \, V(2) \, G_C^{(0)}(2, 1')
$$

(4.110)

and equivalently for higher order terms. The term where the external points are paired, giving rise to a disconnected or unlinked diagram with a vacuum diagram contribution, clearly vanishes owing to the integration along both the forward and return parts of the contour, giving two terms differing only by a minus sign.

The generic component in a diagram, the first order term, is graphically represented by the diagram

$$
G_C^{(1)}(\mathbf{x}, \tau; \mathbf{x}', \tau') \quad = \quad \underset{\mathbf{x}\tau}{\bullet} \!\!\xleftarrow{\hspace{1cm}}\!\! \underset{\mathbf{x}_1 \tau_1}{\times} \!\!\xleftarrow{\hspace{1cm}}\!\! \underset{\mathbf{x}'\tau'}{\bullet}
$$

(4.111)

where a cross has been introduced to symbolize the interaction of the particles with the scalar potential

$$
\underset{\mathbf{x}\tau}{\xrightarrow{\quad\times\quad}} \quad \equiv \quad V(\mathbf{x}, \tau)
$$

(4.112)

and a thin line is used to represent the zeroth-order or free thermal equilibrium contour-ordered Green's function

$$
\underset{\mathbf{x}\tau}{\bullet} \xleftarrow{\quad R \quad} \underset{\mathbf{x}'\tau'}{\bullet} \quad \equiv \quad G_C^{(0)}(\mathbf{x}, \tau; \mathbf{x}', \tau')
$$

(4.113)

in order to distinguish it from the contour-ordered Green's function in the presence of the potential V, the full contour-ordered Green's function

$$\equiv \quad G_C(\mathbf{x}, \tau; \mathbf{x}', \tau') \qquad (4.114)$$

depicted as a thick line.

With this dictionary or stenographic rules, the analytical form, Eq. (4.110), is obtained from the diagram, Eq. (4.111), since integration is implied over the variables of the internal points where interaction with the potential takes place. The only difference to equilibrium standard Feynman diagrammatics is that internal integrations are not over time or imaginary time, but over the contour variable.

The second-order expression in perturbation theory leads to two terms giving identical contributions, since interchange of pairs of fermion operators introduces no factor of -1. The resulting factor of two exactly cancels the factor of two originating from the expansion of the exponential in Eq. (4.109). This feature repeats for the higher-order terms, and for particles interacting with a scalar potential $V(\mathbf{x}, t)$, we have the following diagrammatic representation of the contour-ordered Green's function:

$$G_C(\mathbf{x}, \tau; \mathbf{x}', \tau') = \quad\quad = \quad\quad$$

$$+ \quad\quad + \quad\quad$$

$$+ \quad\quad + \quad\cdots\ , \qquad (4.115)$$

where all ingredients now represent contour quantities according to the above dictionary.

Exercise 4.7. Show that for a particle coupled to a scalar potential $V(\mathbf{x}, t)$, the infinite series

$$G(1, 1') = G_0(1, 1') + \int d\mathbf{x}_2 \int_c d\tau_2\, G_0(1, 2)\, V(\mathbf{x}_2, \tau_2)\, G_0(2, 1') + \cdots \qquad (4.116)$$

by iteration can be captured in the Dyson equation

$$G(1, 1') = G_0(1, 1') = \int d\mathbf{x}_2 \int_c d\tau_2\, G_0(1, 2)\, V(\mathbf{x}_2, \tau_2)\, G(2, 1')\,, \qquad (4.117)$$

which has the diagrammatic representation

$$= \quad + \quad\quad .\quad (4.118)$$

If in Eq. (4.117) we operate with the inverse free contour ordered Green's function which satisfies (recall Exercise 4.1 on page 89)

$$\int d\mathbf{x}_2 \int_c d\tau_2 \, G_0^{-1}(1,2) \, G_0(2,1') = \delta(1-1') \tag{4.119}$$

we obtain

$$\int d\mathbf{x}_2 \int_c d\tau_2 \, (G_0^{-1}(1,2) - V(2)) \, G(2,1') = \delta(1-1') . \tag{4.120}$$

As expected, the coupling to a classical field can be accounted for by adding the potential term to the free Hamiltonian. The δ-function contains, besides products in δ-functions in the degrees of freedom, the contour variable δ-function specified in Eq. (4.45). We shall write the equation, absorbing the potential in the inverse propagator, in condensed matrix notation

$$(G_0^{-1} \boxtimes G)(1,1') = \delta(1-1') = (G \boxtimes G_0^{-1})(1,1') \tag{4.121}$$

where $\boxtimes$ signifies matrix multiplication in the spatial variable (and possible internal degrees of freedom) and contour time variables. The latter, adjoint, equation corresponds to the choice of iterating from the left instead of the right.

4.4.2 Particles coupled to a stochastic field

If the potential $V(\mathbf{x}, t)$ of the previous section is treated as a stochastic Gaussian random variable (with zero mean), the diagrams in perturbation theory, Eq. (4.115), will be turned into the diagrams for the averaged Green's function according to the prescription: pair together pairwise potential crosses in all possible ways and substitute for the paired crosses the Gaussian correlator of the stochastic variable. For the lowest order contribution to the averaged contour ordered Green's function we thus have the diagram

$$\langle G_C^{(2)}(1,1') \rangle \quad = \qquad \qquad \qquad \tag{4.122}$$

where the outermost labels 1 and 1' as well as the internal labels 2 and 3 are suppressed, and the following notation has been introduced for the correlator:

$$= \quad \langle V(\mathbf{x},\tau) \, V(\mathbf{x}',\tau') \rangle . \tag{4.123}$$

If the stochastic variable is taken as time independent, $V(\mathbf{x})$, we cover the case of particles in a random impurity potential (treated in the Born approximation), and the correlator, the impurity correlator, is given by.

$$\langle V(\mathbf{x})V(\mathbf{x}') \rangle = n_{\mathrm{i}} \int d\mathbf{r}\, V_{\mathrm{imp}}(\mathbf{x} - \mathbf{r})\, V_{\mathrm{imp}}(\mathbf{x}' - \mathbf{r}) \,. \tag{4.124}$$

where $V_{\mathrm{imp}}(\mathbf{x})$ is the potential created at position $\mathbf{x}$ by a single impurity at the origin, and n_{i} is their concentration.[34]

4.4.3 Interacting fermions and bosons

The next level of complication is the important case of interacting fermions and bosons. Let us look at the generic fermion–boson interaction, Eq. (2.71), or equivalently, the jellium electron–phonon interaction, and let the ψ-field denote the fermi field in the Green's function we are looking at

$$G(1,1') = -i\,\mathrm{Tr}\left(\rho_0\, T_C\left(e^{-i\int_C d\tau\, H_{H_0}^{(i)}(\tau)}\, \psi_{H_0}(1)\, \psi_{H_0}^\dagger(1')\right)\right) \,. \tag{4.125}$$

Here the contour C is either the real-time contour of Figure 4.5, or the general contour of Figure 4.4.[35]

Expanding the exponential we get terms in increasing order of the coupling constant. The zeroth-order term just gives the free or thermal equilibrium contour-ordered Green's function, say for fermions, Eq. (4.57). The term linear in the phonon or boson field vanishes as discussed in Section 4.3.3, and we consider the second-order term[36]

$$G^{(2)}(1,1') = -i\,\mathrm{Tr}\left(\rho_0 T_C\left(\frac{(-i)^2}{2!}\int_C d\tau_3\, H_{H_0}^{(i)}(\tau_3)\int_C d\tau_2\, H_{H_0}^{(i)}(\tau_2)\, \psi_{H_0}(1)\,\psi_{H_0}^\dagger(1')\right)\right)$$

$$= \frac{i}{2!}\, g^2 \int_C d\tau_3 d\tau_2 \int d\mathbf{x}_3 d\mathbf{x}_2\, \mathrm{Tr}\left(e^{-\beta H_0} T_C\left(\psi_{H_0}^\dagger(\mathbf{x}_3,\tau_3)\psi_{H_0}(\mathbf{x}_3,\tau_3)\phi_{H_0}(\mathbf{x}_3,\tau_3)\right.\right.$$

$$\times\ \left.\left. \psi_{H_0}^\dagger(\mathbf{x}_2,\tau_2)\,\psi_{H_0}(\mathbf{x}_2,\tau_2)\,\phi_{H_0}(\mathbf{x}_2,\tau_2)\,\psi_{H_0}(1)\,\psi_{H_0}^\dagger(1')\right)\right) \,. \tag{4.126}$$

The expression has the form of a string of fermi and bose operators subdued to the contour-ordering operation and thermally weighted by the Hamiltonian for the free fields which is Gaussian. The trace over these independent degrees of freedom splits

---

[34] For details on quenched disorder and impurity averaging see Chapter 3 of reference [1].

[35] For the general contour of Figure 4.4, we should recall the cancelation of the disconnected diagrams against the vacuum diagrams of the denominator. However, the uninitiated reader need not worry about this by adopting the closed real-time contour. For the general case, the proof of cancelation can be consulted in Section 9.5.2.

[36] The use of states with a non-definite number of bosons, as useful in the theory of Bose–Einstein condensation, will be discussed in Section 10.6.

into a product of two separate traces containing only fermi or bose fields weighted by their respective free field Hamiltonians, $H_0 = H_f^{(0)} + H_b^{(0)}$. Higher-order terms in the expansion have the same form, they just contain strings with a larger number of fields. In perturbation theory the task is to evaluate such terms, or rather first break them down into Gaussian products as accomplished by Wick's theorem.

Consider the expression in Eq. (4.126), and perform the following choice of pairings: the creation fermi field indexed by the external label $1'$, $\psi_{H_0}^\dagger(1')$, is paired with the annihilation field associated with an internal point whose creation field is paired with the annihilation field associated with the other internal point, thereby fixing the final fermion pairing. Since the internal points represents dummy integrations this kind of choice gives rise to two identical expressions, an observation that can be used to cancel the factorial factor, $1/2!$, originating from the expansion of the exponential function in Eq. (4.125). The string of boson or phonon fields contains only two fields simply leading to the appearance of their contour-ordered thermal average. The considered second-order expression from the Wick decomposition for the contour-ordered fermion Green's functions thus becomes

$$G_C^{(2)}(1,1') \rightarrow ig^2 \int d\mathbf{x}_3 \int_C d\tau_3 \int d\mathbf{x}_2 \int_C d\tau_2 \, G_G^{(0)}(1,3) \, G_C^{(0)}(3,2) \, D_C^{(0)}(3,2) \, G_C^{(0)}(2,1') \,.$$

$$(4.127)$$

The presence of the imaginary unit in Eq. (5.26) is the result of one lacking factor of $-i$ for our convention of Green's functions: two factors of $-i$ are provided by the interaction and one provided externally in the definition of the Green's function.

The next step is then to visualize these unwieldy expressions arising in perturbation theory in terms of diagrams and a few stenographic rules, the Feynman rules. The considered second-order term in the coupling constant, Eq. (4.127), can be represented by the first diagram in Figure 4.6.

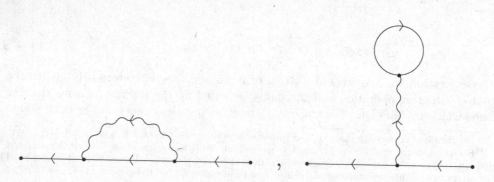

Figure 4.6 Lowest order fermion–boson diagrams.

Here the straight line represents the free or thermal equilibrium contour ordered fermion Green's function and the wavy line represents the thermal equilibrium contour-ordered boson Green's function:

$$\underset{x\tau}{\bullet}\wwwwwwwwwww\underset{x'\tau'}{\bullet} \equiv D_C^{(0)}(\mathbf{x}, \tau; \mathbf{x}', \tau') \qquad (4.128)$$

i.e.

$$D_C^{(0)}(1, 1') = -i\,\mathrm{tr}_b\left(\rho_b^{(0)}\, T_c(\phi_{H_0}(1)\,\phi_{H_0}^\dagger(1'))\right) = -i\,\langle T_c(\phi_{H_0}(1)\,\phi_{H_0}^\dagger(1'))\rangle \qquad (4.129)$$

as tr_b denotes the trace with respect to the boson degrees of freedom and $\rho_b^{(0)}$ is the thermal equilibrium statistical operator for the free bosons. As a Feynman rule, each vertex carries a factor of the coupling constant.

Another decomposition according to Wick's theorem of the second-order expression in Eq. (4.126) corresponds to when the fermi field indexed by the external label $1'$, $\psi_{H_0}^\dagger(1')$, is paired with the annihilation field associated with an internal point and the creation field of that vertex is paired with the field corresponding to the external point 1, thereby fixing the final fermion pairing, and again giving rise to two identical expressions, which in this case are the expression

$$G_C^{(2)}(1, 1') \to -ig^2 \int dx_3 \int_C d\tau_3 \int dx_2 \int_C d\tau_2\, G_C^{(0)}(1, 2)\, G_C^{(0)}(2, 1')\, D_C^{(0)}(3, 2)\, G_C^{(0)}(3, 3)\,. \qquad (4.130)$$

The corresponding expression can, according to the above dictionary for Feynman diagrams, be represented by the second diagram in Figure 4.6. We note the relative minus sign compared with the term represented by the first diagram in Figure 4.6 that reflects a general feature, which in diagrammatic terms can be stated as the Feynman rule: associated with a closed loop of fermion propagators is a factor of minus one.

The considered expressison corresponding to the second diagram in Figure 4.6 contains the fermion contour-ordered Green's function taken at equal contour times, $G_C^{(0)}(\mathbf{x}, \tau; \mathbf{x}, \tau)$, and therefore needs interpretation. Recalling that the annihilation field occurs to the right of the creation field originally in the interaction Hamiltonian, and labeling the contour variable of the latter by τ', we then have for the contour variables of these fields $\tau \overset{c}{<} \tau'$, and the propagator closing on itself represents the G-lesser Green's function, $G_0^<(\mathbf{x}, \tau; \mathbf{x}, \tau)$, corresponding to the density of the fermions. This is indicated by the direction of the arrow on the propagator closing on itself in the second diagram in Figure 4.6.

The final option for pairings in the Wick decomposition of the second-order expression in Eq. (4.126) corresponds to pairing the fermi creation field indexed by the external label $1'$, $\psi_{H_0}^\dagger(1')$, with the annihilation field indexed by the external label 1, $\psi_{H_0}(1)$. The pairings of the fermi fields labeled by the internal points can again be done in a two-fold way, and the corresponding expression arises

$$G_C^{(2)}(1,1') \rightarrow -ig^2 G_C^{(0)}(1,1') \int d\mathbf{x}_3 \int_C d\tau_3 \int d\mathbf{x}_2 \int_C d\tau_2 \, G_C^{(0)}(3,2) \, D_C^{(0)}(3,2) \, G_C^{(0)}(2,3) \,,$$

(4.131)

which can be represented by the diagram depicted in Figure 4.7.

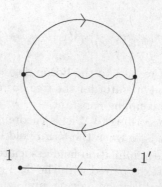

Figure 4.7 Unlinked or second-order vacuum diagram contribution to G_C.

The vacuum bubble gives a vanishing overall factor owing to forward and return contour integrations canceling each other for the case of the real-time closed contour.

The expression corresponding to the second diagram in Figure 4.6 vanishes for the case of electron–phonon interaction as it contains an overall factor that vanishes. Letting $\mathbf{x}, \tau$ represent the internal point where the fermi propagator closes on itself (representing the quantity $G_0^<(\mathbf{x},\tau;\mathbf{x},\tau)$, the free fermionic density which is independent of the variables), the term involving the phonon propagator then becomes

$$\int d\mathbf{x} \, D_0(\mathbf{x},\tau;\mathbf{x}',\tau') = 0$$

(4.132)

since the integrand is the divergence of a function with a vanishing boundary term.[37]

The second-order contribution in the electron–phonon coupling to the contour-ordered electron Green's function is thus represented by the diagram depicted in Figure 4.8.

---

[37]Thus the theory does not contain any so-called tadpole diagrams, which is equivalent to the vanishing of the average of the phonon field. In the Sommerfeld treatment of the Coulomb interaction in a pure metal, tadpole or Hartree diagrams are also absent, though for a different reason. They are canceled by the interaction with the homogeneous background charge (recall Exercise 2.12 on page 44).

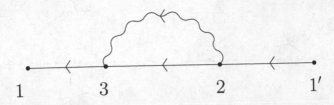

Figure 4.8 Second-order contribution to G_C for the electron–phonon interaction.

We observe the usual Feynman rule expressing the superposition principle: integrate over all internal space points (and sum over all internal spin degrees of freedom) and integrate over the internal contour time variable associated with each vertex. In addition we have the Feynman rule: only topologically different diagrams appear; interchange of internal dummy integration variables has been traded with the factorial from the exponential function.

The next non-vanishing term will, according to Wick's theorem for a string of bose fields, be the fourth order term for the fermion–boson coupling, and the expression

$$G_C^{(2)}(1,1') = -i\frac{(-i)^4}{4!}\text{Tr}\left(-\rho_0 T_C \left(\int_C d2 \dots \int_C d5 \, H_{H_0}^{(i)}(2) \, H_{H_0}^{(i)}(3)\right.\right.$$

$$\times \quad \left.\left. H_{H_0}^{(i)}(4) \, H_{H_0}^{(i)}(5) \, \psi_{H_0}(1) \, \psi_{H_0}^\dagger(1')\right)\right) \tag{4.133}$$

needs to be Wick de-constructed. To get the diagrammatic expression for this term plot down four dots on a piece of paper representing the four internal points in the fourth-order perturbative expression; label them 2, 3, 4 and 5, and the two *external* states, 1 and 1'. Attach at each internal dot, or vertex, a wiggly stub and incoming and outgoing stubs representing the three field operators for each interaction Hamiltonian. To get connected diagrams (the unlinked diagrams again vanish owing to the vanishing of vacuum bubbles) we proceed as follows. The *external* field $\psi_{H_0}^\dagger(1')$ can be paired with any of the fermi annihilation fields associated with the internal points, giving rise to four identical contributions since the internal points represent dummy integration variables. The creation field emerging from this point can be paired with annihilation fields at the remaining three vertices, giving rise to three identical contributions, and the creation field emerging from this vertex has two options: either connecting to one of the two remaining internal vertices or to the external point. In both cases, two identical terms arise, thereby canceling the overall factor from the expansion of the exponential function, 1/4!, in Eq. (4.133). In the latter case, the factor of two occurs because of the two-fold way of pairing the boson fields, and this latter term is thus, according to the above dictionary, represented by the last diagram in Figure 4.9. Pairing the boson fields for the former case gives three different contributions as represented by the first three topologically different diagrams in Figure 4.9.

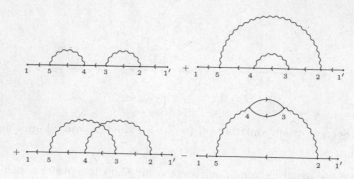

Figure 4.9 Fourth-order diagrams in the coupling constant.

The first three diagrams in Figure 4.9 are solely the result of emission and absorption of phonons by the electron or bosons by fermions in general. The last diagram is the signature of the presence of the Fermi sea: a phonon can cause electron–hole excitations, or in QED a photon can cause electron–positron pair creation. From the boson point of view, such bubble-diagrams with additional decorations are basic, the generic boson self-energy diagram, the self-energy being a quantity we introduce in the next section.

Exercise 4.8. Obtain by brute force application of Wick's theorem for the fermi and phonon field strings the corresponding Feynman diagrams for the fermi propagator to sixth order in the fermion–boson coupling.

Exercise 4.9. Obtain by brute force application of Wick's theorem for the fermi and phonon field strings the corresponding Feynman diagrams for the phonon propagator to second order in the coupling.

The feature that the total combinatorial choice factor cancels the factorial factor, $1/n!$, originating from the expansion of the exponential function is quite general. For the Nth order term

$$-i\frac{(-i)^N}{N!}\mathrm{Tr}\Big(e^{-\beta H_0}T_C\Big(\int_C d2\ldots\int_C d(N+1)\,H^{(i)}_{H_0}(2)\cdots H^{(i)}_{H_0}(N+1)\psi_{H_0}(1)\psi^{\dagger}_{H_0}(1')\Big)\Big)$$

$$(4.134)$$

all connected combinations that differ only by permutations of $H^{(i)}_{H_0}$ give identical contributions, thus canceling the factor $1/N!$ in front, and as a consequence only topologically different diagrams appear. This has a very important consequence for diagrammatics, viz. that it allows separating off sub-parts in a diagram and summing them separately. We shall shortly return to this in the next section, and in much more detail in Chapter 9.[38]

---

[38]We note that the diagrammatic structure of amplitudes for quantum processes can be captured in the two options: to interact or not to interact! The resulting Feynman diagrams being all the topologically different ones constructable by the vertices and propagators of the theory. We shall take this Shakespearian point of view as the starting point when we construct the general diagrammatic and formal structure of quantum field theories in Chapter 9.

The diagrammatic representation of the perturbative expansion of the electron Green's function for the case of electron–phonon interaction, or in general the fermion Green's function for fermion–boson interaction, thus becomes

$$(4.135)$$

In the perturbative expression for the contour-ordered Green's function for the case of electron–phonon interaction, each interaction contains one phonon field operator, a fermion creation and annihilation field, all with the same contour time. The Feynman diagrammatics is thus characterized by a vertex with incoming and outgoing fermi lines and a phonon line, a three-line vertex.

The totality of diagrams can be captured by the following tool-box and instructions. With the diagrammatic ingredients, an electron propagator line, a phonon propagator line and the electron–phonon vertex construct all possible topologically connected diagrams. This is Wick's theorem in the language of Feynman diagrams. We recall that, whenever an odd number of fermi fields are interchanged, a minus sign appears. Diagrammatically this can be incorporated by the additional sign rule: for each loop of fermi propagator lines in a diagram a minus sign is attributed. Accompanying this are the additional Feynman rules, which for our choices become the following. In addition to the usual rule of the superposition principle: sum over all internal labels, our conventions leads for fermion–boson interaction to the additional Feynman rule: a diagram containing n boson lines is attributed the factor $i^n g^{2n}(-1)^F$, where F is the number of closed loops formed by fermion propagators.

4.5   The self-energy

We have so far only derived diagrammatic formulas from formal expressions. Now we shall argue directly in the diagrammatic language to generate new diagrammatic expressions from previous ones, and thereby diagrammatically derive new equations.

In order to get a grasp of the totality of diagrams for the contour-ordered Green's function or propagator we shall use their topology for classification. We introduce the one-particle irreducible (1PI) propagator, corresponding to all the diagrams that can not be cut in two by cutting an internal particle line. In the following example

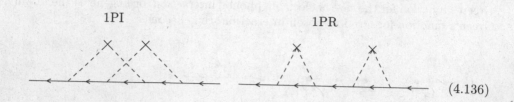

$$(4.136)$$

the first diagram is one-particle irreducible, 1PI, whereas the second is one-particle reducible, 1PR. Here we have used the diagrammatics for the impurity-averaged propagator in a Gaussian random field instead of the analogous diagrammatics for the electron–boson or electron–phonon interaction to illustrate that the arguments are topological and valid for any type of interaction and its diagrammatics.[39]

Amputating the external propagator lines of the one-particle irreducible diagrams (below displayed for the impurity-averaged propagator), we obtain the self-energy:

$$\Sigma(1,1')\ \equiv$$

$$(4.137)$$

consisting, by construction, of all amputated diagrams that can not be cut in two by cutting one bare propagator line.

---

[39]For a detailed discussion of the impurity-averaged propagator, which is of interest in its own right, we refer to Chapter 3 in reference [1].

4.5. The self-energy

We can now go on and uniquely classify all diagrams of the (impurity-averaged) propagator according to whether they can be cut in two by cutting an internal particle line at only one place, or at only two, three, etc., places. By construction we uniquely exhaust all the possible diagrams for the propagator (the subscript is a reminder that we are considering the contour-ordered Green's function, but we leave it out from now on)

$$G_C(1,1') \;=\; \xleftarrow{\hspace{2cm}}$$

$$=\; \xleftarrow{\hspace{1cm}} \;+\; \xleftarrow{\hspace{0.5cm}} \Sigma \xleftarrow{\hspace{0.5cm}}$$

$$+\; \xleftarrow{} \Sigma \xleftarrow{} \Sigma \xleftarrow{}$$

$$+\; \xleftarrow{} \Sigma \xleftarrow{} \Sigma \xleftarrow{} \Sigma \xleftarrow{}$$

$$+\; \cdots \;. \tag{4.138}$$

By iteration, this equation is seen to be equivalent to the equation[40]

$$\xleftarrow{\hspace{1cm}} \;=\; \xleftarrow{\hspace{1cm}} \;+\; \xleftarrow{} \Sigma \xleftarrow{} \tag{4.139}$$

and we obtain the Dyson equation

$$G(1,1') \;=\; G_0(1,1') + \int d\mathbf{x}_3 \int_C d\tau_3 \int d\mathbf{x}_2 \int_C d\tau_2 \; G_0(1,3)\,\Sigma(3,2)\,G(2,1') \tag{4.140}$$

---

[40] In the last term we can interchange the free and full propagator, because iterating from the left generates the same series as iterating from the right.

which we can write in matrix notation

$$G = G_0 + G_0 \boxtimes \Sigma \boxtimes G \tag{4.141}$$

where $\boxtimes$ signifies matrix multiplication in the spatial variable (and possible internal degrees of freedom) and contour time variables. Arguing on the topology of the diagrams has reorganized them and we have obtained a new type of equation.[41]

4.5.1 Non-equilibrium Dyson equations

The standard topological arguments of the previous section for diagrams organizes them into irreducible sub-parts and we obtained the Dyson equation, Eq. (4.141). We could of course have iterated Eq. (4.138) from the other side to obtain

$$G = G_0 + G \boxtimes \Sigma \boxtimes G_0 . \tag{4.142}$$

For an equilibrium state the two equations are redundant, since time convolutions by Fourier transformation become simple products for which the order of factors is irrelevant (as discussed in detail in Section 5.6). However, in a non-equilibrium state, the two equations contain different information and subtracting them is a useful way of expressing the non-equilibrium dynamics as we shall exploit in Chapter 7.

Introduce the inverse of the free contour-ordered Green's function, Eq. (4.141),

$$(G_0^{-1} \boxtimes G_0)(1,1') = \delta(1-1') = (G_0 \boxtimes G_0^{-1})(1,1') , \tag{4.143}$$

where

$$G_0^{-1}(1,1') = G_0^{-1}(1)\,\delta(1-1') \tag{4.144}$$

and

$$G_0^{-1}(1) = \left(i\hbar \frac{\partial}{\partial \tau_1} - h(1) \right) , \tag{4.145}$$

where h denotes the single-particle Hamiltonian for the theory under consideration. The two non-equilibrium Dyson equations, Eq. (4.141) and Eq. (5.69), can then be expressed through operating with the inverse free contour-ordered Green's function from the left

$$(G_0^{-1} - \Sigma) \boxtimes G = \delta(1-1') \tag{4.146}$$

and from the right

$$G \boxtimes (G_0^{-1} - \Sigma) = \delta(1-1') . \tag{4.147}$$

These two non-equilibrium Dyson equations will prove useful in Chapter 7 where quantum kinetic equations are considered.[42]

By operating with the inverse of the free propagator, the explicit appearance of the free propagator (or rather the non-interacting propagator since, possible external fields can be included) has been removed. We can in fact remove its presence completely by expressing the self-energy in terms of skeleton diagrams where only the full propagator appears, and we now turn to these.

---

[41]The self-energy is just one out of the infinitely many one-particle irreducible vertex functions which occur in a quantum field theory. Their significance will become clear when, in Chapter 10, we encounter the usefulness of the effective action.

[42]If one, by the end of the day, in the Dyson equations uses the lowest-order approximation for the self-energy, this whole venture into the diagrammatic jungle is hardly worthwhile since a civilized Golden Rule calculation suffices.

4.5.2   Skeleton diagrams

So far we have only a perturbative description of the self-energy; i.e. we have a representation of the self-energy as a functional of the free contour-ordered Green's function and the impurity correlator, $\Sigma[G_0]$. For the case of fermion–boson interaction the self-energy is a functional of both types of free contour-ordered Green's functions, $\Sigma[G_0, D_0]$. The self-energy, Σ, can in naive perturbation theory be described as the sum of diagrams that can not be cut in two by cutting only one internal free propagator line. In a realistic description of a physical system, we always need to invoke the specifics of the problem in order to implement a controlled approximation. To this end we must study the actual correlations in the system, and it is necessary to have the self-energy expressed in terms of the full propagator. Coherent quantum processes correspond to an infinite repetition of bare processes, and the diagrammatic approach is precisely useful for capturing this feature, as irreducible re-summations are easily described diagrammatically. In order to achieve a description of the self-energy in terms of the full propagator, let us consider the perturbative expansion of the self-energy.

For any given self-energy diagram in the perturbative expansion, Eq. (4.137), we also encounter self-energy diagrams with all possible self-energy decorations on internal lines; for example, for the case of particles in a random potential:

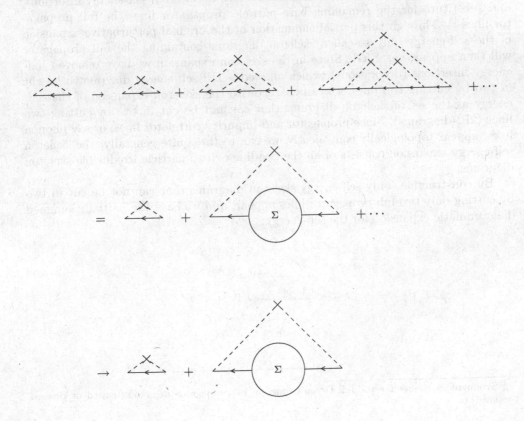

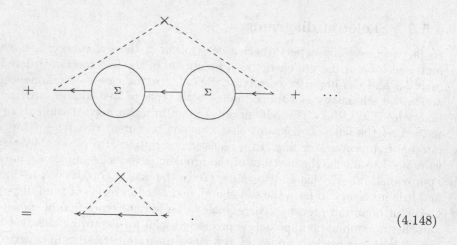

$$+ \qquad\qquad\qquad\qquad + \quad \cdots$$

$$= \qquad\qquad\qquad\qquad . \qquad\qquad\qquad\qquad\qquad (4.148)$$

We can uniquely classify all these self-energy decorations in the perturbative expansion according to whether the particle line can be cut into two, three, or more pieces by cutting particle lines (the step indicated by the second arrow in Eq. (4.148)). We can therefore partially sum the self-energy diagrams according to the unique prescription: for a given self-energy diagram, remove all internal self-energy insertions, and substitute for the remaining bare particle propagator lines the full propagator lines.[43] Through this partial summation of the original perturbative expansion of the self-energy only so-called skeleton diagrams containing the full propagator will then appear, i.e. $\Sigma[G]$. Since in the skeleton expansion we have removed self-energy insertions (decorations), which allowed a 1PI self-energy diagram to be cut in two by cutting two lines, we can characterize the skeleton expansion of the self-energy as the set of skeleton diagrams that can not be cut in two by cutting two lines (2PI-diagrams). Since propagator and impurity correlator lines, or say phonon lines, appear topologically equivalently, we can restate quite generally: the skeleton self-energy expansion consists of all the two-line or two-particle irreducible skeleton diagrams.

By construction, only self-energy skeleton diagrams that can not be cut in two by cutting only two full propagator lines appear, and we have the partially summed diagrammatic expansion for the self-energy:

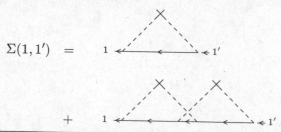

$$\Sigma(1,1') =$$

$$+$$

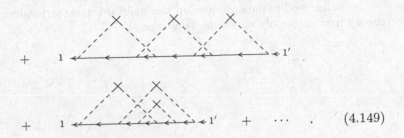

$$+ \quad 1 \qquad \qquad \qquad 1' \quad + \quad \cdots \quad . \qquad (4.149)$$

The partial summation of diagrams is unique, since the initial and final impurity correlator lines are attached internally in different ways in each class of summed diagrams. No double counting of diagrams thus takes place owing to the different topology of the skeleton self-energy diagrams, and all diagrams in the perturbative expansion of the self-energy, Eq. (4.137), are by construction contained in the skeleton diagrams of Eq. (4.149).

What has been achieved by the partial summation, where each diagram corresponds to an infinite sum of terms in perturbation theory, is that the self-energy is expressed as a functional of the exact propagators or full Green's functions

$$\Sigma(1,1') = \Sigma_{(1,1')}[G,D] \ . \tag{4.150}$$

We can continue this topological classification, and introduce the higher-order vertex functions; however, we defer this until Chapter 9.

Exercise 4.10. Draw the rest, in Eq. (4.149), of the four skeleton self-energy diagrams with three impurity correlators.

Exercise 4.11. Draw the skeleton self-energy diagrams for fermion–boson interaction to fourth order in the coupling.

4.6   Summary

We have presented the formalism needed for treating general non-equilibrium situations. The closed time path formalism was shown to facilitate a convenient and compact treatment of non-equilibrium statistical Green's functions. Perturbation theory valid for non-equilibrium states turned out in standard fashion, reflecting a general Wick theorem for closed time path strings of operators, and the Feynman diagrams for the contour ordered Green's functions become of standard form. For the reader with knowledge of equilibrium theory the good news is thus that the general non-equilibrium formalism is formally equivalent to the equilibrium theory if one elevates time to the contour level. For the reader not familiar with equilibrium theory the good news is rejoice: knowledge of equilibrium theory is not needed, since the equilibrium case is just a special simple case of the presented general theory. However, the apparatus of the closed time path formalism needs a physical interpretation, we need to get back to real time. In the next chapter we shall introduce the real-time technique and develop the diagrammatic structure of non-equilibrium theory in a physically appealing language.

5

Real-time formalism

The contour-ordered Green's function considered in the previous chapter was ideal for discussing general closed time path properties such as the perturbative diagrammatic structure for non-equilibrium states. However, the contour-ordered Green's function lacks physical transparency and does not appeal to intuition.[1] We need a different approach, which brings quantities back to real time. To accomplish this we introduce a representation where forward and return parts of the closed time path are ordered by numbers, specifying the position of a contour time by two indices, $i = 1, 2$. Next is the diagrammatic perturbation theory in the real-time technique then formulated in a fashion where the aspects of non-equilibrium states emerge in the physically most appealing way. In particular, we shall construct the representation where spectral properties and quantum statistics show up on a different footing in the diagrams. Lastly, we consider the connection to the imaginary-time treatment of non-equilibrium states, and establish its equivalence to the real-time approach propounded in this chapter.[2]

5.1 Real-time matrix representation

To let our physical intuition come into play; we need to get from contour times back to real times. This is achieved by labeling the forward and return contours of the closed time path, depicted in Figure 4.5, by numbers, specifying the position of a contour time by an index. The forward contour we therefore label c_1 and the return contour c_2, i.e. a contour time variable gets tagged by the label 1 or 2 specifying its belonging to forward or return contour, respectively.[3]

The contour ordered Green's function is by this tagging mapped onto a 2×2-

---

[1] As the imaginary time Green's function discussed in Section 5.7.1 does not appeal to intuition.

[2] In this chapter we follow the exposition given in references [3] and [9].

[3] Instead of labeling the two branches by 1 and 2, one can also label them by $\pm$ as in the original works of Schwinger [5] and Keldysh [10]. However, when stating Feynman rules, numbers are convenient for labeling.

matrix in the dynamical index or Schwinger–Keldysh space

$$G_C(1,1') \to \hat{G}(1,1') \equiv \begin{pmatrix} \hat{G}_{11}(1,1') & \hat{G}_{12}(1,1') \\ \hat{G}_{21}(1,1') & \hat{G}_{22}(1,1') \end{pmatrix} \qquad (5.1)$$

according to the prescription: the ij-component in the matrix Green's function $\hat{G}$ is $G_c(1,1')$ for 1 lying on c_i and $1'$ lying on c_j, $i,j = 1,2$. The times appearing in the components of the matrix Green's function are now standard times, $1 = (\mathbf{x}, t_1)$, and we can identify them in terms of our previously introduced Green's functions, the real-time Green's functions introduced in Section 3.3. The matrix structure reflects the essence in the real-time formulation of non-equilibrium quantum statistical mechanics due to Schwinger [5]: letting the quantum dynamics do the doubling of the degrees of freedom necessary for describing non-equilibrium states.[4]

The 11-component of the matrix in Eq. (5.1) becomes

$$\hat{G}_{11}(1,1') = -i\langle T(\psi(\mathbf{x}_1,t_1)\,\psi^\dagger(\mathbf{x}_{1'},t_{1'}))\rangle = -i\langle T(\psi(1)\,\psi^\dagger(1'))\rangle, \qquad (5.2)$$

the time-ordered Green's function, where $\psi(\mathbf{x},t) = \psi_{\mathcal{H}}(\mathbf{x},t)$ is the field in the full Heisenberg picture for the species of interest. Contour ordering on the forward contour is just usual time-ordering.

Analogously, the 21-component becomes

$$\hat{G}_{21}(1,1') = G^>(1,1') = -i\langle \psi(1)\,\psi^\dagger(1')\rangle, \qquad (5.3)$$

i.e. *G-greater*, and the 12-component is *G-lesser*

$$\hat{G}_{12}(1,1') = G^<(1,1') = \mp i\,\langle \psi^\dagger(1')\,\psi(1)\rangle, \qquad (5.4)$$

where upper and lower signs refer to bose and fermi fields, respectively, and the 22-component is the anti-time-ordered Green's function

$$\hat{G}_{22}(1,1') = \tilde{G}(1,1') = -i\langle \tilde{T}(\psi(1)\,\psi^\dagger(1'))\rangle. \qquad (5.5)$$

We note that the time-ordered and anti-time-ordered Green's functions can be expressed in terms of *G-greater* and *G-lesser*, recall Eq. (3.64), and

$$\hat{G}_{11}(1,1') = \theta(t_1 - t_{1'})\,G^>(1,1') + \theta(t_{1'} - t_1)\,G^<(1,1'). \qquad (5.6)$$

The matrix Green's function in Eq. (5.1) can therefore be expressed in terms of the real-time Green's functions introduced in Section 3.3

$$\hat{G}(1,1') = \begin{pmatrix} G(1,1') & G^<(1,1') \\ G^>(1,1') & \tilde{G}(1,1') \end{pmatrix}. \qquad (5.7)$$

The way of representing the information in the contour-ordered Green's function as in Eq. (5.1) or equivalently Eq. (5.7) is respectable as, for example, the matrix

---

[4]The thermo-field approach to non-equilibrium theory also employs a doubling of the degrees of freedom (see, for example, reference [11]), but in our view not in as physically appealing way as does the real-time version of the closed time path formulation.

is anti-hermitian with transposition meaning interchange of all arguments including that of the dynamical index (note the importance of the sign convention for handling fermi fields under ordering operations). For real bosons the matrix is real and symmetric. However, when it comes to understanding non-equilibrium contributions from various processes, as described by Feynman diagrams, the present form of the matrix Green's function lacks physical transparency, and offers no basis for developing intuition. We shall therefore eventually transform to a different matrix form, and as a final act liberate ourselves from the matrix outfit altogether.

Let us now establish the Feynman rules in the real-time technique for the matrix Green's function in the dynamical index or Schwinger–Keldysh space.

5.2 Real-time diagrammatics

Instead of having the diagrammatics represent the perturbative expansion of the contour Green's function as in the previous chapter, we shall map the diagrams to the real-time domain where eventually a proper physical interpretation of the diagrams can be obtained.

5.2.1 Feynman rules for a scalar potential

We start with the simplest kind of coupling, that of particles interacting with an external classical field. For particles interacting with a scalar potential $V(\mathbf{x}, t)$, we have the diagrammatic expansion of the contour ordered Green's function depicted in Eq. (4.115) on page 105. The first-order diagram corresponded to the term

$$G_C^{(1)}(1, 1') = \int d\mathbf{x}_2 \int_C d\tau_2 \, G_C^{(0)}(1, 2) \, V(2) \, G_C^{(0)}(2, 1') . \tag{5.8}$$

Parameterizing the real-time contour we have

$$\int_C d\tau_2 = \int_{-\infty}^{\infty} dt + \int_{\infty}^{-\infty} dt = \int_{-\infty}^{\infty} dt - \int_{-\infty}^{\infty} dt \tag{5.9}$$

and the first order term for the matrix ij-component becomes

$$\hat{G}_{ij}^{(1)}(1, 1') = \int d\mathbf{x}_2 \int_{-\infty}^{\infty} dt_2 \, \hat{G}_{i1}^{(0)}(1, 2) \, V(2) \, \hat{G}_{1j}^{(0)}(2, 1')$$

$$- \int d\mathbf{x}_2 \int_{-\infty}^{\infty} dt_2 \, \hat{G}_{i2}^{(0)}(1, 2) \, V(2) \, \hat{G}_{2j}^{(0)}(2, 1') . \tag{5.10}$$

Introducing in Schwinger–Keldysh or dynamical index space the matrix

$$\hat{V}_{ij}(1) = V(1) \, \tau_{ij}^{(3)} \tag{5.11}$$

proportional to the third Pauli-matrix

$$\tau^{(3)} = \begin{pmatrix} 1 & 0 \\ 0 & -1 \end{pmatrix} \tag{5.12}$$

we have

$$\hat{G}_{ij}^{(1)}(1,1') = \int d\mathbf{x}_2 \int_{-\infty}^{\infty} dt_2 \, \hat{G}_{ik}^{(0)}(1,2) \, \hat{V}_{kk'}(2) \, \hat{G}_{k'j}^{(0)}(2,1') \, , \tag{5.13}$$

where summation over repeated Schwinger–Keldysh or dynamical indices are implied. Instead of treating individual indexed components, the condensed matrix notation is applied and the matrix equation becomes

$$\hat{G}^{(1)} = \hat{G}^{(0)} \otimes \hat{V} \hat{G}^{(0)} = \hat{G}^{(0)} \hat{V} \otimes \hat{G}^{(0)} \, , \tag{5.14}$$

where $\otimes$ signifies matrix multiplication in the spatial variable (as well as possible internal degrees of freedom) and the real time, for the latter integration from minus to plus infinity of times. For the components of the free equilibrium matrix Green's function, $\hat{G}^{(0)}$, we have, according to Section 3.4, explicit expressions.

We introduce a diagrammatic notation for this real-time matrix Green's function contribution

$$\hat{G}^{(1)}(\mathbf{x}_1, t_1; \mathbf{x}_{1'}, t_{1'}) \quad = \quad \begin{array}{ccc} \bullet & \times & \bullet \\ \mathbf{x}_1 t_1 & \mathbf{x}_2 t_2 & \mathbf{x}_{1'} t_{1'} \end{array} \tag{5.15}$$

The diagram has the same form as the one depicted in Eq. (4.111) for the contour Green's function, but is now interpreted as an equation for the matrix propagator in Schwinger–Keldysh space: each line now represents the free matrix Green's function, $\hat{G}^{(0)}$, and the cross represents the matrix for the potential coupling, Eq. (5.11). We get the extra Feynman rule characterizing the non-equilibrium technique: matrix multiplication over internal dynamical indices is implied.

For the coupling to the scalar potential, the higher-order diagrams are just repetitions of the basic first-order diagram, and we can immediately write down the expression for the matrix propagator for a diagram of arbitrary order. Re-summation of diagrams to get the Dyson equation, as discussed in Section 4.5, is trivial for coupling to external classical fields, giving

$$\hat{G} = \hat{G}^{(0)} + \hat{G}^{(0)} \otimes \hat{V} \hat{G} \, , \quad \hat{G} = \hat{G}^{(0)} + \hat{G} \hat{V} \otimes \hat{G}^{(0)} \, , \tag{5.16}$$

where the potential can be placed on either side of the convolution symbol.

According to Eq. (3.65), Eq. (3.66) and Eq. (3.67), the free equilibrium matrix Green's function, $\hat{G}^{(0)}$, satisfies

$$G_0^{-1}(1) \, \hat{G}_0(1,1') = \tau^{(3)} \delta(1-1') \, , \tag{5.17}$$

where $G_0^{-1}(1)$ is given by Eq. (3.69) for the case of coupling to both a scalar and a vector potential. Since we want the inverse matrix Green's function operating on the free equilibrium matrix Green's function to produce the unit matrix in all variables including the dynamical index, it can be accomplished by either of the objects carrying the third Pauli matrix, $\tau^{(3)}$. For example, introducing the matrix representation

$$\check{G}(1,1') \equiv \tau^{(3)} \hat{G}(1,1') = \begin{pmatrix} \hat{G}_{11}(1,1') & \hat{G}_{12}(1,1') \\ -\hat{G}_{21}(1,1') & -\hat{G}_{22}(1,1') \end{pmatrix} \tag{5.18}$$

we then have

$$G_0^{-1}(1)\,\check{G}_0(1,1') = \underline{1}\,\delta(1-1'),\tag{5.19}$$

where the unit matrix $\underline{1}$ in the dynamical index space will often be denoted by 1 and often left out when operating on a matrix in the dynamical index space. Introducing the inverse free matrix Green's function

$$G_0^{-1}(1,1') = G_0^{-1}(1)\,\delta(1-1')\,\underline{1}\tag{5.20}$$

we have

$$(G_0^{-1} \otimes \check{G}_0)(1,1') = \underline{1}\,\delta(1-1') = (\check{G}_0 \otimes G_0^{-1})(1,1').\tag{5.21}$$

We can therefore rewrite the Dyson equations for real-time matrix Green's function in the forms

$$((G_0^{-1} - V) \otimes \check{G})(1,1') = \underline{1}\,\delta(1-1')\tag{5.22}$$

and

$$(\check{G} \otimes (G_0^{-1} - V))(1,1') = \underline{1}\,\delta(1-1').\tag{5.23}$$

We note that in the matrix representation, Eq. (5.18), the coupling to a scalar potential is a scalar, i.e. proportional to the unit matrix in Schwinger–Keldysh space

$$\check{G}^{(1)} = \check{G}^{(0)} \otimes \check{V}\,\check{G}^{(0)},\tag{5.24}$$

where

$$\check{V}_{ij}(1) = V(1)\,\delta_{ij}\;,\quad \check{V}(1) = V(1)\,\underline{1}.\tag{5.25}$$

The matrix representation introduced in Eq. (5.18) serves the purpose of absorbing the minus signs associated with the return contour into the third Pauli matrix.

5.2.2 Feynman rules for interacting bosons and fermions

For a three-line type vertex, such as in the case of fermion–boson interaction or electron–phonon interaction, more complicated coupling matrices appear in the dynamical index or Schwinger–Keldysh space than for the case of coupling to an external field. For illustration of the matrix structure in the dynamical index space it suffices to consider the generic boson–fermion coupling in Eq. (2.71). As noted in Section 2.4.3, this is also equivalent to considering the electron–phonon interaction in the jellium model where the electrons couple only to longitudinal compressional charge configurations of the ionic lattice, the longitudinal phonons. Our interest is to display the dynamical index structure of propagators and vertices; later these can be sprinkled with whatever additional indices they deserve to be dressed with: species index, spin, color, flavor, Minkowski, phonon branch, etc.

In the expression for the lowest-order perturbative contribution to the contour ordered Green's function, Eq. (4.127), we parameterize the two real-time contours according to Eq. (5.9). In Schwinger–Keldysh space this term then becomes

$$\hat{G}_{ij}^{(1)} = ig^2\hat{G}_{ii'}^{(0)} \otimes \hat{\gamma}_{i'l'}^{k}\,\hat{G}_{l'l}^{(0)}\,\hat{D}_{kk'}^{(0)}\,\hat{\tilde{\gamma}}_{lj'}^{k'} \otimes \hat{G}_{j'j}^{(0)}\tag{5.26}$$

or equivalently for the components of the lowest order self-energy matrix components

$$\hat{\Sigma}_{ij}^{(1)} = ig^2 \hat{\gamma}_{il'}^{k} \, \hat{G}_{l'l}^{(0)} \, \hat{D}_{kk'}^{(0)} \, \hat{\bar{\gamma}}_{lj}^{k'} \, , \tag{5.27}$$

where the third rank tensors representing the phonon absorption and emission vertices are identical

$$\hat{\gamma}_{ij}^{k} = \delta_{ij} \, \tau_{jk}^{(3)} = \hat{\bar{\gamma}}_{ij}^{k} \, . \tag{5.28}$$

The third rank vertex tensors vanish unless electron and phonon indices are identical, reflecting the fact that the fields in a vertex correspond to the same moment in contour time. The presence of the imaginary unit in Eq. (5.26) is the result of one lacking factor of $-i$ for our convention of Green's functions: two factors of $-i$ are provided by the interaction and one provided externally in the definition of the Green's functions. Such features are collected in one's own private choice of Feynman rules.

In the present representation, Eq. (5.1), instead of thinking in terms of the diagrammatic *matrix* representation one can visualize the *components* diagrammatically, and we would have diagrams with Green's functions attached to either of the forward or return parts of the contour. It can be useful once to draw these kind of diagrams, but eventually we shall develop a form of diagram representation without reference to the contour but instead to the distinct different physical properties represented by the retarded and kinetic Green's functions of Section 3.3.2.

The vertices, Eq. (5.28), are diagonal in the fermion, i.e. lower Schwinger–Keldysh indices since the two fermi field operators carry the same time variable. The boson field attached to that vertex has of course the same time variable, but the other bose field it is paired with can have a time variable residing on either the forward or backward path, giving the possibilities of ± 1 as reflected in the matrix elements of the third Pauli matrix.

The diagrammatic representation of the matrix Green's function, Eq. (5.26), is displayed in Figure 5.1, where straight and wiggly lines represent fermion and boson matrix Green's functions, or the free electron and free phonon matrix Green's functions, respectively, and the vertices represent the third rank tensors specified in Eq. (5.28).

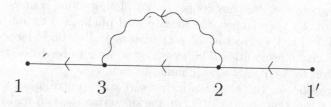

Figure 5.1 Diagrammatic representation of the matrix Green's function G for fermion–boson interaction.

In the matrix representation specified by Eq. (5.18), the diagram represents (using $\delta_{ij} = \tau_{ik}^{(3)} \tau_{kj}^{(3)}$),

$$\check{G}_{ij}^{(1)} = ig^2 \check{G}_{ii'}^{(0)} \otimes \tilde{\gamma}_{i'l'}^k \check{G}_{l'l}^{(0)} \check{D}_{kk'}^{(0)} \tilde{\tilde{\gamma}}_{lj'}^{k'} \otimes \check{G}_{j'j}^{(0)}, \tag{5.29}$$

where the absorption vertex is

$$= \tilde{\gamma}_{ij}^k = \hat{\gamma}_{ii'}^{k'} \tau_{i'j}^{(3)} \tau_{k'k}^{(3)} = \delta_{ij} \tau_{jk}^{(3)} \tag{5.30}$$

and the emission vertex is

$$= \tilde{\tilde{\gamma}}_{ij}^k = \hat{\tilde{\gamma}}_{ii'}^k \tau_{i'j}^{(3)} = \delta_{ij} \delta_{jk}. \tag{5.31}$$

In this representation the absorption and emission vertices thus differ.
In terms of the lowest order matrix self-energy, Eq. (5.29) becomes

$$\check{G}^{(1)} = \check{G}^{(0)} \otimes \check{\Sigma}^{(1)} \otimes \check{G}^{(0)}. \tag{5.32}$$

5.3 Triagonal and symmetric representations

Since only two components of the matrix Green's function, Eq. (5.1), are independent it can be economical to remove part of this redundancy. In the original article of Keldysh [10], one component was eliminated by the linear transformation, the $\pi/4$-rotation in Schwinger–Keldysh space,

$$\hat{G} \rightarrow L\hat{G}L^\dagger, \tag{5.33}$$

where the orthogonal matrix, $L^\dagger = L$, is

$$L = \frac{1}{\sqrt{2}}(1 - i\tau^{(2)}) = \frac{1}{\sqrt{2}} \begin{pmatrix} 1 & -1 \\ 1 & 1 \end{pmatrix} \tag{5.34}$$

i.e. 1 denotes the 2×2 unit matrix and $\tau^{(2)}$ is the second Pauli matrix

$$\tau^{(2)} = \begin{pmatrix} 0 & -i \\ i & 0 \end{pmatrix}. \tag{5.35}$$

Using the following identities (recall Section 3.3)

$$G^R(1, 1') = \hat{G}_{11}(1, 1') - \hat{G}_{12}(1, 1') = \hat{G}_{21}(1, 1') - \hat{G}_{22}(1, 1') \tag{5.36}$$

and

$$G^A(1,1') = \hat{G}_{11}(1,1') - \hat{G}_{21}(1,1') = \hat{G}_{12}(1,1') - \hat{G}_{22}(1,1') \tag{5.37}$$

and

$$G^K(1,1') = \hat{G}_{21}(1,1') + \hat{G}_{12}(1,1') = \hat{G}_{11}(1,1') + \hat{G}_{22}(1,1') \tag{5.38}$$

and

$$0 = \hat{G}_{11}(1,1') - \hat{G}_{12}(1,1') - \hat{G}_{21}(1,1') + \hat{G}_{22}(1,1') \tag{5.39}$$

the linear transformation, the $\pi/4$-rotation in Schwinger–Keldysh space Eq. (5.33), amounts to[5]

$$\begin{pmatrix} \hat{G}_{11} & \hat{G}_{12} \\ \hat{G}_{21} & \hat{G}_{22} \end{pmatrix} \to \begin{pmatrix} 0 & G^A \\ G^R & G^K \end{pmatrix} \tag{5.40}$$

where the retarded, advanced and the Keldysh or kinetic Green's functions all were introduced in Section 3.3.2.

For real bosons or phonons, the matrix

$$D = \begin{pmatrix} 0 & D^A(\mathbf{x},t,\mathbf{x}',t') \\ D^R(\mathbf{x},t,\mathbf{x}',t') & D^K(\mathbf{x},t,\mathbf{x}',t') \end{pmatrix} \tag{5.41}$$

is real and symmetric, regarded as a matrix in all its arguments, i.e. including its dynamical indices which at this level amounts to the interchange R ↔ A. This symmetric form is the useful representation, the symmetric representation, needed when functional methods are employed, as discussed in Chapters 9 and 10.

In condensed matter physics a representation in terms of triagonal matrices is often used, originally introduced by Larkin and Ovchinnikov [12]. To obtain this triagonal representation, the $\pi/4$-rotation in Schwinger–Keldysh space is performed on the matrix Green's function in Eq. (5.18)

$$G = L\check{G}L^\dagger \tag{5.42}$$

and the triagonal matrix is obtained[6]

$$G = \begin{pmatrix} G^R & G^K \\ 0 & G^A \end{pmatrix}. \tag{5.43}$$

Not only are these representations economical, they are also appealing from a physical point of view as G^R and G^K contain distinctly different information: the spectral function has the information about the quantum states of a system, the energy spectrum, and the kinetic Green's function, G^K, has the information about

---

[5]The alternative not to work with matrices at this stage, but instead base the description on the *G-greater* and *G-lesser* Green's functions is discussed in detail in Section 5.7. This choice emerges if one starts from the so-called imaginary-time formalism, as we shall discuss. We shall eventually abandon the matrices and interpret diagrams directly in terms of the three types of Green's functions and two simple rules for their behavior at vertices, the real rules: the RAK-rules.

[6]No confusion with the notation for the time-ordered Green's function should arise.

the occupation of these states for non-equilibrium situations as discussed in Section 3.4.

The identity in Eq. (5.39) is of the type that guarantees that vacuum diagrams lead to vanishing contributions.

Exercise 5.1. Consider free phonons in thermal equilibrium at temperature T, and show that their matrix Green's function in the triagonal representation

$$D^{(0)} = \begin{pmatrix} D_0^R & D_0^K \\ 0 & D_0^A \end{pmatrix} \qquad (5.44)$$

has components that in terms of the momentum and energy variables or equivalently wave vector and frequency variables are specified by

$$D_0^R(\mathbf{k}, \omega) = (D_0^A(\mathbf{k}, \omega))^* = \frac{-\omega_{\mathbf{k}}^2}{\omega_{\mathbf{k}}^2 - (\omega + i\delta)^2} \qquad (5.45)$$

and

$$D_0^K(\mathbf{k}, \omega) = (D^R(\mathbf{k}, \omega) - D^A(\mathbf{k}, \omega)) \coth \frac{\hbar\omega}{2kT}, \qquad (5.46)$$

where $\omega_{\mathbf{k}} = c|\mathbf{k}|$ is the linear dispersion relation for the longitudinal phonons, c being the longitudinal sound velocity.

5.3.1 Fermion–boson coupling

Let us consider what happens to the fermion–boson interaction or electron–phonon interaction dynamical index vertices when transforming to the triagonal matrix representation, i.e. let us find the tensors for the vertices. To obtain the coupling matrices for the fermion–boson interaction in this representation we transform all matrix Green's functions according to

$$G_{ij}^{(1)} = L_{ii'} \check{G}_{i'j'}^{(1)} L_{j'j}^\dagger \qquad (5.47)$$

and similarly for the phonon Green's function, and inserting the identity according to[7]

$$\delta_{ij} = L_{ii'}^\dagger L_{i'j} \qquad (5.48)$$

the absorption vertex becomes

$$\gamma_{ij}^k = L_{ii'} \check{\gamma}_{i'j'}^{k'} L_{j'j}^\dagger L_{k'k}^\dagger \quad = \qquad (5.49)$$

and the emission vertex becomes

---

[7]From this it immediately follows that the coupling matrix for a scalar field in the triagonal representation is the unit matrix in Schwinger–Keldysh space.

$$\tilde{\gamma}_{ij}^k = L_{ii'} L_{kk'} \, \tilde{\gamma}_{i'j'}^{k'} \, L_{j'j}^\dagger \quad = \qquad\qquad (5.50)$$

and simple calculation gives for the vertices

$$\gamma_{ij}^1 = \tilde{\gamma}_{ij}^2 = \frac{1}{\sqrt{2}} \, \delta_{ij} \qquad\qquad (5.51)$$

and

$$\gamma_{ij}^2 = \tilde{\gamma}_{ij}^1 = \frac{1}{\sqrt{2}} \, \tau_{ij}^{(1)} \,, \qquad\qquad (5.52)$$

where $\tau_{ij}^{(1)}$ is the first Pauli matrix

$$\tau^{(1)} = \begin{pmatrix} 0 & 1 \\ 1 & 0 \end{pmatrix} . \qquad\qquad (5.53)$$

The fermion–boson vertices can be considered basic as two-particle interaction can also be formulated in terms of them, as discussed in Section 5.3.2. The above four types of vertices thus represent the additional dressing of vertices needed for describing non-equilibrium situations. In Section 5.4 we shall describe the physical significance of the dynamical index structure of the vertices in the symmetric or triagonal representations.

The diagrammatic representation is the same irrespective of the matrix representation used, only the matrices and tensors vary. The diagram displayed in Figure 5.1 represents in the triagonal representation the string of matrices

$$G_{ij}^{(1)}(1,1') = ig^2 G_{ii'}^{(0)}(1,3) \otimes \gamma_{i'l'}^k \, G_{l'l}^{(0)}(3,2) \, D_{kk'}^{(0)}(3,2) \, \tilde{\gamma}_{lj'}^{k'} \otimes G_{j'j}^{(0)}(2,1') \,, \quad (5.54)$$

where straight and wiggly lines represent the free fermion and boson matrix Green's functions, or the free electron and free phonon matrix Green's functions, respectively, in the triagonal representation, and the vertices are specified in Eq. (5.51) and Eq. (5.52).

The virtues of the triagonal representation are that the coupling matrix for a classical field is the unit matrix in Schwinger–Keldysh space, and both the matrix Green's function and matrix self-energies are triagonal matrices, as we show in Section 5.5,

$$\Sigma = \begin{pmatrix} \Sigma^R & \Sigma^K \\ 0 & \Sigma^A \end{pmatrix} , \qquad\qquad (5.55)$$

making operative the property that triagonal matrix structure is invariant with respect to matrix multiplication.

5.3.2 Two-particle interaction

Another important interaction we will encounter is the two-body or two-particle interaction, say Coulomb electron–electron interaction. The ready-made form for perturbative expansion of the contour ordered Green's function becomes, for the case of two-particle interaction,

$$G(1,1') = \mathrm{Tr}\left(\rho_0 T_C\left(S\,\psi_{H_0}(1)\,\psi^\dagger_{H_0}(1')\right)\right),\qquad(5.56)$$

where

$$S = e^{-i\int_C d\tau_1 \int dx_1 \int_C d\tau_2 \int dx_2\,\psi^\dagger_{H_0}(\mathbf{x}_1,\tau_1)\,\psi^\dagger_{H_0}(\mathbf{x}_2,\tau_2)U(\mathbf{x}_2,\tau_2;\mathbf{x}_1,\tau_1)\psi_{H_0}(\mathbf{x}_2,\tau_2)\,\psi_{H_0}(\mathbf{x}_1,\tau_1)}$$

$$(5.57)$$

and for an instantaneous interaction

$$U(\mathbf{x}_2,\tau_2;\mathbf{x}_1,\tau_1) = V(\mathbf{x}_2,\mathbf{x}_1)\,\delta(\tau_2-\tau_1),\qquad(5.58)$$

where, $V(\mathbf{x}_2,\mathbf{x}_1)$ is for example the Coulomb interaction, and the contour delta function of Eq. (4.45) appears. In the *hat*-representation, the two-body interaction will thus get the matrix representation

$$\hat{U}(\mathbf{x}_2,t_2;\mathbf{x}_1,t_1) = \tau^{(3)}\,\delta(t_2-t_1)\,V(\mathbf{x}_2,\mathbf{x}_1).\qquad(5.59)$$

The basic vertex for two-particle interaction is thus the one depicted in Figure 5.2, where the wiggly line represents the matrix two-particle interaction specified in Eq. (5.59).

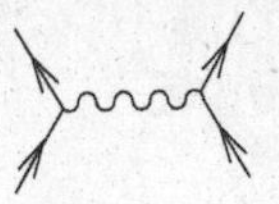

Figure 5.2 Two-particle interaction vertex.

However, the basic vertex for two-particle interaction can be interpreted as two separate vertices in terms of the action of the real-time dynamical indices, and can be formulated identically to the case of electron–boson or electron–phonon interaction. Although $\hat\gamma_{e-ph}$, Eq. (5.28), of course is capable of coupling the upper and lower branch it is of no importance since such terms vanish since $\hat{U}$ is diagonal. One is thus free to choose either of the forms

$$\hat\gamma^k_{ij} \propto \delta_{ij}\,\tau^{(3)}_{jk} \quad \text{or} \quad \hat\gamma^k_{ij} \propto \delta_{ij}\,\delta_{jk}\qquad(5.60)$$

the former choice making the separated two-particle or electron–electron interaction vertices identical in the dynamical indices to the case of fermion–boson or electron–phonon interaction.

Exercise 5.2. The wavy line in Figure 5.2, representing the two-body interaction, can be assigned an arbitrary direction, which then in turn can be put to use in accounting for the momentum flow in the Feynman diagrams for two-body interactions.

Assuming the interaction in Eq. (5.58) is translational invariant and instantaneous, its Fourier transform becomes independent of the energy variable, $U(\mathbf{q}, \omega) = V(\mathbf{q})$.

Show that, for the two-body interaction, the following Feynman rule applies in the momentum-energy variables. At both vertices in the basic interaction appearing in diagrams, Figure 5.2, the out-going electron momentum and energy variables equals the in-coming electron variables plus, for the case of momentum, the amount carried by the interaction line, counted with a plus or minus sign determined by convention by the arbitrarily assigned direction of the interaction wavy line. As a result, of course, the total out-going electron momenta and energies equals the in-coming ones in Figure 5.2.

Exercise 5.3. Obtain the matrix equations corresponding to the two lowest-order terms in the electron–electron interaction for the electron matrix Green's function corresponding to the diagrams in Figure 5.3.

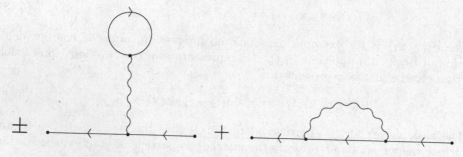

Figure 5.3 Lowest-order two-particle interaction diagrams.

These correspond to the following self-energies.

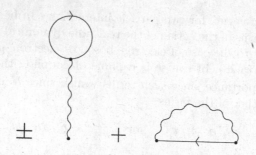

Figure 5.4 Lowest-order two-particle interaction self-energy diagrams.

These are the Hartree and Fock terms.[8]

---

[8] In order for all diagrams to appear with a plus sign it is customary to bury fermionic quantum statistical minus signs in the Feynman rule: each closed loop of fermi propagators is assigned a minus sign.

Exercise 5.4. Apply Wick's theorem to obtain the result that, to second order in the electron–phonon interaction, the diagrams for the electron matrix Green's function are given by the diagrams corresponding to the first three self-energy diagrams in Figure 5.5.

Exercise 5.5. Apply Wick's theorem to obtain the connected diagrams for the fermion matrix Green's function to second order in the two-particle interaction corresponding to the self-energy diagrams depicted in Figure 5.5.

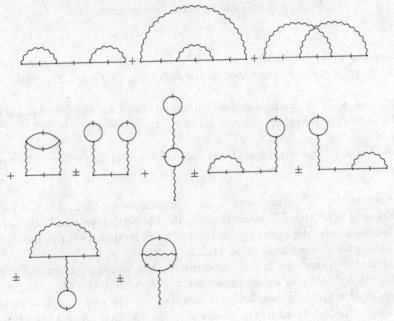

Figure 5.5 Second-order two-particle interaction self-energy diagrams.

5.4   The real rules: the RAK-rules

The matrix structure of the contour ordered Green's function was studied in the previous sections, and the proper choice of representation, that of Section 5.3, was governed by the split of information carried by the various matrix components, spectral properties and quantum statistics. The matrix structure of the basic interaction vertices should also be interpreted and will give rise to efficient rules in terms of our preferred labeling of propagators. Going through the functioning of the dynamical indices of vertices and the various possibilities for propagator attachments, leads to the observation that the diagrammatic rules significant for describing non-equilibrium states need not be formulated in terms of the individual dynamical or Schwinger–Keldysh indices of the vertices, but can with profit be formulated in terms of the labels of the three different types of propagators entering in the non-equilibrium description R, A and K. Consider, for example, the basic fermion–boson diagram depicted in Figure 5.6.

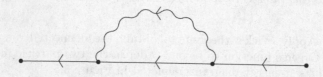

Figure 5.6 Basic fermion–boson diagram.

The boson propagator can be either D^R, D^A or D^K, and the non-equilibrium diagrammatic rules can now be stated as the following two rules, the *real rules*.

For the case of D^A a change in the dynamical index for the fermion takes place only at the **A**bsorption vertex and vice versa for the case of D^R.

For the case of D^K *no* change in the dynamical fermion index takes place at *either* of the vertices.

The effect of the D^K component is thus analogous to that of a Gaussian distributed classical field with D^K as correlator, an observation we shall take advantage of when discussing the dephasing properties of the electron–electron interaction on the weak localization effect in Section 11.3.2.

To analyze the dynamical index structure for the propagator given by the diagram in Figure 5.6, we can for example use the fact that the G_{21} component for the fermion matrix Green's function vanishes, i.e. we use the triagonal representation, and one immediately scans the diagram by in addition using identities such as $G^R(1, 1') D^A(1, 1') = 0$, and obtains for the corresponding self-energy components (adapting here the Feynman rule of absorbing the factor ig^2 into the phonon propagator)

$$\Sigma^R(1, 1') = \frac{1}{2}\left(D^R(1, 1') G^K(1, 1') + D^K(1, 1') G^R(1, 1')\right) \tag{5.61}$$

and

$$\Sigma^A(1, 1') = \frac{1}{2}\left(D^A(1, 1') G^K(1, 1') + D^K(1, 1') G^A(1, 1')\right) \tag{5.62}$$

and

$$
\begin{aligned}
\Sigma^K(1, 1') &= \frac{1}{2}(D^R(1, 1') G^R(1, 1') + D^A(1, 1') G^A(1, 1') + D^K(1, 1') G^K(1, 1')) \\
&= \frac{1}{2}((G^R(1, 1') - G^A(1, 1'))(D^R(1, 1') - D^A(1, 1'))) \\
&\quad + \frac{1}{2} D^K(1, 1') G^K(1, 1').
\end{aligned}
\tag{5.63}
$$

Equivalent to an external Gaussian distributed classical field, the D^K component does not sense the quantum statistics of the fermions for the case of retarded and advanced quantities, but of course carries the information of the quantum statistics of the bosons. Contrarily, the D^R and D^A components introduce the G^K component carrying the information of the quantum statistics of the fermions, the non-equilibrium distribution of the fermions.

The choice of the arrow on the boson Green's function in Figure 5.6 is of course arbitrary, the opposite one corresponding to the interchange $D^R(1, 1') \rightarrow (D^A(1', 1))^*$, the complex conjugation being irrelevant for a real boson field, say for phonons.

We have finally arrived at a convenient and complete physical interpretation of the dynamical index that reflects the need for doubling the degrees of freedom to describe non-equilibrium states.

5.5   Non-equilibrium Dyson equations

The standard topological arguments for partial summation of Feynman diagrams, as presented in Section 4.5.2, organizes them into one-particle irreducible sub-parts and two-particle irreducible self-energy skeleton diagrams, and we arrived at the Dyson equation, Eq. (4.141), where the self-energy is expressed in terms of the full propagators. When the corresponding equation for contour ordered quantities are lifted to the real time matrix representation we obtain the matrix Dyson equation

$$\hat{G} = \hat{G}_0 + \hat{G}_0 \otimes \tau^{(3)} \hat{\Sigma} \tau^{(3)} \otimes \hat{G}, \tag{5.64}$$

where the $\tau^{(3)}$-matrices absorb the minus signs from the return part of the closed time path, or equivalently

$$\check{G} = \check{G}_0 + \check{G}_0 \otimes \check{\Sigma} \otimes \check{G}. \tag{5.65}$$

In the triagonal representation, the three equations in the matrix Dyson equation

$$G = G_0 + G_0 \otimes \Sigma \otimes G \tag{5.66}$$

take the forms

$$G^{R(A)} = G_0^{R(A)} + G_0^{R(A)} \otimes \Sigma^{R(A)} \otimes G^{R(A)} \tag{5.67}$$

and, for the kinetic Green's function,

$$G^K = G_0^K + G_0^R \otimes \Sigma^R \otimes G^K + G_0^R \otimes \Sigma^K \otimes G^A + G_0^K \otimes \Sigma^A \otimes G^A. \tag{5.68}$$

The matrix self-energy, Σ, can in naive perturbation theory be described as the sum of diagrams that can not be cut in two by cutting only one internal free propagator line, and is from this point of view a functional of the free matrix Green's functions, $\Sigma = \Sigma[G_0, D_0]$. As discussed in Section 4.5.2, the self-energy can also be thought of as a functional of the full matrix Green's function, $\Sigma = \Sigma[G, D]$, and is then the sum of all the skeleton self-energy diagrams, i.e. the diagrams which can not be cut in two be cutting only *two* full propagator lines. It is the latter representation that is useful in the Dyson equation.

Equivalently, by iterating from the left gives the matrix Dyson equation

$$G = G_0 + G \otimes \Sigma \otimes G_0. \tag{5.69}$$

For an equilibrium state the two equations are redundant, since time convolutions by Fourier transformation become simple products for which the order of factors is irrelevant. However, in a non-equilibrium state, the two matrix equations contain different information and subtracting them is a useful way of expressing the non-equilibrium dynamics, and we shall exploit this in Chapters 7 and 8.

Since the transformation of the real-time matrix self-energy is identical to the one for the matrix Green's function we get, analogously to the equations from Eq. (5.36) to Eq. (5.39), and therefore for the components of the self-energy matrix,

$$\Sigma^R = \hat{\Sigma}_{11} - \hat{\Sigma}_{12} = \hat{\Sigma}_{21} - \hat{\Sigma}_{22} \tag{5.70}$$

$$\Sigma^A = \hat{\Sigma}_{11} - \hat{\Sigma}_{21} = \hat{\Sigma}_{12} - \hat{\Sigma}_{22} \tag{5.71}$$

$$\Sigma^K = \hat{\Sigma}_{11} + \hat{\Sigma}_{22} = \hat{\Sigma}_{12} + \hat{\Sigma}_{21} \tag{5.72}$$

$$0 = \hat{\Sigma}_{11} - \hat{\Sigma}_{12} + - \hat{\Sigma}_{21} + \hat{\Sigma}_{22}. \tag{5.73}$$

By construction

$$\hat{\Sigma}_{11}(\mathbf{x}_1, t_1, \mathbf{x}_{1'}, t_{1'}) = \begin{cases} \hat{\Sigma}_{12}(\mathbf{x}_1, t_1, \mathbf{x}_{1'}, t_{1'}) & t_{1'} > t_1 \\ \hat{\Sigma}_{21}(\mathbf{x}_1, t_1, \mathbf{x}_{1'}, t_{1'}) & t_1 > t_{1'} \end{cases} \tag{5.74}$$

and

$$\hat{\Sigma}_{22}(\mathbf{x}_1, t_1, \mathbf{x}_{1'}, t_{1'}) = \begin{cases} \hat{\Sigma}_{21}(\mathbf{x}_1, t_1, \mathbf{x}_{1'}, t_{1'}) & t_{1'} > t_1 \\ \hat{\Sigma}_{12}(\mathbf{x}_1, t_1, \mathbf{x}_{1'}, t_{1'}) & t_1 > t_{1'} \end{cases} \tag{5.75}$$

and the matrix self-energy has in the triagonal representation the same triagonal form as the matrix Green's function

$$\Sigma = \begin{pmatrix} \Sigma^R & \Sigma^K \\ 0 & \Sigma^A \end{pmatrix}. \tag{5.76}$$

Exercise 5.6. Introducing

$$\Sigma^<(\mathbf{x}_1, t_1, \mathbf{x}_{1'}, t_{1'}) = \hat{\Sigma}_{12}(\mathbf{x}_1, t_1, \mathbf{x}_{1'}, t_{1'}) \tag{5.77}$$

and

$$\Sigma^>(\mathbf{x}_1, t_1, \mathbf{x}_{1'}, t_{1'}) = \hat{\Sigma}_{21}(\mathbf{x}_1, t_1, \mathbf{x}_{1'}, t_{1'}) \tag{5.78}$$

show that we have, identically to the relationships for the Green's functions, the relation for the retarded self-energy

$$\Sigma^R(\mathbf{x}, t, \mathbf{x}', t') = \theta(t - t') \left(\Sigma^>(\mathbf{x}, t, \mathbf{x}', t') - \Sigma^<(\mathbf{x}, t, \mathbf{x}', t') \right) \tag{5.79}$$

and advanced self-energy

$$\Sigma^A(\mathbf{x}, t, \mathbf{x}', t') = -\theta(t' - t) \left(\Sigma^>(\mathbf{x}, t, \mathbf{x}', t') - \Sigma^<(\mathbf{x}, t, \mathbf{x}', t') \right) \tag{5.80}$$

and for the kinetic component

$$\Sigma^K(\mathbf{x}, t, \mathbf{x}', t') = \Sigma^>(\mathbf{x}, t, \mathbf{x}', t') + \Sigma^<(\mathbf{x}, t, \mathbf{x}', t') . \tag{5.81}$$

Show that the components of the self-energy matrix satisfies

$$\Sigma^A(\mathbf{x}, t, \mathbf{x}', t') = \left(\Sigma^R(\mathbf{x}', t', \mathbf{x}, t)\right)^* \tag{5.82}$$

and

$$\Sigma^K(\mathbf{x}, t, \mathbf{x}', t') = \left(\Sigma^K(\mathbf{x}', t', \mathbf{x}, t)\right)^* . \tag{5.83}$$

Exercise 5.7. Show that in the case where the matrix Green's function is represented in symmetric form

$$G = \begin{pmatrix} 0 & G^A \\ G^R & G^K \end{pmatrix} \tag{5.84}$$

the matrix self-energy has the form

$$\Sigma = \begin{pmatrix} \Sigma^K & \Sigma^R \\ \Sigma^A & 0 \end{pmatrix} . \tag{5.85}$$

We shall not at present take the diagrammatics beyond the self-energy to higher-order vertices, since in the following chapters only the Dyson equation is needed. In Chapter 9 we shall study diagrammatics in their full glory.

From the equation of motion for the free Green's function (or fields) we then get, for the matrix Green's function, the equations of motion

$$(i\partial_{t_1} - h(1))G(1, 1') = \delta(1 - 1') + (\Sigma \otimes G)(1, 1') \tag{5.86}$$

and

$$(i\partial_{t_{1'}} - h^*(1'))G(1, 1') = \delta(1 - 1') + (G \otimes \Sigma)(1, 1') \tag{5.87}$$

or introducing the inverse free Green's function

$$G_0^{-1}(1, 1') = (i\partial_{t_1} - h(1)) \, \delta(1 - 1') \tag{5.88}$$

the two equations can be expressed through operating with the inverse free matrix Green's function from the left

$$(G_0^{-1} - \Sigma) \otimes G = \delta(1 - 1') \tag{5.89}$$

and from the right

$$G \otimes (G_0^{-1} - \Sigma) = \delta(1 - 1') . \tag{5.90}$$

These two non-equilibrium Dyson equations will prove useful in Chapter 7 where quantum kinetic equations are considered.

The matrix equation, Eq. (5.89), comprises the three coupled equations for $G^{R,A,K}$

$$(G_0^{-1} - \Sigma^{R(A)}) \otimes G^{R(A)} = \delta(1 - 1') \tag{5.91}$$

and

$$G_0^{-1} \otimes G^K = \Sigma^R \otimes G^K + \Sigma^K \otimes G^A . \tag{5.92}$$

Analogously, from Eq. (5.90), we obtain

$$G^{R(A)} \otimes (G_0^{-1} - \Sigma^{R(A)}) = \delta(1 - 1') \tag{5.93}$$

and

$$G^K \otimes G_0^{-1} = G^R \otimes \Sigma^K + G^K \otimes \Sigma^A . \tag{5.94}$$

Exercise 5.8. Show that, subtracting the left and right Dyson equations for G^K, the resulting equation can be written in the form

$$[G_0^{-1} \overset{\otimes}{,} G^K]_- - \frac{1}{2}[\Sigma^R + \Sigma^A \overset{\otimes}{,} G^K]_- - \frac{1}{2}[\Sigma^K \overset{\otimes}{,} G^R + G^A]_-$$

$$= -\frac{1}{2}[\Sigma^K \overset{\otimes}{,} (G^R - G^A)]_+ + \frac{1}{2}[(\Sigma^R - \Sigma^A) \overset{\otimes}{,} G^K]_+ . \tag{5.95}$$

If at the end of the day, one makes the lowest-order approximation for the self-energy (as often done!), introducing the Green's function formalism and diagrammatics is of course ridiculous as final results follow from Fermi's Golden Rule.[9] A virtue of the real-time formalism and its associated Feynman diagrams is that nontrivial approximations can be established using the diagrammatic estimation technique, and higher-order correlations studied systematically, as we shall consider in the following chapters, not least in chapter 10.[10]

Before studying applications of the real-time technique we shall make obsolete one version of the imaginary-time formalism, viz. the too pervasive Matsubara technique. The general imaginary-time formalism has virtues for special Euclidean field theory purposes as well as for expedient proofs establishing conserving approximations. After the discussion of the equilibrium Dyson equation in the next section, we demonstrate the equivalence of the imaginary-time formalism to the closed time path formulation and the real-time technique introduced in this chapter.

5.6 Equilibrium Dyson equation

In equilibrium all quantities depend only on time differences, and for translational invariant situations also only on spatial differences, and convolutions are by Fourier transformation turned into products. In terms of the self-energy we therefore have for the retarded Green's function the equilibrium Dyson equation[11]

$$G^R(\mathbf{p}, E) = G_0^R(\mathbf{p}, E) + G_0^R(\mathbf{p}, E)\,\Sigma^R(E, \mathbf{p})\,G^R(\mathbf{p}, E) \tag{5.96}$$

---

[9]In the same vein, if one employs a mean-field approximation, introducing the formalism of quantum field theory seems excessive. This point of view was taken in references [1] and [13].

[10]For a discussion of the diagrammatic estimation technique see chapter 3 of reference [1].

[11]We recall the result of Section 3.4, that in thermal equilibrium all the various Green's functions can be expressed in terms of, for example, the (imaginary part of the) retarded Green's function.

which we immediately solve to get

$$G^{\mathrm{R}}(\mathbf{p}, E) = \frac{1}{G_0^{-1}(\mathbf{p}, E) - \Sigma^{\mathrm{R}}(E, \mathbf{p})} = \frac{1}{E - \epsilon_{\mathbf{p}} - \Sigma^{\mathrm{R}}(E, \mathbf{p})} . \tag{5.97}$$

The retarded self-energy determines the analytic structure of the retarded Green's function, i.e. the location of the poles of the analytically continued retarded Green's function onto the second Riemann sheet through the branch cut along the real axis (recall Section 3.4), the generic situation being that of a simple pole. For given momentum value the simple pole is located at $E = E_1 + iE_2$, determined by $E_1 = \epsilon_{\mathbf{p}} + \Re e \, \Sigma(E, \epsilon_{\mathbf{p}})$ and $E_2 = \Im m \, \Sigma(E, \epsilon_{\mathbf{p}})$, and as

$$G^{\mathrm{R}}(\mathbf{p}, t) = \int_{-\infty}^{\infty} dE \, e^{-iEt} \, G^{\mathrm{R}}(\mathbf{p}, E) \tag{5.98}$$

the imaginary part of the self-energy thereby determines the temporal exponential decay of the Green's function, i.e. the lifetime of (in the present case) momentum states. The effect of interactions are clearly to give momentum states a finite lifetime.

For the Fourier transform of Eq. (5.97) we get (in three spatial dimensions for the prefactor to be correct)

$$G_E^{\mathrm{R}}(\mathbf{x} - \mathbf{x}') = \frac{-m}{2\pi\hbar^2} \frac{e^{\frac{i}{\hbar}|\mathbf{x}-\mathbf{x}'|\sqrt{2m(E-\Sigma^{\mathrm{R}}(E, p_E\hat{\mathbf{p}}))}}}{|\mathbf{x} - \mathbf{x}'|} \tag{5.99}$$

where p_E is the solution of the equation $p_E = \sqrt{2m(E - \Sigma^{\mathrm{R}}(E, p_E\hat{\mathbf{p}}))}$. Interactions will thus provide a finite spatial and temporal range of the Green's function.

For the case of electrons, say in a metal, the advanced Green's function likewise describes the attenuation of the holes.

Exercise 5.9. Show that the spectral function in equilibrium is given by (using now the grand canonical ensemble)

$$A(E, \mathbf{p}) = \frac{\Gamma(E, \mathbf{p})}{\left(E - \xi_{\mathbf{p}} - \Re e\Sigma^{\mathrm{R}}(E, \mathbf{p})\right)^2 + \left(\frac{\Gamma(E, \mathbf{p})}{2}\right)^2} \tag{5.100}$$

where

$$\Re e\Sigma(E, \mathbf{p}) \equiv \frac{1}{2}\left(\Sigma^{\mathrm{R}}(E, \mathbf{p}) + \Sigma^{\mathrm{A}}(E, \mathbf{p})\right) \tag{5.101}$$

and

$$\Gamma(E, \mathbf{p}) \equiv i\left(\Sigma^{\mathrm{R}}(E, \mathbf{p}) - \Sigma^{\mathrm{A}}(E, \mathbf{p})\right) . \tag{5.102}$$

We note that the sum-rule satisfied by the spectral weight function, Eq. (3.89), sets limitation on the dependence of the self-energy on the energy variable. The general features of interaction is to broaden the peak in the spectral weight function and to shift, renormalize, energies.

Exercise 5.10. Show that for bosons in equilibrium at temperature T, their self-energy components satisfy the fluctuation–dissipation relations

$$\Sigma^K(E, \mathbf{p}) = \left(\Sigma^R(E, \mathbf{p}) - \Sigma^A(E, \mathbf{p}) \right) \coth \frac{E}{2kT} \qquad (5.103)$$

and for fermions

$$\Sigma^K(E, \mathbf{p}) = \left(\Sigma^R(E, \mathbf{p}) - \Sigma^A(E, \mathbf{p}) \right) \tanh \frac{E}{2kT} . \qquad (5.104)$$

5.7   Real-time versus imaginary-time formalism

Although we shall mainly use the real-time technique presented in this chapter throughout, it is useful to be familiar with the equivalent imaginary-time formalism in view of the vast amount of literature where this method has been employed. Or more importantly to realize the link between the imaginary-time formalism and the Martin–Schwinger–Abrikosov–Gorkov–Dzyaloshinski–Eliashberg–Kadanoff–Baym–Langreth analytical continuation procedure. In the classic textbooks of Kadanoff and Baym [14] and Abrikosov, Gorkov and Dzyaloshinski [15] on non-equilibrium statistical mechanics, the imaginary-time formalism introduced by Matsubara [16] and Fradkin [17] and Martin and Schwinger [18] was used. Being then a Euclidean field theory it possesses nice convergence properties. However, it lacks appeal to intuition.

5.7.1   Imaginary-time formalism

The workings of the imaginary-time formalism are based on the mathematical formal resemblance of the Boltzmann statistical weighting factor in the equilibrium statistical operator $\rho \propto e^{-H/kT}$ and the evolution operator $\hat{U} \propto e^{-iHt/\hbar}$ for an isolated system. The imaginary time Green's function

$$\mathcal{G}(\mathbf{x}, \tau; \mathbf{x}', \tau') \equiv -\mathrm{Tr}\left(e^{-\frac{H - \mu N}{kT}} T_\tau(\psi(\mathbf{x}, \tau)\, \tilde{\psi}(\mathbf{x}', \tau')) \right) \qquad (5.105)$$

is defined in terms of field operators depending on *imaginary* time according to (we suppress all other degrees of freedom than space)

$$\psi(\mathbf{x}, \tau) = e^{\frac{1}{\hbar}\tau(H - \mu N)}\, \psi(\mathbf{x})\, e^{-\frac{1}{\hbar}\tau(H - \mu N)} \qquad (5.106)$$

and

$$\tilde{\psi}(\mathbf{x}, \tau) = e^{\frac{1}{\hbar}\tau(H - \mu N)}\, \psi^\dagger(\mathbf{x})\, e^{-\frac{1}{\hbar}\tau(H - \mu N)} , \qquad (5.107)$$

where $\psi(\mathbf{x})$ is the field operator in the Schrödinger picture, and T_τ provides the imaginary time ordering (with the usual minus sign involved for an odd number of interchanges of fermi fields). The τs involved are *real* variables, the use of the word imaginary refers to the transformation $t \to -i\tau$ in which case the time-ordered real-time Green's function, Eq. (4.10), transforms into the imaginary-time Green's function (more about this shortly). Note that $\psi(\mathbf{x}, \tau)$ and $\tilde{\psi}(\mathbf{x}, \tau)$ are not each others adjoints. Knowledge of the imaginary-time Green's function allows the calculation of thermodynamic average values.

The imaginary-time single-particle Green's function respects the Kubo–Martin–Schwinger boundary conditions, for example

$$\mathcal{G}(\mathbf{x}, \tau; \mathbf{x}', 0) = \pm \mathcal{G}(\mathbf{x}, \tau; \mathbf{x}', \beta), \tag{5.108}$$

owing to the cyclic invariance property of the trace (the notation $\beta = \hbar/kT$ is used). The periodic boundary condition is for bosons, and the anti-periodic boundary condition is for fermions (the identical consideration in connection with the fluctuation–dissipation theorem was discussed in Section 3.4, and is further discussed in Section 6.5). We note the crucial role of the (grand) canonical ensemble as elaborated in Section 3.4.

In its simple equilibrium applications in statistical mechanics, thermodynamics, or in linear response theory, the involved imaginary-time Green's functions are expressed in terms of a single so-called Matsubara frequency

$$\mathcal{G}(\mathbf{x}, \tau; \mathbf{x}', \tau') = \frac{1}{\beta} \sum_{\omega_n} e^{-i\omega_n (\tau - \tau')} \mathcal{G}(\mathbf{x}, \mathbf{x}'; \omega_n), \tag{5.109}$$

where $\omega_n = 2n\pi/\beta$ for bosons and $\omega_n = (2n+1)\pi/\beta$ for fermions, respectively, $n = 0, \pm 1, \pm 2, \dots$. Equilibrium or thermodynamic properties and linear transport coefficients can therefore be expressed in terms of only one Matsubara frequency, and the analytical continuation to obtain them from the imaginary-time Green's functions is trivial, say the retarded Green's function is obtained by $G^{\mathrm{R}}(\mathbf{x}, \mathbf{x}'; \omega) = \mathcal{G}(\mathbf{x}, \mathbf{x}'; i\omega_n \rightarrow \omega + i0^+)$ as the two functions coincide according to $G^{\mathrm{R}}(i\omega_n) = \mathcal{G}(\omega_n)$ for $\omega_n > 0$.

The imaginary-time Green's functions can also be used to study non-equilibrium states by letting the external potential depend on the imaginary time. The Matsubara technique is then a bit cumbersome, but can be used to derive exact equations, say, the Dyson equation for real-time Green's functions. In fact this was the method used originally to study non-equilibrium superconductivity in the quasi-classical approximation [19].[12] However, for general non-equilibrium situations, the necessary analytical continuation in arbitrarily many Matsubara frequencies becomes nontrivial (and are usually left out of textbooks), and are more involved than using the real-time technique. Furthermore, when approximations are made, the real-time results obtained upon analytical continuation can be spurious. However, the main disadvantage of the imaginary-time formalism is that it lacks physical transparency. We shall therefore not discuss it further in the way it is usually done in textbooks, but use a contour formulation to show its equivalence to the real-time formalism.[13]

---

[12]Amazingly, the non-equilibrium theory of superconductivity was originally obtained using the Matsubara technique [19], as, I guess, the imaginary-time formalism was in rule at the Landau Institute. A plethora of papers and textbooks have perpetrated the use of the imaginary-time formalism. It is *the* contestant to be the most important *frozen accident* in the evolution of non-equilibrium theory. Let's iron out unfortunate fluctuations of the past! Its proliferation also testifies to the fact that idiosyncratically written papers, such as the seminal paper of Schwinger [5], can be a long time in germination.

[13]The imaginary-time formalism can be useful for special purpose applications such as diagram-

5.7.2   Imaginary-time Green's functions

The imaginary-time Green's functions are profitably interpreted as contour-ordered Green's function, viz. on an imaginary-time contour. First we note, that the *times* entering the imaginary-time Green's function can be interpreted as contour times. Choosing the times in the time ordered Green's function in Eq. (3.61), instead to lie on the contour starting at, say, t_0 and ending down in the lower complex time plane at $t_0 - i\beta$, the appendix contour c_a in Figure 4.4, turns the expression Eq. (3.61) into the equation for the imaginary-time Green's function, Eq. (5.105). This observation, by the way, gives the standard Feynman diagrammatics for the imaginary-time Green's function since Wick's theorem involving the appendix contour is a trivial corollary of the general Wick's theorem of Section 4.3.3. We can thus, for example, immediately write down the non-equilibrium Dyson equation for the imaginary-time Green's function, t_1 and $t_{1'}$ lying on the appendix contour c_a. Considering the case where the non-equilibrium situation is the result of a time-dependent potential, V, the Dyson equation for the imaginary-time Green's functions or appendix contour-ordered Green's function is

$$G(1,1') \;=\; G_0(1,1') + \sum_{\sigma_3 \sigma_2} \int d\mathbf{x}_3 \int_{c_a} d\tau_3 \int d\mathbf{x}_2 \int_{c_a} d\tau_2 \, G_0(1,3)\, \Sigma(3,2)\, G(2,1')$$

$$+ \; \sum_{\sigma_2} \int d\mathbf{x}_2 \int_{c_a} d\tau_2 \, G_0(1,2)\, V(2)\, G(2,1') \,. \tag{5.110}$$

The appendix contour interpretation of the imaginary-time Green's function is in certain situations more expedient than the real-time formulation when it for example comes to diagrammatically proving exact relationships, since it has fewer diagrams than the real-time approach if unfolding its matrix structure is needed. It should thus be used in such situations, but then it is preferable not to use the Matsubara frequency technique, but instead stick to the appendix contour formalism.

In practice we need to know how to analytically continue imaginary-time quantities to real-time functions, say for the imaginary-time Dyson equation, or for terms appearing in perturbative expansions of imaginary-time quantities. Instead of turning to standard textbook imaginary-time formalism, the Matsubara technique, it is in view of the above preferable to go to the appendix contour ordered Green's functions, and perform the analytical continuation from there. In fact, this analytical continuation procedure becomes equivalent to the analytical continuation of the contour ordered functions in the general contour formalism, for example the transition from the contour-ordered Green's function to real-time Green's functions, which we now turn to discuss. We illustrate this in the next section by performing the proceduce for the Dyson equation.

The Boltzmann factor can guarantee analyticity of the Green's function for times on the appendix contour c_a and indeed in the whole strip corresponding to translating the real time t_0 (recall Exercise 3.21 on page 74). The appendix contour can therefore

---

matic proofs of conservation laws, i.e. for proving exact relations. The imaginary-time formalism is seen to be a simple corollary of the closed time path formalism.

be deformed into the contour depicted in Figure 4.4 on page 91, landing the original times on the imaginary appendix contour onto the real axis producing the contour-ordered Green's function. This provides the analytical continuation from imaginary times to the real times of interest. We now turn to discuss the procedure in detail.

5.7.3 Analytical continuation procedure

By using the closed time path approach, the analytical continuation procedure is automated, and the equations of motion for the real-time correlation functions are obtained without the irrelevant detour into Matsubara frequencies.

The procedure to obtain the real time Dyson equation from either the imaginary-time formalism, i.e. from Eq. (5.110), or from the general contour formalism Dyson equation, Eq. (4.141), is in fact the same. In, for example, the perturbative expansion of the contour-ordered Green's function or in the Dyson equation, we encounter objects integrated over the contour depicted in Figure 4.4 on page 91, and we need to obtain the corresponding formulae in terms of real-time functions, and as we demonstrate now this is equivalent to analytical continuation.

Since space (and spin) and contour-time variables play different roles in the following argument, in fact only the contour-time variables play a role, we separate space and contour-time matrix notations

$$(A \times B)(1, 1') \equiv \sum_{\sigma_2} \int d\mathbf{x}_2 \, A(1, 2) \, B(2, 1') \tag{5.111}$$

and

$$(A \Box B)(1, 1') \equiv \int_c d\tau_2 \, A(1, 2) \, B(2, 1') \,. \tag{5.112}$$

Here the contour-time integration could refer to the imaginary time appendix contour c_a, or the contour c_i, stretching from t_0 through t_1 and $t_{1'}$ and back again to t_0 (or all the way to infinity and back) and finally along the appendix contour to $t_0 - i\beta$, the contour depicted in Figure 4.4. The latter contour is obtained from the appendix contour by the allowed analytical continuation procedure as discussed at the end of the previous section. We shall not be interested in initial correlations, and can therefore let the initial time protrude to the far past, $t_0 \to -\infty$, and the contour in Figure 4.4 becomes the real-time closed contour, C, depicted in Figure 4.5. In the case of analytical continuation from the imaginary-time appendix contour we shall also eventually let the real-time t_0 protrude to the far past. Everything in the following, however, would be equally correct if we stick to the general contour, c_i, depicted in Figure 4.4, allowing treating general initial states and therefore including the completely general non-equilibrium problem. We would then just in addition to integrations over the closed time path, have terms with integrations over the imaginary time appendix contour c_a.

Consider the case where the non-equilibrium situation is the result of a time-dependent potential, V. The Dyson equation for the imaginary-time Green's function is then given in Eq. (5.110), and for the contour-ordered Green's function we analogously have the equation

$$G_C(1,1') = G_C^{(0)}(1,1') + \sum_{\sigma_3\sigma_2} \int d\mathbf{x}_3 \int_C d\tau_3 \int d\mathbf{x}_2 \int_C d\tau_2\, G_C^{(0)}(1,3)\, \Sigma(3,2)\, G_C(2,1')$$

$$+ \sum_{\sigma_2} \int d\mathbf{x}_2 \int_C d\tau_2\, G_C^{(0)}(1,2)\, V(2)\, G_C(2,1')\,, \qquad (5.113)$$

which can be written on the form (dropping the contour reminder subscript)

$$G = G_0 + G_0 \boxtimes \Sigma \boxtimes G + G^{(0)}\, V \boxtimes G\,, \qquad (5.114)$$

where Σ denotes the self-energy for the problem of interest.

We thus encounter explicitly contour matrix-multiplication, or multiplication in series, a term of the form

$$C(\tau_1,\tau_{1'}) = \int_c d\tau\, A(\tau_1,\tau)\, B(\tau,\tau_{1'}) = (A\square B)(\tau_1,\tau_{1'})\,, \qquad (5.115)$$

where A and B are functions of the contour variable, and the involved contour could be any of the three one can encounter as discussed above. Degrees of freedom are suppressed since they play no role in the following demonstration that contour integrations can be turned into integrations over the real-time axis. To accomplish this we recall that the functions $C^<(\tau_1,\tau_{1'})$ and $C^>(\tau_1,\tau_{1'})$ are analytic functions in the strips $0 < \Im m(\tau_1 - \tau_{1'}) < \beta$ and $-\beta < \Im m(\tau_1 - \tau_{1'}) < 0$, respectively.

Let us demonstrate the analytical continuation procedure for the case of $C^<$. A *lesser* quantity means by the general prescription, Eq. (4.41), that the contour time τ_1 appears earlier than the contour time $\tau_{1'}$, whatever contour is involved, i.e. we have chosen the relationship $\tau_1 \underset{c}{<} \tau_{1'}$ (irrespective of the numerical relationship of their corresponding real time values in the case of the real-time contour). Exploiting analyticity, the contour C or c_i or the imaginary time contour c_a is deformed into the contour $c_1 + c_{1'}$ depicted in Figure 5.7.[14]

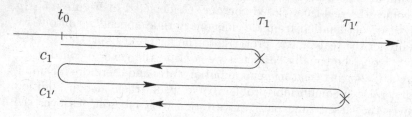

Figure 5.7 Deforming either of the contours C or c_i or c_a into the contour built by the contours c_1 and $c_{1'}$.

---

[14]Starting the ascent to real times, we essentially follow Langreth [20].

The expression in Eq. (5.115), for the chosen contour ordering, therefore becomes

$$C^<(1,1') = \int_{c_1} d\tau \, A(\tau_1, \tau) \, B(\tau, \tau_{1'}) + \int_{c_{1'}} d\tau \, A(\tau_1, \tau) \, B(\tau, \tau_{1'})$$

$$= \int_{c_1} d\tau \, A(\tau_1, \tau) \, B^<(\tau, \tau_{1'}) + \int_{c_{1'}} d\tau \, A^<(\tau_1, \tau) \, B(\tau, \tau_{1'}) \quad (5.116)$$

and in the last equality, we have used the fact that on contour c_1 we have $\tau \overset{<}{\underset{c}{}} \tau_{1'}$, and on contour $c_{1'}$ we have $\tau \overset{>}{\underset{c}{}} \tau_1$. In the event of including initial correlations, or rather staying with the general exact equation, the additional term with integration over the appendix contour should be retained in the above equation.

Splitting in forward and return contour parts we have (as a consequence of the contour positioning of the times on the contour parts in question as indicated to the right)

$$C^<(1,1') = \int_{\overrightarrow{c_1}} d\tau \, A^>(\tau_1, \tau) \, B^<(\tau, \tau_{1'}) \qquad \overrightarrow{c_1} : \tau \overset{<}{\underset{c}{}} \tau_1$$

$$+ \int_{\overleftarrow{c_1}} d\tau \, A^<(\tau_1, \tau) \, B^<(\tau, \tau_{1'}) \qquad \overleftarrow{c_1} : \tau \overset{>}{\underset{c}{}} \tau_1$$

$$+ \int_{\overrightarrow{c_{1'}}} d\tau \, A^<(\tau_1, \tau) \, B^<(\tau, \tau_{1'}) \qquad \overrightarrow{c_{1'}} : \tau \overset{<}{\underset{c}{}} \tau_{1'}$$

$$+ \int_{\overleftarrow{c_{1'}}} d\tau \, A^<(\tau_1, \tau) \, B^>(\tau, \tau_{1'}) \qquad \overleftarrow{c_{1'}} : \tau \overset{>}{\underset{c}{}} \tau_{1'}. \quad (5.117)$$

Parameterizing the contours in terms of the real time variable, as in Eq. (4.35), and noting that the external contour variables, τ_1 and $\tau_{1'}$, now can be identified by their corresponding values on the real time axis, gives ($t_0 \to -\infty$)

$$C^<(1,1') = \int_{-\infty}^{t_1} dt \, (A^>(t_1, t) - A^<(t_1, t)) \, B^<(t, t_{1'})$$

$$+ \int_{-\infty}^{t_{1'}} dt \, A^<(t_1, t)(B^<(t, t_{1'}) - B^>(t, t_{1'})) \quad (5.118)$$

and thereby

$$C^<(1,1') = \int\limits_{-\infty}^{\infty} dt\, \theta(t_1 - t)(A^>(t_1,t) - A^<(t_1,t))\, B^<(t,t_{1'})$$

$$+ \int\limits_{-\infty}^{\infty} dt\, A^<(t_1,t)\, \theta(t_{1'} - t)(B^<(t,t_{1'}) - B^>(t,t_{1'}))\,. \qquad (5.119)$$

Introducing the retarded function

$$A^R(1,1') = \theta(t_1 - t_{1'})\,(A^>(1,1') - A^<(1,1')) \qquad (5.120)$$

and the advanced function

$$A^A(1,1') = \theta(t_{1'} - t_1)\,(A^>(1,1') - A^<(1,1')) \qquad (5.121)$$

we have the real-time rule for multiplication in series

$$C^< = A^R \circ B^< + A^< \circ B^A\,, \qquad (5.122)$$

where $\circ$ symbolizes matrix multiplication in real time, i.e. integration over the internal real-time variable from minus infinity to plus infinity of times.

Analogously one shows

$$C^> = A^R \circ B^> + A^> \circ B^A\,. \qquad (5.123)$$

We shall also need an expression for C^R, and from Eq. (5.122) and Eq. (5.123) we get

$$C^R(t_1,t_{1'}) = \theta(t_1 - t_{1'})\,((A^R \circ B^>)(t_1,t_{1'}) + (A^> \circ B^A)(t_1,t_{1'})$$

$$- (A^R \circ B^<)(t_1,t_{1'}) + (A^< \circ B^A)(t_1,t_{1'}))$$

$$= \theta(t_1 - t_{1'})((A^R \circ (B^> - B^<))(t_1,t_{1'}) + ((A^> - A^<) \circ B^A)(t_1,t_{1'}))\,.$$

$$(5.124)$$

Expressing retarded and advanced functions according to Eq. (5.120) and Eq. (5.121) gives

$$C^R(t_1,t_{1'}) = \theta(t_1 - t_{1'}) \left(\int\limits_{-\infty}^{t_1} dt\, (A^>(t_1,t) - A^<(t_1,t))\, (B^>(t,t_{1'}) - B^<(t,t_{1'})) \right.$$

$$\left. + \int\limits_{-\infty}^{t_{1'}} dt\, (A^>(t_1,t) - A^<(t_1,t))\, (B^>(t,t_{1'}) - B^<(t,t_{1'})) \right)$$

$$= \theta(t_1 - t_{1'}) \int_{t_{1'}}^{t_1} dt \, (A^>(t_1, t) - A^<(t_1, t)) \, (B^>(t, t_{1'}) - B^<(t, t_{1'})) \, .$$

$$(5.125)$$

Using the fact that $t_{1'} < t < t_1$ (as otherwise both left- and right-hand sides vanish) we obtain

$$C^R = A^R \circ B^R \, . \qquad (5.126)$$

Analogously one arrives at

$$C^A = A^A \circ B^A \, . \qquad (5.127)$$

By using Eq. (5.122), Eq. (5.123), Eq. (5.126) and Eq. (5.127) we find, owing to the associative property of the composition, that the analytical continuation of the contour quantity

$$D = A \square B \square C \qquad (5.128)$$

becomes

$$D^{\gtrless} = A^R \circ B^R \circ C^{\gtrless} + A^R \circ B^{\gtrless} \circ C^A + A^{\gtrless} \circ B^A \circ C^A \qquad (5.129)$$

and, by induction, Eq. (5.129) generalizes to an arbitrary number of functions multiplied in series. Note that the retarded and advanced functions appear to the left and right, respectively, of the *greater* and *lesser* Green's functions.

Employing the analytical continuation procedure, one can thus from the imaginary-time Green's function formalism arrive with equal ease at the real-time non-equilibrium Dyson equations of Section 5.5.

The only other ingredient encountered in perturbation expansions is the product of contour-ordered Green's functions of the form

$$C(\tau_1, \tau_{1'}) = A(\tau_1, \tau_{1'}) \, B(\tau_1, \tau_{1'}) \, , \qquad (5.130)$$

multiplication in *parallel*. This occurs, for example, for pair-creation or electron-hole excitations, etc., or for a self-energy diagram for example for fermion–boson interaction, in which case one might prefer to have the same sequence of the contour variables in all of the functions as in the self-energy insertion of the diagram in Figure 5.6 (this is, of course, a matter of taste).

Following the above procedure we immediately get for the analytical continuation of multiplication in parallel

$$C^{\gtrless}(t_1, t_{1'}) = A^{\gtrless}(t_1, t_{1'}) \, B^{\gtrless}(t_1, t_{1'}) \, . \qquad (5.131)$$

With these tools at hand, we can turn any exact imaginary-time formula, or any diagram in the perturbative expansion of the imaginary-time Green's function or a contour ordered quantity, say the contour-ordered Green's functions of Section 4.4, into products of real-time Green's functions. This automatic mechanical continuation to real times is much preferable than to do it in the Matsubara frequencies. With this at hand a very effective way of studying non-equilibrium states in the real time formalism is available, as discussed in the classic text [14], and whether using this or the other three-fold representation is a matter of taste. However, the *G-greater* and *G-lesser* Green's functions are quantum statistically quantities of the same nature, whereas in the representation introduced in Section 5.3, the Green's functions carry distinct information.

5.7.4 Kadanoff–Baym equations

As an example of using the analytical continuation procedure we shall, from the Dyson equation for the imaginary-time Green's function in Eq. (5.110) (or the general contour-ordered Green's function, Eq. (5.114)), obtain the equations of motion for the physical correlation functions, the *lesser* and *greater* Green's functions on the real-time axis. Let us therefore consider the Dyson equation for the imaginary-time Green's function or the general contour-ordered Green's function of the form

$$G_0^{-1} \boxtimes G = \delta(1 - 1') + \Sigma \boxtimes G, \qquad (5.132)$$

where external fields are included in G_0^{-1}. Applying the rule for multiplication in series gives

$$G_0^{-1} \otimes G^{\lessgtr} = \Sigma^{R} \otimes G^{\lessgtr} + \Sigma^{\lessgtr} \otimes G^{A} \qquad (5.133)$$

and similarly for the right-hand Dyson equation

$$G^{\lessgtr} \otimes G_0^{-1} = G^{R} \otimes \Sigma^{\lessgtr} + G^{\lessgtr} \otimes \Sigma^{A}. \qquad (5.134)$$

Subtracting the left and right Dyson equations gives

$$[G_0^{-1} \overset{\otimes}{,} G^{\lessgtr}]_- = \Sigma^{R} \otimes G^{\lessgtr} - G^{\lessgtr} \otimes \Sigma^{A} + \Sigma^{\lessgtr} \otimes G^{A} - G^{R} \otimes \Sigma^{\lessgtr} \qquad (5.135)$$

which can be rewritten

$$[G_0^{-1} \overset{\otimes}{,} G^{\lessgtr}]_- - \frac{1}{2}[\Sigma^{R} + \Sigma^{A} \overset{\otimes}{,} G^{\lessgtr}]_- - \frac{1}{2}[\Sigma^{\lessgtr} \overset{\otimes}{,} G^{R} + G^{A}]_-$$

$$= -\frac{1}{2}[\Sigma^{<} \overset{\otimes}{,} G^{>}]_+ + \frac{1}{2}[\Sigma^{>} \overset{\otimes}{,} G^{<}]_+. \qquad (5.136)$$

These two equations, the Kadanoff–Baym equations, can be used as basis for considering quantum kinetics.

We recall that the kinetic Green's function is specified according to $G^{K} = G^{>} + G^{<}$, and note that adding the two equations in Eq. (5.136) we recover Eq. (5.95). We note that the equations, Eq. (5.129), satisfied by G^{K} are satisfied by both $G^{>}$ and $G^{<}$, or rather we should appreciate the observation that their equations are identical with respect to splitting into retarded and advanced Green's functions.

Since the equations for $G^{\lessgtr}$ mixes, through for example self-energies according to Eq. (5.131), it is economical to work instead solely with G^{K}. However, there can be special circumstances where the advantage is reversed, for example when discussing the dynamics of a tunnel junction. One should note that the quantum statistics of particle species manifests itself quite differently depending on which type of kinetic propagator one chooses to employ.

5.8 Summary

We have presented the real-time formalism necessary for treating non-equilibrium situations. For the reader not familiar with equilibrium theory the good news is that equilibrium theory is just an especially simple case of the presented general theory. In the real-time formulation of the properties of non-equilibrium states the dynamics is used to provide a doubling of the degrees of freedom, and one encounters at least two types of Green's functions. To get a physically transparent representation, we introduced the real-time matrix representation of the contour-ordered Green's functions to describe non-equilibrium states. This allowed us to represent matrix Green's function perturbation theory in terms of Feynman diagrams in a standard fashion. We introduced the physical representation corresponding to the two Green's functions representing the spectral and quantum statistical properties of a system. We then showed that the matrix notation can be broken down into two simple rules for the universal vertex structure in the dynamical indices. This allowed us to formulate the non-equilibrium aspects of the Feynman diagrams directly in terms of the various matrix Green's function components, R, A, K, establishing the real rules. In this way we were able to express how the different features of the spectral and quantum statistical properties enter into the diagrammatic representation of non-equilibrium processes. We ended the chapter by showing the equivalence of the imaginary-time and the closed time path and the real-time formalisms, all formally identical, and transformed into each other by analytical continuation. In the rest of the book we shall demonstrate the versatility of the real-time technique. Before constructing the functional formulation of quantum field theory from its Feynman diagrams, and show that its classical limit can be used to study classical statistical dynamics, as done in the last chapter, in the next three chapters we demonstrate various applications of the real-time formalism to the study of quantum dynamics.

6

Linear response theory

There exists a regime of overlap between the equilibrium and non-equilibrium behavior of a system, the non-equilibrium behavior of weakly perturbed states. When a system is perturbed ever so slightly, its response will be linear in the perturbation, say the current of the conduction electrons in a metal will be proportional to the strength of the applied electric field. This regime is called the linear response regime, and though the system is in a non-equilibrium state all its characteristics can be inferred from the properties of its equilibrium state. In the next chapter we shall go beyond the linear regime by showing how to obtain quantum kinetic equations. The kinetic-equation approach to transport is a general method, and allows in principle nonlinear effects to be considered. However, in many practical situations one is interested only in the linear response of a system to an external force. The linear response limit is a tremendous simplification in comparison with general non-equilibrium conditions, and is the subject matter of this chapter. In particular the linear response of the density and current of an electron gas are discussed. The symmetry properties of response functions, and the fluctuation–dissipation theorem are established. Lastly we demonstrate how correlation functions can be measured in scattering experiments, as illustrated by considering neutron scattering from matter. Needless to say, in measurements of (say) the current in a macroscopic body, far less information in the current correlation function is probed.

6.1   Linear response

In this section we consider the response of an arbitrary property of a system to a general perturbation. The Hamiltonian consists of two parts:

$$H = H_0 + H'_t, \tag{6.1}$$

where H_0 governs the dynamics in the absence of the perturbation H'_t.

For the expectation value of a quantity A for a system in state ρ we have

$$A(t) = \text{Tr}(\rho(t) A) = \text{Tr}(U(t, t') \rho(t') U^\dagger(t, t') A). \tag{6.2}$$

Expanding the time-evolution operator to linear order in the applied perturbation we get

$$U(t,t') = U_0(t,t') - U_0(t,t_r) \frac{i}{\hbar} \int_{t'}^{t} d\bar{t}\, H_I'(\bar{t})\, U_0^\dagger(t',t_r) + \mathcal{O}((H_t')^2)\,, \qquad (6.3)$$

where the perturbation is in the interaction picture with respect to H_0

$$H_I'(t) = U_0^\dagger(t,t_r) H_t'\, U_0(t,t_r)\,. \qquad (6.4)$$

For the statistical operator we thus have the perturbative expansion in terms of the perturbation

$$\rho(t) = \rho_0(t) + \rho_1(t) + \mathcal{O}((H_t')^2)\,, \qquad (6.5)$$

where

$$\rho_0(t) = U_0(t,t_i)\, \rho_i\, U_0^\dagger(t,t_i) = U_0(t,t_r)\, \rho_0(t_r)\, U_0^\dagger(t,t_r) \qquad (6.6)$$

and the linear correction in the applied potential is given by

$$\rho_1(t) = \frac{i}{\hbar} U_0(t,t_i)\, \rho_i\, U_0(t_i,t_r) \int_{t_i}^{t} d\bar{t}\, H_I'(\bar{t})\, U_0^\dagger(t,t_r)$$

$$- \frac{i}{\hbar} U_0(t,t_r) \int_{t_i}^{t} d\bar{t}\, H_I'(\bar{t})\, U_0^\dagger(t_i,t_r)\, \rho_i\, U_0^\dagger(t,t_i)\,. \qquad (6.7)$$

We have assumed that prior to time t_i, the applied field is absent, and the system is in state ρ_i. For the expectation value we then get to linear order

$$A(t) = \mathrm{Tr}(\rho_0(t)\, A) + \frac{i}{\hbar} \int_{t_i}^{t} d\bar{t}\, \mathrm{Tr}(\rho_0(t_r)\, [H_I'(\bar{t}), A_I(t)])\,. \qquad (6.8)$$

So far the statistical operator at the reference time has been arbitrary; however, typically we shall assume the state prior to the application of the perturbation is the thermal equilibrium state of the system.

We first discuss the density response to an external scalar potential, and afterwards the current response to a vector potential.

6.1.1 Density response

In this section we consider the density response to an applied external field. The external field is represented by the potential $V(\mathbf{x}, t)$, and the Hamiltonian consists of two parts:

$$\mathcal{H}(t) = H + H_{V(t)}\,, \qquad (6.9)$$

where H governs the dynamics in the absence of the applied potential, and the applied potential couples to the density of the system as specified by the operator, Eq. (2.28),

$$H_{V(t)} = \int d\mathbf{x}\, n(\mathbf{x})\, V(\mathbf{x}, t)\,. \qquad (6.10)$$

The density will adjust to the applied potential, and according to Eq. (6.8) the deviation from equilibrium is to linear order

$$\delta n(\mathbf{x}, t) = n(\mathbf{x}, t) - n_0(\mathbf{x}, t) = \int d\mathbf{x}' \int_{t_i}^{\infty} dt' \, \chi(\mathbf{x}, t; \mathbf{x}', t') \, V(\mathbf{x}', t') , \qquad (6.11)$$

where

$$n_0(\mathbf{x}, t) = \mathrm{Tr}(\rho_0(t) \, n(\mathbf{x})) \qquad (6.12)$$

is the density in the absence of the potential, and the linear density response can be specified in the various ways by the density–density response function:[1]

$$
\begin{aligned}
\chi(\mathbf{x}, t; \mathbf{x}', t') &= -\frac{i}{\hbar} \theta(t - t') \mathrm{Tr}(\rho_0(t_r)[n(\mathbf{x}, t), n(\mathbf{x}', t')]) \\
&\equiv -\frac{i}{\hbar} \theta(t - t') \langle [n(\mathbf{x}, t), n(\mathbf{x}', t')] \rangle_0 \\
&= -\frac{i}{\hbar} \theta(t - t') \mathrm{Tr}(\rho_0(t_r)[\delta n(\mathbf{x}, t), \delta n(\mathbf{x}', t')]) \\
&\equiv \chi^{\mathrm{R}}(\mathbf{x}, t; \mathbf{x}', t') .
\end{aligned}
\qquad (6.13)
$$

The density operator is in the interaction picture with respect to H

$$n(\mathbf{x}, t) = e^{iH(t - t_r)} \, n(\mathbf{x}) \, e^{-iH(t - t_r)} \qquad (6.14)$$

and we have introduced the density deviation operator $\delta n(\mathbf{x}, t) \equiv n(\mathbf{x}, t) - n_0(\mathbf{x}, t)$. The retarded density response function appears in Eq. (6.11) in respect of causality; i.e. a change in the density at time t can occur only as a cause of the applied potential prior to that time.

Before the external potential is applied we assume a stationary state with respect to the unperturbed Hamiltonian H, and the initial state is described by a statistical operator of the form

$$\rho_i = \rho_i(H) = \sum_\lambda \rho_\lambda |\lambda\rangle\langle\lambda| \qquad (6.15)$$

where the $|\lambda\rangle$s are the eigenstates of H,

$$H |\lambda\rangle = \epsilon_\lambda |\lambda\rangle \qquad (6.16)$$

and $\rho_\lambda = \rho_i(\epsilon_\lambda)$ is the probability for finding the unperturbed system with energy ϵ_λ. The unperturbed statistical operator is then time independent, $\rho_0(t) = \rho_i$, and the equilibrium density profile is time independent, $n_0(\mathbf{x}, t) = n_0(\mathbf{x}) = \mathrm{Tr}(\rho_i \, n(\mathbf{x}))$.

---

[1] A response function is a *retarded* Green's function. Our preferred choice of Green's functions, for which we developed diagrammatic non-equilibrium perturbation theory, is thus the proper physical choice.

The response function will then only depend on the time difference:

$$\chi(\mathbf{x}, t; \mathbf{x}', t') = \chi(\mathbf{x}, \mathbf{x}'; t - t')$$

$$= -\frac{i}{\hbar}\theta(t - t')\sum_{\lambda\lambda'}(\rho_\lambda - \rho_{\lambda'})\langle\lambda|n(\mathbf{x})|\lambda'\rangle\langle\lambda'|n(\mathbf{x}')|\lambda\rangle e^{\frac{i}{\hbar}(\epsilon_\lambda - \epsilon_{\lambda'})(t - t')}. \qquad (6.17)$$

In linear response, each Fourier component contributes additively, so without loss of generality we just need to seek the response at one driving frequency, say ω,

$$V(\mathbf{x}, t) = V_\omega(\mathbf{x})\, e^{-i\omega t}. \qquad (6.18)$$

For any ω in the the upper half plane, $\Im m\,\omega > 0$, the applied potential vanishes in the far past, $V(t \to -\infty) = 0$, and the state of the system in the far past becomes smoothly independent of the applied potential. For ω real we are thus interested in the analytic continuation from the upper half plane of the frequency-dependent response function.

Since we shall be interested in steady-state properties, the time integration in Eq. (6.11) can be performed by letting the arbitrary initial time t_i be taken in the remote past. By letting t_i approach minus infinity, transients are absent, and there is then only a linear density response at the driving frequency

$$\delta n(\mathbf{x}, t) = n(\mathbf{x}, t) - n_0(\mathbf{x}) = \delta n(\mathbf{x}, \omega)\, e^{-i\omega t}. \qquad (6.19)$$

We obtain for the Fourier transform of the linear density response

$$\delta n(\mathbf{x}, \omega') = \delta n(\mathbf{x}, \omega)\, \delta(\omega - \omega'), \qquad (6.20)$$

where

$$\delta n(\mathbf{x}, \omega) = \int d\mathbf{x}'\, \chi(\mathbf{x}, \mathbf{x}'; \omega)\, V_\omega(\mathbf{x}') \qquad (6.21)$$

and

$$\chi(\mathbf{x}, \mathbf{x}'; \omega) = \sum_{\lambda\lambda'}\frac{\rho_\lambda - \rho_{\lambda'}}{\epsilon_\lambda - \epsilon_{\lambda'} + \hbar\omega + i0}\langle\lambda|n(\mathbf{x})|\lambda'\rangle\langle\lambda'|n(\mathbf{x}')|\lambda\rangle \qquad (6.22)$$

is the Fourier transform of the time-dependent linear response function for a steady state. The positive infinitesimal stems from the theta function; i.e. causality causes the response function $\chi_\omega \equiv \chi_\omega^R$ to be an analytic function in the upper half plane.

If the Hamiltonian H describes the dynamics of independent particles, the linear response function becomes

$$\chi(\mathbf{x}, \mathbf{x}'; t - t') = -\frac{i}{\hbar}\theta(t - t')\sum_{\lambda\lambda'}(\rho_\lambda - \rho_{\lambda'})e^{\frac{i}{\hbar}(\epsilon_\lambda - \epsilon_{\lambda'})(t - t')}\psi_\lambda^*(\mathbf{x})\psi_{\lambda'}(\mathbf{x})\psi_{\lambda'}^*(\mathbf{x}')\psi_\lambda(\mathbf{x}')$$

$$(6.23)$$

where $\psi_\lambda(\mathbf{x}) = \langle\mathbf{x}|\lambda\rangle$ now denotes the energy eigenfunction of a particle corresponding to the energy eigenvalue ϵ_λ, and ρ_λ.the probability for its occupation. For the Fourier transform we have

$$\chi(\mathbf{x}, \mathbf{x}'; \omega) = \sum_{\lambda\lambda'}\frac{\rho_\lambda - \rho_{\lambda'}}{\epsilon_\lambda - \epsilon_{\lambda'} + \hbar\omega + i0}\psi_\lambda^*(\mathbf{x})\psi_{\lambda'}(\mathbf{x})\psi_{\lambda'}^*(\mathbf{x}')\psi_\lambda(\mathbf{x}'). \qquad (6.24)$$

Looking ahead to Eq. (D.22), we can express the Fourier transform of the density response function·in terms of the single particle spectral function (see Appendix D)

$$\chi(\mathbf{x},\mathbf{x}';\omega) = \int\limits_{-\infty}^{\infty} \frac{dE}{2\pi} \int\limits_{-\infty}^{\infty} \frac{dE'}{2\pi} \frac{\rho_i(E') - \rho_i(E)}{E' - E + \hbar\omega + i0} A(\mathbf{x},\mathbf{x}';E) A(\mathbf{x}',\mathbf{x},E'). \quad (6.25)$$

Introducing the propagators for a single particle instead of the spectral functions

$$A(\mathbf{x},\mathbf{x}';E) = i\left[G^R(\mathbf{x},\mathbf{x}',E) - G^A(\mathbf{x},\mathbf{x}',E)\right] \quad (6.26)$$

we have expressed the response function in terms of the single-particle propagators, quantities we know how to handle well, as we have developed the diagrammatic perturbation theory for them.[2]

6.1.2 Current response

In this section, we shall discuss the linear current response. We shall specifically discuss the electric current response to an applied time-dependent electric field represented by a vector potential **A**:[3]

$$\mathbf{E} = -\frac{\partial \mathbf{A}}{\partial t}, \quad (6.27)$$

thereby in view of the preceding we have covered the general case of coupling to an electromagnetic field.

Inserting the expression for the current density operator, Eq. (2.47), into the linear response formula, Eq. (6.8), and recalling the perturbation, Eq. (2.51), the average current density becomes to linear order

$$\mathbf{j}(\mathbf{x},t) = \mathrm{Tr}(\rho_0(t)\,\mathbf{j}_{A(t)}(\mathbf{x},t)) = \frac{i}{\hbar} \int\limits_{t_i}^{t} d\bar{t}\, \mathrm{Tr}(\rho_0(t_r)[\mathbf{j}_p(\mathbf{x},t), H_{A(\bar{t})}]), \quad (6.28)$$

where $\mathbf{j}_p(\mathbf{x},t)$ is just the paramagnetic part of the current density operator in the interaction picture with respect to H.

To linear order in the external electric field we therefore see that the current density

$$j_\alpha(\mathbf{x},t) = \mathrm{Tr}(\rho_0(t)\,j_\alpha^P(\mathbf{x})) + \sum_\beta \int d\mathbf{x}' \int\limits_{t_i}^{\infty} dt'\, Q_{\alpha\beta}(\mathbf{x},t;\mathbf{x}',t') A_\beta(\mathbf{x}',t') \quad (6.29)$$

is determined by the current response function

$$Q_{\alpha\beta}(\mathbf{x},t;\mathbf{x}',t') = K_{\alpha\beta}(\mathbf{x},t;\mathbf{x}',t') - \frac{e^2\rho_0(\mathbf{x},\mathbf{x},t)}{m} \delta_{\alpha\beta}\, \delta(\mathbf{x}-\mathbf{x}')\, \delta(t-t'), \quad (6.30)$$

---

[2]If the particles have coupling to other degrees of freedom the propagators are still operators with respect to these, and a trace with respect to these degrees of freedom should be performed, as discussed in Section 6.2.

[3]The case of representing the electric field as the gradient of a scalar potential can be handled with an equal amount of labor and the treatments are equivalent by gauge invariance.

where we have introduced the current-current response function

$$K_{\alpha\beta}(\mathbf{x}, t; \mathbf{x}', t') = \frac{i}{\hbar}\theta(t - t') \,\text{Tr}(\rho_0(t_r) \,[j_\alpha^p(\mathbf{x}, t), j_\beta^p(\mathbf{x}', t')])$$

$$\equiv \frac{i}{\hbar}\theta(t - t')\langle [j_\alpha^p(\mathbf{x}, t), j_\beta^p(\mathbf{x}', t')]\rangle_0 \qquad (6.31)$$

and $\text{Tr}(\rho_0(t)j_\alpha^p(\mathbf{x}))$ is a possible current density in the absence of the field. Here we shall not consider superconductivity or magnetism, and can therefore in the following assume that this term vanishes.

Assuming that we have a stationary state with respect to the unperturbed Hamiltonian before the external field is applied, the response function depends only on the relative time

$$K_{\alpha\beta}(\mathbf{x}, t; \mathbf{x}', t') = \frac{i}{\hbar}\theta(t - t') \sum_{\lambda\lambda'} (\rho_\lambda - \rho_{\lambda'})\langle\lambda|j_\alpha^p(\mathbf{x})\lambda'\rangle\langle\lambda'|j_\beta^p(\mathbf{x}')|\lambda\rangle e^{\frac{i}{\hbar}(\epsilon_\lambda - \epsilon_{\lambda'})(t - t')}.$$

$$(6.32)$$

In linear response each frequency contributes additively so we just need to seek the response at one driving frequency, say ω,

$$\mathbf{A}(\mathbf{x}, t) = \mathbf{A}(\mathbf{x}, \omega)\,e^{-i\omega t}. \qquad (6.33)$$

The time integration in Eq. (6.29) can then be performed by letting the arbitrary initial time, t_i, be taken in the remote past (letting t_i approach minus infinity), and we only get a current response at the driving frequency

$$j_\alpha(\mathbf{x}, t) = j_\alpha(\mathbf{x}, \omega)\,e^{-i\omega t}. \qquad (6.34)$$

For the Fourier transform of the current density we then have

$$j_\alpha(\mathbf{x}, \omega) = + \sum_\beta \int d\mathbf{x}'\, Q_{\alpha\beta}(\mathbf{x}, \mathbf{x}'; \omega)A_\beta(\mathbf{x}', \omega) + \mathcal{O}(\mathbf{E}^2), \qquad (6.35)$$

where

$$Q_{\alpha\beta}(\mathbf{x}, \mathbf{x}'; \omega) = K_{\alpha\beta}(\mathbf{x}, \mathbf{x}'; \omega) - \frac{\rho_0(\mathbf{x}, \mathbf{x})e^2}{m}\delta_{\alpha\beta}\,\delta(\mathbf{x} - \mathbf{x}') \qquad (6.36)$$

and

$$K_{\alpha\beta}(\mathbf{x}, \mathbf{x}'; \omega) = \sum_{\lambda\lambda'} \frac{\rho_{\lambda'} - \rho_\lambda}{\epsilon_\lambda - \epsilon_{\lambda'} + \hbar\omega + i0} \,\langle\lambda|j_\alpha^p(\mathbf{x})\lambda'\rangle\,\langle\lambda'|j_\beta^p(\mathbf{x}')|\lambda\rangle. \qquad (6.37)$$

For the case of a single particle, the paramagnetic current density matrix element is given by

$$\langle\lambda|\mathbf{j}^p(\mathbf{x})|\lambda'\rangle = \frac{e\hbar}{2im}\psi_\lambda^*(\mathbf{x})\left(\overrightarrow{\frac{\partial}{\partial\mathbf{x}}} - \overleftarrow{\frac{\partial}{\partial\mathbf{x}}}\right)\psi_{\lambda'}(\mathbf{x}), \qquad (6.38)$$

where the arrows indicate whether differentiating to the left or right. For a system of independent particles, the response function then becomes

$$
K_{\alpha\beta}(\mathbf{x},\mathbf{x}';\omega) = \left(\frac{e\hbar}{m}\right)^2 \int_{-\infty}^{\infty} \frac{dE}{2\pi} \int_{-\infty}^{\infty} \frac{dE'}{2\pi} \frac{\rho_{\mathrm{i}}(E') - \rho_{\mathrm{i}}(E)}{E' - E + \hbar\omega + i0}
$$

$$
\times \quad [G^{\mathrm{R}}(\mathbf{x},\mathbf{x}';E) - G^{\mathrm{A}}(\mathbf{x},\mathbf{x}';E)]
$$

$$
\times \quad \overleftrightarrow{\nabla}_{x_\alpha} \overleftrightarrow{\nabla}_{x'_\beta} [G^{\mathrm{R}}(\mathbf{x}',\mathbf{x},E') - G^{\mathrm{A}}(\mathbf{x}',\mathbf{x},E')] . \tag{6.39}
$$

We have introduced the abbreviated notation

$$
\overleftrightarrow{\nabla}_{\mathbf{x}} = \frac{1}{2}\left(\frac{\overrightarrow{\partial}}{\partial \mathbf{x}} - \frac{\overleftarrow{\partial}}{\partial \mathbf{x}}\right) \tag{6.40}
$$

for the differential operator associated with the current vertex in the position representation.

In the expression for the current response kernel we can perform one of the energy integrations, and exploiting the analytical properties of the propagators half of the terms are seen not to contribute, and we obtain for the current response function for an electron gas (the factor of 2 accounts for spin)

$$
K_{\alpha\beta}(\mathbf{x},\mathbf{x}',\omega) = -2\left(\frac{e\hbar}{m}\right)^2 \int_{-\infty}^{\infty} \frac{dE}{2\pi} f_0(E)\Big(A(\mathbf{x},\mathbf{x}';E) \overleftrightarrow{\nabla}_{x_\alpha}\overleftrightarrow{\nabla}_{x'_\beta} G^{\mathrm{A}}(\mathbf{x}',\mathbf{x};E-\hbar\omega)
$$

$$
+ \quad G^{\mathrm{R}}(\mathbf{x},\mathbf{x}';E+\hbar\omega) \overleftrightarrow{\nabla}_{x_\alpha}\overleftrightarrow{\nabla}_{x'_\beta} A(\mathbf{x}',\mathbf{x};E)\Big) . \tag{6.41}
$$

Gauge invariance implies a useful expression for the longitudinal part of the current response function, i.e. the current response to a longitudinal electric field, $\nabla \times \mathbf{E} = \mathbf{0}$, viz.[4]

$$
K_{\alpha\beta}(\mathbf{x},\mathbf{x}';\omega = 0) = \frac{e^2 \rho_0(\mathbf{x},\mathbf{x},\omega=0)}{m} \delta_{\alpha\beta}\, \delta(\mathbf{x}-\mathbf{x}') \tag{6.42}
$$

and the longitudinal part of the current response function can be written in the form

$$
Q_{\alpha\beta}(\mathbf{x},\mathbf{x}';\omega) = K_{\alpha\beta}(\mathbf{x},\mathbf{x}';\omega) - K_{\alpha\beta}(\mathbf{x},\mathbf{x}';\omega=0) . \tag{6.43}
$$

We can therefore express the longitudinal current density response solely in terms of the paramagnetic response function

$$
j_\alpha(\mathbf{x},\omega) = \sum_\beta \int d\mathbf{x}'\, [K_{\alpha\beta}(\mathbf{x},\mathbf{x}';\omega) - K_{\alpha\beta}(\mathbf{x},\mathbf{x}';\omega=0)]\, A_\beta(\mathbf{x}',\omega) . \tag{6.44}
$$

---

[4]For a detailed discussion see chapter 7 of reference [1], and for its relation to the causal and dissipative character of linear response see appendix E of reference [1].

6.1.3   Conductivity tensor

Expressing the current density in terms of the electric field

$$j_\alpha(\mathbf{x}, \omega) = \sum_\beta \int d\mathbf{x}' \, \sigma_{\alpha\beta}(\mathbf{x}, \mathbf{x}'; \omega) \, E_\beta(\mathbf{x}', \omega) + \mathcal{O}(\mathbf{E}^2) \tag{6.45}$$

introduces the conductivity tensor,

$$\sigma_{\alpha\beta}(\mathbf{x}, \mathbf{x}'; \omega) = \frac{Q_{\alpha\beta}(\mathbf{x}, \mathbf{x}'; \omega)}{i\omega} \tag{6.46}$$

or, equivalently for the longitudinal part,

$$\sigma_{\alpha\beta}(\mathbf{x}, \mathbf{x}', \omega) = \frac{K_{\alpha\beta}(\mathbf{x}, \mathbf{x}', \omega) - K_{\alpha\beta}(\mathbf{x}, \mathbf{x}', \omega = 0)}{i\omega}. \tag{6.47}$$

We note that the conductivity tensor is analytic in the upper half plane as causality demands, and as a consequence the real and imaginary parts are related through principal value integrals, Kramers–Kronig relations,

$$\Re\sigma_{\alpha\beta}(\mathbf{x}, \mathbf{x}', \omega) = \frac{1}{\pi} P \int\limits_{-\infty}^{\infty} d\omega' \, \frac{\Im m \, \sigma_{\alpha\beta}(\mathbf{x}, \mathbf{x}'; \omega')}{\omega' - \omega} \tag{6.48}$$

and

$$\Im m \, \sigma_{\alpha\beta}(\mathbf{x}, \mathbf{x}', \omega) = -\frac{1}{\pi} P \int\limits_{-\infty}^{\infty} d\omega' \, \frac{\Re\sigma_{\alpha\beta}(\mathbf{x}, \mathbf{x}'; \omega')}{\omega' - \omega}. \tag{6.49}$$

The time average of the response function, $K_{\alpha\beta}(\mathbf{x}, \mathbf{x}'; \omega = 0)$, is a real function, and we have (for ω real)

$$\Re\sigma_{\alpha\beta}(\mathbf{x}, \mathbf{x}'; \omega) = \Re\left(\frac{-i}{\omega} K_{\alpha\beta}(\mathbf{x}, \mathbf{x}'; -\omega)\right) = \frac{1}{\omega} \Im m \, K_{\alpha\beta}(\mathbf{x}, \mathbf{x}'; \omega). \tag{6.50}$$

The real part of the conductivity tensor for an electron gas is according to Eq. (6.41) given by

$$\Re\sigma_{\alpha\beta}(\mathbf{x}, \mathbf{x}', \omega) = \frac{1}{\pi}\left(\frac{e}{m}\right)^2 \int\limits_{-\infty}^{\infty} dE \, \frac{f_0(E) - f_0(E + \hbar\omega)}{\omega}$$

$$\times \ [G^{\mathrm{R}}(\mathbf{x}, \mathbf{x}'; E + \hbar\omega) - G^{\mathrm{A}}(\mathbf{x}, \mathbf{x}'; E + \hbar\omega)]$$

$$\times \ \overleftrightarrow{\nabla}_{x_\alpha} \overleftrightarrow{\nabla}_{x'_\beta} \ [G^{\mathrm{R}}(\mathbf{x}', \mathbf{x}; E) - G^{\mathrm{A}}(\mathbf{x}', \mathbf{x}; E)]. \tag{6.51}$$

In the case where the electron gas in the absence of the applied field is in the thermal state, only electrons occupying levels in the thermal layer around the Fermi surface contribute to the real part of the longitudinal conductivity, as expected.

6.1.4   Conductance

Often we are interested only in the total average current through the system (S denotes a cross-sectional surface through the system)

$$I(\omega) = \int_S d\mathbf{s} \cdot \mathbf{j}(\mathbf{x}, \omega) \tag{6.52}$$

and a proper description is in terms of the conductance, the inverse of the resistance. Let us consider a hypercube of volume L^d, and choose the surface S perpendicular to the direction of the current flow, say, the α-direction. In terms of the conductivity we have (where ds_α denotes the infinitesimal area on the surface S):

$$I_\alpha(\omega) = \sum_\beta \int_S ds_\alpha \int d\mathbf{x}' \, \sigma_{\alpha\beta}(\mathbf{x}, \mathbf{x}', \omega) \, E_\beta(\mathbf{x}', \omega) \,. \tag{6.53}$$

Since the current, by particle conservation, is independent of the position of the cross section we get

$$I_\alpha(\omega) = L^{-1} \int d\mathbf{x} \, \mathbf{j}(\mathbf{x}, \omega) = L^{-1} \sum_\beta \int d\mathbf{x} \int d\mathbf{x}' \, \sigma_{\alpha\beta}(\mathbf{x}, \mathbf{x}', \omega) \, E_\beta(\mathbf{x}', \omega) \,. \tag{6.54}$$

For the case of a spatially homogeneous external field in the β-direction, $E_\alpha(\mathbf{x}, \omega) = \delta_{\alpha\beta} \, E(\omega)$, we have in terms of the applied voltage across the system, $V_\beta(\omega) = E(\omega) \, L$,

$$I_\alpha(\omega) = G_{\alpha\beta}(\omega) \, V_\beta(\omega) \tag{6.55}$$

where we have introduced the conductance tensor

$$G_{\alpha\beta}(\omega) = L^{-2} \int d\mathbf{x} \int d\mathbf{x}' \, \sigma_{\alpha\beta}(\mathbf{x}, \mathbf{x}', \omega) \tag{6.56}$$

the inverse of the resistance tensor.

For a translational invariant state, the conductance and conductivity are related according to

$$G_{\alpha\beta}(L) = L^{d-2} \sigma_{\alpha\beta}(L) \,. \tag{6.57}$$

6.2   Linear response of Green's functions

The linear response of physical quantities can also, for a many-body system, conveniently be expressed in terms of the linear response of the single-particle Green's function as it specifies average quantities. For example, the average current density can be expressed in terms of the kinetic component of the matrix Green's function (recall Eq. (3.83)), and we are therefore interested in its linear response. We represent the external electric field $\mathbf{E}$ by a time-dependent vector potential $\mathbf{A}$ according to

Eq. (6.27), and not by a scalar potential; the two cases can be handled with an equal amount of labor and are equivalent by gauge invariance. According to Eq. (2.51), the linear coupling to the vector potential is through the coupling to the current operator. The linear correction to the Green's function for this perturbation is thus represented by the diagram depicted in Figure 6.1.

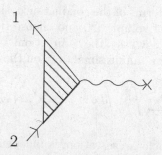

Figure 6.1 Linear response diagram for a propagator.

The vertex in Figure 6.1 consists of the diagrams produced by inserting the vector potential coupling into any electron line in any diagram for the Green's function in question. For our interest, the Green's function is the kinetic or Keldysh propagator, as the labels 1 and 2 allude to in Figure 6.1, referring to the triagonal representation of the matrix Green's function. The resulting propagator to linear order in the vector potential is denoted by δG^{K}.

To get the current density, say for electrons, we insert the kinetic Green's function into the current density formula, Eq. (3.83), and assuming that the current density vanishes in the absence of the field we obtain

$$\mathbf{j}(\mathbf{x},t) = \frac{e\hbar}{2m}\left(\frac{\partial}{\partial\mathbf{x}} - \frac{\partial}{\partial\mathbf{x}'}\right)\delta G^{\mathrm{K}}(\mathbf{x},t,\mathbf{x}',t)\bigg|_{\mathbf{x}'=\mathbf{x}} + i\frac{e^2}{m}\mathbf{A}(\mathbf{x},t)\,G^{\mathrm{K}}(\mathbf{x},t,\mathbf{x},t)\,. \quad (6.58)$$

Recall, if the particles are coupled to other degrees of freedom, the propagators are still operators with respect to these, and a trace with respect to these degrees of freedom should be performed, resulting in the presence of vertex corrections corresponding to insertion of the external vertex into all propagators.

According to the expression for the linear correction to the Green's function, as depicted in Figure 6.1, the propagator to linear order in the vector potential is

$$\delta G^{\mathrm{K}}(1,1') = \frac{ie\hbar}{2m}\mathrm{Tr}\left(\tau^1\int d2\,\mathbf{A}(2)\cdot\left(\frac{\partial}{\partial\mathbf{x}_2} - \frac{\partial}{\partial\mathbf{x}_{2'}}\right)G(1,2')\,G(2,1')\bigg|_{2'=2}\right), \quad (6.59)$$

if in the trace Tr we include the trace over interactions as well as meaning trace over Schwinger–Keldysh indices. The matrix τ^1 in Schwinger–Keldysh space insures that the kinetic component is projected out at the measuring vertex.

For the conductivity tensor we then get in the triagonal matrix representation of the Green's functions

$$\sigma_{\alpha\beta}(\omega) = \frac{e^2\hbar}{2\pi\omega}\text{Tr}\left(\tau^1\frac{1}{V}\sum_{\mathbf{PP'}}\int_{-\infty}^{\infty}dE\, p_\alpha p'_\beta\, G(\mathbf{p}_+,\mathbf{p}'_+,E+\omega)\, G(\mathbf{p}'_-,\mathbf{p}_-,E)\right) - \frac{ne^2}{i\omega m}\,, \tag{6.60}$$

where n is, for example, the density of electrons, and $\mathbf{p}_\pm = \mathbf{p}\pm\mathbf{q}$ and $\mathbf{p}'_\pm = \mathbf{p}'\pm\mathbf{q}$, $\mathbf{q}$ being the wave vector of the electric field. Here we have directly arrived at expressing the response function, the conductivity, in terms of the single-particle propagators, quantities we know how to handle well, as we have developed the diagrammatic perturbation theory for them.[5]

Transport coefficients are thus represented by Feynman diagrams of the form depicted in Figure 6.2, an infinite sum of diagrams captured in the conductivity diagram.

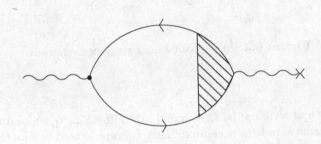

Figure 6.2 Linear response or conductivity diagram.

Different types of linear response coefficients correspond to the action of different single-particle operators at the excitation and measuring vertices. The excitation vertex is proportional to the unit matrix in the dynamical or Schwinger–Keldysh index in the triagonal representation and the measuring vertex is attributed in the dynamical indices the first Pauli matrix in order that the trace over the dynamical indices picks out the kinetic component. As we discuss next, for the case of fermions the high-energy contribution from the propagator term exactly cancels the diamagnetic term in Eq. (6.60).

Let us consider the case where the electrons only interact with a random potential,

---

[5]If the particles have coupling to other degrees of freedom the propagators are still operators with respect to these, and a trace with respect to these degrees of freedom should be performed.

in which case the conductivity becomes

$$\sigma_{\alpha\beta}(\omega) = \frac{e^2\hbar}{2\pi\omega}\text{Tr}\left(\tau^1\int dE\,\mathbf{p}\cdot\mathbf{p}'\,\langle G(\mathbf{p},\mathbf{p}',E+\omega)\,G(\mathbf{p}',\mathbf{p},,E)\rangle\right) - \frac{ne^2}{i\omega m}\,,\quad(6.61)$$

where the bracket means average with respect to the random potential. The quantity to be impurity averaged is thus the product of two Green's functions, as depicted in Figure 6.3.

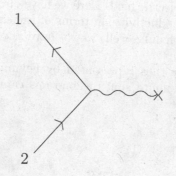

Figure 6.3 Propagator linear response diagram.

Denoting the first term on the right in Eq. (6.61) by $K_{\alpha\beta}(\mathbf{q},\omega)$, and unfolding the trace in the dynamical indices it can explicitly be represented by the three terms[6]

$$K_{\alpha\beta}(\mathbf{q},\omega) = K_{\alpha\beta}^{\text{RA}}(\mathbf{q},\omega) + K_{\alpha\beta}^{\text{RR}}(\mathbf{q},\omega) + K_{\alpha\beta}^{\text{AA}}(\mathbf{q},\omega)\,,\qquad(6.62)$$

where

$$K_{\alpha\beta}^{\text{RA}}(\mathbf{q},\omega) = \frac{i}{\pi}\left(\frac{e}{m}\right)^2\frac{1}{V}\sum_{\mathbf{pp}'}\int_{-\infty}^{\infty}dE\,(f_0(E) - f_0(E+\hbar\omega))$$

$$\times\quad p_\alpha p'_\beta\langle G^{\text{R}}(\mathbf{p}_+,\mathbf{p}'_+;E+\hbar\omega)\,G^{\text{A}}(\mathbf{p}'_-,\mathbf{p}_-;E)\rangle\qquad(6.63)$$

and

$$K_{\alpha\beta}^{\text{RR}}(\mathbf{q},\omega) = -\frac{i}{\pi}\left(\frac{e}{m}\right)^2\frac{1}{V}\sum_{\mathbf{pp}'}p_\alpha p'_\beta\int_{-\infty}^{\infty}dE\,f_0(E)$$

$$\times\quad\langle G^{\text{R}}(\mathbf{p}_+,\mathbf{p}'_+;E+\hbar\omega)\,G^{\text{R}}(\mathbf{p}'_-,\mathbf{p}_-;E)\rangle\qquad(6.64)$$

---

[6]Actually, the terms should be kept together under the momentum summation for reasons of convergence. For clarity of presentation, however, we write the three terms separately.

and

$$K_{\alpha\beta}^{AA}(\mathbf{q},\omega) = \frac{i}{\pi}\left(\frac{e}{m}\right)^2 \frac{1}{V}\sum_{\mathbf{p}\mathbf{p}'} p_\alpha p'_\beta \int_{-\infty}^{\infty} dE\, f_0(E)$$

$$\times \langle G^A(\mathbf{p}_+,\mathbf{p}'_+;E)\,G^A(\mathbf{p}'_-,\mathbf{p}_-;E-\hbar\omega)\rangle. \qquad (6.65)$$

In the first term integrations are limited to the Fermi surface and can be easily performed. The two other terms are regular, having the poles of the product of Green's functions in the same half plane. The leading-order contribution from these terms cancels exactly the density terms giving for a degenerate electron gas the conductivity tensor

$$\sigma_{\alpha\beta}(\mathbf{q},\omega) = \frac{e^2 v_F^2}{\pi}\int_{-\infty}^{\infty} dE\, \frac{f_0(E)-f_0(E+\hbar\omega)}{\omega}$$

$$\times \frac{1}{V}\sum_{\mathbf{p}\mathbf{p}'} \hat{p}_\alpha \hat{p}'_\beta \langle G^R(\mathbf{p}_+,\mathbf{p}'_+;E+\hbar\omega)\,G^A(\mathbf{p}'_-,\mathbf{p}_-;E)\rangle. \qquad (6.66)$$

The apparent singular ω-dependence is thus canceled which is no accident but, as noted earlier, a consequence of gauge invariance.[7]

We shall make use of this formula in Chapter 11, where we discuss weak localization.

Exercise 6.1. The classical conductivity of a disordered conductor corresponds to including only the ladder diagrams for the impurity averaged vertex function, i.e. all the diagrams of order $(\hbar/p_F l)^0$,

$$(6.67)$$

Analytically we have that the three-point vector vertex in the ladder approximation, $\mathbf{\Gamma}^L$, satisfies the equation

$$\mathbf{\Gamma}_E^L(\mathbf{p},\mathbf{q},\omega) = \mathbf{p} + n_i \int \frac{d\mathbf{p}'}{(2\pi\hbar)^3}\, |V_{\mathrm{imp}}(\mathbf{p}-\mathbf{p}')|^2 G^R(\mathbf{p}'_+,E_+)G^A(\mathbf{p}'_-,E)\,\mathbf{\Gamma}_E^L(\mathbf{p}',\mathbf{q},\omega). \qquad (6.68)$$

---

[7]For details of the calculations regarding this point we refer the reader to chapter 8 of reference [1].

Consider the normal skin effect, where the wavelength of the electric field is much larger than the mean free path, $q\, l \ll 1$.[8] We can therefore set the wave vector of the electric field $\mathbf{q}$ equal to zero in the propagators, and thereby in the vertex function as its scale of variation consequently is the Fermi wave vector $k_{\mathrm{F}} = p_{\mathrm{F}}/\hbar$.

Show that the classical conductivity is given by

$$\sigma(\omega) = \frac{\sigma_0}{1 - i\omega\tau_{\mathrm{tr}}} \quad , \quad \sigma_0 = \frac{ne^2\tau_{\mathrm{tr}}}{m} \tag{6.69}$$

where $\tau_{\mathrm{tr}} \equiv \tau_{\mathrm{tr}}(\epsilon_{\mathrm{F}})$ is the transport relaxation time

$$\frac{\hbar}{\tau_{\mathrm{tr}}(\epsilon_{\mathrm{F}})} = 2\pi n_i N_0 \int \frac{d\hat{\mathbf{p}}'_{\mathrm{F}}}{4\pi}\, |V_{\mathrm{imp}}(\mathbf{p}_{\mathrm{F}} - \mathbf{p}'_{\mathrm{F}})|^2 (1 - \hat{\mathbf{p}}_{\mathrm{F}} \cdot \hat{\mathbf{p}}'_{\mathrm{F}})\,. \tag{6.70}$$

In the following we shall assume that, prior to applying the perturbation, the system is in its thermal equilibrium state. It is therefore of importance for the relevance of linear response theory to verify that this is a stable state, i.e. weak fields do not perturb a system out of this state, as will be shown in Section 6.4. But first we establish the general properties of response functions satisfied in thermal equilibrium states.

6.3 Properties of response functions

Response functions must satisfy certain relationships. In order to be specific, we illustrate these relationships by considering the current response function. We have already noted that causality causes the response function to be analytic in the upper half ω-plane. The current response function therefore has the representation in terms of the current spectral function, $\Im m K_{\alpha\beta}$ (the current response function vanishes in the limit of large ω),

$$K_{\alpha\beta}(\mathbf{x}, \mathbf{x}'; \omega) = \int_{-\infty}^{\infty} \frac{d\omega'}{\pi}\, \frac{\Im m K_{\alpha\beta}(\mathbf{x}, \mathbf{x}'; \omega')}{\omega' - \omega - i0}\,. \tag{6.71}$$

Since $K_{\alpha\beta}(\mathbf{x}, t; \mathbf{x}', t')$ contains a commutator of hermitian operators multiplied by the imaginary unit, it is real, and we have the property of the response function (ω real)

$$[K_{\alpha\beta}(\mathbf{x}, \mathbf{x}'; \omega)]^* = K_{\alpha\beta}(\mathbf{x}, \mathbf{x}'; -\omega) \tag{6.72}$$

and the real part of the response function is even[9]

$$\Re e K_{\alpha\beta}(\mathbf{x}, \mathbf{x}'; -\omega) = \Re e K_{\alpha\beta}(\mathbf{x}, \mathbf{x}'; \omega) \tag{6.73}$$

---

[8] For an applied field of wavelength much shorter than the mean free path, $q \gg 1/l$, the corrections to the bare vertex can be neglected.

[9] In the presence of a magnetic field $\mathbf{B}$, we must also reverse the direction of the field, for example, $\Im m K_{\alpha\beta}(\mathbf{x}, \mathbf{x}'; \omega, \mathbf{B}) = -\Im m K_{\alpha\beta}(\mathbf{x}, \mathbf{x}'; -\omega, -\mathbf{B})$.

and the imaginary part is odd

$$\Im m\, K_{\alpha\beta}(\mathbf{x}, \mathbf{x}'; -\omega) = -\Im m K_{\alpha\beta}(\mathbf{x}, \mathbf{x}'; \omega). \tag{6.74}$$

From the spectral representation, Eq. (6.37), we have

$$\Im m K_{\alpha\alpha}(\mathbf{x}, \mathbf{x}; \omega) = \pi \sum_{\lambda\lambda'} \rho(\epsilon_\lambda) |\langle\lambda| j_\alpha^p(\mathbf{x})|\lambda'\rangle|^2 \left(\delta(\epsilon_\lambda - \epsilon_{\lambda'} - \hbar\omega) - \delta(\epsilon_\lambda - \epsilon_{\lambda'} + \hbar\omega)\right).$$
$$\tag{6.75}$$

For the thermal equilibrium state, where

$$\rho(\epsilon_{\lambda'}) = \rho(\epsilon_\lambda)\, e^{\frac{\epsilon_\lambda - \epsilon_{\lambda'}}{kT}} \tag{6.76}$$

we then obtain[10]

$$\Im m\, K_{\alpha\alpha}(\mathbf{x}, \mathbf{x}; \omega) = \pi \left(1 - e^{-\hbar\omega/kT}\right) \sum_{\lambda\lambda'} \rho(\epsilon_\lambda) |\langle\lambda| j_\alpha^p(\mathbf{x})|\lambda'\rangle|^2\, \delta(\epsilon_\lambda - \epsilon_{\lambda'} + \hbar\omega).$$
$$\tag{6.77}$$

For a state where the probability distribution, $\rho(\epsilon_\lambda)$, is a decreasing function of the energy, such as in the case of the thermal equilibrium state, the imaginary part of the diagonal response function is therefore positive for positive frequencies

$$\Im m\, K_{\alpha\alpha}(\mathbf{x}, \mathbf{x}; \omega \geq 0) \geq 0. \tag{6.78}$$

For the imaginary part of the diagonal part of the response function K we therefore have[11]

$$\omega\, \Im m K_{\alpha\alpha}(\mathbf{x}, \mathbf{x}, \omega) \geq 0. \tag{6.79}$$

From the spectral representation, Eq. (6.71), we then find that the real part of the response function at zero frequency is larger than zero, $\Re e K_{\alpha\alpha}(\mathbf{x}, \mathbf{x}, \omega = 0) > 0$. The diagonal part of the real part of the response function is therefore positive for small frequencies. Since for large frequencies, the integral in Eq. (6.71) is controlled by the singularity in the denominator, and as $\Im m K_{\alpha\alpha}(\mathbf{x}, \mathbf{x}, \omega)$ is a decaying function, the real part of the response function is negative for large frequencies, eventually approaching zero.

6.4 Stability of the thermal equilibrium state

In this section we shall show that the thermal equilibrium state is stable; i.e. manipulating the system by coupling its physical properties to a weak classical field that vanishes in the past and future can only increase the energy of the system. The average energy of a system is

$$E(t) = \langle H(t)\rangle = \text{Tr}(\rho(t)H(t)). \tag{6.80}$$

---

[10]Note that the nature of the discussion is general; we already encountered the similar one for the spectral weight function in Section 3.4.

[11]We stress the important role played by the canonical ensemble.

The rate of change of the expectation value for the energy (the term appearing when differentiating the statistical operator with respect to time vanishes, as seen by using the von Neumann equation, Eq. (3.14), and the cyclic property of the trace),

$$
\frac{dE}{dt} = \mathrm{Tr}\left(\rho(t)\frac{dH}{dt}\right) = -\int d\mathbf{x}\,\mathrm{Tr}(\rho(t)\,\mathbf{j}_t(\mathbf{x}))\cdot\dot{\mathbf{A}}(\mathbf{x},t)
\tag{6.81}
$$

has the perturbation expansion in the time-dependent external field, $\mathbf{A}$,

$$
\frac{dE}{dt} = \frac{-i}{\hbar}\sum_{\alpha\beta}\int d\mathbf{x}\int d\mathbf{x}'\int_{t_i}^{t}dt'\,\dot{A}_\alpha(\mathbf{x},t)\,\langle[j_\alpha^{\mathrm{p}}(\mathbf{x},t),j_\beta^{\mathrm{p}}(\mathbf{x}',t')]\rangle_0\,A_\beta(\mathbf{x}',t')
$$

$$
-\int d\mathbf{x}\,\langle\mathbf{j}_t(\mathbf{x})\rangle_0\cdot\dot{\mathbf{A}}(\mathbf{x},t)+\mathcal{O}(A^3)
$$

$$
= -\sum_{\alpha\beta}\int d\mathbf{x}\int d\mathbf{x}'\int_{t_i}^{\infty}dt'\,\dot{A}_\alpha(\mathbf{x},t)\,Q_{\alpha\beta}(\mathbf{x},t;\mathbf{x}',t')\,A_\beta(\mathbf{x}',t')
$$

$$
-\int d\mathbf{x}\,\langle\mathbf{j}_t(\mathbf{x})\rangle_0\cdot\dot{\mathbf{A}}(\mathbf{x},t)+\mathcal{O}(A^3)\,.
\tag{6.82}
$$

The dot signifies differentiation with respect to time. We recall that the equilibrium current, $\langle\mathbf{j}_t(\mathbf{x})\rangle_0$, is in fact time independent.

An external field therefore performs, in the time span between t_i and t_f, the work

$$
W \equiv E(t_f) - E(t_i)
$$

$$
= -\sum_{\alpha\beta}\int_{t_i}^{t_f}dt\int d\mathbf{x}\int d\mathbf{x}'\int_{t_i}^{\infty}dt'\,\dot{A}_\alpha(\mathbf{x},t)\,K_{\alpha\beta}(\mathbf{x},t;\mathbf{x}',t')\,A_\beta(\mathbf{x}',t')
$$

$$
= \sum_{\alpha\beta}\int_{t_i}^{t_f}dt\int d\mathbf{x}\int d\mathbf{x}'\int_{t_i}^{\infty}dt'\,A_\alpha(\mathbf{x},t)\,\frac{dK_{\alpha\beta}(\mathbf{x},t;\mathbf{x}',t')}{dt}\,A_\beta(\mathbf{x}',t')
$$

$$
+ \mathcal{O}(A^3)\,.
\tag{6.83}
$$

In the first equality we have noticed that the diamagnetic term in the response function Q does not contribute. For the second equality we have assumed that the vector potential vanishes in the past and in the future (i.e. the time average of the electric field is zero), so that the boundary terms vanish, and we observe that in that case there is no linear contribution; to linear order the energy of the system is unchanged.

For an isotropic system we have

$$K_{\alpha\beta}(\mathbf{x}, \mathbf{x}', \omega) = K(\mathbf{x}, \omega)\, \delta(\mathbf{x} - \mathbf{x}')\, \delta_{\alpha\beta} \tag{6.84}$$

and we obtain, in view of Eq. (6.79), the result that the mean change in energy of the system to second order is positive

$$\Delta E \equiv W = \int d\mathbf{x} \int_{-\infty}^{\infty} \frac{d\omega}{2\pi}\, \omega\, \Im m\, K(\mathbf{x}, \omega)\, \mathbf{A}(\mathbf{x}, \omega) \cdot \mathbf{A}^*(\mathbf{x}, -\omega) \geq 0. \tag{6.85}$$

Interacting weakly with the physical quantities of a system in thermal equilibrium through a classical field, which vanishes in the past and in the future, can thus only lead to an increase in the energy of the system; the energy never decreases. The thermodynamic equilibrium state is thus a stable state.[12]

In the case of a monochromatic field

$$\mathbf{A}(\mathbf{x}, t) = \frac{1}{2}\left(\mathbf{A}(\mathbf{x}, \omega)e^{-i\omega t} + \mathbf{A}^*(\mathbf{x}, \omega)e^{i\omega t}\right) = \Re e\left(\mathbf{A}(\mathbf{x}, \omega)e^{-i\omega t}\right) \tag{6.86}$$

we have for the mean rate of change of the energy to second order in the applied field, $T \equiv 2\pi/\omega$,

$$\overline{\frac{dE_\omega}{dt}}^T \equiv \frac{1}{T}\int_0^T dt\, \frac{dE}{dt} = \frac{-i}{4\hbar}\frac{1}{T}\sum_{\alpha\beta}\int_0^T dt \int_{t_1}^t dt' \int d\mathbf{x} \int d\mathbf{x}'\, \dot{A}_\alpha(\mathbf{x}, t)$$

$$\times\ \langle[j_\alpha^P(\mathbf{x}, t), j_\beta^P(\mathbf{x}', t')]\rangle_0\, A_\beta(\mathbf{x}', t') \tag{6.87}$$

as the diamagnetic term averages in time to zero. Turning the field on in the far past, $t_i \to -\infty$, we have in terms of the response function

$$\overline{\frac{dE_\omega}{dt}}^T = \frac{-i\omega}{4}\sum_{\alpha\beta}\int d\mathbf{x} \int d\mathbf{x}'\, A_\alpha^*(\mathbf{x}, \omega)\left(K_{\alpha\beta}(\mathbf{x}, \mathbf{x}', \omega) - K_{\beta\alpha}(\mathbf{x}', \mathbf{x}, -\omega)\right) A_\beta(\mathbf{x}', \omega)$$

$$= \frac{\omega}{2}\sum_{\alpha\beta}\int d\mathbf{x} \int d\mathbf{x}'\, A_\alpha^*(\mathbf{x}, \omega)\, \Im m K_{\alpha\beta}(\mathbf{x}, \mathbf{x}', \omega)\, A_\beta(\mathbf{x}', \omega). \tag{6.88}$$

We can, according to Eq. (6.74), rewrite the average work performed by the external field in the form

$$\overline{\frac{dE_\omega}{dt}}^T = \sum_\lambda \hbar\omega\, \rho(\epsilon_\lambda)\left(P_\lambda(\hbar\omega) - P_\lambda(-\hbar\omega)\right), \tag{6.89}$$

---

[12]It is important to stress the crucial role of the canonical (or grand canonical) ensemble for the validity of Eq. (6.79).

where

$$
P_\lambda(\hbar\omega) \; = \; \frac{2\pi}{\hbar} \sum_{\lambda'} \left| \frac{1}{2} \int d\mathbf{x} \, \langle\lambda'| \, \mathbf{j}_\mathrm{p}(\mathbf{x}) \cdot \mathbf{A}(\mathbf{x},\omega)|\lambda\rangle \right|^2 \delta(\epsilon_\lambda - \epsilon_{\lambda'} + \hbar\omega) \tag{6.90}
$$

is Fermi's Golden Rule expression for the probability for the transition from state λ to any state λ' in which the system absorbs the amount $\hbar\omega$ of energy from the external field, and $P_\lambda(-\hbar\omega)$ is the transition probability for emission of the amount $\hbar\omega$ of energy to the external field. The equation for the change in energy is thus a master equation for the energy, and we infer that the energy exchange between a system and a classical field oscillating at frequency ω takes place in lumps of magnitude $\hbar\omega$.

At each frequency we have for the average work done on the system by the external field:

$$
\overline{\frac{dE_\omega}{dt}}^T \; = \; \frac{1}{2} \sum_{\alpha\beta} \int d\mathbf{x} \int d\mathbf{x}' \; E_\alpha^*(\mathbf{x},\omega) \, \Re e \sigma_{\alpha\beta}(\mathbf{x},\mathbf{x}',\omega) \, E_\beta(\mathbf{x}',\omega) \,, \tag{6.91}
$$

where we have utilized Eq. (6.50) to introduce the real part of the conductivity tensor.

For a translational invariant system we have

$$
\sigma_{\alpha\beta}(\mathbf{x},\mathbf{x}',\omega) \; = \; \sigma_{\alpha\beta}(\mathbf{x}-\mathbf{x}',\omega) \; = \; \frac{1}{V} \sum_{\mathbf{q}} e^{i\mathbf{q}\cdot(\mathbf{x}-\mathbf{x}')} \, \sigma_{\alpha\beta}(\mathbf{q},\omega) \tag{6.92}
$$

and we get for each wave vector

$$
E_\alpha(\mathbf{x},\omega) = E_\alpha(\mathbf{q},\omega) \, e^{i\mathbf{q}\cdot\mathbf{x}} \tag{6.93}
$$

the contribution

$$
\overline{\frac{dE_{\mathbf{q}\omega}}{dt}}^T \; = \; \frac{V}{2} \sum_{\alpha\beta} E_\alpha^*(\mathbf{q},\omega) \, \Re e \sigma_{\alpha\beta}(\mathbf{q},\omega) \, E_\beta(\mathbf{q},\omega) \,. \tag{6.94}
$$

Each harmonic contributes additively, and we get for the average energy absorption for arbitrary spatial dependence of the electric field the expression

$$
\overline{\frac{dE_\omega}{dt}}^T \; = \; \frac{V}{2} \sum_{\alpha\beta} \sum_{\mathbf{q}} E_\alpha^*(\mathbf{q},\omega) \, \Re e \sigma_{\alpha\beta}(\mathbf{q},\omega) \, E_\beta(\mathbf{q},\omega) \,. \tag{6.95}
$$

For an isotropic system the conductivity tensor is diagonal

$$
\sigma_{\alpha\beta}(\mathbf{x},\mathbf{x}',\omega) \; = \; \delta_{\alpha\beta} \, \sigma(\mathbf{x}-\mathbf{x}',\omega) \tag{6.96}
$$

and we have

$$
\overline{\frac{dE_\omega}{dt}}^T \; = \; \frac{V}{2} \sum_{\alpha} |E_\alpha(\mathbf{q},\omega)|^2 \, \Re e \sigma_{\alpha\alpha}(\mathbf{q},\omega) \,. \tag{6.97}
$$

For the spatially homogeneous field case, $E_\alpha(\mathbf{q} \neq \mathbf{0}, \omega) = 0$, we then obtain

$$\overline{\frac{dE_\omega}{dt}}^T = \frac{V}{2} \sum_\alpha |E_\alpha^*(\mathbf{q} = \mathbf{0}, \omega)|^2 \sum_{\mathbf{q}} \Re e \, \sigma_{\alpha\alpha}(\mathbf{q}, \omega) . \tag{6.98}$$

Since

$$\frac{1}{V} \sum_{\mathbf{q}} \Re e \, \sigma_{\alpha\alpha}(\mathbf{q}, \omega) = \Re e \, \sigma_{\alpha\alpha}(\mathbf{x}, \mathbf{x}, \omega) = \frac{1}{\omega} \Im m K_{\alpha\alpha}(\mathbf{x}, \mathbf{x}, \omega) \geq 0 \tag{6.99}$$

we obtain the result that, for a system in thermal equilibrium, the average change in energy can only be *increased* by interaction with a weak periodic external field[13]

$$\overline{\frac{dE_\omega}{dt}}^T \geq 0 . \tag{6.100}$$

The thermal state is stable against a weak periodic perturbation.[14]

Considering the isotropic d.c. case we get directly from Eq. (6.91) the familiar Joule heating expression for the energy absorbed per unit time in a resistor biased by voltage $\mathcal{U}$

$$\overline{\frac{dE}{dt}}^T = \frac{1}{2} G \mathcal{U}^2 = \frac{1}{2} R I^2 , \tag{6.101}$$

where R is the resistance, the inverse conductance, $R \equiv G^{-1}$, and we have used the fact that in the d.c. case the imaginary part of the conductance tensor vanishes.

The absorbed energy of a system in thermal equilibrium interacting with an external field is dissipated in the system, and we thus note that $\Re e \, \sigma$, or equivalently $\Im m \, K$, describes the dissipation in the system.

6.5 Fluctuation–dissipation theorem

The most important hallmark of linear response is the relation between equilibrium fluctuations and dissipation. We shall illustrate this feature by again considering the current response function; however, the argument is equivalent for any correlation function. We introduce the current correlation function in the thermal equilibrium state

$$\tilde{K}_{\alpha\beta}^{(j)}(\mathbf{x}, t; \mathbf{x}', t') \equiv \frac{1}{2} \langle \{\delta j_\alpha^{\mathrm{p}}(\mathbf{x}, t), \delta j_\beta^{\mathrm{p}}(\mathbf{x}', t')\} \rangle_0 , \tag{6.102}$$

where

$$\delta j_\alpha^{\mathrm{p}}(\mathbf{x}, t) \equiv j_\alpha^{\mathrm{p}}(\mathbf{x}, t) - \langle j_\alpha^{\mathrm{p}}(\mathbf{x}, t) \rangle_0 \tag{6.103}$$

is the deviation from a possible equilibrium current, $\langle \mathbf{j}^{\mathrm{p}}(\mathbf{x}, t) \rangle_0$, which in fact is independent of time. However, for notational simplicity we assume in the following that

---

[13]In fact, from the positivity of $\Re e \, \sigma_{\alpha\alpha}(\mathbf{q}, \omega)$ for arbitrary wave vector we find that the conclusion is valid for arbitrary spatially varying external field.

[14]Since this result is valid at any frequency, we again obtain the result that a system in thermal equilibrium is stable.

the equilibrium current density vanishes. By taking the anti-commutator, we have symmetrized the correlation function, and since the current operator is hermitian, the correlation function is a real function.

Since the statistical average is with respect to the equilibrium state (for an arbitrary Hamiltonian H), we have on account of the cyclic property of the trace

$$
\begin{aligned}
K_{\alpha\beta}^{>}(\mathbf{x}, t; \mathbf{x}', t') &\equiv \operatorname{Tr}\left(e^{-H/kT} j_\alpha^{\mathrm{p}}(\mathbf{x}, t)\, j_\beta^{\mathrm{p}}(\mathbf{x}', t')\right) \equiv \langle j_\alpha^{\mathrm{p}}(\mathbf{x}, t)\, j_\beta^{\mathrm{p}}(\mathbf{x}', t')\rangle_0 \\
&= \operatorname{Tr}\left(e^{-H/kT} j_\beta^{\mathrm{p}}(\mathbf{x}', t')\, j_\alpha^{\mathrm{p}}(\mathbf{x}, t + i\hbar/kT)\right) \\
&= K_{\alpha\beta}^{<}(\mathbf{x}, t + i\hbar/kT; \mathbf{x}', t')
\end{aligned}
\tag{6.104}
$$

as we define

$$
K_{\alpha\beta}^{<}(\mathbf{x}, t; \mathbf{x}', t') \equiv \operatorname{Tr}\left(e^{-H/kT} j_\beta^{\mathrm{p}}(\mathbf{x}', t')\, j_\alpha^{\mathrm{p}}(\mathbf{x}, t)\right) \equiv \langle j_\beta^{\mathrm{p}}(\mathbf{x}', t')\, j_\alpha^{\mathrm{p}}(\mathbf{x}, t)\rangle_0 \, . \tag{6.105}
$$

We note the crucial role played by the assumption of a (grand) canonical ensemble.

We assume the canonical ensemble average exists for all real times t and t', and consequently $K^{<}$ is an analytic function in the region $0 < \Im m(t - t') < \hbar/kT$, and $K^{>}$ is analytic in the region $-\hbar/kT < \Im m(t - t') < 0$. For the Fourier transforms we therefore obtain the relation

$$
K_{\alpha\beta}^{>}(\mathbf{x}, \mathbf{x}'; \omega) = e^{-\hbar\omega/kT}\, K_{\alpha\beta}^{<}(\mathbf{x}, \mathbf{x}'; \omega) \, . \tag{6.106}
$$

We observe the following relation of the commutator to the retarded and advanced correlation functions

$$
\begin{aligned}
K_{\alpha\beta}^{>}(\mathbf{x}, t; \mathbf{x}', t') - K_{\alpha\beta}^{<}(\mathbf{x}, t; \mathbf{x}', t') &= \langle [j_\alpha^{\mathrm{p}}(\mathbf{x}, t), j_\beta^{\mathrm{p}}(\mathbf{x}', t')]\rangle_0 \\
&= -i\hbar \left(K_{\alpha\beta}^{\mathrm{R}}(\mathbf{x}, t; \mathbf{x}', t') - K_{\alpha\beta}^{\mathrm{A}}(\mathbf{x}, t; \mathbf{x}', t')\right) \, ,
\end{aligned}
\tag{6.107}
$$

where we have introduced the advanced correlation function

$$
K_{\alpha\beta}^{\mathrm{A}}(\mathbf{x}, t; \mathbf{x}', t') = -\frac{i}{\hbar}\, \theta(t' - t)\, \langle [j_\alpha^{\mathrm{p}}(\mathbf{x}, t), j_\beta^{\mathrm{p}}(\mathbf{x}', t')]\rangle_0 \tag{6.108}
$$

corresponding to the retarded one appearing in the current response, Eq. (6.31),

$$
K_{\alpha\beta}^{\mathrm{R}}(\mathbf{x}, t; \mathbf{x}', t') \equiv K_{\alpha\beta}(\mathbf{x}, t; \mathbf{x}', t') \, . \tag{6.109}
$$

Since the response function involves the commutator of two hermitian operators we immediately verify that (for ω real)

$$
K_{\alpha\beta}^{\mathrm{R(A)}}(\mathbf{x}, \mathbf{x}', -\omega) = [K_{\alpha\beta}^{\mathrm{R(A)}}(\mathbf{x}, \mathbf{x}', \omega)]^{*} \, . \tag{6.110}
$$

Analogous to Eq. (6.105) we have for the correlation function, the anti-commutator,

$$
\langle \{j_\alpha^{\mathrm{p}}(\mathbf{x}, t), j_\beta^{\mathrm{p}}(\mathbf{x}', t')\}\rangle_0 = K_{\alpha\beta}^{>}(\mathbf{x}, t; \mathbf{x}', t') + K_{\alpha\beta}^{<}(\mathbf{x}, t; \mathbf{x}', t') \, . \tag{6.111}
$$

Using Eq. (6.105) we can rewrite

$$\tilde{K}_{\alpha\beta}^{(j)}(\mathbf{x}, \mathbf{x}', \omega) = \frac{1}{2} K_{\alpha\beta}^{>}(\mathbf{x}, \mathbf{x}', \omega) \left(1 + e^{\hbar\omega/kT}\right)$$

$$= \left(\frac{1}{2}\left(K_{\alpha\beta}^{>}(\mathbf{x}, \mathbf{x}', \omega) + K_{\alpha\beta}^{<}(\mathbf{x}, \mathbf{x}', \omega)\right) - \frac{1}{2}\left(K_{\alpha\beta}^{>}(\mathbf{x}, \mathbf{x}', \omega) - K_{\alpha\beta}^{<}(\mathbf{x}, \mathbf{x}', \omega)\right)\right)$$

$$\times \frac{1}{2}\left(1 + e^{\hbar\omega/kT}\right) \qquad (6.112)$$

and thereby[15]

$$\tilde{K}_{\alpha\beta}^{(j)}(\mathbf{x}, \mathbf{x}', \omega) = \left(K_{\alpha\beta}^{R}(\mathbf{x}, \mathbf{x}', \omega) - K_{\alpha\beta}^{A}(\mathbf{x}, \mathbf{x}', \omega)\right) \frac{\hbar}{2i} \coth\frac{\hbar\omega}{2kT}. \qquad (6.113)$$

Using Eq. (6.113), and noting that for omega real (we establish this as a consequence of time-reversal invariance in the next section)

$$K_{\alpha\beta}^{A}(\mathbf{x}, \mathbf{x}', \omega) = [K_{\alpha\beta}^{R}(\mathbf{x}, \mathbf{x}', \omega)]^{*} \qquad (6.114)$$

we then get the relation between the correlation function and the imaginary part of the response function

$$\tilde{K}_{\alpha\beta}^{(j)}(\mathbf{x}, \mathbf{x}', \omega) = \hbar \coth\frac{\hbar\omega}{2kT} \Im m K_{\alpha\beta}(\mathbf{x}, \mathbf{x}', \omega). \qquad (6.115)$$

We have established the relationship between the imaginary part of the linear response function, governing according to Eq. (6.88) the dissipation in the system, and the equilibrium fluctuations, the fluctuation–dissipation theorem.[16]

According to the fluctuation–dissipation theorem we can express the change in energy of a system in an external field of frequency ω, Eq. (6.88), in terms of the current fluctuations

$$\overline{\frac{dE_{\omega}}{dt}}^{T} = \frac{1}{2\hbar\omega \coth\frac{\hbar\omega}{2kT}} \sum_{\alpha\beta} \int d\mathbf{x} \int d\mathbf{x}' \, E_{\alpha}^{*}(\mathbf{x}, \omega) \, \tilde{K}_{\alpha\beta}^{j}(\mathbf{x}, \mathbf{x}', \omega) E_{\beta}(\mathbf{x}', \omega). \qquad (6.116)$$

For the current fluctuations we have (recall Eq. (6.50))

$$\frac{1}{2}\langle\{j_{\alpha}^{P}(\mathbf{x}, \omega), j_{\beta}^{P}(\mathbf{x}', -\omega)\}\rangle = \tilde{K}_{\alpha\beta}^{(j)}(\mathbf{x}, \mathbf{x}', \omega)$$

---

[15]If we introduced

$$K_{\alpha\beta}^{K}(\mathbf{x}, t; \mathbf{x}', t') = \frac{i}{\hbar}\langle\{\delta j_{\alpha}^{P}(\mathbf{x}, t), \delta j_{\beta}^{P}(\mathbf{x}', t')\}\rangle_{0} = 2\frac{i}{\hbar}\tilde{K}_{\alpha\beta}^{(j)}(\mathbf{x}, t; \mathbf{x}', t')$$

we would be in accordance with the standard notation of the book.

[16]Formally the fluctuation–dissipation theorem expresses the relationship between a commutator and anti-commutator canonical equilibrium average. The fluctuation–dissipation relation, Eq. (6.115), is also readily established by comparing the spectral representation of the imaginary part of the retarded current response function, Eq. (6.37), with that of $\tilde{K}^{(j)}$. The fluctuation–dissipation relationship expresses that the system is in equilibrium and described by the canonical ensemble.

$$= \hbar\omega \coth \frac{\hbar\omega}{2kT} \, \Re e \, \sigma_{\alpha\beta}(\mathbf{x}, \mathbf{x}', \omega) \qquad (6.117)$$

and the equal-time current fluctuations are specified by

$$\tilde{K}_{\alpha\beta}^{(j)}(\mathbf{x}, t; \mathbf{x}', t) = \int_{-\infty}^{\infty} \frac{d\omega}{2\pi} \, \tilde{K}_{\alpha\beta}^{(j)}(\mathbf{x}, \mathbf{x}', \omega) = \hbar \int_{-\infty}^{\infty} \frac{d\omega}{2\pi} \, \coth \frac{\hbar\omega}{2kT} \, \Im m K_{\alpha\beta}(\mathbf{x}, \mathbf{x}', \omega)$$

$$(6.118)$$

and Eq. (6.79) guarantees the positivity of the equal-time and space current density fluctuations.

In a macroscopic description we have a local relationship between field and current density, Ohm's law,

$$j_\alpha(\mathbf{x}, \omega) = \sigma_{\alpha\beta}(\mathbf{x}, \omega) \, E_\beta(\mathbf{x}, \omega) \qquad (6.119)$$

or equivalently

$$\sigma_{\alpha\beta}(\mathbf{x}, \mathbf{x}', \omega) = \sigma_{\alpha\beta}(\mathbf{x}, \omega) \, \delta(\mathbf{x} - \mathbf{x}') \, . \qquad (6.120)$$

The equilibrium current density fluctuations at point $\mathbf{x}$ are then specified by

$$\langle j_\alpha^2 \rangle_{\mathbf{x}\omega} \equiv \frac{1}{2V} \int d(\mathbf{x} - \mathbf{x}') \, \langle \{ j_\alpha^{\mathrm{p}}(\mathbf{x}, \omega), j_\alpha^{\mathrm{p}}(\mathbf{x}', -\omega) \} \rangle_0$$

$$= \frac{1}{V} \, \tilde{K}_{\alpha\alpha}^{(j)}(\mathbf{x}, \omega) = \frac{1}{V} \, \hbar\omega \coth \frac{\hbar\omega}{2kT} \, \Re e \, \sigma_{\alpha\alpha}(\mathbf{x}, \omega) \, . \qquad (6.121)$$

We note that the factor

$$\frac{\hbar\omega}{2} \coth \frac{\hbar\omega}{2kT} = \hbar\omega \left(n(\omega) + \frac{1}{2} \right) \qquad (6.122)$$

is the average energy of a harmonic oscillator, with frequency ω, in the thermal state. The average energy consists of a thermal contribution described by the Bose function, and a zero-point quantum fluctuation contribution.

In the high-temperature limit where relevant frequencies are small compared to the temperature, $\hbar\omega \ll kT$, we get for the current fluctuations in a homogeneous conductor with conductivity σ, Johnson noise,

$$\langle j_\alpha^2 \rangle_\omega = \frac{2kT\sigma}{V} \qquad (6.123)$$

independent of the specific nature of the conductor.

In the linear response treatment we have assumed the field fixed, and studied the fluctuations in the current density. However, fluctuations in the current (or charge) density gives rise to fluctuations in the electromagnetic field as well. As an example of using the fluctuation–dissipation theorem we therefore turn the point of view around, and using Ohm's law obtain that the (longitudinal) electric field fluctuations are given by

$$\langle E_\alpha^2 \rangle_{\mathbf{x}\omega} = \frac{1}{|\sigma(\mathbf{x}, \omega)|^2} \, \langle j_\alpha^2 \rangle_{\mathbf{x}\omega} \, . \qquad (6.124)$$

According to Eq. (6.121) we then obtain for the (longitudinal) electric field fluctuations

$$\langle E_\alpha^2 \rangle_{\mathbf{x}\omega} = \frac{1}{V}\, \hbar\omega \coth \frac{\hbar\omega}{2kT}\, \frac{\Re e\sigma_{\alpha\alpha}(\mathbf{x},\omega)}{|\sigma_{\alpha\alpha}(\mathbf{x},\omega)|^2}. \tag{6.125}$$

In the high-temperature limit, $\hbar\omega \ll kT$, we have for the (longitudinal) electric field fluctuations, Nyquist noise,

$$\langle E_\alpha^2 \rangle_\omega = \frac{2kT}{\sigma V}. \tag{6.126}$$

6.6 Time-reversal symmetry

Hermitian operators will by suitable phase choice have a definite sign under time reversal: position and electric field have positive sign, and velocity and magnetic field have negative sign.[17] The following considerations can be performed for any pair of operators (see Exercise 6.2), but we shall for definiteness consider the current operator, and show that Eq. (6.114) is a consequence of time-reversal invariance.

In case the Hamiltonian is time-reversal invariant,

$$\left(T\,[j_\alpha^{\mathrm{P}}(\mathbf{x},t), j_\beta^{\mathrm{P}}(\mathbf{x}',t')]\, T^\dagger \right)^\dagger = [T\,j_\beta^{\mathrm{P}}(\mathbf{x}',t')\, T^\dagger, T\,j_\alpha^{\mathrm{P}}(\mathbf{x},t)\, T^\dagger]$$

$$= -\,[j_\alpha^{\mathrm{P}}(\mathbf{x},-t), j_\beta^{\mathrm{P}}(\mathbf{x}',-t')] \tag{6.127}$$

and

$$\langle\psi|[j_\alpha^{\mathrm{P}}(\mathbf{x},t), j_\beta^{\mathrm{P}}(\mathbf{x}',t')]|\psi\rangle = \langle T\psi|T\,[j_\alpha^{\mathrm{P}}(\mathbf{x},t), j_\beta^{\mathrm{P}}(\mathbf{x}',t')]^\dagger T^\dagger|T\psi\rangle, \tag{6.128}$$

where $|T\psi\rangle$ is the time-reversed state of $|\psi\rangle$. Consequently,

$$\mathrm{Tr}(\rho(H)[j_\alpha^{\mathrm{P}}(\mathbf{x},-t), j_\beta^{\mathrm{P}}(\mathbf{x}',-t')]) = -\mathrm{Tr}(\rho(H)[j_\alpha^{\mathrm{P}}(\mathbf{x},t), j_\beta^{\mathrm{P}}(\mathbf{x}',t')]) \tag{6.129}$$

and we therefore find that time-reversal invariance implies

$$K_{\alpha\beta}^{\mathrm{R}}(\mathbf{x},\mathbf{x}';\omega) = \left[K_{\alpha\beta}^{\mathrm{A}}(\mathbf{x},\mathbf{x}';\omega)\right]^* \tag{6.130}$$

i.e. we have established Eq. (6.114).

Exercise 6.2. Consider two physical quantities represented by the operators $A_1(\mathbf{x},t)$ and $A_2(\mathbf{x},t)$, which transform under time reversal according to

$$T\,A_i(\mathbf{x},t)\,T^\dagger = s_i\,A_i(\mathbf{x},-t) \quad,\quad s_i = \pm 1,\ i = 1,2. \tag{6.131}$$

Show that when the Hamiltonian is invariant under time reversal, the response function

$$A_{ij}(\mathbf{x},\mathbf{x}',t-t') \equiv \mathrm{Tr}(\rho(H)[A_i(\mathbf{x},t), A_j(\mathbf{x}',t')]) \tag{6.132}$$

---

[17]For a discussion of time-reversal symmetry we refer the reader to chapter 2 of reference [1].

satisfies the relations

$$A_{ij}(\mathbf{x}, \mathbf{x}', t - t') = -s_i s_j A_{ij}(\mathbf{x}, \mathbf{x}', t' - t) = s_i s_j A_{ij}(\mathbf{x}', \mathbf{x}, t - t') \qquad (6.133)$$

and thereby[18]

$$A_{ij}(\mathbf{x}, \mathbf{x}', \omega) = -s_i s_j A_{ij}(\mathbf{x}, \mathbf{x}', -\omega) = s_i s_j A_{ij}(\mathbf{x}', \mathbf{x}, \omega) . \qquad (6.134)$$

6.7   Scattering and correlation functions

In measurements on macroscopic bodies only very crude information of the microscopic state is revealed. For example, in a measurement of the current only the conductance is revealed and not any of the complicated spatial structure of the conductivity. To reveal the whole structure of a correlation function takes a more individualized source than that provided by a battery. It takes a particle source such as the one used in a scattering experiment, using for example neutrons from a spallation source.

In this section we shall consider transport of particles (neutrons, photons, etc.) through matter. To be specific we consider the scattering of slow neutrons by a piece of matter. A neutron interacts with the nuclei of the substance (all assumed identical). The interaction potential is short ranged, and we take for the interaction with the nucleus at position $\mathbf{R}_N$[19]

$$V(\mathbf{r}_n - \mathbf{R}_N) = a\, \delta(\mathbf{r}_n - \mathbf{R}_N) . \qquad (6.135)$$

We have thus neglected the spin of the nuclei (or consider the case of spin-less bosons).[20] For the interaction of a neutron with the nuclei of the substance we then have

$$V(\mathbf{r}_n) = \sum_N V(\mathbf{r}_n - \mathbf{R}_N) = a \sum_N \delta(\mathbf{r}_n - \mathbf{R}_N) . \qquad (6.136)$$

The interaction is weak, and the scattering can be treated in the Born approximation. For the transition rate between initial and final states we then have

$$\Gamma_{\text{fi}} = \frac{2\pi}{\hbar} \, |\langle f| \sum_N V(\hat{\mathbf{r}}_n - \hat{\mathbf{R}}_N)|i\rangle|^2 \, \delta(E_{\text{f}} - E_{\text{i}}) . \qquad (6.137)$$

For simplicity we assume that the states of the substance can be labeled solely by their energy

$$|i\rangle \;\doteq\; |\mathbf{p}', E_{\text{S}}^{(i)}\rangle \;=\; |\mathbf{p}'\rangle|E_{\text{S}}^{(i)}\rangle , \qquad\qquad |f\rangle \;\doteq\; |\mathbf{p}, E_{\text{S}}^{(f)}\rangle \;=\; |\mathbf{p}\rangle|E_{\text{S}}^{(f)}\rangle , \qquad (6.138)$$

---

[18] If the Hamiltonian contains a term coupling to a magnetic field, the symmetry of the correlation function is $A_{ij}(\mathbf{x}, \mathbf{x}', \omega, \mathbf{B}) = -s_i s_j A_{ij}(\mathbf{x}, \mathbf{x}', -\omega, -\mathbf{B}) = s_i s_j A_{ji}(\mathbf{x}', \mathbf{x}, \omega, -\mathbf{B})$.

[19] We thus exclude the possibility of any nuclear reaction taking place.

[20] However, it is precisely the magnetic moment of the neutron that makes it an ideal tool to investigate the magnetic properties of matter. The subject of neutron scattering is thus vast, and for a general reference we refer the reader to reference [21].

where the initial and final energies are

$$E_f = E_S^{(f)} + \frac{\mathbf{p}^2}{2m_n} , \qquad E_i = E_S^{(i)} + \frac{\mathbf{p}'^2}{2m_n} \qquad (6.139)$$

and m_n is the mass of the neutron. We introduce the energy transfer from the neutron to the material

$$\hbar\omega = \frac{\mathbf{p}'^2}{2m_n} - \frac{\mathbf{p}^2}{2m_n} = E_S^{(f)} - E_S^{(i)} \qquad (6.140)$$

and we have for the transition probability per unit time

$$\Gamma_{fi} = \frac{2\pi}{\hbar} |\langle \mathbf{p}, E_S^{(f)} | a \sum_N \delta(\hat{\mathbf{r}}_n - \hat{\mathbf{R}}_N) | \mathbf{p}', E_S^{(i)} \rangle|^2 \, \delta(E_S^{(f)} - E_S^{(i)} - \hbar\omega) . \qquad (6.141)$$

Since the interaction is inelastic, the differential cross section of interest, $d^2\sigma/d\hat{\mathbf{p}} \, d\epsilon$, is the fraction of incident neutrons with momentum $\mathbf{p}'$ being scattered into a unit solid angle $d\hat{\mathbf{p}}$ with energy in the range between ϵ and $\epsilon + d\epsilon$. Noting that

$$\Delta\mathbf{p} = p^2 \, dp \, d\hat{\mathbf{p}} = m_n \, p \, d\epsilon \, d\hat{\mathbf{p}} \qquad (6.142)$$

we obtain for the inelastic differential cross section for neutron scattering off the substance

$$\frac{d^2\sigma}{d\hat{\mathbf{p}} \, d\epsilon} = \frac{m_n^2 L^6}{(2\pi\hbar)^3} \frac{p}{p'} \frac{2\pi}{\hbar} |\langle \mathbf{p}, E_S^{(f)} | a \sum_N \delta(\hat{\mathbf{r}}_n - \hat{\mathbf{R}}_N) | \mathbf{p}', E_S^{(i)} \rangle|^2$$

$$\times \quad \delta(E_S^{(f)} - E_S^{(i)} - \hbar\omega) \qquad (6.143)$$

which we can express as

$$\frac{d^2\sigma}{d\hat{\mathbf{p}} \, d\epsilon} = \frac{m_n^2 a^2}{(2\pi\hbar)^3} \frac{p}{p'} \frac{2\pi}{\hbar} \int d\mathbf{x} \int d\mathbf{x}' \, e^{-\frac{i}{\hbar}(\mathbf{x}-\mathbf{x}')\cdot(\mathbf{p}-\mathbf{p}')} \int_{-\infty}^{\infty} \frac{d(t-t')}{2\pi\hbar} \, e^{-i(t-t')\omega}$$

$$\times \quad \langle E_S^{(f)} | n(\mathbf{x},t) | E_S^{(i)} \rangle \langle E_S^{(i)} | n(\mathbf{x}',t') | E_S^{(f)} \rangle , \qquad (6.144)$$

where $n(\mathbf{x},t)$ is the density operator for the nuclei of the material in the Heisenberg picture with respect to the substance Hamiltonian $\hat{H}_S$.

Exercise 6.3. Show that, for scattering off a single heavy nucleus, $M \gg m_n$, we have for the total cross section

$$\sigma = \int_{4\pi} d\hat{\mathbf{p}} \int_0^{\infty} d\epsilon \, \frac{d^2\sigma}{d\hat{\mathbf{p}} \, d\epsilon} = 4\pi \left(\frac{m_n \, a}{2\pi\hbar^2}\right)^2 . \qquad (6.145)$$

In the scattering experiment we know only the probability distribution for the initial state of the material, which we shall assume to be the thermal equilibrium state

$$\rho_S = \sum_\lambda |E_S(\lambda)\rangle\, P(E_S(\lambda))\, \langle E_S(\lambda)| , \tag{6.146}$$

where

$$P(E_S(\lambda)) = \frac{e^{-E_S(\lambda)/kT}}{Z_S} , \quad H_S\, |E_S(\lambda)\rangle = E_S(\lambda)\, |E_S(\lambda)\rangle . \tag{6.147}$$

For the transition rate weighted over the thermal mixture of initial states of the substance we have

$$\Gamma_{\mathbf{f}\mathbf{p}'} \equiv \sum_\lambda P(E_S^{(i)}(\lambda))\, \Gamma_{\mathrm{fi}}$$

$$= \frac{m_n^2 a^2}{(2\pi\hbar)^3}\, \frac{p}{p'}\, \frac{2\pi}{\hbar} \sum_\lambda P(E_S(\lambda)) \int d\mathbf{x} \int d\mathbf{x}' \int\limits_{-\infty}^{\infty} \frac{d(t-t')}{2\pi\hbar}\, e^{-i(t-t')\omega}$$

$$\times\; e^{-\frac{i}{\hbar}(\mathbf{x}-\mathbf{x}')\cdot(\mathbf{p}-\mathbf{p}')}\langle E_S^{(f)}|n(\mathbf{x},t)|E_S(\lambda)\rangle\, \langle E_S(\lambda)|n(\mathbf{x}',t')|E_S^{(f)}\rangle \tag{6.148}$$

and we obtain for the weighted differential cross section (we use the same notation)

$$\frac{d^2\sigma}{d\hat{\mathbf{p}}\,d\epsilon} = \sum_\lambda P(E_S^{(i)}(\lambda))\, \frac{d^2\sigma}{d\hat{\mathbf{p}}\,d\epsilon}$$

$$= \frac{m_n^2 a^2}{(2\pi\hbar)^3}\, \frac{p}{p'}\, \frac{2\pi}{\hbar} \sum_\lambda P(E_S(\lambda)) \int d\mathbf{x} \int d\mathbf{x}' \int\limits_{-\infty}^{\infty} \frac{d(t-t')}{2\pi\hbar}\, e^{-i(t-t')\omega}$$

$$\times\; e^{-\frac{i}{\hbar}(\mathbf{x}-\mathbf{x}')\cdot(\mathbf{p}-\mathbf{p}')}\langle E_S^{(f)}|n(\mathbf{x},t)|E_S(\lambda)\rangle\langle E_S(\lambda)|n(\mathbf{x}',t')|E_S^{(f)}\rangle . \tag{6.149}$$

Furthermore, in the experiment the final state of the substance is not measured, and we must sum over all possible final states of the substance, and we obtain finally for the observed differential cross section (we use the same notation)

$$\frac{d^2\sigma}{d\hat{\mathbf{p}}\,d\epsilon} = \frac{m_n^2 a^2}{(2\pi\hbar)^3}\, \frac{p}{p'}\, \frac{2\pi}{\hbar} \int d\mathbf{x} \int d\mathbf{x}'\, e^{-\frac{i}{\hbar}(\mathbf{x}-\mathbf{x}')\cdot(\mathbf{p}'-\mathbf{p})} \int\limits_{-\infty}^{\infty} \frac{d(t-t')}{2\pi\hbar}\, e^{i(t-t')\omega}$$

$$\times\; \langle n(\mathbf{x},t)\, n(\mathbf{x}',t')\rangle , \tag{6.150}$$

where the bracket denotes the weighted trace with respect to the state of the substance

$$\langle n(\mathbf{x},t)\, n(\mathbf{x}',t')\rangle \equiv \mathrm{tr}_S(\rho_S\, n(\mathbf{x},t)\, n(\mathbf{x}',t')) . \tag{6.151}$$

We thus obtain the formula

$$\frac{d^2\sigma}{d\hat{p}\,d\epsilon} = \frac{p}{p'}\frac{m_n^2\,a^2}{(2\pi\hbar)^3\hbar^2}\,V\,S(\mathbf{q},\omega) \tag{6.152}$$

where $S(\mathbf{q},\omega)$ is the Fourier transform of the space-time density correlation function $S(\mathbf{x},t;\mathbf{x}',t') \equiv \langle n(\mathbf{x},t)\,n(\mathbf{x}',t')\rangle$, and $\hbar\mathbf{q} \equiv \mathbf{p}' - \mathbf{p}$ and $\hbar\omega$ is the momentum and energy transfer from the neutron to the substance. We note that $S(-\mathbf{q},-\omega) = S(\mathbf{q},\omega)$. This correlation function is often referred to as the dynamic structure factor.[21] The dynamic structure factor gives the number of density excitations of the system with a given energy and momentum. A scattering experiment is thus a measurement of the density correlation function.

Exercise 6.4. Show that, for a target consisting of a single nucleus of mass M in the thermal state, the dynamic structure factor is given by

$$S(\mathbf{q},\omega) = \frac{1}{V}\sqrt{\frac{2\pi M}{kTq^2}}\,e^{-\frac{M}{2kTq^2}\left(\omega - \frac{\hbar q^2}{2M}\right)}. \tag{6.153}$$

For the differential cross section of a Boltzmann gas of N non-interacting nuclei we have according to Eq. (6.153)

$$\frac{d^2\sigma}{d\hat{p}\,d\epsilon} = N\,\frac{m_n^2\,a^2}{(2\pi\hbar)^3}\frac{p}{p'}\frac{2\pi}{\hbar}\,S(\mathbf{q},\omega)$$

$$= N\,\frac{m_n^2\,a^2}{(2\pi\hbar)^3}\frac{p}{p'}\frac{2\pi}{\hbar}\int_{-\infty}^{\infty} dt\, e^{-i\omega t}e^{-\frac{q^2}{2M}\left(t^2 kT - i\hbar t\right)}. \tag{6.154}$$

Exercise 6.5. Show that the limiting behavior of the total cross section for a Boltzmann gas is

$$\sigma = \int_{4\pi} d\hat{p}\int_0^{\infty} d\epsilon\,\frac{d^2\sigma}{d\hat{p}\,d\epsilon}$$

---

[21]We here follow the conventional notation, although in the standard notation of this chapter we have $S(\mathbf{x},\mathbf{x}',\omega) = \chi^>(\mathbf{x},\mathbf{x}',\omega)$. According to the fluctuation–dissipation theorem, the structure function is related to the density response function according to $S(\mathbf{x},\mathbf{x}',\omega) = 2\hbar n(\omega)\,\Im m\,\chi(\mathbf{x},\mathbf{x}',\omega)$.

$$
= \begin{cases} 4\pi N \left(\frac{m_n a}{2\pi\hbar^2}\right)^2 \dfrac{2}{\sqrt{\pi}\left(1 + \frac{m_n}{M}\right)^2 \sqrt{\frac{M\,p'^2}{2m_n^2 kT}}} & \text{for} \qquad \frac{M p'^2}{2m_n^2 kT} \ll 1 \\[4ex] 4\pi N \left(\frac{m_n a}{2\pi\hbar^2}\right)^2 \dfrac{1}{\left(1 + \frac{m_n}{M}\right)^2} & \text{for} \qquad \frac{M p'^2}{2m_n^2 kT} \gg 1 \ . \end{cases}
\tag{6.155}
$$

The divergent result for low energies is caused by the almost vanishing flux of incoming neutrons being scattered by the moving nuclei in the gas, and in the opposite limit we recover the result for scattering off N free and non-interacting nuclei.

For a discussion of the liquid–gas transition, and the phenomenon of critical opalescence we refer the reader to chapter 7 of reference [1].

6.8   Summary

The non-equilibrium states of a system which allows a description with sufficient accuracy by taking into account only the linear response occupies an especially simple regime. In fact, the non-equilibrium properties of such states could be completely understood in terms of the fluctuations characterizing the equilibrium state. Since the equilibrium state possesses universal properties, so does the dissipative regime of ever so slight perturbations, a feature with many important practical consequences. In Chapter 11 we shall return to study the linear response functions, the transport coefficients or conductivities. In particular we shall study the electrical conductivity of a disordered conductor in the quantum regime and take into account nonlinear effects in an applied magnetic field. To discuss such intricacies we shall express transport coefficients in terms of Green's functions and thereby have the powerful method of Feynman diagrams at our disposal. The density response function is for a system of charged particles equivalent to the effective interaction as density fluctuations are the source of the interaction. The effective interaction in a disordered conductor is discussed in Chapter 11. In the next two chapters, we shall study general non-equilibrium states, and universal properties are in general completely lost.

7

Quantum kinetic equations

In this chapter, the quantum field theoretic method will be used to derive quantum kinetic equations. The classical limit can be established, and quantum corrections can be studied systematically. Of importance is the fact that the treatment allows us to assess the validity regime of the kinetic equations by diagrammatic estimates. The quasi-classical Green's function technique is introduced. It will allow us to go beyond classical kinetics and, for example, to discuss renormalization effects due to interactions in a controlled approximation. Thermo-electric effects, being depending on particle–hole asymmetry, are not tractable in the quasi-classical technique and are dealt with on a separate basis.[1]

7.1   Left–right subtracted Dyson equation

In a non-equilibrium situation, the fluctuation–dissipation relation is no longer valid, and the kinetic propagator, G^{K}, is no longer specified by the spectral function and the quantum statistics of the particles, as for example in Eq. (3.116). To derive quantum kinetic equations the left and right matrix Dyson equation's, Eq. (5.66) and Eq. (5.69), are subtracted giving

$$[G_0^{-1} - \Sigma \overset{\otimes}{,} G]_- = 0, \tag{7.1}$$

the left–right subtracted Dyson equation. The reason behind this trick will soon become clear. Here we have again used $\otimes$ to signify matrix multiplication in the spatial and time variables, and introduced notation stressing the matrix multiplication structure in these variables

$$[A \overset{\otimes}{,} B]_- = A \otimes B - B \otimes A \quad , \quad [A \overset{\otimes}{,} B]_+ = A \otimes B + B \otimes A, \tag{7.2}$$

the latter anti-commutation notation to be employed immediately also. The general quantum kinetic equation is obtained by taking the kinetic or Keldysh component,

---

[1]This chapter, as well and the following chapter, follows the exposition given in references [3] and [9].

the off-diagonal component, of equation Eq. (7.1) giving

$$[G_0^{-1} - \Re\Sigma \overset{\otimes}{,} G^K]_- - [\Sigma^K \overset{\otimes}{,} \Re G]_- = \frac{i}{2}[\Sigma^K \overset{\otimes}{,} A]_+ - \frac{i}{2}[\Gamma \overset{\otimes}{,} G^K]_+ \quad (7.3)$$

where we have introduced the spectral weight function

$$A(1, 1') \equiv i(G^R(1, 1') - G^A(1, 1')) \quad (7.4)$$

and[2]

$$\Re G(1, 1') \equiv \frac{1}{2}(G^R(1, 1') + G^A(1, 1')) \quad (7.5)$$

and similarly for the self-energies

$$\Gamma(1, 1') \equiv i(\Sigma^R(1, 1') - \Sigma^A(1, 1')) \quad (7.6)$$

and

$$\Re\Sigma(1, 1') \equiv \frac{1}{2}(\Sigma^R(1, 1') + \Sigma^A(1, 1')) . \quad (7.7)$$

The way we have grouped the self-energy combinations in Eq. (7.3) appears at the moment rather arbitrary (compare this also with Section 5.7.4). Recall that A and Γ can be expressed as $A = i(G^> - G^<)$ and $\Gamma = i(\Sigma^> - \Sigma^<)$ and appear on the right side, whereas $\Re\Sigma$ and $\Re G$ are of a different nature. We shall later understand the physics involved in this difference of appearance of the self-energies: those on the left describe renormalization effects, i.e. effects of virtual processes, whereas those on the right describe real dissipative collision processes. The presence of the self-energy entails one having to deal with a complicated set of equations for an infinite hierarchy of the correlation functions, the starting equation being the Dyson equation. Of course, the general quantum kinetic equation is useless in practice unless an approximate expression for the self-energy is available.

Notice that in equilibrium, say at temperature T, the exact quantum kinetic equation is an empty statement since the Green's functions are related according to the fluctuation–dissipation relation, which for the case of fermions reads[3]

$$G^K(E, \mathbf{p}) = \left(G^R(E, \mathbf{p}) - G^A(E, \mathbf{p})\right) \tanh \frac{E}{2kT} \quad (7.8)$$

and consequently

$$\Sigma^K(E, \mathbf{p}) = \left(\Sigma^R(E, \mathbf{p}) - \Sigma^A(E, \mathbf{p})\right) \tanh \frac{E}{2kT} . \quad (7.9)$$

As a consequence, the two terms on the right in Eq. (7.3) cancel each other and the terms on the left are trivially zero in an equilibrium state since the convolution $\otimes$ in this case is commutative.

In a non-equilibrium situation, the fluctuation–dissipation relation is no longer valid. Since a Green's function is a traced quantity, a closed set of equations can

---

[2]The choice of notation reflects that in the Wigner or mixed coordinates, A and $\Re G$ will be purely real functions as shown in Exercise 7.1 on page 182.

[3]Displayed for simplicity for the case of a translational invariant state.

not be obtained, and one gets complicated equations for an infinite hierarchy of the correlation functions. If one, preferably by some controlled approximation, can break the hierarchy, usually at most at the two-particle correlation level, one obtains quantum kinetic equations, i.e. equations having the form of kinetic equations, but which contain quantum features which are not included in the classical Boltzmann equation [22]. One of the earliest applications of the non-equilibrium Green's function technique was to derive such kinetic equation [10] [14], though owing to their complicated structure they leave in general little progress in their solution by analytical means. However, as we shall see, combined with the diagrammatic estimation technique, the enterprise has the virtue of giving access to quantitative criteria for the validity of the so prevalently used Boltzmann equation, and thus not just the unquantified statement of lowest-order perturbation theory.

We now embark on the manipulations leading to a form of the quantum kinetic equation resembling classical kinetic equations. This is done by introducing Wigner coordinates.

7.2   Wigner or mixed coordinates

To derive quantum kinetic equations resembling the form of classical kinetic equations, we introduce the mixed or Wigner coordinates

$$\mathbf{R} = \frac{\mathbf{x}_1 + \mathbf{x}_{1'}}{2} \quad , \quad \mathbf{r} = \mathbf{x}_1 - \mathbf{x}_{1'} \tag{7.10}$$

and time variables[4]

$$T = \frac{t_1 + t_{1'}}{2} \quad , \quad t = t_1 - t_{1'} \tag{7.11}$$

in order to separate the variables, $(\mathbf{r}, t)$, describing the microscopic properties, governed by the characteristics of the system, from the variables, $(\mathbf{R}, T)$, describing the macroscopic properties, governed by the non-equilibrium features of the state under consideration, say as a result of the presence of an applied potential. To implement this separation of variables, we Fourier transform all functions with respect to the relative coordinates, say for a Green's function

$$G(X, p) \equiv \int dx \, e^{-ipx} \, G(X + x/2, X - x/2) \tag{7.12}$$

where the abbreviated notation has been introduced

$$X = (T, \mathbf{R}) \quad , \quad x = (t, \mathbf{r}) \tag{7.13}$$

and

$$p = (E, \mathbf{p}) \quad , \quad xp = -Et + \mathbf{p} \cdot \mathbf{r} . \tag{7.14}$$

We then express the current and density in terms of the mixed variables. The average charge density, Eq. (3.54), becomes (the factor of two is from the spin of the

---

[4]No danger of confusion with the notation for the temperature should occur.

particles, say electrons)

$$\rho(\mathbf{R}, T) = -2ie \int \frac{d\mathbf{p}}{(2\pi)^3} \int_{-\infty}^{\infty} dE \, G^<(E, \mathbf{p}, \mathbf{R}, T) \tag{7.15}$$

and the average electric current density in the presence of a vector potential $\mathbf{A}$, Eq. (3.57), becomes in terms of the mixed variables

$$\mathbf{j}(\mathbf{R}, T) = -\frac{e}{m} \int \frac{d\mathbf{p}}{(2\pi)^3} \int_{-\infty}^{\infty} dE \, (\mathbf{p} - e\mathbf{A}(\mathbf{R}, T)) \, G^<(E, \mathbf{p}, \mathbf{R}, T) . \tag{7.16}$$

Since

$$G^< = \frac{1}{2} G^K + \frac{i}{2} A \tag{7.17}$$

the current and density can also be expressed in terms of the kinetic Green's function

$$\mathbf{j}(\mathbf{R}, T) = -\frac{e}{m} \int \frac{d\mathbf{p}}{(2\pi)^3} \int_{-\infty}^{\infty} dE \, (\mathbf{p} - e\mathbf{A}(\mathbf{R}, T)) \, G^K(E, \mathbf{p}, \mathbf{R}, T) \tag{7.18}$$

and for the density (up to a state independent constant)

$$\rho(\mathbf{R}, T) = -2ie \int \frac{d\mathbf{p}}{(2\pi)^3} \int_{-\infty}^{\infty} dE \, G^K(E, \mathbf{p}, \mathbf{R}, T) . \tag{7.19}$$

Exercise 7.1. Show that, for an arbitrary non-equilibrium state, retarded and advanced Green's functions in the mixed coordinates are related according to

$$(G^R(\mathbf{R}, T, \mathbf{p}, E))^* = G^A(\mathbf{R}, T, \mathbf{p}, E) . \tag{7.20}$$

As a consequence, the spectral function in the mixed coordinates is a real function, and

$$(\Sigma^R(\mathbf{R}, T, \mathbf{p}, E))^* = \Sigma^A(\mathbf{R}, T, \mathbf{p}, E) . \tag{7.21}$$

Note that in the Wigner coordinates, the spectral function is twice the imaginary part of the advanced Green's function.

Exercise 7.2. Show for an arbitrary non-equilibrium state the spectral representation in the mixed coordinates, the Kramers–Kronig relations,

$$G^{R(A)}(X, p) = \int_{-\infty}^{\infty} \frac{dE'}{-2\pi i} \frac{G^R(X, p') - G^A(X, p')}{E - E' \, (\pm) \, i0}$$

$$= \int_{-\infty}^{\infty} \frac{dE'}{2\pi} \frac{A(X, p')}{E - E' \, (\pm) \, i0} , \qquad p' \equiv (\mathbf{p}, E') . \tag{7.22}$$

We now show that a convolution $C = A \otimes B$ in the mixed coordinates is given by

$$(A \otimes B)(X, p) = e^{\frac{i}{2}(\partial_X^A \partial_p^B - \partial_p^A \partial_X^B)} A(X, p) B(X, p),$$

(7.23)

where

$$\partial_X^A = (-\partial_T, \nabla_{\mathbf{R}}), \quad \partial_p^A = (-\partial_E, \nabla_{\mathbf{p}})$$

(7.24)

and

$$\partial_X^A \partial_p^B \equiv -\frac{\partial^A}{\partial T} \frac{\partial^B}{\partial E} + \frac{\partial^A}{\partial \mathbf{R}} \cdot \frac{\partial^B}{\partial \mathbf{p}}$$

(7.25)

and the upper index refers to the function operated on. Let us here for clarity distinguish quantities in the mixed coordinates by a tilde

$$\tilde{C}(X, x) \equiv C(X + x/2, X - x/2) = C(x_1, x_{1'}).$$

(7.26)

Consider the convolution

$$C(x_1, x_{1'}) \equiv \int dx_2 \, A(x_1, x_2) B(x_2, x_{1'}),$$

(7.27)

which in mixed coordinates becomes

$$\tilde{C}(X, x) \equiv \int dx_2 \, A(X + x/2, x_2) B(x_2, X - x/2)$$

$$= \int dx_2 \, \tilde{A}\left(\frac{1}{2}(X + x/2 + x_2), X + x/2 - x_2\right)$$

$$\tilde{B}\left(\frac{1}{2}(x_2 + X - x/2), x_2 - (X - x/2)\right).$$

(7.28)

Making the shift of variable

$$x_2 \rightarrow x_2 - (X - x/2)$$

(7.29)

eliminates the X-dependence in the variable at the relative coordinate place, giving

$$\tilde{C}(X, x) = \int dx_2 \, \tilde{A}(X + x_2/2, x - x_2) \tilde{B}(X - x/2 + x_2/2, x_2).$$

(7.30)

In the mixed coordinates we have

$$C(X, p) = \int dx \, e^{-ixp} \int dx_2 \, \tilde{A}(X + x_2/2, x - x_2) \tilde{B}(X - x/2 + x_2/2, x_2)$$

$$= \int dx \, e^{-ixp} \int dx_2 \int \frac{dp'}{(2\pi)^4} e^{-ip'(x - x_2)} A(X + x_2/2, p')$$

$$\times \int \frac{dp''}{(2\pi)^4} e^{-ip'' x_2} B(X - x/2 + x_2/2, p''),$$

(7.31)

where in the last equality the integrand has been expressed in the mixed coordinates. Performing a Taylor expansion and partial integrations then leads to Eq. (7.23).

In particular, for the case of interest of slowly varying perturbations, which corresponds to the lowest-order Taylor expansion, the convolution becomes

$$(A \otimes B)(X, p) = A(X, p) \, B(X, p) + \frac{i}{2} \left(\partial_X A(X, p) \right) \partial_p B(X, p)$$

$$- \frac{i}{2} \left(\partial_p A(X, p) \right) \partial_X B(X, p) . \tag{7.32}$$

In the mixed coordinates, the operator part of the inverse Green's function, G_0^{-1} of Eq. (3.68), becomes a simple multiplicative factor

$$G_0^{-1}(E, \mathbf{p}, \mathbf{R}, T) = E - \xi_{\mathbf{p}} - V(\mathbf{R}, T) , \tag{7.33}$$

where $V(\mathbf{R}, T)$ is an applied potential, and $\xi_{\mathbf{p}} = \epsilon_{\mathbf{p}} - \mu$ is the single-particle energy measured from the chemical potential, and for quadratic dispersion, such as the case for the free electron model, $\epsilon_{\mathbf{p}} = \mathbf{p}^2/2m$.

7.3 Gradient approximation

To make progress towards an intelligible and tractable equation, one assumes that the spatial and temporal inhomogeneity is weak, inducing only slow variations in Green's functions and self-energies.[5] In the following we assume the non-equilibrium state is induced by an applied potential, $V(\mathbf{R}, T)$, which is a slowly varying function of its variables compared to the characteristic scales of equilibrium Green's functions and self-energies.[6] We shall, for example, have a degenerate Fermi system in mind, say conduction electrons in a metal, where the characteristic scales are the Fermi energy and momentum. This allows for the approximation where only lowest-order terms in the variation is kept, the so-called gradient approximation. In this approximation we thus have

$$[A \overset{\otimes}{,} B]_+ \ \rightarrow \ 2A(X, p) \, B(X, p) \tag{7.34}$$

and

$$-i[A \overset{\otimes}{,} B]_- \ \rightarrow \ [A, B]_{\mathbf{p}} , \tag{7.35}$$

where

$$[A, B]_{\mathbf{p}} = \partial_X^A A \, \partial_p^B B - \partial_p^A A \, \partial_X^B B$$

$$= \left(\partial_E^A \partial_T^B - \partial_T^A \partial_E^B - \nabla_{\mathbf{p}}^A \cdot \nabla_{\mathbf{R}}^B + \nabla_{\mathbf{R}}^A \cdot \nabla_{\mathbf{p}}^B \right) A(X, p) \, B(X, p) , \tag{7.36}$$

---

[5] If one is interested only in the linear response, such an assumption is not needed, but the gradient approximation allows, in principle, inclusion of all the nonlinear effects of a slightly inhomogeneous perturbation.

[6] The coupling to a vector potential will be handled in Section 7.6.

and the subscript p on the bracket signifies its resemblance to the Poisson bracket of classical mechanics.

In the gradient approximation, the quantum kinetic equation, Eq. (7.3), becomes

$$[G_0^{-1} - \Re e\Sigma, G^K]_p - [\Sigma^K \overset{\otimes}{,} \Re e G]_p = i\Sigma^K A - i\Gamma G^K. \tag{7.37}$$

The first term on the left-hand side becomes, in the gradient approximation,

$$[G_0^{-1} \overset{\otimes}{,} G^K]_- \rightarrow [G_0^{-1} \overset{\otimes}{,} G^K]_p$$

$$= \partial_T G^K(E, \mathbf{p}, \mathbf{R}, T) + \partial_E G^K(E, \mathbf{p}, \mathbf{R}, T) \partial_T V(\mathbf{R}, T)$$

$$+ \nabla_{\mathbf{R}} G^K(E, \mathbf{p}, \mathbf{R}, T) \cdot \nabla_{\mathbf{p}} \xi_{\mathbf{p}} - \nabla_{\mathbf{p}} G^K(E, \mathbf{p}, \mathbf{R}, T) \cdot \nabla_{\mathbf{R}} V(\mathbf{R}, T). \tag{7.38}$$

In fact, the first term is always exact, and so is the third term for the case of quadratic dispersion.[7] We note that they are identical in form to the driving terms in the Boltzmann equation, whereas the last term on the right, which also appears in the Boltzmann equation, here is valid only in the gradient approximation, i.e. the magnitude of the characteristic wave vector of the potential, q, is small compared with the characteristic wave vector of the system, which in the case of degenerate fermions is the Fermi wave vector, $q < k_F$ (usually no restriction at all for transport situations in degenerate Fermi systems). The second term on the right looks strange in the Boltzmann context, but we shall soon integrate the equation over E, upon which this term disappears.

Since in equilibrium a Poisson bracket vanishes, the kinetic equation reduces to

$$0 = \Sigma^K(E, \mathbf{p}) A(E, \mathbf{p}) - \Gamma(E, \mathbf{p}) G^K(E, \mathbf{p}) \tag{7.39}$$

and this identity can be interpreted as the statement of determining the equilibrium distribution function as the one for which the right-hand side, the collision integral, vanishes.

7.3.1 Spectral weight function

To make further progress we study the spectral weight function. The equation of motion for the spectral weight function is obtained by subtracting the diagonal components of Eq. (7.1), giving

$$[G_0^{-1} - \Re e\Sigma \overset{\otimes}{,} A]_- - [\Gamma \overset{\otimes}{,} \Re e G]_- = 0. \tag{7.40}$$

In the gradient approximation, the non-equilibrium spectral function satisfies (according to Eq. (7.40)) the equation

$$[E - \xi_{\mathbf{p}} - V(\mathbf{R}, T) - \Re e\Sigma^R, A]_p + [\Re e G^R, \Gamma]_p = 0. \tag{7.41}$$

---

[7]The first term is not dependent on the gradient approximation, but as usual is exact, simply owing to the equation being first order in time, and similarly for the second term for the case of quadratic dispersion.

We note that

$$A(E, \mathbf{p}, \mathbf{R}, T) = \frac{\Gamma(E, \mathbf{p}, \mathbf{R}, T)}{\left(E - \xi_{\mathbf{p}} - V(\mathbf{R}, T) - \Re\Sigma^{\mathrm{R}}(E, \mathbf{p}, \mathbf{R}, T)\right)^2 + \left(\frac{\Gamma(E, \mathbf{p}, \mathbf{R}, T)}{2}\right)^2} \quad (7.42)$$

solves Eq. (7.41) since, because $[A, B]_{\mathrm{p}} = -[B, A]_{\mathrm{p}}$, and noting that

$$\Re\left(G^{\mathrm{R}}(E, \mathbf{p}, \mathbf{R}, T)\right)^{-1} = E - \xi_{\mathbf{p}} - V(\mathbf{R}, T) - \Re\Sigma^{\mathrm{R}}(E, \mathbf{p}, \mathbf{R}, T), \quad (7.43)$$

the left-hand side of equation Eq. (7.41) can then be rewritten in the form

$$-i\left[\Re\left(G^{\mathrm{R}}\right)^{-1} - \frac{i}{2}\Gamma, \left(\Re\left(G^{\mathrm{R}}\right)^{-1} - \frac{i}{2}\Gamma\right)^{-1}\right]_{\mathrm{p}}$$

$$+ \ i\left[\Re\left(G^{\mathrm{R}}\right)^{-1} + \frac{i}{2}\Gamma, \left(\Re\left(G^{\mathrm{R}}\right)^{-1} + \frac{i}{2}\Gamma\right)^{-1}\right]_{\mathrm{p}}, \quad (7.44)$$

which vanishes, since for any function F, we have $[A, F(A)]_{\mathrm{p}} = 0$. In the far past, where the system is assumed undisturbed, i.e. V vanishes, the presented solution, Eq. (7.42), reduces to the equilibrium spectral function

$$A(E, \mathbf{p}) = \frac{\Gamma(E, \mathbf{p})}{\left(E - \xi_{\mathbf{p}} - \Re\Sigma(E, \mathbf{p})\right)^2 + (\Gamma(E, \mathbf{p})/2)^2}, \quad (7.45)$$

which in this case can be obtained directly from Eq. (7.40). The solution Eq. (7.42) is therefore the sought solution since it satisfies the correct initial condition.

Adding the left and right Dyson equations for the retarded non-equilibrium Green's function, and performing the expansion within the gradient approximation, Eq. (7.34), we similarly obtain the result

$$G^{\mathrm{R}}(E, \mathbf{p}, \mathbf{R}, T) = \frac{1}{G_0^{-1}(E, \mathbf{p}, \mathbf{R}, T) - \Sigma^{\mathrm{R}}(E, \mathbf{p}, \mathbf{R}, T)}$$

$$= \frac{1}{E - \xi_{\mathbf{p}} - V(\mathbf{R}, T) - \Sigma^{\mathrm{R}}(E, \mathbf{p}, \mathbf{R}, T)}, \quad (7.46)$$

and similarly for the advanced Green's function.

7.3.2 Quasi-particle approximation

If the interaction is weak the self-energies are small, and the spectral weight function is a peaked function in the variable E, in fact in the absence of interactions according to Eq. (7.42)

$$A(E, \mathbf{p}, \mathbf{R}, T) = 2\pi \, \delta(E - \xi_{\mathbf{p}} - V(\mathbf{R}, T)) \quad (7.47)$$

and therefore is G^K also a peaked function in the variable E. We first consider this so-called quasi-particle approximation.[8] In Section 7.5 we will consider the case of strong electron–phonon interaction and the spectral weight can not be approximated by a delta function, and a different approach to obtaining a kinetic equation must be developed.

The reason for subtracting the left and right Dyson equations is that the term linear in E in G_0^{-1} then disappears, thereby, in view of Eq. (7.47), allowing the equation, Eq. (7.37), to be integrated with respect to this variable giving

$$(\partial_T + \nabla_{\mathbf{p}}\xi_{\mathbf{p}} \cdot \nabla_{\mathbf{R}} - \nabla_{\mathbf{R}} V(\mathbf{R}, T) \cdot \nabla_{\mathbf{p}}) h(\mathbf{p}, \mathbf{R}, T)$$

$$= \Sigma^K(E = \xi_{\mathbf{p}} + V(\mathbf{R}, T), \mathbf{p}, \mathbf{R}, T)$$

$$- \Gamma(E = \xi_{\mathbf{p}} + V(\mathbf{R}, T), \mathbf{p}, \mathbf{R}, T) h(\mathbf{p}, \mathbf{R}, T), \qquad (7.48)$$

where we have introduced the distribution function

$$h(\mathbf{p}, \mathbf{R}, T) = -\int_{-\infty}^{\infty} \frac{dE}{2\pi i} G^K(E, \mathbf{p}, \mathbf{R}, T). \qquad (7.49)$$

The two self-energy terms on the left in Eq. (7.37) must be neglected in this approximation since they are by assumption small and in addition multiplied by the characteristic frequency, ω_0, of the external potential which is small compared with the characteristic frequency of the system, which in the case of degenerate fermions is the Fermi energy, $\omega_0 \ll \epsilon_F$. In the event that the left–right subtracted Dyson equation allows for integrating over E, equal time quantities appear, and the distribution function is of the Wigner type, and is related similarly to densities and currents.[9]

In equilibrium the distribution function is for fermions given by

$$h_0(\mathbf{p}) = \tanh \frac{\xi_{\mathbf{p}}}{2kT} \qquad (7.50)$$

in which case the sum of the two terms on the right in Eq. (7.48) vanish. We shall now focus on the terms on the right-hand side of equation Eq. (7.48), and realize they describe collisions and dissipative effects.

Since the equation for the Green's function is not closed we will eventually have to make an approximation that cuts off the hierarchy of correlations. For states not too far from equilibrium, this can be done at the level of self-energies if, for example, vertex corrections can be shown to be small in some parameter, viz. the one characterizing the equilibrium approximation. To this end we recall the usefulness of the diagrammatic estimation technique.

---

[8]This is of course a most unfortunate choice of labeling used in the literature. The physical implication of the approximation simply being that in between collisions, the particle motion is that of a free particle.

[9]For a discussion of the Wigner function see chapter 4 of reference [1].

7.4   Impurity scattering

We now start to consider interactions of relevance, and begin with the simplest case; that of impurity scattering. In the clean limit where impurity scattering say of electrons in a metal or semiconductor is weak, so that any tendency to localization in a three-dimensional sample can be neglected, i.e. $\epsilon_F \tau \gg \hbar$,[10] diagrams with crossing of impurity lines can be neglected, and the impurity self-energy is[11]

$$\Sigma(E, \mathbf{p}, \mathbf{R}, T) \quad \equiv \quad \begin{array}{c} \mathbf{p}E \longleftarrow \overset{\times}{\underset{\mathbf{p}'ERT}{\triangle}} \longrightarrow \mathbf{p}E \end{array} \qquad (7.51)$$

corresponding to the analytical expression for the real-time matrix self-energy

$$\Sigma(\mathbf{p}, E, \mathbf{R}, T) = n_i \int \frac{d\mathbf{p}'}{(2\pi\hbar)^3} \, |V_{\mathrm{imp}}(\mathbf{p} - \mathbf{p}')|^2 \, G(\mathbf{p}', E, \mathbf{R}, T) \,. \qquad (7.52)$$

For the kinetic component of the self-energy we have

$$\Sigma^{K}(\mathbf{p}, E, \mathbf{R}, T) = n_i \int \frac{d\mathbf{p}'}{(2\pi)^3} \, |V_{\mathrm{imp}}(\mathbf{p} - \mathbf{p}')|^2 \, G^{K}(\mathbf{p}', E, \mathbf{R}, T) \qquad (7.53)$$

and

$$\Gamma(\mathbf{p}, E, \mathbf{R}, T) \quad = \quad i(\Sigma^{R}(\mathbf{p}, E, \mathbf{R}, T) \; - \; \Sigma^{A}(\mathbf{p}, E, \mathbf{R}, T))$$

$$= \quad n_i \int \frac{d\mathbf{p}'}{(2\pi)^3} \, |V_{\mathrm{imp}}(\mathbf{p} - \mathbf{p}')|^2 \, A(\mathbf{p}', E, \mathbf{R}, T) \,. \qquad (7.54)$$

Since we work to lowest order in the impurity concentration, n_i, the spectral weight should be replaced by the delta function expression, and we obtain

$$(\partial_T + \nabla_{\mathbf{p}} \xi_{\mathbf{p}} \cdot \nabla_{\mathbf{R}} - \nabla_{\mathbf{R}} V(\mathbf{R}, T) \cdot \nabla_{\mathbf{p}}) \, h(\mathbf{p}, \mathbf{R}, T) = I^{(1)}[f] \qquad (7.55)$$

where the right side, the electron-impurity collision integral, is

$$I^{(1)}[f] = -2\pi n_i \int \frac{d\mathbf{p}'}{(2\pi)^3} \, |V_{\mathrm{imp}}(\mathbf{p} - \mathbf{p}')|^2 \, \delta(\xi_{\mathbf{p}} - \xi_{\mathbf{p}'})(h(\mathbf{p}, \mathbf{R}, T) - h(\mathbf{p}', \mathbf{R}, T)) \,. \quad (7.56)$$

We have arrived at the classical kinetic equation describing the motion of a particle in a weakly disordered system, the Boltzmann equation for a particle in a random

---

[10]In a strictly one-dimensional sample localization is typically dominant and in a two-dimensional sample it is important at low enough temperatures. The first quantum correction to this classical limit, the weak localization effect, is discussed in Chapter 11.

[11]For a detailed description of the standard impurity average Green's function technique and diagrammatic estimation, we refer the reader to reference [1], where also inclusion of multiple impurity scattering is shown to be equivalent to the considered Born approximation by inclusion of the t-matrix.

potential. The derived equation is called a kinetic equation because the collision integral is not a functional in time (or space), i.e. local in both the space and time variables, and a functional only with respect to the momentum variable. The only difference signaling we are considering the degenerate electron gas is the quantum statistics, which dictates the distribution function to respect the Pauli principle, i.e the equilibrium distribution is specified by Eq. (7.50).

The weak-disorder kinetic equation for a particle in a random potential is of course immediately obtained from classical mechanics, granted a stochastic treatment of the impurity scattering, giving the collision integral

$$I_t^{(1)}[f] = -\sum_{\mathbf{p}'} \{W(\mathbf{p}',\mathbf{p})f(\mathbf{p},t) - W(\mathbf{p},\mathbf{p}')f(\mathbf{p}',t)\}, \qquad (7.57)$$

where $W(\mathbf{p}',\mathbf{p})$ is the classical transition rate between momentum states, the classical scattering cross section. In classical mechanics the distribution function concept is unproblematic because we can simultaneously specify position and momentum, and the terms on the left-hand side of Eq. (7.55) are simply the streaming terms in phase space for the situation in question.

In the quantum case we have, in the Born approximation for the transition rate between momentum states,

$$\begin{aligned} W(\mathbf{p}',\mathbf{p}) &= \frac{2\pi n_{\mathrm{i}}}{\hbar V}\,|V_{\mathrm{imp}}(\mathbf{p}-\mathbf{p}')|^2\,\delta(\epsilon_{\mathbf{p}'}-\epsilon_{\mathbf{p}}) \\[2mm] &= \frac{2\pi}{\hbar}n_{\mathrm{i}}V\,|\langle\mathbf{p}|V_{\mathrm{imp}}(\hat{\mathbf{x}})|\mathbf{p}'\rangle|^2\,\delta(\epsilon_{\mathbf{p}'}-\epsilon_{\mathbf{p}})\,. \end{aligned} \qquad (7.58)$$

We note that in the Born approximation we always have $W(\mathbf{p}',\mathbf{p}) = W(\mathbf{p},\mathbf{p}')$.[12]

We note that the expression $W(\mathbf{p}',\mathbf{p})$ in Eq. (7.58) is Fermi's Golden Rule expression for the transition probability per unit time from momentum state $\mathbf{p}$ to momentum state $\mathbf{p}'$ (or vice versa) caused by the scattering off an impurity, times the number of impurities. The two terms in the collision integral thus have a simple interpretation because they describe the scattering in and out of a momentum state. For example, the first term in the collision integral of the Boltzmann equation, Eq. (7.56), is a loss term, and gives the rate of change of occupation of a phase space volume due to the scattering of an electron from momentum $\mathbf{p}$ to momentum $\mathbf{p}'$ by the random potential. The probability per unit time of being scattered out of the phase space volume around $\mathbf{p}$, and into a volume around $\mathbf{p}'$, is the product of three probabilities: (the probability that an electron is in that phase space volume to be available for scattering) $\times$ (the transition probability per unit time for the transition from state $\mathbf{p}$ to $\mathbf{p}'$) $\times$ (the probability that there is an impurity in the space volume to scatter). Similarly we have the interpretation of the other term as a scattering-in term.

The obtained equation is a quasi-classical equation because, in between collisions with impurities, the electrons move along straight lines just as in classical mechan-

---

[12] In general, potential scattering is time-reversal invariant, and we always have $W(\mathbf{p}',\mathbf{p}) = W(-\mathbf{p},-\mathbf{p}')$. If, in addition, the potential is invariant with respect to space inversion, we have $W(\mathbf{p}',\mathbf{p}) = W(-\mathbf{p}',-\mathbf{p})$, and thereby $W(\mathbf{p}',\mathbf{p}) = W(\mathbf{p},\mathbf{p}')$.

ics, but the scattering cross section is the quantum mechanical one.[13] Besides the inherent quantum statistics, this is the only quantum feature surviving in the weak disorder limit, $\hbar/\epsilon_F \tau \ll 1$, where τ is the characteristic time scale for the dynamics, the momentum relaxation time, soon to be discussed. The presented diagrammatic method for deriving transport equations is capable of going beyond the Markov process described by the classical kinetic equation, to include quantum effects. One can construct a kinetic equation determining the first quantum correction, the weak localization effect, but it is easier to employ linear response theory as described in Chapter 11.

Let us study the simplest non-equilibrium situation where the distribution is out of momentum equilibrium for only a single momentum state on the Fermi surface

$$f_{\mathbf{p}'}(t) = f_0(\epsilon_{\mathbf{p}'}) + \delta f_{\mathbf{p}}(t)\, \delta_{\mathbf{p},\mathbf{p}'} \tag{7.59}$$

and we assume no external fields. The Boltzmann equation then reduces to

$$\frac{\partial \delta f_{\mathbf{p}}(t)}{\partial t} = -\frac{\delta f_{\mathbf{p}}}{\tau_{\mathbf{p}}} . \tag{7.60}$$

whose solution describes the exponential relaxation to equilibrium

$$f_{\mathbf{p}}(t) = f_0(\epsilon_{\mathbf{p}}) + \delta f_{\mathbf{p}}(t=0)\, e^{-t/\tau_{\mathbf{p}}} \tag{7.61}$$

and the momentum relaxation time (which for the considered isotropic Fermi surface does not depend on the direction of the momentum)

$$\frac{1}{\tau} = \sum_{\mathbf{p}'(\neq \mathbf{p})} W_{\mathbf{p}',\mathbf{p}} \tag{7.62}$$

is seen to be identical to the imaginary part of the retarded self-energy for $E = \epsilon_F$

$$\frac{1}{\tau} = \frac{1}{\tau(\epsilon_F)} = 2\pi n_i \int \frac{d\mathbf{p}'}{(2\pi)^3}\, |V_{\mathrm{imp}}(\mathbf{p}_F - \mathbf{p}')|^2\, \delta(\epsilon_{\mathbf{p}'} - \epsilon_F) . \tag{7.63}$$

We noted above that the collision integral rendered the kinetic equation a stochastic equation for the momentum, Pauli's master equation. In the case where $\tau(\mathbf{p})$ can be considered independent of the momentum $\mathbf{p}$, τ is the phenomenological parameter of the Drude theory of conduction, and $\Delta t/\tau(\mathbf{p})$ is, according to Eq. (7.61), the probability that an electron with momentum $\mathbf{p}$ in the time span Δt will suffer a collision with total loss of momentum direction memory. Such an assumption is not valid in the quantum mechanical description as the scattering of a wave sets up correlations that can not lead to a total memory loss in general, as we shall discuss in detail in Chapter 11.

One might miss Pauli blocking factors in the expression for the collision integral, Eq. (7.56), but they need not, as just shown, appear in the considered case of potential

---

[13]If we go beyond the considered Born approximation and include multiple scattering, we encounter the exact cross section for scattering off an impurity as expressed by the t-matrix. For a discussion see chapter 3 in reference [1].

scattering. If one uses the Kadanoff–Baym form of the kinetic equation, Eq. (5.136), Pauli blocking factors would then appear in intermediate results. Another lesson to learn is that the form of the appearance of the quantum statistics, here the Fermi–Dirac distribution function or other forms, depends on the type of Green's functions one employs; a case in question is our choice leading to the distribution function in Eq. (7.49) and Eq. (7.50).

For the sole purpose of obtaining the weak-disorder kinetic equation, the use of quantum field theoretic methods and Feynman diagrams is hardly necessary. However, it allows us in a simple way to assess the validity criterion for the classical kinetic description, and to go beyond the classical limit and study quantum corrections. In view of the neglected diagrams, the validity of the Boltzmann equation requires $\hbar/\epsilon_F \tau \ll 1$, or equivalently $p_F \gg \hbar/l$, where $l = v_F \tau$ is the mean free path.[14] In addition for the gradient approximation to be valid, the characteristic frequency and wave vector of the perturbation must satisfy the weak restrictions $\hbar\omega < \epsilon_F, q < k_F$. There can be some satisfaction in *deriving* the Boltzmann equation, in particular to establish validity criteria, i.e. to establish the Landau criterion and not instead the devastating for applications Peierls criterion, $\hbar\omega < kT$, which an argument based on a simple quasi-particle picture would suggest. But for the sake of deriving classical kinetic equations, the venture into quantum field theory is over-kill. The more so, that in practice it is difficult to go beyond the linear regime systematically and study nonlinear effects. However, there exists a successful technique that leads to an exception to this state of affairs, viz. the so-called quasi-classical Green's function technique. We consider this technique applied in the normal state in Section 7.5, and its even more important application to superconductivity will be studied in Chapter 8.

Exercise 7.3. Show that the continuity equation is obtained by integrating the kinetic equation, Eq. (7.55), with respect to the momentum variable.

For a discussion of the classical Boltzmann transport coefficients for a degenerate Fermi system, electrical and thermal conductivities, we refer the reader to chapter 5 of reference [1]. Here we just note that, for the case of a time-independent electric field, the solution to the Boltzmann equation, Eq. (7.56), to linear order is immediately obtained giving for the conductivity, σ_0, the Boltzmann result

$$\sigma_0 = \frac{ne^2 \tau_{tr}}{m} \tag{7.64}$$

where $\tau_{tr} \equiv \tau_{tr}(\epsilon_F)$ is the transport relaxation time in the Born approximation

$$\frac{\hbar}{\tau_{tr}(\epsilon_F)} = 2\pi n_i N_0 \int \frac{d\hat{\mathbf{p}}_F'}{4\pi} |V_{imp}(\mathbf{p}_F - \mathbf{p}_F')|^2 (1 - \hat{\mathbf{p}}_F \cdot \hat{\mathbf{p}}_F') . \tag{7.65}$$

The appearance of the transport time expresses the simple fact that small angle scattering is not effective in degrading the current. For isotropic scattering the momentum and transport relaxation times are identical, as each scattering direction is

---

[14] This so-called Landau criterion is not sufficient for the applicability of the Boltzmann equation in low-dimensional systems, $d \leq 2$. This is a subject we shall discuss in detail in Chapter 11.

weighted equally. The transport relaxation time is the characteristic time a particle can travel before the direction of its velocity is randomized.

Exercise 7.4. Show that the retarded impurity self-energy, Eq. (7.51), in equilibrium and for $|E - \epsilon_F| \ll \epsilon_F$ and $|\mathbf{p} - \mathbf{p}_F| \ll p_F$ just becomes the constant

$$\Sigma^{\mathrm{R}}(E, \mathbf{p}) = -i\frac{\hbar}{2\tau} \qquad (7.66)$$

where

$$\frac{\hbar}{\tau} = 2\pi n_i N_0 \int \frac{d\hat{\mathbf{p}}'_F}{4\pi} \, |V_{\mathrm{imp}}(\mathbf{p}_F - \mathbf{p}'_F)|^2 \,. \qquad (7.67)$$

For later use, we end this section on dynamics due to impurity scattering by considering Boltzmannian motion and its large scale features, Brownian motion.

7.4.1   Boltzmannian motion in a random potential

In later chapters we shall discuss quantum corrections to classical transport. However, in many cases we often still need to know only the classical kinetics of the particle motion. We therefore take this opportunity to discuss the Boltzmannian motion of a particle scattered by impurities, although we shall not need these results before we discuss destruction of phase coherence due to electron–phonon interaction in Chapter 11. The Boltzmann theory is a stochastic description of the classical motion of a particle in a weakly disordered potential. At each instant the particle has attributed a probability for a certain position and velocity (or momentum). In the absence of external fields the Boltzmann equation for a particle in a random potential has the form

$$\frac{\partial f(\mathbf{x}, \mathbf{p}, t)}{\partial t} + \mathbf{v} \cdot \frac{\partial f(\mathbf{x}, \mathbf{p}, t)}{\partial \mathbf{x}} = -\int \frac{d\mathbf{p}'}{(2\pi\hbar)^3} \, W(\mathbf{p}, \mathbf{p}') \left[f(\mathbf{x}, \mathbf{p}, t) - f(\mathbf{x}, \mathbf{p}', t) \right], \quad (7.68)$$

where we have introduced the notation $\mathbf{v} = \mathbf{v_p} = \mathbf{p}/m$ for the particle velocity.

The Boltzmann equation is first order in time (the state of a particle is completely determined in classical mechanics by specifying its position and momentum), and the solution for such a Markov process can be expressed in terms of the conditional probability F for the particle to have position $\mathbf{x}$ and momentum $\mathbf{p}$ at time t given it had position $\mathbf{x}'$ and momentum $\mathbf{p}'$ at time t'

$$f(\mathbf{x}, \mathbf{p}, t) = \int \frac{d\hat{\mathbf{p}}'}{4\pi} \int d\mathbf{x}' \, F(\mathbf{x}, \mathbf{p}, t; \mathbf{x}', \mathbf{p}', t') \, f(\mathbf{x}', \mathbf{p}', t') \,. \qquad (7.69)$$

For elastic scattering only the direction of momentum can change, and consequently we need only integrate over the direction of the momentum. In the absence of external fields the motion in between scattering events is along straight lines, and the conditional probability describes how the particle by impurity scattering, is thrown between different straight-line segments, i.e. a Boltzmannian path.

We define the Boltzmann propagator as the conditional probability for the initial condition that it vanishes for times $t < t'$, the retarded Green's function for the Boltzmann equation. The equation obeyed by the Boltzmann propagator is thus, assuming isotropic scattering,

$$
\left(\frac{\partial}{\partial t} + \mathbf{v_p} \cdot \frac{\partial}{\partial \mathbf{x}} + \frac{1}{\tau}\right) F(\mathbf{p}, \mathbf{x}, t; \mathbf{p}', \mathbf{x}', t') - \frac{1}{\tau} \int \frac{d\hat{\mathbf{p}}}{4\pi} F(\mathbf{p}, \mathbf{x}, t; \mathbf{p}', \mathbf{x}', t')
$$

$$
= \hat{\delta}(\hat{\mathbf{p}} - \hat{\mathbf{p}}') \, \delta(\mathbf{x} - \mathbf{x}') \, \delta(t - t') , \tag{7.70}
$$

where $\hat{\delta}$ is the spherical delta function

$$
\int \frac{d\hat{\mathbf{p}}'}{4\pi} \hat{\delta}(\hat{\mathbf{p}} - \hat{\mathbf{p}}') f(\mathbf{p}') = f(\mathbf{p}) . \tag{7.71}
$$

The equation for the Boltzmann propagator is solved by Fourier transformation, and we obtain

$$
F(\mathbf{p}, \mathbf{x}, t; \mathbf{p}', \mathbf{x}', t') = \int \frac{d\mathbf{q} \, d\omega}{(2\pi)^4} \, e^{i\mathbf{q} \cdot (\mathbf{x} - \mathbf{x}') - i\omega(t - t')} \, F(\mathbf{p}, \mathbf{p}'; \mathbf{q}, \omega) , \tag{7.72}
$$

where

$$
F(\mathbf{p}, \mathbf{p}'; \mathbf{q}, \omega) = \frac{1}{-i\omega + \mathbf{p} \cdot \mathbf{q}/m + 1/\tau} \left(\frac{1/\tau}{-i\omega + \mathbf{p}' \cdot \mathbf{q}/m + 1/\tau} I(q, \omega) + \hat{\delta}(\hat{\mathbf{p}} - \hat{\mathbf{p}}')\right) \tag{7.73}
$$

and

$$
I(q, \omega) = \frac{ql}{ql - \arctan ql/(1 - i\omega\tau)} , \tag{7.74}
$$

where $l = v\tau$ is the mean free path.

We note, by direct integration, the property

$$
F(\mathbf{x}, \mathbf{p}, t; \mathbf{x}', \mathbf{p}', t') = \int \frac{d\hat{\mathbf{p}}''}{4\pi} \int d\mathbf{x}'' \, F(\mathbf{x}, \mathbf{p}, t; \mathbf{x}'', \mathbf{p}'', t'') \, F(\mathbf{x}'', \mathbf{p}'', t''; \mathbf{x}', \mathbf{p}', t') \tag{7.75}
$$

the signature of a Markov process.[15] This property will be utilized in Section 11.3.1 in the calculation of the dephasing rate in weak localization due to electron–phonon interaction.

7.4.2 Brownian motion

If we are interested only in the long-time and large-distance behavior of the particle motion, $|\mathbf{x} - \mathbf{x}'| \gg l$, $t - t' \gg \tau$, the wave vectors and frequencies of importance in

---

[15] For a Markov process, the *future* is independent of the *past* when the *present* is known, i.e. the causality principle of classical physics in the context of a stochastic dynamic system, here the process in question is Boltzmannian motion.

the Boltzmann propagator, Eq. (7.73), satisfy $ql, \omega\tau \ll 1$, and we obtain the diffusion approximation

$$I(q,\omega) \simeq \frac{1/\tau}{-i\omega + D_0 q^2} , \qquad (7.76)$$

where $D_0 = vl/3$ is the diffusion constant in the considered case of three dimensions (and isotropic scattering). By Fourier transforming we find that, in the diffusion approximation, the dependence on the magnitude of the momentum (velocity) in the momentum directional averaged Boltzmann propagator appears only through the diffusion constant, $t > t'$,

$$D(\mathbf{x}, t; \mathbf{x}, t') \equiv \int \frac{d\hat{\mathbf{p}} d\hat{\mathbf{p}}'}{(4\pi)^2} F(\mathbf{p}, \mathbf{x}, t; \mathbf{p}', \mathbf{x}', t') = \int \frac{d\mathbf{q} d\omega}{(2\pi)^4} \frac{e^{i\mathbf{q}\cdot(\mathbf{x}-\mathbf{x}')-i\omega(t-t')}}{-i\omega + D_0 q^2}$$

$$= \frac{e^{-(\mathbf{x}-\mathbf{x}')^2/4D_0(t-t')}}{(4\pi D_0(t-t'))^{d/2}} . \qquad (7.77)$$

This diffusion propagator describes the diffusive or Brownian motion of the particle, the conditional probability for the particle to diffuse from point $\mathbf{x}'$ to $\mathbf{x}$ in time span $t - t'$, described by the one parameter, the diffusion constant. The absence of the explicit appearance of the magnitude of the velocity reflects the fact that the local velocity is a meaningless quantity in Brownian motion.

Exercise 7.5. Show that

$$\langle \mathbf{x}^2 \rangle_{t,\mathbf{x}',t'} \equiv \int d\mathbf{x} \, \mathbf{x}^2 \, D(\mathbf{x}, t; \mathbf{x}', t') = \mathbf{x}'^2 + 2dD_0(t-t') , \qquad (7.78)$$

where the d on the right-hand side is the spatial dimension.

If we are interested only in the long-time and large-distance behavior of the Boltzmannian motion we can, as noted above, get a simplified description of the classical motion of a particle in a random potential. We are thus not interested in the zigzag Boltzmannian trajectories, but only in the smooth large-scale behavior. It is instructive to relate the large-scale behavior to the velocity (or momentum) moments of the distribution function, and the corresponding physical quantities, density and current density. Expanding the distribution function on spherical harmonics

$$f(\mathbf{x}, \mathbf{p}, t) = f_0(\epsilon_\mathbf{p}, \mathbf{x}, t) + \mathbf{p} \cdot \mathbf{f}(\epsilon_\mathbf{p}, \mathbf{x}, t) + \cdots \qquad (7.79)$$

we have the particle current density given in terms of the first moment

$$\mathbf{j}(\mathbf{x}, t) = \frac{1}{m} \int \frac{d\mathbf{p}}{(2\pi\hbar)^3} \, \mathbf{p} \, \mathbf{p} \cdot \mathbf{f}(\epsilon_\mathbf{p}, \mathbf{x}, t) = \frac{1}{3m} \int \frac{d\mathbf{p}}{(2\pi\hbar)^3} \, p^2 \, \mathbf{f}(\epsilon_\mathbf{p}, \mathbf{x}, t) \qquad (7.80)$$

and the density given in terms of the zeroth moment

$$n(\mathbf{x}, t) = \int \frac{d\mathbf{p}}{(2\pi\hbar)^3} \, f_0(\epsilon_\mathbf{p}, \mathbf{x}, t) . \qquad (7.81)$$

Taking the spherical average

$$\langle \ldots \rangle \equiv \int \frac{d\hat{\mathbf{p}}}{4\pi} \ldots \tag{7.82}$$

of the force-free Boltzmann equation, Eq. (7.68), we obtain the zeroth moment equation

$$\frac{\partial f_0(\epsilon_{\mathbf{p}}, \mathbf{x}, t)}{\partial t} + \frac{p^2}{3m} \nabla_{\mathbf{x}} \cdot \mathbf{f}(\epsilon_{\mathbf{p}}, \mathbf{x}, t) = 0 . \tag{7.83}$$

Integrating this equation with respect to momentum gives the continuity equation

$$\frac{\partial n(\mathbf{x}, t)}{\partial t} + \nabla_{\mathbf{x}} \cdot \mathbf{j}(\mathbf{x}, t) = 0 . \tag{7.84}$$

This result is of course independent of whether external fields are present or not. This is seen directly from the Boltzmann equation by integrating with respect to momentum as we have the identity

$$\int \frac{d\hat{\mathbf{p}}}{4\pi} I_{\mathbf{x}, \mathbf{p}, t}[f] = 0 \tag{7.85}$$

simply reflecting that the collision integral respects particle conservation.

Taking the first moment of the Boltzmann equation, $\langle \mathbf{p} \ldots \rangle$,

$$\int \frac{d\hat{\mathbf{p}}}{4\pi} \mathbf{p} \left(\frac{\partial f(\mathbf{x}, \mathbf{p}, t)}{\partial t} + \mathbf{v_p} \cdot \frac{\partial f(\mathbf{x}, \mathbf{p}, t)}{\partial \mathbf{x}} - I_{\mathbf{x}, \mathbf{p}, t}[f] \right) = 0 \tag{7.86}$$

we obtain the first moment equation

$$\frac{p^2}{3} \left(\frac{\partial}{\partial t} + \frac{1}{\tau(\epsilon_{\mathbf{p}})} \right) \mathbf{f}(\mathbf{x}, \mathbf{p}, t) + \frac{p^2}{3m} \frac{\partial f_0(\mathbf{x}, \mathbf{p}, t)}{\partial \mathbf{x}} = 0 , \tag{7.87}$$

where we have repeatedly used the angular average formulas

$$\int \frac{d\hat{\mathbf{p}}}{4\pi} p_\alpha p_\beta = \frac{p^2}{3} \delta_{\alpha\beta} , \qquad \int \frac{d\hat{\mathbf{p}}}{4\pi} p_\alpha p_\beta p_\gamma = 0 . \tag{7.88}$$

We have thus reduced the kinetic equation to a closed set of equations relating the two lowest moments of the distribution function, f_0 and $\mathbf{f}$, and we get the equation satisfied by the zeroth moment f_0:

$$\left(\frac{\partial}{\partial t} + \frac{1}{\tau(\epsilon_{\mathbf{p}})} \right) \frac{\partial f_0(\mathbf{x}, \mathbf{p}, t)}{\partial t} - \frac{p^2}{3m^2} \triangle_{\mathbf{x}} f_0(\mathbf{x}, \mathbf{p}, t) = 0 . \tag{7.89}$$

In a metal the derivatives of the zeroth harmonic of the distribution function for the conduction electrons, $\partial_t f_0(\epsilon_{\mathbf{p}}, \mathbf{x}, t)$ and $\triangle_{\mathbf{x}} f_0(\epsilon_{\mathbf{p}}, \mathbf{x}, t)$, are peaked at the Fermi energy, and we can use the approximations

$$\int \frac{d\mathbf{p}}{(2\pi\hbar)^3} p^2 \triangle_{\mathbf{x}} f_0(\epsilon_{\mathbf{p}}, \mathbf{x}, t) \simeq p_F^2 \int \frac{d\mathbf{p}}{(2\pi\hbar)^3} \triangle_{\mathbf{x}} f_0(\epsilon_{\mathbf{p}}, \mathbf{x}, t) \tag{7.90}$$

and

$$\int \frac{d\mathbf{p}}{(2\pi\hbar)^3} \left(\frac{\partial}{\partial t} + \frac{1}{\tau(\epsilon_{\mathbf{p}})}\right) \frac{\partial f_0(\epsilon_{\mathbf{p}}, \mathbf{x}, t)}{\partial t} \simeq \left(\frac{\partial}{\partial t} + \frac{1}{\tau}\right) \frac{\partial n(\mathbf{x}, t)}{\partial t} , \tag{7.91}$$

where as usual $\tau \equiv \tau(\epsilon_{p_F})$. Assuming only low-frequency oscillations in the density, $\omega\tau \ll 1$,

$$\left|\frac{\partial^2 n}{\partial t^2}\right| \ll \frac{1}{\tau}\left|\frac{\partial n}{\partial t}\right| \tag{7.92}$$

and we obtain from Eq. (7.89) the continuity equation on diffusive form

$$\left(\frac{\partial}{\partial t} - D_0 \triangle_{\mathbf{x}}\right) n(\mathbf{x}, t) = 0 . \tag{7.93}$$

Since $\nabla_{\mathbf{x}} f_0(\epsilon_{\mathbf{p}}, \mathbf{x}, t)$ is peaked at the Fermi energy, we can use the approximation

$$\int \frac{d\mathbf{p}}{(2\pi\hbar)^3} p^2 \nabla_{\mathbf{x}} f_0(\epsilon_{\mathbf{p}}, \mathbf{x}, t) \simeq p_F^2 \int \frac{d\mathbf{p}}{(2\pi\hbar)^3} \nabla_{\mathbf{x}} f_0(\epsilon_{\mathbf{p}}, \mathbf{x}, t) \tag{7.94}$$

and assuming only low-frequency current oscillations

$$\left|\frac{\partial \mathbf{j}(\mathbf{x}, t)}{\partial t}\right| \ll \frac{1}{\tau}|\mathbf{j}(\mathbf{x}, t)| \tag{7.95}$$

we obtain from the first moment equation, Eq. (7.87), the diffusion expression for the current density

$$\mathbf{j}(\mathbf{x}, t) = -D_0 \frac{\partial n(\mathbf{x}, t)}{\partial \mathbf{x}} . \tag{7.96}$$

If we assume that the particle is absent prior to time t', at which time the particle is created at point $\mathbf{x}'$, the diffusion equation, Eq. (7.93), gets a source term, and we obtain for the conditional probability or diffusion propagator $D(\mathbf{x}, t; \mathbf{x}', t')$

$$n(\mathbf{x}, t) = \int d\mathbf{x}' \, D(\mathbf{x}, t; \mathbf{x}', t') \, n(\mathbf{x}', t') \tag{7.97}$$

the equation

$$\left(\frac{\partial}{\partial t} - D_0 \triangle_{\mathbf{x}}\right) D(\mathbf{x}, t; \mathbf{x}', t') = \delta(\mathbf{x} - \mathbf{x}') \, \delta(t - t') \tag{7.98}$$

with the initial condition

$$D(\mathbf{x}, t; \mathbf{x}', t') = 0 , \quad \text{for} \quad t < t' . \tag{7.99}$$

We can solve the equation for the diffusion propagator, the retarded Green's function for the diffusion equation, by referring to the solution of the free particle Schrödinger Green's function equation, Eq. (C.24), and letting $it \to t$, and $\hbar/2m \to D_0$, and we obtain

$$D(\mathbf{x}, t; \mathbf{x}', t') = \theta(t - t') \frac{e^{-\frac{(\mathbf{x}-\mathbf{x}')^2}{4D_0(t-t')}}}{(4\pi D_0(t - t'))^{d/2}} . \tag{7.100}$$

Exercise 7.6. Show that the Diffuson or diffusion propagator has the path integral representation

$$D(\mathbf{x}, t; \mathbf{x}', t') = \int_{\mathbf{x}_{t'}=\mathbf{x}'}^{\mathbf{x}_t=\mathbf{x}} \mathcal{D}\mathbf{x}_{\bar{t}}\; e^{-S_{\mathcal{E}}[\mathbf{x}_{\bar{t}}]} = \int_{\mathbf{x}_{t'}=\mathbf{x}'}^{\mathbf{x}_t=\mathbf{x}} \mathcal{D}\mathbf{x}_{\bar{t}}\; e^{-\int_{t'}^{t} d\bar{t}\, L_{\mathcal{E}}(\dot{\mathbf{x}}_{\bar{t}})} \qquad (7.101)$$

where the Euclidean action $S_{\mathcal{E}}[\mathbf{x}_{\bar{t}}]$ is specified by the Euclidean Lagrangian

$$L_{\mathcal{E}}(\dot{\mathbf{x}}_t) = \frac{\dot{\mathbf{x}}_t^2}{4D_0}. \qquad (7.102)$$

The probability density of diffusive paths is therefore given by

$$P_D[\mathbf{x}_{\bar{t}}] \equiv e^{-S_{\mathcal{E}}[\mathbf{x}_{\bar{t}}]} = e^{-\int_{t'}^{t} d\bar{t}\, \frac{\dot{\mathbf{x}}_{\bar{t}}^2}{4D_0}}. \qquad (7.103)$$

Note that the velocity entering the above Wiener measure is not the local velocity but the velocity averaged over Boltzmannian paths.[16]

Exercise 7.7. Show that, for a diffusing particle, we have the Gaussian property for the characteristic function

$$< e^{i\mathbf{q}\cdot(\mathbf{x}(t)-\mathbf{x}(t'))} >_D = \frac{\int \mathcal{D}\mathbf{x}_{\bar{t}}\, P_D[\mathbf{x}_{\bar{t}}]\, e^{i\mathbf{q}\cdot(\mathbf{x}(t)-\mathbf{x}(t'))}}{\int \mathcal{D}\mathbf{x}_{\bar{t}}\, P_D[\mathbf{x}_{\bar{t}}]} = e^{-D_0 q^2 |t-t'|}. \qquad (7.104)$$

Exercise 7.8. Consider the Diffuson or diffusion propagator specified by the ladder diagrams

---

$$+ \quad \begin{array}{c} \mathbf{p}+ \\ \\ \times \qquad \times \\ \\ \mathbf{p}_-' \end{array} \qquad + \cdots \Bigg) . \tag{7.105}$$

Show that for $ql, \omega\tau \ll 1$, $E \simeq \epsilon_{\mathrm{F}}$, the Diffuson exhibits the diffusion pole

$$D(\mathbf{q}, \omega) \equiv \tau u^{-2} D_E(\mathbf{q}, \omega) = \frac{1}{-i\omega + D_0 q^2} , \tag{7.106}$$

where $D_0 = v_{\mathrm{F}}^2 \tau / d$ is the diffusion constant in d dimensions.

7.5   Quasi-classical Green's function technique

When particles interact there can be strong dependence of the self-energy on the energy variable E, as in the case of electron–phonon interaction in strong coupling materials, say as in a metal such as lead, which is the type of system we for example shall have in mind in this section. The employed quasi-particle approximation Eq. (7.47) is not valid and the structure in the spectral weight, Eq. (7.45), must be respected, leaving no chance of simplicity by integrating over the energy variable E, i.e. of obtaining equations involving only equal-time Green's functions.

There exists a consistent and self-contained approximation scheme for a degenerate Fermi system, valid for a wide range of phenomena, that does not employ the restrictive quasi-particle approximation. It is called the quasi-classical approximation.[17] The electron–phonon interaction can lead to an important structure in the self-energy, i.e. in $\Re e\Sigma$ and Γ, as a function of the variable E. In contrast, as noted by Migdal, the momentum dependence is very weak as a consequence of the phonon energy being small compared with the Fermi energy [25]. The spectral weight function thus becomes a peaked function of the momentum, and we shall exploit this peaked character.

The left–right subtraction trick dismissed the strong linear E-dependence in the inverse propagator G_0^{-1}, and similarly its strong momentum dependence, its $\xi_{\mathbf{p}}$-dependence, $\xi_{\mathbf{p}} = \epsilon_{\mathbf{p}} - \mu$. It therefore allows, when there is only weak momentum dependence of the self-energy, i.e. short-range effective interaction, which is typically the case for electronic interactions, integration over the variable $\xi_{\mathbf{p}}$, so-called ξ-integration. The peaked character of the spectral weight in the variable ξ will, in conjunction with multiplying other quantities, restrict their momentum dependence to the Fermi surface. We shall therefore consider the ξ-integrated Green's function

---

[17]This scheme was first applied by Prange and Kadanoff in their treatment of transport phenomena in the electron–phonon system [23]. It was later extended to describe transport in superfluid systems by Eilenberger [24], the topic of the next chapter.

or quasi-classical Green's function[18]

$$g(\mathbf{R}, \hat{\mathbf{p}}, t_1, t_{1'}) = \frac{i}{\pi} \int d\xi\, G(\mathbf{R}, \mathbf{p}, t_1, t_{1'}) .$$ (7.107)

We note that care should be exercised with respect to ξ-integration, since the integrand is not well behaved for large values of ξ, falling off only as $1/\xi$. The ξ-integration should be understood in the following sense of deforming the integration contour as depicted in Figure 7.1: the ξ-integration is split into a low- and high-energy contribution, and only the low-energy contribution is important in the kinetic equation since high-energy contributions do not contribute.

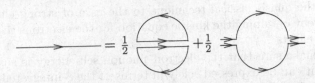

Figure 7.1 Splitting in high- and low-energy contributions.

The semicircles are specified by a cut-off energy E_c, which is chosen much larger than the Fermi energy. The remaining high-energy contribution to the Green's function does not depend on the non-equilibrium state, i.e. it is a constant, and therefore drops out of the left–right subtracted Dyson equation. We immediately return to this point again when expressing physical quantities, such as average currents and densities in terms of the quasi-classical Green's function, and later in Section 8.3 to provide a less formal and more physical understanding of the quasi-classical Green's function.

Let us first determine measurable quantities in terms of the quasi-classical Green's function, say density and currents, in the presence of an electromagnetic field $(\mathbf{A}, \varphi)$. The charge density becomes, in terms of the quasi-classical Green's function,

$$\rho(\mathbf{R}, T) = -\frac{1}{2} e N_0 \int \frac{d\hat{\mathbf{p}}}{4\pi} \int dE\, g^{\mathrm{K}}(E, \hat{\mathbf{p}}, \mathbf{R}, T) - 2e^2 N_0\, \varphi(\mathbf{R}, T) ,$$ (7.108)

where N_0 is the density of states at the Fermi energy, and the current density is given by

$$\mathbf{j}(\mathbf{R}, T) = -\frac{1}{2} e N_0 \int \frac{d\hat{\mathbf{p}}}{4\pi} \int dE\, \mathbf{v}_{\mathrm{F}}\, g^{\mathrm{K}}(E, \hat{\mathbf{p}}, \mathbf{R}, T) .$$ (7.109)

---

[18]Here and in the following, we assume for simplicity a spherical Fermi surface. For a general Fermi surface one decomposes according to

$$\frac{d\mathbf{p}}{(2\pi)^3} = \frac{d\xi_{\hat{\mathbf{p}}}}{v_{\mathrm{F}}} \frac{ds_{\mathrm{F}}}{(2\pi)^3} ,$$

where $d\xi_{\hat{\mathbf{p}}}/v_{\mathrm{F}}$ is the length of the momentum increment measured from the Fermi surface in the directions $\pm\hat{\mathbf{p}}$, and ds_{F} is the corresponding Fermi surface area element.

The ξ-integration does not respect the proper order of integrations, momentum integration being last because of the convergence property of the Green's function. We thus encounter the high energy contribution to the density, the second term on the right in Eq. (7.108), whereas in the current density the high-energy contribution cancels the term proportional to the vector potential, the so-called diamagnetic term. We observe, as discussed in Section 6.2, that as far as the high-energy contributions are concerned, the analysis and their calculation is equivalent to their appearance in linear response expressions.[19] The high-energy contributions also follow from the gauge transformation properties of the Green's function.

7.5.1   Electron–phonon interaction

Here we apply the quasi-classical technique to the case of strong electron–phonon interaction, thereby obtaining the kinetic equation for the electrons that includes the renormalization effects.

Let us first make sure that the electron–phonon self-energy is susceptible to ξ-integration, i.e. it can be expressed solely in terms of the ξ-integrated Green's function or quasi-classical Green's function. Migdal's theorem states that the electron–phonon self-energy diagrams for the electron Green's function where phonon lines cross are small in the parameter ω_D/ϵ_F, where ω_D is the typical phonon energy, i.e. vertex corrections are negligible [25].[20] In this approximation, which indeed is a good one in metals, with an accuracy of order 1%, the electronic self-energy is represented by a single skeleton diagram as depicted in Figure 7.2.

Figure 7.2 Electron–phonon self-energy.

or analytically for the lowest order in ω_D/ϵ_F contribution to the electron self-energy

$$\Sigma_{ij}^{(e-ph)}(1,1') = ig^2\,\gamma_{ii'}^k\,G_{i'j'}(1,1')\,D_{kk'}(1,1')\,\tilde{\gamma}_{j'j}^{k'}\,. \tag{7.110}$$

In the mixed coordinates with respect to the spatial coordinates we get (suppressing

---

[19] A detailed discussion of this is given in chapters 7 and 8 of reference [1].

[20] The demonstration of Migdal's theorem is quite analogous to that of crossing impurity diagrams being small. Crossing lines result in propagators having restriction on the momentum range for which they provide a large contribution. Contributions from such diagrams thus become small owing to phase space restrictions. In the case of electron–phonon interaction, the range is set by the typical phonon energy. For details on diagram estimation see chapter 3 of reference [1].

the arguments on the left $(\mathbf{R}, \mathbf{p}, t_1, t_{1'}))$[21]

$$\Sigma_{ij}^{(\text{e}-\text{ph})} = ig^2 \gamma_{ii'}^k \int \frac{d\mathbf{p}'}{(2\pi)^3} \, G_{i'j'}(\mathbf{R}, \mathbf{p}', t_1, t_{1'}) \, D_{kk'}(\mathbf{R}, \mathbf{p} - \mathbf{p}', t_1, t_{1'}) \, \tilde{\gamma}_{j'j}^{k'}. \quad (7.111)$$

The momentum integration can be split into integrations over angular (or in general Fermi surface) and length of the momentum measured from the Fermi surface

$$\int \frac{d\mathbf{p}}{(2\pi)^3} = \int d\xi \, N(\xi) \int \frac{d\hat{\mathbf{p}}}{4\pi} = N_0 \int d\xi \int \frac{d\hat{\mathbf{p}}}{4\pi} \quad (7.112)$$

and the last equality is valid when particle–hole symmetry applies.[22] Using the fact that the Debye energy is small compared with the Fermi energy,[23] the various electron Green's function are tied to the Fermi surface, and we obtain the electron–phonon matrix self-energy expressed in terms of the quasi-classical matrix electron Green's function

$$\sigma_{ij}^{(\text{e}-\text{ph})}(\mathbf{R}, \hat{\mathbf{p}}, t_1, t_{1'}) = \frac{\lambda}{4} \gamma_{ii'}^k \int d\hat{\mathbf{p}}' g_{i'j'}(\mathbf{R}, \hat{\mathbf{p}}', t_1, t_{1'}) \, D_{kk'}(\mathbf{R}, p_{\text{F}}(\hat{\mathbf{p}} - \hat{\mathbf{p}}'), t_1, t_{1'}) \tilde{\gamma}_{j'j}^{k'} \,,$$
$$(7.113)$$

where $\lambda = g^2 N_0$ is the dimensionless electron–phonon coupling constant. The matrix components of the matrix self-energy are therefore

$$\sigma_{\text{e}-\text{ph}}^{\text{R(A)}}(\mathbf{R}, \hat{\mathbf{p}}, t_1, t_{1'}) = \frac{\lambda}{8} \int d\hat{\mathbf{p}}' \left(g^{\text{K}}(\mathbf{R}, \hat{\mathbf{p}}', t_1, t_{1'}) D^{\text{R(A)}}(\mathbf{R}, p_{\text{F}}(\hat{\mathbf{p}} - \hat{\mathbf{p}}'), t_1, t_{1'}) \right.$$

$$\left. + \; g^{\text{R(A)}}(\mathbf{R}, \hat{\mathbf{p}}', t_1, t_{1'}) D^{\text{K}}(\mathbf{R}, p_{\text{F}}(\hat{\mathbf{p}} - \hat{\mathbf{p}}'), t_1, t_{1'}) \right) \quad (7.114)$$

and

$$\sigma_{\text{e}-\text{ph}}^{\text{K}}(\mathbf{R}, \hat{\mathbf{p}}, t_1, t_{1'}) = \frac{\lambda}{8} \int d\hat{\mathbf{p}}' (g^{\text{R}}(\mathbf{R}, \hat{\mathbf{p}}', t_1, t_{1'}) D^{\text{R}}(\mathbf{R}, p_{\text{F}}(\hat{\mathbf{p}} - \hat{\mathbf{p}}'), t_1, t_{1'})$$

$$+ \; g^{\text{A}}(\mathbf{R}, \hat{\mathbf{p}}', t_1, t_{1'}) D^{\text{A}}(\mathbf{R}, p_{\text{F}}(\hat{\mathbf{p}} - \hat{\mathbf{p}}'), t_1, t_{1'})$$

$$+ \; g^{\text{K}}(\mathbf{R}, \hat{\mathbf{p}}', t_1, t_{1'}) D^{\text{K}}(\mathbf{R}, p_{\text{F}}(\hat{\mathbf{p}} - \hat{\mathbf{p}}'), t_1, t_{1'})) \quad (7.115)$$

or equivalently

$$\sigma_{\text{e}-\text{ph}}^{\text{K}} = \frac{\lambda}{8} \int d\hat{\mathbf{p}}' \left((g^{\text{R}} - g^{\text{A}})(D^{\text{R}} - D^{\text{A}}) + g^{\text{K}} D^{\text{K}} \right) \quad (7.116)$$

---

[21] In the case of impurity scattering, the self-energy is expressed in terms of the quasi-classical Green's function according to

$$\sigma_{\text{imp}}(E, \mathbf{R}, T) = -\frac{i}{2\tau} \int \frac{d\hat{\mathbf{p}}'}{4\pi} \, g(\hat{\mathbf{p}}', E, \mathbf{R}, T) \,,$$

where the high-energy cut-off is provided by the momentum dependence of the impurity potential, providing necessary convergence.

[22] Or rather, owing to this step the quasi-classical approximation is unable to account for effects due to particle–hole asymmetry.

[23] Or equivalently, the sound velocity is small compared with the Fermi velocity.

since

$$g^{\mathrm{R}}(t_1, t_{1'}) \, D^{\mathrm{A}}(t_1, t_{1'}) \; = \; 0 \; = \; g^{\mathrm{A}}(t_1, t_{1'}) \, D^{\mathrm{R}}(t_1, t_{1'}) \,. \tag{7.117}$$

Utilizing the peaked character of the electron spectral weight function in the ξ-variable, the momentum dependence of the self-energy can be neglected, and the left–right subtracted Dyson equations, Eq. (7.1), can be integrated with respect to ξ, giving the quantum kinetic equation

$$\left[g_0^{-1} + i\Re\sigma \,\overset{\circ}{,}\, g^{\mathrm{K}} \right]_- \; = \; 2i\sigma^{\mathrm{K}} \; - \; i\left[(\sigma^{\mathrm{R}} - \sigma^{\mathrm{A}}) \,\overset{\circ}{,}\, g^{\mathrm{K}} \right]_+ , \tag{7.118}$$

where

$$g_0^{-1}(\mathbf{R}, \hat{\mathbf{p}}, t_1, t_{1'}) = g_0^{-1}(\mathbf{R}, \hat{\mathbf{p}}, t_1) \, \delta(t_1 - t_{1'}) \tag{7.119}$$

and

$$g_0^{-1}(\mathbf{R}, \hat{\mathbf{p}}, t_1) = \partial_{t_1} + \mathbf{v}_{\mathrm{F}} \cdot (\nabla_{\mathbf{x}_1} - ie\mathbf{A}(\mathbf{R}, t_1)) + ie\phi(\mathbf{R}, t_1) - \frac{e^2}{2m}\mathbf{A}^2(\mathbf{R}, t_1). \tag{7.120}$$

Here $\mathbf{v}_{\mathrm{F}} = \mathbf{p}_{\mathrm{F}}/m$, the Fermi velocity, specifies the Fermi surface direction, and $\circ$ implies matrix multiplication in the time variable. We have considered the case where, say, the electrons in a metal are subject to electromagnetic fields.

From the spectral representation

$$G^{\mathrm{R(A)}}(E, \mathbf{p}, \mathbf{R}, T) \; = \; \int_{-\infty}^{\infty} \frac{dE'}{2\pi} \, \frac{A(E', \mathbf{p}, \mathbf{R}, T)}{E - E' \, \underset{(\pm)}{} \, i0} \tag{7.121}$$

it follows that ξ-integrating $\Re eG$ gives a state independent constant and the last term on the left-hand side in Eq. (7.3) vanishes upon ξ-integration.

The form of g_0^{-1} follows from the following observation where for definiteness we focus on the scalar potential term. First transform to the mixed spatial coordinates

$$\varphi(\mathbf{x}_1, t_1) \, G(\mathbf{x}_1, t_1, \mathbf{x}_{1'}, t_{1'}) \; = \; \varphi(\mathbf{R} + \mathbf{r}/2, t_1) \, G(\mathbf{R}, \mathbf{r}, t_1, t_{1'}) \,. \tag{7.122}$$

Since the Green's function $G(\mathbf{R}, \mathbf{r})$ is a wildly oscillating function in the relative coordinate $\mathbf{r}$, the function is essentially zero when $r \gg k_{\mathrm{F}}^{-1}$, and since we shall assume that the scalar potential is slowly varying on the atomic length scale we have

$$\varphi(\mathbf{x}_1, , t_1) \, G(\mathbf{x}_1, t_1, \mathbf{x}_{1'}, t_{1'}) \; \simeq \; \varphi(\mathbf{R}) \, G(\mathbf{R}, \mathbf{r}, t_1, t_{1'}). \tag{7.123}$$

It can be instructive to perform the equivalent argument on the Fourier-transformed product giving (being irrelevant for the manipulations, the time variables are suppressed)

$$\varphi(\mathbf{R} + \mathbf{r}/2) \, G(\mathbf{R}, \mathbf{r}) \; = \; \int d\mathbf{k} d\mathbf{P} d\mathbf{p} \; e^{i(\mathbf{R}\cdot\mathbf{P} + \mathbf{r}\cdot\mathbf{p})} \, \varphi(\mathbf{k}) \, G(\mathbf{P} - \mathbf{k}, \mathbf{p} - \mathbf{k}/2) \,, \tag{7.124}$$

where the shifts of variables, $\mathbf{P} + \mathbf{k} \to \mathbf{P}$ and $\mathbf{P} + \mathbf{k}/2 \to \mathbf{p}$, have been performed. The quasi-classical approximation consists of the weak assumption that the external perturbation only has Fourier components for wave vectors small compared with the

Fermi wave vector, $k \ll k_F$, so that $G(\mathbf{P}-\mathbf{k}, \mathbf{p}-\mathbf{k}/2) \simeq G(\mathbf{P}-\mathbf{k}, \mathbf{p})$, again leading to the stated result, Eq. (7.123). In the quasi-classical approximation the effect of the Lorentz force is lost, and for a perturbing electric field we might as well transform to a gauge where the vector potential is absent. This is the price paid for the quasi-classical approximation, which is less severe in the superconducting state, and we will return to effects of the Lorentz force in the normal state in Section 7.6. However, we note that the influence on the phase of the Green's function is fully incorpotated in the quasi-classical approximation, a fact we shall exploit when considering the weak localization effect in Chapter 11.

A simplification which arises in the normal state, and should be contrasted with the more complicated situation in the superconducting state to be discussed in Section 8.2.3, is the lack of structure in the ξ-integrated retarded and advanced Green's functions

$$g^{R(A)}(\mathbf{R}, \hat{\mathbf{p}}, t_1, t_{1'}) = \genfrac{}{}{0pt}{}{+}{(-)} \, \delta(t_1 - t_{1'}) \quad , \quad g^{R(A)}(\mathbf{R}, \hat{\mathbf{p}}, E, T) = \genfrac{}{}{0pt}{}{+}{(-)} \, 1 \qquad (7.125)$$

and they thus contain no information since particle–hole asymmetry effects are neglected, i.e. the variation of the density of states through the Fermi surface is neglected. This fact leaves the quantum kinetic equation, Eq. (7.118), together with the self-energy expressions a closed set of equations for g^K.

We emphasize again that in obtaining the quasi-classical equation of motion only the degeneracy of the Fermi system is used, restricting the characteristic frequency and wave vectors to modestly obey the restrictions

$$q \ll k_F \quad , \quad \omega \ll \epsilon_F . \qquad (7.126)$$

These criteria are well satisfied for transport phenomena in degenerate Fermi systems.

In contrast to the performed approximation for the convolution in space due to the degeneracy of the Fermi system, there is in general no simple approximation for the convolution in the time variables. Two different approximation schemes are immediately available: one consists of linearization with respect to a perturbation such as an electric field, allowing frequencies restricted only by the Fermi energy, $\omega < \epsilon_F$, to be considered, but of course restricted to weak fields. The other assumes perturbations to be sufficiently slowly varying in time that a lowest-order expansion in the time derivative is valid

$$[A \, \stackrel{\circ}{,} \, B]_- \simeq \partial_E^A A \, \partial_T^B B - \partial_T^A A \, \partial_E^B B . \qquad (7.127)$$

For definiteness, we shall employ the second scheme here. In order to reduce the general quantum kinetic equation, Eq. (7.118), to a simpler looking transport equation, we introduce the mixed coordinates with respect to the temporal coordinates and perform the *gradient* expansion in these variables giving

$$((1 - \partial_e \Re e\sigma)\partial_T + \partial_T \Re e\sigma \, \partial_E + \mathbf{v}_F \cdot \nabla_\mathbf{R} + e\partial_T \varphi \, \partial_E) g^K = I_{e-ph} , \qquad (7.128)$$

where the collision integral is

$$I_{e-ph} = 2i\sigma^K - \gamma g^K , \qquad (7.129)$$

where

$$\gamma = i(\sigma^R - \sigma^A).\tag{7.130}$$

The two terms in the collision integral constitute the scattering-in and scattering-out terms, respectively. According to Eq. (7.114) and Eq. (7.116) they are determined by (space and time variables suppressed)

$$\gamma(E,\hat{\mathbf{p}}) = -\pi \int \frac{d\hat{\mathbf{p}}'}{4\pi} \int dE'\, \mu(p_F(\hat{\mathbf{p}} - \hat{\mathbf{p}}'), E - E') \left(\coth \frac{E' - E}{2T} - h(E', \hat{\mathbf{p}}') \right)$$

$$\tag{7.131}$$

and

$$i\sigma^K_{e-ph}(E,\hat{\mathbf{p}}) = -\pi \int \frac{d\hat{\mathbf{p}}'}{4\pi} \int dE'\, \mu(p_F(\hat{\mathbf{p}} - \hat{\mathbf{p}}'), E - E') \left(h(E', \hat{\mathbf{p}}') \coth \frac{E' - E}{2T} - 1 \right),$$

$$\tag{7.132}$$

where we have introduced the distribution function

$$h(E, \hat{\mathbf{p}}, \mathbf{R}, T) = \frac{1}{2} g^K(E, \hat{\mathbf{p}}, \mathbf{R}, T)\tag{7.133}$$

and

$$\mu(\mathbf{q}, E) = \frac{iN_0 |g_{\mathbf{q}}|^2}{2\pi} \left(D^R(E, \mathbf{q}) - D^A(E, \mathbf{q}) \right)\tag{7.134}$$

is the Eliashberg spectral weight function. Here we have allowed for a more general longitudinal electron–phonon coupling than the jellium model. The coupling is denoted $g_{\mathbf{q}}$, corresponding to momentum transfer $\mathbf{q}$. The connection to the jellium model is $|g_{\mathbf{q}}| = g\sqrt{\omega_{\mathbf{q}}/2}$, where $\omega_{\mathbf{q}} = c\,\mathbf{q}$ is the energy of a phonon with momentum $\mathbf{q}$, c being the sound velocity.

We have further assumed that the phonons are in thermal equilibrium at temperature T,[24] and have therefore used the fluctuation–dissipation relation for bosons, Eq. (5.103),

$$D^K(E, \mathbf{p}) = \left(D^R(E, \mathbf{p}) - D^A(E, \mathbf{p}) \right) \coth \frac{E}{2kT}.\tag{7.135}$$

We note that the variables in the distribution function are quite different from that of the classical Boltzmann equation for electron–phonon interaction, which is the Wigner coordinates $(\mathbf{p}, \mathbf{R}, T)$. Here an energy variable and a position on the Fermi surface appear separately (besides space and time). This feature of the quasi-classical equation reflects the fact that we do not rely on a definite relation between

---

[24]This is not necessary, but would otherwise lead to the requirement of considering the kinetic equation for the phonons also. For typical transport situations in a metal, the approximation, viz. considering the phonons a heat reservoir, is applicable.

the energy and momentum variables as is the case in the quasi-particle approximation of Section 7.3.2.

Introducing the Fermi and Bose type distribution functions

$$f(E, \hat{\mathbf{p}}, \mathbf{R}, T) = \frac{1}{2}\left(1 - h(E, \hat{\mathbf{p}}, \mathbf{R}, T)\right) \tag{7.136}$$

and

$$n(E) = -\frac{1}{2}\left(1 - \coth\frac{E}{2kT}\right) \tag{7.137}$$

the collision integral takes the more familiar form

$$I_{\text{e-ph}} = -2\pi \int \frac{d\hat{\mathbf{p}}'}{4\pi} \int dE' \; \mu(p_{\text{F}}(\hat{\mathbf{p}} - \hat{\mathbf{p}}'), E - E') \, R^{E\hat{\mathbf{p}}}_{E'\hat{\mathbf{p}}'} \,, \tag{7.138}$$

where

$$R^{E\hat{\mathbf{p}}}_{E'\hat{\mathbf{p}}'} = (1 + n(E - E'))f(E, \hat{\mathbf{p}})(1 - f(E', \hat{\mathbf{p}}'))$$

$$- \; n(E - E')(1 - f(E, \hat{\mathbf{p}}))f(E', \hat{\mathbf{p}}') \,. \tag{7.139}$$

Finally, introducing a gauge invariant distribution function by the substitution

$$f(E) \;\rightarrow\; f(E - \varphi(\mathbf{R}, T)) \tag{7.140}$$

we obtain the quantum kinetic equation

$$((1 - \partial_E \Re e\sigma)\partial_T + \partial_T \Re e\sigma \, \partial_E + \mathbf{v}_{\text{F}} \cdot (\nabla_{\mathbf{R}} + e\mathbf{E}(\mathbf{R}, T)) \, \partial_E) f = I_{\text{e-ph}}[f] \,, \tag{7.141}$$

where $\mathbf{E}(\mathbf{R}, T) = -\nabla_{\mathbf{R}}\varphi(\mathbf{R}, T)$ is the perturbing electric field. We note that the self-energy terms on the right-hand side in the kinetic equation describe collision processes, and we now turn to show that the self-energies on the left describe renormalization effects, in particular mass renormalization due to the electron–phonon interaction.

From the kinetic equation, Eq. (7.141), Prange and Kadanoff [23] drew the conclusion that many-body effects can be seen only in time-dependent transport properties and that static transport coefficients, such as d.c. conductivity and thermal conductivity are correctly given by the usual Boltzmann results. However, there is a restriction to the generality of this statement, viz. that in deriving the quasi-classical equation of motion particle–hole symmetry was assumed. Within the quasi-classical scheme, all thermoelectric coefficients therefore vanish, and no conclusion can be drawn about many-body effects on the thermo-electric properties. In Section 7.6.1, we shall by not employing the quasi-classical scheme consider how thermo-electric properties do get renormalized by the electron–phonon interaction.

7.5.2   Renormalization of the a.c. conductivity

As an example of electron–phonon renormalization of time-dependent transport coefficients we shall consider the a.c. conductivity in the frequency range $\omega\tau_{e-ph} \gg 1$,
where $1/\tau_{e-ph}$ is the clean-limit electron–phonon scattering rate for an electron on
the Fermi surface

$$
\frac{1}{\tau(\epsilon_F, T)} =
\begin{cases}
\frac{7\pi\zeta(3)}{2}\lambda\,\frac{(kT)^3}{\hbar(p_F c)^2} & kT \ll 2p_F c \\[4mm]
2\pi\lambda\frac{kT}{\hbar} & kT \gg 2p_F c
\end{cases}
\tag{7.142}
$$

ζ being Riemann's zeta function.[25] For definiteness we also consider the temperature
to be low compared to the Debye temperature θ_D. In this high-frequency limit the
collision integral can be neglected and the linearized kinetic equation takes the simple
form

$$
(1 - \partial_E \Re e\sigma)\partial_T h + \partial_T \Re e\sigma\,\partial_E h_0 + e\mathbf{v}_F \cdot \mathbf{E}(T)\,\partial_E h_0 = 0
\tag{7.143}
$$

for a spatially homogeneous electric field. Except for a real constant, just renormal-
izing the chemical potential, we have according to the Feynman rules

$$
\Re e\sigma = \frac{1}{2}N_0 \int\frac{d\hat{\mathbf{p}}'}{4\pi}\int dE'\,|g_{\mathbf{p}_F-\mathbf{p}_F'}|^2\,h(E',\hat{\mathbf{p}}',\mathbf{R},T)\,\Re e D(\mathbf{p}_F - \mathbf{p}_F', E - E')\,,
\tag{7.144}
$$

where $g_{\mathbf{p}_F-\mathbf{p}_F'}$ denotes the electron–phonon coupling, and

$$
\Re e D = \frac{1}{2}(D^R + D^A)\,.
\tag{7.145}
$$

For an applied monochromatic field, $\mathbf{E}(t) = \mathbf{E}_0\exp\{-i\omega t\}$, the solution can be
sought in the form

$$
h_1 = ae\mathbf{E}\cdot\mathbf{v}_F\,\partial_E\,h_0\,,
\tag{7.146}
$$

where the constant a remains to be determined. Inserting Eq. (7.146) into the kinetic
equation we obtain

$$
a = \frac{1}{-i\omega(1 + \lambda^*)}\,,
\tag{7.147}
$$

where

$$
\lambda^* = 2N_0 \int\frac{d\hat{\mathbf{p}}'}{4\pi}\int dE'\,\frac{|g_{\mathbf{p}_F-\mathbf{p}_F'}|^2}{\omega_{\mathbf{p}_F-\mathbf{p}_F'}}\,(1 - \hat{\mathbf{p}}\cdot\hat{\mathbf{p}}')\,.
\tag{7.148}
$$

The current can now be evaluated and we obtain for the frequency dependence
of the conductivity

$$
\sigma(\omega) = \frac{ne^2}{-i\omega m_{\text{opt}}}
\tag{7.149}
$$

---

[25] For a calculation of the collision rate see Exercise 8.8 on page 237 and Section 11.3.1, and, for
example, chapter 10 of reference [1].

where the optical mass is renormalized according to

$$m_{\mathrm{opt}} = m(1 + \lambda^*) \tag{7.150}$$

a result originally obtained by Holstein using a different approach, viz. linear response theory [26]. We note that it is the non-equilibrium electron contribution to the real part of the self-energy that makes the optical mass renormalization different from the specific heat mass renormalization, $m \to (1 + \lambda)m$ (see also the result of Exercise 8.8 on page 237).

As a consequence of electron–phonon interaction, the physically observed mass of the electron is not the mass or band structure effective mass of the electron, but it has been changed owing to the interaction.[26] Furthermore, we note that the magnitude of the mass renormalization depends on how the system is probed, the optical mass being different from the specific heat mass.

7.5.3   Excitation representation

The quasi-classical theory leads to equations which are more general than the Boltzmann equation, and the kinetic equation looks quite different. We have shown that the basic variables, besides space and time, are the energy variable and the momentum position on the Fermi surface. Although the electron–phonon interaction does not permit the quasi-particle approximation *a priori*, we recapitulate the derivation of reference [23] showing that it is still possible to cast the electron–phonon transport theory into the standard Landau–Boltzmann form. We start by defining a quasi-particle energy $E_{\mathbf{p}}$, which is defined implicitly by (we suppress the space-time variables and use the short notation $E_{\mathbf{p}} = E(\mathbf{p}, \mathbf{R}, T)$)

$$E_{\mathbf{p}} = \xi_{\mathbf{p}} + \Re e\sigma(E_{\mathbf{p}} + e\varphi(\mathbf{R}, T), \hat{\mathbf{p}}, \mathbf{R}, T) \tag{7.151}$$

thereby satisfying the equations

$$\nabla_{\mathbf{p}} E_{\mathbf{p}} = Z_{\mathbf{p}} \nabla_{\mathbf{p}} \xi_{\mathbf{p}} \tag{7.152}$$

and

$$\nabla_{\mathbf{R}} E_{\mathbf{p}} = Z_{\mathbf{p}} (e\nabla_{\mathbf{R}} \varphi \, \partial_E \Re e\sigma + \nabla_{\mathbf{R}} \Re e\sigma) \Big|_{E = E_{\mathbf{p}} + e\varphi(\mathbf{R}, T)} \tag{7.153}$$

and

$$\partial_T E_{\mathbf{p}} = Z_{\mathbf{p}} (e\partial_T \, \varphi \partial_E \Re e\sigma + \partial_T \Re e\sigma) \Big|_{E = E_{\mathbf{p}} + e\varphi(\mathbf{R}, T)} \tag{7.154}$$

where in Eq. (7.151), assuming for simplicity a spherical Fermi surface, any angular dependence of the real part of the self-energy has been neglected, and the so-called

---

[26] Thus interaction causes renormalization of observable quantities. This point of view is the rationale for avoiding the ubiquitous infinities occurring in quantum field theories such as QED, and being taken to an extreme since there the unobservable bare mass (and the bare coupling constant, the bare electron charge) is taken, it turns out, to be infinite in order to provide the finite and accurate predictions of QED by phenomenologically introducing the observed mass and charge.

wave-function renormalization constant

$$Z_{\mathbf{p}} = (1 - \partial_E \Re e\sigma)^{-1}\Big|_{E=E_{\mathbf{p}}+e\varphi(\mathbf{R},T)} \tag{7.155}$$

has been introduced.

The energy variable E in the kinetic equation is now set equal to $E_{\mathbf{p}} + e\varphi$ and we introduce the distribution function (again suppressing the space-time variables)

$$n_{\mathbf{p}} = f(E, \hat{\mathbf{p}}, \mathbf{R}, T)\Big|_{E=E_{\mathbf{p}}+e\varphi(\mathbf{R},T)} \tag{7.156}$$

Using the relations

$$\nabla_{\mathbf{p}} n = \nabla_{\mathbf{p}} E_{\mathbf{p}}(\partial_E f)\Big|_{E=E_{\mathbf{p}}+e\varphi(\mathbf{R},T)} \tag{7.157}$$

and

$$\nabla_{\mathbf{R}} n = (\nabla_{\mathbf{R}} f + \nabla_{\mathbf{R}}(E_{\mathbf{p}} + e\varphi(\mathbf{R},T))\partial_E f)\Big|_{E=E_{\mathbf{p}}+e\varphi(\mathbf{R},T)} \tag{7.158}$$

and

$$\partial_T n = (\partial_T f + \partial_T(E_{\mathbf{p}} + e\varphi(\mathbf{R},T))\partial_E f)\Big|_{E=E_{\mathbf{p}}+e\varphi(\mathbf{R},T)} \tag{7.159}$$

and Eqs. (7.152–7.154), we obtain the kinetic equation of the form

$$Z_{\mathbf{p}}^{-1}(\partial_T + \nabla_{\mathbf{p}} E_{\mathbf{p}} \cdot \nabla_{\mathbf{R}} - \nabla_{\mathbf{R}}(E_{\mathbf{p}} + \varphi(\mathbf{R},T)) \cdot \nabla_{\mathbf{p}}) \, n(\mathbf{p}, \mathbf{R}, T) = \tilde{I}_{\text{e-ph}} \tag{7.160}$$

with the electron–phonon collision integral

$$\tilde{I}_{\text{e-ph}} = -\frac{2\pi}{N_0} \int \frac{d\mathbf{p}'}{(2\pi)^3} \, Z_{\mathbf{p}'} \, \tilde{\mu}(\mathbf{p} - \mathbf{p}') \, R_{\hat{\mathbf{p}}'}^{\hat{\mathbf{p}}} \,, \tag{7.161}$$

where

$$R_{\hat{\mathbf{p}}'}^{\hat{\mathbf{p}}} = (1 + N(E_{\mathbf{p}} - E_{\mathbf{p}'}))n_{\mathbf{p}}(1 - n_{\mathbf{p}'}) - N(E_{\mathbf{p}} - E_{\mathbf{p}'})(1 - n_{\mathbf{p}})n_{\mathbf{p}'} \tag{7.162}$$

and

$$\tilde{\mu}(\mathbf{p} - \mathbf{p}') = \frac{iN_0|g_{\mathbf{p}-\mathbf{p}'}|^2}{2\pi}(D^{\text{R}}(\mathbf{p} - \mathbf{p}', E_{\mathbf{p}} - E_{\mathbf{p}'}) - D^{\text{A}}(\mathbf{p} - \mathbf{p}', E_{\mathbf{p}} - E_{\mathbf{p}'})) \,. \tag{7.163}$$

In transforming the collision integral we have utilized the substitution

$$N_0 \int \frac{d\hat{\mathbf{p}}}{4\pi} \int dE \;\to\; N_0 \int \frac{d\hat{\mathbf{p}}}{4\pi} \int d\xi_{\mathbf{p}} \frac{dE_{\mathbf{p}}}{d\xi_{\mathbf{p}}} \;\to\; \int \frac{d\mathbf{p}}{(2\pi)^3} Z_{\mathbf{p}} \,. \tag{7.164}$$

Since the sound velocity is much smaller than the Fermi velocity, the phonon damping is negligible, and the phonon spectral weight function has delta function character

$$\tilde{\mu}(\mathbf{p} - \mathbf{p}') = N_0|g_{\mathbf{p}-\mathbf{p}'}|^2(\delta(E_{\mathbf{p}} - E_{\mathbf{p}'} - \omega_{\mathbf{p}-\mathbf{p}'}) - \delta(E_{\mathbf{p}} - E_{\mathbf{p}'} + \omega_{\mathbf{p}-\mathbf{p}'})) \,. \tag{7.165}$$

The kinetic equation can then be written in the final form

$$(\partial_T + \nabla_{\mathbf{p}} E_{\mathbf{p}} \cdot \nabla_{\mathbf{R}} - \nabla_{\mathbf{R}}(E_{\mathbf{p}} + \varphi(\mathbf{R}, T)) \cdot \nabla_{\mathbf{p}}) \, n(\mathbf{p}, \mathbf{R}, T) = I_{\text{e-ph}} , \qquad (7.166)$$

where the electron–phonon collision integral is

$$I_{\text{e-ph}} = -2\pi \int \frac{d\mathbf{p}'}{(2\pi)^3} \, Z_{\mathbf{p}} Z_{\mathbf{p}'} |g_{\mathbf{p}-\mathbf{p}'}|^2 \, R_{\hat{\mathbf{p}}'}^{\hat{\mathbf{p}}} \, (\delta(E_{\mathbf{p}} - E_{\mathbf{p}'} - \omega_{\mathbf{p}-\mathbf{p}'}).$$

$$- \, \delta(E_{\mathbf{p}} - E_{\mathbf{p}'} + \omega_{\mathbf{p}-\mathbf{p}'})) . \quad (7.167)$$

This has the form of the familiar Landau–Boltzmann equation, except for the fact that the transition matrix elements are renormalized.

We stress that only the quasi-classical approximation was used to derive the above kinetic equation. In particular, we have not assumed any relation between the lifetime of a electron in a momentum state at the Fermi surface and the temperature. This would have been necessary for invoking a quasi-particle description in order to justify the existence of long-lived electronic momentum states. It has thus been established from microscopic principles that the validity of the Landau–Boltzmann description of the electron–phonon system is determined not by the Peierls criterion (stating the upper bound is not the Fermi energy but the temperature), but by the Landau criterion

$$\frac{\hbar}{\tau(\epsilon_{\text{F}}, T)} \ll \epsilon_{\text{F}} . \qquad (7.168)$$

This is of importance for the validity of the Boltzmann description of transport in semiconductors, for which the Peierls criterion would be detrimental.

7.5.4   Particle conservation

That an approximation for the quasi-classical Green's function respects conservation laws, say particle number conservation, is not in general as easily stated as for the microscopic Green's function. We therefore establish it here explicitly. The collision integral, Eq. (7.167), has the invariant

$$\int \frac{d\mathbf{p}}{(2\pi)^3} \, I_{\text{e-ph}} = 0 , \qquad (7.169)$$

which we shall see expresses the conservation of the number of particles, here the electrons in question. Integrating the kinetic equation, Eq. (7.166), with respect to momentum we obtain the continuity equation

$$\partial_T n + \nabla_{\mathbf{R}} \cdot \mathbf{j} = 0 , \qquad (7.170)$$

where

$$n(\mathbf{R}, T) = 2 \int \frac{d\mathbf{p}}{(2\pi)^3} \, n(\mathbf{p}, \mathbf{R}, T) \qquad (7.171)$$

and

$$\mathbf{j}(\mathbf{R}, T) = 2 \int \frac{d\mathbf{p}}{(2\pi)^3} \nabla_{\mathbf{p}} E_{\mathbf{p}} \, n(\mathbf{p}, \mathbf{R}, T) \tag{7.172}$$

are the Landau–Boltzmann expressions for the density and current density and the factor of two accounts for the spin of the electron.

In order to establish that these are indeed the correctly identified densities (in the excitation representation), we should connect one of them with the microscopic expression. Assuming that $|e\varphi| \ll \epsilon_F$, the microscopic expression for the density, Eq. (7.108), is (suppressing space-time variables in quantities, here in φ)

$$n(\mathbf{R}, T) = -2N_0 \int \frac{d\hat{\mathbf{p}}}{4\pi} \int_{-\infty}^{\infty} dE \, f(E + e\varphi, \mathbf{p}) \,. \tag{7.173}$$

In order to compare the density expression in the particle representation with the excitation representation we transform Eq. (7.171) to the particle representation

$$n(\mathbf{R}, T) = 2 \int \frac{d\mathbf{p}}{(2\pi)^3} \, n(\mathbf{p}, \mathbf{R}, T) = 2N_0 \int \frac{d\hat{\mathbf{p}}}{4\pi} \int_{-\infty}^{\infty} dE \, (1 - \partial_E \Re e\sigma) \, f(E, \hat{\mathbf{p}}). \tag{7.174}$$

Since Eq. (7.173) and Eq. (7.174) appear to be different, Eq. (7.172) is also transformed to the particle representation

$$2 \int \frac{d\mathbf{p}}{(2\pi)^3} \nabla_{\mathbf{p}} E_{\mathbf{p}} \, n(\mathbf{p}, \mathbf{R}, T) = 2N_0 \int \frac{d\hat{\mathbf{p}}}{4\pi} \int_{-\infty}^{\infty} dE \, \mathbf{v}_F \, f(E, \hat{\mathbf{p}}) \,. \tag{7.175}$$

Comparing the expression in Eq. (7.175) to that of Eq. (7.109), we observe that it is identical to the quasi-classical current-density expression. The only possibility for the above-mentioned apparent discrepancy not to lead to a violation of the continuity equation is the existence of the identity

$$\partial_T \int d\hat{\mathbf{p}} \int_{-\infty}^{\infty} dE \, f(E, \hat{\mathbf{p}}) \, \partial_E \Re e\sigma = 0 \tag{7.176}$$

which we now prove. Inserting the expression from Eq. (7.144) into the left side of Eq. (7.176) we are led to consider

$$\int d\hat{\mathbf{p}} \int dE \int d\hat{\mathbf{p}}' \int dE' \, |g_{\mathbf{p}_F - \mathbf{p}_F'}|^2 \, (\Re e D(\mathbf{p}_F - \mathbf{p}_F', E - E')$$

$$\partial_T f(E, \hat{\mathbf{p}}) \, \partial_{E'} f(E', \hat{\mathbf{p}}') - \partial_E f(E, \hat{\mathbf{p}}) \, \partial_T f(E', \hat{\mathbf{p}}')) = 0 \tag{7.177}$$

which by interchanging the variables $E, \hat{\mathbf{p}}$ and $E', \hat{\mathbf{p}}'$ is seen to vanish, and the identity Eq. (7.176) is thus established. We have thus established that the approximations made do not violate particle conservation.

7.5.5 Impurity scattering

For electrons interacting with impurities in a conductor, the self-energy is given by the diagram in Eq. (7.51), $\epsilon_F \tau \gg \hbar$, and we can immediately implement the quasi-classical approximation. The equation for the kinetic component of the quasi-classical Green's function in the presence of an electric field becomes

$$(\partial_T + \mathbf{v}_F \cdot \nabla_\mathbf{R} + e\partial_T \varphi \, \partial_E) \, g^K = -\frac{1}{\tau} g^K(E, \hat{\mathbf{p}}, \mathbf{R}, T) + \int \frac{d\hat{\mathbf{p}}}{4\pi} \, g^K(E, \hat{\mathbf{p}}, \mathbf{R}, T) \,, \tag{7.178}$$

where for simplicity we have assumed that the momentum dependence of the impurity potential can be neglected.

In the diffusive limit the quasi-classical kinetic Green's function will be almost isotropic, and an expansion in spherical harmonics needs to keep only the s- and p-wave parts

$$g^K(E, \hat{\mathbf{p}}, \mathbf{R}, T) = g_s^K(E, \mathbf{R}, T) + \hat{\mathbf{p}} \cdot \mathbf{g}_p^K(E, \mathbf{R}, T) \tag{7.179}$$

and

$$|\hat{\mathbf{p}} \cdot \mathbf{g}_p^K| \ll |g_s^K| \,. \tag{7.180}$$

Inserting into the kinetic equation we get the relation

$$\mathbf{g}_p^K(E, \mathbf{R}, T) = -l \nabla_\mathbf{R} \, g_s^K(E, \mathbf{R}, T) \tag{7.181}$$

and using the expressions for the current and density, Eq. (7.108) and Eq. (7.109), we obtain their relationship

$$\mathbf{j}(\mathbf{R}, T) = -D_0 \nabla_\mathbf{R} \rho(\mathbf{R}, T) + \sigma_0 \mathbf{E}(\mathbf{R}, T) \,, \tag{7.182}$$

where we have used the Einstein relation, $\sigma_0 = 2e^2 N_0 D_0$, relating conductivity and the diffusion constant.

In the absence of the electric field, the kinetic equation becomes the diffusion equation for the s-wave component

$$(\partial_T - D_0 \nabla_\mathbf{R}^2) \, g_s^K(E, \mathbf{R}, T) = 0 \,. \tag{7.183}$$

Exercise 7.9. Show that by introducing the distribution function

$$h(\mathbf{p}, \mathbf{R}, T) = \frac{1}{2} g^K(E = \xi_\mathbf{p} + e\varphi(\mathbf{R}, T), \hat{\mathbf{p}}, \mathbf{R}, T) \tag{7.184}$$

the kinetic equation assumes the standard Boltzmann form, Eq. (7.55).

7.6 Beyond the quasi-classical approximation

The importance of the quasi-classical description is the very weak restrictions for its applicability. However, it has two severe limitations. It relies on the assumption of particle–hole symmetry and is thus unable to treat thermo-electric effects, and since momenta are tied to the Fermi surface the effect of the Lorentz force is lost and the quasi-classical Green's function technique is unable to describe magneto-transport. In this section we shall show how these restrictions can be avoided following previous works of Langreth [27] and Altshuler [28]. As an example, in Section 7.6.1 we consider thermo-electric effects in a magnetic field, the Nernst–Ettingshausen effect.

A distribution function is introduced according to

$$G^{\mathrm{K}} = G^{\mathrm{R}} \otimes h - h \otimes G^{\mathrm{A}},\tag{7.185}$$

which upon insertion into the quantum kinetic equation, Eq. (7.3), and by use of the equations of motion for the retarded and advanced Green's functions, and the property that the composition $\otimes$ is associative leads to the equation

$$G^{\mathrm{R}} \otimes B - B \otimes G^{\mathrm{A}} = 0,\tag{7.186}$$

where

$$B[h] = [G_0^{-1} - \Re e\Sigma \overset{\otimes}{,} h]_- + \frac{1}{2}[\Gamma \overset{\otimes}{,} h]_+ - i \Sigma^{\mathrm{K}}.\tag{7.187}$$

In the gradient approximation we then have

$$(G^{\mathrm{R}} - G^{\mathrm{A}})B + [B, \Re eG]_{\mathrm{p}} = 0.\tag{7.188}$$

Inserting the solution of the equation

$$(G^{\mathrm{R}} - G^{\mathrm{A}})B = 0\tag{7.189}$$

into Eq. (7.188), we observe that the second term on the left in Eq. (7.187) has the form of a double Poisson bracket and thus should be dropped in the gradient approximation. The quantum kinetic equation therefore takes the form

$$B[h] = 0\tag{7.190}$$

and expressions in Eq. (7.187) should be evaluated in the gradient approximation.

Since the introduced distribution function is not gauge invariant, we shall not succeed in obtaining an appropriate kinetic equation with the usual expression for the Lorentz force unless the kinetic momentum is introduced instead of the canonical one.[27] Performing a gradient expansion of the term in Eq. (7.187) containing G_0^{-1}, we obtain in the mixed or Wigner coordinates

$$-i[G_0^{-1} \overset{\otimes}{,} h]_{\mathrm{p}} = [E - e\varphi - \xi_{\mathbf{p}-e\mathbf{A}}, h]_{\mathrm{p}},\tag{7.191}$$

where $(\varphi, \mathbf{A})$ are the potentials describing the electromagnetic field.

Within the gradient approximation a gauge-invariant distribution function $\tilde{h}$ can thus be introduced

$$\tilde{h}(\Omega, \mathbf{P}, \mathbf{R}, T) = h(E, \mathbf{p}, \mathbf{R}, T)\tag{7.192}$$

defined by the change of variables

$$\mathbf{P} = \mathbf{p} - e\mathbf{A}(\mathbf{R}, T), \quad \Omega = E - e\varphi(\mathbf{R}, T).\tag{7.193}$$

---

[27]Describing the kinetics in the momentum representation assumes that we are not in the quantum limit where Landau level quantization is of importance, $\hbar\omega_c \ll kT$.

We observe the identity (now indicating the variables involved in the Poisson brackets by subscripts)

$$[A, B]_{\mathbf{p}, E} = [\tilde{A}, \tilde{B}]_{\mathbf{P}, \Omega} + e\mathbf{E} \cdot (\partial_\Omega \tilde{A} \nabla_{\mathbf{P}} \tilde{B} - \partial_\Omega \tilde{B} \nabla_{\mathbf{P}} \tilde{A}) + e\mathbf{B} \cdot (\nabla_{\mathbf{P}} \tilde{A} \times \nabla_{\mathbf{P}} \tilde{B}) , \qquad (7.194)$$

where $\mathbf{E} = -\nabla\varphi - \partial_T \mathbf{A}$ and $\mathbf{B} = \nabla \times \mathbf{A}$ are the electric and magnetic fields, respectively, and $\tilde{A}$ and $\tilde{B}$ are related to A and B by equations analogous to Eq. (7.192). Using this identity, the following driving terms then appear in the gradient approximation

$$-i[G_0^{-1} - \Re e\Sigma, h]_{\mathbf{p}, E} = [\Omega - \xi_{\mathbf{P}} - \Re e\tilde{\Sigma}, h]_{\mathbf{P}, \Omega}$$

$$+ \quad e\mathbf{E} \cdot ((1 - \partial_\Omega \Re e\tilde{\Sigma}) \nabla_{\mathbf{P}} \tilde{h}) + \mathbf{v}^* \partial_\Omega \tilde{h}) + e\mathbf{v}^* \times \mathbf{B} \cdot \nabla_{\mathbf{P}} \tilde{h} , \qquad (7.195)$$

where we have introduced[28]

$$\mathbf{v}^* = \nabla_{\mathbf{P}}(\xi_{\mathbf{P}} + \Re e\tilde{\Sigma}(\Omega, \mathbf{P}, \mathbf{R}, T)) . \qquad (7.196)$$

As a result of the transformation Eq. (7.193), the kinematic and not the canonical momentum enters the kinetic equation, and a gauge invariant kinetic equation is obtained as desired.

We could equally well have obtained the kinetic equation on gauge invariant form by choosing to introduce the mixed representation according to

$$G(X, p) \equiv \int dx e^{-i\mathbf{r} \cdot (\mathbf{p} + e\mathbf{A}(X)) + it(E + e\varphi(X))} G(X, x) \qquad (7.197)$$

whereupon, in accordance with Eq. (7.194), the Poisson bracket can be expressed as

$$[A, B]_{\mathbf{p}, E} = \partial_E A \{\partial_T + \mathbf{u} \cdot \nabla_{\mathbf{R}} + (e\mathbf{E} \cdot \mathbf{u} - (\partial_E A)^{-1} \partial_T A) \partial_E$$

$$+ \quad (e\mathbf{E} + e\mathbf{v} \times \mathbf{B} + (\partial_E A)^{-1} \nabla_{\mathbf{R}} A) \cdot \nabla_{\mathbf{p}}\} B \qquad (7.198)$$

with

$$\mathbf{u} = (\partial_E A)^{-1} \nabla_{\mathbf{p}} A . \qquad (7.199)$$

The kinetic equation thus takes the form

$$\{(1 - \partial_E \Re e\Sigma) \partial_T \quad + \quad \partial_T \Re e\Sigma \, \partial_E + \mathbf{v}^* \cdot (\nabla_{\mathbf{R}} + e\mathbf{E} \, \partial_E)$$

$$+ \quad (e\mathbf{E} + e\mathbf{v}^* \times \mathbf{B}) \cdot \nabla_{\mathbf{p}}\} h = I[h] \qquad (7.200)$$

where the collision integral is given by

$$I[h] = i\Sigma^K - \Gamma h . \qquad (7.201)$$

---

[28] As long as inter-band transitions can be neglected, band structure effects can be included as shown in reference [3].

Exercise 7.10. Consider the case of an instantaneous two-particle interaction between fermions, $V(\mathbf{x})$, such as Coulomb interaction between electrons,

$$U^R(\mathbf{x}, t, \mathbf{x}', t') = V(\mathbf{x} - \mathbf{x}')\,\delta(t - t') = U^A(\mathbf{x}, t, \mathbf{x}', t') \qquad (7.202)$$

and $U^K(\mathbf{x}, t, \mathbf{x}', t') = 0$. The Hartree–Fock self-energy skeleton diagrams, the diagrams in Figure 5.4, do not contribute to the collision integral owing to the instantaneous character of the interaction. The lowest-order self-energy skeleton diagrams contributing to the collision integral are thus specified by the third and fourth diagrams in Figure 5.5.

Show that the corresponding electron–electron collision integral becomes

$$I_{e-e}[f] = -2\pi \int d\mathbf{p}_1 d\mathbf{p}_2 d\mathbf{p}_3 \, (U^R(\mathbf{p} - \mathbf{p}))^2 \, \delta(\mathbf{p} + \mathbf{p}_2 - \mathbf{p}_1 - \mathbf{p}_3)$$

$$\times \quad \delta(\xi_\mathbf{p} + \xi_{\mathbf{p}_2} - \xi_{\mathbf{p}_1} - \xi_{\mathbf{p}_3})$$

$$\times \quad (f_\mathbf{p} f_{\mathbf{p}_2}(1 - f_{\mathbf{p}_1})(1 - f_{\mathbf{p}_3}) - (1 - f_\mathbf{p})(1 - f_{\mathbf{p}_2})f_{\mathbf{p}_1} f_{\mathbf{p}_3})) , \qquad (7.203)$$

where $f_\mathbf{p}$ is the electron distribution function which in equilibrium reduces to the Fermi function. If one uses the the real-time formulation in terms of the Green's functions G^{RAK}, the canceling terms $f_\mathbf{p} f_{\mathbf{p}_1} f_{\mathbf{p}_2} f_{\mathbf{p}_3}$ do not appear explicitly but have to be added and subtracted.

Show that the decay of a momentum or energy state for the above collision integral is given by the following energy relaxation rate

$$\frac{1}{\tau_{e-e}(\mathbf{p})} = -2\pi \int d\mathbf{p}_1 d\mathbf{p}_2 d\mathbf{p}_3 \, (U^R(\mathbf{p} - \mathbf{p})^2 \delta(\mathbf{p} + \mathbf{p}_2 - \mathbf{p}_1 - \mathbf{p}_3)$$

$$\times \quad \delta(\xi_{\mathbf{p}_1} + \xi_{\mathbf{p} + \mathbf{p}_2 - \mathbf{p}_1} - \xi_\mathbf{p} - \xi_{\mathbf{p}_2})$$

$$\times \quad (f_{\mathbf{p}_2}(1 - f_{\mathbf{p}_1})(1 - f_{\mathbf{p}_3}) + f_{\mathbf{p}_3}(1 - f_{\mathbf{p}_2})f_\mathbf{p}) , \qquad (7.204)$$

where the short notation has been introduced for the Fermi function, $f_\mathbf{p} = f_0(\xi_\mathbf{p})$.

Assume that the interaction is due to screened Coulomb interaction

$$(U^R(\mathbf{p}))^2 = \left| \frac{V(\mathbf{p})}{\epsilon(\mathbf{p})} \right|^2 = \left| \frac{\frac{e^2}{\epsilon_0}}{\hbar^{-2}\mathbf{p}^2 + \kappa_s^2} \right|^2 , \qquad (7.205)$$

where $\kappa_s^2 = 2N_0 e^2/\epsilon_0$ is the screening wave vector.

Show that the electron–electron collision rate for an electron on the Fermi surface has the temperature dependence

$$\frac{1}{\tau_{e-e}(T)} = \begin{cases} \dfrac{\pi^2 e^2}{32\epsilon_0 v_F^2 \kappa_s \hbar^3} (kT)^2 & \kappa_s \ll k_F \\[2em] \dfrac{\pi^3}{16} \dfrac{(kT)^2}{\hbar \epsilon_F} & \kappa_s \gg k_F . \end{cases} \qquad (7.206)$$

The life time is seen to be determined by the phase-space restriction owing to Pauli's exclusion principle. The long lifetime of excitations near the Fermi surface due to the exclusion principle is the basis of Landau's phenomenological Fermi liquid theory of strongly interacting degenerate fermions, and its microscopic Green's function foundation.

7.6.1 Thermo-electrics and magneto-transport

As an example of electron–phonon renormalization of a static transport coefficient, we consider the Nernst–Ettingshausen effect, viz. the high-field Nernst–Ettingshausen coefficient, which relates the current density to the vector product of the temperature gradient and the magnetic field. For now, we shall neglect any momentum dependence of the self-energy. The system is driven out of equilibrium by a temperature gradient. The magnetic field is assumed to satisfy the condition

$$\gamma \ll \omega_c , \tag{7.207}$$

where $\omega_c = |e|B/m$ is the Larmor or cyclotron frequency and γ is the collision rate. The collision integral can then be neglected, and the kinetic equation reduces to

$$(\mathbf{v} \cdot \nabla_R + e(\mathbf{v} \times \mathbf{B}) \cdot \nabla_\mathbf{p}) h = 0 . \tag{7.208}$$

In the gradient approximation, the electric current density is according to Eq. (7.16)

$$\mathbf{j}(\mathbf{R}, T) = -e \int \frac{d\mathbf{p}}{(2\pi)^3} \int_{-\infty}^{\infty} dE \, \mathbf{v} \, (Ah - [\Re G, h]_{\mathbf{p}E}) . \tag{7.209}$$

According to Eq. (7.207) and Eq. (7.208), the last term vanishes since

$$[\Re G, h]_{\mathbf{p}E} = -\nabla_\mathbf{p} \Re G \cdot \nabla_R h + e\mathbf{B} \cdot (\nabla_\mathbf{p} \Re G \times \nabla_\mathbf{p} h)$$

$$= -\frac{\partial \Re G}{\partial \xi} (\mathbf{v} \cdot \nabla_R h + e(\mathbf{v} \times \mathbf{B}) \cdot \nabla_\mathbf{p} h) = 0 . \tag{7.210}$$

Inserting the solution of Eq. (7.208)

$$h = h_0 - \frac{|\nabla T|}{eBT} p_y E \frac{\partial h_0}{\partial E} . \tag{7.211}$$

into the current expression and performing a Sommerfeld expansion gives

$$j = \frac{(1+\lambda)S_0}{B^2} \nabla T \times B \quad , \quad \lambda = -\frac{\partial \Re \Sigma}{\partial E} \bigg|_{E=0, \mathbf{p}=\mathbf{p}_F} \tag{7.212}$$

where S_0 is the free electron entropy which in a degenerate electron gas is identical to the specific heat. In the jellium model one has $\lambda = g^2 N_0$. Thus the enhancement of the high-field thermo-electric current is seen to be identical to the enhancement of the equilibrium specific heat.

Taking into account a possible momentum dependence of the self-energy leads to non-equilibrium contributions to the spectral weight function which, however, are difficult to calculate. A calculation within the context of Landau–Boltzmann Fermi-liquid theory leads to the appearance of two $\nabla_{\mathbf{p}}\Re e\Sigma$-dependent terms that exactly cancel each other, thus suggesting the above result to be generally valid [9].

Thermopower measurements agree with the calculated mass enhancement according to Eq. (7.212), see references [29, 30].

7.7 Summary

In this chapter the quantum kinetic equation approach to transport using the real-time approach has been considered. The examples studied were condensed matter systems, but the approach is useful in application to many physical systems, say in nuclear physics in connection with nuclear reactions and heavy ion collisions, as discussed for example in reference [31]. We have also realized the difficulties involved in describing general non-equilibrium states. Since no universality of much help is available in guiding approximations, cases must be dealt with on an individual basis. Here the use of the skeleton diagrammatic representation of the self-energy, just as for equilibrium states, can be a powerful tool to assess controlled approximations in nontrivial expansion parameters as we demonstrated for the case of electron–phonon interaction. This allowed establishing, for example, that the classical Landau–Boltzmann equation has a much wider range of applicability than to be expected *a priori*. The general problem is the vast amount of information encoded in the one-particle Green's functions, truncated objects with boundless information of correlations expressed by higher-order Green's functions. It is therefore necessary to eliminate the information in the equations of motion which do not influence the studied properties, to get rid of any excess information. The quasi-classical Green's function technique being such a successful scheme when it comes to understand the transport properties of metals, except for effects depending on particle–hole asymmetry such as thermo-electric effects. The quasi-classical Green's function technique allowed analytical calculation of mass renormalization effects typical of interactions in quantum systems, and are in general susceptible to numerical treatment.[29] The quasi-classical Green's function technique is the basic tool for studying non-equilibrium properties of the low-temperature superconducting state, a topic we turn to in the next chapter. In fact, the quasi-classical Green's function technique is a corner stone for describing many quantum phenomena in condensed matter, being the systematic starting point for treating quantum corrections to classical kinetics, and we shall exploit this to our advantage when discussing the weak localization effect in Chapter 11.

---

[29]Despite brave efforts, little progress has, to my knowledge, been made using numerics to extend solutions of the *general* quantum kinetic equation to include higher than second-order correlations. This field will undoubtedly be studied in the future using numerics.

8

Non-equilibrium superconductivity

Superconductivity was discovered in 1911 by H. Kamerlingh Onnes. Having succeeded in liquefying helium, transition temperature 4.2 K, this achievement in cryogenic technology was used to cool mercury to the man-made temperature that at that time was closest to absolute zero. He reported the observation that mercury at 4.2 K abruptly entered a new state of matter where the electrical resistance becomes vanishingly small. This extraordinary phenomenon, coined superconductivity, eluted a microscopic understanding until the theory of Bardeen, Cooper and Schrieffer in 1957 (BCS-theory).[1] The mechanism responsible for the phase transition from the normal state to the superconducting state at a certain critical temperature is that an effective *attractive* interaction between electrons makes the normal ground state unstable. As far as conventional or low-temperature superconductors are concerned, the attraction between electrons follows from the form of the phonon propagator, Eq. (5.45), viz. that the electron–phonon interaction is attractive for frequencies less than the Debye frequency, and in fact can overpower the screened Coulomb repulsion between electrons, leading to an effective attractive interaction between electrons.[2] The original BCS-theory was based on a bold ingenious guess of an approximate ground state wave function and its low-energy excitations describing the essentials of the superconducting state. Later the diagrammatic Green's function technique was shown to be useful to describe more generally the properties of superconductors, such as under conditions of spatially varying magnetic fields and especially for general non-equilibrium conditions.

In terms of Green's functions and the diagrammatic technique, the transition from the normal state to the superconducting state shows up as a singularity in the

---

[1]For an important review of the attempts to understand the phenomena of superconductivity and its truly defining state characteristic, the Meissner-effect, i.e. the expulsion of a magnetic field from a piece of material in the superconducting state, we refer the reader to the article by Bardeen [32], written on the brink of the monumental discovery of the theoretical understanding of the new state of matter discovered almost half a century earlier.

[2]In high-temperature superconductors, the attractive interaction is not caused by the ionic background fluctuations but by spin fluctuations.

effective interaction vertex. The effect of a particular class of scattering processes in the normal state drives the singularity. In diagrammatic terms certain vertex corrections, capturing the effect of the particular scattering process, corresponding to re-summation of an infinite class of diagrams, become singular. In the case of superconductivity, the particle–particle ladder self-energy vertex corrections, a typical member of which is depicted in Figure 8.1, where the wiggly line represents the effective attractive electron-electron interactions (in the simplest model simply the electron–phonon interaction) becomes divergent in the normal state, signalling a phase transition at a critical temperature T_c.

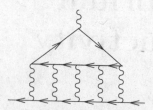

Figure 8.1 Cooper instability diagram.

Although the set of diagrams according to Migdal's theorem by diagrammatic estimation is formally of the order of $\hbar\omega_D/\epsilon_F$, where $\hbar\omega_D$ is the Debye energy, which is typically two orders of magnitude smaller than the Fermi energy, the particle–particle ladder sums up a geometric series to produce a denominator which by vanishing produces a singularity.[3] In the simplest, longitudinal-only electron–phonon model, the critical temperature is given by (see Exercise 8.3 on page 221)

$$kT_c \simeq \hbar\omega_D \, e^{-1/\lambda} \,, \tag{8.1}$$

where $\lambda = N_0 g^2$ is the dimensionless electron–phonon coupling constant in the jellium model (recall Section 7.5.1). We note that the critical temperature is non-analytic in the coupling constant, precisely such non-perturbative effects are captured by re-summation of an infinite class of diagrams. The singularity signals a transition between two states, leading at zero temperature to a ground state that is very different from the normal ground state, and in general at temperatures below the critical one to properties astoundingly different from those of the normal state.

The signifying feature of the superconducting state is, as stressed by Yang [33], that it possesses off-diagonal long-range order, i.e. for pair-wise far away separated

---

[3]The story goes that Landau delayed the publication of Migdal's result for several years, because it is in blatant contradiction to the existence of superconductivity (mediated by phonons). Nowadays we are familiar with the status of diagrammatic estimates such as Migdal's theorem (as discussed in Section 7.5.1). They are not immune to the existence of singularities in certain infinite re-summations of a particular set of diagrams. The situation is formally quite analogous to the singularity involved in Anderson's metal–insulator transition. In revealing the physics in this case, diagrammatic techniques are also useful, as we shall discuss in Chapter 11.

spatial arguments, the two-particle correlation function is non-vanishing

$$\lim_{|\mathbf{x}_1,\mathbf{x}_2-\mathbf{x}_3,\mathbf{x}_4|\to\infty} \langle \psi_\alpha^\dagger(\mathbf{x}_4)\,\psi_\beta^\dagger(\mathbf{x}_3)\,\psi_\gamma(\mathbf{x}_2)\,\psi_\delta(\mathbf{x}_1)\rangle \neq 0\,, \tag{8.2}$$

i.e. when the spatial arguments $\mathbf{x}_1$ and $\mathbf{x}_2$ are chosen arbitrarily far away from the spatial arguments $\mathbf{x}_3$ and $\mathbf{x}_4$, the two-particle correlation function nevertheless stays non-vanishing, contrary to the case of the normal state. An order parameter function, $\Delta_{\gamma\delta}(\mathbf{x},\mathbf{x}')$, expressing this property, can therefore be introduced according to

$$\lim_{|\mathbf{x}_1,\mathbf{x}_2-\mathbf{x}_3,\mathbf{x}_4|\to\infty} \langle \psi_\alpha^\dagger(\mathbf{x}_4)\,\psi_\beta^\dagger(\mathbf{x}_3)\,\psi_\gamma(\mathbf{x}_2)\,\psi_\delta(\mathbf{x}_1)\rangle = \Delta_{\alpha\beta}^*(\mathbf{x}_4,\mathbf{x}_3)\,\Delta_{\gamma\delta}(\mathbf{x}_1,\mathbf{x}_2) \tag{8.3}$$

and we speak of BCS-pairing.

8.1 BCS-theory

In this section we consider the BCS-theory, but shall not go into any details of BCS-ology since instead we shall use the Green's function technique to describe and calculate properties of the superconducting state.[4] The part of the interaction responsible for the instability is captured by keeping in the Hamiltonian only the so-called pairing interaction. In a conventional and clean superconductor, pairing takes place between momentum and spin states $(\mathbf{p},\uparrow)$ and $(-\mathbf{p},\downarrow)$, each others time-reversed states,[5] and we encounter orbital s-wave and spin-singlet pairing and the BCS-Hamiltonian becomes[6]

$$H_{\text{pairing}} = \sum_{\mathbf{p},\sigma} \epsilon_{\mathbf{p}}\, c_{\mathbf{p}\sigma}^\dagger\, c_{\mathbf{p}\sigma} + \sum_{\mathbf{p}\mathbf{p}'} V_{\mathbf{p}\mathbf{p}'}\, c_{\mathbf{p}\uparrow}^\dagger\, c_{-\mathbf{p}\downarrow}^\dagger\, c_{-\mathbf{p}'\downarrow}\, c_{\mathbf{p}'\uparrow}\,, \tag{8.4}$$

where the effective attractive interaction $V_{\mathbf{p}\mathbf{p}'}$ is only non-vanishing for momentum states in the tiny region around the Fermi surface set by the Debye energy, ω_D, for the case where the attraction is caused by electron–phonon interaction. The parameters specifying the boldly guessed BCS-ground state[7]

$$|\text{BCS}\rangle = \prod_{\mathbf{p}} (u_{\mathbf{p}} + v_{\mathbf{p}}\, c_{\mathbf{p}\uparrow}^\dagger\, c_{-\mathbf{p}\downarrow}^\dagger)\, |0\rangle \tag{8.5}$$

---

[4] The properties of the BCS-state are described in numerous textbooks, e.g. reference [34].

[5] In a disordered superconductor, pairing takes place between an exact impurity eigenstate and its time reversed eigenstate.

[6] Other types of pairing occur in Nature. In ^{3}He p-wave pairing occurs, and high-temperature superconductors have d-wave pairing.

[7] The BCS-ground state is seen to be a state that is not an eigenstate of the total number operator, i.e. it does not describe a state with a definite number of electrons (recall Exercise 1.7 on page 20). For massless bosons, such as photons, a number-violating state is not an unphysical state, but for an assembly of fermions having a finite chemical potential and interactions obeying particle conservation it certainly is, and only the enormous explanatory power of the BCS-theory makes it decent to use a formulation that violates the most sacred of conservation laws. In other words, the superconducting state can also be described in terms that do not violate gauge invariance such as when staying fully in the electron–phonon model, but the BCS-theory correctly describes the off-diagonal long-range order, and is a very efficient way for incorporating and calculating the order parameter, characterizing the superconducting state, and its consequences. Quantum field theory is therefore also convenient, but the superconducting state can be described without its use and instead formulated in terms of the one- and two-particle density matrices.

are then obtained by the criterion of minimizing the average energy in the grand canonical ensemble, i.e. the average value of $\langle \text{BCS}|\text{H}_{\text{pairing}} - \mu \text{N}|\text{BCS}\rangle$, the pairing Hamiltonian with energies measured from the chemical potential, which at zero temperature is the Fermi energy, $\xi_{\mathbf{p}} = \epsilon_{\mathbf{p}} - \epsilon_{\text{F}}$. This leads to a gap in the single-particle spectrum close to the Fermi surface. We shall not dwell on BCS-ology as we soon introduce the mean-field approximation at the level of Green's functions, and instead offer it as exercises.

Exercise 8.1. Assume $u_{\mathbf{p}}$ and $v_{\mathbf{p}}$ real so that (recall Exercise 1.7 on page 20) the angle $\phi_{\mathbf{p}}$ parameterizes the amplitudes, $u_{\mathbf{p}} = \sin \phi_{\mathbf{p}}$ and $v_{\mathbf{p}} = \cos \phi_{\mathbf{p}}$. Show that

$$\langle \text{BCS}|\text{H}_{\text{pairing}} - \epsilon_{\text{F}}\text{N}|\text{BCS}\rangle = \sum_{\mathbf{p},\sigma} \xi_{\mathbf{p}} (1 + \cos 2\phi_{\mathbf{p}}) + \frac{1}{4} \sum_{\mathbf{p}\mathbf{p}'} V_{\mathbf{p}\mathbf{p}'} \sin 2\phi_{\mathbf{p}} \sin 2\phi_{\mathbf{p}'}$$

$$(8.6)$$

resulting in the minimum condition of the average grand canonical energy to be

$$2\xi_{\mathbf{p}} \tan 2\phi_{\mathbf{p}} = \sum_{\mathbf{p}'} V_{\mathbf{p}\mathbf{p}'} \sin 2\phi_{\mathbf{p}'} . \qquad (8.7)$$

Using simple geometric relations, $2u_{\mathbf{p}}v_{\mathbf{p}} = \sin 2\phi_{\mathbf{p}}$ and $v_{\mathbf{p}}^2 - u_{\mathbf{p}}^2 = \cos 2\phi_{\mathbf{p}}$, and introducing the quantities $\Delta_{\mathbf{p}} = -\sum_{\mathbf{p}'} V_{\mathbf{p}\mathbf{p}'} u_{\mathbf{p}'}v_{\mathbf{p}'}$ and $E_{\mathbf{p}} = \sqrt{\xi_{\mathbf{p}}^2 + \Delta_{\mathbf{p}}^2}$, show that the minimum condition becomes the self-consistency condition

$$\Delta_{\mathbf{p}} = -\frac{1}{2} \sum_{\mathbf{p}'} V_{\mathbf{p}\mathbf{p}'} \frac{\Delta_{\mathbf{p}'}}{E_{\mathbf{p}'}} \qquad (8.8)$$

for the BCS-energy gap in the excitation spectrum.

Exercise 8.2. Besides the normal state solution, $\Delta_{\mathbf{p}} = 0$, for an attractive interaction the self-consistency condition, Eq. (8.8) has a nontrivial solution, $\Delta_{\mathbf{p}} \neq 0$. Assuming, as dictated by electron–phonon interaction, that the interaction is attractive only in a tiny region around the Fermi energy set by the Debye energy, ω_{D}, the interaction is modeled by a constant attraction in this region, $V_{\mathbf{p}\mathbf{p}'} = -V \theta(\omega_{\text{D}} - |\xi_{\mathbf{p}}|) \theta(\omega_{\text{D}} - |\xi_{\mathbf{p}'}|)$. Show that in this model the self-consistency equation has the solution $\Delta_{\mathbf{p}} = -\Delta \theta(\omega_{\text{D}} - |\xi_{\mathbf{p}}|)$, where the constant Δ, the energy gap, is determined by (the prime indicates that the summation is restricted)

$$1 = \frac{V}{2} \sum_{\mathbf{p}}' \frac{1}{\sqrt{\xi_{\mathbf{p}}^2 + \Delta^2}} , \qquad (8.9)$$

which for weak coupling, $N_0 V \ll 1$ (N_0 being the density of momentum states of the electron gas at the Fermi energy), gives

$$\Delta \simeq 2\hbar\omega_{\text{D}} \, e^{-1/N_0 V} . \qquad (8.10)$$

Show that in this model

$$\langle BCS| H_{\text{pairing}} - \epsilon_F N |BCS\rangle = -\frac{\Delta^2}{V} + \sum_{\mathbf{p}} \left(\xi_{\mathbf{p}} - \frac{\xi_{\mathbf{p}}^2}{E_{\mathbf{p}}} \right) \qquad (8.11)$$

and thereby that the energy difference per unit volume between the state with $\Delta \neq 0$ and the normal state, where states up to the Fermi surface are filled according to Eq. (1.105), is given by $-N_0 \Delta^2/2$.[8] The state with $\Delta \neq 0$ is thus favored as the ground state by the pairing interaction.

Exercise 8.3. Introduce new operators by the Bogoliubov–Valatin transformation[9]

$$\gamma_{\mathbf{p}\uparrow}^{\dagger} = u_{\mathbf{p}} c_{\mathbf{p}\uparrow}^{\dagger} - v_{\mathbf{p}}^* c_{-\mathbf{p}\downarrow} \quad , \quad \gamma_{-\mathbf{p}\downarrow}^{\dagger} = u_{\mathbf{p}} c_{-\mathbf{p}\downarrow}^{\dagger} + v_{\mathbf{p}}^* c_{\mathbf{p}\uparrow} \qquad (8.12)$$

and their adjoints, leaving them canonical as the normalization condition, $|u_{\mathbf{p}}|^2 + |v_{\mathbf{p}}|^2 = 1$, is insisted, assuring the anti-commutation relations

$$\{\gamma_{\mathbf{p}\uparrow}, \gamma_{\mathbf{p}'\uparrow}^{\dagger}\} = \delta_{\mathbf{p}\mathbf{p}'} = \{\gamma_{\mathbf{p}\downarrow}, \gamma_{\mathbf{p}'\downarrow}^{\dagger}\} \qquad (8.13)$$

as well as

$$\{\gamma_{\mathbf{p}\uparrow}, \gamma_{\mathbf{p}'\downarrow}^{\dagger}\} = 0 = \{\gamma_{\mathbf{p}\downarrow}, \gamma_{\mathbf{p}'\downarrow}\} \quad , \quad \{\gamma_{\mathbf{p}\uparrow}, \gamma_{\mathbf{p}'\uparrow}\} = 0 = \{\gamma_{\mathbf{p}\uparrow}, \gamma_{\mathbf{p}'\downarrow}\} . \qquad (8.14)$$

Show that $H_{\text{pairing}} - \epsilon_F N$ is diagonalized by the transformation to the Hamiltonian, up to an irrelevant constant term,

$$H_{\text{pairing}} - \epsilon_F N = \sum_{\mathbf{p}} E_{\mathbf{p}} (\gamma_{\mathbf{p}\uparrow}^{\dagger} \gamma_{\mathbf{p}\uparrow} + \tilde{\gamma}_{\mathbf{p}\downarrow}^{\dagger} \tilde{\gamma}_{\mathbf{p}\downarrow}) , \qquad (8.15)$$

provided $2\xi_{\mathbf{p}} u_{\mathbf{p}} v_{\mathbf{p}} + (v_{\mathbf{p}}^2 - u_{\mathbf{p}}^2)\Delta_{\mathbf{p}} = 0$, where $\Delta_{\mathbf{p}}$ satisfies the self-consistency equation Eq. (8.8) (assuming for simplicity real amplitudes). Equivalently, noting the coefficients can be chosen real,

$$u_{\mathbf{p}}^2 = \frac{1}{2} \left(1 + \frac{\xi_{\mathbf{p}}}{E_{\mathbf{p}}} \right) \quad , \quad v_{\mathbf{p}}^2 = \frac{1}{2} \left(1 - \frac{\xi_{\mathbf{p}}}{E_{\mathbf{p}}} \right) \quad , \quad u_{\mathbf{p}} v_{\mathbf{p}} = \frac{\Delta_{\mathbf{p}}}{2E_{\mathbf{p}}} . \qquad (8.16)$$

This provides a general description of the BCS-Hamiltonian in terms of free fermionic quasi-particles with energy dispersion $E_{\mathbf{p}} = \sqrt{\xi_{\mathbf{p}}^2 + \Delta_{\mathbf{p}}^2}$, and an energy gap in the spectrum has appeared.

Show the $|BCS\rangle$-state is the vacuum state for the γ-operators, $\gamma_{\mathbf{p}}|BCS\rangle = 0$.

At finite temperatures Pauli's exclusion principle for the BCS-quasi-particles, which is equivalent to the anti-commutation properties of the γ-operators, gives that

---

[8]This so-called condensation energy is typically seven orders of magnitude smaller than the average Coulomb energy, and for the pairing Hamiltonian to make sense it is implicitly assumed that the Coulomb energy for an electron is the same in the two states, which the success of the BCS-theory then indicates.

[9]Recall the particle–hole symmetry of the BCS-state discussed in Exercise 2.8. on page 39.

at temperature T the probability of occupation of energy state $E_{\mathbf{p}}$ is given by the Fermi function

$$\langle \gamma_{\mathbf{p}\uparrow}^{\dagger} \gamma_{\mathbf{p}\uparrow} \rangle = \frac{1}{e^{E_{\mathbf{p}}/kT} + 1} = \langle \gamma_{-\mathbf{p}\downarrow}^{\dagger} \gamma_{-\mathbf{p}\downarrow} \rangle \, . \qquad (8.17)$$

Show consequently that the energy gap is temperature dependent as determined self-consistently by the gap equation

$$\Delta_{\mathbf{p}} = -\frac{1}{2} \sum_{\mathbf{p}'} V_{\mathbf{p}\mathbf{p}'} \frac{\Delta_{\mathbf{p}'}}{E_{\mathbf{p}'}} \tanh\left(\frac{E_{\mathbf{p}'}}{2kT}\right) \, . \qquad (8.18)$$

Show in the simple model considered in the previous exercise, that the energy gap vanishes at the critical temperature, T_{c}, given by

$$kT_{\mathrm{c}} \simeq \hbar\omega_{\mathrm{D}} \, e^{-1/N(0)V} \, . \qquad (8.19)$$

The BCS-theory is a mean field self-consistent theory with anomalous terms as specified by the off-diagonal long-range order. The effective Hamiltonian of the superconducting state can therefore also be arrived at by the following argument. The effective two-body interaction is short ranged, of the order of the Fermi wavelength, the inter-atomic distance, and can be approximated by the effective local two-body interaction, a delta potential characterized by a coupling strength γ (in the electron–phonon model γ is the square of the electron–phonon coupling constant, $\gamma = g^2$). The attractive two-body interaction term then becomes

$$V = -\frac{1}{2}\gamma \sum_{\alpha,\beta} \int d\mathbf{x} \, \psi_{\alpha}^{\dagger}(\mathbf{x}) \, \psi_{\beta}^{\dagger}(\mathbf{x}) \, \psi_{\beta}(\mathbf{x}) \, \psi_{\alpha}(\mathbf{x}) \, , \qquad (8.20)$$

assuming a spin-independent interaction. This is of course still a hopelessly complicated many-body problem. The BCS-theory is a self-consistent theory where the interaction term is substituted according to

$$V \to -\frac{1}{2}\gamma \sum_{\alpha,\beta} \int d\mathbf{x} \, (\Delta_{\alpha\beta}^{*}(\mathbf{x},\mathbf{x}) \, \psi_{\beta}(\mathbf{x}) \, \psi_{\alpha}(\mathbf{x}) \, + \, \Delta_{\beta\alpha}(\mathbf{x},\mathbf{x}) \psi_{\alpha}^{\dagger}(\mathbf{x}) \, \psi_{\beta}^{\dagger}(\mathbf{x})) \qquad (8.21)$$

a manageable quadratic form, however with anomalous terms. The implicit assumption for a self-consistent theory is thus that the fluctuations in the states of interest of the difference between the two operators in Eq. (8.20) and Eq. (8.21) are small. This is analogous to the Hartree–Fock treatment of the electron–electron interaction in the normal state. These normal terms should also be considered, but in a conventional superconductor such as a metal like tin, these effects lead to only a tiny renormalization of the electron mass, and we can think of them as included through the dispersion relation. In a strongly interacting degenerate Fermi system such as ^{3}He, these interactions need to be taken into account and must be dealt with in terms of Landau's Fermi liquid theory, a quasi-particle description (for details see reference

[35] and for the application of the quasi-classical Green's function technique see reference [36]). One should be aware that the BCS-approximation is quite a bold move since the BCS-Hamiltonian breaks a sacred conservation law, viz. particle number conservation, or equivalently, gauge invariance is spontaneously broken.[10]

For conventional superconductors we encounter orbital s-wave and spin-singlet pairing where the interaction part of the Hamiltonian is

$$V_{\text{BCS}} = -\gamma \int d\mathbf{x} \, (\Delta^*(\mathbf{x}) \, \psi_\uparrow(\mathbf{x}) \, \psi_\downarrow(\mathbf{x}) + \Delta(\mathbf{x}) \, \psi_\downarrow^\dagger(\mathbf{x}) \, \psi_\uparrow^\dagger(\mathbf{x})) \qquad (8.22)$$

as the superconducting order parameter is[11]

$$\Delta(\mathbf{x}) = \langle \psi_\uparrow(\mathbf{x}) \, \psi_\downarrow(\mathbf{x}) \rangle . \qquad (8.23)$$

Of importance is the feature of self-consistency, i.e. the bracket means average with respect to the order-parameter dependent BCS-Hamiltonian

$$H_{\text{BCS}} = \sum_{\alpha=\downarrow,\uparrow} \int d\mathbf{x} \, \psi_\alpha^\dagger(\mathbf{x}) \left(\frac{1}{2m} \left(\frac{\hbar}{i} \frac{\partial}{\partial \mathbf{x}} - e\mathbf{A}(\mathbf{x}, t) \right)^2 - \mu \right) \psi_\alpha(\mathbf{x})$$

$$- \gamma \int d\mathbf{x} \, (\Delta^*(\mathbf{x}) \, \psi_\uparrow(\mathbf{x}) \, \psi_\downarrow(\mathbf{x}) + \Delta(\mathbf{x}) \, \psi_\downarrow^\dagger(\mathbf{x}) \, \psi_\uparrow^\dagger(\mathbf{x})) \qquad (8.24)$$

and Eq. (8.24) and Eq. (8.23) thus represent a complicated set of coupled equations. We have placed the superconductor in an electromagnetic field represented by a vector potential which, except for weak fields or for temperatures near the critical temperature, through self-consistency leads to unquenchable analytic intractabilities. Only for simple and highly symmetric situations can the order parameter be specified *a priori*, thereby opening up for analytical tractability.

In the Heisenberg picture, the equation of motion governed by the BCS Hamiltonian is for the spin-up electron field component

$$i\frac{\partial \psi_\uparrow(\mathbf{x}, t)}{\partial t} = \left(\frac{1}{2m} (-i\nabla_\mathbf{x} - e\mathbf{A}(\mathbf{x}, t))^2 - \mu \right) \psi_\uparrow(\mathbf{x}, t) + \gamma\Delta(\mathbf{x}, t) \, \psi_\downarrow^\dagger(\mathbf{x}, t) \quad (8.25)$$

and for the spin-down adjoint component

$$-i\frac{\partial \psi_\downarrow^\dagger(\mathbf{x}, t)}{\partial t} = \left(\frac{1}{2m} (i\nabla_\mathbf{x} - e\mathbf{A}(\mathbf{x}, t))^2 - \mu \right) \psi_\downarrow^\dagger(\mathbf{x}, t) - \gamma\Delta^*(\mathbf{x}, t) \, \psi_\uparrow(\mathbf{x}, t). \quad (8.26)$$

The BCS-Hamiltonian therefore leads to a set of coupled equations of motion for the single-particle time-ordered Green's function

$$G(\mathbf{x}, t; \mathbf{x}', t') = -i\langle T(\psi_\uparrow(\mathbf{x}, t) \, \psi_\uparrow^\dagger(\mathbf{x}', t')) \rangle \qquad (8.27)$$

---

[10]In the electron–phonon model, the Hamiltonian is gauge invariant.
[11]In the case of p-wave or d-wave pairing, the order parameter has additional spin dependence.

and the anomalous or Gorkov Green's function

$$\overline{F}(\mathbf{x}, t; \mathbf{x}', t') = -i\langle T(\psi_\downarrow^\dagger(\mathbf{x}, t)\, \psi_\uparrow^\dagger(\mathbf{x}', t'))\rangle, \qquad (8.28)$$

viz. the Gorkov equations[12]

$$\left(i\frac{\partial}{\partial t} - \frac{1}{2m}(-i\nabla_\mathbf{x} - e\mathbf{A}(\mathbf{x}, t))^2 + \mu\right) G(\mathbf{x}, t, \mathbf{x}', t') + \gamma\Delta(\mathbf{x}, t)\,\overline{F}(\mathbf{x}, t, \mathbf{x}', t')$$

$$= \delta(\mathbf{x} - \mathbf{x}')\delta(t - t') \qquad (8.29)$$

and

$$\left(-i\frac{\partial}{\partial t_1} - \frac{1}{2m}(i\nabla_{\mathbf{x}_1} - e\mathbf{A}(\mathbf{x}_1, t_1))^2 + \mu\right)\overline{F}(1, 1') + \gamma\Delta^*(1)\,G(1, 1') = 0, \quad (8.30)$$

where in the latter equation we have introduced the usual condensed notation. The spin labeling of the functions is irrelevant since no spin-dependent interactions, such as spin flip interactions due to magnetic impurities, are presently included and spin up and down are therefore equivalent, except for the singlet feature of the anomalous Green's function as we consider s-wave pairing. The order parameter is specified by the equal space and time anomalous Green's function

$$\Delta^*(\mathbf{x}, t) = i\,\overline{F}(\mathbf{x}, t^+; \mathbf{x}, t) = \langle\psi_\downarrow^\dagger(\mathbf{x}, t)\psi_\uparrow^\dagger(\mathbf{x}, t)\rangle. \qquad (8.31)$$

When the effect of pairing is taken into account, the Feynman diagrammatics in the electron–phonon or BCS-model is modified by the presence of lines describing the additional channel due to the non-vanishing of the anomalous Green's function. However, as the order parameter is small compared with the Fermi energy in a conventional superconductor (as well as in superfluid He-3), this new scale is irrelevant for diagram estimation, and Migdal's theorem is then again valid (as first noted by Eliashberg [37]). The peaked structure at the Fermi momentum of the Green's functions thus remains as in the normal state, and the argument for the validity of the Migdal approximation now becomes identical for the superfluid case once it is based on the correct ground state, i.e. the anomalous self-energy terms are included. A theory of strong coupling superconductivity, Eliashberg's theory, is thus available of which the BCS-theory is the weak coupling limit, $kT_c \ll \hbar\omega_D$, in accordance with Eq. (8.1). It is convenient to collect the equations of motion for the normal and anomalous Green's functions into a single matrix equation of motion, and this is done by introducing the Nambu field, by which the BCS-Hamiltonian is turned into a quadratic form. Furthermore, we shall introduce the contour ordered and not just the time ordered Green's functions in order to describe the non-equilibrium states of a superconductor.

---

[12]Had we used the canonical ensemble, the chemical potential would be absent in Eq. (8.29), and since

$$\overline{F}_N(\mathbf{x}, t^+; \mathbf{x}', t) = -i\langle N + 2|\psi_\downarrow^\dagger(\mathbf{x}, t)\,\psi_\uparrow^\dagger(\mathbf{x}', t)|N\rangle = e^{2\frac{i}{\hbar}\mu t}\overline{F}_\mu(\mathbf{x}, t; \mathbf{x}', t)$$

the term $-2\mu\,\overline{F}_N(\mathbf{x}, t; \mathbf{x}', t')$ would appear on the left in Eq. (8.30).

8.1.1 Nambu or particle–hole space

In order to write the BCS-Hamiltonian, Eq. (8.24), in standard quadratic form of a field, we introduce with Nambu the pseudo-spinor field

$$\Psi(1) \equiv \left(\begin{array}{c} \psi_\uparrow(1) \\ \psi_\downarrow^\dagger(1) \end{array} \right) \equiv \left(\begin{array}{c} \Psi_1(1) \\ \Psi_2(1) \end{array} \right), \tag{8.32}$$

where, introducing condensed notation, $1 \equiv (t_1, \mathbf{x}_1)$ comprises the spatial variable and in the Heisenberg picture also the time variable. The adjoint Nambu field is

$$\Psi^\dagger(1) \equiv (\psi_\uparrow^\dagger(1), \psi_\downarrow(1)) \equiv (\Psi_1^\dagger(1), \Psi_2^\dagger(1)), \tag{8.33}$$

where in the last definition we have introduced the matrix notation for the Nambu or particle–hole space.

The BCS-Hamiltonian is

$$H_{\text{BCS}} = \int d\mathbf{x}_1 \left(\sum_{\sigma=\uparrow,\downarrow} \psi_\sigma^\dagger(1)h(1)\psi_\sigma(1) + \Delta(1)\psi_\uparrow^\dagger(1)\psi_\downarrow^\dagger(1) + \Delta^*(1)\psi_\downarrow(1)\psi_\uparrow(1)) \right), \tag{8.34}$$

where

$$h(1) = \frac{1}{2m}\left(-i\nabla_{\mathbf{x}_1} - e\mathbf{A}(\mathbf{x}_1, t_1)\right)^2 + e\phi(\mathbf{x}_1, t_1) - \mu \tag{8.35}$$

and the presence of coupling of the electrons to a classical electromagnetic field has been included.

Consider the quadratic form in terms of the Nambu field

$$H = \int d\mathbf{x}_1 \; \Psi^\dagger(1) \left(\begin{array}{cc} h(1) & \Delta(1) \\ \Delta^*(1) & -h^*(1) \end{array} \right) \Psi(1), \tag{8.36}$$

where $h^*(1)$ denotes complex conjugate of the single-particle Hamiltonian

$$h^*(1) = \frac{1}{2m}\left(i\nabla_{\mathbf{x}_1} - e\mathbf{A}(\mathbf{x}_1, t_1)\right)^2 + e\phi(\mathbf{x}_1, t_1) - \mu. \tag{8.37}$$

The off-diagonal terms are identical to the ones in the BCS-Hamiltonian, but only the first of the diagonal terms

$$\int d\mathbf{x}_1 (\psi_\uparrow^\dagger(1)h(1)\psi_\uparrow(1) - \psi_\downarrow(1)h^*(1)\psi_\downarrow^\dagger(1)) \tag{8.38}$$

gives the corresponding kinetic energy term. In the second term partial integrations are performed, giving

$$\int d\mathbf{x}_1 \; \psi_\downarrow(\mathbf{x}_1, t_1)h^*(1)\psi_\downarrow^\dagger(1) = \int d\mathbf{x}_1 \; h(\mathbf{x}_{1'}, t_1)\psi_\downarrow(1')\psi_\downarrow^\dagger(1)\Big|_{\mathbf{x}_{1'}=\mathbf{x}_1}. \tag{8.39}$$

Using the equal-time anti-commutation relations for the electron fields produces the wanted order of the operators but an additional delta function

$$\int d\mathbf{x}_1 \; h(\mathbf{x}_{1'}, t_1) \left(\delta(\mathbf{x}_1 - \mathbf{x}_{1'}) - \psi_\downarrow^\dagger(1)\psi_\downarrow(1') \right)\Big|_{\mathbf{x}_{1'}=\mathbf{x}_1}, \tag{8.40}$$

which, however, is just a state independent (infinite) constant that has no influence on the dynamics and can be dropped. We have thus shown that the BCS-Hamiltonian can equivalently be written in terms of the Nambu field as

$$H_{\mathrm{BCS}} = \int d\mathbf{x}_1 \, \Psi^\dagger(1) \begin{pmatrix} h(1) & \Delta(1) \\ \Delta^*(1) & -h^*(1) \end{pmatrix} \Psi(1). \tag{8.41}$$

Two by two (2×2) matrices are introduced in Nambu space according to

$$\Psi(1) \triangle \Psi^\dagger(1') \;\equiv\; \begin{pmatrix} \psi_\uparrow(1) \\ \psi_\downarrow^\dagger(1) \end{pmatrix} (\psi_\uparrow^\dagger(1'), \psi_\downarrow(1'))$$

$$=\; \begin{pmatrix} \psi_\uparrow(1)\psi_\uparrow^\dagger(1') & \psi_\uparrow(1)\psi_\downarrow(1') \\ \psi_\downarrow^\dagger(1)\psi_\uparrow^\dagger(1') & \psi_\downarrow^\dagger(1)\psi_\downarrow(1') \end{pmatrix} \tag{8.42}$$

or, in Nambu index notation,

$$(\Psi(1) \triangle \Psi^\dagger(1'))_{ij} \equiv \Psi_i(1) \, \Psi_j^\dagger(1') \,. \tag{8.43}$$

For the opposite sequence we define the $2{\times}2$-matrix

$$(\Psi^\dagger(1) \triangle \Psi(1'))_{ij} \;\equiv\; \Psi_j^\dagger(1) \, \Psi_i(1')$$

$$=\; \begin{pmatrix} \psi_\uparrow^\dagger(1)\psi_\uparrow(1') & \psi_\downarrow(1)\psi_\uparrow(1') \\ \psi_\uparrow^\dagger(1)\psi_\downarrow^\dagger(1') & \psi_\downarrow(1)\psi_\downarrow^\dagger(1') \end{pmatrix} \,. \tag{8.44}$$

The Nambu field is seen to satisfy the canonical anti-commutation rules

$$[\Psi(\mathbf{x}) \stackrel{\triangle}{,} \Psi^\dagger(\mathbf{x}'))]_+ = \delta(\mathbf{x} - \mathbf{x}') \, \underline{1} \tag{8.45}$$

and

$$[\Psi^\dagger(\mathbf{x}) \stackrel{\triangle}{,} \Psi^\dagger(\mathbf{x}'))]_+ = \underline{0} = [\Psi(\mathbf{x}) \stackrel{\triangle}{,} \Psi(\mathbf{x}'))]_+ \tag{8.46}$$

where $\underline{1}$ and $\underline{0}$ are the unit and zero matrices in Nambu space, respectively.

The contour-ordered Green's function is defined in particle–hole or Nambu space according to

$$\underline{G}(1,1') = -i\langle T_{c_t}(\Psi_\mathcal{H}(1) \triangle \Psi_\mathcal{H}^\dagger(1')) \rangle \,, \tag{8.47}$$

where the subscript indicates the field is in the Heisenberg picture. For $t_1 \stackrel{>}{c} t_{1'}$ the contour-ordered Green's function becomes

$$\underline{G}^>(1,1') \;=\; -i\langle (\Psi_\mathcal{H}(1) \triangle \Psi_\mathcal{H}^\dagger(1')) \rangle$$

$$=\; -i \begin{pmatrix} \langle \psi_\uparrow(1)\psi_\uparrow^\dagger(1') \rangle & \langle \psi_\uparrow(1)\psi_\downarrow(1') \rangle \\ \langle \psi_\downarrow^\dagger(1)\psi_\uparrow^\dagger(1') \rangle & \langle \psi_\downarrow^\dagger(1)\psi_\downarrow(1') \rangle \end{pmatrix} \,. \tag{8.48}$$

In Nambu index notation the *greater* Green's function simply becomes

$$\underline{G}_{ij}^>(1,1') = -i\langle \Psi_i(1) \, \Psi_j^\dagger(1') \rangle. \tag{8.49}$$

The *lesser* Green's function is then

$$\underline{G}^{<}(1,1') = i\langle(\Psi_{\mathcal{H}}^{\dagger}(1')\triangle\Psi_{\mathcal{H}}(1))\rangle$$

$$= i \begin{pmatrix} \langle\psi_{\uparrow}^{\dagger}(1')\psi_{\uparrow}(1)\rangle & \langle\psi_{\downarrow}(1')\psi_{\uparrow}(1)\rangle \\ \langle\psi_{\uparrow}^{\dagger}(1')\psi_{\downarrow}^{\dagger}(1)\rangle & \langle\psi_{\downarrow}(1')\psi_{\downarrow}^{\dagger}(1)\rangle \end{pmatrix} \tag{8.50}$$

and in matrix notation

$$\underline{G}^{<}_{ij}(1,1') = -i\langle\Psi_i^{\dagger}(1)\,\Psi_j(1')\rangle\,. \tag{8.51}$$

To acquaint ourselves with Nambu space we consider the dynamics of the Nambu field governed by the BCS-Hamiltonian. In the presence of classical electromagnetic fields, the free one-particle Hamiltonian is

$$H_0(t) = \int d\mathbf{x}\,\Psi_{H_0}^{\dagger}(\mathbf{x},t)\,\underline{h}(\mathbf{x},t)\,\Psi_{H_0}(\mathbf{x},t)\,, \tag{8.52}$$

where

$$\underline{h}(1) = \begin{pmatrix} h(1) & 0 \\ 0 & -h^*(1) \end{pmatrix} \equiv h_{ij}(1)\,. \tag{8.53}$$

From the equations of motion for the free Nambu field

$$i\partial_{t_1}\Psi_i(1) = h_{ij}(1)\Psi_j(1) \tag{8.54}$$

and

$$i\partial_{t_1}\Psi_i^{\dagger}(1) = -h_{ij}^*(1)\Psi_j^{\dagger}(1)\,, \tag{8.55}$$

where the fields are in the Heisenberg picture with respect to $H_0(t)$, the equations of motion for the free Nambu Green's functions become

$$(i\partial_{t_1} - \underline{h}(1))\underline{G}_0^{>}(1,1') = 0 \tag{8.56}$$

and

$$i\partial_{t_{1'}}\,\underline{G}_0^{>}(1,1') = -\underline{G}_0^{>}(1,1')\,\overset{\leftarrow}{\underline{h}}{}^*(1')\,, \tag{8.57}$$

where the arrow indicates that the spatial differential operator operates to the left. Identical equations of motion are obtained for $\underline{G}_0^{<}(1,1')$.

The presence of the pairing interaction then leads to the appearance of a self-energy which is purely off-diagonal in Nambu space

$$\underline{\Sigma}_{\text{BCS}} = \begin{pmatrix} 0 & \Delta(1) \\ \Delta^*(1) & 0 \end{pmatrix}\,. \tag{8.58}$$

In order to get more symmetric equations we perform the transformation

$$\underline{G}(1,1') \to \tau_3\,\underline{G}(1,1') \equiv G\,, \tag{8.59}$$

where τ_3 denotes the third Pauli matrix in Nambu space. The equations of motion for the free Nambu Green's functions then become

$$i\tau_3 \partial_{t_1} G_0^{\gtrless}(1,1') = h_N(1) G_0^{\gtrless}(1,1') \tag{8.60}$$

and

$$i\tau_3 \partial_{t_{1'}} G_0^{\gtrless}(1,1') = -G_0^{\gtrless}(1,1') \overleftarrow{h}_N^{*}(1'), \tag{8.61}$$

where

$$h_N(1) = \underline{\underline{h}}(1)\tau_3 = -\frac{1}{2m}\overrightarrow{\partial}^2(1) + e\phi(1) - \mu \tag{8.62}$$

and

$$\overrightarrow{\partial}(1) = \nabla_{\mathbf{x}_1} - ie\tau_3 \mathbf{A}(\mathbf{x}_1, t_1). \tag{8.63}$$

The BCS-self-energy, describing the pairing interaction, then becomes

$$\Sigma_{\mathrm{BCS}} = \begin{pmatrix} 0 & -\Delta(1) \\ \Delta^{*}(1) & 0 \end{pmatrix}. \tag{8.64}$$

Introducing the Nambu field facilitates the description of the particle–hole coherence in a superconductor. Next we introduce the real-time formalism for describing non-equilibrium states as discussed in Chapter 5, here for the purpose of describing non-equilibrium superconductivity. For a superconductor this means adding to the Nambu-indices of Green's functions the additional Schwinger–Keldysh or dynamical indices.

Exercise 8.4. Show that, in equilibrium, the retarded Nambu Green's function has the form (unit matrices in Nambu space are suppressed)

$$G^R(E, \mathbf{p}) = (E\tau_3 - \xi(\mathbf{p}) - \Sigma^R(E, \mathbf{p}))^{-1}. \tag{8.65}$$

In a strong coupling superconductor the self-energy has, according to the electron–phonon model, the form

$$\Sigma^R(E) = (1 - Z^R(E))E\tau_3 - i\Phi^R(E)\tau_1, \tag{8.66}$$

and show as a consequence that the retarded Nambu Green's function becomes

$$G^R(E, \mathbf{p}) = \frac{-E\,Z^R(E)\,\tau_3 - \xi(\mathbf{p}) - i\Phi^R(E)\,\tau_1}{(\xi(\mathbf{p}))^2 - E^2(Z^R(E))^2 + (\Phi^R(E))^2}. \tag{8.67}$$

8.1.2   Equations of motion in Nambu–Keldysh space

The contour-ordered Green's function in Nambu space is defined according to

$$G_C(1,1') = -i\tau_3 \langle T_{c_t}(\Psi_{\mathcal{H}}(1)\triangle\Psi_{\mathcal{H}}^{\dagger}(1'))\rangle \tag{8.68}$$

and is mapped into real-time dynamical or Schwinger–Keldysh space according to the usual rule, Eq. (5.1),

$$G_C(1,1') \rightarrow G(1,1') \equiv \begin{pmatrix} G_{11}(1,1') & G_{12}(1,1') \\ -G_{21}(1,1') & -G_{22}(1,1') \end{pmatrix}, \tag{8.69}$$

where the Schwinger–Keldysh components now are Nambu matrices

$$G_{11}(1,1') = -i\,\tau_3\,\langle T(\Psi_{\mathcal{H}}(1)\,\triangle\,\Psi_{\mathcal{H}}^\dagger(1'))\rangle \tag{8.70}$$

and *G-lesser*

$$G_{12}(1,1') = G^<(\mathbf{x}_1,t_1,\mathbf{x}_{1'},t_{1'}) = i\,\tau_3\,\langle\psi_{\mathcal{H}}^\dagger(\mathbf{x}_{1'},t_{1'})\,\triangle\,\psi_{\mathcal{H}}(\mathbf{x}_1,t_1)\rangle \tag{8.71}$$

and *G-greater*

$$G_{21}(1,1') = G^>(\mathbf{x}_1,t_1,\mathbf{x}_{1'},t_{1'}) = -i\,\tau_3\,\langle\psi_{\mathcal{H}}(\mathbf{x}_1,t_1)\,\triangle\,\psi_{\mathcal{H}}^\dagger(\mathbf{x}_{1'},t_{1'})\rangle \tag{8.72}$$

and

$$G_{22}(1,1') = -i\,\tau_3\,\langle\tilde{T}(\Psi_{\mathcal{H}}(1)\,\triangle\,\Psi_{\mathcal{H}}^\dagger(1'))\rangle \tag{8.73}$$

and the Pauli matrix appears because of the convention, Eq. (8.59).

The information contained in the various Schwinger–Keldysh components of the matrix Green's function is rather condensed and it can be useful to have explicit expressions for two independent components, say *G-lesser* and *G-greater*, from which all other relevant Green's functions can be constructed.

Exercise 8.5. Show that, in terms of the electron field, we have

$$G^>(1,1') = -i\tau_3\begin{pmatrix} \langle\psi_\uparrow(1)\,\psi_\uparrow^\dagger(1')\rangle & \langle\psi_\uparrow(1)\,\psi_\downarrow(1')\rangle \\ \langle\psi_\downarrow^\dagger(1)\,\psi_\uparrow^\dagger(1')\rangle & \langle\psi_\downarrow^\dagger(1)\,\psi_\downarrow(1')\rangle \end{pmatrix} \tag{8.74}$$

and

$$G^<(1,1') = i\tau_3\begin{pmatrix} \langle\psi_\uparrow^\dagger(1')\,\psi_\uparrow(1)\rangle & \langle\psi_\downarrow(1')\,\psi_\uparrow(1)\rangle \\ \langle\psi_\uparrow^\dagger(1')\,\psi_\downarrow^\dagger(1)\rangle & \langle\psi_\downarrow(1')\,\psi_\downarrow^\dagger(1)\rangle \end{pmatrix}. \tag{8.75}$$

The matrix Green's function on triagonal form

$$G = \begin{pmatrix} G^{\mathrm{R}} & G^{\mathrm{K}} \\ 0 & G^{\mathrm{A}} \end{pmatrix} \tag{8.76}$$

has, according to the construction in Section 5.3, the retarded and advanced components

$$G^{\mathrm{R}}(\mathbf{x},t,\mathbf{x}',t') = -i\theta(t-t')\,\tau_3\,\langle[\psi_{\mathcal{H}}(\mathbf{x},t)\stackrel{\triangle}{,}\psi_{\mathcal{H}}^\dagger(\mathbf{x}',t')]_+\rangle$$

$$= \theta(t-t')\,(G^>(\mathbf{x},t,\mathbf{x}',t') - G^<(\mathbf{x},t,\mathbf{x}',t')) \tag{8.77}$$

and

$$G^{\mathrm{A}}(\mathbf{x},t,\mathbf{x}',t') = i\theta(t'-t)\,\tau_3\,\langle[\psi_{\mathcal{H}}(\mathbf{x},t)\stackrel{\triangle}{,}\psi_{\mathcal{H}}^\dagger(\mathbf{x}',t')]_+\rangle$$

$$= -\theta(t'-t)\,(G^>(\mathbf{x},t,\mathbf{x}',t') - G^<(\mathbf{x},t,\mathbf{x}',t')) \tag{8.78}$$

and the Keldysh or kinetic Green's function

$$G^K(\mathbf{x}, t, \mathbf{x}', t') = -i\tau_3 \langle [\psi_{\mathcal{H}}(\mathbf{x}, t) \overset{\triangle}{,} \psi_{\mathcal{H}}^\dagger(\mathbf{x}', t')]_- \rangle$$

$$= G^>(\mathbf{x}, t, \mathbf{x}', t') + G^<(\mathbf{x}, t, \mathbf{x}', t') \,. \tag{8.79}$$

Exercise 8.6. Write down the retarded, advanced and kinetic components of the Nambu Green's function in terms of the electron field.

The equation of motion, the non-equilibrium Dyson equation, for the matrix Green's function becomes

$$(G_0^{-1} \otimes G)(1, 1') = \delta(1 - 1') + (\Sigma \otimes G)(1, 1') \tag{8.80}$$

and for the conjugate equation

$$(G \otimes G_0^{-1})(1, 1') = \delta(1 - 1') + (G \otimes \Sigma)(1, 1') \,, \tag{8.81}$$

where the inverse free matrix Green's function in Nambu–Keldysh space

$$G_0^{-1}(1, 1') = G_0^{-1}(1)\,\delta(1 - 1') \tag{8.82}$$

is specified in the triagonal representation by

$$G_0^{-1}(1) = (i\tau^{(3)}\,\partial_{t_1} - h(1)) \,, \tag{8.83}$$

and

$$\tau^{(3)} = \begin{pmatrix} \tau_3 & 0 \\ 0 & \tau_3 \end{pmatrix} \tag{8.84}$$

is the 4×4-matrix, diagonal in Keldysh indices and τ_3 the third Pauli matrix in Nambu space. Here

$$h(1) = -\frac{1}{2m}\partial^2(1) + e\phi(1) - \mu \,, \tag{8.85}$$

where

$$\partial(1) = \nabla_{\mathbf{x}_1} - ie\tau^{(3)}\mathbf{A}(\mathbf{x}_1, t_1) \,. \tag{8.86}$$

Written out in components the matrix equation, Eq. (8.80), this gives

$$G_0^{-1}(1)\,G^{R(A)}(1, 1') = \delta(1 - 1') + (\Sigma^{R(A)} \otimes G^{R(A)})(1, 1') \tag{8.87}$$

and

$$G_0^{-1}(1)\,G^K(1, 1') = (\Sigma^R \otimes G^K)(1, 1') + (\Sigma^K \otimes G^A)(1, 1') \,. \tag{8.88}$$

Subtracting the left and right Dyson equations, Eq. (8.80) and Eq. (8.81), we obtain an equation identical in form to that of the normal state, Eq. (7.1), an equation for the spectral weight function and a quantum kinetic equation of the form Eq. (7.3). However, they are additionally matrix equations in Nambu space. Generally they are too complicated to be analytically tractable. It is therefore of importance that the quasi-classical approximation works for the superconducting state, at least excellently in low temperature superconductors where the superconducting coherence length, $\xi_0 = \hbar v_F / \pi \Delta$, is much longer than the Fermi wavelength. In other words, the small distance information in the above equations is irrelevant and should be removed. After discussing the gauge transformation properties of the Nambu Green's functions, we turn to describe the quasi-classical theory of non-equilibrium superconductors which precisely does that.[13]

8.1.3 Green's functions and gauge transformations

The field operator representing a charged particle transforms according to

$$\psi(\mathbf{x}, t) \rightarrow \psi(\mathbf{x}, t)\, e^{ie\Lambda(\mathbf{x}, t)} \equiv \tilde{\psi}(\mathbf{x}, t) \tag{8.89}$$

under the gauge transformation

$$\varphi(\mathbf{x}, t) \rightarrow \varphi(\mathbf{x}, t) + \frac{\partial \Lambda(\mathbf{x}, t)}{\partial t} \quad , \quad \mathbf{A}(\mathbf{x}, t) \rightarrow \mathbf{A}(\mathbf{x}, t) - \nabla_{\mathbf{x}} \Lambda(\mathbf{x}, t) . \tag{8.90}$$

The probability and current density of the particles will be invariant to this shift; quantum mechanics is gauge invariant.

The matrix Green's function therefore transforms according to

$$
\tilde{G}^{<}(1, 1') = i \begin{pmatrix} e^{ie(\Lambda(1) - \Lambda(1'))} \langle \psi_\uparrow^\dagger(1') \psi_\uparrow(1) \rangle & e^{ie(\Lambda(1) + \Lambda(1'))} \langle \psi_\downarrow(1') \psi_\uparrow(1) \rangle \\ -e^{-ie(\Lambda(1) + \Lambda(1'))} \langle \psi_\uparrow^\dagger(1') \psi_\downarrow^\dagger(1) \rangle & -e^{-ie(\Lambda(1) - \Lambda(1'))} \langle \psi_\downarrow(1') \psi_\downarrow^\dagger(1) \rangle \end{pmatrix}
$$

$$
= e^{ie\Lambda(1)\tau_3}\, G^{<}(1, 1')\, e^{-ie\Lambda(1')\tau_3} \tag{8.91}
$$

and similarly for

$$
\tilde{G}^{>}(1, 1') = i \begin{pmatrix} e^{ie(\Lambda(1) - \Lambda(1'))} \langle \psi_\uparrow(1) \psi_\uparrow^\dagger(1') \rangle & e^{ie(\Lambda(1) + \Lambda(1'))} \langle \psi_\uparrow(1) \psi_\downarrow(1') \rangle \\ -e^{-ie(\Lambda(1) + \Lambda(1'))} \langle \psi_\downarrow^\dagger(1) \psi_\uparrow^\dagger(1') \rangle & -e^{-ie(\Lambda(1) - \Lambda(1'))} \langle \psi_\downarrow^\dagger(1) \psi_\downarrow(1') \rangle \end{pmatrix}
$$

$$
= e^{ie\Lambda(1)\tau_3}\, G^{>}(1, 1')\, e^{-ie\Lambda(1')\tau_3} . \tag{8.92}
$$

The other Green's functions in Nambu space, $G^{R,A,K}$, are linear combinations of $G^{\lessgtr}$, and therefore transform similarly. The gauge transformation then transforms the matrix Green's function in Keldysh space according to

$$
\tilde{G}(1, 1') = e^{ie\Lambda(1)\,\tau^{(3)}}\, G(1, 1')\, e^{-ie\Lambda(1')\,\tau^{(3)}} . \tag{8.93}
$$

The flexibility of gauge transformations allows one to choose potentials that minimize the temporal variation of the order parameter, such as facilitating transformation to the gauge where the order parameter is real, the real Δ gauge, where the

---

[13]The technique has also been used to derive kinetic equations for quasi-one-dimensional conductors with a charge-density wave resulting from the Peierls instability [38].

phase of the order parameter vanishes, $\chi = 0$. This is achieved by choosing the gauge transformation

$$-e\varphi \;\rightarrow\; \Phi = \frac{1}{2}\dot{\chi} - e\varphi \tag{8.94}$$

and

$$-\frac{e}{m}\mathbf{A} \;\rightarrow\; \mathbf{v}_{\mathrm{s}} = -\frac{1}{2m}(\nabla\chi + 2e\mathbf{A}) \tag{8.95}$$

introducing the gauge-invariant quantities, the superfluid velocity, $\mathbf{v}_{\mathrm{s}}$, and the electro-chemical potential, Φ, of the condensate or Cooper pairs.

8.2   Quasi-classical Green's function theory

The superconducting state introduces the additional energy scale of the order parameter, which in the BCS-case equals the energy gap in the excitation spectrum. In a conventional superconductor, as well as in superfluid He-3, this scale is small compared with the Fermi energy. The peaked structure at the Fermi momentum of the Green's functions thus remains as in the normal state, and the arguments for the superfluid case that brings us from the left and right Dyson equations, Eq. (8.80) and Eq. (8.81), to the subtracted Dyson equation for the quasi-classical Green's function are thus identical to those of Section 7.5 for the normal state, and we obtain the matrix equation, the Eilenberger equations,

$$[g_0^{-1} + i\sigma \overset{\circ}{,} g]_- \;=\; 0\,, \tag{8.96}$$

which gives the three coupled equations for $g^{\mathrm{R(A)}}$ and g^{K} where

$$g_0^{-1}(\hat{\mathbf{p}}, \mathbf{R}, t_1, t_{1'}) \;=\; g_0^{-1}(\hat{\mathbf{p}}, \mathbf{R}, t_1)\,\delta(t_1 - t_{1'}) \tag{8.97}$$

and[14]

$$g_0^{-1}(\hat{\mathbf{p}}, \mathbf{R}, t_1) \;=\; \tau^{(3)}\partial_{t_1} + \mathbf{v}_F \cdot (\nabla_{\mathbf{R}} - ie\tau^{(3)}\mathbf{A}(\mathbf{R}, t_1)) + ie\phi(\mathbf{R}, t_1) \tag{8.98}$$

and the ξ-integrated or quasi-classical four by four (4×4) matrix Green's function

$$g(\mathbf{R}, \hat{\mathbf{p}}, t_1, t_{1'}) \;=\; \frac{i}{\pi} \int d\xi\, G(\mathbf{R}, \mathbf{p}, t_1, t_{1'}) \tag{8.99}$$

is defined in the same way as in Section 7.5, capturing the low-energy behavior of the Green's functions.[15] The equations for $g^{\mathrm{R(A)}}$ determines the spectral densities and the equation for g^{K} is the quantum kinetic equation.[16]

---

[14]The $\mathbf{A}^2$-term is smaller in the quasi-classical expansion parameter $\lambda_F/\xi_0 \sim \Delta/E_F$, the ratio of the Fermi wavelength and the superconducting coherence length, $e^2\mathbf{A}^2/m \sim ev_F A\lambda_F/\xi_0$.

[15]In Section 8.3, the quasi-classical Green's functions will be introduced not by ξ-integration but by considering the spatial behavior of the Green's functions on the scale much larger than the inter-atomic distance.

[16]We follow the exposition given in reference [3] and reference [9].

Writing out for the components, we have for the spectral components

$$[g_0^{-1} + i\sigma^{R(A)} \, \overset{\circ}{,} \, g^{R(A)}]_- = 0 \qquad (8.100)$$

and for the kinetic component the quantum kinetic equation

$$[g_0^{-1} + i\Re e\sigma \, \overset{\circ}{,} \, g^K]_- = 2i\sigma^K - i[(\sigma^R - \sigma^A) \, \overset{\circ}{,} \, g^K]_+ \, . \qquad (8.101)$$

The self-energy comprises the effective electron–electron interaction, impurity scattering and electron spin-flip scattering due to magnetic impurities. The impurity scattering is in the weak disorder limit described by the self-energy[17]

$$\sigma_{\mathrm{imp}}(\hat{\mathbf{p}}, \mathbf{R}, t_1, t_{1'}) = -i\pi n_i N_0 \int \frac{d\hat{\mathbf{p}}'}{4\pi} \, |V_{\mathrm{imp}}(\hat{\mathbf{p}} \cdot \hat{\mathbf{p}}')|^2 \, g(\mathbf{R}, \hat{\mathbf{p}}', t_1, t_{1'}) \qquad (8.102)$$

quite analogous to that of the normal state, Eq. (7.51), except that the Green's function in addition to the real-time dynamical index structure is a matrix in Nambu space.

Even a small amount of magnetic impurities can, owing to their breaking of time reversal symmetry and consequent disruption of the coherence of the superconducting state, have a drastic effect on the properties of a superconductor, leading to the phenomena of gap-less superconductivity, and an amount of a few percent can destroy superconductivity completely [39]. We therefore include spin-flip scattering of electrons, which in contrast to normal impurities leads to pair-breaking and the quite different physics just mentioned. We assume that the positions and spin-states of the magnetic impurities are random, and owing to the latter assumption we can limit the analysis to the last term in Eq. (2.25).[18] In terms of the Nambu field the scattering off the magnetic impurities then becomes

$$V_{\mathrm{sf}} \rightarrow \sum_a \int d\mathbf{x} \, u(\mathbf{x} - \mathbf{x}_a) \, S_a^z \, \Psi^\dagger(\mathbf{x}) \Psi(\mathbf{x}) \, , \qquad (8.103)$$

where as usual in the Nambu formalism, an infinite constant has been dropped. The spin-flip self-energy has the additional feature, compared to the impurity scattering, of averaging over the random spin orientations of $S_a^z S_a^z$. Assuming that all impurities have the same spin, $S_a = S$, the averaging gives the factor $S(S+1)/3$ and the spin-flip self-energy then becomes

$$\sigma_{\mathrm{sf}}(\hat{\mathbf{p}}, \mathbf{R}, t_1, t_{1'}) = -i\pi n_{\mathrm{magn.imp.}} N_0 S(S+1) \int \frac{d\hat{\mathbf{p}}'}{4\pi} \, |u(\hat{\mathbf{p}} \cdot \hat{\mathbf{p}}')|^2 \, \tau^{(3)} \, g(\mathbf{R}, \hat{\mathbf{p}}', t_1, t_{1'}) \, \tau^{(3)}$$
$$(8.104)$$

where $n_{\mathrm{magn.imp}}$ is the concentration of magnetic impurities. Since the exchange interaction is weak, only s-wave scattering needs to be taken into consideration, and

---

[17]The weak disorder limit refers to $\hbar/\epsilon_F \tau \ll 1$, and the neglect of localization effects, but we could of course trivially include multiple scattering by introducing the t-matrix instead of the impurity potential. For a discussion see for example chapter 3 of reference [1].

[18]Magnetic impurity scattering was discussed in Exercise 2.5 on page 37, and for example in chapter 11 of reference [1].

the spin-flip self-energy becomes

$$\sigma_{sf}(\hat{\mathbf{p}}, \mathbf{R}, t_1, t_{1'}) = -\frac{i}{2\tau_s} \int \frac{d\hat{\mathbf{p}}'}{4\pi} \, \tau^{(3)} \, g(\mathbf{R}, \hat{\mathbf{p}}', t_1, t_{1'}) \, \tau^{(3)} \,, \qquad (8.105)$$

where the spin-flip scattering time is

$$\frac{1}{\tau_s} = 2\pi n_{mag.imp} N_0 S(S+1) \int \frac{d\hat{\mathbf{p}}'}{4\pi} \, |u(\hat{\mathbf{p}} \cdot \hat{\mathbf{p}}')|^2 \,. \qquad (8.106)$$

When inelastic effects are of interest, they are for example described by the electron–phonon interaction through the self-energy whose matrix components of the matrix self-energy are

$$\sigma_{e-ph}^{R(A)}(\mathbf{R}, \hat{\mathbf{p}}, t_1, t_{1'}) = \frac{\lambda}{8} \int d\hat{\mathbf{p}}' (g^K(\mathbf{R}, \hat{\mathbf{p}}', t_1, t_{1'}) \, D^{R(A)}(\mathbf{R}, p_F(\hat{\mathbf{p}} - \hat{\mathbf{p}}'), t_1, t_{1'})$$

$$+ \; g^{R(A)}(\mathbf{R}, \hat{\mathbf{p}}', t_1, t_{1'}) \, D^K(\mathbf{R}, p_F(\hat{\mathbf{p}} - \hat{\mathbf{p}}'), t_1, t_{1'})) \qquad (8.107)$$

and

$$\sigma_{e-ph}^K(\mathbf{R}, \hat{\mathbf{p}}, t_1, t_{1'}) = \frac{\lambda}{8} \int d\hat{\mathbf{p}}' (g^R(\mathbf{R}, \hat{\mathbf{p}}', t_1, t_{1'}) D^R(\mathbf{R}, p_F(\hat{\mathbf{p}} - \hat{\mathbf{p}}'), t_1, t_{1'})$$

$$+ \; g^A(\mathbf{R}, \hat{\mathbf{p}}', t_1, t_{1'}) D^A(\mathbf{R}, p_F(\hat{\mathbf{p}} - \hat{\mathbf{p}}'), t_1, t_{1'})$$

$$+ \; g^K(\mathbf{R}, \hat{\mathbf{p}}', t_1, t_{1'}) D^K(\mathbf{R}, p_F(\hat{\mathbf{p}} - \hat{\mathbf{p}}'), t_1, t_{1'})) \qquad (8.108)$$

or

$$\sigma_{e-ph}^K = \frac{\lambda}{8} \int d\hat{\mathbf{p}}' \, ((g^R - g^A)(D^R - D^A) + g^K D^K) \qquad (8.109)$$

since

$$g^R(t_1, t_{1'}) \, D^A(t_1, t_{1'}) = 0 = g^A(t_1, t_{1'}) \, D^R(t_1, t_{1'}) \,. \qquad (8.110)$$

The difference of the self-energies in comparison with the normal state is that the electron quasi-classical propagators are now matrices in Nambu space.

Currents and densities are in the quasi-classical description, just as in the normal state, split into low- and high-energy contributions. The charge density becomes, in terms of the quasi-classical Green's function,

$$\rho(\mathbf{R}, T) = -2eN_0 \left(e \varphi(\mathbf{R}, T) + \frac{1}{8} \int \frac{d\hat{\mathbf{p}}}{4\pi} \int_{-\infty}^{\infty} dE \, \mathrm{Tr}(g^K(E,, \hat{\mathbf{p}}, \mathbf{R}, T)) \right) \,, \quad (8.111)$$

where N_0 is the density of states at the Fermi energy, and Tr denotes the trace with respect to Nambu or particle–hole space. The current density is given by

$$\mathbf{j}(\mathbf{R}, T) = -\frac{eN_0 v_F}{4} \int \frac{d\hat{\mathbf{p}}}{4\pi} \int_{-\infty}^{\infty} dE \, \hat{\mathbf{p}} \, \mathrm{Tr}(\tau_3 \, g^K(E,, \hat{\mathbf{p}}, \mathbf{R}, T)) \,. \qquad (8.112)$$

The order parameter is specified in terms of the off-diagonal component of the quasi-classical kinetic propagator according to (absorbing the coupling constant)

$$\Delta(\mathbf{R}, T) = -\frac{i\lambda}{8} \int \frac{d\hat{\mathbf{p}}}{4\pi} \int dE \, \text{Tr}((\tau_1 - i\tau_2) \, g^K(E, \hat{\mathbf{p}}, \mathbf{R}, T)) \,. \tag{8.113}$$

Exercise 8.7. Show that the quasi-classical retarded Nambu Green's function in the thermal equilibrium state is

$$g^R(E) = \frac{E \, Z^R(E) \, \tau_3 + i\Phi^R(E) \, \tau_1}{\sqrt{E^2 (Z^R(E))^2 - (\Phi^R(E))^2}} \,. \tag{8.114}$$

In the strong coupling case the order parameter is

$$\Delta = \frac{\Phi^R(E)}{Z^R(E)} \,. \tag{8.115}$$

8.2.1 Normalization condition

In the superconducting state the retarded and advanced quasi-classical propagators do not reduce to scalars (times the unit matrix in Nambu space) as in the normal state, and the quantum kinetic matrix equation, Eq. (8.96), constitutes a complicated coupled set of equations describing the states (as specified by $g^{R(A)}$) and their occupation (as described by g^K). Since the quantum kinetic equation, Eq. (8.96), is homogeneous and the time convolution associative, the whole hierarchy $g \circ g$, $g \circ g \circ g, \ldots$, are solutions if g itself is a solution. A normalization condition to cut off the hierarchy is therefore needed. For a translationally invariant state in thermal equilibrium it follows from Exercise 8.7 (or see the explicit expressions obtained in Section 8.2.3) that $(g^R(E))^2$ and $(g^A(E))^2$ equal the unit matrix in Nambu space, $(g^R(E))^2 = 1 = (g^A(E))^2$, and the fluctuation–dissipation relation, $g^K(E) = (g^R(E)) - g^A(E)) \tanh(E/2T)$, then guarantees that the 21-component in Schwinger–Keldysh indices of $g \circ g$ vanishes, $g^R(E) g^K(E) + g^K(E) g^A(E) = 0$. Since the quantum kinetic equation, Eq. (8.101), is first order in the spatial variable, the solution is uniquely specified by boundary conditions. Since a non-equilibrium state can spatially join up smoothly with the thermal equilibrium state we therefore anticipate the general validity of the normalization condition

$$(g \circ g)(t_1 - t_{1'}) = \delta(t_1 - t_{1'}) \,. \tag{8.116}$$

The function $g \circ g$ is thus a trivial solution to the kinetic equation, but contains the important information of normalization. Section 8.3 provides a detailed proof of the normalization condition.

The three coupled equations for the quasi-classical propagators $g^{R,A,K}$ in equations Eq. (8.101) and Eq. (8.100) constitute, together with the normalization condition, the powerful quasi-classical theory of conventional superconductors. Writing out the components in the normalization condition we have

$$g^{R(A)} \circ g^{R(A)} = \delta(t_1 - t_{1'}) \tag{8.117}$$

and

$$g^R \circ g^K + g^K \circ g^A = 0 \,. \tag{8.118}$$

8.2.2   Kinetic equation

The normalization condition, Eq. (8.118), is solved by representing the kinetic Green's function in the form

$$g^{\mathrm{K}} = g^{\mathrm{R}} \circ h - h \circ g^{\mathrm{A}}, \tag{8.119}$$

where h so far is an arbitrary matrix distribution function in particle–hole space. The existence of such a representation is provided by the normalization condition, Eq. (8.117) and Eq. (8.118), as the choice

$$h = \frac{1}{4}(g^{\mathrm{R}} \circ g^{\mathrm{K}} - g^{\mathrm{K}} \circ g^{\mathrm{A}}) \tag{8.120}$$

solves Eq. (8.119). This choice is by no means unique, in fact the substitution

$$h \rightarrow h + g^{\mathrm{R}} \circ k + k \circ g^{\mathrm{A}} \tag{8.121}$$

leads to the same g^{K} for arbitrary k.[19]

Using the equation of motion for $g^{\mathrm{R(A)}}$, Eq. (8.201), and the fact that the time convolution composition $\circ$ is associative, the kinetic equation, Eq. (8.101), is brought to the form for the distribution matrix

$$g^{\mathrm{R}} \circ B[h] - B[h] \circ g^{\mathrm{A}} = 0, \tag{8.122}$$

where

$$B[h] = \sigma^{\mathrm{K}} + h \circ \sigma^{\mathrm{A}} - \sigma^{R} \circ h + [g_0^{-1} \overset{\circ}{,} h]_- . \tag{8.123}$$

The quasi-classical equations are integral equations with respect to the energy variable, and only in special cases, such as at temperatures close to the critical temperature, are they amenable to analytical treatment. However, they can be solved numerically and provide a remarkably accurate description of non-equilibrium phenomena in conventional superconductors. The quantum kinetic equation is thus a powerful tool to obtain a quantitative description of non-equilibrium properties of superconductors.

Before we unfold the information contained in the quantum kinetic equation we consider the equation for the spectral densities or generalized densities of states, Eq. (8.100), as they are input for solving the kinetic equation.

8.2.3   Spectral densities

The equation of motion for the retarded and advanced propagators in Eq. (8.96) becomes

$$[g_0^{-1} + i\sigma^{\mathrm{R(A)}} \overset{\circ}{,} g^{\mathrm{R(A)}}]_- = 0. \tag{8.124}$$

In the static case, we note in general that it follows from Eq. (8.124) that $g^{\mathrm{R(A)}}$ is traceless, so that

$$g^{\mathrm{R(A)}} = \alpha^{\mathrm{R(A)}} \tau_3 + \beta^{\mathrm{R(A)}} \tau_1 + \gamma^{\mathrm{R(A)}} \tau_2 . \tag{8.125}$$

---

[19]A choice making the resemblance between the Boltzmann equation and Eq. (8.122) immediate in the quasi-particle approximation has been introduced in reference [40].

The quantities $\alpha^{R(A)}$, $\beta^{R(A)}$ and $\gamma^{R(A)}$ denote generalized densities of states.

We need to consider only one set of generalized densities of states since from the equality

$$G^{R(A)}(1,1') = \tau_3 \left(G^{R(A)}(1,1')\right)^\dagger \tau_3 \tag{8.126}$$

it follows in general that

$$\alpha^A = -(\alpha^R)^* \quad , \quad \beta^A = (\beta^R)^* \quad , \quad \gamma^A = (\gamma^R)^* . \tag{8.127}$$

In a translationally invariant state of a superconductor in thermal equilibrium, the spectral densities depend only on the energy variable, E, and the real and imaginary parts of the spectral densities are even and odd functions, respectively. In general, the equations for the spectral functions have to be solved numerically, for which they are quite amenable, and they then serve as input information in the quantum kinetic equation.

To elucidate the information contained in Eq. (8.124), we solve it in equilibrium and take the BCS-limit, obtaining

$$g^{R(A)} = \alpha^{R(A)} \tau_3 + \beta^{R(A)} \tau_1 \tag{8.128}$$

as

$$g^R(E) = \frac{E \tau_3 + i\Delta \tau_1}{\sqrt{E^2 - \Delta^2}} . \tag{8.129}$$

Splitting in real and imaginary parts

$$\alpha^{R(A)} = \overset{+}{(-)} N_1(E) + i R_1(E) \quad , \quad \beta^{R(A)} = N_2(E) \overset{+}{(-)} i R_2(E) , \tag{8.130}$$

where

$$N_1(E) = \frac{|E|}{\sqrt{E^2 - \Delta^2}} \Theta(E^2 - \Delta^2) \tag{8.131}$$

is the density of states of BCS-quasi-particles, and

$$N_2(E) = \frac{\Delta}{\sqrt{\Delta^2 - E^2}} \Theta(\Delta^2 - E^2) \tag{8.132}$$

and

$$R_1(E) = -\frac{E}{\Delta} N_2(E) \quad , \quad R_2 = \frac{\Delta}{E} N_1(E) \tag{8.133}$$

with Δ being the BCS-energy gap.

Exercise 8.8. Show that in the weak coupling limit, the equilibrium electron–phonon self-energy is specified by (recall the notation of Exercise 8.4 on page 228)

$$\Re Z^R(E) = 1 + \lambda \tag{8.134}$$

and

$$\Im m(E Z^R(E)) = \frac{1+\lambda}{2\tau(E)} \equiv \frac{1}{2\tau_{\text{in}}} , \tag{8.135}$$

where $\lambda = g^2 N_0$ is the dimensionless electron–phonon coupling constant and the inelastic electron–phonon collision rate is given by

$$\frac{1}{\tau(E)} = \frac{\lambda\pi}{4(cp_{\rm F})^2} \int_{-\infty}^{\infty} dE'\, N_1(E') \frac{(E'-E)|E'-E|\cosh\frac{E}{2T}}{\sinh\frac{(E'-E)}{2T}\cosh\frac{E'}{2T}}. \tag{8.136}$$

For temperatures close to the transition temperature, $\Delta \ll T$, the rate becomes equal to that of the normal state and we obtain for the collision rate for an electron on the Fermi surface

$$\frac{1}{\tau(E=0)} = \frac{\lambda\pi}{(cp_{\rm F})^2} \int_0^{\infty} dE\, \frac{E^2}{\sinh\frac{E}{T}} = \frac{7\pi}{2}\zeta(3)\frac{\lambda T^3}{(cp_{\rm F})^2} \tag{8.137}$$

where ζ is Riemann's zeta function.[20]

We note that, in the electron–phonon model, the superconductor is always gapless as the interaction leads to pair breaking and smearing of the spectral densities. The inelastic collision rate is finite, the pair-breaking parameter, and N_1 is nonzero for all energies.

8.3   Trajectory Green's functions

A physically transparent approach to the quasi-classical Green's function theory of superconductivity revealing the physical content of ξ-integration and providing a general proof of the important normalization condition was given by Shelankov, and we follow in this section the presentation of reference [40]. The quasi-classical theory for a superconductor is based on the existence of a small parameter, viz. that all relevant length scales of the system: the superconducting coherence length, $\xi_0 = \hbar v_{\rm F}/\pi\Delta$, and the impurity mean free path, $l = v_{\rm F}\tau$, are large compared with the microscopic length scale of a degenerate Fermi system, the inverse of the Fermi momentum, $p_{\rm F}^{-1}$, the inter-atomic distance, $k_{\rm F}^{-1}/\xi_0 \ll 1$ (throughout we set $\hbar = 1$). In addition, the length scale for the variation of the external fields, $\lambda_{\rm external}$, as well as the order parameter are smoothly varying functions on this atomic length scale.

The 4×4 matrix Green's function (matrix with respect to both Nambu and Schwinger–Keldysh index) can be expressed through its Fourier transform

$$G(\mathbf{x}_1, \mathbf{x}_2, t_1, t_2) = \int \frac{d\mathbf{p}}{(2\pi)^3}\, e^{i\mathbf{p}\cdot\mathbf{r}}\, G(\mathbf{p}, \mathbf{R}, t_1, t_2), \tag{8.138}$$

where on the right-hand side the spatial Wigner coordinates, the relative, $\mathbf{r} = \mathbf{x}_1 - \mathbf{x}_2$, and center of mass coordinates, $\mathbf{R} = (\mathbf{x}_1 + \mathbf{x}_2)/2$, have been introduced. For a degenerate Fermi system, we recall from Chapter 7 that the Green's functions are peaked at the Fermi surface, and for distances $r \gg p_{\rm F}^{-1}$ the exponential is in general rapidly oscillating and we can make use of the identity

$$\frac{e^{i\mathbf{p}\cdot\mathbf{r}}}{2\pi i} = \frac{e^{-ipr}}{pr}\delta(\hat{\mathbf{p}} + \hat{\mathbf{r}}) - \frac{e^{ipr}}{pr}\delta(\hat{\mathbf{p}} - \hat{\mathbf{r}}), \tag{8.139}$$

---

[20]The electron–phonon collision rate can be modified owing to the presence of disorder, as we will discuss in Section 11.3.1.

where a hat on a vector denotes as usual the unit vector in the direction of the vector. Thus for $r \gg p_F^{-1}$ the matrix Green's function can be expressed in the form (suppressing here the time coordinates since they are immaterial for the following)

$$G(\mathbf{x}_1, \mathbf{x}_2) = -\frac{m}{2\pi} \frac{e^{ip_F|\mathbf{x}_1 - \mathbf{x}_2|}}{|\mathbf{x}_1 - \mathbf{x}_2|} g_+(\mathbf{x}_1, \mathbf{x}_2) + \frac{m}{2\pi} \frac{e^{-ip_F|\mathbf{x}_1 - \mathbf{x}_2|}}{|\mathbf{x}_1 - \mathbf{x}_2|} g_-(\mathbf{x}_1, \mathbf{x}_2), \quad (8.140)$$

where, assuming $|\mathbf{x}_1 - \mathbf{x}_2| \gg p_F^{-1}$,

$$g_\pm(\mathbf{x}_1, \mathbf{x}_2) = \frac{i}{2\pi} \int_{-\infty}^{\infty} v_F \, d(p - p_F) \, e^{\pm i(p - p_F)|\mathbf{x}_1 - \mathbf{x}_2|} \, G(\pm p\hat{\mathbf{r}}, \mathbf{R}) \quad (8.141)$$

and the rapid convergence of the integrand limits the integration over the length of the momentum to the region near the Fermi surface.

The equations of motion for the slowly varying functions, $g_\pm$, are obtained by substituting into the (left) Dyson equation, which gives

$$\pm i v_F \hat{\mathbf{r}} \cdot \nabla_{\mathbf{x}_1} g_\pm(\mathbf{x}_1, \mathbf{x}_2) + H(\pm\hat{\mathbf{r}}, \mathbf{x}_1) \circ g_\pm(\mathbf{x}_1, \mathbf{x}_2) = 0, \quad (8.142)$$

where (re-introducing briefly the time variables)

$$H(\mathbf{n}, \mathbf{x}, t_1, t_2) = \left(i\tau_3 \frac{\partial}{\partial t_1} - e\phi(\mathbf{x}, t_1) + e v_F \tau_3 \mathbf{n} \cdot \mathbf{A}(\mathbf{x}, t_1) \right) \delta(t_1 - t_2)$$

$$- \Sigma(\mathbf{n}, \mathbf{x}, t_1, t_2) \quad (8.143)$$

and we have used the fact that the components of the matrix self-energy are peaked for small spatial separations, $|\mathbf{x}_1 - \mathbf{x}_2| \lesssim p_F^{-1}$, i.e. slowly varying functions of the momentum as discussed in Section 7.5, and

$$\Sigma(\mathbf{n}, \mathbf{x}, t_1, t_2) = \int d\mathbf{r} \, e^{ip_F \mathbf{n} \cdot \mathbf{r}} \, \Sigma(\mathbf{x} + \mathbf{r}/2, \mathbf{x} - \mathbf{r}/2, t_1, t_2). \quad (8.144)$$

The circle in Eq. (8.142) denotes, besides integration with respect to the internal time, an additional matrix multiplication with respect to Nambu and dynamical indices. Since $|\mathbf{x}_1 - \mathbf{x}_2| \gg p_F^{-1}$, the second spatial derivative is negligible because the envelope functions, $g_\pm$, are slowly varying, and consequently the differentiation acts only along the straight line connecting the space points in question, the classical trajectory connecting the points. Only the influence of the external fields on the phase of the propagator is thus included and the effects of the Lorentz force are absent, as expected in the quasi-classical Green's function technique. Thermo-electric and other particle–hole symmetry broken effects are also absent just as in the normal state as discussed in Chapter 7.

Specifying a linear trajectory by a position, $\mathbf{R}$, and its direction, $\mathbf{n}$, the positions on the linear trajectory, $\mathbf{r}$, can be specified by the distance, y, from the position $\mathbf{R}$

$$\mathbf{r} = \mathbf{R} + y\mathbf{n}. \quad (8.145)$$

For the propagator on the trajectory we then have

$$g_\pm(\mathbf{n}, \mathbf{R}, y_1, y_2) = g_\pm(\mathbf{R} + y_1 \mathbf{n}, \mathbf{R} + y_2 \mathbf{n}) \quad (8.146)$$

and we introduce the matrix Green's function on the trajectory

$$g(\mathbf{n}, \mathbf{R}, y_1, y_2) \equiv \begin{cases} g_+(\mathbf{R} + y_1\mathbf{n}, \mathbf{R} + y_2\mathbf{n}) & y_1 > y_2 \\ \\ g_-(\mathbf{R} + y_1\mathbf{n}, \mathbf{R} + y_2\mathbf{n}) & y_1 < y_2. \end{cases} \tag{8.147}$$

Then, according to Eq. (8.141), and again with $|y_1 - y_2| \gg p_F^{-1}$,

$$g(\mathbf{n}, \mathbf{R}, y_1, y_2) = \frac{i}{2\pi} \int_{-\infty}^{\infty} v_F \, d(p - p_F) \, e^{\pm i(p - p_F)(y_1 - y_2)} \, G(p\mathbf{n}, \mathbf{R} + (y_1 + y_2)\mathbf{n}/2)$$

$$\tag{8.148}$$

and we observe that the trajectory Green's function describes the propagation of particles with momentum value p_F along the direction $\mathbf{n}$, and satisfies according to Eq. (8.142), for $|y_1 - y_2| \gg p_F^{-1}$, the equation

$$iv_F \frac{\partial}{\partial y_1} g(y_1, y_2) + H(\mathbf{n}, y_1) \circ g(y_1, y_2) = 0, \tag{8.149}$$

where the notation

$$g(y_1, y_2) \equiv g(\mathbf{n}, \mathbf{R}, y_1, y_2) \tag{8.150}$$

has been introduced. Equation (8.149) is incomplete as we have no information at the singular point, $y_1 = y_2$. Forming the quantity

$$g(y + \delta, y) - g(y - \delta, y) = \frac{-v_F}{\pi} \int_{-\infty}^{\infty} d(p - p_F) \, G(p\mathbf{n}, \mathbf{R} + \mathbf{n}(y + \delta/2)) \, \sin((p - p_F)\delta)$$

$$\tag{8.151}$$

and assuming $\xi_0 \gg \delta \gg p_F^{-1}$, we can neglect the dependence in the center of mass coordinate on δ, and as the contribution from the momentum integration comes from the regions far from the Fermi surface in the limit of vanishing δ, we can insert the normal state Green's functions to obtain (recall Eq. (7.125))

$$g(y + \delta, y) - g(y - \delta, y) = \delta(t_1 - t_2), \tag{8.152}$$

where the unit matrix in Nambu–Keldysh space has been suppressed on the right-hand side, and $\delta \ll \xi_0, \lambda_{\text{external}}$. This result can be included in the equation of motion, Eq. (8.149), as a source term, and we obtain the quasi-classical equation of motion

$$iv_F \frac{\partial}{\partial y_1} g(y_1, y_2) + H(\mathbf{n}, y_1) \circ g(y_1, y_2) = iv_F \, \delta(y_1 - y_2). \tag{8.153}$$

Together with the similarly obtained conjugate equation

$$-iv_F \frac{\partial}{\partial y_2} g(y_1, y_2) + g(y_1, y_2) \circ H(\mathbf{n}, y_2) = iv_F \, \delta(y_1 - y_2) \tag{8.154}$$

we have the equations determining the non-equilibrium properties of a low-temperature superconductor.

Exercise 8.9. Show that the retarded, advanced and kinetic components of the trajectory Green's function satisfy the relations

$$g^{\mathrm{R}}(\mathbf{n}, \mathbf{R}, y_1, t_1, y_2, t_2) = -\tau_3 \left(g^{\mathrm{A}}(\mathbf{n}, \mathbf{R}, y_2, t_2, y_1, t_1) \right)^\dagger \tau_3 \qquad (8.155)$$

and

$$g^{\mathrm{K}}(\mathbf{n}, \mathbf{R}, y_1, t_1, y_2, t_2) = \tau_3 \left(g^{\mathrm{K}}(\mathbf{n}, \mathbf{R}, y_2, t_2, y_1, t_1) \right)^\dagger \tau_3 \qquad (8.156)$$

and for spin-independent dynamics

$$g^{\mathrm{R}}(\mathbf{n}, \mathbf{R}, y_1, t_1, y_2, t_2) = \tau_1 \left(g^{\mathrm{A}}(-\mathbf{n}, \mathbf{R}, y_1, t_2, y_2, t_1) \right)^{\mathrm{T}} \tau_1 \qquad (8.157)$$

and

$$g^{\mathrm{K}}(\mathbf{n}, \mathbf{R}, y_1, t_1, y_2, t_2) = \tau_1 \left(g^{\mathrm{K}}(-\mathbf{n}, \mathbf{R}, y_1, t_2, y_2, t_1) \right)^{\mathrm{T}} \tau_1 . \qquad (8.158)$$

From the quasi-classical equations of motion, Eq. (8.153) and Eq. (8.154), it follows that for $y_1 \neq y_2$

$$\frac{\partial}{\partial y} \left(g(y_1, y) \circ g(y, y_2) \right) = 0 \qquad (8.159)$$

and the function $g(y_1, y) \circ g(y, y_2)$ jumps to constant values at the fixed positions y_1 and y_2. Since we know the jumps of g we get

$$g(y_1, y) \circ g(y, y_2) = \begin{cases} g(y_1, y_2) & y_1 > y > y_2 \\ 0 & y \notin [y_1, y_2] \\ -g(y_1, y_2) & y_1 < y < y_2 \end{cases} \qquad (8.160)$$

where the value zero follows from the decay of the Green's function as a function of the spatial variable as the positions in the quasi-classical Green's function satisfy the constraint $|y_1 - y| \gg l$, and in a disordered conductor the Green's function decays according to $g(y_1, y) \propto \exp\{|y_1 - y|/2l\}$, where l is the impurity mean free path (recall Exercise 7.4 on page 192).

Introducing the *coinciding* argument trajectory Green's functions (suppressing the time variables)

$$g_\pm(\mathbf{n}, \mathbf{r}) \equiv \lim_{\delta \to 0} g_{(\pm)}(\mathbf{n}, \mathbf{R}, y \pm \delta, y) \qquad (8.161)$$

we observe that their left–right subtracted Dyson equations of motion according to Eq. (8.153) and Eq. (8.154) are

$$\pm i v_{\mathrm{F}} \cdot \nabla_{\mathbf{r}} \, g_\pm + H \circ g_\pm - g_\pm \circ H = 0 \qquad (8.162)$$

and according to Eq. (8.152) and Eq. (8.160) they satisfy the relations

$$g_\pm \circ g_\pm = \pm g_\pm \qquad (8.163)$$

and

$$g_\pm \circ g_\mp = 0 = g_\mp \circ g_\pm \qquad (8.164)$$

and

$$g_+ - g_- = 1\,, \qquad (8.165)$$

where 1 is the unit matrix in Nambu–Keldysh space.

The quantity $g(\mathbf{n}, \mathbf{r}) = g_+(\mathbf{n}, \mathbf{r}) + g_-(\mathbf{n}, \mathbf{r})$ therefore satisfies the equation of motion

$$i\mathbf{v}_F \cdot \nabla_\mathbf{r}\, g + H \circ g - g \circ H = 0 \qquad (8.166)$$

and, according to Eq. (8.163), Eq. (8.164) and Eq. (8.165), the normalization condition

$$g \circ g = 1\,. \qquad (8.167)$$

The equation of motion is the same as that for the ξ-integrated Green's function and the above analysis provides an explicit procedure for the ξ-integration as

$$g(\mathbf{n}, \mathbf{r}) = \frac{i}{\pi} \lim_{\delta \to 0} \int_{-\infty}^{\infty} v_F\, d(p - p_F)\, G(p\mathbf{n}, \mathbf{R})\, \cos((p - p_F)\delta)\,. \qquad (8.168)$$

The integral is convergent when δ is finite and independent of δ for $\delta \ll \xi_0$. The dropping of the high-energy contributions in the ξ-integration procedure is in this procedure made explicit by the small distance cut-off.

The quantum effects included in the quantum kinetic equation for g^K is thus the particle–hole coherence due to the pairing interaction whereas the kinetics is classical.

8.4 Kinetics in a dirty superconductor

A characteristic feature of a solid is that it contains imperfections, generally referred to as impurities. Typically superconductors thus contain impurities, and of relevance is a dirty superconductor. The kinetics in a disordered superconductor will be diffusive. In the dirty limit where the mean free path is smaller than the coherence length, or $kT_c < \hbar/\tau$, the integral equation with respect to the ordinary impurity scattering, i.e. the non-spin-flip impurity scattering, can then be reduced to a much simpler differential equation of the diffusive type.[21] We therefore return to the coupled equations for the quasi-classical propagators $g^{R,A,K}$, Eq. (8.96), supplemented by the normalization condition, Eq. (8.116).

In the dirty limit, the Green's function will be almost isotropic, and an expansion in spherical harmonics needs only keep the s- and p-wave parts

$$g(\hat{\mathbf{p}}, \mathbf{R}, t_1, t_{1'}) = g_s(\mathbf{R}, t_1, t_{1'}) + \hat{\mathbf{p}} \cdot \mathbf{g}_p(\mathbf{R}, t_1, t_{1'}) \qquad (8.169)$$

and

$$|\hat{\mathbf{p}} \cdot \mathbf{g}_p(\mathbf{R}, t_1, t_{1'})| \ll |g_s(\mathbf{R}, t_1, t_{1'})|\,. \qquad (8.170)$$

---

[21] Quite analogous to deriving the diffusion equation from the Boltzmann equation as discussed in Sections 7.4.2 and 7.5.5.

The self-energy is then

$$\sigma(\hat{\mathbf{p}}, \mathbf{R}, t_1, t_{1'}) = \sigma_{\mathrm{s}}(\mathbf{R}, t_1, t_{1'}) + \hat{\mathbf{p}} \cdot \boldsymbol{\sigma}_{\mathrm{p}}(\mathbf{R}, t_1, t_{1'}) , \qquad (8.171)$$

where

$$\hat{\mathbf{p}} \cdot \boldsymbol{\sigma}_{\mathrm{p}}(\mathbf{R}, t_1, t_{1'}) = -i\pi n_{\mathrm{i}} N_0 \int \frac{d\hat{\mathbf{p}}'}{4\pi} \, |V_{\mathrm{imp}}(\hat{\mathbf{p}} \cdot \hat{\mathbf{p}}')|^2 \, \hat{\mathbf{p}}' \cdot \mathbf{g}_{\mathrm{p}}(\mathbf{R}, t_1, t_{1'}) \qquad (8.172)$$

and

$$\sigma_{\mathrm{s}} = -\frac{i}{2\tau} g_{\mathrm{s}} + \sigma_{\mathrm{s}}' , \qquad (8.173)$$

where

$$\sigma_{\mathrm{s}}' = -\frac{i}{2\tau_{\mathrm{s}}} \tau_3 \, g_{\mathrm{s}} \, \tau_3 + \sigma_{\mathrm{s}}^{\mathrm{e-ph}} \qquad (8.174)$$

contains the effects of spin-flip and electron–phonon scattering.

Performing the angular integration gives

$$\boldsymbol{\sigma}_{\mathrm{p}} = \frac{-i}{2} \left(\frac{1}{\tau} - \frac{1}{\tau_{\mathrm{tr}}} \right) \mathbf{g}_{\mathrm{p}} , \qquad (8.175)$$

where τ_{tr} is the impurity transport life time determining the normal state conductivity

$$\frac{1}{\tau_{\mathrm{tr}}} = 2\pi n_{\mathrm{i}} N_0 \int \frac{d\hat{\mathbf{p}}'}{4\pi} \, |V_{\mathrm{imp}}(\hat{\mathbf{p}} \cdot \hat{\mathbf{p}}')|^2 \, (1 - \hat{\mathbf{p}} \cdot \hat{\mathbf{p}}') . \qquad (8.176)$$

The inverse propagator has exactly the form

$$g_0^{-1} = g_{0_{\mathrm{s}}}^{-1} + \hat{\mathbf{p}} \cdot \mathbf{g}_{0_{\mathrm{p}}}^{-1} , \qquad (8.177)$$

where

$$g_{0_{\mathrm{s}}}^{-1} = (\tau_3 \, \partial_{t_1} + ie\phi(\mathbf{R}, t_1)) \, \delta(t_1 - t_{1'}) \qquad (8.178)$$

and

$$\mathbf{g}_{0_{\mathrm{p}}}^{-1} = v_{\mathrm{F}} \, \boldsymbol{\partial} \quad , \quad \boldsymbol{\partial} = (\nabla_{\mathbf{R}} - ie\tau_3 \mathbf{A}(\mathbf{R}, t_1)) \, \delta(t_1 - t_{1'}) . \qquad (8.179)$$

The kinetic equation in the dirty limit can be split into even and odd parts with respect to $\hat{\mathbf{p}}$

$$[g_{0_{\mathrm{s}}}^{-1} + i\sigma_{\mathrm{s}}' \, \overset{\circ}{,} \, g_{\mathrm{s}}]_- + \frac{1}{3} v_{\mathrm{F}} [\boldsymbol{\partial} \, \overset{\circ}{,} \, \mathbf{g}_{\mathrm{p}}]_- = 0 \qquad (8.180)$$

and

$$\frac{1}{2\tau_{\mathrm{tr}}} [g_{\mathrm{s}} \, \overset{\circ}{,} \, \mathbf{g}_{\mathrm{p}}]_- + v_F [\boldsymbol{\partial} \, \overset{\circ}{,} \, g_{\mathrm{s}}]_- = 0 . \qquad (8.181)$$

Using the s- and p-wave parts of the normalization condition gives

$$g_{\mathrm{s}} \circ g_{\mathrm{s}} = \delta(t_1 - t_{1'}) \qquad (8.182)$$

and

$$[g_{\mathrm{s}} \, \overset{\circ}{,} \, \mathbf{g}_{\mathrm{p}}]_+ = \mathbf{0} \qquad (8.183)$$

and we get

$$\mathbf{g}_p = -l\, g_s \circ [\boldsymbol{\partial} \, \overset{\circ}{,} \, g_s]_- \, , \tag{8.184}$$

where $l = v_F \tau_{\mathrm{tr}}$ is the impurity mean free path.

Upon inserting into Eq. (8.180), an equation for the isotropic part of the quasi-classical Green's function is obtained, the Usadel equation [41],

$$[g_{0s}^{-1} + i\sigma_s' - D_0\, \boldsymbol{\partial} \circ g_s \circ \boldsymbol{\partial} \, \overset{\circ}{,} \, g_s]_- = 0 \, . \tag{8.185}$$

We have obtained a kinetic equation which is local in space, an equation for the quasi-classical Green's function for coinciding spatial arguments. This equation is the starting point for considering general non-equilibrium phenomena in a dirty superconductor.

Exercise 8.10. Show that the current density in the dirty limit takes the form

$$\mathbf{j}(\mathbf{R}, T) = \frac{eN_0 D_0}{4} \int_{-\infty}^{\infty} dE \, \mathrm{Tr}(\tau_3 \, (g_s^R \circ \boldsymbol{\partial} \circ g_s^K + g_s^K \circ \boldsymbol{\partial} \circ g_s^A)) \, , \tag{8.186}$$

which by using the Einstein relation, $\sigma_0 = 2e^2 N_0 D_0$, can be expressed in terms of the conductivity of the normal state.

8.4.1 Kinetic equation

In the dirty limit, the kinetic equation

$$g^R \circ B[h] - B[h] \circ g^A = 0 \tag{8.187}$$

is specified by

$$B[h] = (g_0^R)^{-1} \circ h - h \circ (g_0^A)^{-1} - i\sigma_{\mathrm{e-ph}}^K$$

$$- D_0\, \boldsymbol{\partial} \circ g^R \circ [\boldsymbol{\partial} \, \overset{\circ}{,} \, h]_- - D_0\, [\boldsymbol{\partial} \, \overset{\circ}{,} \, h]_- \circ g^A \circ \boldsymbol{\partial} \, , \tag{8.188}$$

where

$$\left(g_0^{R(A)}\right)^{-1} = -iE\tau_3 + ie\varphi(\mathbf{r}, t) + \hat{\Delta} + i\sigma_{\mathrm{e-ph}}'^{R(A)} + \frac{1}{2\tau_s}\tau_3\, g^{R(A)}\, \tau_3 \, . \tag{8.189}$$

Inelastic effects are included through the electron–phonon interaction.

In the low frequency limit, the problem simplifies, and we discuss this case in order to show how the matrix distribution function enters the collision integral. For superconducting states close to the transition temperature, the Ginzburg–Landau regime, the component γ is negligible, as discussed in the next section, and the distribution matrix h can be chosen diagonal in Nambu space

$$h = h_1 1 + h_2 \tau_3 \, . \tag{8.190}$$

We then perform a Taylor expansion in Eq. (8.187), and linearize the equation with respect to $h_1 - h_0$ and h_2. To expose the kinetic equations satisfied by the distribution functions we multiply the kinetic equation with Pauli matrices and take the trace in particle–hole space, in fact for the present case we take the trace of the equation and the trace of the equation multiplied by τ_3, and obtain the two coupled equations for the distribution functions

$$N_1 \dot{h}_1 + R_2 \Re\dot{\Delta}\partial_E h_1 + 2R_2 \Im m\Delta\, h_2 - D_0 \nabla_{\mathbf{R}} \cdot M_1(E, E)\, \nabla_{\mathbf{R}} h_1$$

$$+ \; D_0 \nabla_{\mathbf{R}} \cdot (\nabla_{\mathbf{R}} h_2 + \dot{\mathbf{p}}_s \partial_E h_0) - 4N_2 R_2\, \mathbf{p}_s \cdot (\nabla_{\mathbf{R}} h_2 + \dot{\mathbf{p}}_s \partial_E h_0)$$

$$= \; K_1[h_1] \tag{8.191}$$

and

$$N_1(\dot{h}_2 + \dot{\Phi}\,\partial_E h_0) + 2N_2 \Re e\Delta\, h_2 - N_2 \Im m\dot{\Delta}\,\partial_E h_0 - 4D_0 N_2 R_2\, \mathbf{p}_s \cdot \nabla_{\mathbf{R}} h_1$$

$$- \; D_0 \nabla_{\mathbf{R}} \cdot M_2(E, E)\,(\nabla_{\mathbf{R}} h_2 + \dot{\mathbf{p}}_s \partial_E h_0) = K_2[h_2]\,, \tag{8.192}$$

where the collision integrals are given by, $i = 1, 2$,

$$K_i[h_i] \;=\; -\pi \int_{-\infty}^{\infty} dE'\, \mu(E - E')\, M_i(E, E') \frac{h_i(E) \cosh^2\left(\frac{E}{2T}\right) - h_i(E') \cosh^2 \frac{E'}{2T}}{\sinh \frac{E - E'}{2T} \cosh \frac{E}{2T} \cosh \frac{E'}{2T}}\,,$$

$$\tag{8.193}$$

where

$$M_i(E, E') \;=\; \begin{cases} N_1(E)\, N_1(E') + R_2(E)\, R_2(E') & i = 1 \\ N_1(E)\, N_1(E') + N_2(E)\, N_2(E') & i = 2 \end{cases} \tag{8.194}$$

and μ is the Fermi surface average of the function in Eq. (7.134), the Eliashberg function, $\alpha^2 F(E - E') = \mu(E - E')$,

$$\mu(E - E') \;=\; \frac{i\lambda}{2\pi} \int \frac{d\hat{\mathbf{p}}'}{4\pi}\, (D^{\mathrm{R}}(\hat{\mathbf{p}}' \cdot \hat{\mathbf{p}}, E' - E) - D^{\mathrm{A}}(\hat{\mathbf{p}}' \cdot \hat{\mathbf{p}}, E' - E))\,, \tag{8.195}$$

or in general the Fermi surface weighted average of the phonon spectral weight function and the momentum-dependent coupling function (recall Eq. (7.134)).

Together with the expressions for charge and current density and Maxwell's equations, the kinetic equations for the distribution functions supplemented with the equations for the generalized densities of states and the order parameter equation, constitute a complete description of a dirty conventional superconductor in the low-frequency limit.

Exercise 8.11. Show that in the Debye model of lattice vibrations, the Eliashberg function becomes

$$\mu(E) \;=\; \frac{\lambda}{4(cp_{\mathrm{F}})^2}\, E|E|\, \theta(\omega_{\mathrm{D}} - |E|)\,. \tag{8.196}$$

8.4.2   Ginzburg–Landau regime

In this section we shall derive the time-dependent Ginzburg–Landau equation for the order parameter.[22] First we further reduce the equation determining the components of the spectral part of the Usadel equation

$$[g_0^{-1} + i\sigma'^{R(A)} - D_0\, \partial \circ g^{R(A)} \circ \partial \,\mathring{,}\, g^{R(A)}]_- = 0 \qquad (8.197)$$

by considering the case where temporal non-equilibrium is slow.

We shall treat the pairing effect (contained in $\Re\,\sigma_{e-ph}^{R(A)}$) in the BCS-approximation and approximate the electronic damping by the equilibrium expression Eq. (8.136). Then the retarded (advanced) electron–phonon self-energy reduces to

$$\sigma_{e-ph}^{R(A)} = \left(-\lambda E \underset{(+)}{\bar{(\,)}} \frac{i}{2\tau_{in}}\right)\tau_3 - i\hat{\Delta}\,, \qquad (8.198)$$

where τ_{in} is the inelastic electron–phonon scattering time, and the gap matrix is

$$\hat{\Delta} = \begin{pmatrix} 0 & \Delta \\ \Delta^* & 0 \end{pmatrix}, \qquad (8.199)$$

where (from now on we drop the s-wave index)

$$\Delta = -\frac{i\lambda}{8}\int_{-\omega_D}^{\omega_D} dE\, \mathrm{Tr}\left((\tau_1 - i\tau_2)g^K\right) \qquad (8.200)$$

is the order parameter.

We assume that the characteristic non-equilibrium frequency, ω, satisfies $\omega < \Delta, T, 1/\tau$. We can then make a temporal gradient expansion in Eq. (8.197) and obtain to lowest order for the off-diagonal components

$$\frac{1}{2}D_0(\alpha\mathcal{D}^2(\beta - i\gamma) - (\beta - i\gamma)\nabla_{\mathbf{R}}^2\alpha)^{R(A)}$$

$$= \left(\left(-iE_{(-)}^{+}\frac{i}{2\tau_{in}}\right)(\beta - i\gamma) - \Delta\alpha + \frac{i}{\tau_s}\alpha(\beta - i\gamma) - \dot{\Phi}\,\partial_E(\beta - i\gamma)\right)^{R(A)} \qquad (8.201)$$

and

$$\frac{1}{2}D_0(\alpha\mathcal{D}^{*2}(\beta + i\gamma) - (\beta + i\gamma)\nabla_{\mathbf{R}}^2\alpha)^{R(A)}$$

$$= \left(\left(-iE_{(-)}^{+}\frac{i}{2\tau_{in}}\right)(\beta + i\gamma) - \Delta^*\alpha + \frac{i}{\tau_s}\alpha(\beta + i\gamma) + \dot{\Phi}\,\partial_E(\beta + i\gamma)\right)^{R(A)}, \qquad (8.202)$$

where

$$\mathcal{D} = \nabla_{\mathbf{R}} - 2ie\mathbf{A} \qquad (8.203)$$

<hr />

[22]We essentially follow reference [42] and reference [43].

is the gauge co-variant derivative. · We note that all time-dependent terms cancel except the one involving the electro-chemical potential of the condensate. Together with the normalization condition

$$(\alpha^{\mathrm{R}})^2 + (\beta^{\mathrm{R}})^2 + (\gamma^{\mathrm{R}})^2 = 1 \qquad (8.204)$$

these equations determine the generalized densities of states. In view of Eq. (8.155), say only the retarded components needs to be evaluated, and in the following we therefore leave out the superscript.

Assuming the superconductor is in the Ginzburg–Landau regime where the temperature is close to the critical temperature, $\Delta(T) \ll T$, we can iterate Eq. (8.202) starting with the density of states for the normal state, i.e. $\alpha \to 1$, and neglect spatial variations. We then obtain to a first approximation

$$\beta + i\gamma = \frac{\Delta^*}{-iE + 1/2\tau_{\mathrm{in}} + 1/\tau_{\mathrm{s}}} \qquad (8.205)$$

and similarly for $\beta - i\gamma$. Then using the normalization condition, Eq. (8.204), gives the first order correction to α. In the next iteration we then obtain

$$\beta + i\gamma = \frac{\Delta^*}{-iE + 1/2\tau_{\mathrm{in}} + 1/\tau_{\mathrm{s}}}$$

$$+ \frac{i}{2} \left(\frac{D_0 \mathcal{D}^* \Delta^*}{(-iE + 1/2\tau_{\mathrm{in}} + 1/\tau_{\mathrm{s}})^2} + \frac{(-iE + 1/2\tau_{\mathrm{in}})|\Delta|^2 \Delta^*}{(-iE + 1/2\tau_{\mathrm{in}} + 1/\tau_{\mathrm{s}})^4} \right) \qquad (8.206)$$

and similarly for $\beta - i\gamma$. It follows from these equations that γ is smaller than the other components by the amount Δ/T, and can be neglected in the Ginzburg–Landau regime.

The distribution matrix in Nambu space we assume to be of the form

$$h = h_0 + h_1 + h_2 \tau_3 . \qquad (8.207)$$

Making the slow frequency gradient expansion of the kinetic propagator, Eq. (8.119), and keeping only linear terms in the distribution functions h_1 and h_2 we obtain

$$g^{\mathrm{K}} = h_0(g^{\mathrm{R}} - g^{\mathrm{A}}) - \frac{i}{2}[h_0, g^{\mathrm{R}} + g^{\mathrm{A}}]_{\mathrm{p}}$$

$$+ h_1(g^{\mathrm{R}} - g^{\mathrm{A}}) + h_2(g^{\mathrm{R}}\tau_3 - \tau_3 g^{\mathrm{A}}) , \qquad (8.208)$$

where the Poisson bracket is with respect to time and energy variables. In the expression for the order parameter, Eq. (8.113), we therefore obtain

$$\Delta(\mathbf{R}, T) = -\frac{\lambda}{4} \int_{-\infty}^{\infty} dE \, \left(h_0(\beta - \beta^*) - \frac{i}{2}[h_0, (\beta + \beta^*)]_{\mathrm{p}} \right.$$

$$\left. + h_1(\beta - \beta^*) - h_2(\beta + \beta^*) \right) . \qquad (8.209)$$

Using the known pole structure of $h_0 = \tanh E/2T$, terms involving this function can be evaluated by the residue theorem, and we arrive at the time-dependent Ginzburg–Landau equation for the order parameter

$$(A - B|\Delta|^2 - C(\partial_T - D_0 \mathcal{D}^2) + \chi)\,\Delta \;=\; 0\,, \tag{8.210}$$

where

$$\chi \;=\; \frac{1}{\Delta}\int\limits_{-\infty}^{\infty} dE\, R_2\, h_1 \tag{8.211}$$

is Schmid's control function, controlling the magnitude of the order parameter, and the coefficients can be expressed in terms of the poly-gamma-functions (ψ being the di-gamma-function)

$$A \;=\; \ln\frac{T_c}{T} + \psi(1/2 + \rho T/T_c) - \psi(1/2 + \rho) \tag{8.212}$$

and

$$B \;=\; -\frac{1}{(4\pi T)^2}\left(\psi^{(2)}(1/2 + \rho) + \frac{1}{3}\rho_s\,\psi^{(3)}(1/2 + \rho)\right) \tag{8.213}$$

and

$$C \;=\; -\frac{1}{4\pi T}\left(\psi^{(1)}(1/2 + \rho)\right)\,, \tag{8.214}$$

where $\rho = \rho_s + \rho_{in}$

$$\rho_s \;=\; \frac{1}{2\pi\tau_s T}\,,\qquad \rho_{in} \;=\; \frac{1}{4\pi\tau_{in}T} \tag{8.215}$$

and we have used the relation for the transition temperatures in the presence and absence of pair-breaking mechanisms

$$\ln\frac{T_{c0}}{T_c} \;=\; \psi(1/2 + \rho T/T_c) - \psi(1/2)\,. \tag{8.216}$$

Evaluating the coefficients gives the time-dependent Ginzburg–Landau equation

$$\frac{\pi}{8T_c}\,\dot\Delta(\mathbf{x},t) \;=\; -\left(\frac{T - T_c}{T_c} + \frac{7\zeta(3)}{8\pi^2}\frac{\Delta^2}{T_c^2} + \xi^2(0)\,(4m^2\mathbf{v}_s^2 - \nabla_\mathbf{x}^2) + \chi\right)\Delta(\mathbf{x},t)\,, \tag{8.217}$$

where $\xi^2(0) = \pi D_0/8T_c$ is the coherence length in the dirty limit.

In the normal state close to the transition temperature, there will be superconducting fluctuations in the order parameter. In that case, the first term in the time-dependent Ginzburg–Landau equation, Eq. (8.217), dominates and the thermal fluctuations of the order parameter decays with the relaxation time

$$\tau_R^N \;=\; \frac{\pi}{8}\frac{1}{|T - T_c|}\,. \tag{8.218}$$

In the superconducting state the relaxation of the order parameter is, according to Eq. (8.211), determined by the non-equilibrium distribution of the quasi-particles, which in turn is influenced by the time dependence of the order parameter. In the spatially homogeneous situation where h_2 vanishes, the kinetic equation, Eq. (8.191), reduces to

$$N_1 \dot{h}_1 + R_2 \dot{\Delta} \partial_E h_0 = -\frac{N_1}{\tau_{\text{in}}} h_1 . \qquad (8.219)$$

Calculating the control function gives

$$\chi = \frac{\pi}{4T_c} \tau_{\text{in}} \dot{\Delta} \qquad (8.220)$$

and according to the time-dependent Ginzburg–Landau equation the relaxation time for the order parameter, $\Delta \tau_{\text{in}} \gg 1$, is

$$\tau_R = \frac{\pi^3}{7\zeta(3)} \frac{T_c}{\Delta} \tau_{\text{in}} . \qquad (8.221)$$

Experimental observation of the relaxation of the magnitude of the order parameter can been achieved by driving the superconductor out of thermal equilibrium by a laser pulse [44].

8.5 Charge imbalance

Under non-equilibrium conditions in a superconductor a difference in the electro-chemical potential between the condensate and the quasi-particles can exist, reflecting the finite rate of conversion between supercurrent and normal current. For example, charge imbalance occurs when charge from a normal metal is injected into a super-conductor in a tunnel junction. As an application of the theory of non-equilibrium superconductivity, we shall consider the phenomenon of charge imbalance generated by the combined presence of a supercurrent and a temperature gradient. We shall limit ourselves to the case of temperatures close to the critical temperature where analytical results can be obtained.[23]

The charge density is in the real Δ-gauge, recall Section 8.1.3,

$$\rho = 2eN_0 \left(\Phi + \int_{-\infty}^{\infty} dE \, N_1(E) \, f_2(E) \right) , \qquad (8.222)$$

where the condensate electro-chemical potential in general is $\Phi = \dot{\chi}/2 - e\varphi$, χ being the phase of the order parameter. We have introduced distribution functions related to the original ones according to $h_1 = 1 - 2f_1$ and $h_2 = -2f_2$.[24] We could insert instead the full distribution function, $f = f_1 + f_2$, in Eq. (8.222) as f_1 is an odd function, and thereby observe that the charge of the quasi-particles described by the distribution function is the elementary charge, the full electronic charge.

---

[23]We essentially follow the presentation of reference [45]. For general references to charge imbal-ance in superconductors, as well as other non-equilibrium phenomena, we refer to the articles in reference [42].

[24]In reference [46], they are referred to as the longitudinal and transverse distribution functions.

The strong Coulomb force suppresses charge fluctuations, but it is possible to have a charge imbalance between the charge carried by the condensate of correlated electrons and the charge carried by quasi-particles

$$Q^* = 2eN_0 \int_{-\infty}^{\infty} dE\, N_1(E)\, f_2(E)\,. \tag{8.223}$$

The presence of a temperature variation, $T(\mathbf{r}) = T + \delta T(\mathbf{r})$ creates a non-equilibrium distribution in the thermal mode

$$\delta f = f_1 - f_0 = -\frac{E}{T}\frac{\partial f_0}{\partial E}\delta T \tag{8.224}$$

where f_0 is the Fermi function. The presence of a supercurrent, $\mathbf{p}_s = m\mathbf{v}_s = -(\nabla\chi + 2e\mathbf{A})/2$, couples via the kinetic equation for the charge mode the thermal and the charge mode. For a stationary situation with f_2 homogeneous in space we have according to Eq. (8.192)

$$2N_2\Delta f_2 - 4D_0 N_2 R_2\, \mathbf{p}_s\cdot\nabla f_1 = K_2[f_2]\,. \tag{8.225}$$

The first term on the left gives rise to conversion between the supercurrent and the current carried by quasi-particles, while the second term is a driving term proportional to $\mathbf{v}_s\cdot\nabla T$. Close to the transition temperature, T_c, the collision integral is dominated by energies in the region $E' \simeq T$. In this energy regime we have $\alpha \simeq 1$ and $\tilde{\beta}, \tilde{\gamma} \simeq \eta$ where we have introduced the notation $\eta = \Delta(T)/T$ for the small parameter of the problem. The collision integral then becomes proportional to the inelastic collision rate

$$K_2[f_2] = -\frac{N_1}{\tau(E)}\left(f_2 + \frac{\partial f_0}{\partial E}\int_{-\infty}^{\infty} dE\, N_1(E)\, f_2(E)\right)\,. \tag{8.226}$$

The last term is proportional to the charge imbalance, and we get the following kinetic equation

$$\left(2\Delta\tau(E)N_2 + N_1\right)f_2 + Q^*\frac{N_1}{2N_0}\frac{\partial f_0}{\partial E} = -4mD_0\frac{E}{T}\frac{\partial f_0}{\partial E}\tau(E)N_2 R_2\, \mathbf{v}_s\cdot\nabla T. \tag{8.227}$$

Integrating with respect to the energy variable gives

$$\frac{Q^*}{2N_0} = -\frac{4mD_0}{T}\frac{A}{1-B}\,\mathbf{v}_s\cdot\nabla T\,, \tag{8.228}$$

where

$$A = -\int_{-\infty}^{\infty} dE\,\frac{\partial f_0}{\partial E}\frac{N_1 N_2 R_2 E\tau(E)}{N_1 + 2\Delta\tau(E)N_2} \tag{8.229}$$

and

$$B = - \int\limits_{-\infty}^{\infty} dE \, \frac{\partial f_0}{\partial E} \, \frac{N_1^2}{N_1 + 2\Delta\tau(E)N_2} \, . \tag{8.230}$$

Assuming weak pair-breaking the quantities can be evaluated. To zeroth order in η we have $B \simeq 1$ as $N_1 \simeq 1$ and $N_2 \simeq 0$. From the structure of the densities of states it is apparent that the main correction contribution comes from the energy range Δ up to a few Δ. We can therefore use the high-energy expansion

$$N_2 = \frac{\Delta\Gamma}{E^2 + \Gamma^2} \, , \tag{8.231}$$

where

$$\Gamma = \frac{1}{2\tau(E)} + \frac{1}{\tau_s} + \frac{1}{2} D_0(\mathbf{p}_s^2 - \Delta^{-1}\nabla^2\Delta) \tag{8.232}$$

is the pair-breaking parameter. In the limit of weak pair-breaking, $\Gamma \ll \Delta$, and Δ is small as we assume the temperature is close to the critical temperature, so that $\Delta(\tau(E)\Gamma)^{1/2} \ll T$, and we get

$$B = 1 - \frac{\pi\Delta}{4T} (2\tau(E)\Gamma)^{1/2} \, . \tag{8.233}$$

In the BCS-limit, A is logarithmically divergent due to the singular behavior of the density of states, but the pair-breaking smears out the singularity and gives a logarithmic cut-off at $\ln(4\Delta/\Gamma)$, and we have

$$A = \frac{\Delta}{8T} \left(\ln \frac{4\Delta}{\Gamma} + 2(2\tau(E)\Gamma - 1)^{1/2} \arctan((2\tau(E)\Gamma - 1)^{1/2}) \right) \, . \tag{8.234}$$

In the limit where electron–phonon interaction provides the main pair breaking mechanism, $\Gamma \sim 1/2\tau_E$, the charge imbalance thus becomes

$$Q^* = \frac{2}{3\pi} \frac{p_F l}{T} 2N_0 (\mathbf{v}_s \cdot \nabla T) \ln(8\Delta\tau(E)) \, . \tag{8.235}$$

For a discussion of the experimental observation of charge imbalance we refer the reader to reference [47].

8.6   Summary

In this chapter we have considered non-equilibrium superconductivity. By using the quasi-classical Green's technique, a theory with an accuracy in the 1% range was constructed that were able to describe the non-equilibrium states of a conventional superconductor. This is a rather impressive achievement bearing in mind that a superconductor is a messy many-body system. In general one obtains coupled equations for spectral densities, non-equilibrium distribution functions and the order parameter, which of course in general are inaccessible to analytic treatment, but which can be handled by numerics. The versatility of the quasi-classical Green's technique to understand non-equilibrium phenomena in superfluids is testified by the wealth of results obtained using it. For the reader interested in non-equilibrium superconductivity, we give the general references where further applications can be trailed [48] [49].

9

Diagrammatics and generating functionals

At present, the only general method available for gaining knowledge from the fundamental principles about the dynamics of a system is the perturbative study. According to Feynman, as described in Chapter 4, instead of formulating quantum theory in terms of operators,[1] the canonical formulation, for calculational purposes quantum dynamics can conveniently be formulated in terms of a few simple stenographic rules, the Feynman rules for propagators and interaction vertices.

In Chapters 4 and 5, we showed how to arrive at the Feynman rules of diagrammatic perturbation theory for *non-equilibrium* states starting from the Hamiltonian defining the theory. The feature of non-equilibrium states, originally carried by the dynamical indices, could be expressed in terms of two simple universal vertex rules for the RAK-components of the matrix Green's functions. We are thus well acquainted with diagrammatics even for the description of non-equilibrium situations. However, for the situations studied using the quantum kinetic equations in Chapters 7 and 8, only the Dyson equation was needed, i.e. the self-energy, the 2-state one-particle irreducible amputated Green's function. No need for higher-order vertex functions was required, and the full flourishing diagrammatics was not put into action. In this chapter we shall proceed the other way around. We shall show that the diagrammatics of a physical theory, including the description of non-equilibrium states, can be obtained by simply stating quantum dynamics, the superposition principle, as the two exclusive options for a particle: *to interact or not to interact*! From this simple Shakespearean approach we shall construct the Feynman diagrammatics of non-equilibrium dynamics. Thus starting with bare propagators and vertices defining a physical theory, and constructing its dynamics in diagrammatic perturbation theory, we then show how to capture all of the diagrammatics in terms of a single functional differential equation. In this way we shall by simple topological arguments for diagrams construct the generating functional approach to quantum field theory of non-equilibrium states. The corresponding analytic generating functional technique

---

[1]Or equivalently for that matter in terms of path or functional integrals as we discuss in the next chapter.

is originally due to Schwinger [50]. In the next chapter, we shall then follow Feynman and instead of describing the dynamics of a theory in terms of differential equations, describe its corresponding representation in terms of path integrals. These analytical condensed techniques shall prove very powerful when unraveling the content of a field theory. The methods were originally developed to study equilibrium state properties, in fact strings of field operators evaluated in the vacuum state as relevant to the Green's function's of QED, and later taken over to study equilibrium properties of many-body systems. In the following we shall develop these methods for general non-equilibrium states.

A point we wish to stipulate is that diagrammatics and the equivalent functional methods are a universal language of physics with applications ranging from high to low energies: from particle physics over solid state physics even to classical stochastic physics and soft condensed matter physics, as we shall exemplify in the following chapters. In the next chapter, we shall eventually use the effective action approach to study Bose–Einstein condensation, viz. the properties of a trapped Bose gas. In Chapter 12, we shall consider classical statistical dynamics, classical Langevin dynamics, where the fluctuations are caused by the stochastic nature of the Langevin force, a problem which, interestingly enough, mathematically is formally equivalent to a quantum field theory.

9.1 Diagrammatics

According to the Feynman rules, the quantum theory of particle dynamics is defined by its bare propagators and vertices, specifying the possible transmutations of particles and thereby describing how any given particle configuration can be propagated into another one. In the standard model, the elementary particles consist of *matter* constituents: leptons (electron, electron neutrino and their heavier cousins) and quarks (up and down and their heavier cousins, three families in all), all spin-1/2 particles and therefore fermions, and the *force* carriers which are bosons and mediate interaction through their exchange between the matter constituents, or realize transmutation of particles through decay. The electro-weak force is mediated by the photon and the heavy vector bosons, and the strong force is mediated by gluons. In condensed matter physics, electromagnetism or simply the Coulomb interaction is the relevant interaction; typically the interactions of electrons with photons, phonons, magnons and other electrons are of chief interest. In statistical physics, thermal as well as quantum fluctuations are of interest but the diagrammatics are the same, even for non-equilibrium states, the emphasis of this book. In equilibrium statistical mechanics, thermodynamics, thermal and not quantum fluctuations are often of chief importance, and the use of diagrams are also of great efficiency, for example in understanding phase transitions. In Chapter 12 we demonstrate the usefulness of Feynman diagrams even in the context of *classical* physics, viz. in the context of classical stochastic dynamics. At the level of diagrammatics there is no essential difference in the treatment of different physical systems and different types of fluctuations, and all cases will here be dealt with in a unified description.

A generic particle physics experiment consists of colliding particles in certain states and at a later time detecting the resulting debris of particle content in their respective states, or rather reconstructing these since typically the particle content of interest has long ceased to exist once the detector signals are recorded.[2] To any possible outcome only a probability P can according to quantum mechanics be attributed. To each possible process (a final configuration of particles *given* an initial one) is thus associated a (*conditional*) probability P. The probability for a certain process occurring, is according to the fundamental principle of quantum mechanics, specified by a probability *amplitude* A, a *complex* number, giving the probability for the process as the absolute square of the probability amplitude[3]

$$P = |A|^2 . \tag{9.1}$$

In order not to clutter diagrams and equations with indices, a compound label is introduced

$$1 \equiv (s_1, \mathbf{x}_1, t_1, \sigma_1, \ldots) \tag{9.2}$$

for a complete specification of a particle state and it thus includes: species type s, space and time coordinates $(\mathbf{x}, t)$, internal (spin, flavor, color) degrees of freedom $\sigma, \ldots$, or say in discussing superconductivity a Nambu index. Most importantly, since we also allow for non-equilibrium situations the index 1 includes a dynamical or Schwinger–Keldysh index in addition, or equivalently we let the temporal coordinate t become a contour time τ on the contour depicted in Figure 4.4 or Figure 4.5. However, we shall for short refer to the labeling 1 as the *state* label. Instead of the position, the complementary momentum representation is of course more often used owing to calculational advantages or experimental relevance, say in connection with particle scattering, but for the present exposition one might advantageously have the more intuitive position representation in mind.

We now embark on constructing the dynamics of a non-equilibrium quantum field theory in terms of diagrams, i.e. stating the laws of nature in terms of the propagators of species and their vertices of interaction.

9.1.1 Propagators and vertices

Feynman has given us a lucid way of representing and calculating probability amplitudes in terms of diagrams. In this framework a theory is defined in terms of the particles it describes, their propagators and their possible interactions. Each particle is attributed a free or bare propagator, $G_{12}^{(0)}$, the probability amplitude to freely propagate between the states in question, say spin states and space-time points $(\mathbf{x}_1, \sigma_1, t_1)$ and $(\mathbf{x}_2, \sigma_2, t_2)$. The corresponding free propagator or Green's function,

---

[2]Indeed any dynamics of particles can be viewed as caused by collisions, i.e. interactions, and the following diagrammatic discussion is valid for any in-put/out-put kind of machinery. The dynamics need not be dictated by the laws of physics for diagrammatics to work, it can be the result of any mechanism of choice, say a random walk. The diagrammatic approach can therefore also be used to study statistical mechanics models, and for the brave perhaps models of evolution or climate, or the stock market for the greedy.

[3]For almost 100 years, no mechanics beyond these probabilities has been found despite many brave attempts. Furthermore, we stress the weird quantum feature that the probabilities have to be calculated through the more fundamental *amplitudes*, which are the true carriers of the dynamics of the theory.

the amplitude for no interaction, is represented diagrammatically by a line as shown in Figure 9.1, where a dot signifies a state label.

Figure 9.1 Diagrammatic notation for bare propagators.

In the context of quantum theory, the propagator or Green's function is the conditional probability amplitude for the event 1 to take place *given* that event 2 has taken place. All states have equal status and the bare propagators are symmetric functions of the state labels, $G_{12}^{(0)} = G_{21}^{(0)}$. The free propagator is species specific

$$G_{12}^{(0)} \propto \delta_{s_1 s_2} , \tag{9.3}$$

a free particle can not change its identity.[4]

In the treatment of non-equilibrium states in the real-time technique, the real-time forward and return contour matrix representation, Eq. (5.1), or better the economical and more physical symmetric representation of the bare propagator, should thus be used, the latter having the following additional matrix structure in the dynamical or Schwinger–Keldysh indices:

$$G_0 = \begin{pmatrix} 0 & G_0^A \\ G_0^R & G_0^K \end{pmatrix} . \tag{9.4}$$

The bare vertices describe the possible interactions allowed to take place, and generic examples, the three- and four-line attachment or connector vertices, are displayed in Figure 9.2.

Figure 9.2 Diagrammatic notation for bare vertices.

---

[4]If, say, there is no spin dynamics then $G_{12}^{(0)} \propto \delta_{\sigma_1 \sigma_2}$. Sometimes it is convenient to include in the free propagator the change in the internal degrees of freedom of the particle; for example, if the spin of the particle is coupled to an external magnetic field. The chosen notation is seen to be capable of dealing with any kind of dynamics.

Without risking confusion, we have in accordance with standard notation also used a single dot in connection with vertices (and, say, not a triangle with three attached dots or a box with four attached dots), and here the dot does not specify a single state label but several, as specified by the protruding stubs to which propagators can be attached. The rationale for this is that quantum field theories are local in time, so that at least all time labels of the propagators meeting at a vertex are identical. In the 3-connector vertex, the single dot with its three protruding stubs thus represents three state labels where propagators can be attached and they can all be different. The form of the vertices, as specified by the indices, describes how particle species are transmuted into other particle species or how a particle changes its quantum numbers owing to interaction. The numerical value of the vertex, the amplitude for the process specified by g, the coupling constant or charge, gives the strength of the process.

The two ingredients, propagators and vertices, are the only building blocks for constructing the Feynman diagrams. In condensed matter physics, the corresponding amplitudes represented by the propagators and vertices are the only ones needed to specify the theory. These numbers are taken from experiment, for example from the measured values of the mass and charge of the electron. However, in relativistic particle physics they are only bare parameters, i.e. rendered unobservable quantities owing to the presence of interactions. For example, the value of the mass entering a bare propagator is a quantity unreachable by experiment (i.e. has no manifestation in the world of facts) since it corresponds to the non-existent situation where the particle is not allowed to interact. The interaction causes the mass to change, and in order to make contact with experiment the knowledge of the measured masses (and charges) must be introduced into the theory through the scheme of renormalization.[5] The expressions for the bare propagators are known *a priori*, since they are specified by the space-time symmetry, and the forms of the vertices are given by the symmetry of the theory, but their numerical values must be taken from comparison with experiment.[6]

In elementary particle physics, only the two types of vertices displayed in Figure 9.2 occur, the 4-connector vertex being relevant only for the gluon–gluon coupling. The 3-connector vertex is ubiquitous, for example describing electron–photon interaction or pair creation such as in QED. In fact, in QED, the theory restricted to the multiplet of electron and its anti-particle, the positron, and the photon, the vertex is nonzero for various species combinations, describing both electron or positron emission or absorption of a photon, or pair creation or destruction. In condensed matter physics, the 3-connector could for example describe electron–phonon, electron–electron or electron–magnon interactions, as discussed in Section 2.4. In statistical physics, where the propagators describe both thermal and quantum fluctuations and

---

[5]Of course, the interactions encountered in condensed matter physics in the same manner lead to renormalization of, say, the electron mass, as we have calculated in Section 7.5.2. However, this is a finite amount on top of the infinitely renormalized bare mass. Usually this is an effect of only a few percent of the electron mass, except in for example the case of heavy fermion systems.

[6]In relativistic quantum theory the forms of the propagators are specified by Lorentz invariance. For a massive particle the propagator or Green's function is specified by its bare mass and the type of particle in question. Also the *form* of the interaction can be obtained from the symmetry and Lorentz invariance of the theory, whereas the strength of the coupling constants are phenomenological parameters, i.e. they are obtained by comparison of theory and experiment.

for example effects of quenched disorder, vertices of arbitrary complexity can occur.[7] In the theory of phase transitions, which is an equilibrium theory, the diagrams describe transitions, i.e. thermal fluctuations, between the possible states of the order parameter relevant to the transition and critical phenomenon in question. However, we shall frame the arguments in the appealing particle representation, but since arguments are about the topological character of diagrams the formalism applies to any representation and any type of fluctuations and thus to any kind of field theory.

9.1.2   Amplitudes and superposition

Consider an amplitude $A_{1234...N}$ specified by N external states, an N-state amplitude. It could, for example, describe the transition probability amplitude for collision of two particles in states 1 and 2, respectively, to end up in a particle configuration described by the states $3, 4, ..., N$, or the decay of a particles in state 1 into particles in states $2, 3, ..., N$, etc. This general conditional probability amplitude is represented by the N-state diagram shown in Figure 9.3.[8]

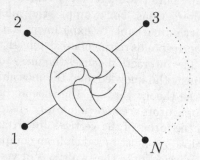

Figure 9.3 Diagrammatic notation for the N-external-state amplitude $A_{1234...N}$.

Specifying any amplitude is done by following *the laws of Nature*, quantum dynamics, which at the diagrammatic level of bare propagators and vertices is the basic rule that a particle has two options: *to interact or not to interact!*[9] The probability amplitude for a given process, characterized by the fixed initial and final state labels, is then construed as represented by the multitude of topologically different diagrams that can be constructed using the building blocks of the theory, viz. all the topolog-

---

[7]A case in question within the context of *classical* stochastic phenomena will be discussed in Chapter 12. The simplest vertex, a two-line vertex, is of course also relevant, viz. describing a particle interacting with an external classical field, but it is trivial to include, as will become clear shortly and we leave it implicit in the discussion for the moment.

[8]In statistical mechanics the diagrams can represent probabilities directly, say transitions between configurations of the order parameter.

[9]The former option is evident since otherwise the particle would live undetected, devoid of influence. The latter option is required by the fact that not all particles can interact directly.

ically different diagrams that the vertices and bare propagators allow. Examples of diagrams for the 4-state amplitude are shown in Figure 9.4 for the theory defined by having only a 4-connector vertex.

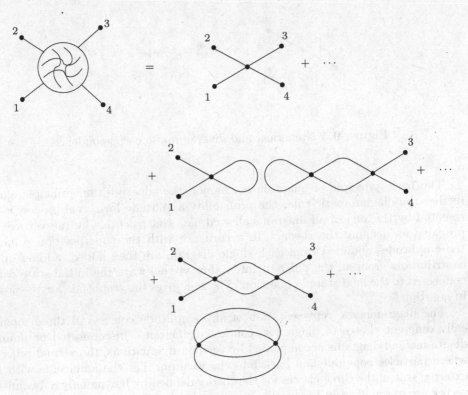

Figure 9.4 Generic types of diagrams.

The numerical value represented by a diagram is obtained by multiplying together the amplitudes for each component, propagators and vertices, constituting the diagram,[10] and in accordance with the superposition principle summation occurs over all internal labels, adding up all the alternative ways the process can be effected, for example summation over all the alternative space-time points where interaction could take place is performed.[11] The first diagram on the right in Figure 9.4 thus

---

[10]This rule is often left implicit, but represents the multiplication rule of quantum mechanics: that amplitudes for events effected in a sequence should be multiplied in order to get the amplitude for the sequence of events. The expression of causality in quantum mechanics.

[11]Only topologically different diagrams appear, interchanging the labeling of interaction points, i.e. permutation of vertices, are not additionally counted. This is precisely how the diagrammatics of (non-equilibrium) quantum field theory turned out as discussed in Chapter 4; the important

represents the analytical expression as displayed in Figure 9.5, and we have introduced the convention that repeated indices are summed over, or as we shall say state labels appearing twice are contracted.

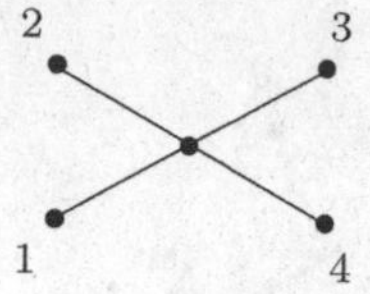

$$= \quad G^{(0)}_{11'}\, G^{(0)}_{22'}\, g_{1'2'3'4'}\, G^{(0)}_{3'3}\, G^{(0)}_{4'4}$$

Figure 9.5 Numerical and diagrammatic correspondence.

The basic principle of quantum mechanics, the superposition principle, entails further the diagrammatic rule: the probability amplitude for a real process is represented by the *sum* of all diagrams allowed, i.e. constructable by the vertices and propagators defining the theory. In accordance with the superposition principle, the amplitudes obtained from each single diagram are then added, adding up the contributions from all the different internal or virtual ways the initial state can be connected to the final state in question. The sum gives the amplitude for the process in question.

The diagrammatic representation of any amplitude consists of three topologically different classes of diagrams: connected diagrams, disconnected or unlinked diagrams, and diagrams accompanied by vacuum fluctuations, the virtual processes where particles pops out and back into the vacuum. For the amplitude with four external states, the three classes for the theory defined by having only a 4-connector vertex are exemplified in Figure 9.4.

The last diagram in Figure 9.4 represents the type of diagrams where a diagram (here a connected one) appears together with a vacuum fluctuation diagram. Vacuum diagrams close onto themselves, no propagator lines end up on the external states, and they appear as unlinked diagrams. According to the multiplication rule, the two amplitudes represented by the two sub-parts of the total diagram are multiplied together to get the total amplitude represented by the diagram. The first and second diagrams on the right in Figure 9.4 are of the connected and disconnected type, respectively. These diagrams, according to the general rule of diagram construction, can also be accompanied by any vacuum fluctuations constructable. The symbol $+\cdots$ in the figure summarizes envisioning *all diagrams constructable with the vertices and propagators defining the theory*. The total class of diagrams is thus an infinite myriad with infinite repetitions.

The totality of all diagrams can thus (with the help of our most developed sense)

---

feature that the factorial provided by the expansion of the exponential function is canceled by this redundancy.

be envisioned perturbatively. However, this is of little use unless only trivial lowest order perturbation theory needs to be invoked. One approach to a more powerful diagrammatic representation is by using topological arguments to partially re-sum the diagrammatic perturbation expansion in terms of effective vertices and the full 2-state propagators, i.e. in terms of so-called skeleton diagrams.[12] In the next section, we shall first pursue the hierarchal option on our way to this goal, expressing any N-state amplitude in terms of amplitudes with different numbers of external states.

Before embarking on deriving the fundamental diagrammatic equation, we introduce the inverse propagator. The inverse of the free or bare propagator is specified by the (partial differential) equation satisfied by the free propagator

$$(G_0^{-1})_{1\bar{1}} \, G_{\bar{1}2}^{(0)} = \delta_{12} \tag{9.5}$$

or since the propagator is symmetric in its labels

$$(G_0^{-1})_{1\bar{1}} \, G_{\bar{1}2}^{(0)} = \delta_{12} = G_{1\bar{1}}^{(0)} \, (G_0^{-1})_{\bar{1}2} \, . \tag{9.6}$$

We have written the equation satisfied by the free propagator in matrix notation, in terms of an integral operator as summation over repeated indices is implied.[13] For later use we introduce diagrammatic notation for the inverse free propagators as depicted in Figure 9.6.

$$(G_0^{-1})_{11'} \quad = \quad \overset{1}{\bullet}\!\!\!-\!\!\!\diagup\!\!\!-\!\!\!\overset{1'}{\bullet}$$

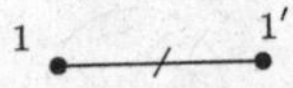

Figure 9.6 Diagrammatic notation for the inverse free propagator.

Using the basic diagrammatic rule: to interact or not, we shall start obtaining diagrammatic identities relating amplitudes, and eventually express these diagrammatic relations in terms of differential equations.

9.1.3   Fundamental dynamic relation

To get started on a systematic categorization of the plethora of diagrams, let us first consider the case where one particle is not allowed to interact and let us separate out its state to appear on the left in the diagram specifying the amplitude in question as depicted in Figure 9.7. Since *not* interacting is an option even for a particle capable of interacting, this seemingly irrelevant case of a completely non-interacting particle is a first step in the general deconstruction of an N-state amplitude into amplitudes with less external states, and allows furthermore a comment on the quantum statistics of identical particles.

---

[12]This was performed in Section 4.5.2, starting with the canonical formalism.

[13]The inverse free contour-ordered Green's function encountered in Section 4.4.1, or the inverse free matrix Green's function of Section 5.2.1, stipulating the additional matrix structure in the dynamical indices, had integral kernels typically consisting of differential operators operating on the delta function.

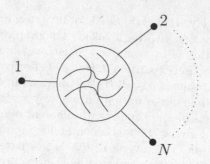

Figure 9.7 General N-state diagram.

Since the particle in state 1 is assumed not to interact, its only option is to propagate directly to a final state, and the amplitude $A_{1234\ldots N}$ can in this case be expressed in terms of the amplitude which has two external states less according to the basic rule: *everything can happen on the way between* the $(N-2)$ other final states, and the diagrammatic equation displayed in Figure 9.8 is obtained.

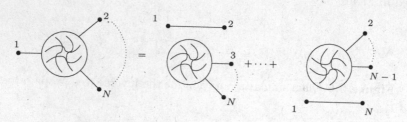

Figure 9.8 Diagrams for the non-interacting particle labeled by 1.

The N-state amplitude is in this case represented by the amplitudes specified by $(N-2)$ external states, i.e. $A_{23\ldots N}$ *without* the index M labeling the state where the propagator starting in state 1 ends up. If $s_M \neq s_1$, the process is not allowed since a non-interacting particle can not propagate to a different species state, and this feature is faithfully respected by the diagrammatics, since then the corresponding propagator according to Eq. (9.3) vanishes, $G_{1M}^{(0)} = 0$, and the contribution from the corresponding diagram vanishes since by the multiplication rule the bare propagator amplitude multiplies the adjacent $(N-2)$-state amplitude.

The quantum statistics of identical particles introduces minus signs when two identical fermions interchange states and the amplitudes are symmetric upon interchange of bosons, say

$$A_{213\ldots N} = \pm A_{123\ldots N} \tag{9.7}$$

where the upper (lower) sign is for bosons (fermions), respectively.

For the case of non-interacting identical particles, only free propagation and effects of the quantum statistics of the particles are involved as displayed in Figure 9.9.

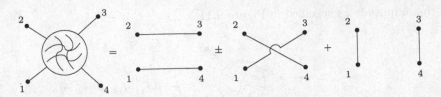

Figure 9.9 The 4-state diagrams for two non-interacting identical particles.

In the following we consider first bosons, in which case the amplitude functions are symmetric upon interchange of pairs of external state labels. The features of antisymmetry for fermions are then added.[14] The symmetry property of amplitudes forces the vertices to be symmetric in their indices, e.g. for the 3-vertex $g_{213} = g_{123}$, etc.

Returning to the diagram for the general N-state amplitude and respecting the other option for the particle in state 1, *to interact*, gives the additional first two diagrams as depicted on the right in Figure 9.10 for the case of a theory with three- and four-line connector vertices. The equation relating amplitudes as depicted in Figure 9.10 is the fundamental dynamic equation of motion in the diagrammatic language (for the case of three- and four-line vertices but trivially generalized).

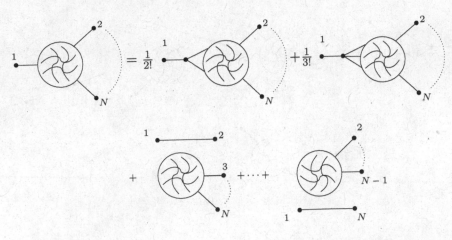

Figure 9.10 Fundamental dynamic equation for three- and four-vertex interactions.

The option of interaction through the 3-state vertex is for the N-external-state amplitude expressed in terms of the amplitude $A_{233...N+1}^{(N+1)}$ with $(N+1)$ external states, where two internal propagators are contracted at the vertex. This leads to the first diagram on the right in Figure 9.10, representing according to the Feynman

---

[14]In diagrammatics the essential is the topology of a diagram, and the interpretation of diagrams for the case of fermions is by the end of the day the same as for bosons except for the rule that a relative minus sign must be assigned to a diagram for each closed loop of fermion propagators.

rules the amplitude as specified in Figure 9.11.

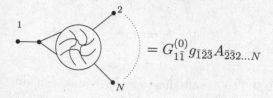

$$= G_{1\bar{1}}^{(0)} g_{\bar{1}2\bar{3}} A_{\bar{2}\bar{3}2...N}$$

Figure 9.11 Diagram and corresponding analytical expression.

Repeated state labels are summed over in accordance with the superposition principle. Similarly for diagrams with higher-order vertices in Figure 9.10 displayed for a theory with an additional 4-attachment vertex.

Although combinatorial prefactors are an abomination in diagrammatics we have in accordance with custom introduced them in Figure 9.10 by hand, the convention being: an N-line vertex carries an explicit prefactor $1/(N-1)!$, the reason being to be relieved at a different junction as immediately to be revealed. Consider a theory with only a 3-attachment vertex, and follow the further adventures of one of the particles emanating from the interaction vertex according to its two options, interact or not, as depicted in Figure 9.12.

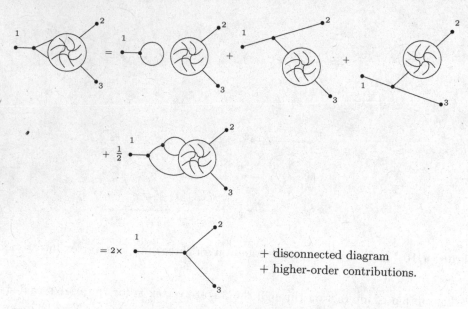

+ disconnected diagram
+ higher-order contributions.

Figure 9.12 Further adventures of a particle line emanating at a vertex.

The upper row of diagrams on the right in Figure 9.12 corresponds to the option of not interacting. In lowest order in the interaction, the second and third diagrams on the right give the same contribution. The inserted combinatorial factor in Figure 9.10 is thus the device to make the bare vertex diagram (here a 3-vertex) appear with no combinatorial factor. In a theory with only a 3-attachment vertex, the inserted combinatorial factor appearing with the vertex in Figure 9.10, thus makes the diagrammatic expansion of the 3-state amplitude start out with the lowest-order connected diagram, the bare vertex 3-state amplitude, carrying no additional factor as depicted in Figure 9.13.

Figure 9.13 Lowest-order connected 3-state diagram for a 3-vertex theory.

A similar function has the combinatorial factor inserted in front of the 4-vertex diagram in Figure 9.10.

9.1.4   Low order diagrams

Let us now familiarize ourselves with the Feynman rules and derive the expressions of lowest-order diagrammatic perturbation theory. The reader not interested in entering into this infinite forest of diagrams can skip the next few pages and go straight to the next sections where more powerful methods are developed. These will allow us systematically to generate the jungle of diagrams. However, for the adventurous reader let us see what kind of diagrams will emerge when we apply the simple law of dynamics, *to interact or not to interact!* A lesson to be learned from this is that although the basic rule is as simple as it possibly can be, in this brute force generation of diagrams one can easily miss a diagram, something history has proved over and again. The functional methods we shall consider shortly are able to capture the complete diagammatics in a simple way and in this way are able to help us in ensuring against mistakes.

We can now in any diagram follow the further possible options of any particle line emanating at a vertex, interact or not, and in this way unfold order by order the infinite total canopy in the jungle of diagrams constituting perturbation theory. For example, consider the 2-state amplitude (or two-point or 2-state propagator or Green's function) and a theory with the option of interaction only through the 3-attachment vertex. The two options for dynamics then generate the diagrams depicted in Figure 9.14.

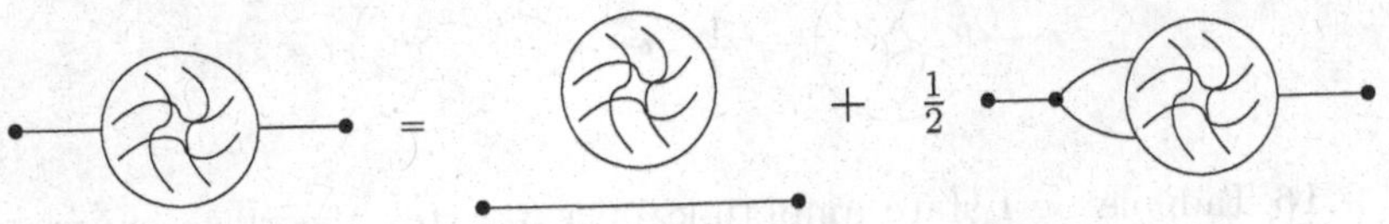

Figure 9.14 Interaction or not option for the 2-state amplitude.

A new diagrammatic entity enters in the first diagram on the right in Figure 9.14, the sum of all vacuum diagrams. The first diagram on the right in Figure 9.14 represents the product of two quantities, the *bare* 2-state amplitude, the bare propagator, *times* the amplitude resulting from the sum of all vacuum diagrams: free propagation accompanied by vacuum fluctuations, and nothing further is to be revealed diagrammatically in this part. The second diagram on the right corresponds to the option of interaction (in QED it could represent photon absorption or emission by electrons and positrons or pair creation). We note the general structure emerging in this way for the 2-state amplitude: the appearance of the bare 2-state amplitude and the appearance of a higher-order amplitude, here the 3-state amplitude.

Next we concentrate on the second diagram on the right in Figure 9.14, and explore the options, *interact or not*, of one of the lines emanating from the vertex and obtain the diagrams depicted in Figure 9.15.

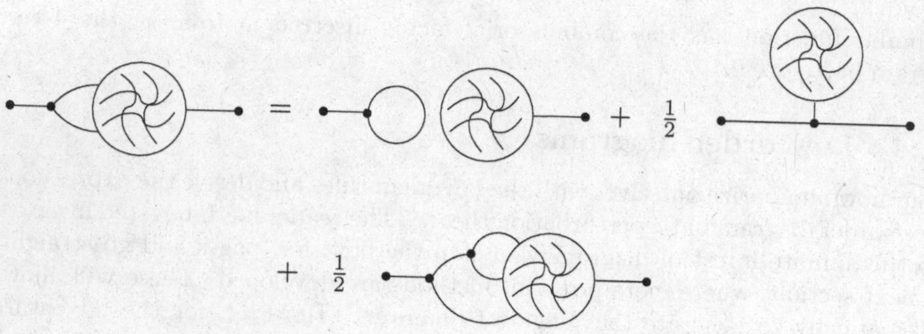

Figure 9.15 Diagrams generated by particle emanating at the vertex.

The first two diagrams on the right in Figure 9.15 correspond to the option of not interacting, viz. either propagating freely back to the vertex or freely to the external state. The last diagram encompasses the option of interacting, exposing one more vertex in our 3-state vertex theory.

The 1-state amplitude appearing in the first and second diagram on the right in Figure 9.15 (as a disconnected and connected piece, respectively), the tadpole diagram, can in a 3-vertex theory be expressed in terms of the 2-state amplitude contracted at the vertex as depicted in Figure 9.16, since the only option for the line is to interact (the option of not interacting was already exhausted in the first diagram in Figure 9.14).

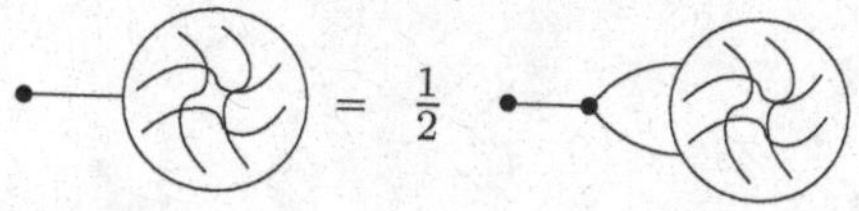

Figure 9.16 Tadpole or 1-state amplitude in a 3-vertex theory is expressable in terms of the vertex and the 2-state amplitude contracted at the vertex.

Inserting into the second diagram on the right in Figure 9.16 the expression for the 2-state amplitude specified by the expression in Figure 9.14 gives in a three-line-vertex theory the diagrammatic equation for the tadpole depicted in Figure 9.17.

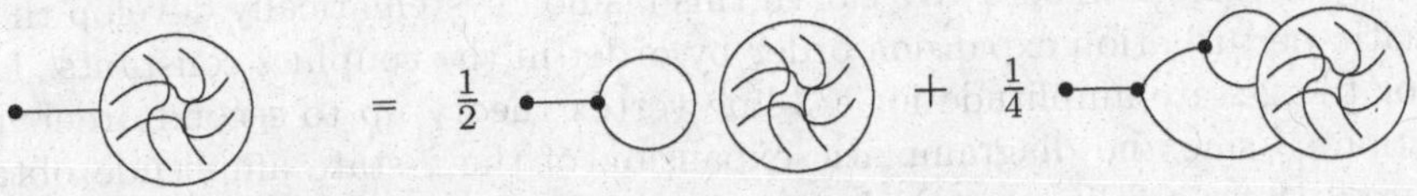

Figure 9.17 Tadpole equation for a three-line-vertex theory.

The 1-state diagram, the tadpole, has thus been expressed in terms of the bare tadpole *times* the amplitude representing the sum of all the vacuum diagrams plus a higher correlation amplitude, here the 3-state amplitude contracted at vertices according to the second diagram on the right in Figure 9.17.

Exercise 9.1. Obtain the diagrammatic equation for the tadpole if a 4-line vertex is also included in the theory.

Let us now further expose interactions in the 2-state amplitude in Figure 9.14. Insert the diagrammatic expansion of the tadpole in Figure 9.17 into the first diagram on the right in Figure 9.15, and then substitute the resulting expression for the second diagram on the right in Figure 9.14, and further explore the options for particle lines emanating from vertices, interaction or not. This gives the diagrammatic expansion of the 2-state amplitude depicted in Figure 9.18.

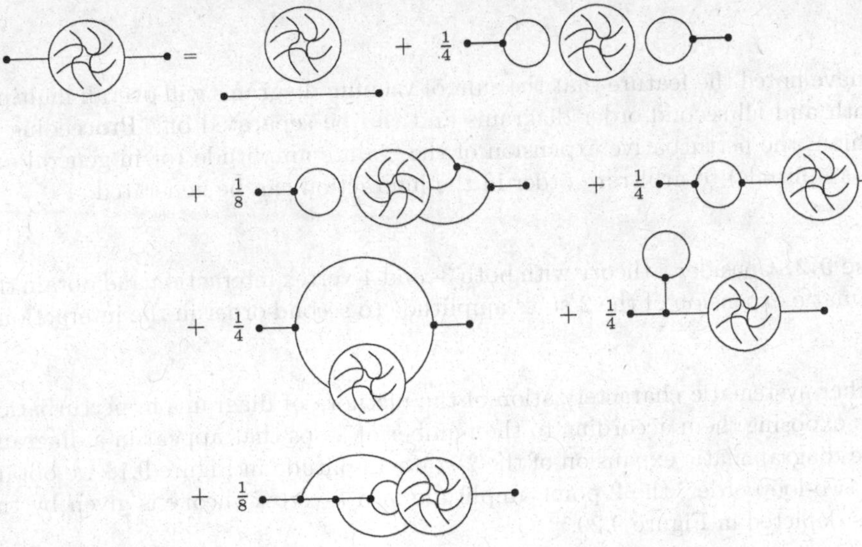

Figure 9.18 The 2-state amplitude equation for a 3-line-vertex theory exposed to second order in the coupling.

In this way an amplitude is expressed in terms of higher-order amplitudes appearing as the vertices launch propagator lines into states represented by amplitudes of ever higher state numbers. We can in this fashion systematically develop the diagrammatic perturbation expansion order by order in the coupling constants. Let us do it for the 2-state amplitude for a 3-line vertex theory up to second order in the interaction. Using the diagrammatic expansion of the 2-state amplitude obtained in Figure 9.18 for a 3-line vertex theory, the diagrammatic expansion of the 2-state amplitude to second order in the 3-vertex can now explicitly be identified by neglecting effects higher than second order. The 2-state amplitude to second order in the coupling thus has the diagrammatic expansion depicted in Figure 9.19.

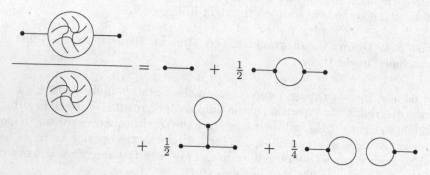

Figure 9.19 The 2-state amplitude to second order for a 3-vertex theory.

We have noted the feature that the sum of vacuum diagrams will overall multiply the zeroth and all second-order diagrams and can be separated off. Proceeding in this fashion, the perturbative expansion of the 2-state amplitude (or in general any N-state amplitude) to arbitrary order in the interaction can be generated.

Exercise 9.2. Consider a theory with both 3- and 4-vertex interaction and obtain the diagrammatic expansion of the 2-state amplitude to second order in the interactions.

Another systematic characterization of the plethora of diagrams in perturbation theory is exposing them according to the number of loops that appear in a diagram. From the diagrammatic expansion of the 2-state amplitude in Figure 9.18 we obtain that, to two-loop order, the 2-point amplitude in a 3-vertex theory is given by the diagrams depicted in Figure 9.20.[15]

---

[15]This type of expansion, the loop expansion, will give rise to a powerful systematic approximation scheme as discussed in Section 10.4. In quantum field theory it corresponds to a power series expansion in $\hbar$, the number of loops in a diagram corresponds to the power in $\hbar$, and is thus a way systematically to include quantum fluctuations.

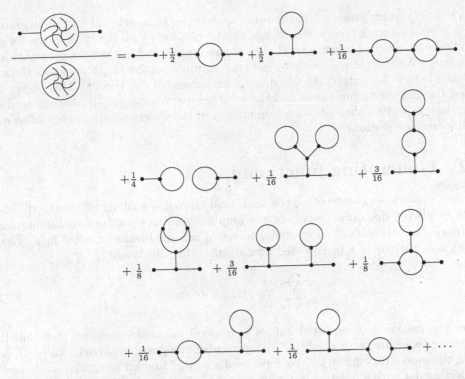

Figure 9.20 The 2-state amplitude to two-loop order for a 3-line-vertex theory.

In low order perturbation theory, we have noticed the feature that the sum of all vacuum diagrams separates off, and we show in the Section 9.5 that all amplitudes can be expressed in terms of their corresponding connected amplitude times, the amplitude representing *the sum of all the vacuum diagrams*.[16]

Exercise 9.3. Consider a theory with both 3- and 4-vertex interaction and obtain the diagrammatic expansion of the 2-state amplitude to one-loop order.

Exercise 9.4. Consider a theory with both 3- and 4-vertex interaction and obtain the diagrammatic expansion of the 2-state amplitude to two-loop order.

In this section we have proceeded from simplicity, the simple rules of diagrammatics, to complexity, the multitude of systematically generated diagrams by the simple law of dynamics, to interact or not to interact. However, this scheme soon

---

[16] From the canonical version of non-equilibrium perturbation theory considered in Chapter 4, we know that *the sum of all the vacuum diagrams* is an irrelevant number to the theory, in fact just one. But in standard zero-temperature formulation and finite temperature imaginary-time formulation of perturbation theory they appear, and to include these cases we include them in the diagrammatic discussion. Vacuum diagrams can be of use in their own right as discussed and taken advantage of in Chapters 10 and 12.

gets messy; just try your luck in the previous exercise to muscle out all the diagrams for a 3- plus 4-vertex theory. In order not to be blinded by all the trees in the forest we shall now proceed to get a total view of the jungle, and in this way we return to simplicity. We shall introduce an object that contains all the amplitudes of a theory and the vehicle for extracting any desired amplitude of the theory. This object is called the generating functional and the vehicle for revealing amplitudes will be differentiation, and we shall obtain a formulation of the diagrammatic theory in terms of differential equations.

9.2   Generating functional

We now embark on constructing the analytical theory describing efficiently the totality of all the diagrams describing the amplitudes, the quantities containing the information of the theory. The complete set of all amplitudes possible in a given theory can conveniently be collected into a generating functional

$$Z[J] = \sum_{N=0}^{\infty} \frac{1}{N!} A_{12...N} J_1 J_2 \cdots J_N , \qquad (9.8)$$

where summation over repeated indices is implied, or as we shall say state labels appearing twice are contracted.[17] The function of the possible particle states, J, is called the source (or current).[18] We have used a square bracket to remind us that we are dealing not with a function but a functional.[19] The expansion coefficients are the amplitudes of the theory. Here the generating functional or generator is considered to generate all the probability *amplitudes* of the quantum field theory in question.[20] In the diagrammatic approach, the $(N = 0)$-term, the value of $Z[J = 0]$, shall by definition be taken to be the amplitude representing the sum of all the vacuum diagrams of the theory in question.[21]

---

[17]For the continuous parts of the compound state label index the summation is actually integration, summation over small volumes. We shortly elaborate on this, but for simplicity we let this feature be implicit using matrix contraction for convolution.

[18]The source functions not only as a *source* for particles, but also as a *sink*, i.e. particle lines not only emanate from the source but can also terminate there, a feature we bury in the indices and need not display explicitly in the diagrammatics.

[19]A functional maps a *function*, here J, into a number.

[20]Actually, quantum field theory requires the substitution $J \to iJ$, but for convenience we leave out at this stage the imaginary unit since it is irrelevant for the ensuing discussion. The imaginary unit is fully installed in Chapter 10.

[21]In a $T = 0$ quantum field theory, the sum of all vacuum diagrams equals according to the Gell-Mann–Low theorem, Eq. (4.20), a phase factor of modulus one. In the closed time path formulation, which we shall always have in mind, the sum of all vacuum diagrams are by construction equal to one. The $(N = 0)$-term can therefore be set equal to one, i.e. giving the normalization condition $Z[J = 0] = 1$. Since our interest is the real-time treatment of non-equilibrium situations, the closed time path guarantees the even stronger normalization condition of the generator, viz. $Z[J] = 1$, provided that the sources on the two parts of the closed time path are taken as identical. When calculating physical quantities, the sum of all vacuum diagrams in fact drops out as an overall factor, a feature we have already encountered in low order perturbation theory in the previous section. However, vacuum diagrams can in themselves be a useful calculational device, a feature we shall employ when employing the effective action approach in Chapter 10 and Chapter 12. In

This way of collecting all the data of a theory into a single object, the generator of the theory, is indeed quite general. In equilibrium statistical mechanics the generating functional will be the partition function in the presence of an external field, the source (recall the general relation between quantum theory and thermodynamics as discussed in Section 1.1 (there displayed explicitly only for the simplest case of a single particle, the general case being obtained straightforwardly). The construction of the generating functional is also analogous to how the probabilities in a classical stochastic theory are collected into a generating function that generates the probabilities of interest of the stochastic variable (in that case the $(N = 0)$-term is one by normalization). In that context the generating function is usually reserved to denote the generator of the moments of the probability distribution involving a Fourier transformation of the probability distribution. This avenue we shall also take advantage of in the context of quantum field theory of non-equilibrium states when we introduce functional integration in Chapter 10.

Since the values of the source function J in different states are independent, varying the magnitude of the source for a given state influences only the source for the state in question and we have for such a variation (a formal discussion of the involved functional differentiation is given in the next section)

$$\frac{\delta J_M}{\delta J_1} = \delta_{1M} , \tag{9.9}$$

i.e. the Kronecker function which vanishes unless $1 = M$. Differentiating the generating functional with respect to the source function J and subsequently setting $J = 0$ therefore generates the amplitudes of the theory of interest, for example

$$\frac{\delta^N Z[J]}{\delta J_1 \delta J_2 \dots \delta J_N}\bigg|_{J=0} = A_{12\dots N} , \tag{9.10}$$

where the factorial in Eq. (9.8) is canceled by the same number of equal terms appearing due to the symmetry, Eq. (9.7), of the probability amplitude.[22] In the particle picture of quantum field theory the function J acts as a source for creating or absorbing a particle in the state specified by its argument.

For continuous variables, such as space and (contour or for real forward and return) time, the summation in Eq. (9.8) is actually short for integration, and we encounter instead of the Kronecker function, Eq. (9.10), Dirac's delta function,[23] say in the spatial variable

$$\frac{\delta J_{\mathbf{x}}}{\delta J_{\mathbf{x}'}} = \delta(\mathbf{x} - \mathbf{x}'). \tag{9.11}$$

However, this feature will in our notation be kept implicit for continuous variables.

We have used the symbol δ to designate that the type of differentiation we have in mind is *functional* differentiation, the *strength* of the source is varied for given state label.

---

the present chapter, the starting point is diagramatics and for that reason the $(N = 0)$-term is by definition taken to be the sum of all the vacuum diagrams.

[22] We first discuss the Bose case, the Fermi case needs the introduction of Grassmann numbers, as discussed in Section 9.4.

[23] For a discussion of Dirac's delta function we refer to appendix A of reference [1].

Thus functional generation of the amplitudes is achieved by functional differentiation. We therefore dwell for a moment on the mathematical rules of functional differentiation. However, in the intuitive approach of this chapter, we could in view of Eq. (9.9) simply *define* functional differentiation as the sorcery: cutting open the contraction of the source and amplitude, thereby exposing the state.

9.2.1   Functional differentiation

Functional differentiation maps a functional, $F[J]$, into a function according to the limiting procedure

$$\frac{\delta F[J]}{\delta J(x)} = \lim_{\varepsilon \to 0} \frac{F[J(x') + \varepsilon \delta(x - x')] - F[J]}{\varepsilon} . \tag{9.12}$$

More precisely, into a function of x and in general still a functional of J. Since we shall be dealing with functionals which have Taylor expansions, i.e. have a perturbation expansion in terms of the source, an equivalent definition is

$$F[J + \delta J] - F[J] = \int dx \, \frac{\delta F[J]}{\delta J(x)} \, \delta J(x) + \mathcal{O}(\delta J^2) . \tag{9.13}$$

The functional derivative measures the change in the functional due to an infinitesimal change in the magnitude of the *function* at the argument in question.

The operational definition of Dirac's delta function

$$J(x) = \int dx' \, \delta(x - x') \, J(x') \tag{9.14}$$

is thus seen to be identical to the functional derivative specified in Eq. (9.11) if in Eq. (9.12) or Eq. (9.13) we choose F to be the functional

$$F[J] = J(x) \tag{9.15}$$

for fixed x, or returning to our index notation $F[J] = J_x$.

For the functional defined by the integral

$$F[J] \equiv \int dx \, f(x) \, J(x) \tag{9.16}$$

we get for the functional derivative

$$\frac{\delta F[J]}{\delta J(x)} = f(x) \tag{9.17}$$

exposing the kernel.

As regards the discrete degrees of freedom we have instead of Eq. (9.13)

$$F[J + \Delta J] - F[J] = \sum_{\sigma_1 = -j}^{j} \frac{\Delta F[J]}{\Delta J_{\sigma_1}} \, \Delta J_{\sigma_1} \tag{9.18}$$

and if we choose F to be the functional

$$F[J] = J_{\sigma_1} \tag{9.19}$$

the functional derivative becomes

$$\frac{\Delta J_{\sigma_1}}{\Delta J_{\sigma_{1'}}} = \delta_{\sigma_1, \sigma_{1'}} \tag{9.20}$$

i.e. the Kronecker part in Eq. (9.9). The δ on the right-hand side in Eq. (9.9) is thus a product of delta and Kronecker functions in the continuous respectively discrete variables.

As usual in theoretical physics, to be in command of formal manipulations one needs only to be in command of the exponential function. In the context of functional differentiation, we note that the functional differential equation

$$\frac{\delta F[J]}{\delta J(x)} = \frac{\delta G[J]}{\delta J(x)} F[J] \tag{9.21}$$

has the solution

$$F[J] = e^{G[J]} , \tag{9.22}$$

which is proved directly using the expansion of the exponential function or follows from the chain rule for functional differentiation

$$\frac{\delta}{\delta g(x)} f(G[g]) = \frac{\delta G[g]}{\delta g(x)} \frac{\partial f(G)}{\partial G} \tag{9.23}$$

for arbitrary functional G and function f.[24]

Of particular importance is the case

$$F[J] \equiv e^{\int dx\, f(x)\, J(x)} , \tag{9.24}$$

where f is an arbitrary function, and in this case we get for the functional derivative

$$\frac{\delta F[J]}{\delta J(x)} = f(x)\, F[J] . \tag{9.25}$$

Exercise 9.5. Standard rules for differentiation applies to functional differentiation. Verify for example the rule

$$\frac{\delta}{\delta f(x)}(F[f]G[f]) = \left(\frac{\delta}{\delta f(x)} F[f] \right) G[f] + F[g] \left(\frac{\delta}{\delta f(x)} G[f] \right) \tag{9.26}$$

and the functional Taylor series expansion

$$F[f_1 + f_2] = e^{\int dx\, f_2(x)\, \frac{\delta}{\delta f_1(x)}} F[f_1] . \tag{9.27}$$

---

[24]In equations Eq. (9.12) and Eq. (9.13) we deviate from our general notation that capital letters represent functionals whereas lower capital letters denote functions.

9.2.2 From diagrammatics to differential equations

We shall now show how to capture the whole diagrammatics in a single functional differential equation. We introduce the diagrammatic notation for the generating functional, Z, displayed in Figure 9.21.

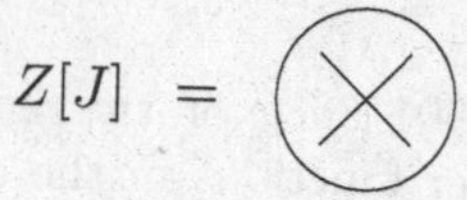

$$Z[J] \;=\; \bigotimes$$

Figure 9.21 Diagrammatic notation for the generating functional.

According to the definition of the generating function in terms of amplitudes and sources, Eq. (9.8), we have the relation as shown in Figure 9.22.

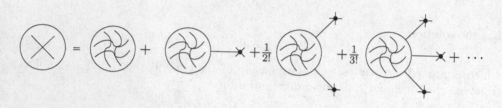

Figure 9.22 Diagrammatic representation of the generating functional.

The first term on the right of Figure 9.22 is the sum of all vacuum diagrams and independent of the source, and we have introduced the diagrammatic notation that a cross designates the source, the label of the source being that of the state indicated by the corresponding dot as shown in Figure 9.23.

$$\times = J$$

Figure 9.23 Diagrammatic notation for the source in the state indicated by the dot.

We have thus introduced the new diagrammatic feature that a particle line, as dictated by the generating functional, can end up on a source. The propagator dot and the corresponding source dot are thus shared in accordance with the convention that the corresponding state label are contracted, i.e. repeated indices are summed, integrated, over in accordance with the definition in Eq. (9.8).[25]

If the source is not set to zero after differentiation

$$A_{12\ldots N}[J] \;=\; \frac{\delta^N Z[J]}{\delta J_1 \delta J_2 \cdots \delta J_N} \tag{9.28}$$

---

[25]The dot was also used in connection with the vertices, and another reason for this is that in fact a vertex is a generalization of a source, generating multi-particle states.

we generate a new quantity, the amplitude in the presence the source. A source dependent amplitude is a function of the state labels exposed by the labels of the sources with respect to which the generating functional is differentiated as well as a functional of the source.

For the non-interacting theory in the presence of the source, the amplitude $A_1[J]$ is represented by the diagram depicted in Figure 9.24.

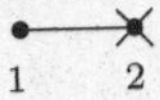

$$\bullet\!\!\longrightarrow\!\!\times = G_{12}^{(0)} J_2$$
$$12$$

Figure 9.24 Diagrammatic representation of the amplitude A_1 for a free theory in the presence of the source.

We now turn to show how to express in terms of functional differential equations, all the diagrammatic equations relating amplitudes, as exemplified in Figure 9.10, and derived by the simple diagrammatic rule: to interact or not. This is achieved by first expressing the fundamental dynamic diagrammatic equation displayed in Figure 9.10, in terms of a differential equation for the generating functional.

The first derivative of the generating functional generates according to its definition the terms

$$\frac{\delta Z[J]}{\delta J_1} = A_1 + A_{1\bar{2}} J_{\bar{2}} + \frac{1}{2} A_{1\bar{2}\bar{3}} J_{\bar{2}} J_{\bar{3}} + \frac{1}{3!} A_{1\bar{2}\bar{3}\bar{4}} J_{\bar{2}} J_{\bar{3}} J_{\bar{4}} + \cdots . \quad (9.29)$$

Differentiating the generating functional with respect to the source of a certain label removes this source, corresponding diagrammatically to removing a cross, and exposes this state in a bare propagator as each source dependent amplitude thus no longer ends up on this source but in the corresponding particle state, a particle is launched.[26] An external state, no longer contracted with the source, is thus exposed in each of the diagrams on the right-hand side in Figure 9.22 representing the generating functional, viz. the state with the label of the source with respect to which we differentiate. We therefore introduce the diagrammatic notation for the first derivative of the generating functional, the 1-state amplitude in the presence of the source, where a state on a free propagator line extrudes from the generating functional as depicted in Figure 9.25.

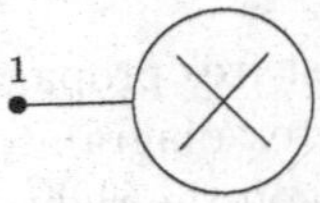

$$\bullet\!\!-\!\!\!\bigotimes \equiv \frac{\delta Z}{\delta J_1}$$
$$1$$

Figure 9.25 Diagram representing the first derivative.

The cross in the diagram in Figure 9.25 is there to remind us that the first

---

[26]Or terminated as kept track of for convenience by yet an index in the collective index, and not as in Chapter 4 by an arrow.

derivative, the source dependent 1-state amplitude function, is still a functional of the source.

The equation for the first derivative of the generating function, Eq. (9.29), can therefore be expressed diagrammatically as depicted in Figure 9.26.

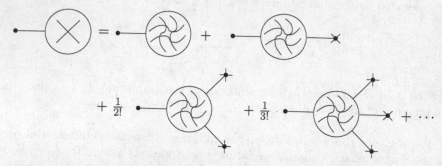

Figure 9.26 Diagrammatic expansion of the 1-state amplitude in the presence of the source.

Let us consider a 3-vertex theory. The first diagram on the right in Figure 9.26, the tadpole, is then given by the diagram in Figure 9.16, i.e. specified by the vertex and the 2-state amplitude. The second diagram on the right in Figure 9.26 can according to the two options of the external state line, interact or not, be split into the two diagrams on the right-hand side depicted in Figure 9.27. For the latter option the exposed state propagates directly to the source as depicted in the first diagram on the right.

Figure 9.27 Interaction or not options for the 1-source term.

The structure of the above equation is: free propagation to the source times the sum of vacuum diagrams plus exposed vertex diagram.

Similarly for the 2-source diagram on the right in Figure 9.26 we get the options as depicted in Figure 9.28. The factor of two appearing in front multiplying the first term on the right is the result of the option that when non-interacting the external state line can end up on either of the two sources and the two diagrams specify the same number.

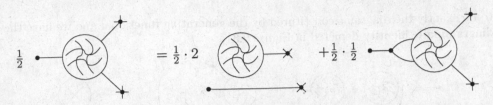

Figure 9.28 Interaction or not options for the 2-source term.

The structure of the above equation is: free propagation to the source times the 1-state amplitude contracted on the source (one integer lower-state amplitude than the one on the left) plus exposed vertex diagram.

Similarly for the 3-source diagram we have the options as depicted in Figure 9.29 (and we have equivalent for the further higher-numbered source diagrams in Figure 9.26).

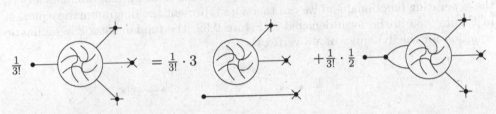

Figure 9.29 Interaction or no interaction options for the 3-source term.

If we collect the resulting diagrams into their two different types: those with the amplitude factor of free propagation to the source and those with an exposed vertex, the diagrammatic equation for the first derivative of the generating functional becomes the one depicted in Figure 9.30.

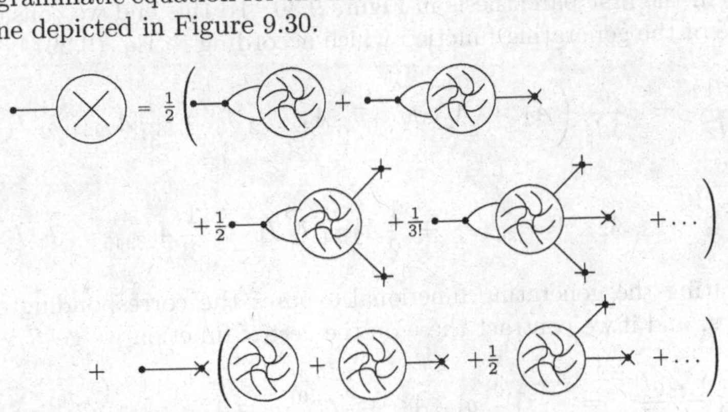

Figure 9.30 First derivative equation for a 3-vertex theory.

The sum of the diagrams in the parenthesis in the last line of Figure 9.30 are seen to be exactly the diagrams constituted by the generating functional and we have the diagrammatic identity depicted in Figure 9.31.

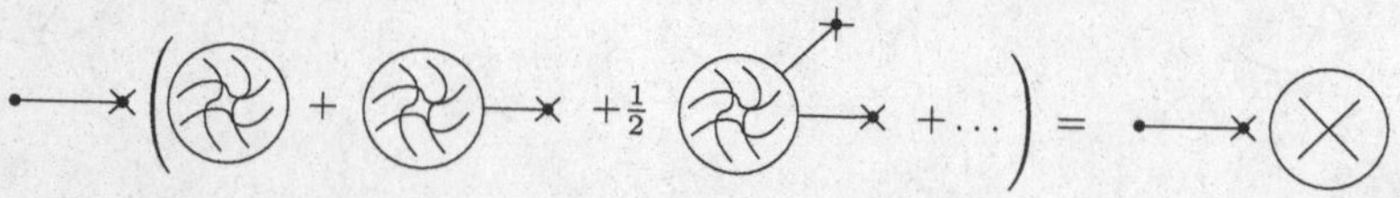

Figure 9.31 Propagation to the source times generating functional part.

The systematics of the prefactors of the diagrams in the parenthesis in Figure 9.31 are easily identified through their generation: the term with N sources getting the prefactor $1/N!$.

The diagrams in the first parenthesis in Figure 9.30 can also be expressed in terms of the generating functional. They all start out with the launched propagator entering the 3-vertex whose two other stubs either exposes lines in the sum of vacuum diagrams, or the 1-state amplitude contracted on the source, or the 2-state amplitude contracted on the source, etc. These latter parts thus sum up diagrammatically to the generating functional and we can therefore represent the diagrammatic equation in Figure 9.30 in the form depicted in Figure 9.32, the fundamental diagrammatic equation for the dynamics of a 3-vertex theory.

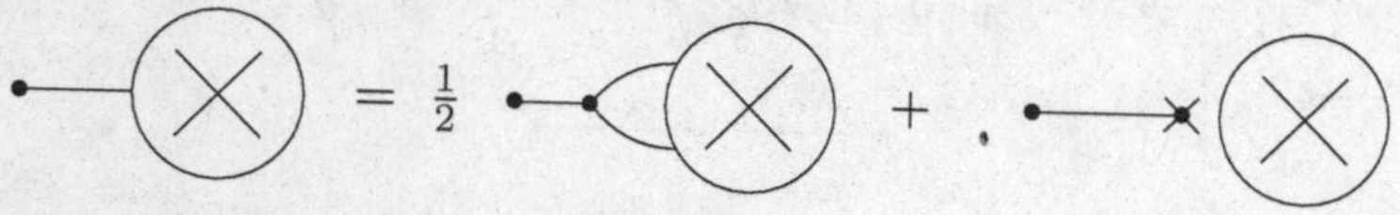

Figure 9.32 Fundamental diagrammatic equation for the 1-state amplitude, in the presence of the source, for a 3-vertex theory.

Next we wish to identify the analytical expression corresponding to the first diagram on the right in Figure 9.32, or equivalently, the analytical expression for the diagrams in the first parenthesis in Figure 9.30. To this end we consider the second derivative of the generating function which according to Eq. (9.29) becomes

$$
\frac{\delta^2 Z[J]}{\delta J_3 \delta J_2} = \frac{\delta}{\delta J_3}\left(A_2 + A_{2\bar{2}}\, J_{\bar{2}} + \frac{1}{2} A_{2\bar{2}\bar{3}}\, J_{\bar{2}} J_{\bar{3}} + \frac{1}{3!} A_{2\bar{2}\bar{3}\bar{4}}\, J_{\bar{2}} J_{\bar{3}} J_{\bar{4}} + \ldots \right)
$$

$$
= A_{23} + A_{23\bar{3}}\, J_{\bar{3}} + \frac{1}{2} A_{23\bar{3}\bar{4}}\, J_{\bar{3}} J_{\bar{4}} + \frac{1}{3!} A_{23\bar{3}\bar{4}\bar{5}}\, J_{\bar{3}} J_{\bar{4}} J_{\bar{5}} + \ldots \; . \tag{9.30}
$$

Differentiating the generating functional exposes the corresponding state labels of amplitudes, and if we contract these on the vertex function we get[27]

$$
\frac{1}{2} G^{(0)}_{1\bar{1}}\, g_{\bar{1}23}\, \frac{\delta^2 Z[J]}{\delta J_3 \delta J_2} = \frac{1}{2} G^{(0)}_{1\bar{1}}\, g_{\bar{1}23} A_{23} + \frac{1}{2} G^{(0)}_{1\bar{1}}\, g_{\bar{1}23} A_{23\bar{3}} J_{\bar{3}} + \frac{1}{2} G^{(0)}_{1\bar{1}}\, g_{\bar{1}23}\, \frac{1}{2} A_{23\bar{3}\bar{4}}\, J_{\bar{3}} J_{\bar{4}}
$$

---

[27]The vertices can thus be viewed as internal sources for creation and annihilation of particles, a point we shall exploit later.

$$+ \ \frac{1}{2} G_{1\bar{1}}^{(0)} \ g_{\bar{1}23} \frac{1}{3!} A_{23\bar{3}\bar{4}\bar{5}} \ J_{\bar{3}} \ J_{\bar{4}} \ J_{\bar{5}} \ + \ ... \tag{9.31}$$

i.e. exactly the analytical expression corresponding to the diagrams in the first parenthesis on the right in Figure 9.30. The correct factorial prefactors are generated term by term, the term with N sources getting the prefactor $1/N!$, and all terms have an overall factor $1/2$ since they were generated by a 3-line vertex theory. The diagrams in the first parenthesis in Figure 9.30 are thus represented in terms of differentiating the generating functional twice. We have thus derived diagrammatically the fundamental analytical equation, the Dyson–Schwinger equation, obeyed by the generating functional for a 3-vertex theory[28]

$$\frac{\delta Z[J]}{\delta J_1} = G_{1\bar{1}}^{(0)} \left(\frac{1}{2} g_{\bar{1}23} \frac{\delta^2}{\delta J_3 \delta J_2} + J_{\bar{1}} \right) Z[J] \,. \tag{9.32}$$

Just as in diagrammatics, the ingredients here are the bare propagators and vertices, but now instead of the diagrammatic rule of dynamics, to interact or not, we have instead free propagation to the source and differentiations with respect to the source.

The two lines protruding out of the generator in the first diagram on the right in Figure 9.32 has thus the same operational meaning as in Figure 9.25: it signifies differentiation with respect to the source, here where the labels with which the differentiation takes place are contracted at the vertex.[29]

By introducing the generating functional, the diagrammatic equations for amplitudes in the presence of a source can be represented by a differential equation, so far we have achieved it for the 1-state amplitude, but the game can be continued by taking further derivatives. The functional differential equation, Eq. (9.32), will thus be the fundamental dynamic equation for a 3-vertex theory.

The power of the generating functional technique is that all the relations existing between the amplitudes in a theory, as expressed by the diagrammatic equation in Figure 9.10, are contained in the fundamental functional differential equation, of the type Eq. (9.32) (or Eq. (9.34) the analogous equation for a theory with an additional four-line vertex, or quite generally for a theory with an arbitrary number of vertices). This is quite a compression of the information contained in the set of diagrammatic relations between amplitudes that has been achieved here. From the fundamental differential equation we can obtain all the diagrammatic equations relating amplitudes by functional differentiation. All the diagrammatic equations are thus equivalently representable by differential equations. For example, for the 2-state amplitude or Green's function in a 3-vertex theory we obtain by differentiating with respect to the source on both sides in Eq. (9.32)

$$A_{11'} = \frac{\delta}{\delta J_{1'}} \ G_{1\bar{1}}^{(0)} \left(\frac{1}{2} g_{\bar{1}23} \frac{\delta^2}{\delta J_3 \delta J_2} + J_{\bar{1}} \right) Z[J] \Big|_{J=0}$$

$$= G_{1\bar{1}}^{(0)} \left(\frac{1}{2} g_{\bar{1}23} \frac{\delta^3}{\delta J_3 \delta J_2 \delta J_{1'}} + \delta_{\bar{1}1'} \right) Z[J] \Big|_{J=0} \,, \tag{9.33}$$

---

[28]The generating functional approach to quantum field theory was championed by Schwinger [50].
[29]We note that the vertex in the equation in Figure 9.32 is acting like a 3-particle source .

which is the functional representation of the diagrammatic equation depicted in Figure 9.14.

We now have two ways of interpreting the two lines entering the generating functional in the first diagram on the right in Figure 9.32, either in the diagrammatic language options of *interact or not*, or as two functional differentiations of the generating functional.

Exercise 9.6. Obtain by diagrammatic reasoning for a theory with both 3- and 4-vertex interaction the Dyson–Schwinger equation (letting $\delta/\delta J \to \delta/i\delta J$ for proper quantum field theory notation, for details see Section 10.2.1)

$$\frac{\delta Z[J]}{\delta J_1} = G^{(0)}_{1\bar{1}} \left(J_{\bar{1}} + \frac{1}{2} g_{\bar{1}23} \frac{\delta^2 Z[J]}{\delta J_3 \delta J_2} + \frac{1}{3!} g_{\bar{1}234} \frac{\delta^3 Z[J]}{\delta J_4 \delta J_3 \delta J_2} \right) Z[J] \qquad (9.34)$$

satisfied by the generating functional.

For a non-interacting, free, quantum field theory we can solve Eq. (9.32) immediately (with $\delta/\delta J \to \delta/i\delta J$ for proper quantum field theory notation) and obtain for the generator of the free theory

$$Z_0[J] = e^{\frac{i}{2} J_1 G^{(0)}_{1\bar{1}} J_{\bar{1}}} . \qquad (9.35)$$

The overall multiplying constant equals one in accordance with the normalization $Z_0[J = 0] = 1$, the sum of all the vacuum diagrams are equal to one. For the free theory this follows trivially, the only vacuum diagram being the one where the free propagator closes on itself, and since the equal time propagator by nature of being a the conditional probability amplitude it satisfies $G_0(\mathbf{x}, t; \mathbf{x}', t;) = \delta(\mathbf{x} - \mathbf{x}')$, leaving the vacuum diagram equal to one. Since our interest is the real-time description of non-equilibrium situations, the closed time path formalism guarantees the even stronger normalization condition of the generator, $Z[J] = 1$, provided that the source on the two parts of the closed time path are taken as identical. We note that the free closed time path generator is unity, $Z_0[J] = 1$, if the sources on the two contour parts are identical, $J_+ = J_-$, in view of the identity Eq. (5.39). We have

$$Z_0[J] = e^{iW_0[J]} , \qquad W_0 = \frac{1}{2} J_1 G^{(0)}_{1\bar{1}} J_{\bar{1}} \qquad (9.36)$$

and W_0 vanishes, $W_0[J] = 0$, if the sources on the two contour parts are identical, $J_+ = J_-$.

In the closed time path formalism, the free Green's function entering Eq. (9.35) is the free contour-ordered Green's function. If we introduce the two parts of the closed contour explicitly and the notation $J_\pm$ for the source on the forward and return parts, respectively, the components of the matrix Green's function of Eq. (5.1) appears multiplied by the respective sources and integrations are over real time.

If we wish to express the generator in the physical or symmetric matrix Green's function representation, we should rotate the real-time sources by $\pi/4$ to give

$$\begin{pmatrix} J_1 \\ J_2 \end{pmatrix} = \frac{1}{\sqrt{2}} \begin{pmatrix} 1 & -1 \\ 1 & 1 \end{pmatrix} \begin{pmatrix} J_+ \\ J_- \end{pmatrix} \qquad (9.37)$$

as well as the Green's functions, and we obtain, suppressing variables other than the time,

$$W_0[J] = \int\limits_{-\infty}^{\infty} dt \int\limits_{-\infty}^{\infty} dt' (J_2(t)\, G_0^R(t,t')\, J_1(t') + J_1(t)\, G_0^A(t,t')\, J_2(t') + J_2(t)\, G_0^K(t,t')\, J_2(t')).$$

$$(9.38)$$

By choosing properly the real-time dynamical indices of the sources, we can by differentiation generate the various real-time propagators, G_0^{RAK}.

9.3 Connection to operator formalism

In Chapter 4 we showed how to derive the Feynman diagrammatics for non-equilibrium situations starting from the canonical formulation in terms of quantum fields, i.e. we started from the equations of motion for the contour or real-time Green's functions, describing the interactions in the system, and ended up with their diagrammatic representation in terms of perturbation theory. In this chapter, we have started from diagrammatics and have obtained the equation of motion for the contour or real-time Green's functions in terms of the generating functional. We can also make the direct connection back to the quantum fields by expressing the contour or real-time generating functional in terms of them according to[30]

$$Z[J] = \left\langle T_c\, e^{i \int_c dxd\tau\, \phi(\mathbf{x},\tau)\, J(\mathbf{x},\tau)} \right\rangle = \mathrm{Tr}(\rho(H) T_c\, e^{i \int_c dxd\tau\, \phi(\mathbf{x},\tau)\, J(\mathbf{x},\tau)}) \qquad (9.39)$$

since for example the two-point Green's function (modulo the imaginary unit) is then specified in terms of the, for simplicity, scalar quantum field operator, $\phi(\mathbf{x}, \tau)$, on the multi-particle space according to

$$\langle T_c(\phi(\mathbf{x},\tau)\,\phi(\mathbf{x}',\tau'))\rangle = -\frac{\delta Z[J]}{\delta J_{\mathbf{x}',\tau'}\,\delta J_{\mathbf{x},\tau}}\bigg|_{J=0}. \qquad (9.40)$$

In Eq. (9.39) the contour is the closed time path depicted in Figure 4.5, as we have a non-equilibrium situation in mind, and we have in Eq. (9.40) generated the contour ordered 2-state Green's function. Introducing the two parts of the closed contour, the matrix Green's function of Eq. (5.1) emerges. If we wish to generate the components of the symmetric or physical matrix Green's function, Eq. (5.41), we should rotate the fields and sources according to Eq. (9.37) (recall Eq. (9.38)).

Exercise 9.7. Consider the case of a self-coupled bose field as described by the potential $V(\phi)$. Show that the generating functional can be expressed in terms of the free generator according to

$$Z[J] = e^{-iV\left(\frac{\delta}{i\delta J(\mathbf{x},\tau)}\right)} Z_0[J]. \qquad (9.41)$$

---

[30]For the case of zero temperature the generator is the vacuum-to-vacuum amplitude in the presence of coupling to the source, $Z[J] = \langle 0| T_c\, e^{i \int_c dxd\tau\, \phi(\mathbf{x},\tau)\, J(\mathbf{x},\tau)}|0\rangle$.

We have considered, as above, the case of a real or hermitian bose field. If we considered spin-less bosons, say sodium atoms at low temperatures where their internal degrees of freedom can be neglected, we would have for the generating functional

$$Z[\eta, \eta^*] = \left\langle T_c \, e^{i \int_c dx dt \, \psi(\mathbf{x},t) \, \eta(\mathbf{x},t) + \psi^\dagger(\mathbf{x},t) \, \eta^*(\mathbf{x},t)} \right\rangle, \qquad (9.42)$$

where now the source is not as above a real function, but a doublet of complex c-number functions. We note the important feature of the closed time path formalism that the generator equals one if the sources are identical on the upper and lower parts of the contour.

9.4   Fermions and Grassmann variables

For the case of fermions, the sources must be anti-commuting numbers, so-called Grassmann variables, in order to respect the antisymmetry property of Green's functions or amplitudes in general. In quantum field theory, we shall always be concerned with a Grassmann algebra consisting of an even number of generators

$$\{\eta_\alpha, \eta_\beta\} = 0, \qquad (9.43)$$

where $\alpha, \beta = 1, 2, \ldots, 2n$. All possible products (ordered by convention $\alpha < \beta$, etc.), the set $\{1, \eta_\alpha, \eta_\alpha \eta_\beta, \ldots, \eta_\alpha \cdots \eta_\nu\}$, constitute a basis for the Grassmann algebra, which in addition is a vector space over the complex numbers of dimension 2^{2n} since any generator can either be included or not in a product. Since we consider an even number of generators, they can be grouped in pairs, so-called conjugates, and renamed, η_α and η_α^*, i.e. now $\alpha = 1, \ldots, n$. The conjugation property is endowed with the properties: $(\eta_\alpha^*)^* = \eta_\alpha$, and $(c \eta_\alpha)^* = c^* \eta_\alpha$ and $(\eta_\alpha \cdots \eta_\beta)^* = \eta_\beta^* \cdots \eta_\alpha^*$.

Each of the variables satisfies its Grassmann or exterior algebra, and owing to the anti-commutation relation, which implies $\eta^2 = 0$, the highest polynomial to be built is thus linear

$$f(\eta) = c_0 + c_1 \eta, \qquad (9.44)$$

the monomial, where the coefficients c_0 and c_1 are arbitrary complex numbers. Similarly for a pair η and η^*

$$f(\eta, \eta^*) = c_0 + c_1 \eta + c_2 \eta^* + c_3 \eta \eta^*. \qquad (9.45)$$

The linear space of functions of conjugate variables being four-dimensional.

As a consequence of the anti-commutation relation,

$$e^{\eta + \eta^*} = 1 + \eta + \eta^*. \qquad (9.46)$$

Exercise 9.8. Show that for pairs with different labels, say $\eta_1 \eta_1^*$ and η_2, η_2^* they commute and powers vanish, i.e.

$$[\eta_1 \eta_1^*, \eta_2, \eta_2^*] = 0, \quad (\eta_1 \eta_1^*)^2 = 0. \qquad (9.47)$$

Show that

$$e^{\sum\limits_{\alpha=1}^{n} \eta_\alpha \eta_\alpha^*} = \prod_{\alpha=1}^{n} e^{\eta_\alpha \eta_\alpha^*} = \prod_{\alpha=1}^{n} (1 + \eta_\alpha \eta_\alpha^*) . \tag{9.48}$$

Differentiation, symbolized by the operator $\partial/\partial\eta$, is the linear operation defined by first anti-commuting the variable next to the operation, giving for example

$$\frac{\partial}{\partial\eta}(\eta^* \eta) = \frac{\partial}{\partial\eta}(-\eta\eta^*) = -\eta^* . \tag{9.49}$$

For the function in Eq. (9.45) we thus get the derivatives

$$\frac{\partial}{\partial\eta} f(\eta, \eta^*) = c_1 + c_2 \eta^* \quad , \quad \frac{\partial}{\partial\eta^*} f(\eta, \eta^*) = c_2 - c_3 \eta \tag{9.50}$$

and

$$\frac{\partial}{\partial\eta^*} \frac{\partial}{\partial\eta} f(\eta, \eta^*) = c_3 = -\frac{\partial}{\partial\eta} \frac{\partial}{\partial\eta^*} f(\eta, \eta^*) . \tag{9.51}$$

Differentiations with respect to a pair of Grassmann variables thus anti-commute.

For the case of fermions, the sources we encounter must satisfy the algebra of anti-commuting variables, say the anti-commutation relations

$$\eta(\mathbf{x}, \tau) \, \eta^*(\mathbf{x}', \tau') = -\eta^*(\mathbf{x}', \tau') \, \eta(\mathbf{x}, \tau) \tag{9.52}$$

and we have the generating functional

$$Z[\eta, \eta^*] = \left\langle T_c e^{i \int d\mathbf{x} \int_c d\tau \, (\psi(\mathbf{x},\tau)\eta(\mathbf{x},\tau) + \psi^\dagger(\mathbf{x},\tau)\eta^*(\mathbf{x},\tau))} \right\rangle \tag{9.53}$$

generating for example the two-point fermion contour ordered Green's function, or propagator, according to

$$G(\mathbf{x}, \tau; \mathbf{x}', \tau') = i \frac{\delta^2 Z[\eta, \eta^*]}{\delta\eta^*(\mathbf{x}', \tau') \, \delta\eta(\mathbf{x}, \tau)} = -i\langle T_c(\psi(\mathbf{x}, \tau) \, \psi^\dagger(\mathbf{x}', \tau'))\rangle \tag{9.54}$$

the anti-commutation of the fields under the contour ordering being respected since derivatives with respect to Grassmann variables anti-commute.

Instead of the equality

$$\left[\frac{\delta}{\delta J(\mathbf{x}, \tau)} , J(\mathbf{x}', \tau') \right] = \delta(\mathbf{x} - \mathbf{x}') \, \delta(\tau - \tau') \tag{9.55}$$

valid for bosonic sources, we thus have for fermions

$$\left\{ \frac{\delta}{\delta\eta(\mathbf{x}, \tau)} , \eta(\mathbf{x}', \tau') \right\} = \delta(\mathbf{x} - \mathbf{x}') \, \delta(\tau - \tau') = \left\{ \frac{\delta}{\delta\eta^*(\mathbf{x}, \tau)} , \eta^*(\mathbf{x}', \tau') \right\} \tag{9.56}$$

and the following combinations of differentiations anti-commuting

$$\left\{ \frac{\delta}{\delta\eta(\mathbf{x}, \tau)} , \frac{\delta}{\delta\eta^*(\mathbf{x}', \tau')} \right\} = 0 = \left\{ \frac{\delta}{\delta\eta(\mathbf{x}, \tau)} , \frac{\delta}{\delta\eta(\mathbf{x}', \tau')} \right\} . \tag{9.57}$$

The topological arguments of Section 9.2.2 are unchanged for the case of including also fermions, and we obtain for example the fundamental dynamical equation for the case of electrons interacting through Coulomb interaction, V,

$$\frac{\delta Z[\eta, \eta^*]}{\delta \eta_1} = G_{1\bar{1}}^{(0)} \left(V_{\bar{1}234} \frac{\delta^3}{\delta \eta_4 \delta \eta_3^* \delta \eta_2^*} + \eta_{\bar{1}}^* \right) Z[\eta, \eta^*] \tag{9.58}$$

or for coupled fermions and bosons for example

$$\frac{\delta Z[J, \eta, \eta^*]}{\delta \eta_1} = G_{1\bar{1}}^{(0)} \left(g_{\bar{1}23} \frac{\delta^3}{\delta J_3 \delta \eta_2^*} + \eta_{\bar{1}}^* \right) Z[J, \eta, \eta^*] \tag{9.59}$$

and similar for the other source derivatives. Here we have for once made the species labeling of the sources explicit.[31]

For a non-interacting, free, quantum field theory we can immediately solve the corresponding Eq. (9.32), and obtain the generator of the free theory for fermions, recall Eq. (9.48),

$$Z_0[\eta^*, \eta] = e^{\frac{i}{2} \eta_{\bar{1}}^* G_{1\bar{1}}^{(0)} \eta_{\bar{1}}} . \tag{9.60}$$

9.5　Generator of connected amplitudes

We now show how to express the generator of all amplitudes, connected and disconnected, in terms of a less redundant quantity, the generator of connected amplitudes. Their relation is provided simply by the exponential function

$$Z[J] = e^{W[J]} , \quad W[J] = \ln Z[J] . \tag{9.61}$$

We shall first provide an intuitive demonstration arguing only at the diagrammatic level, and then give the general combinatorial proof.

9.5.1　Source derivative proof

The diagrams collected in the generating functional Z contain redundancy, viz. the presence of disconnected diagrams.[32] Say, for an 8-state amplitude there will a diagram which is the product of the first diagram on the right in Figure 9.4 multiplied by itself, describing processes which do not interfere. Furthermore, there is the redundancy of disconnected vacuum diagrams, the blobs of particles in and out of the vacuum. The disconnected diagrams, we now show, quite generally can be factored out of any N-state diagram. By this procedure the generator will be expressed in terms of the generator of only connected diagrams. It turns out that it is the exponential function which relates these two quantities. The presence of disconnected

---

[31] For any theory whose diagrammatics we derived in Chapter 4, we know the vertices and we can now immediately write down the fundamental functional differential equation satisfied by the generating functional.

[32] For the diagrammatics we encountered in Chapter 4, all physical quantities were *ab initio* expressed in terms of connected diagrams owing to using the closed time path or contour formulation.

diagrams is equivalent to processes that do not interfere with each other. The physical content of expressing the theory only in terms of connected diagrams is profound, viz. it is possible to describe a subsystem without bothering about the rest of the Universe with which it does not interact. This is in accordance with all experimental experience: processes separated far enough in space do not influence each other. We have in the diagrammatic approach stated the laws of Nature in terms of diagrammatic rules and we now show that the feature of having to deal only with connected diagrams is built in implicitly.[33]

Let us go back to the equation for the first derivative of the generator, the diagrammatic equation depicted in Figure 9.26. For the first diagram on the right, the tadpole or 1-state amplitude, we have in general the diagrammatic relationship depicted in Figure 9.33.

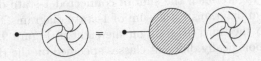

Figure 9.33 Tadpole and connected tadpole relation.

In Figure 9.33, the hatched circle denotes the sum of all *connected* tadpole or 1-state amplitude diagrams. The diagrammatic argument for the validity of this relation is that since the external particle line has no option of ending on an external state it must enter into a vertex, thereby creating connected diagrams, and any such can be accompanied by any vacuum side *show*.

The class of diagrams contained in the second term on the right in Figure 9.26 can be split topologically into the two distinct classes depicted in Figure 9.34.

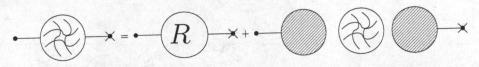

Figure 9.34 One-source road diagrams and disconnected diagrams.

Here the first diagram contains all the diagrams where we can follow at least one set of connected lines from the external state to the source, there is a *road* from the external state to the source. The second diagram is the sum of diagrams where there is no road from the initial state to the source, the external state and the source are disconnected. Then the two propagator lines must enter connected diagrams which

---

[33]In the canonical derivation of quantum field theory diagrammatics of chapter 4, the cancellation of the disconnected diagrams follows from the same argument as given in this and the next subsections, or the observation was superfluous in the close time path formulation as they occurred in multiples with opposite signs.

can be accompanied by any vacuum diagram. In the road diagram, disconnected diagrams must be vacuum diagrams. In the *road* diagram the disconnected vacuum bubbles can therefore be split off and the *connected* road diagram appears, as depicted in the first diagram on the right in Figure 9.35.

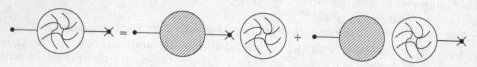

Figure 9.35 Splitting off the sum of vacuum diagrams in the road diagram.

To get the second set of diagrams on the right in Figure 9.35, we have used the relation depicted in Figure 9.33, the sum of connected 1-state diagrams multiplied by the sum of vacuum diagrams is the sum of 1-state diagrams.

Next we go on to the third diagram on the right in Figure 9.26. It can be split uniquely into the topologically different classes specified on the right in Figure 9.36.

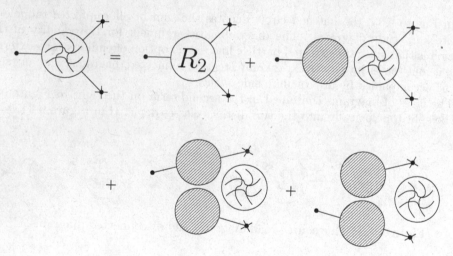

Figure 9.36 Road diagram classification.

Here the first diagram on the right comprises all the diagrams with roads from the external state to *both* sources, the second diagram all the diagrams with no roads from the external state to the sources, and the last two diagrams comprise all the diagrams with roads to only one of the sources. Clearly, this groups the diagrams uniquely into topologically different classes. In the road diagram the vacuum part splits off from the connected road diagram to both sources and we get the relation depicted in Figure 9.37.

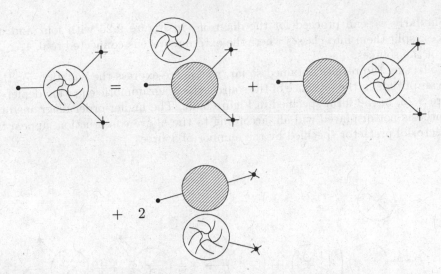

Figure 9.37 Splitting off the vacuum diagrams in the road diagram in Figure 9.36.

Here the factor of two appears in front of the last diagram because the last two diagrams in Figure 9.36 give identical contributions, and we have again used the fact that the sum of connected 1-state diagrams multiplied by the sum of vacuum diagrams is the sum of 1-state diagrams.

For the class of diagrams contained in the third term on the right in Figure 9.26 for the 1-state amplitude in the presence of the source, we can again split them uniquely into different topological classes: the set where the external state is connected to all the three sources, or to two or only one or none, i.e. the external state is disconnected from the sources, and we obtain the relation depicted in Figure 9.38.

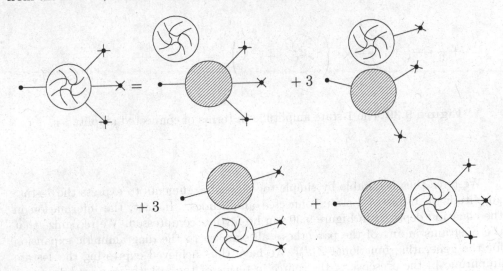

Figure 9.38 Road diagram classification.

Similarly we can proceed for the diagrams in Figure 9.26 with four and more sources: split them into classes where the external state is connected to 0, 1, 2, 3, 4, etc., of the sources.

Collecting the results obtained so far, we can re-express the equation for the 1-state amplitude in the presence of the source, the diagrammatic equation depicted in Figure 9.26, in the form specified in Figure 9.39. The higher-order diagrams in the parenthesis not displayed will all, according to the above construction, appear with the factorial prefactor specified by the number of sources.

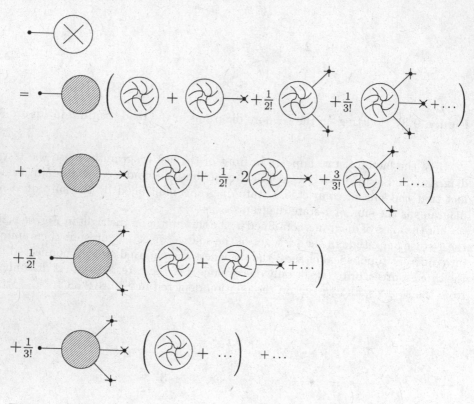

Figure 9.39 The 1-state amplitude in terms of connected amplitudes.

We have thus been able by simple topological arguments to express the 1-state amplitude in terms of 1-state connected amplitudes. In fact, the information in the equation depicted in Figure 9.39 can be further compressed. We recognize that the diagrams in any of the parenthesis all sum up to the diagrammatic expansion for the generating functional $Z[J]$. We have thus achieved expressing the 1-state amplitude in the presence of the source in terms of the 1-state *connected* diagrams in the presence of the source and the generator $Z[J]$ as depicted in Figure 9.40

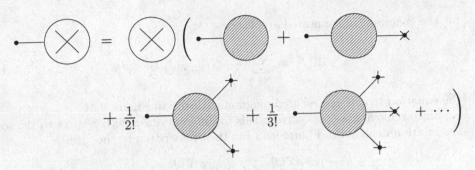

Figure 9.40 First derivative diagrammatic equation.

On the right we see the 1-state connected diagrams in the presence of the source. We shall therefore introduce the generator of connected diagrams, and we introduce a hatched circle with a cross as the diagrammatic notation for the generator of connected diagrams as depicted in Figure 9.41.

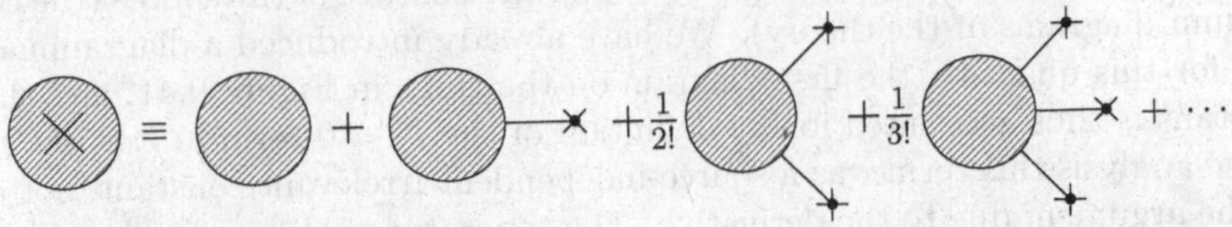

Figure 9.41 Generator of connected Green's functions.

The first term on the right in Figure 9.41 comprises the sum of all connected vacuum diagrams. Removing a cross in the connected generator specified in Figure 9.41 by functional differentiation exposes the sum of 1-state connected diagrams in the presence of the source, etc.

The diagrammatically derived equation depicted in Figure 9.40 can then be rewritten in the form depicted in Figure 9.42.

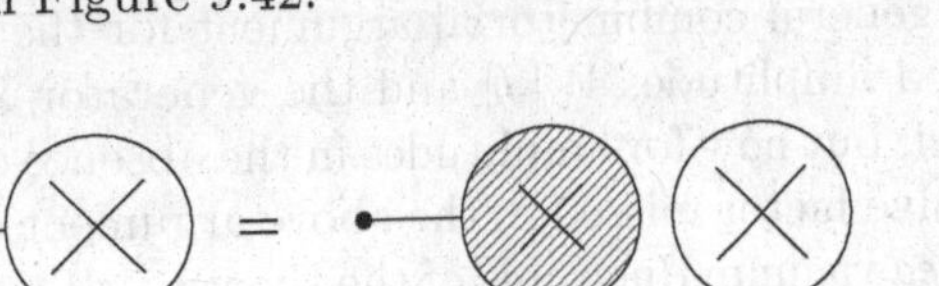

Figure 9.42 Relation between the derivatives of the generator and connected generator.

We introduce the notation $G_{12...N}$ for the amplitude represented by the N-state *connected* diagrams, the hatched circle with N external states, and have analytically for the generator of connected amplitudes

$$W[J] = \sum_{N=0}^{\infty} \frac{1}{N!} \, G_{12\ldots N} \, J_1 J_2 \cdots J_N \,, \tag{9.62}$$

the equation that is represented diagrammatically in Figure 9.41.

Since removing a cross corresponds to differentiation with respect to the source, the relation depicted in Figure 9.42 can then be written in the form

$$\frac{\delta Z[J]}{\delta J_1} = Z[J] \, \frac{\delta W[J]}{\delta J_1} \,. \tag{9.63}$$

We immediately solve equation Eq. (9.63), by the above analysis up to an undetermined multiplicative constant, and obtain

$$Z[J] = e^{W[J]} \,. \tag{9.64}$$

The overall multiplicative factor will in the following subsection be determined to be the sum of all *connected* vacuum diagrams in the absence of the source (the connected vacuum diagrams of the theory). We have already introduced a diagrammatic notation for this quantity, the first diagram on the right in Figure 9.41, and the overall constant is thus accounted for by definition of the $N = 0$-term in Eq. (9.62). In the above analysis this term was a source-independent irrelevant constant, not captured by the argument due to the derivative. The generator of all amplitudes, $Z[J]$, is thus equal to the exponential of the generator of only connected amplitudes. The simple structure of the combinatorial factors in the definition of the exponential function is thus enough at the level of generators to express the relationship between the connected diagrams and all the diagrams, including disconnected diagrams.

Inversely we have[34]

$$W[J] = \ln Z[J] \,. \tag{9.65}$$

9.5.2 Combinatorial proof

We now give the general combinatorial argument for the relation between the generator of connected amplitudes $W[J]$ and the generator $Z[J]$, again arguing at the diagrammatic level, but now for amplitudes in the absence of sources. This will fix the overall multiplicative factor missed in the above argument to be determined to be the sum of all *connected* vacuum diagrams of the theory.[35] This is achieved by the following observation. Any N-state amplitude can be classified according to its connected and disconnected sub-diagrammatic topological feature of its external attachments.

---

[34] In thermodynamics $Z[J]$ represents the partition function and $W[J]$ represents the free energy, and we have the diagrammatics necessary for a field theoretic approach to critical phenomena, and the renormalization group. In probability theory, $Z[J]$ is the characteristic function, the generator of moments and $W[J]$ is the generator of cumulants. In a quantum field theory we should restore the imaginary unit, $iW[J] = \ln Z[J]$.

[35] The argument also shows that for the time-ordered Green's function defined in terms of the field operators, Eq. (4.21) (or the contour-ordered Green's function, Eq. (4.50)), the denominator exactly cancels the separated off vacuum diagrams in the numerator.

For example, for the 3-state amplitude we get the topological classification as depicted on the right in Figure 9.43 (skipping for clarity the overall factor representing the sum of all vacuum diagrams in the absence of the source accompanying each of the diagrams on the right in Figure 9.43).

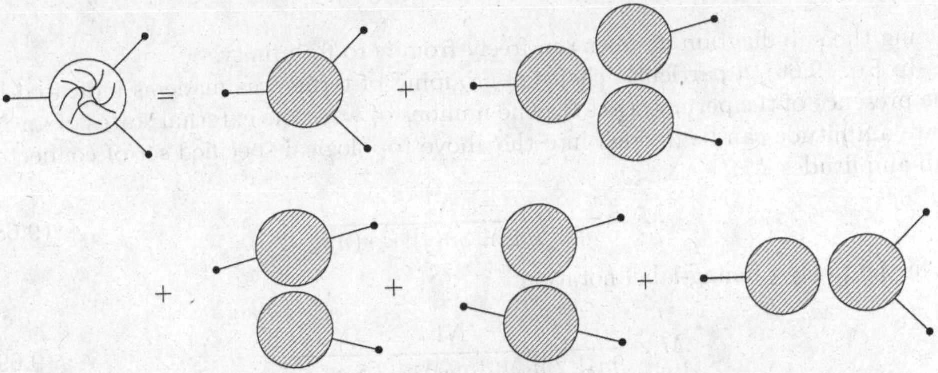

Figure 9.43 The 3-state diagrams in terms of connected diagrams.

The general combinatorial proof of the relationship between the generator of all amplitudes, the $A_{1...N}$s, and the generator of connected amplitudes, the $G_{1...N}$s, now proceeds. Any N-state amplitude $A_{1...N}$ is a sum over all the possible products of *connected* sub-amplitudes (multiplied by the overall sum of vacuum in the absence of the source which we keep implicit). Any N-state amplitude can thus be divided into its 1-state connected sub-amplitude parts (say m_1 in all, $m_1 \geq 0$), multiplying its 2-state connected amplitudes (say m_2 in all),..., and its n-state connected amplitudes (say m_n in all), and we have (suppressing on the right the overall multiplicative factor representing the sum of vacuum diagrams)

$$A^{(N)}_{1,2,...,N} \quad \rightarrow \quad {\sum_{\{m_n\}}}' G^{(1)}_{P_1} \cdots G^{(1)}_{P_{m_1}} \, G^{(2)}_{P_{m_1+1}, P_{m_1+2}} \cdots G^{(2)}_{P_{m_1+m_2}, m_1+m_2+1} \cdots$$

$$\cdots \, G^{(n)}_{P_{m_1+\cdots+(n-1)m_{n-1}+1},...,P_{m_1+...+m_{n-1}+n}} \cdots G^{(n)}_{P_{N-n},...,P_N} \,, \qquad (9.66)$$

where G denotes a connected amplitude, and the arrow indicates that a particular choice of external state labels has been chosen as indicated by the permutation P of the N labels. By construction the numbers specifying the sub-amplitudes satisfy a constraint, the relation $m_1 + 2m_2 + \cdots + nm_n = N$, since we consider the N-state amplitude, the case of N external states. The prime on the summation indicates that for each N the sum is over only sets of sub-amplitude labeling values that satisfy the constrain. Some of these ms are by construction zero; for example for, say, the 4-state amplitude there is the combination ($m_1 = 1, m_2 = 0, m_3 = 1, m_4 = 0$) describing the diagram with one 1-state connected diagram multiplying a 3-state connected diagram. Clearly, $m_{N+n} = 0$ for $n \geq 1$. Introducing the notation $m_p = 0$ to mean that there is no connected sub-amplitude with p external states we can write the constrain

$$\sum_{m_p=0}^{\infty} p\, m_p = N \tag{9.67}$$

letting the sub-diagram number run freely from zero to infinity.

In Eq. (9.66), a particular choice of grouping of terms was made as indicated by the presence of the permutation P. The number of ways the external states of an N-state amplitude can be divided into the above topological specified set of connected sub-amplitudes is

$$M \equiv \frac{N!}{m_1!(2m_2)!(3m_3)! \cdots (nm_n)!} \tag{9.68}$$

or in the freely running-label notation

$$M \equiv \frac{N!}{m_1!(2m_2)!(3m_3)! \cdots (\infty m_\infty)!}, \tag{9.69}$$

where ∞m_∞ simply indicates that for high enough external state labeling number, say beyond L, we have $(L+n)m_{L+n} = 0$ for any $n \geq 1$ owing to the constraint, Eq. (9.67). In the generating functional where the N-state amplitude is contracted with N external sources, all of these terms have identical value.

Within each subset of sub-amplitudes, for example the product of 2-state diagrams, the labels defining the external states could have been paired differently giving $(2m_2)!/((m_2)!(2!)^{m_2})$ differently chosen sub-amplitudes which when contracted with the sources give the same value. For the set of 3-state sub-amplitudes there are analogously $(3m_3)!/((m_3)!(3!)^{m_3})$ possible choices giving identical contribution, etc. For the N-state amplitude contracted with the N sources, we then have

$$\frac{1}{N!} A_{1,\ldots,N}^{(N)} J_1 \cdots J_N = \sum_{\{m_n\}} \frac{1}{m_1!} (G_1 J_1)^{m_1} \frac{1}{m_2!} \left(\frac{1}{2!} G_{12} J_1 J_2\right)^{m_2} \cdots \frac{1}{m_L!} \left(\frac{1}{L!} G_{1\ldots L} J_1 \cdots J_L\right)^{m_L}. \tag{9.70}$$

The generator Z can therefore be expressed in terms of *connected* amplitudes according to (the $N=0$-term, the sum of all vacuum diagrams in the absence of the source, will be dealt with shortly)

$$Z[J] = \sum_{N=0}^{\infty} \frac{1}{N!} A_{1,2,\ldots,N}^{(N)} J_1 J_2 \cdots J_N$$

$$= \sum_{N=0}^{\infty} \sum_{\{m_n\}} \frac{1}{m_1!} (G_1 J_1)^{m_1} \frac{1}{m_2!} \left(\frac{1}{2!} G_{12} J_1 J_2\right)^{m_2} \cdots \frac{1}{m_L!} \left(\frac{1}{L!} G_{12\ldots L} J_1 J_2 \cdots J_L\right)^{m_L}. \tag{9.71}$$

This can be rewritten

$$Z[J] = \sum_{N=0}^{\infty} \sideset{}{'}\sum_{\{m_n\}} \frac{1}{m_1!}(G_1 J_1)^{m_1} \frac{1}{m_2!}(\frac{1}{2!}G_{12}J_1 J_2)^{m_2} \cdots \frac{1}{m_L!}(\frac{1}{L!}G_{12\ldots L}J_1 J_2 \cdots J_L)^{m_L}$$

$$= \sum_{m_1, m_2, \ldots = 0}^{\infty} \frac{1}{m_1!}(G_1 J_1)^{m_1} \frac{1}{m_2!}(\frac{1}{2!}G_{12}J_1 J_2)^{m_2} \cdots \frac{1}{m_n!}(\frac{1}{n!}G_{12\ldots n}J_1 J_2 \cdots J_n)^{m_n} \cdots$$

$$(9.72)$$

where the last summation runs freely over all m_ls so clearly any term in the first sum is present once in the second sum, and any term in the second sum is unique. Any term in the double sum is also unique and contains any term in the sum on the right with the freely running summation and we have argued for the validity of the last equality sign in Eq. (9.72).

In the discussion we suppressed the multiplicative factor representing the sum of all the vacuum diagrams. To get the correct formula for $Z[J]$, we should thus in Eq. (9.72) interpret the term with all m_ps equal to zero, $m_1 = 0 = m_2 = m_3$, as the sum of all the vacuum diagrams or rather as unity since we should remember the overall multiplicative factor we left out of the argument representing the sum of all the vacuum diagrams, $Z[J = 0]$, connected and disconnected. We shall now obtain the expression for $Z[J = 0]$ in terms of the sum of *connected* vacuum diagrams. The combinatorial argument runs equivalent to the above. A vacuum diagram with disconnected parts classifies itself into connected vacuum parts characterized according to the number of vertices in the connected diagrams: a product of products of connected diagrams with one, two, etc., vertices. The constraint and the combinatorics will then be the same as above, N now characterizing the total number of vertices in the vacuum diagrams in question, and we end up with the terms on the right-hand side of Eq. (9.72) except for the absence of the source and the Gs now having the meaning of connected vacuum diagrams with the possible different numbers of vertices. We have thus shown that the sum of all the vacuum diagrams is given by the exponential of the sum of all *connected* vacuum diagrams. Note that the term contributing the unit term to this exponential function is provided by the vacuum contribution for the option of not interacting, the contribution of the free theory as discussed at the end of Section 9.2.2. Diagrammatically we have thus identified that the first diagram in Figure 9.41 represents the term $W[J = 0]$, the sum of all connected vacuum diagrams.

We therefore get

$$Z[J] = e^{W[J=0]} e^{G_1 J_1} e^{\frac{1}{2!}G_{12}J_1 J_2} \cdots e^{\frac{1}{n!}G_{12\ldots n}J_1 J_2 \cdots J_n} \cdots , \qquad (9.73)$$

where $W[J = 0]$ denotes the first term on the right in the definition of the generator of connected diagrams in Figure 9.41, and thereby

$$Z[J] = e^{W[J]} , \qquad (9.74)$$

where $W[J]$ is given by the expression in Eq. (9.62), since the G-amplitudes above were, by construction, the connected ones.

9.5.3 Functional equation for the generator

By construction $W[J]$ is the generator of connected amplitudes or Green's functions

$$G_{12\ldots N} = \frac{\delta^N W[J]}{\delta J_1 \delta J_2 \cdots \delta J_N}\bigg|_{J=0}. \tag{9.75}$$

For the first derivative of the generator of connected Green's functions we get according to the defining equation, Eq. (9.62), the trivial equation

$$\frac{\delta W[J]}{\delta J_{\bar{1}}} = G_{\bar{1}} + \sum_{N=1}^{\infty} \frac{1}{N!} G_{\bar{1}12\ldots N} J_1 J_2 \cdots J_N, \tag{9.76}$$

which has the diagrammatical form depicted in Figure 9.44.

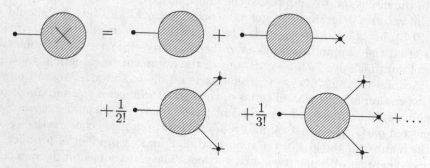

Figure 9.44 First derivative of the generator of connected amplitudes.

The equation Eq. (9.76), displayed diagrammatically in Figure 9.44, has no reference to the content of the theory in question, but expresses only the polynomial structure of the generator (of connected amplitudes) in terms of the source. To get the theory into play we shall use the fundamental dynamic equation, Eq. (9.32), and the established relation, Eq. (9.74). Since Z is related to W by the exponential function, all equations for Z can be turned into equations for W (the exponential function is the one which when differentiated brings back itself). Inserting Eq. (9.74) into the fundamental equation, Eq. (9.32), or rather Eq. (9.34) for the 3- plus 4-vertex theory, and using Eq. (9.22), we get

$$\frac{\delta W[J]}{\delta J_1} = G_{12}^{(0)} \left(J_2 + \frac{1}{2} g_{234} \left(\frac{\delta^2 W[J]}{\delta J_4 \delta J_3} + \frac{\delta W[J]}{\delta J_3} \frac{\delta W[J]}{\delta J_4} \right) \right.$$

$$\left. + \frac{1}{3!} g_{2345} \left(\frac{\delta^3 W[J]}{\delta J_5 \delta J_4 \delta J_3} + 3 \frac{\delta W[J]}{\delta J_3} \frac{\delta^2 W[J]}{\delta J_5 \delta J_4} + \frac{\delta W[J]}{\delta J_3} \frac{\delta W[J]}{\delta J_4} \frac{\delta W[J]}{\delta J_5} \right) \right) \tag{9.77}$$

the fundamental functional differential equation for the generator of connected amplitudes (here for the case of a 3- plus 4-vertex theory).

In diagrammatic notation we therefore have the equation depicted in Figure 9.45.

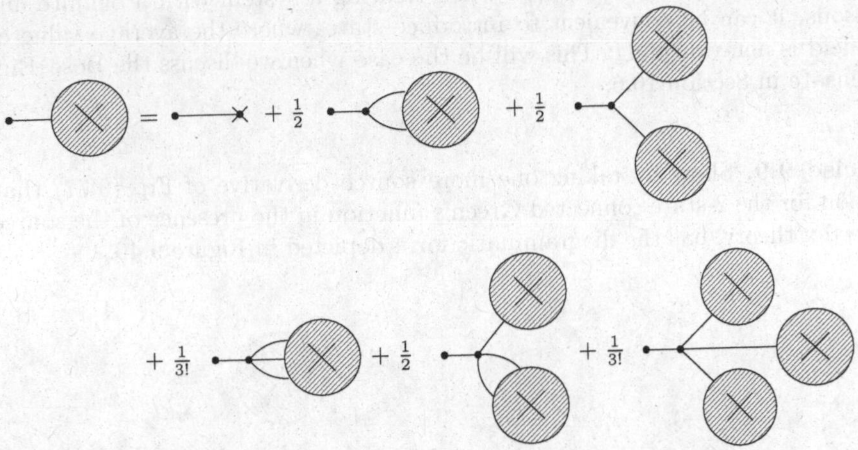

Figure 9.45 Fundamental equation for the generator of connected Green's function for a 3- plus 4-vertex theory.

In deriving the equation depicted diagrammatically in Figure 9.45 we reversed our previous order of first deriving equations by the diagrammatic rule, to interact or not, and instead used the fundamental functional differential equation, Eq. (9.32), whereby the propagator lines emerging from vertices into connected Green's functions represents functional differentiations. We could of course also immediately arrive at the equation in Figure 9.45 diagrammatically, the options for entering into connected diagrams for say propagators emerging from the 4-vertex is either into a 3-state diagram, or 2- and 1-state diagrams. The prefactor of the next to last diagram on the right in Figure 9.45 is caused by the three identical diagrams with the appearance of the 2-state diagram.

We have thus expressed the 1-state connected Green's function, in the presence of the source, in terms of higher-order *connected* Green's function and the free propagators and vertices of the theory. By taking further derivatives in Eq. (9.77) we can obtain the differential equation satisfied by any connected Green's function and immediately write down its diagrammatic analog. We have thus made the full circle back to the canonically derived non-equilibrium Feynman diagrammatics of Chapters 4 and and 5, where the diagrams represented averages of the quantum fields, but now we are armed in addition with the powerful tool of a functional formulation of non-equilibrium quantum field theory.

The amplitudes which in this chapter were defined in terms of the diagrams are thus for the case of quantum field theory the expectation values of products of the quantum fields of the theory. For example, the 2-state connected amplitude is the 2-point Green's function (normal or anomalous for the superconducting state as dictated by the Nambu index), etc. The 1-state amplitude is the average value of the quantum field. The 1-state amplitude, the tadpole, thus vanishes for a state with a definite number of particles, but can be non-vanishing for, for example, photons in a coherent state. However, even when treating a system with a definite number of bosons, it can be convenient to introduce states where the average value of the bose field is non-vanishing. This will be the case when we discuss the Bose–Einstein condensate in Section 10.6.

Exercise 9.9. Show by taking one more source derivative of Eq. (9.77) that the equation for the 2-state connected Green's function in the presence of the source, for a 3-vertex theory, has the diagrammatic form depicted in Figure 9.46.

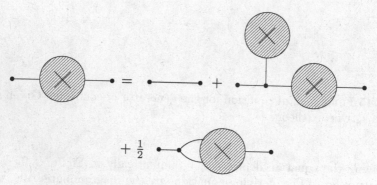

Figure 9.46 Equation for the 2-state connected Green's function for a 3-vertex theory.

Then argue instead diagrammatically from the equation in Figure 9.45 to obtain the above equation. One then learns to appreciate the skill of differentiation.

9.6   One-particle irreducible vertices

In order to get a handle of the totality of connected diagrams we shall further exploit their topology for classification. We introduce the concept of one-particle irreducible diagrams (1PI-diagrams). All diagrams can then be classified uniquely by the topological property: they can be cut in two by cutting zero (1PI), one, two, etc., internal bare lines. This will lead to the appearance of the one-particle irreducible vertices, and to the important formulation of the theory in terms of the effective action.

Consider the 1-state connected Green's function in the presence of the source, i.e. the derivative of the generator of connected Green's functions

$$\varphi_1 = \frac{\delta W[J]}{\delta J_1} . \tag{9.78}$$

We shall refer to this function as the field.[36] Besides being a function of the state exposed by differentiation, the field is also a functional of the source, $\varphi_1 = \varphi_1[J]$. We shall leave this feature implicit. However, in the diagrammatic notation we shall keep the source dependence explicit, through the cross, as we introduce the diagrammatic notation depicted in Figure 9.47 for the field.

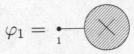

Figure 9.47 Diagrammatic representation of the 1-state connected Green's function or average field, the tadpole.

The state label of the field, exposed by differentiating the generator of connected Green's functions, launches a free propagator which in its further propagation has two options. The trivial one is where it propagates directly to the source in accordance with the first diagram on the right in Figure 9.45, this option being represented by the first diagram on the right in Figure 9.48. The other option corresponds to interaction and the corresponding diagrams can be uniquely classified topologically into distinct classes as follows: the exposed state where a propagator is launched can enter into a connected diagrammatic structure which has the property that it can not be cut in two by cutting only *one* internal bare propagator line, i.e. excepting the launched propagator. By definition such diagrams must not end on the source, and these diagrams are thus a subset of the set of diagrams described by the first diagram on the right in Figure 9.44, and are referred to as one-particle irreducible diagrams, 1PI-diagrams. Diagrammatically this set of diagrams is represented by the second diagram on the right in Figure 9.48. The next option is that the launched propagator enters into a one-particle irreducible diagrammatic part and emerges into a diagrammatic part such that the total diagram can be cut into two parts by cutting one internal line at exactly *one* or *two* or *three*, etc., places, all of these lines therefore emerging into the 1-state connected Green's function in the presence of the source, the field.[37] Diagrammatically these sets of diagrams are therefore represented by the third, etc., diagrams on the right in Figure 9.48 (combinatorial factors are inherited from our convention, here expressed in the starting equation depicted in Figure 9.44). The 1-state connected Green's function, the tadpole, is thus represented in terms of the one-particle irreducible vertices with attached tadpoles as depicted in Figure 9.48.

---

[36]Or average field or classical field. The reason for this terminology will become clear in the next chapter (or by comparison with the diagrammatic representation of the canonical operator formalism). The diagrammatic structure of the theories considered are identical to those of the quantum field theories we studied in Chapter 4. Therefore, interpreting the diagrammatic theory as a quantum field theory, the 1-state amplitude is the average value of the quantum field (in the presence of the source). For photons in a coherent state it describes the classical state of the electromagnetic field.

[37]Two (or more) lines can not enter into the same connected diagram, since then it is part of the one-particle irreducible part.

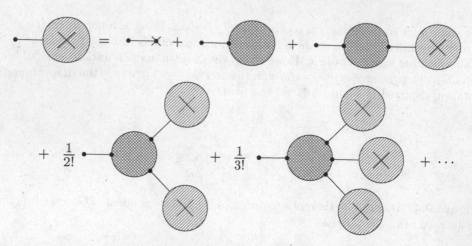

Figure 9.48 The 1P-irreducible vertex representation of the 1-state connected Green's function in the presence of the source.

We could continue and construct the one-particle irreducible vertex representation for any N-state amplitude, but we do not pause for that and relegate it to Section 9.6.2.

The 1-state connected amplitude in the presence of the source which by definition had the diagrammatic expansion depicted in Figure 9.44 and for a 3- plus 4-vertex theory was shown to satisfy the diagrammatic equation depicted in Figure 9.45 has now been organized into a different diagrammatic classification by introducing the one-particle irreducible vertex functions, $\Gamma_{1,2,\ldots,N}$, which diagrammatically are represented by cross-hatched circles with amputated lines protruding and the dots as usual represent the states where lines can end up or emerge from, as shown in Figure 9.49.

$$\Gamma_{12\ldots N} =$$

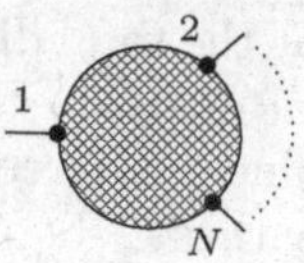

Figure 9.49 One-particle irreducible N-vertex function.

The diagrams on the right in Figure 9.48 correspond to a different re-grouping of the diagrams compared to those on the right-hand side in Figure 9.45, the re-grouping being based on a topological feature easily visually recognizable for any diagram: its 0, 1, 2, etc., irreducibility with respect to internal cutting. According to the topological construction, the one-particle irreducible vertices do not depend on the source J. They are uniquely specified in diagrammatic perturbation theory in terms of the bare vertices and bare propagators (and their topological property of one-particle irreducibility). As we shall see in the next section, they provide yet another way of capturing the content of the diagrammatic perturbation theory. The virtue of the diagrammatic relationship expressed in Figure 9.48 is that no loops appear explicitly, they are all buried in the one-particle irreducible vertices.

The diagrammatic structure of the equation expressed in Figure 9.48 should be stressed: the tadpole in the presence of the source is expressed in terms of tadpoles in the presence of the source attached to 1PI-irreducible vertices, i.e. in terms of so-called tree diagrams, diagrams that become disconnected by cutting just one propagator line. This observation shall be further developed in Section 10.3.

Analytically, the diagrammatic equation in Figure 9.48 reads

$$\varphi_1 = G_{12}^{(0)} \left(J_2 + \Gamma_2 + \Gamma_{23}\,\varphi_3 + \frac{1}{2}\Gamma_{234}\,\varphi_3\,\varphi_4 + \frac{1}{3!}\Gamma_{2345}\,\varphi_3\,\varphi_4\,\varphi_5 + \cdots \right). \quad (9.79)$$

To write Eq. (9.79) in a compact form, we collect the one-particle irreducible vertices into a generator, the generator of the one-particle irreducible vertex functions, the effective action[38]

$$\Gamma[\varphi] \equiv \sum_N^{\infty} \frac{1}{N!}\,\Gamma_{12\ldots N}\,\varphi_1\,\varphi_2\cdots\varphi_N \quad (9.80)$$

so that the one-particle irreducible vertices, or one-particle irreducible amputated Green's functions, are obtained by functional differentiation

$$\Gamma_{12\ldots N} = \left. \frac{\delta^N \Gamma[\varphi]}{\delta\varphi_1 \delta\varphi_2 \cdots \delta\varphi_N} \right|_{\varphi=0}. \quad (9.81)$$

Recall that the one-particle irreducible vertices, $\Gamma_{12\ldots N}$, by construction do not depend on the source, and the field is a function we can vary as it is a functional of the source which is at our disposal to vary.

We can then rewrite Eq. (9.79) as

$$\varphi_1 = G_{12}^{(0)} \left(J_2 + \frac{\delta\Gamma[\varphi]}{\delta\varphi_2} \right). \quad (9.82)$$

We introduce the diagrammatic notation depicted in Figure 9.50 for the effective action, the generator of 1PI-vertices.

$$\Gamma[\varphi] \quad = \qquad \varphi$$

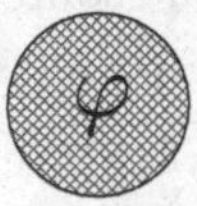

Figure 9.50 Diagrammatic notation for the effective action.

We introduce the diagrammatic notation for the functional derivative of the effective action depicted in Figure 9.51.

---

[38]The effective action is also referred to as the effective potential for the theory. We shall return to the reason for the terminology in Section 9.8. In the next chapter we develop the effective action approach, developing functional integral expressions for the effective action.

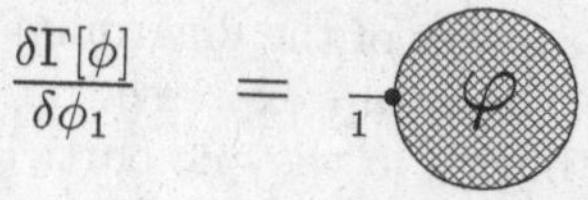

$$\frac{\delta\Gamma[\phi]}{\delta\phi_1} = \underset{1}{\bullet}\!\!\!\!-\!\!\!\bigcirc\!\!\varphi$$

Figure 9.51 Diagrammatic notation for the first derivative of the effective action.

The dot in Figure 9.51 signifies as usual a state label and the functional dependence on the field is made explicit. Similarly, diagrams containing additional dots represent additional functional derivatives with respect to the field, and give, upon setting the field equal to zero, $\varphi = 0$, the one-particle irreducible vertices depicted diagrammatically in Figure 9.49.

Operating on both sides of Eq. (9.79) with the inverse free propagator according to Eq. (9.5) thus gives

$$0 = J_1 + \Gamma_1 + (-G_0^{-1} + \Gamma)_{12}\,\varphi_2 + \frac{1}{2}\Gamma_{123}\,\varphi_2\,\varphi_3 + \frac{1}{3!}\Gamma_{1234}\,\varphi_2\,\varphi_3\,\varphi_4 + \dots \quad (9.83)$$

and we can rewrite Eq. (9.83) in the form (upon absorbing the inverse free propagator in the definition of the 2-state irreducible vertex $(-G_0^{-1} + \Gamma)_{12} \to \Gamma_{12}$):

$$0 = J_1 + \frac{\delta\Gamma[\varphi]}{\delta\varphi_1}. \quad (9.84)$$

Diagrammatically, Eq. (9.84) is represented as depicted in Figure 9.52.

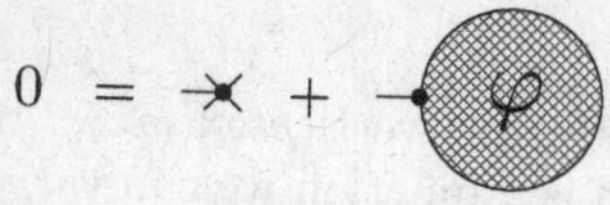

$$0 = \!\!-\!\!\times + \bullet\!\!-\!\!\bigcirc\!\!\varphi$$

Figure 9.52 Source and effective action relationship.

The content of Eq. (9.78) and Eq. (9.84) is that up to an overall constant, the effective action is the functional Legendre transform of the generator of connected Green's functions[39]

$$\Gamma[\varphi] = W[J] - J\,\varphi \quad (9.85)$$

and the Legendre transformation thus determines the overall value, $\Gamma[\varphi = 0]$.

We note that in the absence of the source, $J = 0$, Eq. (9.84) becomes[40]

$$\frac{\delta\Gamma[\varphi]}{\delta\varphi_1} = 0, \quad (9.86)$$

the effective action is stationary with respect to the field. This is an equation stating that the possible values of the field can be sought among the ones which make the effective action stationary.

---

[39] In equilibrium statistical mechanics, the effective action Γ is thus Gibbs potential or free energy, i.e. the (Helmholtz) potential or free energy in the presence of coupling to an external source, a J-reservoir.

[40] In the applications to non-equilibrium situations we consider in Chapter 12, this option is not available as part of the source is an external classical force, the classical force that drives the system out of equilibrium, and we shall employ Eq. (9.84).

9.6.1 Symmetry broken states

Having introduced the effective action according to Eq. (9.80) we are considering the normal state, i.e. we assume that the field vanishes in the absence of the source

$$\varphi_1 = \frac{\delta W[J]}{\delta J_1}\bigg|_{J=0} = 0. \tag{9.87}$$

These, however, are not the only type of states existing in nature, there exist states with spontaneously broken symmetry, i.e. states for which[41]

$$\varphi_1 = \frac{\delta W[J]}{\delta J_1}\bigg|_{J=0} \equiv \varphi_1^{cl} \neq 0. \tag{9.88}$$

We shall consider precisely such a situation and use the formalism presented in this chapter when we discuss Bose–Einstein condensation in Section 10.6. In Chapter 8 we already encountered the generic symmetry broken state, the superconducting state. It can be discussed as well in the present formalism by just allowing the field or order parameter to be a composite object. We discuss this case in Section 10.5 where we in addition to a one-particle source include a two-particle source.

For a symmetry broken state we shall define the effective action according to

$$\Gamma[\varphi] \equiv \sum_N^\infty \frac{1}{N!} \Gamma_{12\ldots N}[\varphi_1^{cl}] (\varphi_1 - \varphi_1^{cl})(\varphi_2 - \varphi_2^{cl}) \cdots (\varphi_N - \varphi_N^{cl}). \tag{9.89}$$

This means that according to Eq. (9.84) we again have

$$\frac{\delta \Gamma[\varphi]}{\delta \varphi_1}\bigg|_{\varphi = \varphi^{cl}} = -J_1. \tag{9.90}$$

The effective action vanishes for vanishing source.

By a shift of variables, $\varphi - \varphi^{cl} \to \varphi$, we can of course rewrite Eq. (9.89) as

$$\Gamma[\varphi] = \sum_N^\infty \frac{1}{N!} \Gamma_{12\ldots N}\, \varphi_1 \varphi_2 \cdots \varphi_N, \tag{9.91}$$

where now the vertices $\Gamma_{12\ldots N} = \Gamma_{12\ldots N}[\varphi_1^{cl} = 0]$ are evaluated in the normal or so-called disordered or symmetric state where the classical field vanishes, $\varphi_1^{cl} = 0$. We thus realize the fundamental importance of the effective action: it allows us to

---

[41]Such states are well-known in equilibrium statistical mechanics, for example from the existence of ferro-magnetism, the appearance below a definite critical temperature of an ordered state with a magnetization in a definite direction despite the rotational invariance of the Hamiltonian. These spontaneously broken symmetry states were first studied in the mean field approximation, the Landau theory, and the full theory of phase transitions, critical phenomena, were obtained by Wilson using field theoretic methods. Superfluid phases are broken symmetry states, and even more fundamentally, the masses of quarks are the result of the Higgs field having a nonzero value.

explore the existence of symmetry broken states by searching for extrema of the effective action, i.e. solutions of

$$\frac{\delta \Gamma[\varphi]}{\delta \varphi_1} = 0 \qquad\qquad (9.92)$$

for which the field is different from zero, $\varphi \neq 0$.

We shall also encounter symmetry broken states created by a simpler mechanism, viz. owing to the presence of an external classical field, but again the effective action approach shall prove useful for such non-equilibrium states, as we elaborate in Chapter 12.

9.6.2   Green's functions and one-particle irreducible vertices

In this section we shall show that since the generator of connected Green's functions and the effective action are related by a Legendre transform we can, by using the functional methods, easily obtain the systematic functional differential equations expressing connected Green's functions in terms of the one-particle irreducible vertex functions. But first let us argue for such equations at the purely diagrammatic level.

The connected 2-state amplitude or Green's function is expressed solely in terms of the 2-state one-particle irreducible vertex according to the diagrammatic expansion as depicted in Figure 9.53.

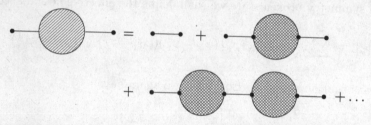

Figure 9.53 Self-energy representation of 2-state propagator.

The reason for this is, that any 2-state diagram is uniquely classified topologically according to whether it can not be cut in two or can be cut in two by cutting an internal particle line at only one place, or at two, three, etc., places. By construction we thus uniquely exhaust all the possible diagrams for the 2-state propagator. The 2-state one-particle irreducible vertex is also called the self-energy.

The diagrammatic equation in Figure 9.53 can be expressed in the form depicted in Figure 9.54, which is seen by iterating the equation in Figure 9.54.

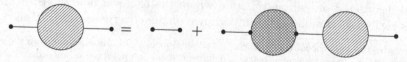

Figure 9.54 Dyson equation for the 2-state Green's function.

The diagrammatic equation depicted in Figure 9.54 corresponds analytically to the equation for the 2-state Green's function expressed in term of the 2-state irreducible vertex, the self-energy (recall we absorbed the inverse free propagator in Γ_{12}, $(-G_0^{-1} + \Gamma)_{12} \to \Gamma_{12}$, i.e. Σ_{12} denotes the one-particle irreducible 2-state vertex)

$$G_{12} = G_{12}^{(0)} + G_{13}^{(0)} \Sigma_{34} G_{42} \tag{9.93}$$

or equivalently the equation

$$G_{12} = G_{12}^{(0)} + G_{13} \Sigma_{34} G_{42}^{(0)} \tag{9.94}$$

by iterating from the other side. We have obtained the non-equilibrium Dyson equations.[42]

We now show how the Dyson equation can be obtained by using the effective action. More importantly, we show that we can use Eq. (9.84) to obtain the equation for the connected 2-state amplitude in the presence of the source where it is expressed in terms of the derivative of the effective action. This will lead to a simple method whereby all amplitudes can be expressed in terms of the 2-state connected amplitude, the full propagator, and one-particle irreducible vertices.

The Legendre transformation, according to Eq. (9.78), gives rise to the relation

$$\frac{\delta}{\delta J_1} = \frac{\delta \varphi_2}{\delta J_1} \frac{\delta}{\delta \varphi_2} = \frac{\delta^2 W[J]}{\delta J_1 \delta J_2} \frac{\delta}{\delta \varphi_2} = G_{12} \frac{\delta}{\delta \varphi_2} . \tag{9.95}$$

Taking the derivative of Eq. (9.84) with respect to the source then gives

$$-\frac{\delta^2 W[J]}{\delta J_2 \delta J_3} \frac{\delta^2 \Gamma[\varphi]}{\delta \varphi_3 \delta \varphi_1} = \delta_{12} . \tag{9.96}$$

Adding and subtracting in the effective action the so-called free term, $\Gamma[\varphi] \equiv -\frac{1}{2}\varphi_1 (G^{(0)})_{12}^{-1} \varphi_2 + \Gamma_i[\varphi]$, i.e. splitting off again the inverse propagator term we previously included in Γ_{12}, so that Γ_i now denotes the original effective action introduced in Eq. (9.80), provides the self-energy

$$\Sigma_{12} = \frac{\delta^2 \Gamma_i[\varphi]}{\delta \varphi_1 \delta \varphi_2} \tag{9.97}$$

in the presence of the source, as expressed through the field. Inserting into Eq. (9.96) gives

$$\frac{\delta^2 \Gamma[\varphi]}{\delta \varphi_1 \delta \varphi_2} = -(G^{(0)})_{12}^{-1} + \frac{\delta^2 \Gamma_i[\varphi]}{\delta \varphi_1 \delta \varphi_2} \tag{9.98}$$

and inserting into Eq. (9.96) gives

$$-\frac{\delta^2 W[J]}{\delta J_2 \delta J_1} \left(-(G^{(0)})_{21}^{-1} + \frac{\delta^2 \Gamma_i[\varphi]}{\delta \varphi_2 \delta \varphi_1} \right) = \delta_{12} . \tag{9.99}$$

---

[42] Recovering the non-equilibrium Dyson equations thus makes contact with quantum field theory studied by canonical means in the previous chapters. For the non-equilibrium states we studied in the previous chapters, we had valid approximate expressions for the self-energy, and did not need to go further into the diagrammatic structure of higher-order vertices.

Matrix multiplying by the bare propagator from the right gives[43]

$$\frac{\delta^2 W[J]}{\delta J_1 \delta J_2} = G_{12}^{(0)} + \frac{\delta^2 W[J]}{\delta J_1 \delta J_3}\frac{\delta^2 \Gamma_i[\varphi]}{\delta\varphi_3 \delta\varphi_4}G_{42}^{(0)},\tag{9.100}$$

which in terms of diagrams has the form depicted in Figure 9.55.

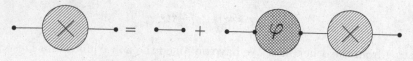

Figure 9.55 Dyson equation for the 2-state Green's function in the presence of the source.

Iterating the equation gives the full propagator

$$\frac{\delta^2 W[J]}{\delta J_1 \delta J_{1'}} = G_{11'}^{(0)} + G_{12}^{(0)}\frac{\delta^2\Gamma_i[\varphi]}{\delta\varphi_2\delta\varphi_3}G_{31'}^{(0)} + G_{12}^{(0)}\frac{\delta^2\Gamma_i[\varphi]}{\delta\varphi_2\delta\varphi_3}G_{34}^{(0)}\frac{\delta^2\Gamma_i[\varphi]}{\delta\varphi_4\delta\varphi_5}G_{51'}^{(0)} + \dots$$

$$(9.101)$$

the analog of the Dyson equation depicted in Figure 9.53, but now for the case where the source is present. The second derivative relationship between the generator of connected Green's functions and the effective action can compactly be rewritten suppressing the matrix indices, i.e. the two state labels occurring upon differentiation are now only indicated by the primes, in the form

$$W''[J] = \frac{1}{G_0^{-1} - \Gamma_i''[\varphi]}\tag{9.102}$$

as we recall the formula for a matrix X

$$\frac{1}{1-X} = 1 + X + X^2 + X^3 + \dots .\tag{9.103}$$

This is the relationship between the full propagator and the self-energy we arrived at earlier by topological classification of diagrams, expressing the connected 2-point Green's function in terms of the self-energy. Here we have constructed the functional analog in terms of a functional differential equation.

By taking further source derivatives of Eq. (9.96), we express the higher-order connected Green's functions in terms of the full propagator and the higher-order one-particle irreducible vertices.

Taking the derivative of Eq. (9.96) with respect to the source and using Eq. (9.84) gives

$$\frac{\delta^3 W[J]}{\delta J_1 \delta J_2 \delta J_3} = \frac{\delta^2 W[J]}{\delta J_1 \delta J_{1'}}\frac{\delta^2 W[J]}{\delta J_2 \delta J_{2'}}\frac{\delta^2 W[J]}{\delta J_3 \delta J_{3'}}\frac{\delta^3\Gamma[\varphi]}{\delta\varphi_{1'}\delta\varphi_{2'}\delta\varphi_{3'}}.\tag{9.104}$$

In terms of diagrams we have for Eq. (9.104) the relation depicted in Figure 9.56.

---

[43]This equation is of course immediately recognized as the Dyson equation, Eq. (4.141), $G_{12} = G_{12}^{(0)} + G_{13}\Sigma_{34}G_{42}^{(0)}$.

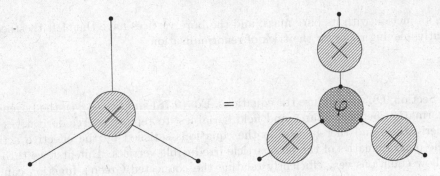

Figure 9.56 Connected 3-state diagram expressed by the 1P-irreducible 3-vertex.

Exercise 9.10. Show by taking further source derivatives of Eq. (9.96) that the equation obtained for the 4-state connected Green's function has the diagrammatic form (for a theory with 3- and 4-connector vertices) depicted in Figure 9.57.

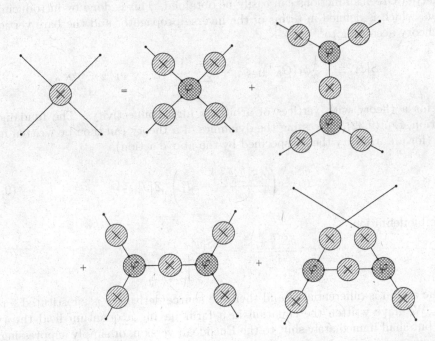

Figure 9.57 Connected 4-state diagram expressed by 1P-irreducible vertices.

If in the above equation we set the source to zero, and thereby the field to zero, instead of encountering quantities depending on the source and field, we will obtain expressions for the connected Green's functions in term of the full 2-point Green's function and the irreducible vertex functions. Since the full 2-point Green's function is the one into which we can feed our phenomenological knowledge of the mass of a particle, these equations are basic for the renormalization procedure. The bare

Green's function with its bare mass, and the bare vertices have thus left the theory explicitly, leaving room for the trick of renormalization.

In Section 9.8, we shall use the equations, Eq. (9.78) and Eq. (9.84), the Legendre transformation between source and field variables, to replace source-derivatives by field-derivatives and thereby obtain the equations satisfied by the effective action and the diagrammatics of the one-particle irreducible vertices. But first we turn to show how equations very efficiently relating the connected Green's function can be generated.

9.7   Diagrammatics and action

In this section we show how the fundamental differential equation for the dynamics, Eq. (9.32), can be turned into an equation from which the relationships between the connected Green's functions can easily be obtained. This is done by introducing the action, which is defined in terms of the inverse propagator and the bare vertices of the theory according to[44]

$$S[\phi] \equiv -\frac{1}{2}\phi_1 (G_0^{-1})_{12}\phi_2 + \sum_N \frac{1}{N!}\, g_{12...N}\, \phi_1\phi_2\cdots\phi_N \,, \qquad (9.105)$$

here for a theory with vertices of arbitrary high connectivity. The fundamental equation, Eq. (9.32), expressing the dynamics of a theory can then be written in the form (for an arbitrary theory specified by the above action)

$$0 = \left(\frac{\delta S[\frac{\hbar}{i}\frac{\delta}{\delta J}]}{\delta \phi_1} + J_1 \right) Z[J] \,, \qquad (9.106)$$

where by definition

$$\frac{\delta S[\frac{\hbar}{i}\frac{\delta}{\delta J}]}{\delta \phi_1} = \frac{\delta S[\phi]}{\delta \phi_1}\Bigg|_{\phi \to \frac{\hbar}{i}\frac{\delta}{\delta J}} \qquad (9.107)$$

i.e. the action is differentiated and then the source derivative is substituted for the field. We have written the equation in a form having a quantum field theory in mind but shall immediately shift to the Euclidean version, or simply suppressing the appearance of $\hbar/i$ by absorbing the factor in the source derivative.

---

[44]At this junction in the generating functional formulation of a quantum field theory the solemnity of the *action* is scarcely noticed, but as just another formal construction. In the next chapter we show how the action in the functional integral formulation of a quantum field theory naturally appears as the fundamental quantity describing the dynamics. The action can also be given a fundamental status in the operator formulation of the generating functional technique (recall Section 3.3), if the dynamics is based on Schwinger's quantum action principle [50]. However, the point of the presentation in this chapter is to base the dynamics directly on diagrams and then by simple topological arguments construct the generating functional technique.

Since the generator of connected diagrams, W, is the logarithm of Z, we have the relation valid for an arbitrary functional F

$$\frac{1}{Z[J]} \frac{\delta}{\delta J_1} (Z[J] F[J]) = \left(\frac{\delta W[J]}{\delta J_1} + \frac{\delta}{\delta J_1} \right) F[J] \qquad (9.108)$$

and by repetition

$$\frac{\delta^N}{\delta J_1 \cdots \delta J_N} Z[J] = Z[J] \left(\frac{\delta W[J]}{\delta J_1} + \frac{\delta}{\delta J_1} \right) \cdots \left(\frac{\delta W[J]}{\delta J_N} + \frac{\delta}{\delta J_N} \right), \qquad (9.109)$$

where operator notation has been used, i.e. the operations are supposed to operate on a functional F.

Since the action is a sum of polynomials we have according to Eq. (9.109)

$$\frac{\delta S[\frac{\delta}{\delta J}]}{\delta \phi_1} Z[J] = Z[J] \frac{\delta S[\frac{\delta W}{\delta J} + \frac{\delta}{\delta J}]}{\delta \phi_1}. \qquad (9.110)$$

The fundamental equation, Eq. (9.106), can thus be written in the form

$$0 = \frac{\delta S[\frac{\delta W}{\delta J} + \frac{\delta}{\delta J}]}{\delta \phi_1} + J_1 . \qquad (9.111)$$

Using the explicit form of the action for an arbitrary theory we have

$$\frac{\delta S[\frac{\delta W}{\delta J} + \frac{\delta}{\delta J}]}{\delta \phi_1} = -(G^{(0)})^{-1}_{12} \left(\frac{\delta W}{\delta J_2} + \frac{\delta}{\delta J_2} \right)$$

$$+ \sum_N \frac{1}{(N-1)!} g_{12 \ldots N} \left(\frac{\delta W}{\delta J_2} + \frac{\delta}{\delta J_2} \right) \cdots \left(\frac{\delta W}{\delta J_N} + \frac{\delta}{\delta J_N} \right)$$

$$(9.112)$$

and using Eq. (9.111) and performing the differentiations and lastly multiply by the bare propagator we immediately recover Eq. (9.77) (for the 3- plus 4-vertex theory).

Having the fundamental equation on the form specified in Eq. (9.111) turns out in practice to be very useful for generating the relations between the connected full Green's functions, and exemplifies the expediency and powerfulness of the generating functional formalism.

9.8 Effective action and skeleton diagrams

In this section, we shall use the equations, Eq. (9.78) and Eq. (9.84), the Legendre transformation between source and field variables, to replace source-derivatives by field-derivatives and thereby obtain the equations obeyed by the effective action. Upon setting the field to zero, $\varphi = 0$, we then obtain the skeleton diagrammatic equations satisfied by the one-particle irreducible vertices. Instead of using topological diagrammatic arguments to obtain the skeleton diagrammatics, we turn to use the generating functional method to achieve the same goal.

On the right side in Eq. (9.112) we can introduce the average field and obtain

$$
\frac{\delta S[\frac{\delta W}{\delta J} + \frac{\delta}{\delta J}]}{\delta \phi_1} = -(G^{(0)})_{12}^{-1} \left(\varphi_2 + \frac{\delta^2 W}{\delta J_2 \delta J_{2'}} \frac{\delta}{\delta \varphi_{2'}} \right)
$$

$$
+ \sum_N \frac{1}{(N-1)!} \, g_{12\ldots N} \left(\varphi_2 + \frac{\delta^2 W}{\delta J_2 \delta J_{2'}} \frac{\delta}{\delta \varphi_{2'}} \right) \cdots \left(\varphi_N + \frac{\delta^2 W}{\delta J_N \delta J_{N'}} \frac{\delta}{\delta \varphi_{N'}} \right) ,
$$

(9.113)

where we in addition have used Eq. (9.95) to substitute the field derivative for the source derivative.

Inserting Eq. (9.111) into Eq. (9.84) and using Eq. (9.78) thus gives the relation between the action and the effective action

$$
\frac{\delta \Gamma[\varphi]}{\delta \varphi_1} = -\frac{\delta S\left[\varphi + W''[J] \frac{\delta}{\delta \varphi}\right]}{\delta \varphi_1} ,
$$

(9.114)

where the right-hand side is short for the right-hand side in Eq. (9.113).

For a 3- plus 4-vertex theory we obtain

$$
\frac{\delta \Gamma[\varphi]}{\delta \varphi_1} = -(G_0^{-1})_{12}\, \varphi_2 + \frac{1}{2} g_{123}\, \varphi_2\, \varphi_3 + \frac{1}{2} g_{123}\, \frac{\delta^2 W[J]}{\delta J_2 \delta J_3}
$$

$$
+ \frac{1}{3!} g_{1234}\, \varphi_2\, \varphi_3\, \varphi_4 + \frac{3}{3!} g_{1234}\, \varphi_4\, \frac{\delta^2 W[J]}{\delta J_2 \delta J_3}
$$

$$
+ \frac{1}{3!} g_{1234}\, \frac{\delta^2 W[J]}{\delta J_2 \delta J_5} \frac{\delta^2 W[J]}{\delta J_3 \delta J_6} \frac{\delta^2 W[J]}{\delta J_4 \delta J_7} \frac{\delta^3 \Gamma[\varphi]}{\delta \varphi_5 \delta \varphi_6 \delta \varphi_7}
$$

(9.115)

as the last term emerges upon noting

$$
\frac{\delta}{\delta \varphi_{2'}} \frac{\delta^2 W[J]}{\delta J_3 \delta J_4} = \frac{\delta J_{2''}}{\delta \varphi_{2'}} \frac{\delta^3 W[J]}{\delta J_{2''} \delta J_3 \delta J_4} = \frac{\delta^2 \Gamma[\varphi]}{\varphi_{2'} \varphi_{2''}} \frac{\delta^3 W[J]}{\delta J_{2''} \delta J_3 \delta J_4}
$$

$$
= \frac{\delta^2 W[J]}{\delta J_3 \delta J_5} \frac{\delta^2 W[J]}{\delta J_4 \delta J_6} \frac{\delta^3 \Gamma[\varphi]}{\delta \varphi_{2''} \delta \varphi_5 \delta \varphi_6} ,
$$

(9.116)

where in obtaining the last equality we have used Eq. (9.96) and Eq. (9.104). The relationship expressed in Eq. (9.116) has the diagrammatic representation depicted in Figure 9.58.

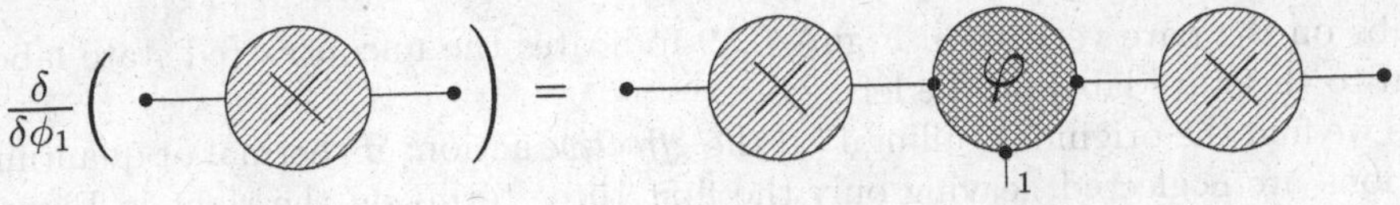

$$\frac{\delta}{\delta\phi_1}\left(\text{—⊗—} \right) = \text{—⊗—φ—⊗—}$$

Figure 9.58 Average field dependence of the propagator.

The implicit dependence of the propagator on the average field, through the source, is thus such that taking the derivative inserts a one-particle irreducible vertex in accordance with the relation depicted in Figure 9.58.

The equation for the first derivative of the effective action, Eq. (9.115), has for a 3- plus 4-vertex theory the diagrammatic representation depicted in Figure 9.59.

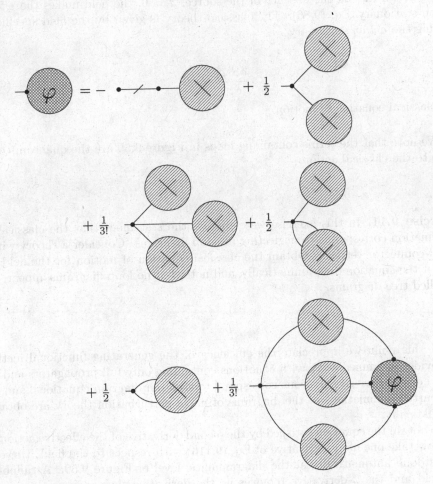

Figure 9.59 Diagrammatic relation for the first derivative of the effective action for a 3- plus 4-vertex theory.

The stubs on the bare vertices in Figure 9.59 indicates the uncontracted state label identical to the state label on the left.

Here we find the origin for calling $\Gamma[\varphi]$ the *effective* action: if thermal or quantum fluctuations are neglected, leaving only the first three terms on the right in Figure 9.59, the (derivative of the) *effective* action reduces to the (derivative of the) action. This corresponds to dropping the W''-terms in Eq. (9.114). In other words, the exact equation of motion for the field, Eq. (9.84), can be obtained from the equation determining the classical field (where S is the action given in Eq. (10.37))

$$0 = J_1 + \frac{\delta S[\phi]}{\delta \phi_1} \tag{9.117}$$

by *substituting* the one-particle irreducible vertices for the bare vertices in the action S. We recall that in the absence of the source, $J = 0$, the field makes the *effective* action stationary, Eq. (9.86). The classical theory is given by the field specified by making the *action* stationary

$$\frac{\delta S[\varphi]}{\delta \varphi_1} = 0 \tag{9.118}$$

the classical equation of motion.

We note that the terms containing loops in Figure 9.59 are the quantum corrections to the classical action.

Exercise 9.11. In this exercise we elaborate the statement that the classical approximation corresponds to neglecting all loop diagrams. Consider a theory with 3- and 4-connector vertices. Obtain the classical equation of motion for the field. Interpret the equation diagrammatically, and note that no loop diagrams appear, only so-called tree diagrams.

At this point we appreciate the efficiency of the generating functional method: it provides us immediately with equations containing only full propagators and vertices, i.e. the derived equations correspond to skeleton diagram equations, and infinite partial summations of the diagrams of naive perturbation theory are obtained automatically.

To obtain the equation satisfied by the second derivative of the effective action, we can now take one more derivative of Eq. (9.116) with respect to the field. However, this is done automatically at the diagrammatic level of Figure 9.59. A tadpole is the field and the φ-derivative removes it; the derivative thus reduces the number of tadpoles present by one. For the field dependence of the propagator we use the relation depicted in Figure 9.58. For the second derivative of the effective action, we thus find that it satisfies the equation depicted in Figure 9.60.

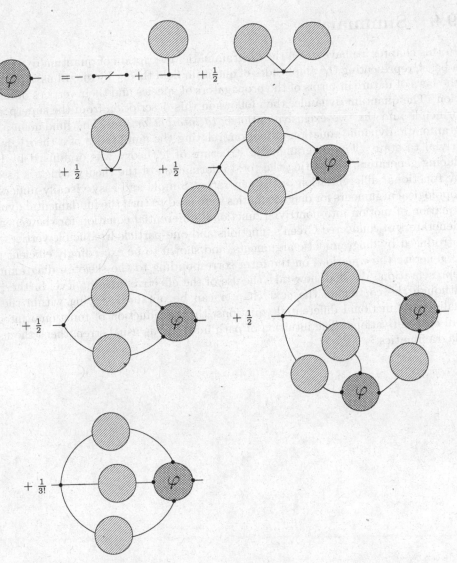

Figure 9.60 Diagrammatic relation for the second derivative of the effective action.

Taking further derivatives, we obtain the equations satisfied by the higher derivatives of the effective action, and upon setting the field to zero, $\varphi = 0$, we obtain the skeleton diagrammatics for the one-particle irreducible vertices.

In the next chapter we shall study the effective action formalism in detail, and give a functional integral evaluation which gives an interpretation of the effective action in terms of vacuum diagrams.

9.9   Summary

In this chapter we have taken the diagrammatic description of quantum dynamics as a basis, representing the amplitudes of quantum field theory by diagrams, and stating the laws of nature in terms of the propagators of species and their vertices of interaction. The quantum dynamics then follows in this description from the superposition principle and the two exclusive options: *to interact or not*. The fundamental diagrammatic dynamic equation of motion, relating the amplitudes of a theory, is then trivial to state. The diagrammatic structure of a theory was organized by introducing generators, encrypting the total information of the theory which is assessed by functional differentiation of the generator. Simple and easy visually understood topological arguments for diagrammatics were used to turn the fundamental dynamic equation of motion into nontrivial functional differential equation for the generator. Generators of connected Green's functions and one-particle irreducible vertices were introduced by diagrammatic arguments, and shown to be exceedingly efficient tools to generate the equations on the form corresponding to the skeleton diagrammatic representation. We shall now take the use of the effective action a level further, and although the content of the next chapter can be obtained staying within the formalism of functional differential equations, the introduction of functional integrals will ease derivations. The intuition of path integrals as usual strengthens the use of diagrammatics.

10

Effective action

In the previous chapter we introduced the one-particle irreducible effective action by collecting the one-particle irreducible vertex functions into a generator whose argument is the field, the one-state amplitude in the presence of the source. The effective action thus generates the one-particle irreducible amputated Green's functions. We shall now enhance the usability of the non-equilibrium effective action by establishing its relationship to the sum of all one-particle irreducible vacuum diagrams. To facilitate this it is convenient to add the final mathematical tool to the arsenal of functional methods, viz. functional integration or path integrals over field configurations. We are then following Feynman and instead of describing the field theory in terms of differential equations, we get its corresponding representation in terms of functional or path integrals. This analytical condensed technique shall prove powerful when unraveling the content of a field theory. The loop expansion of the non-equilibrium effective action is developed, and taken one step further as we introduce the *two-particle* irreducible effective action valid for non-equilibrium states. As an application of the effective action approach, we consider a dilute Bose gas and a trapped Bose–Einstein condensate.

10.1 Functional integration

Functional differentiation has its integral counterpart in functional integration. We shall construct an integration over *functions* and not just numbers as in elementary integration of a function. We approach this infinite-dimensional kind of integration with care (or, from a mathematical point of view, carelessly), i.e. we base it on our usual integration with respect to a single variable and take it to a limit. To deal with any function, $\varphi(\mathbf{x}, t)$, of continuous variables such as space-time, $(\mathbf{x}, t)$, the continuous variables must be discretized, i.e. space-time is divided into a set of small volumes of size Δ covering all or the relevant part of space-time, and the value of the function φ is specified in each such small volume or equivalently on the corresponding mesh of N lattice sites, φ_M, $M = 1, 2, \ldots, N$. This is immediately incorporated into our condensed state label notation

$$1 \equiv (s_1, \mathbf{x}_1, t_1, \sigma_1, \ldots) \tag{10.1}$$

if the space and time variables are now interpreted as discrete. To treat arbitrary non-equilibrium states, a real-time dynamical or Schwinger–Keldysh index is included or the time variable is replaced by the contour time variable for treating general non-equilibrium situations. We shall first consider a real scalar field, and in each cell the field can then take on any real value.

The functional integral of a functional, $F[\varphi]$, of a real function φ, is then defined as the limit[1]

$$\int \mathcal{D}\varphi \, F[\varphi] \equiv \lim_{N \to \infty} \int_{-\infty}^{\infty} \prod_{M=1}^{N} d\varphi_M \, F(\varphi_1, \ldots, \varphi_N) \,. \tag{10.2}$$

The functional integral is a sum over all field configurations.

Shifting each of the integration variables a constant amount, $\varphi_M \to \varphi_M + \varphi_M^{(0)}$, leaves the integrations invariant, and we have the property of a functional integral

$$\int \mathcal{D}\varphi \, F[\varphi] \equiv \int \mathcal{D}\varphi \, F[\varphi + \varphi_0] \,. \tag{10.3}$$

Quantum field theory describes a system with infinitely many degrees of freedom and the functional integral is the infinite dimensional version of the path integral formulation of quantum mechanics, the zero-dimensional quantum field theory, which is discussed in Appendix A.

10.1.1   Functional Fourier transformation

The main functional integral tool will be that of functional Fourier transformation, and to obtain that we recall that usual Fourier transformation of functions is equivalent to the integral representation of Dirac's delta function in terms of the exponential function.[2]

The delta functional, i.e. the functional δ satisfying for any functional F

$$F[J^{(1)}] = \int \mathcal{D}J^{(2)} \, F[J^{(2)}] \, \delta[J^{(1)} - J^{(2)}] \,, \tag{10.4}$$

is construed as a product of delta functions over all the cells, and is constructed as the limit of a product of delta functions, each of which can be represented in terms of its usual integral expression

$$\left(\frac{2\pi}{\Delta}\right)^N \prod_{M=1}^{N} \delta(J_M^{(1)} - J_M^{(2)}) = \int_{-\infty}^{\infty} \prod_{M=1}^{N} d\varphi_M \, e^{i\Delta \, \varphi_M \, (J_M^{(1)} - J_M^{(2)})} \,. \tag{10.5}$$

---

[1]The functional integral over a complex function, a complex field, is defined analogously, involving integration over the real and imaginary parts of the field.

[2]For a discussion of Dirac's delta function and Fourier transformation we refer to Appendix A of reference [1].

For the integration over space and contour time we introduce the notation

$$\varphi J \equiv \int d\mathbf{x} dt \, \varphi(\mathbf{x}, t) \, J(\mathbf{x}, t) = \lim_{N \to \infty} \sum_{M=1}^{N} \Delta \, \varphi_M J_M \,. \tag{10.6}$$

We thus obtain the following functional integral representation of the delta functional

$$\delta[J^{(1)} - J^{(2)}] = \int \mathcal{D}\varphi \, e^{i\varphi(J^{(1)} - J^{(2)})} \,, \tag{10.7}$$

where the normalization factor $\lim_{N \to \infty} (2\pi/\Delta)^N$ has been incorporated into the definition of the functional integral. The delta functional expresses according Eq. (10.4) the identity of two *functions*, i.e. the equality of the two for any value of their argument.

Having the integral representation of the delta functional at hand, Eq. (10.7), we immediately have for the functional Fourier transformation

$$F[\varphi] = \int \mathcal{D}J \, e^{-i\varphi J} F[J] \tag{10.8}$$

the inverse relation

$$F[J] = \int \mathcal{D}\varphi \, e^{iJ\varphi} F[\varphi] \,. \tag{10.9}$$

Functional Fourier transformation is thus the product of ordinary Fourier transforms over each cell.

The mathematical job performed by functional Fourier transformation is, just as in usual Fourier transformation, to change, now *functional*, differential equations into algebraic equations. As far as physics is concerned, the functional integral provides an explicit interpretation, in terms of the superposition principle, of the dynamics of quantum fields, viz. as a sum over all intermediate field configurations leading from an initial to a final state of the field, quite analogous to the path integral in quantum mechanics, the zero-dimensional quantum field theory, as discussed in Appendix A.

10.1.2 Gaussian integrals

The mathematics of quantum mechanics of a single particle resides in the one-dimensional Gaussian integral

$$I(a) = \int\limits_{-\infty}^{\infty} dx \, e^{-\frac{1}{2}ax^2} = \sqrt{\frac{2\pi}{a}} \tag{10.10}$$

or by completing the square

$$\int\limits_{-\infty}^{\infty} dx \, e^{-\frac{1}{2}ax^2 \pm bx} = \sqrt{\frac{2\pi}{a}} \, e^{\frac{b^2}{2a}} \,, \tag{10.11}$$

where the integral is convergent whenever a is not a negative real number, i.e. $I(a)$ is analytic in the complex a-plane except at the branch cut specified by that of the square root. This message holds true for the functional integrals of quantum field theory.

The functional integral is treated as the limit of a multi-dimensional integral and we consider the N-dimensional Gaussian integral

$$I(A;b) = \int\limits_{-\infty}^{\infty} dx_1 \ldots dx_N \, e^{-C(x_1,\ldots,x_N)} \tag{10.12}$$

specified by the quadratic form

$$C(x) = \frac{1}{2} \sum_{M,M'=1}^{N} x_M A_{M,M'} x_{M'} \pm \sum_{M}^{N} b_M x_M = \frac{1}{2} x^T A x \pm b^T x . \tag{10.13}$$

Here x^T denotes the row tuple $x^T = (x_1,\ldots,x_N)$, and x the corresponding column tuple, and similar notation for the N-tuple b. We assume that the matrix A is real, symmetric, $A^T = A$, and positive, so that it can be diagonalized by an orthogonal matrix S, $S^{-1} = S^T$, and $D = S^T A S$ has then only positive diagonal entries d_M. The Jacobian, $|\det S|$, for the transformation $x = Sy$ is thus one, and the integral becomes the elementary integral, Eq. (10.11), occurring N-fold times,

$$
\begin{aligned}
I(A;b) &= \prod_{M=1}^{N} \int\limits_{-\infty}^{\infty} dy_M \, e^{-\frac{1}{2} d_M y_M^2 \pm y_M (S^T b)_M} = \prod_{M=1}^{N} \sqrt{\frac{2\pi}{d_M}} \, e^{\frac{1}{2d_M}(S^T b)_M^2} \\
&= \sqrt{\det(2\pi D^{-1})} \, e^{\sum_{M=1}^{N} \frac{1}{2d_M}(S^T b)_M^2} .
\end{aligned}
\tag{10.14}
$$

Using $(S^T A S)^{-1} = D^{-1}$ to express $A^{-1} = S D^{-1} S^T$, or in terms of matrix elements $(A^{-1})_{MM'} = \sum_{M_1} S_{MM_1} \frac{1}{d_{M_1}} (S^T)_{M_1 M'}$, and using that $\det A = \det D$, we arrive at the expression for the multi-dimensional Gaussian integral

$$I(A;b) = \left(\det\left(\frac{A}{2\pi} \right) \right)^{-1/2} e^{\frac{1}{2} b^T A^{-1} b} . \tag{10.15}$$

Again the result can be generalized by analytical continuation to the case of a complex symmetric matrix, A, with a positive real part, the branch cut in the complex parameter space being specified by the square root of the determinant.

The Gaussian functional integral is then perceived in the limiting sense of Eq. (10.2) and we have[3]

$$\int \mathcal{D}\varphi \, e^{\frac{i}{2} \varphi A \varphi} = \frac{1}{\sqrt{\mathrm{Det} A}} , \tag{10.16}$$

---

[3]Here we have included extra factors in the definition of the path integral, viz. a factor $1/\sqrt{2\pi i}$ for each integration $d\varphi_M$, explaining the absence of $-i$ and 2π in front of A on the right-hand side. The imaginary unit and 2π can thus be shuffled around.

where the limiting procedure introduces the meaning of the functional determinant distinguished by a capital D in Det. Using the identity $\ln \det A = \operatorname{Tr} \ln A$ we have[4]

$$\int \!\! D\varphi \, e^{\frac{i}{2}\varphi A \varphi} = e^{-\frac{1}{2}\operatorname{Tr}\ln A} . \tag{10.17}$$

Similarly, we obtain from the above analysis

$$\int \!\! D\varphi \, e^{\frac{i}{2}\varphi A \varphi + i\varphi J} = e^{-\frac{1}{2}\operatorname{Tr}\ln A} \, e^{-\frac{i}{2}J A^{-1} J} \tag{10.18}$$

or by in the Gaussian integral, Eq. (10.16), shifting the variable, $\varphi \to \varphi + A^{-1}J$.

The generating functional for the free theory, Eq. (9.35), can thus be expressed in terms of a functional integral

$$Z_0[J] = = e^{\frac{1}{2}\operatorname{Tr}\ln G_0^{-1}} \int \!\! D\varphi \, e^{-\frac{i}{2}\varphi G_0^{-1} \varphi + i\varphi J} . \tag{10.19}$$

We have thus made the first connection between functional integrals and the generating functional and thereby to diagrammatics. In the treatment of non-equilibrium states in the real-time technique, the real-time representation in the form Eq. (5.1) or the more economical symmetric representation of the bare propagator should thus be used

$$G_0 = \begin{pmatrix} 0 & G_0^{\mathrm{A}} \\ G_0^{\mathrm{R}} & G_0^{\mathrm{K}} \end{pmatrix} . \tag{10.20}$$

in order to have a symmetric inverse propagator as demanded for the functional integral to be well-defined.

The functional

$$S_0[\varphi] = -\frac{1}{2}\varphi \, G_0^{-1} \, \varphi \tag{10.21}$$

is called the free action, or action for the free theory.[5]

The normalization constant, guaranteeing the normalization of the generator $Z_0[J=0] = 1$, is often left implicit as overall constants of functional integrals have

---

[4]The identity is obvious for a diagonal matrix, and therefore for a diagonalizable matrix which is the case of interest here. The identity follows generally from the product expansion of the exponential function, $\det e^A = \det \lim_{n\to\infty}(I + A/n)^n = \lim_{n\to\infty}(\det(I + A/n))^n = \lim_{n\to\infty}(1 + \operatorname{Tr}A/n) + \mathcal{O}(1/n^2))^n = e^{\operatorname{Tr}A}$. Or, by changing the parameters in a matrix gives for the variation $\ln\det(A+\Delta A) - \ln\det A = \ln\det(I + A^{-1}\Delta A) = \ln(1 + \operatorname{Tr}(A^{-1}\Delta A) + \mathcal{O}((A^{-1}\Delta A)^2)) = \ln(\operatorname{Tr}(A^{-1}\Delta A)) + \mathcal{O}((A^{-1}\Delta A)^2)$, and thereby the sought relation as the overall constant not determined by the variation of the function is fixed by considering the identity matrix as $\det I = 1$ and $\ln I = (\ln(I - (I - I)) = -\sum_{n=1}^{\infty}(I - I)^n/n = 0I$. In connection with functional integrals we thus encounter infinite products, the functional determinant, a highly divergent object, but happily such overall constants have no physical significance.

[5]A convergence factor in the exponent, $-\epsilon\phi^2$, for security, can be assumed absorbed in the inverse free propagator.

no bearing on the physics they describe, resulting in[6]

$$Z_0[J] = = \int \mathcal{D}\varphi \, e^{iS_0[\varphi] + i\varphi J} .$$

(10.22)

Since our interest is the real-time treatment of non-equilibrium situations, the closed time path guarantees the even stronger normalization condition of the generator, $Z_0[J] = 1$, provided that the sources on the two parts of the closed time path are taken to be identical, such as for example is the case for coupling to an external classical field.

To treat functional integration over a complex function, we first consider integration over the real and imaginary parts of an N-tuple with complex entries and have, for the multiple Gaussian integral,

$$\int dz^\dagger dz \, e^{-\frac{1}{2}z^\dagger A z} = \left(\det \left(\frac{A}{2\pi} \right) \right)^{-1} ,$$

(10.23)

where $\dagger$ denotes in addition to transposition complex conjugates, i.e. hermitian conjugation. We note the additional square root power of the determinant in comparison with the Gaussian integral over real variables, Eq. (10.15).

For the case of a complex function $\psi(\mathbf{x}, t)$, the functional integral becomes

$$\int \mathcal{D}\psi^*(\mathbf{x}, t)\mathcal{D}\psi(\mathbf{x}, t) \, F[\psi^*(\mathbf{x}, t), \psi(\mathbf{x}, t)] = \lim_{N, N' \to \infty} \int_{-\infty}^{\infty} \prod_{M, M'=1}^{N, N'} d\psi_M^* d\psi_{M'} \, F[\psi_1^*, \dots, \psi_N]$$

(10.24)

and for the Gaussian integral (shuffling again irrelevant constants)

$$\int \mathcal{D}\psi^*(\mathbf{x}, t) \, \mathcal{D}\psi(\mathbf{x}, t) \, e^{-\frac{i}{2}\psi^* A^{-1} \psi} = \frac{1}{\mathrm{Det} A}$$

(10.25)

where A is a hermitian and positive definite matrix.

Just as for zero-dimensional quantum field theory, i.e. quantum mechanics, where path integrals allow us to write down the solution of the Schrödinger equation in explicit form, so functional integration allows us to write down explicitly the solution of the functional differential equation specifying a quantum field theory as considered in Section 9.2.2. In Section 10.2, we show how this is done by introducing the concept of action and show how it can be used to get a useful functional integral representation of the full theory. But first functional integration over Grassmann variables is introduced in order to cope with fermions.

---

[6]The normalization of the free generator is in the canonical or operator formalism of equilibrium zero temperature quantum field theory the statement that for a quadratic action the addition of the coupling to the source does not produce a transition from the vacuum state.

10.1.3 Fermionic path integrals

To treat a fermionic field theory in terms of path integrals, we shall need to introduce integration over anti-commuting objects. The most general function of a Grassmann variable, η, is (recall Section 9.4) the monomial

$$f(\eta) = c_0 + c_1\eta \tag{10.26}$$

and integration with respect to a Grassmann variable is *defined* as the linear operation

$$\int d\eta \, f(\eta) = c_1 \tag{10.27}$$

or

$$\int d\eta \, 1 = 0 \tag{10.28}$$

and

$$\int d\eta \, \eta = 1 . \tag{10.29}$$

Integration with respect to a Grassmann variable, Berezin integration, is thus identical to differentiation.

We note that the basic formula of integration, that the integral of a total differential vanishes,

$$\int d\eta \, \frac{df(\eta)}{d\eta} = 0 , \tag{10.30}$$

also holds for Berezin integration as $d\eta/d\eta = 1$. The equivalent is true for the conjugate Grassmann variable η^* (recall Section 9.4).

For a general function of two conjugate Grassmann variables, Eq. (9.45), we then have according to the definitions of integration over Grassmann variables

$$- \int d\eta \, d\eta^* \, f(\eta, \eta^*) = \int d\eta^* d\eta \, f(\eta, \eta^*) = c_3 . \tag{10.31}$$

For the basic Gaussian integral for Grassmann fields we have

$$\int d\eta^* d\eta \, e^{i\eta^* A\eta} = (\text{Det} iA) \tag{10.32}$$

as after transforming to diagonal form

$$\int \prod_M d\eta_M^* \, d\eta_M \, e^{i\sum_M \eta_M^* A_{MM}\eta_M} = \int \prod_M d\eta_M^* \, d\eta_M \left(1 + i\sum_M \eta_M^* A_{MM}\eta_M\right)$$

$$= \prod_M iA_{MM} = \text{Det}(iA) , \tag{10.33}$$

where the first equality sign follows from the property $(\eta^*\eta)^2 = 0$ for anti-commuting numbers (recall Section 9.4), and the second equality sign follows from the definition of integration with respect to Grassmann variables. Thus the Gaussian integral over Grassmann variables gives the inverse determinant in comparison with the case of complex functions.[7]

10.2 Generators as functional integrals

In the previous chapter we showed how all the diagrammatics of a theory, non-equilibrium situations included, could be captured in a generating functional, expressing the whole theory in terms of a single differential equation. The Green's functions were obtained by differentiating the generating function, thereby obtaining the equations of motion for all the Green's functions. We now introduce the functional integral expression for the generating functional, thereby obtaining explicit integral representations for the Green's functions, i.e. explicit solutions of the functional differential equations. Needless to say, only the Gaussian integral can be evaluated, and in practice we are back to perturbation theory and diagrams. But the path integral has its particular benefits as we shall explore in this chapter, and is very useful when it comes to exploit the symmetry of a theory.

We now turn to obtain the functional integral expression for the generating functional for the case where interactions are present. Operating with the inverse bare propagator on the fundamental equation for the dynamics, Eq. (9.32), we get according to Eq. (9.5) the functional differential equation

$$(G_0^{-1})_{12}\,\frac{1}{i}\,\frac{\delta Z[J]}{\delta J_2} = \left(\sum_N \frac{1}{(N-1)!}\, g_{12\dots N}\,\left(\frac{1}{i}\right)^{N-1}\,\frac{\delta^{N-1}}{\delta J_N \cdots \delta J_3 \delta J_2} + J_1 \right) Z[J]$$

(10.34)

where we consider a theory with an arbitrary number of vertices.[8]

We introduce the Fourier functional integral representation of the generating functional

$$Z[J] = \int \mathcal{D}\phi\, Z[\phi]\, e^{i\phi J},$$

(10.35)

where for the dummy functional integration variable we use the notation ϕ to distinguish it from the average field considered in the previous chapter for which we used the notation φ.

---

[7]This is the trick behind the use of supersymmetry methods to avoid the denominator problem in the study of quenched disorder [51]. However, the supersymmetry trick has the disadvantage of not being able to cope with the case of interactions. Anyway, we have confessed our preference to avoid the denominator problem by using the real-time technique.

[8]In Eq. (10.34) we performed the shift $\delta/\delta J \to \delta/i\delta J$ for proper quantum field theory notation as dictated by the functional Fourier transform. Details of the transition between Euclidean and Minkowski (contour-time) field theories are stated in the next section.

The functional Fourier transformation turns the fundamental dynamic equation into the form[9]

$$-i\frac{\delta Z[\phi]}{\delta\phi_1} = \left(-(G_0^{-1})_{12}\,\phi_2 + \sum_N \frac{1}{(N-1)!}\,g_{12...N}\,\phi_2\cdots\phi_N\right)Z[\phi]\,. \qquad (10.36)$$

The term on the left originates from the term $J_1 Z[J]$, and results from a functional partial integration.

We refer to ϕ also as the *field*, and it starts out as just a dummy functional integration variable as introduced in Eq. (10.35), but immediately got a life to itself, Eq. (10.36), through the dynamics of the theory.

We then introduce the action (this at a proper place, but recall also Section 9.7)

$$S[\phi] \equiv -\frac{1}{2}\phi_1(G_0^{-1})_{12}\phi_2 + \sum_N \frac{1}{N!}\,g_{12...N}\,\phi_1\phi_2\cdots\phi_N \qquad (10.37)$$

for a theory with vertices of arbitrary high connectivity. The compact matrix notation covers the action being an integral with respect to space-time (or for non-equilibrium situations contour time) and a summation with respect to internal degrees of freedom (and with respect to the real-time dynamical or Schwinger–Keldysh indices if traded for the contour time). We can therefore introduce the Lagrange density

$$S[\phi] = \int d1\,\mathcal{L}(\phi,\phi')\,. \qquad (10.38)$$

We note that the effective action, Eq. (9.80), has the same functional form as the action except that one-particle irreducible vertex functions appear instead of the bare vertices and in the effective action appears the *average* field.

Since the bare propagator is chosen symmetric in all its variables, i.e. in particular with respect to the dynamical indices as we are treating non-equilibrium states, so is its inverse, and Eq. (10.36) can be written on the form

$$\frac{\delta Z[\phi]}{\delta\phi_1} = i\frac{\delta S[\phi]}{\delta\phi_1}\,Z[\phi] \qquad (10.39)$$

and immediately solved (up to an overall constant which can be fixed by comparing with the free theory) as

$$Z[\phi] = e^{iS[\phi]}\,, \qquad (10.40)$$

and we have the path integral representation of the generating functional (up to a source independent normalization factor)[10]

$$Z[J] = \int \mathcal{D}\phi\,e^{iS[\phi]+i\phi\,J}\,. \qquad (10.41)$$

---

[9]We effortlessly interchange functional integration and differentiation, amounting here to functional integration being a linear operation.

[10]A virtue of the path integral formulation is the ease with which symmetries of the action leads to important relations between Green's functions as discussed in Appendix B.

We note that in the path integral formulation of a quantum field theory, the fundamental dynamic equation, Eq. (10.34), can be stated in terms of the basic theorem of integration, the integral of a derivative vanishes

$$0 = \int \mathcal{D}\phi \, \frac{\delta}{\delta\phi} \, e^{iS[\phi]+i\phi J} \, . \tag{10.42}$$

In the treatment of non-equilibrium states in the real-time technique, a symmetric representation of the bare propagator should thus be used, say

$$G_0 = \begin{pmatrix} 0 & G_0^A \\ G_0^R & G_0^K \end{pmatrix}, \tag{10.43}$$

in order for the path integral to be well-defined. Since our interest is the real-time treatment of non-equilibrium situations, the closed time path guarantees the normalization condition of the generator, $Z[J] = 1$, provided that the sources on the two parts of the closed time path are taken as identical.

The action is specified solely in terms of the (inverse) bare propagators and the bare vertices and captures, according to Eq. (10.37), all the information of the theory, just like the diagrammatics and the generating functional technique, but now in a different way through Eq. (10.41). For a scalar boson field theory we thus have a new formulation not in terms of the quantum field, an operator, but in terms of a scalar field ϕ, a real function of space-time. The price paid for having this simpler object appear as the basic quantity is that to calculate the amplitudes of the theory we must perform a functional integral. In this formalism, the superposition principle manifests itself most explicitly as a summation over all intermediate alternative field configurations. For the case of fermions, the role of the real field is taken over by conjugate pairs of Grassmann fields in order to respect the anti-symmetric property of amplitudes for fermions.

The amplitudes of the theory are obtained by differentiating the generating functional with respect to the source, and they now appear in terms of functional integral expressions[11]

$$A_{12...N} = \int \mathcal{D}\phi \, \phi_1 \phi_2 \cdots \phi_N \, e^{iS[\phi]} \, . \tag{10.44}$$

In the functional integral representation of a quantum field theory, the amplitudes are thus moments of the field weighted with respect to the action. We note that in the functional integral representation, the amplitudes are automatically the contour time-ordered amplitudes (or in zero temperature quantum field theory, the time-ordered amplitudes), because of the time slicing involved in the definition of the functional integral, as we also recall from Eq. (A.16) of Appendix A.[12]

For the generator of connected Green's functions

$$i\,W[J] = \ln Z[J] \tag{10.45}$$

---

[11]The appearance of the imaginary unit for one's favorite choice of defining Green's functions are suppressed. As usual they are part of one's private set of Feynman rules.

[12]Normal ordering of interactions on the other hand, has to be enforced by hand.

we then have

$$e^{iW[J]} = \mathcal{N}^{-1} \int \mathcal{D}\phi \, e^{iS[\phi]} \, e^{i\phi \, J} \, , \qquad (10.46)$$

where $\mathcal{N}$ denotes the normalization factor guaranteeing that $W[J]$ vanishes for vanishing source, $W[J = 0] = 0$. Or in the real-time non-equilibrium technique, the generator of connected Green's functions vanishes, $W[J] = 0$ if the source is taken to be equal on the two parts of the contour, $J_- = J_+$.

From the Legendre transform relating the generator of connected Green's functions to the effective action, Eq. (9.85), and the functional integral representation of the generating functional, Eq. (10.41), a functional integral representation of the effective action, the generator of one-particle irreducible vertices, is obtained (reinstating for once $\hbar$)

$$e^{\frac{i}{\hbar}\Gamma[\varphi]} = \int \mathcal{D}\phi \, e^{\frac{i}{\hbar}(S[\phi] + (\phi - \varphi) \, J)} \, , \qquad (10.47)$$

where the normalization factor has been absorbed in the definition of the functional integral.

By inspecting the path integral expression of the generating functional for the theory in question

$$Z[J] = \int \mathcal{D}\phi \, e^{-\frac{i}{2}\varphi \, G_0^{-1} \, \varphi} \, e^{iS_i[\phi]} \, e^{i\phi \, J} \qquad (10.48)$$

one can envisage the perturbation theory diagrams: expand all exponentials except the one containing the inverse free propagator, and perform the Gaussian integrals. We shall do this in Section 10.2.2, but before that we discuss the relationship between the Euclidean and Minkowski versions of field theories.

10.2.1 Euclid versus Minkowski

The exposition in the previous chapter was mostly explicitly for the Euclidean field theory or thermodynamics. We left out the annoying imaginary unit irrelevant to the functioning of the generating functional technique. In that case, the Green's functions are given by

$$A_{12\ldots N} = \int \mathcal{D}\phi \, \phi_1 \, \phi_2 \cdots \phi_N \, e^{S[\phi]} \, , \qquad (10.49)$$

where the action is a real functional specifying in equilibrium statistical mechanics the probability for a given configuration of the field, the Boltzmann factor.

In a quantum field theory, the transformation between source and field is Fourier transformation involving the imaginary unit.[13] Anyone is entitled to deal with this through one's favorite choice of Feynman rules. We followed the standard choice in Section 4.3.2 where we included the imaginary unit in the definition of the Green's

---

[13] For a quantum field theory expressed in the operator formalism, the imaginary unit will also appear through the time evolution operator, recall Section 4.3.2.

functions, recall for example Eq. (4.39) or Eq. (3.61), and we have for the transition between Euclidean, i.e. imaginary-time field theory and real-time quantum field theory the connections

$$G_0 \leftrightarrow -iG_0 \quad , \quad g \leftrightarrow ig \quad , \quad J \leftrightarrow iJ$$

$$S[\phi] \leftrightarrow \frac{i}{\hbar}S[\phi] . \tag{10.50}$$

Equation (9.32) thus transforms into Eq. (10.34).[14]

For a quantum field theory we have for the generating functional ($\hbar$ is later often discarded)

$$Z[J] = \int \mathcal{D}\phi \, Z[\phi] \, e^{\frac{i}{\hbar}\phi J} = \int \mathcal{D}\phi \, e^{\frac{i}{\hbar}S[\phi] + \frac{i}{\hbar}\phi J} \tag{10.51}$$

and Green's functions are generated according to our choice

$$(-i)^{N-1} \left(\frac{\hbar}{i}\right)^N \frac{\delta Z[J]}{\delta J_1 \delta J_2 \cdots \delta J_N}\Bigg|_{J=0} = A_{12...N} . \tag{10.52}$$

We can swing freely between using real-time and imaginary-time formulation, all formal manipulations being analogous.

In the real-time or closed time path technique there is no denominator problem, but otherwise in order to have proper normalization we should write

$$Z[J] = \frac{\int \mathcal{D}\phi \, e^{\frac{i}{\hbar}S[\phi]} \, e^{\frac{i}{\hbar}\phi J}}{\int \mathcal{D}\phi \, e^{\frac{i}{\hbar}S[\phi]}} \tag{10.53}$$

but often such an overall constant are incorporated in the definition of the functional integral.

10.2.2 Wick's theorem and functionals

We now show how perturbation theory falls out very easily from the functional formulation, viz. Wick's theorem becomes a simple matter of differentiation.

We note the relationship

$$e^{iS[\phi]} \, e^{i\phi J} = e^{iS[-i\frac{\delta}{\delta J}]} \, e^{i\phi J} , \tag{10.54}$$

which is immediately obtained by expanding the exponential of the action on the right-hand side and noting that differentiating with respect to the source substitutes the ϕ-variable, and re-exponentiating gives the exponential of the action as on the left-hand side.

---

[14]The form of the propagator also changes when Wick rotating from real to imaginary time, changing the analytical properties of the propagator.

Let us in the action split off the trivial quadratic term

$$S_0[\phi] = -\frac{1}{2}\phi(G^{(0)})^{-1}\phi \qquad (10.55)$$

the free part, and the interaction part of the action, $S = S_0 + S_i$, is in general

$$S_i[\phi] = \sum_N \frac{1}{N!} g_{12\ldots N}\phi_1\phi_2\cdots\phi_N . \qquad (10.56)$$

The functional integral expression for the generating functional

$$Z[J] = \int\!\mathcal{D}\phi\, e^{iS_i[\phi] - \frac{i}{2}\phi(G^{(0)})^{-1}\phi + i\phi J} \qquad (10.57)$$

then, in accordance with Eq. (10.54), becomes

$$Z[J] = e^{iS_i[-i\frac{\delta}{\delta J}]} Z_0[J] , \qquad (10.58)$$

where $Z_0[J]$ is the generating function for the free theory

$$Z_0[J] = e^{\frac{i}{2}JG^{(0)}J} . \qquad (10.59)$$

We have thus achieved expressing the generating functional in terms of the generator of the free theory. Formula Eq. (10.58) expresses the perturbation theory of the theory in a compact form, and in a very different form compared to how in the operator formulation the full theory was expressed in terms of the free theory as we recall from Section 4.3.2. We now unfold this formula and show that it leads to the diagrammatic perturbation theory from which we started out in this chapter, and of course expressions equivalent to the non-equilibrium diagrammatic perturbation theory we derived in the canonical operator formalism in Chapters 4 and 5 by use of Wick's theorem on operator form.

The exponential containing the interaction is then expanded, for example consider a 3-vertex theory for which we get

$$Z[J] = \left(1 + \frac{1}{3!}g_{123}\frac{(-i)^3\delta^3}{\delta J_1\delta J_2\delta J_3} + \frac{1}{2!}\frac{1}{3!}g_{123}\frac{(-i)^3\delta^3}{\delta J_1\delta J_2\delta J_3}\frac{1}{3!}g_{1'2'3'}\frac{(-i)^3\delta^3}{\delta J_{1'}\delta J_{2'}\delta J_{3'}} + \cdots\right)$$

$$\times\; e^{\frac{i}{2}JG^{(0)}J} . \qquad (10.60)$$

A derivative brings down from the exponential a source contracted with a free propagator and another derivative must eliminate this source if the terms are to survive when at the end the source is set to zero. An odd number of differentiations will thus lead to a vanishing expression, and the derivatives must thus group in pairs, and this can be done in all possible ways.

Before arriving at Wick's theorem, we note that the generator can be related to vacuum diagrams. We expand both exponentials multiplied in Eq. (10.58), again considering a 3-vertex theory,

$$Z[J] = \left(1 + \frac{1}{3!}g_{123}\frac{(-i)^3\delta^3}{\delta J_1\delta J_2\delta J_3} + \frac{1}{2!}\frac{1}{3!}g_{123}\frac{(-i)^3\delta^3}{\delta J_1\delta J_2\delta J_3}\frac{1}{3!}g_{1'2'3'}\frac{(-i)^3\delta^3}{\delta J_{1'}\delta J_{2'}\delta J_{3'}} + \cdots\right)$$

$$\times \left(1 + \frac{i}{2}JG^{(0)}J + \frac{1}{2!}\left(\frac{i}{2}JG^{(0)}J\right)^2 + \cdots\right). \tag{10.61}$$

Now operating with the terms, we get strings of differentiations which will attach free propagators to vertices. Setting the source to zero in the end, $J = 0$, we obtain that $Z[J = 0]$ is the sum of all vacuum diagrams constructable from the vertices and propagators of the theory.

Exercise 10.1. Obtain the perturbative expansion of the generating functional at zero source value, $Z[J = 0]$, to fourth order in the coupling constant for a 3-vertex theory and draw the corresponding vacuum diagrams.

The amplitudes of the theory are generated by taking derivatives of the generating functional, for example for the 2-state amplitude we encounter the further derivatives

$$A_{12}[J] = i\frac{\delta^2 Z[J]}{\delta J_1\delta J_2}$$

$$= i\frac{\delta^2}{\delta J_1\delta J_2}\left(1 + \frac{g_{123}}{3!}\frac{(-i)^3\delta^3}{\delta J_1\delta J_2\delta J_3} + \frac{1}{2!}\frac{g_{123}}{3!}\frac{(-i)^3\delta^3}{\delta J_1\delta J_2\delta J_3}\frac{g_{1'2'3'}}{3!}\frac{(-i)^3\delta^3}{\delta J_{1'}\delta J_{2'}\delta J_{3'}} + \cdots\right)$$

$$\times \left(1 + \frac{i}{2}JG^{(0)}J + \frac{1}{2!}\left(\frac{i}{2}JG^{(0)}J\right)^2 + \cdots\right). \tag{10.62}$$

The resulting perturbative expressions for the amplitude upon setting the source to zero are precisely the ones which corresponds to the original diagrammatic definition of the 2-state Green's function, correct factorials and all. This is Wick's theorem expressed in terms of functional differentiation and obtained by using the functional integral representation of the generating functional. The above scheme gives us back by brute force the diagrammatics in terms of free propagators and vertices we started out with. However, as a calculational tool, the procedure becomes quickly quite laborious. Using the generating functional equations of the previous chapter is more efficient, as we demonstrated in Section 9.7.

Exercise 10.2. Obtain the perturbative expansion of the 2-state Green's function, A_{12}, to lowest order in the coupling constants for a 3- plus 4-vertex theory and draw the corresponding diagrams.

Exercise 10.3. Obtain the perturbative expansion of the 2-state Green's function, A_{12}, to fourth order in the coupling constant for a 3-vertex theory and draw the corresponding diagrams.

We now turn to show how Wick's theorem can be formulated in the functional integral approach. The amplitudes or Green's functions of a quantum field theory are in the functional integral representation of the Green's functions specified by averages over the field, such as in Eq. (10.44). Also, in an expansion of the exponential containing the interaction term in Eq. (10.57), such averages will appear, and we encounter arbitrary correlations with a Gaussian weight. Let us therefore first consider the N-dimensional integral

$$I(p_1, \ldots, p_{2N}) = \int_{-\infty}^{\infty} dx_1 \ldots \int_{-\infty}^{\infty} dx_N \, x_{p_1} \cdots x_{p_{2N}} \, e^{-\frac{1}{2}x^T A x} \qquad (10.63)$$

where A denotes the symmetric matrix of Section 10.1.2, and x_{p_M} denotes any variable *picked* from the N-tuple $(x_1, \ldots, x_N)$ and allowed to appear any number of the possible $2N$ times. We have chosen a string of even factors, since the integral vanishes if an odd number of xs occurred, as seen immediately by diagonalizing the quadratic form. The correlation function to be evaluated can be rewritten

$$I(p_1, \ldots, p_{2N}) = i\frac{\partial}{\partial b_{p_1}} \cdots i\frac{\partial}{\partial b_{p_{2N}}} \int_{-\infty}^{\infty} \prod_{M=1}^{N} dx_M \, e^{-\frac{1}{2}x^T A x} \, e^{-ib^T x} \Bigg|_{b=0}$$

$$= i\frac{\partial}{\partial b_{p_1}} \cdots i\frac{\partial}{\partial b_{p_{2N}}} I(A; b) \Bigg|_{b=0} \qquad (10.64)$$

and according to Eq. (10.15)

$$I(p_1, \ldots, p_{2N}) = \left(\det \left(\frac{A}{2\pi} \right) \right)^{-1/2} i\frac{\partial}{\partial b_{p_1}} \cdots i\frac{\partial}{\partial b_{p_{2N}}} e^{-\frac{1}{2}b^T A^{-1} b} \Bigg|_{b=0}. \qquad (10.65)$$

The expression on the right can be evaluated by use of the formula, valid for arbitrary functions f and g,

$$f\left(i\frac{\partial}{\partial b} \right) g(b) = g\left(i\frac{\partial}{\partial c} \right) f(c) \, e^{-ib^T c} \Bigg|_{c=0}, \qquad (10.66)$$

which is immediately proved by the help of Fourier transformation, i.e. ·by showing the formula for plane wave functions. Employing Eq. (10.66) we obtain

$$I(p_1, \ldots, p_{2N}) = \left(\det \left(\frac{A}{2\pi} \right) \right)^{-1/2} e^{\frac{1}{2}\partial_c^T A^{-1} \partial_c} c_{p_1} \cdots c_{p_{2N}} \, e^{-ib^T c} \Bigg|_{c=0}\Bigg|_{b=0}$$

$$= \left(\det \left(\frac{A}{2\pi} \right) \right)^{-1/2} e^{\frac{1}{2}\partial_c^T A^{-1} \partial_c} c_{p_1} \cdots c_{p_{2N}} \Bigg|_{c=0}, \qquad (10.67)$$

where the last equality is obtained as the terms originating from differentiating the exponential eventually vanish when b is set equal to zero. The only surviving term on the right comes from the term in the expansion of the exponential containing $2N$ differentiations giving

$$I(p_1, \ldots, p_{2N}) = \left(\det \left(\frac{A}{2\pi} \right) \right)^{-1/2} \frac{1}{N! 2^N} (\partial_c^{\mathrm{T}} A^{-1} \partial_c)^N \, c_{p_1} \cdots c_{p_{2N}} \bigg|_{c=0} . \quad (10.68)$$

In each of the N double differentiation operators, we must choose pairs in the pick of the factors on the right thereby uniquely exhausting the pick in order to get a non-vanishing result upon setting $c = 0$. Then upon differentiating and setting $c = 0$, a product of N terms of the form $(A^{-1})_{p_i, p_j}$ occurs with the chosen pairings as indices. Permuting which pair is related to which double differentiation operator gives $N!$ identical products. Furthermore, since A is a symmetric matrix so is A^{-1} (transposition and inverting of a matrix are commuting operations) and we obtain

$$I(p_1, \ldots, p_{2N}) = \left(\det \left(\frac{A}{2\pi} \right) \right)^{-1/2} \sum_{\mathrm{a.p.p.}} \prod (A^{-1})_{p_i, p_j} , \quad (10.69)$$

where the sum is over all possible pairings of the indices in the pick $p_1, \ldots, p_{2N}$, without distinction of the ordering within a pair, explaining in addition the canceling of the factor $1/2^N$. The above observation is the equivalent of Wick's theorem.

With the usual convention of absorbing the functional determinant in the definition of the functional integral we get, in accordance with Eq. (10.63) and Eq. (10.69), that the amplitudes of the free theory are obtained according to

$$A_{12 \ldots 2N} = \int \mathcal{D}\phi \, \phi_1 \phi_2 \cdots \phi_{2N} \, e^{-\frac{i}{2} \varphi G_0^{-1} \varphi}$$

$$= \sum_{\mathrm{a.p.p.}} (iG_0^{-1})_{p_1 p_2} (iG_0^{-1})_{p_3 p_4} \cdots (iG_0^{-1})_{p_{2N-1} p_{2N}} . \quad (10.70)$$

By inspecting the path integral expression for the generating functional

$$Z[J] = \int \mathcal{D}\phi \, e^{-\frac{i}{2} \varphi G_0^{-1} \varphi} \, e^{iS_i[\phi]} \, e^{i\phi J} \quad (10.71)$$

one can envisage its perturbation expansion and corresponding Feynman diagrams by this recipe: expand all exponentials except the one containing the inverse free propagator, the Gaussian term, and evaluate the averages according to the above formula, Eq. (10.70). This recipe for functional integration of products of fields weighted by their Gaussian form provides Wick's theorem, but now in the functional or path integral formulation of the field theory. From this observation we can immediately recover the non-equilibrium Feynman diagrammatics of a quantum field theory by expanding the exponential containing the interaction in Eq. (10.41).

The limiting procedure used in Section 10.1 to define functional integration can be made rigorous only for the Euclidean case. For the quantum field theory case, an alternative now offers itself, viz. to define the functional integrals in terms of their, as above, perturbative expansions in the non-Gaussian interaction part.

Exercise 10.4. If the Gaussian part of the integrand in Eq. (10.63) is interpreted as a probability distribution for the random or stochastic variable x, then Eq. (10.69) is the statement that any correlation function of a Gaussian random variable, with zero mean, is expressed in terms of all possible products of the two-point correlation function.

Show that the generating function, i.e. the Fourier transform of the normalized probability distribution

$$P(x) = \left(\det \left(\frac{A}{2\pi} \right) \right)^{-1/2} e^{-\frac{1}{2}x^{\mathrm{T}} A x} , \tag{10.72}$$

is

$$P(k) = e^{-\frac{1}{2}k^{\mathrm{T}} A^{-1} k} . \tag{10.73}$$

Exercise 10.5. Consider a set of independent stochastic variables $\{x_n\}_{n=1,\ldots,N}$, each with arbitrary probability distributions except for zero mean and same finite variance, say σ. Show that the stochastic variable $X = (x_1 + \cdots + x_N)/\sqrt{N}$ will then obey the central limit theorem, i.e. in the limit $N \to \infty$, the stochastic variable X will be Gaussian distributed with variance σ.

Another application of the formula Eq. (10.66), allows us to rewrite Eq. (10.58)

$$Z[J] = e^{iS_i[-i\frac{\delta}{\delta J}]} e^{\frac{i}{2}JG^{(0)}J} = e^{\frac{1}{2}\frac{\delta}{\delta\phi} G_0 \frac{\delta}{\delta\phi}} e^{iS_i[\phi] + i\varphi J} \tag{10.74}$$

thereby giving the following functional integral expression

$$Z[J] = e^{\frac{1}{2}\mathrm{Tr}\ln G_0^{-1}} \int \mathcal{D}\phi \, e^{\frac{1}{2}\frac{\delta}{\delta\phi} G_0 \frac{\delta}{\delta\phi}} e^{iS_i[\phi] + i\varphi J} \Bigg|_{\phi=0} . \tag{10.75}$$

From here we see directly that $Z[J]$ is the sum of all the vacuum diagrams for the theory in question in the presence of the source J. This observation is again the equivalent of Wick's theorem, but here at its most expedient form involving both functional integration and differentiation.

Introducing the generator of connected Green's functions

$$Z[J] = e^{iW[J]} \tag{10.76}$$

and recalling the combinatorial argument of Section 9.5, the above important observation gives that $iW[J]$ is the sum of all the *connected* vacuum diagrams in the presence of the source J.

For the connected Green's functions we then obtain the functional integral expression

$$G_{12\ldots N} = \frac{\int \mathcal{D}\phi \, \phi_1 \phi_2 \cdots \phi_N \, e^{iS[\phi]}}{\int \mathcal{D}\phi \, e^{iS[\phi]}} \equiv \langle \phi_1 \phi_2 \cdots \phi_N \rangle . \tag{10.77}$$

Often the denominator, which cancels all the disconnected contributions in the numerator, is left implicit as a normalization factor in the definition of the functional integral.

For the average or classical field, φ_1, considered in Section 9.6, we thus have the functional integral expression for the Euclidean case

$$\varphi_1 = \frac{\int \mathcal{D}\phi \; \phi_1 \; e^{S[\phi]}}{\int \mathcal{D}\phi \; e^{S[\phi]}} \equiv \langle \phi_1 \rangle \equiv \overline{\phi}_1 \,, \qquad (10.78)$$

the reason for calling $\varphi_1 \equiv G_1$ the *average* field now being obvious.

The diagrammatics obtained by the above procedures are of course naive perturbation theory, expressed in terms of the bare propagators and vertices. A representation which contains the full propagators and the effective vertices is a better representation since it expresses the physics of a particular situation, viz. the state under consideration. This representation can be obtained at the diagrammatic level by topological arguments, leading to the so-called skeleton diagrams as discussed in Section 4.5.2. In Section 9.8, we followed another way and employed the effective action to show how easily the skeleton diagrammatics is obtained from the analytical functional differentiation formalism. In the next section, we show how the partially re-summed perturbation expansion of Green's functions, the skeleton diagrammatic representation, is expressed in the functional integral formalism.

10.3   Generators and 1PI vacuum diagrams

In the previous section we showed that the generating functionals had perturbative expansions corresponding diagrammatically to the sum of all vacuum diagrams expressed in naive perturbation theory. In this section we shall exploit the functional integral representation of a quantum field theory to relate the various generators to classes of one-particle irreducible vacuum diagrams.

We therefore turn to show that the generator of connected Green's functions can be expressed in terms of the effective action and a *restricted* functional integral. A restricted functional integral is a functional integral interpreted in terms of its perturbative expansion or equivalently the corresponding Feynman diagrams, and where only certain topological classes of diagrams are retained. First, we recall the result derived diagrammatically, the relationship displayed in Figure 9.48: that the tadpole, the first derivative of the generator of connected amplitudes, has a diagrammatic expansion in terms of only *tree* diagrams, tadpoles attached to one-particle irreducible vertices. This means that the generator of connected amplitudes, $W[J]$, itself is given by the irreducible vertices attached to tadpoles. This suggests that the generator of connected amplitudes, $W[J]$, can be specified in terms of the effective action, $\Gamma[\phi]$. We now turn to show that it is indeed the case and this in terms of a functional integral where the effective action appears instead of the action and the functional integral is restricted:

$$i\,W[J] = \int_{\mathrm{CTD}} \mathcal{D}\phi \; e^{i\Gamma[\phi]\,+i\phi\,J} \,, \qquad (10.79)$$

where CTD indicates that only the Connected Tree Diagrams should be kept of all the vacuum diagrams generated by the perturbative expansion of the functional integral. Tree diagrams contain no loops, they are contained within the 1PI vertices, and tree diagrams can be cut in two by cutting a single line of a tadpole.

To keep track of the number of loops in the diagrams generated by the unrestricted functional integral in Eq. (10.79), we introduce the parameter a

$$e^{i\tilde{W}_a[J]} = \int D\phi \, e^{ia^{-1}(\Gamma[\phi] + \phi J)} . \qquad (10.80)$$

The vacuum diagrams generated by this functional integral can be characterized as follows. Separate out in the effective action the quadratic term, which according to Eq. (9.83) is the inverse of the full Green's function of the theory multiplied by a^{-1}. Then expand the rest of the exponential and use Wick's theorem according to the previous section, or rather the just derived procedure for Gaussian averaging of products of fields to obtain the perturbative expansion of the functional integral in Eq. (10.80), and its corresponding Feynman diagrams. A Green's function has thus associated a factor a and each of the one-particle irreducible vertices in the rest of the effective action has associated a factor a^{-1} as has the source, which in this context we also refer to as a vertex (on *a par* with Γ_1). A diagram with V vertices (of either kind) and P propagator lines is thus proportional to a^{P-V}. Since the diagrams generated by the path integral in Eq. (10.80) are vacuum diagrams they are loop diagrams, the tree diagrams being those with zero number of loops. Since it takes two *times* two protruding lines from vertices (or one vertex) to form *one* loop, the number of loops L is specified by $L = P - V + 1$, and an L-loop diagram carries an overall factor proportional to a^{L-1}.[15] The theory defined by the functional integral in Eq. (10.80) can thus be described at the diagrammatic level in terms of the diagrams for the theory where a is unity, $a = 1$, according to

$$\tilde{W}_a[J] = \sum_{L=0} a^{L-1} \tilde{W}^{(L)}[J] , \qquad (10.81)$$

where $\tilde{W}^{(L)}[J]$ comprises the sum of connected vacuum diagrams with L loops for the theory defined by the action $\Gamma[\varphi]$, i.e. the theory specified by Eq. (10.80) for the case $a = 1$. We note that the tree diagrams singled out in Eq. (10.79) correspond to the zero loop term $\tilde{W}^{(0)}[J]$.

In the limit of vanishing a, the value of the functional integral, Eq. (10.80), is determined by the field at which the exponent is stationary, denote it φ, according to

$$e^{i\tilde{W}_a[J]} \propto e^{ia^{-1}(\Gamma[\varphi] + \varphi J)} , \qquad (10.82)$$

where the prefactor (a horrendous determinant term) involves the square root of $a(G_0^{-1} - \Sigma)$ and therefore its lowest power is a^0 and will therefore turn out to be

---

[15] This observation gives, for the effective action, a characterization of its diagrammatic structure, and a controlled approximation scheme, the loop expansion. Say for a quantum field theory, the diagrammatic representation of the effective action corresponds to an expansion in $\hbar^L$, where L is the number of loops in a diagram.

harmless when a eventually is set to zero. The stationary field, φ, is determined as the solution of the equation

$$\frac{\delta\Gamma[\varphi]}{\delta\varphi_1} + J_1 = 0 \tag{10.83}$$

thus making contact with the original theory, since this is the equation satisfied by the effective action, Eq. (9.84). According to Eq. (10.81), in the limit of vanishing a we have $\tilde{W}_a[J] \simeq a^{-1}\tilde{W}^{(0)}[J]$, and by taking the logarithm of Eq. (10.82) we get (noting that in this limit, the constant prefactor in Eq. (10.82) gives no contribution)

$$\tilde{W}^{(0)}[J] = \Gamma[\varphi] + \varphi J. \tag{10.84}$$

But according to the Legendre transformation, Eq. (9.85), this implies

$$\tilde{W}^{(0)}[J] = W[J] \tag{10.85}$$

and we have shown the validity of Eq. (10.79). That is, we have shown that the generator of connected Green's functions can be expressed as the sum of all connected tree diagrams where the vertices are one-particle irreducible.[16] In diagrammatic terms, the generator of connected Green's functions, Eq. (10.79), can thus be displayed as depicted in Figure 10.1.

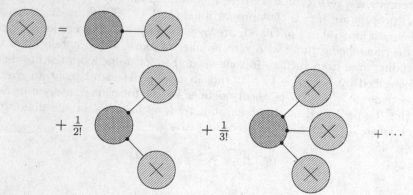

Figure 10.1 The tree diagram expansion of the generator of connected amplitudes in terms of the one-particle irreducible vertices.

The sum of all connected vacuum diagrams in the presence of the source is thus captured by keeping only the tree diagrams if at the same time the bare vertices are exchanged by the one-particle irreducible vertices.

The effective action $\Gamma[\phi_a]$, Eq. (9.80), taken for an arbitrary field value ϕ_a can also be expressed in terms of a restricted functional integral, viz.

$$\Gamma[\phi_a] = \int_{\mathrm{1PICVD}} \mathcal{D}\phi\, e^{iS[\phi+\phi_a]}, \tag{10.86}$$

---

[16]This provides a proof in terms of the functional integral method, that $i\Gamma$ consists of the one-particle irreducible vertices. We already knew this because of its diagrammatic construction according to Section 9.6.

where 1PICVD indicates that in the perturbation expansion, only the *connected one-particle irreducible vacuum diagrams* should be kept of the connected diagrams generated by the perturbative expansion of the path integral, since upon expanding in ϕ_a the prescription on the restricted functional integral generates $\Gamma[\phi_a]$ according to Eq. (9.80). In particular we have shown that $\Gamma[0]$ is the quantity represented by the sum of all connected one-particle irreducible vacuum diagrams for the theory (in the absence of the source).

Since $W[J]$ is related to $\Gamma[\varphi]$ by a Legendre transformation, the above observation for $\Gamma[0]$ corresponds to the statement that $\Gamma[0]$ equals $W[J]$ for the value of the source for which the field $\delta W[J]/\delta J_1$ vanishes. Since the vanishing of $\delta W[J]/\delta J_1$ is equivalently to $\delta Z[J]/\delta J_1$ vanishing, we can state the observation as

$$W[J]\Big|_{\frac{\delta Z[J]}{\delta J_1}=0} = \quad \text{sum of one-particle irreducible connected} \quad\quad (10.87)$$
$$\text{vacuum diagrams (1PICVD).}$$

We shall make use of this observation when we consider the loop expansion of the effective action.

10.4 1PI loop expansion of the effective action

In this section we shall use the path integral representation of the generators to get a useful path integral expression for the effective action which has an explicit diagrammatic expansion. We follow Jackiw, and show how to express the effective action in terms of the one-particle irreducible connected vacuum diagrams for a theory with a shifted action [52].

Consider a field theory specified by the action $S[\phi]$ and the corresponding path integral expression for the generating functional

$$Z[f] = \int \mathcal{D}\phi \, e^{iS[\phi]+if\phi} , \qquad\qquad (10.88)$$

where we have used the notation f for the one-particle source. In fact, in the next chapter, when we consider non-equilibrium phenomena in classical statistical dynamics the source will not be set equal to zero by the end of the day as it will contain the classical force coupled to the classical degree of freedom of interest.

The path integral is invariant with respect to an arbitrary shift of the field, recall Eq. (10.3),

$$\phi \mapsto \phi + \phi_0 \qquad\qquad (10.89)$$

giving for the generating functional

$$Z[f] = \int \mathcal{D}\phi \, e^{iS[\phi+\phi_0]+if(\phi+\phi_0)} = e^{iS[\phi_0]+if\phi_0} Z_1[f] , \qquad\qquad (10.90)$$

where

$$Z_1[f] = \int \mathcal{D}\phi \, e^{i(S[\phi+\phi_0]-S[\phi_0])+if\phi}. \tag{10.91}$$

The subscript on $Z_1[f]$ is not a state label but just discriminates the generator from the original generating functional $Z[f]$. State labels in the functional differentiations are in the following suppressed throughout, and matrix multiplication is implied.

The generator of connected Green's functions then becomes

$$\begin{aligned} W[f] &= -i\ln Z = S[\phi_0] + f\phi_0 - i\ln Z_1[f] \\ &= S[\phi_0] + f\phi_0 + W_1[f]\,, \end{aligned} \tag{10.92}$$

where

$$iW_1[f] = \ln \int \mathcal{D}\phi \, e^{i(S[\phi+\phi_0]-S[\phi_0])+if\phi}. \tag{10.93}$$

To make the so-far arbitrary function ϕ_0 a functional of f, we choose ϕ_0 to be the average field which effects the Legendre transformation to the effective action, $\Gamma[\bar\phi]$, i.e.

$$\phi_0 \equiv \bar\phi = \frac{\delta W[f]}{\delta f}\,, \tag{10.94}$$

where a bar now specifies the average field, $\bar\phi = \varphi$, for visual clarity in the following equations. Recalling that this vice versa gives f implicitly as a functional of $\bar\phi$, $f = f[\bar\phi]$, we have according to Eq. (10.94) and Eq. (10.92)

$$\left(\frac{\delta S[\bar\phi]}{\delta\bar\phi} + f + \frac{\delta W_1}{\delta\bar\phi}\right)\frac{\delta\bar\phi}{\delta f} = 0 \tag{10.95}$$

and thereby, since the second factor on the left is the full Green's function,

$$\frac{\delta S[\bar\phi]}{\delta\bar\phi} + f + \frac{\delta W_1}{\delta\bar\phi} = 0. \tag{10.96}$$

The effective action can, according to Eq. (10.92) and Eq. (10.94), be expressed as

$$\Gamma[\bar\phi] = W[f] - \bar\phi f = S[\bar\phi] + W_1[f] = S[\bar\phi] + W_1[\bar\phi]\,, \tag{10.97}$$

where in the last equality we have been sloppy, using the same notation for W_1 as a functional of the implicit function of $\bar\phi$ as that of f, $W_1[f]$. But by employing Eq. (10.96) in Eq. (10.93) we can in fact eliminate the explicit dependence on f, and get the expression for W_1 as a functional of the average field, $\bar\phi$, as specified by the functional integral

$$iW_1[\bar\phi] = \ln \int \mathcal{D}\phi \, \exp\left\{i(S[\bar\phi+\phi]-S[\bar\phi]) - i\phi\left(\frac{\delta S[\bar\phi]}{\delta\bar\phi} + \frac{\delta W_1}{\delta\bar\phi}\right)\right\}. \tag{10.98}$$

The aim is now to evaluate $W_1[\overline{\phi}]$, or rather to show that it can be expressed in terms of one-particle irreducible connected vacuum graphs. We therefore introduce the generating functional

$$\tilde{Z}[\overline{\phi}; J] = \int \mathcal{D}\phi \, \exp\left\{ i(S[\overline{\phi} + \phi] - S[\overline{\phi}]) - i\phi \frac{\delta S[\overline{\phi}]}{\delta \overline{\phi}} + iJ\phi \right\} \qquad (10.99)$$

for the theory governed by the action

$$\tilde{S}[\overline{\phi}, \phi] = S[\overline{\phi} + \phi] - S[\overline{\phi}] - \phi \frac{\delta S[\overline{\phi}]}{\delta \overline{\phi}}, \qquad (10.100)$$

i.e. the action for the original theory expanded around the average field but keeping only second- and higher-order terms. Correspondingly for the generator of connected Green's functions in this theory we have

$$i\tilde{W}[\overline{\phi}; J] = \ln \tilde{Z}[\overline{\phi}; J] \qquad (10.101)$$

and evidently by comparing Eq. (10.99) and Eq. (10.98)

$$W_1[\overline{\phi}] = \tilde{W}[\overline{\phi}; J]\Big|_{J=-\delta W_1/\delta\overline{\phi}}. \qquad (10.102)$$

We shortly turn to show that for this particular choice of the source as specified in Eq. (10.102), $J = -\delta W_1/\delta\overline{\phi}$, the generator $\tilde{Z}$ vanishes

$$\frac{\delta \tilde{Z}[\overline{\phi}; J]}{\delta J}\Bigg|_{J=-\delta W_1/\delta\overline{\phi}} = 0 \qquad (10.103)$$

or equivalently for the generator of connected Green's functions

$$\frac{\delta \tilde{W}[\overline{\phi}; J]}{\delta J}\Bigg|_{J=-\delta W_1/\delta\overline{\phi}} = 0, \qquad (10.104)$$

i.e. the average field

$$\varphi = \frac{\delta \tilde{W}[\overline{\phi}; J]}{\delta J} \qquad (10.105)$$

vanishes for the theory governed by the action $\tilde{S}$ for the value of the source $J = -\delta W_1/\delta\overline{\phi}$. The statement in Eq. (10.102) thus becomes equivalent to the statement that $W_1[\overline{\phi}]$ is identical to the effective action for the theory governed by $\tilde{S}[\overline{\phi}, \phi]$ for vanishing average field, $\tilde{\Gamma}[\overline{\phi}; \varphi = 0]$. We then use the result of Section 10.3, that in general $\Gamma[\varphi = 0]$ is given by the one-particle irreducible connected vacuum diagrams, or equivalently for the generator of connected Green's functions the expression Eq. (10.87), viz. that $W[f]_{\delta W/\delta f=0}$ consists of the sum of all the one-particle *irreducible* vacuum diagrams. The functional W_1 thus in diagrammatic terms only consists of the sum of all the one-particle *irreducible* vacuum diagrams for the theory governed by $\tilde{S}[\overline{\phi}, \phi]$. These diagrammatic identifications will be exploited shortly.

To establish the validity of Eq. (10.102) we differentiate Eq. (10.98) with respect to the average field $\bar{\phi}$

$$\frac{\delta W_1}{\delta \bar{\phi}} = \frac{1}{Z_1} \int \mathcal{D}\phi \left(\frac{\delta S[\phi + \bar{\phi}]}{\delta \bar{\phi}} - \frac{\delta S[\bar{\phi}]}{\delta \bar{\phi}} - \phi \frac{\delta^2}{\delta \bar{\phi} \, \delta \bar{\phi}} (S[\bar{\phi}] + W_1[\bar{\phi}]) \right)$$

$$\times \quad \exp\left\{ i \left(S[\phi + \bar{\phi}] - S[\bar{\phi}] - \phi \frac{\delta S[\bar{\phi}]}{\delta \bar{\phi}} - \phi \frac{\delta W_1}{\delta \bar{\phi}} \right) \right\}. \tag{10.106}$$

The term originating from the first term in the parenthesis on the right-hand side can be rewritten as

$$\int \mathcal{D}\phi \, \frac{\delta S[\phi + \bar{\phi}]}{\delta \bar{\phi}} \exp\left\{ i \left(S[\phi + \bar{\phi}] - S[\bar{\phi}] - \phi \frac{\delta S[\bar{\phi}]}{\delta \bar{\phi}} - \phi \frac{\delta W_1}{\delta \bar{\phi}} \right) \right\}$$

$$= -i \int \mathcal{D}\phi \, \frac{\delta}{\delta \phi} \exp\left\{ i \left(S[\phi + \bar{\phi}] - S[\bar{\phi}] - \phi \frac{\delta S[\bar{\phi}]}{\delta \bar{\phi}} - \phi \frac{\delta W_1}{\delta \bar{\phi}} \right) \right\}$$

$$+ \int \mathcal{D}\phi \left(\frac{\delta S[\bar{\phi}]}{\delta \bar{\phi}} + \frac{\delta W_1}{\delta \bar{\phi}} \right) \exp\left\{ i \left(S[\phi + \bar{\phi}] - S[\bar{\phi}] - \phi \frac{\delta S[\bar{\phi}]}{\delta \bar{\phi}} - \phi \frac{\delta W_1}{\delta \bar{\phi}} \right) \right\}$$

$$\tag{10.107}$$

and since the first term on the right is an integral of a total derivative it vanishes, giving

$$\frac{\delta W_1}{\delta \bar{\phi}} = \frac{1}{Z_1} \int \mathcal{D}\phi \left\{ \frac{\delta W_1}{\delta \bar{\phi}} - \phi \frac{\delta^2}{\delta \bar{\phi} \delta \bar{\phi}} (S[\bar{\phi}] + W_1[\bar{\phi}]) \right\}$$

$$\times \quad \exp\left\{ i \left(S[\phi + \bar{\phi}] - S[\bar{\phi}] - \phi \frac{\delta S[\bar{\phi}]}{\delta \bar{\phi}} - \phi \frac{\delta W_1}{\delta \bar{\phi}} \right) \right\}. \tag{10.108}$$

The first term on the right in Eq. (10.108) is equal to the term on the left, giving the equation

$$\left(\frac{\delta^2 (S[\bar{\phi}] + W_1[\bar{\phi}])}{\delta \bar{\phi} \, \delta \bar{\phi}} \right) \int \mathcal{D}\phi \, \phi \exp\left\{ i \left(S[\phi + \bar{\phi}] - S[\bar{\phi}] - \phi \frac{\delta S[\bar{\phi}]}{\delta \bar{\phi}} - \phi \frac{\delta W_1}{\delta \bar{\phi}} \right) \right\} = 0.$$

$$\tag{10.109}$$

The first factor

$$\frac{\delta^2}{\delta \bar{\phi} \, \delta \bar{\phi}} \left(S[\bar{\phi}] + W_1[\bar{\phi}] \right) = \frac{\delta^2 \Gamma[\bar{\phi}]}{\delta \bar{\phi} \, \delta \bar{\phi}} \tag{10.110}$$

is according to Eq. (9.95) the inverse Green's function and therefore nonzero, and we have the sought after statement of Eq. (10.103)

$$\int \mathcal{D}\phi \, \phi \exp\left\{ i \left(S[\phi + \bar{\phi}] - S[\bar{\phi}] - \phi \frac{\delta S[\bar{\phi}]}{\delta \bar{\phi}} - \phi \frac{\delta W_1}{\delta \bar{\phi}} \right) \right\} = 0. \tag{10.111}$$

We have thus according to Eq. (10.102) shown that

$$W_1[\bar{\phi}] = \text{sum of all one-particle irreducible connected vacuum} \tag{10.112}$$
$$\text{diagrams (1PICVD) for the theory defined by the}$$
$$\text{action } \tilde{S}[\bar{\phi}, \phi].$$

Dividing the action $\tilde{S}[\bar{\phi}, \phi]$ into its quadratic part and the interaction part

$$\tilde{S}[\bar{\phi}; \phi] = \tilde{S}_0[\bar{\phi}; \phi] + \tilde{S}_{\text{int}}[\bar{\phi}; \phi] \tag{10.113}$$

we have

$$\tilde{S}_0[\bar{\phi}; \phi] = \frac{1}{2}\phi\frac{\delta^2 S[\bar{\phi}]}{\delta\bar{\phi}\delta\bar{\phi}}\phi \equiv \frac{1}{2}\phi\mathcal{D}^{-1}[\bar{\phi}]\,\phi \tag{10.114}$$

and

$$\tilde{S}_{\text{int}}[\bar{\phi}; \phi] = \sum_{N=3}^{\infty}\frac{1}{N!}\frac{\delta^N S[\bar{\phi}]}{\delta\bar{\phi}_1\cdots\delta\bar{\phi}_N}\phi_1\cdots\phi_N\,. \tag{10.115}$$

In the path integral expression for the generator $\tilde{Z}$, Eq. (10.99), the normalization factor

$$\int\mathcal{D}\phi\,e^{\frac{i}{2}\phi\mathcal{D}^{-1}\phi} = \sqrt{i\det\mathcal{D}} \tag{10.116}$$

was kept implicit, but by exposing it the expression for the effective action, Eq. (10.98), can finally be written as

$$\Gamma[\bar{\phi}] = S[\bar{\phi}] - \frac{i}{2}\text{Tr}\ln i\mathcal{D}^{-1}[\bar{\phi}] - i\ln\langle e^{i\tilde{S}_{\text{int}}[\bar{\phi};\phi]}\rangle_{\text{1PICVD}}\,, \tag{10.117}$$

where the last term should be interpreted as

$$\langle e^{i\tilde{S}_{\text{int}}[\bar{\phi};\phi]}\rangle = \int_{\text{1PICVD}}\mathcal{D}\phi\,e^{i\tilde{S}_0[\bar{\phi};\phi]}\,e^{i\tilde{S}_{\text{int}}[\bar{\phi};\phi]} \tag{10.118}$$

and the subscript "1PICVD" indicates the restriction to the one-particle irreducible connected vacuum diagrams resulting from the functional integral. We have explicitly displayed the one-loop contribution, the second term on the right in Eq. (10.117), and consequently we have the normalization

$$\langle 1\rangle_{\text{1PICVD}} = 1. \tag{10.119}$$

The first term on the right in Eq. (10.117), the zero loop or tree approximation, specifies the classical limit, determined by the stationarity of the action, and the second term gives the contribution from the Gaussian fluctuations. The last term, the higher loop contributions, gives the quantum corrections due to interactions, radiative corrections. Reinstating $\hbar$ gives the result that the contribution for a given loop order is proportional to $\hbar$ raised to that power.

For a 3- plus 4-vertex theory, the effective action has the series expansion in terms of one-particle irreducible vacuum diagrams as depicted (explicitly to three loop order) in Figure 10.2, where we have reinstated the notation $\varphi = \bar{\phi}$.

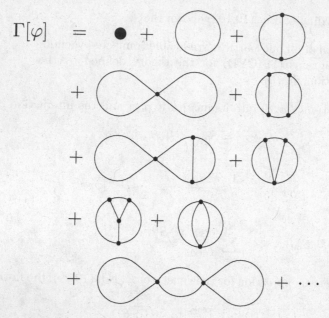

Figure 10.2 The 1PI vacuum diagram expansion of the effective action.

We note that the one-particle reducible diagram depicted in Figure 10.3, which contributes in the sum of vacuum diagrams to $W[J = 0]$, is absent in the series expansion of the effective action.

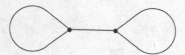

Figure 10.3 One-particle reducible diagram contributing to $W[J = 0]$.

The above functional evaluation of the effective action generates the one-particle irreducible loop expansion in terms of skeleton diagrams, and infinite partial summation of naive perturbation theory diagrams is thus already done. A virtue of the above expansion is that at each loop level for the effective action it contains far fewer diagrams than the naive perturbation expansion.

In Section 10.6, where the effective action approach is applied to a Bose gas, and in Chapter 12, where the theory of classical statistical dynamics is applied to vortex dynamics in a superconductor, we shall need to take the loop expansion to the next level where only two-particle irreducible vacuum diagrams will appear. We therefore first go back to the generating functional technique, but now we will include a two-particle source.

10.5 Two-particle irreducible effective action

The effective action can be taken to the next level in irreducibility in which only two-particle irreducible vertices appear. To construct such a description, we introduce a two-particle source K_{12} in addition to the one-particle source J_1, and a generator of Green's functions where the connected Green's functions of the theory now are contracted on both types of sources, i.e. defined according to the diagrammatic expansion in terms of the two sources as depicted in Figure 10.4.

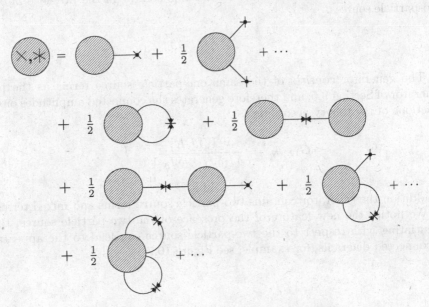

Figure 10.4 Diagrammatic expansion of the generator, $W[J, K]$, in the presence of one- and two-particle sources.

The diagrammatic notation for the two-particle source is thus as displayed in Figure 10.5.

$$K_{12} = \quad {}_1\!\!\!\!\!\!\xrightarrow{\hspace{0.5cm}}\!\!\!\!\!\!{}_2$$

Figure 10.5 Diagrammatic notation for the two-particle source.

The diagrammatic notation for the generator makes explicit the feature that it depends on both a one- and a two-particle source as stipulated in Figure 10.6.

$$W[J, K] \; = \;$$

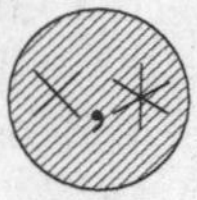

Figure 10.6 Diagrammatic notation for the generator in the presence of one- and two-particle sources.

The generator consists of the same one-particle source terms as the previous generator of Section 9.5, and therefore generates the connected amplitudes or Green's functions of the theory according to

$$G_{12...N} \; = \; \frac{\delta^N W[J, K]}{\delta J_1 \delta J_2 \cdots \delta J_N} \bigg|_{J=0, K=0} . \tag{10.120}$$

In addition the generator contains two-particle source terms and mixed terms.

We notice the new feature of the presence of the two-particle source, that differentiating with respect to the two-particle source can lead to the appearance of disconnected diagrams; for example, see Figure 10.7.

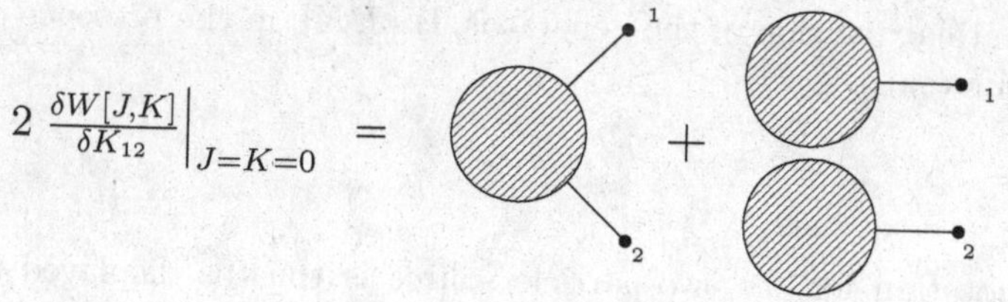

$$2 \frac{\delta W[J, K]}{\delta K_{12}} \bigg|_{J=K=0} \; = \;$$

Figure 10.7 Removing a two-particle source can create disconnected diagrams.

Taking the derivative with respect to the one-particle source

$$\varphi_1 \; = \; \frac{\delta W[J, K]}{\delta J_1} \tag{10.121}$$

we can, analogous to the procedure of Section 9.6, exploit the topological features of diagrams to construct the diagrammatic expansion in terms of two-particle irreducible vertices, as depicted in Figure 10.8. In the following we leave out in the diagrammatic notation the implicit source dependences of quantities.

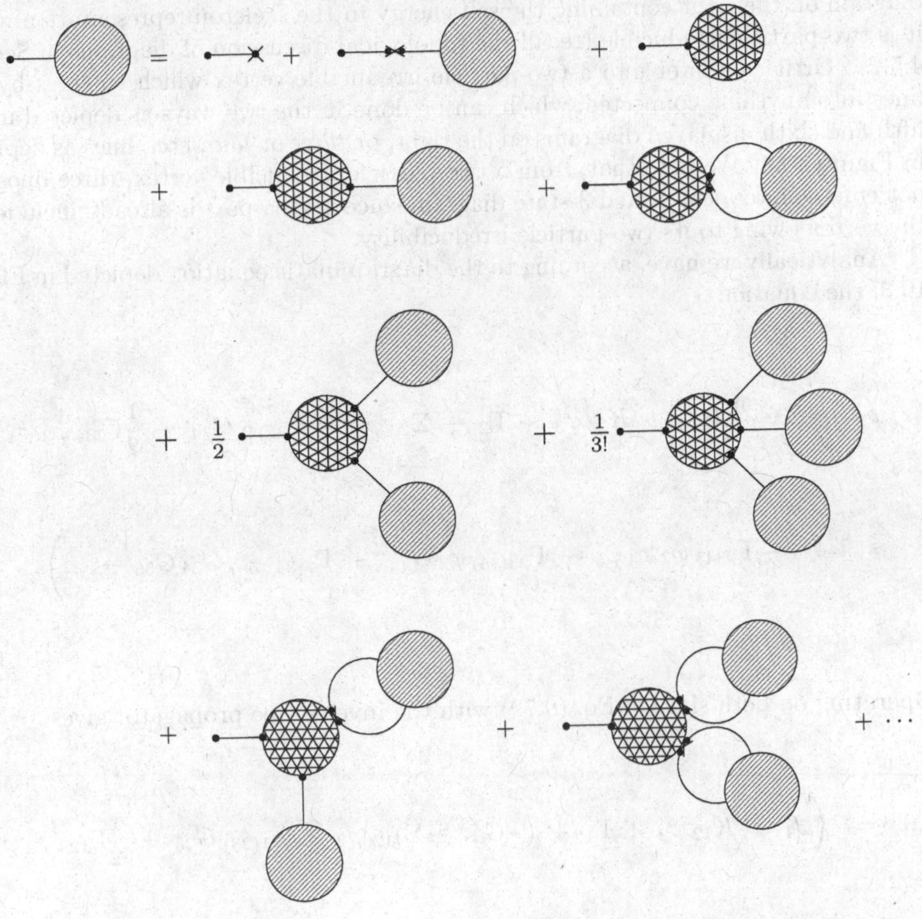

Figure 10.8 Two-particle irreducible expansion of the 1-state Green's function.

The topological arguments for the diagrammatic equation displayed in Figure 10.8 is: the particle state exposed can propagate directly to either a one-particle source or a two-particle source. In the latter case its other state can end up in anything connected, and these two classes of diagrams are depicted as the two first diagrams on the right in Figure 10.8. Or the exposed state can enter into a two-particle irreducible diagram, giving the class of diagrams represented by the third diagram on the right, or into a two-particle reducible diagram. A two-particle irreducible vertex is by definition a vertex diagram which can not be cut in two by cutting only two lines, otherwise it is two-particle reducible. In the case of entering into a two-particle reducible diagram, the exposed state can enter into a two-particle irreducible vertex which emerges by one line into anything connected, accounting for the fourth diagram on the right containing the self-energy in the skeleton representation where it is two-particle irreducible (recall the topological discussion of diagrams in Section 4.5.2). Or it can enter into a two-particle irreducible vertex which emerges by *two* lines into anything connected, which can be done in the two ways as depicted in the fifth and sixth displayed diagrams on the right, or *three* or *four*, etc., lines as depicted in Figure 10.8. We note that, from a two-particle irreducible vertex, three lines can not emerge into a connected 3-state diagram since such a part is already included in the vertex owing to its two-particle irreducibility.

Analytically we have, according to the diagrammatic equation depicted in Figure 10.8, the equation

$$
\varphi_1 \;=\; G_{12}^{(0)} \Big(J_2 \;+\; K_{23}\,\varphi_3 \;+\; \Gamma_2 \;+\; \Sigma_{23}\,\varphi_3 \;+\; \Gamma_{2(34)}\,G_{34} \;+\; \frac{1}{2}\Gamma_{234}\,\varphi_3\,\varphi_4
$$

$$
+\;\; \frac{1}{3!}\Gamma_{2345}\,\varphi_3\,\varphi_4\,\varphi_5 \;+\; \Gamma_{23(45)}\,\varphi_3\,G_{45}^{(0)} \;+\; \Gamma_{2(34)(56)}\,G_{34}\,G_{56} \;+\; \dots \Big).
$$

$$(10.122)$$

Operating on both sides of Eq. (9.79) with the inverse free propagator gives

$$
0 \;=\; \Big(J_1 \;+\; K_{12}\,\varphi_2 \;+\; \Gamma_1 \;+\; (-G_0^{-1}+\Sigma)_{12}\,\varphi_2 \;+\; \Gamma_{1(23)}\,G_{23} \;+\; \frac{1}{2}\Gamma_{123}\,\varphi_2\,\varphi_3
$$

$$
+\;\; \frac{1}{3!}\Gamma_{1234}\,\varphi_2\,\varphi_3\,\varphi_4 \;+\; \Gamma_{12(34)}\,\varphi_2\,G_{34} \;+\; \Gamma_{1(23)(45)}\,G_{23}\,G_{45} \;+\; \dots \Big),
$$

$$(10.123)$$

which corresponds to the diagrammatic equation depicted in Figure 10.9.

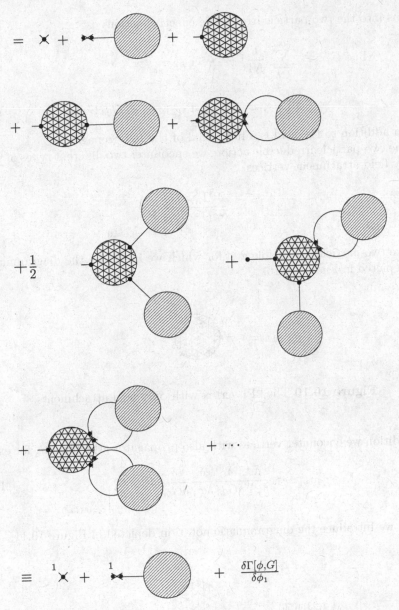

Figure 10.9 Two-particle irreducible vertices and source relation.

The last equality defines the field-derivative of the two-particle irreducible effective action, i.e, just as the diagrams in Figure 9.48 lead to the introduction of the one-particle irreducible effective action, we collect the two-particle irreducible vertex functions into the two-particle irreducible effective action

$$\Gamma[\varphi, G] \;\equiv\; \sum_{N}^{\infty} \frac{1}{N!} \, \Gamma_{12\ldots N} \, \varphi_1 \varphi_2 \cdots \varphi_N$$

$$+\; \Gamma_{1(23)} \, \varphi_1 \, G_{23} \;+\; \Gamma_{1(23)(45)} \, \varphi_1 \, G_{23} \, G_{45} \;+\; \cdots, \qquad (10.124)$$

which in addition to the field is a functional of the full propagators.

In the two-particle irreducible action, we encounter two different types of vertices, viz. only field attachment vertices

$$\Gamma_{12\ldots N} \;=\; \left. \frac{\delta^N \Gamma[\varphi, G]}{\delta \varphi_1 \cdots \delta \varphi_N} \right|_{\phi=0, G=0} \qquad (10.125)$$

which are two-particle irreducible and for which we introduce the diagrammatic notation depicted in Figure 10.10.

$$\Gamma_{12\ldots N} \;=\;$$

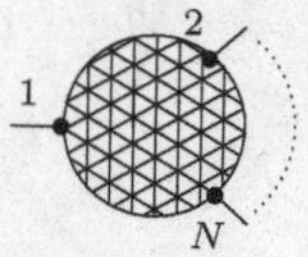

Figure 10.10 The 2PI vertex with only field attachments.

In addition we encounter vertices with also propagator attachments, for example

$$\Gamma_{1(23)4(56)} \;=\; \left. \frac{\delta}{\delta \varphi_1} \frac{\delta}{\delta G_{23}} \frac{\delta}{\delta \varphi_4} \frac{\delta}{\delta G_{56}} \, \Gamma[\varphi, G] \right|_{\phi=0, G=0} \qquad (10.126)$$

for which we introduce the diagrammatic notation depicted in Figure 10.11.

$$\Gamma_{1(23)4(56)} \;=\;$$

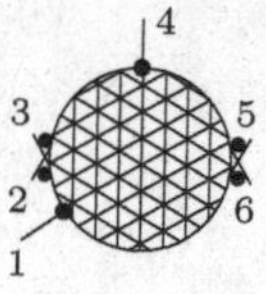

Figure 10.11 Vertex with both field and propagator attachments.

In terms of the two-particle irreducible effective action, we can diagrammatically represent the equation depicted in Figure 10.9 as depicted in Figure 10.12 (redefining $\Gamma_{12} \equiv (-G_0^{-1} + \Sigma)_{12}$).

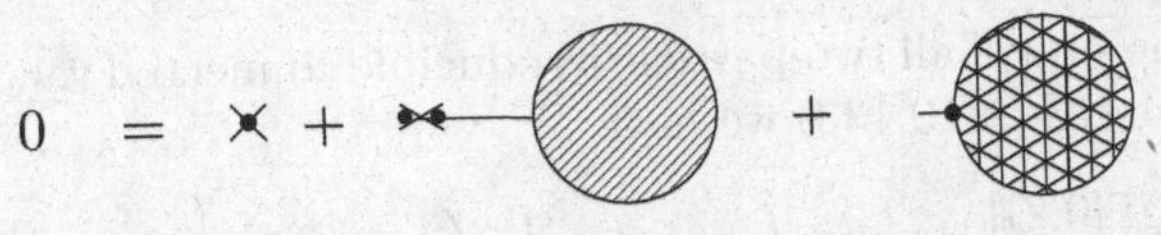

Figure 10.12 Sources and 2PI effective action relation.

Analytically, the diagrammatic relationship depicted in Figure 10.12 is

$$\frac{\delta \Gamma[\varphi, G]}{\delta \varphi_1} = -J_1 - K_{12}\, \varphi_2 \,. \tag{10.127}$$

By diagrammatic construction we have analogously to the one-particle irreducible case, $G_1 \equiv \varphi_1$,

$$G_1 = \frac{\delta W[J, K]}{\delta J_1} \tag{10.128}$$

but now in addition

$$\frac{\delta W[J, K]}{K_{12}} = \frac{1}{2}(G_{12} + \varphi_1\, \varphi_2) \tag{10.129}$$

and these two relationships give implicitly the sources as functions of the field and the full Green's function

$$J = J[\varphi, G] \tag{10.130}$$

and

$$K = K[\varphi, G] \,. \tag{10.131}$$

Since the sources are independent, so are φ and G. We then have the two generators being related by the double Legendre transformation, i.e. with respect to two sources,

$$\Gamma[\varphi, G] = \left(W[J, K] - \varphi J - \frac{1}{2}\varphi K \varphi - \frac{1}{2}GK \right)\Bigg|_{J=J[\varphi, G], K=K[\varphi, G]} \tag{10.132}$$

and we obtain the second relation for the two-particle irreducible effective action and the sources

$$\frac{\delta \Gamma[\varphi, G]}{\delta G_{12}} = -\frac{1}{2}K_{12} \,. \tag{10.133}$$

For $K = 0$ we encounter the usual Legendre transformation and effective action, i.e. $\Gamma[\varphi] = \Gamma[\varphi, G^{(0)}]$ for the value of the Green's function for which

$$\frac{\delta \Gamma[\varphi, G^{(0)}]}{\delta G_{12}^{(0)}} = 0 \,. \tag{10.134}$$

By construction $\Gamma[\varphi, G]$ is the generator, in the field variable φ, of the two-particle irreducible vertices with lines representing the full Green's function, G, and

$\Gamma[\varphi = 0, G]$ is thus the sum of all two-particle irreducible connected vacuum diagrams. Using Eq. (10.132) and Eq. (10.133) we have

$$
\Gamma[0, G] = \mathrm{Tr}\, G \frac{\delta\Gamma[0, G]}{\delta G} - i \ln \int\! \mathcal{D}\phi \; \phi \exp\left\{ i \left(S[\phi] + \phi J^{(0)} - \phi\frac{\delta\Gamma[0, G]}{\delta G}\phi \right) \right\}
$$

$$
+ \quad i \ln \int\! \mathcal{D}\phi \; \phi \exp\left\{ i S_0[\phi] \right\} , \tag{10.135}
$$

where $J^{(0)}$ is the value of the source for which $\delta W[J, K]/\delta J$ vanishes, i.e. tadpoles vanish.

By construction $\Gamma[\varphi, G]$ is the generator with respect to the field, φ, of two-particle irreducible vertex functions. For example, $\delta^2\Gamma[\varphi, G]/\delta\varphi_1\delta\varphi_2$ evaluated at vanishing field, $\varphi = 0$, is the diagrammatic expansion for the inverse two-state Green's function with two-particle reducible diagrams absent and lines representing the full Green's function, i.e.

$$
\frac{\delta^2\Gamma[\varphi, G]}{\delta\varphi_1\delta\varphi_2}\bigg|_{\varphi=0} = G_{12}^{-1} = (G^{(0)} - \Sigma[G])_{12}^{-1} . \tag{10.136}
$$

10.5.1   The 2PI loop expansion of the effective action

In this section we shall take the discussion of Section 10.4 to the next level, the two-particle irreducible (2PI) level and following Cornwall, Jackiw and Tomboulis obtain the expression for the effective action in terms of two-particle irreducible vacuum diagrams [53]. We shall use the path integral representations of the generators to first get a useful path integral expression for the two-particle irreducible effective action which has an explicit diagrammatic expansion. In the two-particle irreducible description of the previous section, physical quantities are expressed in terms of the average field and the full Green's functions. The generating functional with one- and two-particle sources, f and K, corresponding to the diagrammatic expansion in Figure 10.4 is

$$
Z[f, K] = \int\! \mathcal{D}\phi \exp\left\{ iS[\phi] + i\phi f + \frac{i}{2}\phi K\phi \right\} = e^{iW[f, K]} . \tag{10.137}
$$

The normalization constant is chosen so that $Z[f = 0, K = 0] = 1$.

The derivatives of the generating functional generate the average field

$$
\frac{\delta W}{\delta f_1} = \bar{\phi}_1 \tag{10.138}
$$

and the 2-state Green's function according to

$$
\frac{\delta W}{\delta K_{12}} = \frac{1}{2}\left(\bar{\phi}_1\,\bar{\phi}_2 + iG_{12} \right) , \tag{10.139}
$$

where

$$
iG_{12} = \overline{\phi_1\,\phi_2} - \bar{\phi}_1\bar{\phi}_2 \tag{10.140}
$$

and we use for short

$$\overline{\phi_1 \phi_2} = \int \mathcal{D}\phi \, \phi_1 \phi_2 \exp\left\{ iS[\phi] + i\phi f + \frac{i}{2}\phi K \phi \right\} \tag{10.141}$$

for the amplitude A_{12}.

The two-particle irreducible effective action, the double Legendre transform of the generating functional of connected Green's functions, Eq. (10.132)

$$\Gamma[\overline{\phi}, G] = W[f, K] - f\overline{\phi} - \frac{1}{2}\overline{\phi}K\overline{\phi} - \frac{i}{2}GK \tag{10.142}$$

fulfills

$$\frac{\delta\Gamma}{\delta\overline{\phi}} = -f - K\overline{\phi} \tag{10.143}$$

and

$$\frac{\delta\Gamma}{\delta G} = -\frac{i}{2}K . \tag{10.144}$$

The double Legendre transformation can be performed sequentially, i.e. we first define for fixed K

$$\Gamma^K[\overline{\phi}] = \left(W[f, K] - \overline{\phi}f\right)\big|_{\delta W[f,K]/\delta f = \overline{\phi}} \tag{10.145}$$

and then define G according to

$$\frac{\delta\Gamma^K[\overline{\phi}]}{\delta K} = \frac{1}{2}(\overline{\phi}\,\overline{\phi} + iG) \tag{10.146}$$

and the effective action according to

$$\Gamma[\overline{\phi}, G] = \Gamma^K[\overline{\phi}] - \frac{1}{2}\overline{\phi}K\overline{\phi} - \frac{i}{2}GK . \tag{10.147}$$

That the two definitions of the Green's function and the effective action are identical follows from the identity

$$\frac{\delta\Gamma^K[\overline{\phi}]}{\delta K} = \left(\frac{\delta W[f,K]}{\delta f}\frac{\delta f}{\delta K} + \frac{\delta W[f,K]}{\delta K} - \overline{\phi}\frac{\delta f}{\delta K} \right)\bigg|_{\delta W[f,K]/\delta f = \overline{\phi}}$$

$$= \frac{\delta W[f,K]}{\delta K}\bigg|_{\delta W[f,K]/\delta f = \overline{\phi}} . \tag{10.148}$$

Considering K as fixed, $\Gamma^K[\overline{\phi}]$ is the effective action for the theory governed by the action

$$S^K[\phi] = S[\phi] + \frac{1}{2}\phi K \phi. \tag{10.149}$$

We therefore consider the generating functional

$$Z^K[f] = \int \mathcal{D}\phi \, e^{iS^K[\phi] + i\phi f} \tag{10.150}$$

and observe

$$Z^K[f] = Z[f, K] . \tag{10.151}$$

The generating functional of connected Green's functions, for fixed K, is

$$W^K[f] = -i \ln Z^K[f] \tag{10.152}$$

with the corresponding effective action

$$\Gamma^K[\overline{\phi}] = W^K[f] - \overline{\phi} f . \tag{10.153}$$

We can now use the method of functional evaluation of the effective action of Section 10.4 and obtain

$$\Gamma^K[\overline{\phi}] = S^K[\overline{\phi}] + W_1^K[\overline{\phi}] , \tag{10.154}$$

where

$$W_1^K = -i \ln \int \mathcal{D}\phi \, \exp \left\{ i \left(S^K[\phi + \overline{\phi}] - S^K[\overline{\phi}] - \phi \frac{\delta S^K[\overline{\phi}]}{\delta \overline{\phi}} - \phi \frac{\delta W_1^K[\overline{\phi}]}{\delta \overline{\phi}} \right) \right\} . \tag{10.155}$$

Introducing the functional Γ_2 according to the equation

$$\Gamma[\overline{\phi}, G] = S[\overline{\phi}] + \frac{i}{2} \text{Tr} \ln G^{-1} + \frac{i}{2} \text{Tr} \mathcal{D}^{-1}[\overline{\phi}] G + \Gamma_2[\overline{\phi}, G] - \frac{i}{2} \text{Tr} 1 \tag{10.156}$$

with the inverse of the propagator $\mathcal{D}$ defined as

$$\mathcal{D}^{-1}[\overline{\phi}] \equiv \frac{\delta^2 S[\overline{\phi}]}{\delta \overline{\phi} \, \delta \overline{\phi}} \tag{10.157}$$

and using Eqs. (10.147) and (10.154) we have

$$\Gamma_2[\overline{\phi}, G] = -\frac{1}{2} \text{Tr} \left(i \mathcal{D}^{-1}[\overline{\phi}] + K \right) G - \frac{i}{2} \text{Tr} \ln G^{-1} + W_1^K[\overline{\phi}] + \frac{i}{2} \text{Tr} 1 . \tag{10.158}$$

Lastly, we want to show that Γ_2 is the sum of all the two-particle irreducible vacuum graphs in a theory with vertices determined by the action

$$S_{\text{int}}[\phi; \overline{\phi}] = \sum_{N=3}^{\infty} \frac{1}{N!} \frac{\delta^N S[\overline{\phi}]}{\delta \overline{\phi}_1 \cdots \delta \overline{\phi}_N} \phi_1 \cdots \phi_N , \tag{10.159}$$

and propagator lines by the full Green's function G. In order to do so we first eliminate the two-particle source K

$$K = 2i \frac{\delta \Gamma[\overline{\phi}, G]}{\delta G} = G^{-1} - \mathcal{D}^{-1}[\overline{\phi}] + 2i \frac{\delta \Gamma_2[\overline{\phi}, G]}{\delta G} . \tag{10.160}$$

Using Eqs. (10.147), (10.154) and (10.155), the effective action, $\Gamma[\overline{\phi}, G]$, can be rewritten as a functional integral

$$e^{i\Gamma[\overline{\phi}, G]} = \int \mathcal{D}\phi \, \exp \left\{ i \left(S^K[\phi + \overline{\phi}] - S^K[\overline{\phi}] - \phi \frac{\delta S^K[\overline{\phi}]}{\delta \overline{\phi}} - \phi \frac{\delta W_1^K[\overline{\phi}]}{\delta \overline{\phi}} \right) \right\}$$

$$\times \; e^{i(S^K[\overline{\phi}] - \frac{1}{2}\overline{\phi} K \overline{\phi} - \frac{i}{2} G K)} \equiv e^{i(S^K[\overline{\phi}] - \frac{1}{2}\overline{\phi} K \overline{\phi} - \frac{i}{2} G K)} Z_1^K[\overline{\phi}] . \tag{10.161}$$

10.5. Two-particle irreducible effective action

Introducing the generator

$$\tilde{Z}^K[\overline{\phi}, J] = \int \mathcal{D}\phi \, \exp\left\{ i\left((S^K[\phi + \overline{\phi}] - S^K[\overline{\phi}] - \phi \frac{\delta S^K[\overline{\phi}]}{\delta \overline{\phi}} + \phi J \right) \right\} \qquad (10.162)$$

a calculation similar to the one of Section 10.4 gives

$$\left. \frac{\delta \tilde{Z}^K[\overline{\phi}, J]}{\delta J} \right|_{J = -\delta W_1^K / \delta \overline{\phi}} = 0 \,. \qquad (10.163)$$

The average value of ϕ has thus been shown to vanish in the theory governed by the action

$$S^K[\overline{\phi}, \phi] = S^K[\phi + \overline{\phi}] - S^K[\overline{\phi}] - \phi \frac{\delta S^K}{\delta \overline{\phi}} \qquad (10.164)$$

when the source takes the value $J = -\delta W_1^K / \delta \overline{\phi}$. If the generating functional $\tilde{Z}^K[\overline{\phi}, J]$ is multiplied by a factor depending on G and $\overline{\phi}$ the average value of ϕ is still zero. Using Eqs. (10.143), (10.147) and (10.154) we therefore have

$$f + \frac{\delta S^K[\overline{\phi}]}{\delta \overline{\phi}} + \frac{\delta W_1^K[\overline{\phi}]}{\delta \overline{\phi}} = 0 \qquad (10.165)$$

and obtain the following functional integral expression for the two-particle irreducible effective action

$$e^{i\Gamma[\overline{\phi}, G]} = e^{-\frac{i}{2}\overline{\phi}K\overline{\phi} + \frac{1}{2}GK} \int \mathcal{D}\phi \, e^{i(S^K[\phi + \overline{\phi}] + f\phi)} \,. \qquad (10.166)$$

Using Eqs. (10.143) and (10.144) to eliminate the source f

$$f = -\frac{\delta \Gamma}{\delta \overline{\phi}} - 2i \frac{\delta \Gamma}{\delta G} \overline{\phi} \qquad (10.167)$$

we obtain

$$\Gamma[\overline{\phi}, G] - G \frac{\delta \Gamma[\overline{\phi}, G]}{\delta G} = -i \ln \int \mathcal{D}\phi \, e^{iS[\overline{\phi}, G; \phi]} \qquad (10.168)$$

where

$$S[\overline{\phi}, G; \phi] = S[\overline{\phi} + \phi] - \phi \frac{\delta \Gamma[\overline{\phi}, G]}{\delta \overline{\phi}} + i\phi \frac{\delta \Gamma[\overline{\phi}, G]}{\delta G} \phi \,. \qquad (10.169)$$

Differentiating Eq. (10.168) with respect to G we obtain

$$0 = \frac{\delta^2 \Gamma[\overline{\phi}, G]}{\delta G \, \delta G} G - \frac{\delta^2 \Gamma[\overline{\phi}, G]}{\delta G \, \delta \overline{\phi}} \langle \phi \rangle - \frac{\delta^2 \Gamma[\overline{\phi}, G]}{\delta G \, \delta G} i \langle \phi \phi \rangle \,, \qquad (10.170)$$

where the angle brackets denote the average with respect to the action $S[\overline{\phi}, G; \phi]$. The action in Eq. (10.164) with the source term $-\delta W_1^K / \delta \overline{\phi}$ added and the action appearing in Eq. (10.169) differ only by an irrelevant constant, $-S[\overline{\phi}]$, and we can conclude that the average value of the field is zero for the action $S[\overline{\phi}, G; \phi]$, i.e. $\langle \phi \rangle = 0$, and we obtain that

$$G = -i\langle \phi\phi \rangle \qquad (10.171)$$

i.e. G is the full Green's function for the theory governed by the action $S[\overline{\phi}, G; \phi]$.
    Finally we rewrite Eq. (10.160)

$$G^{-1} = \mathcal{D}^{-1}[\overline{\phi}] + K - \Sigma[\overline{\phi}, G] \,, \qquad (10.172)$$

where

$$\Sigma[\overline{\phi}, G] = 2i\frac{\delta\Gamma_2[\overline{\phi}, G]}{\delta G}. \qquad (10.173)$$

Since $\mathcal{D}^{-1}[\overline{\phi}] + K$ is the free inverse Green's function and G^{-1} is the inverse full Green's function for the theory governed by the action in Eq. (10.169), we conclude that Σ is the self-energy, and Eq. (10.172) thereby the Dyson equation. Since the self-energy, Σ, is the sum of one-particle irreducible connected vacuum diagrams, we therefore finally conclude that Γ_2 is given by the sum of two-particle irreducible connected vacuum diagrams.

    We have thus shown that the effective action can be written in the form

$$\Gamma[\overline{\phi}, G] = S[\overline{\phi}] + \frac{i}{2}\mathrm{Tr}\ln G^{-1} + \frac{i}{2}\mathrm{Tr}\mathcal{D}^{-1}[\overline{\phi}]\,G + \Gamma_2[\overline{\phi}, G] - \frac{i}{2}\mathrm{Tr}1, \qquad (10.174)$$

where $\Gamma_2[\overline{\phi}, G]$ is the sum of all two-particle irreducible connected vacuum diagrams in the theory with action $\phi G^{-1}\phi/2 + S_{\mathrm{int}}[\phi : \overline{\phi}]$, i.e.

$$\Gamma_2[\overline{\phi}, G] = -i\ln\langle e^{iS_{\mathrm{int}}[\overline{\phi};\phi]}\rangle_G^{\mathrm{2PI}}, \qquad (10.175)$$

where the superscript and subscript on the angle bracket indicate that the functional integral is restricted to the two-particle irreducible vacuum diagrams and the propagator lines are the full Green's function.

    In general amplitudes or physical quantities can not be calculated exactly, and an approximation scheme must be invoked. If no small dimensionless expansion parameter is available we are at a loss. Furthermore, if non-perturbative effects are prevalent we are left without a general tool to obtain information. To cope with such situations, approximate self-consistent or mean field theories have been useful, although they are uncontrollable as not easily analytically characterized by a small parameter. The effective action approach can be used to systematically study correlations order by order in the loop expansion. It is thus the general starting point for constructing self-consistent approximations. An important feature of the loop expansion is that it is capable of capturing important nonlinearities of a theory. In practice one must at a certain order break the chain of correlations described by the effective action by brute force, a felony we are quite used to in kinetic theory. The rationale behind this scheme working quite well for calculating average properties such as densities and currents is that higher-order correlations average out when interest is in such low-correlation probes. We shall use the effective action approach to study classical statistical dynamics in Chapter 12, but first we apply it in the quantum context, viz. for the study of Bose gases.

10.6 Effective action approach to Bose gases

In this section, the effective action formalism is applied to a gas of bosons.[17] The equations describing the condensate and the excitations are obtained by using the loop expansion for the effective action. For a homogeneous gas, the expansion in terms of the diluteness parameter is identified in terms of the loop expansion. The loop expansion and the limits of validity of the well-known Bogoliubov and Popov equations are examined analytically for a homogeneous dilute Bose gas and numerically for a gas trapped in a harmonic-oscillator potential. The expansion to one-loop order, and hence the Bogoliubov equation, we shall show to be valid for the zero-temperature trapped gas as long as the characteristic length of the trapping potential exceeds the s-wave scattering length.

10.6.1 Dilute Bose gases

The dilute Bose gas has been subject to extensive study for more than half a century, originally in an attempt to understand liquid Helium II, but also as an interesting many-body system in its own right. In 1947, Bogoliubov showed how to describe Bose–Einstein condensation as a state of broken symmetry, in which the expectation values of the field operators are non-vanishing due to the single-particle state of lowest energy being macroscopically occupied, i.e. the annihilation and creation operators for the lowest-energy mode can be treated as c-numbers [55]. In modern terminology, the expectation value of the field operator is the order parameter and describes the density of the condensed bosons. In Bogoliubov's treatment, the physical quantities were expanded in the diluteness parameter $\sqrt{n_0 a^3}$, where n_0 denotes the density of bosons occupying the lowest single-particle energy state, and a is the s-wave scattering length, and Bogoliubov's theory is therefore valid only for homogeneous dilute Bose gases. The inhomogeneous Bose gas was studied by Gross and Pitaevskii, who independently derived a nonlinear equation determining the condensate density [56] [57]. A field-theoretic diagrammatic treatment was applied by Beliaev to the zero-temperature homogeneous dilute Bose gas, showing how to go beyond Bogoliubov's approximation in a systematic expansion in the diluteness parameter $\sqrt{n_0 a^3}$; and also showing how repeated scattering leads to a renormalization of the interaction between the bosons [58, 59]. This renormalization in Beliaev's treatment was a cumbersome issue, where diagrams expressed in terms of the propagator for the non-interacting particles are intermixed with diagrams where the propagator contains the interaction potential. Beliaev's diagrammatic scheme was extended to finite temperatures by Popov and Faddeev [60], and was subsequently employed to extend the Bogoliubov theory to finite temperatures by incorporating terms containing the excited-state operators to lowest order in the interaction potential [61, 62].

A surge of interest in the dilute Bose gas due to the experimental creation of gaseous Bose–Einstein condensates occurred in the mid-1990s [63]. The atomic condensates in the experiments are confined in external potentials, which poses new theoretical challenges; especially, the Beliaev expansion in the diluteness parameter $\sqrt{n_0 a^3}$ is questionable when the density is inhomogeneous. Experiments on trapped

---

[17] In this section we essentially follow reference [54].

Bose gases employ Feshbach resonances to probe the regime of large scattering length, and hence large values of the diluteness parameter. It is therefore of importance to understand the low-density approximations to the exact equations of motion and the corrections thereto. In the following, we shall employ the two-particle irreducible effective action approach, and show that it provides an efficient systematic scheme for dealing with both homogeneous Bose gases and trapped Bose gases. We demonstrate how the effective action formalism can be used to derive the equations of motion for the dilute Bose gas, and more importantly, that the loop expansion can be used to determine the limits of validity of approximations to the exact equations of motion in the trapped case.

10.6.2 Effective action formalism for bosons

A system of spinless non-relativistic bosons is according to Eq. (3.68) and Eq. (10.37) described by the action

$$S[\psi, \psi^\dagger] = \int d\mathbf{r} dt \, \psi^\dagger(\mathbf{r}, t) \left[i \partial_t - h(\mathbf{r}) + \mu \right] \psi(\mathbf{r}, t)$$

$$- \frac{1}{2} \int d\mathbf{r} d\mathbf{r}' dt \, \psi^\dagger(\mathbf{r}, t) \psi^\dagger(\mathbf{r}', t) \, U(\mathbf{r} - \mathbf{r}') \, \psi(\mathbf{r}', t) \psi(\mathbf{r}, t) \,, \quad (10.176)$$

where ψ is the scalar field describing the bosons, and μ the chemical potential. The one-particle Hamiltonian, $h = \mathbf{p}^2/2m + V(\mathbf{r})$, consists of the kinetic term and an external potential, and $U(\mathbf{r})$ is the potential describing the interaction between the bosons. As usual we introduce a matrix notation whereby the field and its complex conjugate are combined into a two-component field $\phi = (\psi, \psi^\dagger) = (\phi_1, \phi_2)$.

The correlation functions of the bose field are obtained from the generating functional

$$Z[\eta, K] = \int \mathcal{D}\phi \exp\left(iS[\phi] + i\eta^\dagger \phi + \frac{i}{2} \phi^\dagger K \phi \right) \quad (10.177)$$

by differentiating with respect to the source $\eta^\dagger = (\eta, \eta^*) = (\eta_1, \eta_2)$. Here $\eta(\mathbf{r}, t)$ denotes a complex scalar field, not a Grassmannn variable, as we are considering bosons. A two-particle source term, K, has been added to the action in the generating functional in order to obtain equations involving the two-point Green's function in a two-particle irreducible fashion as discussed in Section 10.5.1.

The generator of the connected Green's functions is

$$W[\eta, K] = -i \ln Z[\eta, K], \quad (10.178)$$

and the derivative

$$\frac{\delta W}{\delta \eta_i(\mathbf{r}, t)} = \bar{\phi}_i(\mathbf{r}, t) \quad (10.179)$$

gives the average field, $\bar{\phi}$, with respect to the action $S[\phi] + \eta^\dagger \phi + \phi^\dagger K \phi / 2$,

$$\bar{\phi}(\mathbf{r}, t) = \begin{pmatrix} \Phi(\mathbf{r}, t) \\ \Phi^*(\mathbf{r}, t) \end{pmatrix} = \int \mathcal{D}\phi \, \phi(\mathbf{r}, t) \exp\left(iS[\phi] + i\eta^\dagger \phi + \frac{i}{2} \phi^\dagger K \phi \right) = \langle \phi(\mathbf{r}, t) \rangle.$$

$$(10.180)$$

The average field Φ is seen to specify the condensate density and is referred to as the condensate wave function.[18]

The derivative of W with respect to the two-particle source is (recall Figure 10.7)

$$\frac{\delta W}{\delta K_{ij}(\mathbf{r}, t; \mathbf{r}', t')} = \frac{1}{2} \bar{\phi}_i(\mathbf{r}, t) \, \bar{\phi}_j(\mathbf{r}', t') + \frac{i}{2} G_{ij}(\mathbf{r}, t, \mathbf{r}', t') , \qquad (10.181)$$

where G is the full connected two-point matrix Green's function describing the bosons not in the condensate

$$G_{ij}(\mathbf{r}, t, \mathbf{r}', t') = -\frac{\delta^2 W}{\delta \eta_i(\mathbf{r}, t) \, \delta \eta_j(\mathbf{r}', t')}$$

$$= -i \begin{pmatrix} \langle \delta\psi(\mathbf{r}, t) \delta\psi^\dagger(\mathbf{r}', t') \rangle & \langle \delta\psi(\mathbf{r}, t) \delta\psi(\mathbf{r}', t') \rangle \\ \langle \delta\psi^\dagger(\mathbf{r}, t) \delta\psi^\dagger(\mathbf{r}', t') \rangle & \langle \delta\psi^\dagger(\mathbf{r}, t) \delta\psi(\mathbf{r}', t') \rangle \end{pmatrix} , \qquad (10.182)$$

where $\delta\psi(\mathbf{r}, t)$ is the deviation of the field from its mean value, $\delta\psi = \psi - \Phi$. Likewise, we shall write $\phi = \bar{\phi} + \delta\phi$ for the two-component field. We recall that in the path integral representation, averages over fields, such as in Eq. (10.182), are automatically time ordered.

We then introduce the effective action for the bosons, Γ, the generator of the two-particle irreducible vertex functions, through the Legendre transform of the generator of connected Green's functions, W,

$$\Gamma[\bar{\phi}, G] = W[\eta, K] - \eta^\dagger \bar{\phi} - \frac{1}{2} \bar{\phi}^\dagger K \bar{\phi} - \frac{i}{2} \mathrm{Tr} GK . \qquad (10.183)$$

The effective action satisfies according to section 10.5.1 the equations

$$\frac{\delta\Gamma}{\delta\bar{\phi}} = -\eta - K\bar{\phi} , \qquad \frac{\delta\Gamma}{\delta G} = -\frac{i}{2} K . \qquad (10.184)$$

In a physical state where the external sources vanish, $\eta = 0 = K$, the variations of the effective action with respect to the field averages $\bar{\phi}$ and G vanish, yielding the equations of motion

$$\frac{\delta\Gamma}{\delta\bar{\phi}} = 0 \qquad (10.185)$$

and

$$\frac{\delta\Gamma}{\delta G} = 0 . \qquad (10.186)$$

---

[18] Indeed, as pointed out by Penrose and Onsager, Bose–Einstein condensation is associated with off-diagonal long-range order in the two-point correlation function $\lim_{r \to \infty} \langle \psi^\dagger(\mathbf{r}) \psi(\mathbf{0}) \rangle = \langle \psi^\dagger(\mathbf{r}) \rangle \langle \psi(\mathbf{0}) \rangle \neq 0$ [64]. For a conventional description of bosons in terms of field operators we refer to reference [15]. We note that, in the presented effective-action approach, the inherent additional necessary considerations associated with the macroscopic occupation of the ground state in the conventional description is conveniently absent.

According to Section 10.5.1, the effective action can be written in the form

$$\Gamma[\bar\phi, G] = S[\bar\phi] + \frac{i}{2}\operatorname{Tr}\ln G_0 G^{-1} + \frac{i}{2}\operatorname{Tr}(G_0^{-1} - \Sigma^{(1)})G - \frac{i}{2}\operatorname{Tr}1 + \Gamma_2[\bar\phi, G] , \quad (10.187)$$

where G_0 is the non-interacting matrix Green's function,

$$G_0^{-1}(\mathbf{r}, t, \mathbf{r}', t') = -\begin{pmatrix} i\partial_t - h + \mu & 0 \\ 0 & -i\partial_t - h + \mu \end{pmatrix}\delta(\mathbf{r} - \mathbf{r}')\delta(t - t') \quad (10.188)$$

and the matrix

$$\Sigma^{(1)}(\mathbf{r}, t, \mathbf{r}', t') = -\left.\frac{\delta^2 S}{\delta\phi^\dagger(\mathbf{r}, t)\delta\phi(\mathbf{r}', t')}\right|_{\phi=\bar\phi} + G_0^{-1}(\mathbf{r}, t, \mathbf{r}', t') \quad (10.189)$$

will turn out to be the self-energy to one-loop order (see Eq. (10.204)). Using the action describing the bosons, Eq. (10.176), we obtain for the components

$$\Sigma_{ij}^{(1)}(\mathbf{r}, t, \mathbf{r}', t') = \delta(t - t') \, \Sigma_{ij}^{(1)}(\mathbf{r}, \mathbf{r}') \quad (10.190)$$

where

$$\Sigma_{11}^{(1)}(\mathbf{r}, \mathbf{r}') = \delta(\mathbf{r}' - \mathbf{r}) \int d\mathbf{r}'' U(\mathbf{r} - \mathbf{r}'')|\Phi(\mathbf{r}'', t)|^2 + U(\mathbf{r} - \mathbf{r}')\Phi^*(\mathbf{r}', t)\Phi(\mathbf{r}, t)$$

$$(10.191)$$

and

$$\Sigma_{12}^{(1)}(\mathbf{r}, \mathbf{r}') = U(\mathbf{r} - \mathbf{r}')\Phi(\mathbf{r}, t)\Phi(\mathbf{r}', t) \quad (10.192)$$

and

$$\Sigma_{21}^{(1)}(\mathbf{r}, \mathbf{r}') = U(\mathbf{r} - \mathbf{r}')\Phi^*(\mathbf{r}, t)\Phi^*(\mathbf{r}', t) \quad (10.193)$$

and

$$\Sigma_{22}^{(1)}(\mathbf{r}, \mathbf{r}') = \delta(\mathbf{r}' - \mathbf{r}) \int d\mathbf{r}'' U(\mathbf{r} - \mathbf{r}'')|\Phi(\mathbf{r}'', t)|^2 + U(\mathbf{r} - \mathbf{r}')\Phi^*(\mathbf{r}, t)\Phi(\mathbf{r}', t) .$$

$$(10.194)$$

The delta function in the time coordinates reflects the fact that the interaction is instantaneous. Finally, the quantity Γ_2 in Eq. (10.187) is

$$\Gamma_2 = -i \ln\langle e^{iS_{\text{int}}[\bar\phi, \delta\phi]}\rangle_G^{2\text{PI}} , \quad (10.195)$$

where $S_{\text{int}}[\bar\phi, \delta\phi]$ denotes the part of the action $S[\bar\phi + \delta\phi]$ which is higher than second order in $\delta\phi$ in an expansion around the average field. The quantity Γ_2 is conveniently described in terms of the diagrams generated by the action $S_{\text{int}}[\bar\phi, \delta\phi]$, and consists of all the two-particle irreducible vacuum diagrams as indicated by the superscript "2PI", and the diagrams will therefore contain two or more loops. The subscript indicates that propagator lines represent the full Green's function G, i.e. the brackets with subscript G denote the average

$$\langle e^{iS_{\text{int}}[\bar\phi,\delta\phi]}\rangle_G = (\det iG)^{-1/2} \int \mathcal{D}(\delta\phi)\, e^{\frac{i}{2}\delta\phi^\dagger G^{-1}\delta\phi}\, e^{iS_{\text{int}}[\bar\phi,\delta\phi]} . \tag{10.196}$$

The diagrammatic expansion of Γ_2 corresponding to the action for the bosons, Eq. (10.176), is illustrated in Figure 10.13 where the two- and three-loop vacuum diagrams are shown.

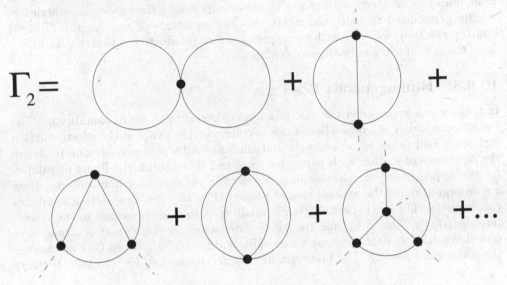

Figure 10.13 Two-loop (upper row) and three-loop vacuum diagrams (lower row) contributing to the effective action.

Since matrix indices are suppressed, the diagrams in Figure 10.13 are to be understood as follows. Full lines represent full boson Green's functions and in the cases where we display the different components explicitly, G_{11} will carry one arrow (according to Eq. (10.182) G_{22} can be expressed in terms of G_{11} and thus needs no special symbol), G_{12} has two arrows pointing inward and G_{21} carries two arrows pointing outward. Dashed lines represent the condensate wave function and can also be decorated with arrows, directed out from the vertex to represent Φ, or directed towards the vertex representing Φ^*. The dots where four lines meet are interaction vertices, i.e. they represent the interaction potential U (which in other contexts will be represented by a wiggly line). When all possibilities for the indices are exhausted, subject to the condition that each vertex has two in-going and two out-going particle lines, we have represented all the terms of Γ_2 to a given loop order. Finally, the expression corresponding to each vacuum diagram should be multiplied by the factor i^{s-2}, where s is the number of loops the diagram contains. In the effective action approach, the appearance of the condensate wave function in the diagrams is automatic, and as noted generally in Section 9.6.1, the approach is well suited to describe broken-symmetry states.

The expansion of the effective action in loop orders was shown in Section 10.3 to be an expansion in Planck's constant. The first term on the right-hand side of Eq. (10.187), $S[\bar{\phi}]$, the zero-loop term, is proportional to $\hbar^0$, and the terms where the trace is written explicitly, the one-loop terms, are proportional to $\hbar^1$. We stress that the effective action approach presented in this chapter is capable of describing arbitrary states, including non-equilibrium situations where the external potential depends on time. Although we in the following in explicit calculations shall limit ourselves to study a Bose gas at zero temperature the theory is straightforwardly generalized to finite temperatures. The equations of motion, Eq. (10.185) and Eq. (10.186), together with the expression for the effective action, Eq. (10.187), form the basis for the subsequent calculations.

10.6.3   Homogeneous Bose gas

In this section we consider the case of a homogeneous Bose gas in equilibrium. The equilibrium theory of a dilute Bose gas is of course well known, but the effective action formalism will prove to be a simple and efficient tool which permits one to derive the equations of motion with particular ease, and to establish the limits of validity for the approximate descriptions often used. For the case of a homogeneous Bose gas in equilibrium, the general theory presented in the previous section simplifies considerably. The single-particle Hamiltonian, h, is then simply equal to the kinetic term, $h(\mathbf{p}) = \mathbf{p}^2/2m \equiv \varepsilon_{\mathbf{p}}$, and the condensate wave function $\Phi(\mathbf{r}, t)$ is a time- and coordinate-independent constant whose value is denoted by $\sqrt{n_0}$, so that n_0 denotes the condensate density. The first term in the effective action, Eq. (10.187), is then

$$S[\Phi] = (\mu n_0 - \frac{1}{2} U_0 n_0^2) \int d\mathbf{r} dt\, 1 \tag{10.197}$$

where

$$U_0 = \int d\mathbf{r}\, U(\mathbf{r}) \tag{10.198}$$

is the zero-momentum component of the interaction potential. For a constant value of the condensate wave function, $\Phi(\mathbf{r}, t) = \sqrt{n_0}$, Eq. (10.194) yields

$$\Sigma^{(1)}(\mathbf{p}) = \begin{pmatrix} n_0(U_0 + U_{\mathbf{p}}) & n_0 U_{\mathbf{p}} \\ n_0 U_{\mathbf{p}} & n_0(U_0 + U_{\mathbf{p}}) \end{pmatrix}. \tag{10.199}$$

Varying, in accordance with Eq. (10.185), the effective action, Eq. (10.187), with respect to n_0 yields the equation for the chemical potential

$$\mu = n_0 U_0 + \frac{i}{2} \int \frac{d^4 p}{(2\pi)^4} [(U_0 + U_{\mathbf{p}})(G_{11}(p) + G_{22}(p)) + U_{\mathbf{p}}(G_{12}(p) + G_{21}(p))]$$

$$- \frac{\delta \Gamma_2}{\delta n_0} \tag{10.200}$$

where the notation for the four-momentum, $p = (\mathbf{p}, \omega)$, has been introduced. The first term on the right-hand side is the zero-loop result, which depends only on the condensate fraction of the bosons. The second term on the right-hand side is the one-loop term which takes the noncondensate fraction of the bosons into account. The term involving the anomalous Green's functions G_{12} and G_{21} will shortly, in Section 10.6.4, be absorbed by the renormalization of the interaction potential. From the last term originate the higher-loop terms, which will be dealt with at the end of this section.

The equation determining the Green's function is obtained by varying the effective action with respect to the matrix Green's function $G(p)$, in accordance with Eq. (10.186), yielding

$$0 = \frac{\delta \Gamma}{\delta G} = -\frac{i}{2} \left(-G^{-1} + G_0^{-1} + \Sigma^{(1)} + \Sigma' \right) , \qquad (10.201)$$

where

$$\Sigma'_{ij} = 2i \frac{\delta \Gamma_2}{\delta G_{ji}} . \qquad (10.202)$$

Introducing the notation for the matrix self-energy

$$\Sigma = \Sigma^{(1)} + \Sigma' \qquad (10.203)$$

Eq. (10.201) is seen to be the Dyson equation

$$G^{-1} = G_0^{-1} - \Sigma . \qquad (10.204)$$

In the context of the dilute Bose gas, this equation is referred to as the Dyson–Beliaev equation.

The Green's function in momentum space is obtained by simply inverting the 2×2 matrix $G_0^{-1}(p) - \Sigma(p)$ resulting in the following components

$$G_{11}(p) = \frac{\omega + \varepsilon_{\mathbf{p}} - \mu + \Sigma_{22}(p)}{D_p} \quad , \quad G_{12}(p) = \frac{-\Sigma_{12}(p)}{D_p} \qquad (10.205)$$

and

$$G_{21}(p) = \frac{-\Sigma_{21}(p)}{D_p} \quad , \quad G_{22}(p) = \frac{-\omega + \varepsilon_{\mathbf{p}} - \mu + \Sigma_{11}(p)}{D_p} \qquad (10.206)$$

all having the common denominator

$$D_p = (\omega + \varepsilon_{\mathbf{p}} - \mu + \Sigma_{22}(p))(\omega - \varepsilon_{\mathbf{p}} + \mu - \Sigma_{11}(p)) + \Sigma_{12}(p)\Sigma_{21}(p). \qquad (10.207)$$

From the expression for the matrix Green's function, Eq. (10.182), it follows that in the homogeneous case its components obey the relationships

$$G_{22}(p) = G_{11}(-p) \quad , \quad G_{12}(-p) = G_{12}(p) = G_{21}(p) . \qquad (10.208)$$

The corresponding relations hold for the self-energy components. We note that the results found for μ and G to zero- and one-loop order coincide with those found in reference [58] to zeroth and first order in the diluteness parameter $\sqrt{n_0 a^3}$. For example, according to Eq. (10.199) we obtain for the components of the matrix Green's function to one loop-order

$$G_{11}^{(1)}(p) = \frac{\omega + \varepsilon_{\mathbf{p}} + n_0 U_{\mathbf{p}}}{\omega^2 - \varepsilon_{\mathbf{p}}^2 - 2n_0 U_{\mathbf{p}} \varepsilon_{\mathbf{p}}} \quad , \quad G_{12}^{(1)}(p) = \frac{-n_0 U_{\mathbf{p}}}{\omega^2 - \varepsilon_{\mathbf{p}}^2 - 2n_0 U_{\mathbf{p}} \varepsilon_{\mathbf{p}}} \,, \quad (10.209)$$

which are the same expressions as the ones in reference [58]. As we shortly demonstrate, the loop expansion for the case of a homogeneous Bose gas is in fact equivalent to an expansion in the diluteness parameter. From Eq. (10.209) we obtain for the single-particle excitation energies to one-loop order

$$E_{\mathbf{p}} = \sqrt{\varepsilon_{\mathbf{p}}^2 + 2n_0 U_{\mathbf{p}} \varepsilon_{\mathbf{p}}} \tag{10.210}$$

which are the well-known Bogoliubov energies [55].

Differentiating with respect to n_0 the terms in Γ_2 corresponding to the two-loop vacuum diagrams gives the two-loop contribution to the chemical potential. Functionally differentiating the same terms with respect to G_{ji} gives the two-loop contributions to the self-energies Σ_{ij}. The diagrams we thus obtain for the chemical potential μ and the self-energy Σ are topologically identical to those found originally by Beliaev [59]; however, the interpretation differs in that the propagator in the vacuum diagrams of Figure 10.13 is the exact propagator, whereas in reference [59] the propagator to one-loop order appears.

In order to establish that the loop expansion for a homogeneous Bose gas is an expansion in the diluteness parameter $\sqrt{n_0 a^3}$, we examine the general structure of the vacuum diagrams comprised by Γ_2. Any diagram of a given loop order differs from any diagram in the preceding loop order by an extra four-momentum integration, the condensate density n_0 to some power k, the interaction potential U to the power $k+1$, and $k+2$ additional Green's functions in the integrand. We can estimate the contribution from these terms as follows. The Green's functions are approximated by the one-loop result Eq. (10.209). The additional frequency integration over a product of $k+2$ Green's functions yields $k+2$ factors of $n_0 U$ (where U denotes the typical magnitude of the Fourier transform of the interaction potential), divided by $2k+3$ factors of the Bogoliubov energy E. The range of the momentum integration provided by the Green's functions is $(mn_0 U)^{1/2}$. The remaining three-momentum integration therefore gives a factor of order $n_0^{-k+1/2} m^{3/2} U^{-k+1/2}$, and provided the Green's functions make the integral converge, the contribution from an additional loop is of the order $(n_0 m^3 U^3)^{1/2}$. This is the case except for the ladder diagrams, in which case the convergence need to be provided by the momentum dependence of the potential. The ladder diagrams will be dealt with separately in the next section where we show that they, through a renormalization of the interaction potential, lead to the appearance of the t-matrix which in the dilute limit is proportional to the s-wave scattering length a and inversely proportional to the boson mass. The renormalization of the interaction potential will therefore not change the estimates performed above, but change only the expansion parameter. Anticipating this change we conclude that the expansion parameter governing the loop expansion is for a homogeneous Bose gas indeed identical to Bogoliubov's diluteness parameter $\sqrt{n_0 a^3}$.

10.6.4 Renormalization of the interaction

Instead of having the interaction potential appear explicitly in diagrams, one should work in the skeleton diagram representation where diagrams are partially summed so that the four-point vertex appears instead of the interaction potential, thus accounting for the repeated scattering of the bosons. In the dilute limit, where the inter-particle distance is large compared to the s-wave scattering length, the ladder diagrams give as usual the largest contribution to the four-point vertex function. The ladder diagrams are depicted in Figure 10.14.

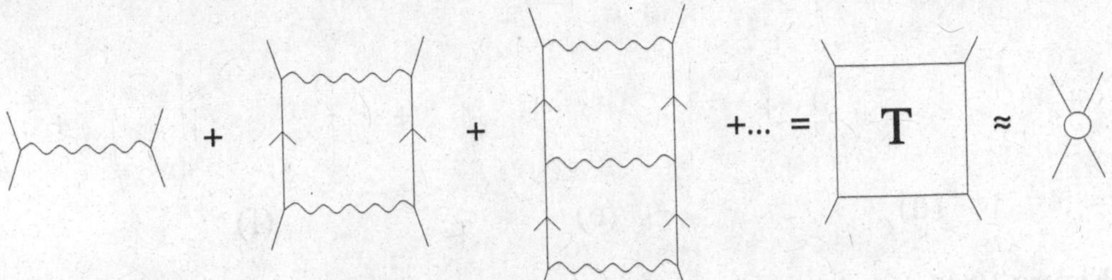

Figure 10.14 Summing all diagrams of the *ladder* type results in the t-matrix, which to lowest order in the diluteness parameter is a momentum-independent constant g, diagrammatically represented by a circle.

On calculating the corresponding integrals, it is found that an extra *rung* in a ladder contributes with a factor proportional not to $\sqrt{n_0 m^3 U^3}$, as was the case for the type of extra loops considered in the previous section, but to $k_0 m U$, where k_0 is the upper momentum cut-off (or inverse spatial range) of the potential, as first noted by Beliaev [59]. The quantity $k_0 m U$ is not necessarily small for the atomic gases under consideration here. Hence, all vacuum diagrams which differ only in the number of ladder rungs that they contain are of the same order in the diluteness parameter, and we have to perform a summation over this infinite class of diagrams. The ladder resummation results in an effective potential $T(p, p', q)$, referred to as the t-matrix and is a function of the two ingoing momenta and the four-momentum transfer. Owing to the instantaneous nature of the interactions, the t-matrix does not depend on the frequency components of the in-going four-momenta, but for notational convenience we display the dependence as $T(p, p', q)$. To lowest order in the diluteness parameter, the t-matrix is independent of four-momenta and proportional to the constant scattering amplitude, $T(0, 0, 0) = 4\pi\hbar^2 a/m = g$. This is illustrated in Figure 10.14, where we have chosen an open circle to represent g. Iterating the equation for the ladder diagrams we obtain the t-matrix equation

$$T(p, p', q) = U_{\mathbf{q}} + i \int d^4 q' \, U_{\mathbf{q}'} \, G_{11}(p+q') \, G_{11}(p-q') \, T(p+q', p-q', q-q'). \quad (10.211)$$

At finite temperatures, the t-matrix takes into account the effects of thermal population of the excited states.

We shall now show how the ladder resummation alters the diagrammatic representation of the chemical potential and the self-energy. In Figure 10.15 are displayed some of the terms up to two-loop order contributing to the chemical potential μ.

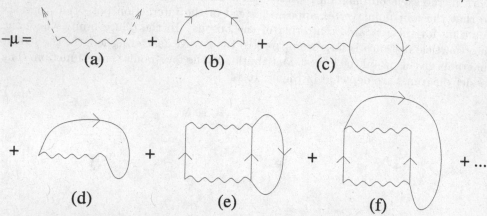

Figure 10.15 Diagrams up to two-loop order contributing to the chemical potential. Only the two-loop diagrams relevant to the resummation of the ladder diagrams are displayed. The two-loop diagrams not displayed are topologically identical to those shown, but differ in the direction of arrows or the presence of anomalous instead of normal propagators.

The first two terms in Eq. (10.200) is represented by diagrams (a)–(d), and the two-loop diagrams (e)–(f) originate from Γ_2. The diagrams labeled (e) and (f) are formally one loop order higher than (c) and (d), but they differ only by containing one additional ladder rung. Hence, the diagrams (c), (d), (e), and (f), and all the diagrams that can be constructed from these by adding ladder rungs, are of the same order in the diluteness parameter $\sqrt{n_0 a^3}$ as just shown above. They are therefore resummed, and as discussed this leads to the replacement of the interaction potential U by the t-matrix.

We note that no ladder counterparts to the diagrams (a) and (b) in Figure 10.15 appear explicitly in the expansion of the chemical potential, since such diagrams are two-particle reducible and are by construction excluded from the effective action Γ_2. However, diagram (b) contains implicitly the ladder contribution to diagram (a). In order to establish this we first simplify the notation by denoting by N_p the numerator of the exact normal Green's function $G_{11}(p)$, which according to Eq. (10.205) is

$$N_p = \omega + \varepsilon_{\mathbf{p}} - \mu + \Sigma_{11}(-p) . \tag{10.212}$$

We then have

$$D_p = N_p N_{-p} - \Sigma_{12}(p)\Sigma_{21}(p) = D_{-p} \tag{10.213}$$

and the contribution from diagram (b) can be rewritten in the form

$$\int d^4 p\, U_{\mathbf{p}}\, G_{12}(p) = \int d^4 p\, U_{\mathbf{p}}\, \frac{\Sigma_{12}(p)}{N_p N_{-p} - \Sigma_{12}(p)\Sigma_{21}(p)}$$

$$= \int d^4p\, U_{\mathbf{p}} \left(\frac{\Sigma_{12}(p)N_p N_{-p}}{D_p^2} - \frac{\Sigma_{12}(p)\Sigma_{21}(p)\Sigma_{12}(p)}{D_p^2} \right)$$

$$= \int d^4p\, U_{\mathbf{p}} \left(\Sigma_{12}(p)G_{11}(p)G_{11}(-p) - \Sigma_{21}(p)G_{12}(p)^2 \right)$$

$$= \int d^4p\, U_{\mathbf{p}} \left(n_0 U_{\mathbf{p}} G_{11}(p)G_{11}(-p) + [\Sigma_{12}(p) - n_0 U_{\mathbf{p}}] \right.$$

$$\left. \times \quad G_{11}(p)G_{11}(-p) - \Sigma_{21}(p)G_{12}(p)^2 \right). \tag{10.214}$$

In Figure 10.16 the last two rewritings are depicted diagrammatically.

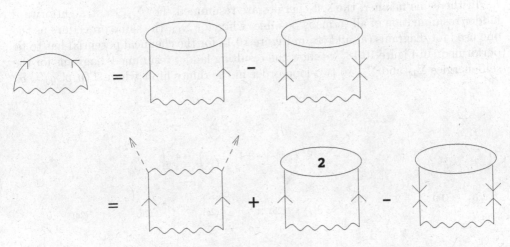

Figure 10.16 Diagrammatic representation of the last two rewritings in Eq. (10.214) which lead to the conclusion that the diagram (b) of Figure 10.15 implicitly contains the ladder contribution to diagram (a). The anomalous self-energy Σ_{12} is represented by an oval with two in-going lines, Σ_{21} is represented by an oval with two outgoing lines, and the sum of the second- and higher-order contributions to Σ_{12} is represented by an oval with the label "2."

We see immediately that the first term on the right-hand side corresponds to the first ladder contribution to diagram (a), and since to one-loop order, $\Sigma_{12}(p) = n_0 U_{\mathbf{p}}$, the other terms in Eq. (10.214) are of two- and higher-loop order. The self-energy in the second term on the right-hand side can be expanded to second loop order, and by iteration this yields all the ladder terms, and the remainder can be kept track of analogously to the way in which it is done in Eq. (10.214). The resulting ladder resummed diagrammatic expression for the chemical potential is displayed in Figure 10.17.

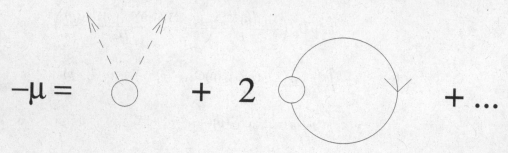

Figure 10.17 The chemical potential to one-loop order after the ladder summation has been performed and the resulting t-matrix been replaced by its expression in the dilute limit, the constant g.

In the same manner, the self-energies are resummed. For Σ_{11}, a straightforward ladder resummation of all terms is possible, while for Σ_{12}, the same procedure as the one used for diagrams (a) and (b) in Figure 10.15 for the chemical potential has to be performed. In Figure 10.18, we show the resulting ladder resummed diagrams for the self-energies Σ_{11} and Σ_{12} to two-loop order in the dilute limit where $T(p, p', q) \approx g$.

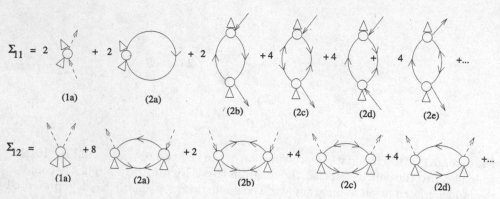

Figure 10.18 Normal Σ_{11} and anomalous Σ_{12} self-energies to two-loop order after the ladder summation has been performed and the resulting t-matrix been replaced by its expression in the dilute limit, the constant g.

In reference [61] a diagrammatic expansion in the potential was performed, which yields to first order the diagram $\Sigma_{11}^{(2a)}$ in Figure 10.18, but not the other two-loop diagrams. This approximation, where the normal self-energy is taken to be $\Sigma_{11} = \Sigma_{11}^{(1a)} + \Sigma_{11}^{(2a)}$, the anomalous self-energy to $\Sigma_{12} = \Sigma_{12}^{(1a)}$, and the diagrams displayed in Figure 10.17 are kept in the expansion of the chemical potential, is referred to as the Popov approximation. Although we showed at the end of Section 10.6.3 that all the two-loop diagrams of Figure 10.18 are of the same order of magnitude in the diluteness parameter $\sqrt{n_0 a^3}$ at zero temperature, the Popov approximation applied at finite temperatures is justified, when the temperature is high enough, $kT \gg gn_0$.

Below, we shall investigate the limits of validity at zero temperature of the Popov approximation in the trapped case.

In this and the preceding section we have shown how the expressions for the self-energies and chemical potential for a homogeneous dilute Bose gas are conveniently obtained by using the effective action formalism, where they simply correspond to working to a particular order in the loop expansion of the effective action. We have established that an expansion in the diluteness parameter is equivalent to an expansion of the effective action in the number of loops. Furthermore, the method provided a way of performing a systematic expansion, and the results are easily generalized to finite temperatures. We now turn to show that the effective action approach provides a way of performing a systematic expansion even in the case of an inhomogeneous Bose gas.

10.6.5 Inhomogeneous Bose gas

We now consider the experimentally relevant case of a Bose gas trapped in an external static potential, thereby setting the stage for the numerical calculations in the next section. In this case, the Bose gas will be spatially inhomogeneous. The effective action formalism is equally capable of dealing with the inhomogeneous gas, in which case all quantities are conveniently expressed in configuration space, as presented in Section 10.6.2. We show that the Bogoliubov and Gross–Pitaevskii theory corresponds to the one-loop approximation to the effective action. The one-loop equations will be exploited further in the next section.

Varying, in accordance with Eq. (10.185), the effective action Γ, Eq. (10.187), with respect to $\Phi^*(\mathbf{r}, t)$, we obtain the equation of motion for the condensate wave function

$$(i\hbar\partial_t - h + \mu)\Phi(\mathbf{r}, t) = g|\Phi(\mathbf{r}, t)|^2 \Phi(\mathbf{r}, t) + 2igG_{11}(\mathbf{r}, t, \mathbf{r}, t)\Phi(\mathbf{r}, t) - \frac{\delta\bar{\Gamma}_2}{\delta\Phi^*(\mathbf{r}, t)}.$$

$$(10.215)$$

To zero-loop order, where only the first term on the right-hand side appears, the equation is the time-dependent Gross–Pitaevskii equation. We have already, as elaborated in the previous section, performed the ladder summation by which the potential is renormalized and the t-matrix appears and substituted its lowest-order approximation in the diluteness parameter, the constant g. Since the t-matrix in the momentum variables is a constant in the dilute limit, it becomes in configuration space a product of three delta functions,

$$T(\mathbf{r}_1, \mathbf{r}_2, \mathbf{r}_3, \mathbf{r}_4) = g\,\delta(\mathbf{r}_1 - \mathbf{r}_4)\,\delta(\mathbf{r}_2 - \mathbf{r}_4)\,\delta(\mathbf{r}_3 - \mathbf{r}_4). \tag{10.216}$$

The quantity $\bar{\Gamma}_2$ is defined as the effective action obtained from Γ_2 by summing the ladder terms whereby U is replaced by the t-matrix, and its diagrammatic expansion is topologically of two-loop and higher order.

The Dyson–Beliaev equation, Eq. (10.204), and the equation determining the condensate wave function, Eq. (10.215), form a set of coupled integro-differential self-consistency equations for the condensate wave function and the Green's function, with the self-energy specified in terms of the Green's function through the effective action according to Eq. (10.202). The Green's function can be conveniently expanded in the amplitudes of the elementary excitations. We write the Dyson–Beliaev equation, Eq. (10.204), in the form

$$\int d\mathbf{r}'' dt'' \left[i\hbar\sigma_3 \partial_t \delta(\mathbf{r} - \mathbf{r}'') \delta(t - t'') + \sigma_3 L(\mathbf{r}, t, \mathbf{r}'', t'') \right] G(\mathbf{r}'', t'', \mathbf{r}', t')$$

$$= \hbar \mathbb{1} \delta(\mathbf{r} - \mathbf{r}') \delta(t - t') , \tag{10.217}$$

where we have introduced the matrix operator

$$L(\mathbf{r}, t, \mathbf{r}', t') = \sigma_3 h \, \delta(\mathbf{r} - \mathbf{r}') \delta(t - t') + \sigma_3 \Sigma(\mathbf{r}, t, \mathbf{r}', t') \tag{10.218}$$

and σ_3 is the third Pauli matrix. Up to one-loop order, the matrix Σ is diagonal in the time and space coordinates and we can factor out the delta functions and write $L(\mathbf{r}, t, \mathbf{r}', t') = \delta(t - t') \, \delta(\mathbf{r} - \mathbf{r}') \, L(\mathbf{r})$, where

$$L(\mathbf{r}) = \begin{pmatrix} h - \mu + 2g|\Phi(\mathbf{r})|^2 & g\Phi(\mathbf{r})^2 \\ -g\Phi^*(\mathbf{r})^2 & -h + \mu - 2g|\Phi(\mathbf{r})|^2 \end{pmatrix}. \tag{10.219}$$

The eigenvalue equation for L are the Bogoliubov equations. The Bogoliubov operator L is not hermitian, but the operator $\sigma_3 L$ is, which renders the eigenvectors of L the following properties. For each eigenvector $\varphi_j(\mathbf{r}) = (u_j(\mathbf{r}), v_j(\mathbf{r}))$ of L with eigenvalue E_j, there exists an eigenvector $\tilde{\varphi}_j(\mathbf{r}) = (v_j^*(\mathbf{r}), u_j^*(\mathbf{r}))$ with eigenvalue $-E_j$. Assuming the Bose gas is in its ground state, the normalization of the positive-eigenvalue eigenvectors can be chosen to be $\langle \varphi_j, \varphi_k \rangle = \delta_{jk}$, where we have introduced the inner product

$$\langle \varphi_j, \varphi_k \rangle = \int d\mathbf{r} \, \varphi_j^\dagger(\mathbf{r}) \sigma_3 \varphi_k(\mathbf{r}) = \int d\mathbf{r} \, (u_j^*(\mathbf{r}) u_k(\mathbf{r}) - v_j^*(\mathbf{r}) v_k(\mathbf{r})). \tag{10.220}$$

It follows that the inner product of the negative-eigenvalue eigenvectors $\tilde{\varphi}$ are

$$\langle \tilde{\varphi}_j, \tilde{\varphi}_k \rangle = \int d\mathbf{r} \, \tilde{\varphi}_j^\dagger(\mathbf{r}) \sigma_3 \tilde{\varphi}_k(\mathbf{r}) = \int d\mathbf{r} \, (v_j(\mathbf{r}) v_k^*(\mathbf{r}) - u_j(\mathbf{r}) u_k^*(\mathbf{r})) = -\delta_{jk} \tag{10.221}$$

and the eigenvectors φ and $\tilde{\varphi}$ are mutually orthogonal, $\langle \varphi_j, \tilde{\varphi}_k \rangle = 0$. By virtue of the Gross–Pitaevskii equation, the vector $\varphi_0(\mathbf{r}) = (\Phi(\mathbf{r}), -\Phi^*(\mathbf{r}))$ is an eigenvector of the Bogoliubov operator L with zero eigenvalue and zero norm. In order to obtain a completeness relation, we must also introduce the vector $\varphi_a(\mathbf{r}) = (\Phi_a(\mathbf{r}), -\Phi_a^*(\mathbf{r}))$ satisfying the relation $L\varphi_a = \alpha\varphi_0$, where α is a constant determined by normalization, $\langle \varphi_0, \varphi_a \rangle = 1$. The resolution of the identity then becomes

$$\sum_j{}' \left(\varphi_j(\mathbf{r})\varphi_j^\dagger(\mathbf{r}') - \tilde{\varphi}_j(\mathbf{r})\tilde{\varphi}_j^\dagger(\mathbf{r}') \right) \sigma_3 + \left(\varphi_a(\mathbf{r})\varphi_0^\dagger(\mathbf{r}') + \varphi_0(\mathbf{r})\varphi_a^\dagger(\mathbf{r}') \right) \sigma_3 = \mathbb{1}\delta(\mathbf{r} - \mathbf{r}')$$

$$\tag{10.222}$$

where the prime on the summation sign indicates that the zero-eigenvalue mode φ_0 is excluded from the sum. Using the resolution of the identity, Eq. (10.222), allows us to invert Eq. (10.217) to obtain the Bogoliubov spectral representation of the Green's function

$$G(\mathbf{r}, \mathbf{r}', \omega) = \hbar \sum_j{}' \left(\frac{1}{-\hbar\omega + E_j} \varphi_j(\mathbf{r})\varphi_j^\dagger(\mathbf{r}') - \frac{1}{-\hbar\omega - E_j} \tilde{\varphi}_j(\mathbf{r})\tilde{\varphi}_j^\dagger(\mathbf{r}') \right). \quad (10.223)$$

It follows from the spectral representation of the Green's function that the eigenvalues E_j are the elementary excitation energies of the condensed gas (here constructed explicitly to one-loop order). Using Eq. (10.223), we can at zero temperature express the non-condensate density or the depletion of the condensate, $n_{\mathrm{nc}} = n - n_0$, in terms of the Bogoliubov amplitudes

$$n_{\mathrm{nc}}(\mathbf{r}) = i \int \frac{d\omega}{2\pi} G_{11}(\mathbf{r}, \mathbf{r}, \omega) = \sum_j{}' |v_j(\mathbf{r})|^2. \quad (10.224)$$

The results obtained in this section form the basis for the numerical calculations presented in the next section.

10.6.6 Loop expansion for a trapped Bose gas

We now turn to determine the validity criteria for the equations obtained to various orders in the loop expansion for the ground state of a Bose gas trapped in an isotropic harmonic potential $V(r) = \frac{1}{2}m\omega_{\mathrm{t}}^2 r^2$. To this end, we shall numerically compute the self-energy diagrams to different orders in the loop expansion.

Working consistently to one-loop order, we need only employ Eq. (10.215) to zero-loop order, providing the condensate wave function, which upon insertion into Eq. (10.219) yields the Bogoliubov operator L to one-loop order, from which the Green's function to one-loop order is obtained from Eq. (10.223). The resulting Green's function is then used to calculate the various self-energy terms numerically. In order to do so, we make the equations dimensionless with the transformations $r = a_{\mathrm{osc}}\tilde{r}$, $\Phi = \sqrt{N_0/a_{\mathrm{osc}}^3}\,\tilde{\Phi}$, $u_j = a_{\mathrm{osc}}^{-3/2}\tilde{u}_j$, $E_j = \hbar\omega_{\mathrm{t}}\tilde{E}_j$, and $g = (\hbar\omega_{\mathrm{t}}a_{\mathrm{osc}}^3/N_0)\tilde{g}$, where $a_{\mathrm{osc}} = \sqrt{\hbar/m\omega_{\mathrm{t}}}$ is the characteristic oscillator length of the harmonic trap, and N_0 is the number of bosons in the condensate.

To zero-loop order, the time-independent Gross–Pitaevskii equation on dimensionless form reads

$$-\frac{1}{2}\nabla_{\tilde{r}}^2\tilde{\Phi} + \frac{1}{2}\tilde{r}^2\tilde{\Phi} + \tilde{g}|\tilde{\Phi}|^2\tilde{\Phi} = \tilde{\mu}\tilde{\Phi}. \quad (10.225)$$

We solve Eq. (10.225) numerically with the steepest-descent method, which has proven to be sufficient for solving the present equation [65]. The result thus obtained for $\tilde{\Phi}$ is inserted into the one-loop expression for the Bogoliubov operator L, Eq. (10.219), in order to calculate the Bogoliubov amplitudes $\tilde{u}_j$ and $\tilde{v}_j$ and the eigenenergies $\tilde{E}_j$. Since the condensate wave function for the ground state, $\tilde{\Phi}$, is real and rotationally symmetric, the amplitudes $\tilde{u}_j, \tilde{v}_j$ in the Bogoliubov equations can be labeled by the two angular momentum quantum numbers l and m, and a radial quantum number n, and we write $\tilde{u}_{nlm}(\tilde{r}, \theta, \phi) = \tilde{u}_{nl}(\tilde{r})Y_{lm}(\theta, \phi)$,

$\tilde{v}_{nlm}(\tilde{r}, \theta, \phi) = \tilde{v}_{nl}(\tilde{r})Y_{lm}(\theta, \phi)$. The resulting Bogoliubov equations are linear and one-dimensional

$$\tilde{L}\tilde{u}_{nl}(\tilde{r}) + \tilde{g}\tilde{\Phi}^2(\tilde{r})\tilde{v}_{nl}(\tilde{r}) = \tilde{E}_{nl}\tilde{u}_{nl}(\tilde{r}) \tag{10.226}$$

and

$$\tilde{L}\tilde{v}_{nl}(\tilde{r}) + \tilde{g}\tilde{\Phi}^2(\tilde{r})\tilde{u}_{nl}(\tilde{r}) = -\tilde{E}_{nl}\tilde{v}_{nl}(\tilde{r}) \tag{10.227}$$

where

$$\tilde{L} = \left(-\frac{1}{2}\frac{1}{\tilde{r}}\frac{\partial^2}{\partial\tilde{r}^2}\tilde{r} + \frac{1}{2}\frac{l(l+1)}{\tilde{r}^2} + \frac{1}{2}\tilde{r}^2 - \tilde{\mu} + 2\tilde{g}\tilde{\Phi}^2(\tilde{r})\right). \tag{10.228}$$

We note that the only parameter in the problem is the dimensionless coupling parameter $\tilde{g} = 4\pi N_0 a/a_{\text{osc}}$. Solving the Bogoliubov equations reduces to diagonalizing the band diagonal $2M \times 2M$ matrix L, where M is the size of the numerical grid. The value of M in the computations was varied between 180 and 240, higher values for stronger coupling, and the grid constant has been chosen to $0.05\,a_{\text{osc}}$ giving a maximum system size of $18\,a_{\text{osc}}$.

In the following, we shall estimate the orders of magnitude and the parameter dependence of the different two- and three-loop self-energy diagrams, and to this end we shall use the one-loop results for the amplitudes $\tilde{u}$, $\tilde{v}$ and the eigenenergies $\tilde{E}$ obtained numerically. When working to two- and three-loop order, one must also consider the corresponding corrections to the approximate t-matrix g. These contributions have been studied in reference [66], and their inclusion will not lead to any qualitative changes of the results.

Let us first compare the one-loop and two-loop contributions to the normal self-energy. The only one-loop term is

$$\Sigma_{11}^{(1a)}(\mathbf{r}, \mathbf{r}', \omega) = 2g|\Phi(\mathbf{r})|^2\delta(\mathbf{r} - \mathbf{r}') = 2gn_0(\mathbf{r})\,\delta(\mathbf{r} - \mathbf{r}'). \tag{10.229}$$

We first compare $\Sigma_{11}^{(1a)}$ with the two-loop term which is proportional to a delta function, i.e. the diagram (2a) in Figure 10.18. We shall shortly compare this diagram to the other two-loop diagrams. For diagram (2a) we have

$$\Sigma_{11}^{(2a)}(\mathbf{r}, \mathbf{r}', \omega) = 2ig\delta(\mathbf{r} - \mathbf{r}')\int\frac{d\omega'}{2\pi}G(\mathbf{r}, \mathbf{r}, \omega') = 2gn_{\text{nc}}(\mathbf{r})\,\delta(\mathbf{r} - \mathbf{r}'). \tag{10.230}$$

The ratio of the two-loop to one-loop self-energy contributions at the point $\mathbf{r}$ is thus equal to the fractional depletion of the condensate at that point. In Figure 10.19 we show the numerically computed dimensionless fractional depletion at the origin, $\tilde{n}_{\text{nc}}(0)/\tilde{n}_0(0)$, where we have introduced the dimensionless notation

$$\tilde{n}_0(\tilde{r}) = |\tilde{\Phi}(\tilde{r})|^2 \quad, \quad \tilde{n}_{\text{nc}}(\tilde{r}) = {\sum_j}' |\tilde{v}_j(\tilde{r})|^2. \tag{10.231}$$

We have chosen to evaluate the densities at the origin, $\mathbf{r} = 0$, in order to avoid a prohibitively large summation over the $l \neq 0$ eigenvectors.

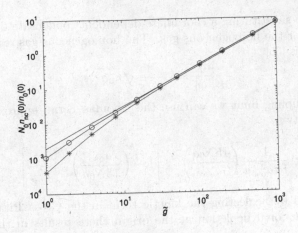

Figure 10.19 Fractional depletion of the condensate $N_0 n_{\rm nc}/n_0$ at the trap center as a function of the dimensionless coupling strength $\tilde{g} = 4\pi N_0 a/a_{\rm osc}$. Asterisks represent the numerical results, circles represent the local-density approximation with the numerically computed condensate density inserted, and the line is the local-density approximation using the Thomas–Fermi approximation for the condensate density.

As apparent from Figure 10.19, the log–log curve has a slight bend initially, but becomes almost straight for coupling strengths $\tilde{g} \gtrsim 100$. A logarithmic fit to the straight portion of the curve gives the relation

$$\frac{\tilde{n}_{\rm nc}(0)}{\tilde{n}_0(0)} \simeq 0.0019\,\tilde{g}^{1.2}. \tag{10.232}$$

When we reintroduce dimensions, the power-law relationship Eq. (10.232) is multiplied by the reciprocal of the number of bosons in the condensate N_0^{-1} because the actual and dimensionless self-energies are related according to

$$\Sigma^{(s)} = \frac{\hbar\omega_t a_{\rm osc}^3}{N_0^{s-1}}\,\tilde{\Sigma}^{(s)}, \tag{10.233}$$

where s denotes the loop order in question. The ratio between different loop orders of the self-energy is thus not determined solely by the dimensionless coupling parameter $\tilde{g} = 4\pi N_0 a/a_{\rm osc}$, but by N_0 and $a/a_{\rm osc}$ separately. We thus obtain for the fractional depletion in the strong-coupling limit, $\tilde{g} \gtrsim 100$,

$$\frac{n_{\rm nc}(0)}{n_0(0)} = \frac{1}{N_0}\frac{\tilde{n}_{\rm nc}(0)}{\tilde{n}_0(0)} \approx 0.041 N_0^{0.2}\left(\frac{a}{a_{\rm osc}}\right)^{1.2}. \tag{10.234}$$

It is of interest to compare our numerical results with approximate analytical results such as those obtained by using the local density approximation (LDA). The

LDA amounts to substituting a coordinate-dependent condensate density in the expressions valid for the homogeneous gas. The homogeneous-gas result for the fractional depletion is [55]

$$\frac{n_{nc}}{n_0} = \frac{8}{3\sqrt{\pi}}\sqrt{n_0 a^3}. \tag{10.235}$$

In the strong-coupling limit we can use the Thomas–Fermi approximation for the condensate density

$$n_0(\mathbf{r}) = \frac{1}{8\pi a_{osc}^2 a}\left(\frac{15 N_0 a}{a_{osc}}\right)^{2/5}\left[1 - \left(\frac{a_{osc}}{15 N_0 a}\right)^{2/5}\frac{r^2}{a_{osc}^2}\right], \tag{10.236}$$

which is obtained by neglecting the kinetic term in the Gross–Pitaevskii equation [67]. For the fractional depletion at the origin there results in the local density approximation

$$\frac{n_{nc}(0)}{n_0(0)} = \frac{(15 N_0)^{1/5}}{3\pi^2\sqrt{2}}\left(\frac{a}{a_{osc}}\right)^{6/5} \tag{10.237}$$

as first obtained in reference [68]. The LDA is a valid approximation when the gas locally resembles that of a homogeneous system, i.e. when the condensate wave function changes little on the scale of the coherence length ξ, which according to the Gross–Pitaevskii equation is $\xi = (8\pi n_0(0)a)^{-1/2}$. For a trapped cloud of bosons in the ground state, its radius R determines the rate of change of the density profile. Since R is a factor $\tilde{g}^{2/5}$ larger than ξ [67], we expect the agreement between the LDA and the exact results to be best in the strong-coupling regime. The fractional depletion of the condensate at the trap center as a function of the dimensionless coupling strength $\tilde{g} = 4\pi N_0 a/a_{osc}$ is shown in Figure 10.19. In Figure 10.19 are displayed both the local-density result Eq. (10.235) with the numerically computed condensate density inserted, and the Thomas–Fermi approximation Eq. (10.237), showing that the LDA indeed is valid when the coupling is strong. Furthermore, inspection of Eq. (10.237) reveals that the LDA coefficient and exponent agree with the numerically found result of Eq. (10.234), which is valid for strong coupling. However, when $\tilde{g} \lesssim 10$, the LDA prediction for the depletion deviates significantly from the numerically computed depletion. Inserting the numerically obtained condensate density into the LDA instead of the Thomas–Fermi approximation is seen not to substantially improve the result, as seen in Figure 10.19.

The relation for the fractional depletion, Eq. (10.234), is in agreement with the results of reference [69], where the leading-order corrections to the Gross–Pitaevskii equation were considered in the one-particle irreducible effective action formalism, employing physical assumptions about the relevant length scales in the problem. These leading-order corrections were found to have the same power-law dependence on N_0 and a/a_{osc}. A direct comparison of the prefactors cannot be made, because the objective of reference [69] was to estimate the higher-loop correction terms to the Gross–Pitaevskii equation and not to the self-energy.

The two-loop term $\Sigma_{11}^{(2a)}$ can, at zero temperature, according to Eq. (10.232) be ignored as long as $\tilde{n}_{nc} \ll \tilde{n}_0$, which is true in a wide, experimentally relevant parameter regime. The one-loop result for the fractional depletion Eq. (10.234) depends very weakly on N_0, so as long as N_0 does not exceed 10^9, which is usually fulfilled in experiments, we can restate the criterion for the validity of Eq. (10.234) into the condition $a \ll a_{\text{osc}}$. In experiments on atomic rubidium and sodium condensates, this condition is fulfilled, except in the instances where Feshbach resonances are used to enhance the scattering length [70].

In Section 10.6.3 we showed that for a homogeneous gas all two-loop diagrams are equally important in the sense that they are all of the same order in the diluteness parameter $\sqrt{n_0 a^3}$. The situation in a trapped system is not so clear, since the density is not constant. We shall therefore compare the five normal self-energy diagrams $\Sigma_{11}^{(2a-e)}$ in Figure 10.18, to see whether they display the same parameter dependence and whether any of the terms can be neglected. In particular, the Popov approximation corresponds to keeping the diagram $\Sigma_{11}^{(2a)}$ but neglects all other two-loop diagrams, and we will now determine its limits of validity at zero temperature. Since diagram (2a) contains a delta function, we shall integrate over one of the spatial arguments of the self-energy terms and keep the other one fixed at the origin, $\mathbf{r} = 0$. We denote by $R^{(j)}$ the ratio between the integrated self-energy terms (j) and (2a),

$$R^{(j)} = \frac{\int d\mathbf{r}\, \Sigma_{11}^{(j)}(0, \mathbf{r}, \omega = 0)}{\int d\mathbf{r}\, \Sigma_{11}^{(2a)}(0, \mathbf{r}, \omega = 0)}. \tag{10.238}$$

In Figure 10.20, we display the ratios $R^{(j)}$ for the different integrated self-energy contributions corresponding to the diagrams where (j) represents (2b) and (2c).

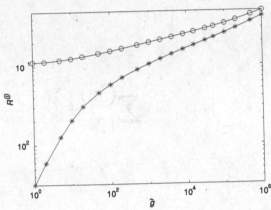

Figure 10.20 Ratio between different two-loop self-energy terms as functions of the dimensionless coupling strength $\tilde{g} = 4\pi N_0 a / a_{\text{osc}}$. Asterisks denote the ratio $R^{(2b)}$ as defined in Eq. (10.238) and circles denote the ratio $R^{(2c)}$. The terms $R^{(2d)}$ and $R^{(2e)}$ are equal and turn out to be equal in magnitude to $R^{(2b)}$, and are not displayed.

The contributions from the diagrams (2d) and (2e) are equal and within our numerical precision turn out to be equal to the contribution from diagram (2c). Furthermore, inspection of the diagrams in Figure 10.18 reveals that when the condensate wave function is real, the anomalous contribution $\Sigma_{12}^{(2a)}$ is equal to $\Sigma_{11}^{(2d)}$, the diagrams $\Sigma_{12}^{(2b)}$ and $\Sigma_{12}^{(2c)}$ are equal to $\Sigma_{11}^{(2c)}$, and $\Sigma_{12}^{(2d)}$ is equal to $\Sigma_{11}^{(2b)}$.

In the parameter regime displayed in Figure 10.20, the contribution from diagram (2a) is larger than the others by approximately a factor of ten, and displays only a weak dependence on the coupling strength. In the weak-coupling limit, $\tilde{g} \lesssim 1$, it is seen that the terms corresponding to diagrams (2b)–(2e) can be neglected as in the Popov approximation, with an error in the self-energy of a few per cent. When the coupling gets stronger, this correction becomes more important. A power-law fit to the ratio $R^{(2c)}$ in the regime where the log–log curve is straight yields the dependence

$$R^{(2c)} \approx 0.065\,\tilde{g}^{0.14}\,, \tag{10.239}$$

which is equal to 0.5 when $\tilde{g} \approx 10^6$; for $\tilde{g}$ greater than this value, the Popov approximation is seen not to be valid. If the ratio between the oscillator length and the scattering length is equal to one hundred, $a_{\mathrm{osc}} = 100a$, the Popov approximation deviates markedly from the two-loop result when N_0 exceeds 10^7, which is often the case experimentally.

In order to investigate the importance of higher-order terms in the loop expansion, we proceed to study the three-loop self-energy diagrams. We have found the number of summations over Bogoliubov levels to be prohibitively large for most three-loop terms; however, we *have* been able to compute the two diagrams $\Sigma_{11}^{(3a)}$ and $\Sigma_{12}^{(3a)}$, displayed in Figure 10.21, for the case where one of the spatial arguments is placed at the origin thereby avoiding a summation over $l \neq 0$ components.

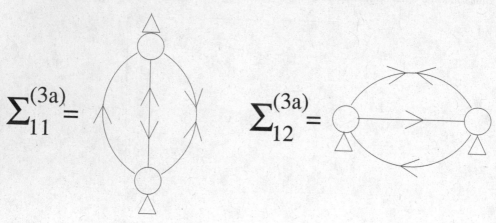

Figure 10.21 Self-energy diagrams to three-loop order which are evaluated numerically.

We compare the diagrams $\Sigma_{11}^{(3a)}$ and $\Sigma_{12}^{(3a)}$ to the two-loop diagrams. As we have seen, diagrams $\Sigma_{11}^{(2b)}$, $\Sigma_{11}^{(2c)}$, and $\Sigma_{11}^{(2d)}$ in Figure 10.18 are similar in magnitude and dependence on $\tilde{g}$, as are of the same order of magnitude and have similar dependence on $\tilde{g}$, and equivalently for the anomalous two-loop diagrams $\Sigma_{12}^{(2a-2d)}$; we have therefore chosen to evaluate only diagrams $\Sigma_{11}^{(2b)}$ and $\Sigma_{12}^{(2a)}$. The results for the ratios $\tilde{\Sigma}_{11}^{(3a)}(0, r, \omega = 0)/\tilde{\Sigma}_{11}^{(2b)}(0, r, \omega = 0)$ and $\tilde{\Sigma}_{12}^{(3a)}(0, r, \omega = 0)/\tilde{\Sigma}_{12}^{(2a)}(0, r, \omega = 0)$, evaluated for different choices of r, are shown in Figure 10.22.

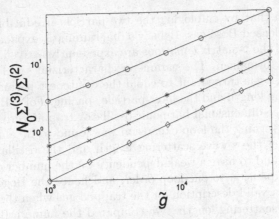

Figure 10.22 Ratio of three-loop to two-loop self-energy diagrams as a function of the dimensionless coupling strength $\tilde{g} = 4\pi N_0 a/a_{\mathrm{osc}}$. Asterisks denote the ratio of the normal self-energy terms $N_0 \Sigma_{11}^{(3a)}/\Sigma_{11}^{(2b)}$ evaluated at the point $(0, a_{\mathrm{osc}}, \omega = 0)$, open circles denote the same ratio evaluated at $(0, 0.5 a_{\mathrm{osc}}, \omega = 0)$, and diamonds denote the same ratio evaluated at $(0, 1.5 a_{\mathrm{osc}}, \omega = 0)$. Crosses denote the ratio of anomalous self-energy terms $N_0 \Sigma_{12}^{(3a)}/\Sigma_{12}^{(2a)}$ at $(0, a_{\mathrm{osc}}, \omega = 0)$.

A linear fit to the log–log plot gives, for the normal terms, the coefficient 0.016 and the exponent 0.76 when $r = 0.5 a_{\mathrm{osc}}$ and the coefficient 0.0029 and the exponent 0.78 when $r = a_{\mathrm{osc}}$, and for the anomalous terms with the choice $r = a_{\mathrm{osc}}$ the coefficient is 0.0015 and the exponent 0.82. Restoring dimensions according to Eq. (10.233) we obtain

$$\frac{\Sigma_{11}^{(3a)}(0, a_{\mathrm{osc}}, \omega = 0)}{\Sigma_{11}^{(2b)}(0, a_{\mathrm{osc}}, \omega = 0)} \approx 0.15 N_0^{-0.2} \left(\frac{a}{a_{\mathrm{osc}}}\right)^{0.8}. \tag{10.240}$$

The ratio between three- and two-loop self-energy terms in the homogeneous case was in Section 10.6.3 found to be proportional to $\sqrt{n_0 a^3}$. A straightforward application of the LDA, substituting the central density $n_0(0)$ for n_0, yields the dependence $\Sigma_{11}^{(3a)}/\Sigma_{11}^{(2b)} \propto N_0^{0.2}(a/a_{\mathrm{osc}})^{1.2}$. This is not in accordance with the numerical result Eq. (10.240) although the self-energies were evaluated at spatial points close to the trap center. The discrepancy between the LDA and the numerical three-loop result is attributed to the fact that we fixed the spatial points in units of a_{osc} while varying the coupling $\tilde{g}$, although the physical situation at the point $r = a_{\mathrm{osc}}$ (and $r = \frac{1}{2} a_{\mathrm{osc}}$ and $r = \frac{3}{2} a_{\mathrm{osc}}$ respectively) varies when $\tilde{g}$ is varied. It is possible that the agreement with the LDA had been better if the length scales had been fixed in units of the actual cloud radius (as given by the Thomas–Fermi approximation) rather than the oscillator length. However, the present calculation agrees fairly well with the LDA as long as the number of atoms in the condensate lies within reasonable bounds. Since $N_0 > 1$ in the condensed state, Eq. (10.240) yields that $\Sigma_{11}^{(3a)} \ll \Sigma_{11}^{(2b)}$ whenever the s-wave scattering length is much smaller than the trap length. We conclude that only when this condition is not fulfilled is it necessary to study diagrams of three-loop order and beyond.

We have shown that by employing the two-particle irreducible effective action approach to a condensed Bose gas, Beliaev's diagrammatic expansion in the diluteness parameter and the t-matrix equations are expediently arrived at with the aid of the effective action formalism. The parameter characterizing the loop expansion for a homogeneous Bose gas turned out to equal the diluteness parameter, the ratio of the s-wave scattering length and the inter-particle spacing. For a Bose gas contained in an isotropic, three-dimensional harmonic-oscillator trap at zero temperature, the small parameter governing the loop expansion was found to be almost proportional to the ratio between the s-wave scattering length and the oscillator length of the trapping potential, and to have a weak dependence on the number of particles in the condensate. The expansion to one-loop order, and hence the Bogoliubov equation, is found to provide a valid description for the trapped gas when the oscillator length exceeds the s-wave scattering length. We compared the numerical results with the local-density approximation, which was found to be valid when the number of particles in the condensate is large compared to the ratio between the oscillator length and the s-wave scattering length. The physical consequences of the self-energy corrections considered are indeed possible to study experimentally by using Feshbach resonances to vary the scattering length. Furthermore, we found that all the self-energy terms of two-loop order are not equally large for the case of a trapped system: in the limit when the number of particles in the condensate is not large compared with the ratio between the oscillator length and the s-wave scattering length, the Popov approximation was shown to be a valid approximation.

10.7 Summary

In this chapter we have considered the effective action. To study its properties and diagrammatic expansions, we introduced the functional integral representations of the generators. We showed how to express the effective action in terms of one-particle and two-particle irreducible loop vacuum diagram expansions. As an application, we applied the two-particle irreducible effective action approach to a condensed Bose gas, and showed that it allows for a convenient and systematic derivation of the equations of motion both in the homogeneous and trapped case. We chose in explicit calculations to apply the formalism to the situation where the temperature was zero, but the formalism is with equal ease capable of dealing with systems at finite temperatures and general non-equilibrium states.

11

Disordered conductors

Quantum corrections to the classical Boltzmann results for transport coefficients in disordered conductors can be systematically studied in the expansion parameter $\hbar/p_F l$, the ratio of the Fermi wavelength and the impurity mean free path, which typically is small in metals and semiconductors. The quantum corrections due to disorder are of two kinds, one being the change in interactions effects due to disorder, and the other having its origin in the tendency to localization. When it comes to an indiscriminate probing of a system, such as the temperature dependence of its resistivity, both mechanisms are effective, whereas when it comes to the low-field magneto-resistance only the weak localization effect is operative, and it has therefore become an important diagnostic tool in material science. We start by discussing the phenomena of localization and (especially weak localization) before turning to study the influence of disorder on interaction effects.

11.1   Localization

In this section the quantum mechanical motion of a particle at zero temperature in a random potential is addressed. In a seminal paper of 1958, P. W. Anderson showed that a particle's motion in a sufficiently disordered three-dimensional system behaves quite differently from that predicted by classical physics according to the Boltzmann theory [71]. In fact, at zero temperature diffusion will be absent, as particle states are localized in space because of the random potential. A sufficiently disordered system therefore behaves as an insulator and not as a conductor. By changing the impurity concentration, a transition from metallic to insulating behavior occurs, the Anderson metal–insulator transition.

In a pure metal, the Bloch or plane wave eigenstates of the Hamiltonian are extended states and current carrying

$$\langle \hat{\mathbf{j}} \rangle_{\text{ext}} \ = \ \int d\mathbf{x} \, \langle \mathbf{p} | \hat{\mathbf{j}}(\mathbf{x}) | \mathbf{p} \rangle \ = \ e \, \mathbf{v}_{\mathbf{p}} \,. \tag{11.1}$$

In a sufficiently disordered system, a typical energy eigenstate has a finite extension, and does not carry any average current

$$\langle \hat{\mathbf{j}} \rangle_{\text{loc}} = \mathbf{0} .\tag{11.2}$$

The last statement is not easily made rigorous, and the phenomenon of localization is quite subtle, a quantum phase transition at zero temperature in a non-equilibrium state.[1]

Astonishing progress in the understanding of transport in disordered systems has taken place since the introduction of the scaling theory of localization [72]. A key ingredient in the subsequent development of the understanding of the transport properties of disordered systems was the intuition provided by diagrammatic perturbation theory. We shall benefit from the physical intuition provided by the developed real-time diagrammatic technique in the present chapter, where it will provide the physical interpretation of the weak localization effect and the diffusion enhancement of interactions. We start by considering the scaling theory of localization.[2]

11.1.1 Scaling theory of localization

We shall consider a macroscopically homogeneous conductor, i.e. one with a spatially uniform impurity concentration, at zero temperature. By macroscopically homogeneous we mean that the impurity concentration on the macroscopic scale, i.e. much larger than the mean free path, is homogeneous. The conductance of a d-dimensional hypercube of linear dimension L is, according to Eq. (6.57), proportional to the conductivity

$$G(L) = L^{d-2}\,\sigma(L) .\tag{11.3}$$

The central idea of the scaling theory of localization is that the conductance rather than the conductivity is the quantity of importance for determining the transport properties of a macroscopic sample. The conductance has dimension of $e^2/\hbar$, independent of the spatial dimension of the sample, and we introduce the dimensionless conductance of a hypercube

$$g(L) \equiv \frac{G(L)}{\frac{e^2}{\hbar}} .\tag{11.4}$$

The one-parameter scaling theory of localization is based on the assumption that the dimensionless conductance solely determines the conductivity behavior of a disordered system. Consider fitting n^d identical blocks of length L, i.e. having the same impurity concentration and mean free path (assumed smaller than the size of the system, $l < L$) into a hypercube of linear dimension nL. The d.c. conductance of the hypercube $g(nL)$ is then related to the conductance of each block, $g(L)$, by

---

[1]For a discussion of wave function localization we refer the reader to chapter 9 of reference [1].
[2]The scaling theory of localization has its inspiration in the original work of Wegner [73] and Thouless [74].

$$g(nL) = f(n, g(L)) \, . \tag{11.5}$$

This is the one-parameter scaling assumption, the conductance of each block solely determines the conductance of the larger block; there is no extra dependence on L or microscopic parameters such as l or λ_F.

For a continuous variation of the linear dimension of a system, the one-parameter scaling assumption results in the logarithmic derivative being solely a function of the dimensionless conductance

$$\frac{d \ln g}{d \ln L} = \beta(g) \, . \tag{11.6}$$

This can be seen by differentiating Eq. (11.5) to get

$$\frac{d \ln g(L)}{d \ln L} = \frac{L}{g} \frac{dg(L)}{dL} = \frac{L}{g} \frac{dg(nL)}{dL} \bigg|_{n=1} = \frac{1}{g} \frac{dg(nL)}{dn} \bigg|_{n=1} = \frac{1}{g} \frac{df(n,g)}{dn} \bigg|_{n=1} \equiv \beta(g(L)) \, . \tag{11.7}$$

The physical significance of the scaling function, β, is as follows. If we start out with a block of size L, with a value of the conductance $g(L)$ for which $\beta(g)$ is positive, then the conductance according to Eq. (11.6) will increase upon enlarging the system, and vice versa for $\beta(g)$ negative. The β-function thus specifies the transport properties at that degree of disorder for a system in the infinite volume limit.

In the limit of weak disorder, large conductance $g \gg 1$, we expect metallic conduction to prevail. The conductance is thus described by classical transport theory, i.e. Ohm's law prevails $G(L) = L^{d-2} \sigma_0$, and the conductivity is independent of the linear size of the system, and we obtain according to Eq. (11.6) the limiting behavior for the scaling function

$$\beta(g) = d - 2 \, , \qquad g \gg 1 \, , \tag{11.8}$$

the scaling function having an asymptotic limit depending only on the dimensionality of the system.

In the limit of strong disorder, small conductance $g \ll 1$, we expect with Anderson [71] that localization prevails, so that the conductance assumes the form $g(L) \propto e^{-L/\xi}$, where ξ is called the localization length, the length scale beyond which the resistance grows exponentially with length.[3] In the low-conductance, so-called strong localization, regime we thus obtain for the scaling function, c being a constant,

$$\beta(g) = \ln g + c \, , \qquad g \ll 1 \, , \tag{11.9}$$

a logarithmic dependence in any dimension.

Since there is no intrinsic length scale to tell us otherwise, it is physically reasonable in this consideration to draw the scaling function as a monotonic non-singular function connecting the two asymptotes. We therefore obtain the behavior of the scaling function depicted in Figure 11.1.

---

[3] At this point we just argue that if the envelope function for a typical electronic wave function is exponentially localized, the conductance will have the stated length dependence, where ξ is the localization length of a typical wave function in the random potential, as it is proportional to the probability for the electron to be at the edge of the sample. For a justification of these statements within the self-consistent theory of localization we refer the reader to chapter 9 of reference [1].

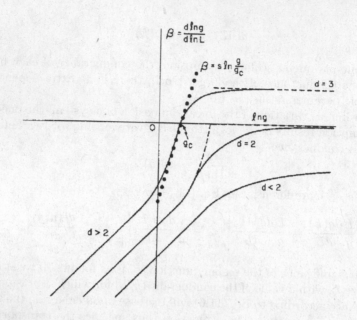

Figure 11.1 The scaling function as function of $\ln g$. Reprinted with permission from E. Abrahams, P. W. Anderson, D. C. Licciardello, and T. V. Ramakrishnan, *Phys. Rev. Lett.*, **42**, 673 (1979). Copyright 1979 by the American Physical Society.

This is precisely the picture expected in three and one dimensions. In three dimensions the unstable fix-point signals the metal–insulator transition predicted by Anderson. The transition occurs at a critical value of the disorder where the scaling function vanishes, $\beta(g_c) = 0$. If we start with a sample with conductance larger than the critical value, $g > g_c$, then upon increasing the size of the sample the conductance increases since the scaling function is positive. In the thermodynamic limit, the system becomes a metal with conductivity σ_0. Conversely, starting with a more disordered sample with conductance less than the critical value, $g < g_c$, upon increasing the size of the system, the conductance will flow to the insulating regime, since the scaling function is negative. In the thermodynamic limit the system will be an insulator with zero conductance. This is the localized state. In one dimension it can be shown exactly, that all states are exponentially localized for arbitrarily small amount of disorder [75, 76, 77, 78], and the metallic state is absent, in accordance with the scaling function being negative. An astonishing prediction follows from the scaling theory in the two-dimensional case where the one-parameter scaling function is also negative. There is no true metallic state in two dimensions.[4]

The prediction of the scaling theory of the absence of a true metallic state in

---

[4]In this day and age, low-dimensional electron systems are routinely manufactured. For example, a two-dimensional electron gas can be created in the inversion layer of an MBE grown GaAs–AlGaAs heterostructure. Two-dimensional localization effects provide a useful tool for probing material characteristics, as we discuss in Section 11.2.

two dimensions was at variance with the previously conjectured theory of *minimal metallic conductivity*. The classical conductivity obtained from the Boltzmann theory has the form, in two and three dimensions $(d = 2, 3)$,[5]

$$\sigma_0 = \frac{e^2}{\hbar} \frac{k_F l}{d\pi^{d-1}} k_F^{d-2} . \tag{11.10}$$

According to Mott [79], the conductivity in three (and two) spatial dimensions should decrease as the disorder increases, until the mean free path becomes of the order of the Fermi wavelength of the electron, $l \sim \lambda_F$. The minimum metallic conductivity should thus occur for the amount of disorder for which $k_F l \sim 2\pi$, and in two dimensions should have the universal value $e^2/\hbar$. Upon further increasing the disorder, the conductivity should discontinuously drop to zero.[6] This is in contrast to the scaling theory, which predicts the conductivity to be a continuous function of disorder. The metal–insulator transition thus resembles a second-order phase transition, a quantum phase transition at zero temperature, in contrast to Mott's first-order conjecture (corresponding to a scaling function represented by the dashed line in Figure 11.1).[7]

The phenomenological scaling theory offers a comprehensive picture of the conductance of disordered systems, and predicts that all states in two dimensions are localized irrespective of the amount of disorder. To gain confidence in this surprising result, one should check the first correction to the metallic limit. We therefore calculate the first quantum correction to the scaling function and verify that it is indeed negative.

11.1.2 Coherent backscattering

In this section we apply the standard diagrammatic impurity Green's function technique to calculate the influence of quenched disorder on the conductivity.[8] In diagrammatic terms, the quantum corrections to the classical conductivity are described by conductivity diagrams, as discussed in Section 6.1.3, where impurity lines connecting the retarded and advanced propagator lines cross. Such diagrams are nominally smaller, determined by the quantum parameter $\hbar/p_F l$, than the classical contribution. The subclass of diagrams, where the impurity lines cross a maximal number of times, is of special importance since their sum exhibits singular behavior. Such a type of diagram is illustrated in Eq. (11.11).

---

[5]In one dimension, the Boltzmann conductivity is $\sigma_0 = 2e^2 l/\pi\hbar$. However, the conclusion to be drawn from the scaling theory is that even the slightest amount of disorder invalidates the Boltzmann theory in one and two dimensions.

[6]In three dimensions in the infinite volume limit, the *conductance* drops to zero at the critical value according to the scaling theory.

[7]The impressive experimental support for the existence of a minimal metallic conductivity in two dimensions is now believed either to reflect the cautiousness one must exercise when attempting to extrapolate measurements at finite temperature to zero temperature, or to invoke a crucial importance of electron–electron interaction in dirty metals even at very low temperatures.

[8]For a detailed description of the standard impurity average Green's function technique we refer the reader to reference [1].

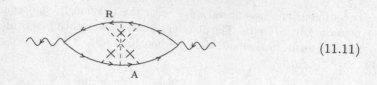

$$(11.11)$$

The maximally crossed diagrams describe the first quantum correction to the classical conductivity, the weak-localization or coherent backscattering effect, a subject we discuss in detail in Section 11.2.

In the frequency and wave vector region of interest, each insertion in a maximally crossed diagram is of order one.[9] Diagrams with maximally crossing impurity lines are therefore all of the same order of magnitude and must accordingly all be summed ($\hbar\mathbf{Q} \equiv \mathbf{p} + \mathbf{p}'$):

$$(11.12)$$

From the maximally crossed diagrams, we obtain analytically, by applying the Feynman rules for conductivity diagrams, the correction to the conductivity of a degenerate Fermi gas, $\hbar\omega, kT \ll \epsilon_{\mathrm{F}}$,[10]

$$\delta\sigma_{\alpha,\beta}(\mathbf{q},\omega) = \left(\frac{e}{m}\right)^2 \frac{\hbar}{\pi} \int \frac{d\mathbf{p}}{(2\pi\hbar)^d} \int \frac{d\mathbf{p}'}{(2\pi\hbar)^d}\, p_\alpha\, p'_\beta\, \tilde{C}_{\mathbf{p},\mathbf{p}'}(\epsilon_{\mathrm{F}},\mathbf{q},\omega)\, G^{\mathrm{R}}(\mathbf{p}_+, \epsilon_{\mathrm{F}} + \hbar\omega)$$

$$\times\quad G^{\mathrm{R}}(\mathbf{p}'_+, \epsilon_{\mathrm{F}} + \hbar\omega) G^{\mathrm{A}}(\mathbf{p}'_-, \epsilon_{\mathrm{F}}) G^{\mathrm{A}}(\mathbf{p}_-, \epsilon_{\mathrm{F}})\,. \tag{11.13}$$

To describe the sum of the maximally crossed diagrams, we have introduced the

---

[9]This is quite analogous to the case of the ladder diagrams important for the classical conductivity, recall Exercise 6.1 on page 163, and for details see chapter 8 of reference [1].

[10]In fact we shall in this section assume zero temperature as we shall neglect any influence on the maximally crossed diagrams from inelastic scattering. Interaction effects will be the main topic of Section 11.3.

so-called Cooperon $\tilde{C}$,[11] corresponding to the diagrams ($\epsilon_F^+ \equiv \epsilon_F + \hbar\omega$):

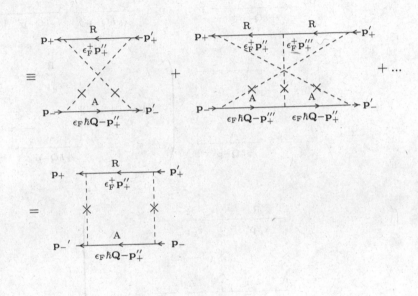

$$\tilde{C}_{\mathbf{p},\mathbf{p}'}(\epsilon_F, \mathbf{q}, \omega) \equiv$$

In the last equality we have twisted the A-line around in each of the diagrams, and by doing so, we of course do not change the numbers being multiplied together.

Let us consider the case where the random potential is delta correlated[12]

$$\langle V(\mathbf{x})V(\mathbf{x}') \rangle = u^2 \, \delta(\mathbf{x} - \mathbf{x}') . \tag{11.15}$$

---

[11] The nickname refers to the singularity in its momentum dependence being for zero total momentum, as is the case for the Cooper pairing correlations resulting in the superconductivity instability as discussed in Chapter 8.

[12] For the case of a short-range potential, the only change being the appearance of the transport time instead of the momentum relaxation time. For details we refer the reader to reference [1].

Since the impurity correlator in the momentum representation then is a constant, u^2, all internal momentum integrations become independent. As a consequence, the dependence of the Cooperon on the external momenta will only be in the combination $\mathbf{p}+\mathbf{p}'$, for which we have introduced the notation $\hbar\mathbf{Q} \equiv \mathbf{p}+\mathbf{p}'$, as well as $\tilde{C}_\omega(\mathbf{p}+\mathbf{p}') \equiv \tilde{C}_{\mathbf{p},\mathbf{p}'}(\epsilon_{\mathrm{F}}, \mathbf{0}, \omega) \equiv \tilde{C}_\omega(\mathbf{Q})$, and we have

$$\tilde{C}_\omega(\mathbf{Q}) = \quad\cdots$$

$$= \quad\cdots$$

$$\equiv \quad\cdots \qquad (11.16)$$

For convenience we have extracted a factor from the maximally crossed diagrams which we shortly demonstrate, Eq. (11.24), is simply the constant u^2 in the relevant parameter regime. We shall therefore also refer to the quantity C as the Cooperon. Diagrammatically we obtain according to Eq. (11.16)

$$
\boxed{C} \;=\; 1 \;+\; \overset{\text{R}}{\underset{\hbar\mathbf{Q}-\mathbf{p}''_+}{\underset{\text{A}}{\times}}}\;\mathbf{p}''_+\;\boxed{C}\;. \tag{11.17}
$$

Analytically the Cooperon satisfies the equation

$$
C_\omega(\mathbf{Q}) = 1 + u^2 \int \frac{d\mathbf{p}''}{(2\pi\hbar)^d}\; G^R(\mathbf{p}''_+, \epsilon_F + \hbar\omega) G^A(\mathbf{p}''_+ - \hbar\mathbf{Q}, \epsilon_F)\, C_\omega(\mathbf{Q})\;. \tag{11.18}
$$

It is obvious that a change in the wave vector of the external field can be compensated by a shift in the momentum integration variable, leaving the Cooperon independent of any spatial inhomogeneity in the electric field, which is smooth on the atomic scale.

The Cooperon equation is a simple geometric series that we can immediately sum

$$
\begin{aligned}
C_\omega(\mathbf{Q}) &= \left(1 + \zeta(\mathbf{Q},\omega) + \zeta^2(\mathbf{Q},\omega) + \zeta^3(\mathbf{Q},\omega) + \dots \right) \\[4pt]
&= 1 + \zeta(\mathbf{Q},\omega)\, C_\omega(\mathbf{Q}) \\[4pt]
&= \frac{1}{1 - \zeta(\mathbf{Q},\omega)}\;,
\end{aligned} \tag{11.19}
$$

where we have for the insertion

$$
\zeta(\mathbf{Q},\omega) = u^2 \int \frac{d\mathbf{p}''}{(2\pi\hbar)^d}\; G^R(\mathbf{p}'', \epsilon_F + \hbar\omega) G^A(\mathbf{p}'' - \hbar\mathbf{Q}, \epsilon_F)\;. \tag{11.20}
$$

Diagrammatically we can express the result

$$
C_\omega(\mathbf{Q}) \;=\; \cfrac{1}{1 \;-\; \overset{\text{R}}{\underset{\epsilon_F\,\hbar\mathbf{Q}-\mathbf{p}''_+}{\underset{\text{A}}{\times}}}\,\epsilon_F^+\,\mathbf{p}''_+}\;. \tag{11.21}
$$

The insertion $\zeta(\mathbf{Q},\omega)$, Eq. (11.20), is immediately calculated for the region of interest, $\omega\tau, Ql \ll 1$, and we have[13]

$$
\zeta(\mathbf{Q},\omega) = 1 + i\omega\tau - D_0\tau Q^2\;. \tag{11.22}
$$

---

[13] For details we refer the reader to [1], where the relation between the Diffuson and its twisted diagrams, the Cooperon, in the case of time-reversal invariance, is also established.

and for the Cooperon

$$C_\omega(\mathbf{Q}) = \frac{\frac{1}{\tau}}{-i\omega + D_0 Q^2}.$$ (11.23)

The Cooperon exhibits singular infrared behavior.[14]

In the singular region the prefactor in Eq. (11.16) equals the constant u^2 as

$$= u^2 \zeta(\mathbf{Q}, \omega) \simeq u^2$$ (11.24)

i.e. in the region of interest we thus have $\tilde{C} = u^2 C$. As far as regards the singular behavior we could equally well have defined the Cooperon by the set of diagrams

as adding a constant to a singular function does not change the singular behavior, and immediately the result of Eq. (11.23) is obtained.

---

[14] The Diffuson, the impurity particle–hole ladder diagrams, also exhibits this singular infrared behavior, which leads to diffusion enhancement of interactions in a disordered conductor as discussed in Section 11.5.

Changing in the conductivity expression, Eq. (11.13), one of the integration variables, $\mathbf{p}' = -\mathbf{p} + \hbar\mathbf{Q}$, we get for the contribution of the maximally crossed diagrams

$$\delta\sigma_{\alpha\beta}(\mathbf{q},\omega) = \left(\frac{e}{m}\right)^2 \frac{\hbar}{\pi} \int \frac{d\mathbf{p}}{(2\pi\hbar)^d} \int' \frac{d\mathbf{Q}}{(2\pi)^d}\, p_\alpha \left(-p_\beta + \hbar Q_\beta\right) \frac{u^2/\tau}{-i\omega + D_0 Q^2}$$

$$\times\; G^{\mathrm{R}}_{\epsilon^+_{\mathrm{F}}}(\mathbf{p}_+)\, G^{\mathrm{R}}_{\epsilon^+_{\mathrm{F}}}(-\mathbf{p}_+ + \hbar\mathbf{Q})\, G^{\mathrm{A}}_{\epsilon_{\mathrm{F}}}(-\mathbf{p}_- + \hbar\mathbf{Q})\, G^{\mathrm{A}}_{\epsilon_{\mathrm{F}}}(\mathbf{p}_-)\,, \qquad (11.26)$$

where the prime on the $\mathbf{Q}$-integration signifies that we need only to integrate over the region $Ql < 1$ from which the large contribution is obtained. Everywhere except in the Cooperon we can therefore neglect $\mathbf{Q}$ as $|\mathbf{p} - \hbar\mathbf{Q}| \sim p \sim p_{\mathrm{F}}$. Assuming a smoothly varying external field on the atomic scale, $q \ll k_{\mathrm{F}}$,[15] we can perform the momentum integration, and obtain to leading order in $\hbar/p_{\mathrm{F}} l$

$$\int \frac{d\mathbf{p}}{(2\pi\hbar)^d}\, p_\alpha p_\beta\, G^{\mathrm{R}}_{\epsilon_{\mathrm{F}}}(\mathbf{p}_+) G^{\mathrm{R}}_{\epsilon_{\mathrm{F}}}(-\mathbf{p}_+)\, G^{\mathrm{A}}_{\epsilon_{\mathrm{F}}}(-\mathbf{p}_-) G^{\mathrm{A}}_{\epsilon_{\mathrm{F}}}(\mathbf{p}_-) = \frac{4\pi\tau^3 N_d(\epsilon_{\mathrm{F}}) p_{\mathrm{F}}^2}{\hbar^3 d} \delta_{\alpha\beta}\,,$$

$$(11.27)$$

where we have also safely neglected the ω dependence in the propagators as for the integration region giving the large contribution, we have $\omega < 1/\tau \ll \epsilon_{\mathrm{F}}/\hbar$.

At zero frequency we have for the first quantum correction to the conductivity of an electron gas

$$\delta\sigma(L) = -\frac{2e^2 D_0}{\pi\hbar} \int' \frac{d\mathbf{Q}}{(2\pi)^d} \frac{1}{D_0 Q^2}\,. \qquad (11.28)$$

In the one- and two-dimensional case the integral diverges for small Q, and we need to assess the lower cut-off.[16] In order to understand the lower cut-off we note that the maximally crossed diagrams lend themselves to a simple physical interpretation. The R-line in the Cooperon describes the amplitude for the scattering sequence of an electron (all momenta being near the Fermi surface as the contribution is otherwise small)

$$\mathbf{p}' \to \mathbf{p}_1 \to \cdots \to \mathbf{p}_N \to \mathbf{p} \simeq -\mathbf{p}' \qquad (11.29)$$

whereas the A-line describes the complex conjugate amplitude for the opposite, i.e. time-reversed, scattering sequence

$$\mathbf{p}' \to -\mathbf{p}_N \to \cdots \to -\mathbf{p}_1 \to \mathbf{p} \simeq -\mathbf{p}' \qquad (11.30)$$

i.e. the Cooperon describes a quantum interference process: the quantum interference between time-reversed scattering sequences. The physical process responsible for the

---

[15] In a conductor a spatially varying electric field will, owing to the mobile charges, be screened. In a metal, say, an applied electric field is smoothly varying on the atomic scale, $q \ll k_{\mathrm{F}}$, and we can set q equal to zero as it appears in combination with large momenta, $p, p' \sim p_{\mathrm{F}}$. For a discussion of the phenomena of screening, we refer the reader to Section 11.5 and chapter 10 of reference [1].

[16] Langer and Neal [80] were the first to study the maximally crossed diagrams, and noted that they give a divergent result at zero temperature. However, in their analysis they did not assess the lower cut-off correctly.

quantum correction is thus coherent backscattering.[17] The random potential acts as sets of mirrors such that an electron in momentum state $\mathbf{p}$ ends up backscattered into momentum state $-\mathbf{p}$. The quantum correction to the conductivity is thus negative as the conductivity is a measure of the initial and final correlation of the velocities as reflected in the factor $\mathbf{p} \cdot \mathbf{p}'$ in the conductivity expression.

The quantum interference process described by the above scattering sequences corresponds in real space to the quantum interference between the two alternatives for a particle to traverse a closed loop in opposite (time-reversed) directions as depicted in Figure 11.2.[18]

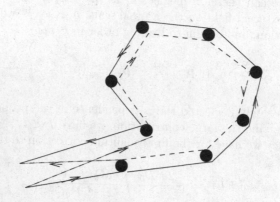

Figure 11.2 Coherent backscattering process.

We are considering the phenomenon of conductivity, where currents through connecting leads are taken in and out of a sample, say, at opposing faces of a hypercube. The maximal size of a loop allowed to contribute to the coherent backscattering process is thus the linear size of the system, as we assume that an electron reaching the end of the sample is irreversibly lost to the environment (leads and battery).[19] For a system of linear size L we then have for the quantum correction to the conductivity

$$\delta\sigma(L) = -\frac{2e^2 D_0}{\pi\hbar} \int_{1/L}^{1/l} \frac{d\mathbf{Q}}{(2\pi)^d} \frac{1}{D_0 Q^2} . \tag{11.31}$$

---

[17]The coherent backscattering effect was considered for light waves in 1968 [81]. It is amusing that a quantitative handling of the phenomena had to await the study of the analogous effect in solid-state physics, and the diagrammatic treatment of electronic transport in metals a decade later. Here we reap the benefits of employing the proper physical representation of Green's functions in the diagrammatic non-equilibrium perturbation theory, leading directly to a physical interpretation of the summed sub-class of diagrams.

[18]We will take advantage of this all-important observation of the physical origin of the quantum correction to the conductivity (originally expressed in references [82, 83]) in Section 11.2, where the real space treatment of weak localization is done in detail.

[19]An electron is assumed never to reenter from the leads phase coherently, and the Cooperon equation should be solved with the boundary condition that the Cooperon vanishes on the lead boundaries, thereby cutting off the singularity. For details we refer the reader to chapter 11 of reference [1].

Performing the integral in the two-dimensional case gives for the first quantum correction to the dimensionless conductance[20]

$$\delta g(L) = -\frac{1}{\pi^2} \ln \frac{L}{l} . \tag{11.32}$$

We note that the first quantum correction to the conductivity indeed is negative, describing the precursor effect of localization. For the asymptotic scaling function we then obtain

$$\beta(g) = -\frac{1}{\pi^2 g} , \qquad g \gg 1 , \tag{11.33}$$

and the first quantum correction to the scaling function is thus seen to be negative in concordance with the scaling picture.

Exercise 11.1. Show that, in dimensions one and three, the first quantum correction to the dimensionless conductance is

$$\delta g(L) = \begin{cases} -\frac{1}{\pi^2}(1 - \frac{l}{L}) & d = 1 \\ \\ -\frac{1}{\pi^3}(\frac{L}{l} - 1) & d = 3 \end{cases} \tag{11.34}$$

and thereby for the scaling function to lowest order in $1/g$

$$\beta(g) = (d - 2) - \frac{a}{g} , \tag{11.35}$$

where

$$a = \begin{cases} \frac{2}{\pi^2} & d = 1 \\ \\ \frac{1}{\pi^3} & d = 3 . \end{cases} \tag{11.36}$$

We can introduce the length scale characterizing localization, the localization length, qualitatively as follows: for a sample much larger than the localization length, $L \gg \xi$, the sample is in the localized regime and we have $g(L) \simeq 0$. To estimate the localization length, we equate it to the length for which $g(\xi) \simeq g_0$, i.e. the length scale, where the scale-dependent part of the conductance is comparable to the Boltzmann conductance. The lowest-order perturbative estimate based on Eq. (11.32) and Eq. (11.34) gives in two and one dimensions the localization lengths $\xi^{(2)} \simeq l \exp \pi k_F l/2$ and $\xi^{(1)} \simeq l$, respectively.

The one-parameter scaling hypothesis has been shown to be valid for the average conductance in the model considered above [73]. Whether the one-parameter scaling picture for the disorder model studied is true for higher-order cumulants of

---

[20]The precise magnitudes of the cut-offs are irrelevant for the scaling function in the two-dimensional case, as a change can produce only the logarithm of a constant in the dimensionless conductance.

the conductance, $\langle g^n \rangle$, is a difficult question that seems to have been answered in the negative in reference [84]. However, a different question is whether deviations from one-parameter scaling are observable, in the sense that a sample has to be so close to the metal–insulator transition that real systems cannot be made homogeneous enough. Furthermore, electron–electron interaction can play a profound role in real materials invalidating the model studied, and leaving room for a metal–insulator transition in low-dimensional systems.[21]

We can also calculate the zero-temperature frequency dependence of the first quantum correction to the conductivity for a sample of large size, $L \gg \sqrt{D_0/\omega} \equiv L_\omega$. From Eq. (11.26) we have

$$\delta\sigma_{\alpha\beta}(\omega) = \delta\sigma(\omega)\,\delta_{\alpha\beta}\,, \tag{11.37}$$

where

$$\delta\sigma(\omega) = -\frac{2e^2 D_0}{\hbar\pi} \int\limits_0^{1/l} \frac{d\mathbf{Q}}{(2\pi)^d} \frac{1}{-i\omega + D_0 Q^2}\,. \tag{11.38}$$

Calculating the integral, we get for the frequency dependence of the quantum correction to the conductivity in, say, two dimensions [86]

$$\frac{\delta\sigma(\omega)}{\sigma_0} = -\frac{1}{\pi k_F l}\,\ln\frac{1}{\omega\tau}\,. \tag{11.39}$$

We note that for the perturbation theory to remain valid the frequency can not be too small, $\omega\tau \simeq 1$.

The quantum correction to the conductivity in two dimensions is seen to be universal

$$\delta\sigma(\omega) = -\frac{1}{2\pi^2}\frac{e^2}{\hbar}\,\ln\frac{1}{\omega\tau}\,. \tag{11.40}$$

Let us calculate the first quantum correction to the current density response to a spatially homogeneous electric pulse

$$\delta\mathbf{j}(t) = \delta\sigma(t)\,\mathbf{E}_0\,, \tag{11.41}$$

where

$$\delta\sigma(t) = -\frac{2e^2 D_0}{\hbar\pi} \int\limits_{-\infty}^{\infty} \frac{d\omega}{2\pi}\, e^{-i\omega t} \int\limits_{1/L}^{1/l} \frac{d\mathbf{Q}}{(2\pi)^d} \frac{1}{-i\omega + D_0 Q^2} = -\frac{2e^2 D_0}{\hbar\pi} \int\limits_{1/L}^{1/l} \frac{d\mathbf{Q}}{(2\pi)^d}\, e^{-iD_0 Q^2 t}$$
$$\tag{11.42}$$

which in the two-dimensional case becomes

$$\delta\sigma(t) = \frac{e^2}{2\pi^2 \hbar t}\left(e^{-\frac{t}{2\tau}} - e^{-\frac{D_0 t}{L^2}}\right)\,. \tag{11.43}$$

---

[21] For a review on interaction effects, see for example [85].

After the short time τ the classical contribution and the above quantum contribution in the direction of the force on the electron dies out, and an *echo* in the current due to coherent backscattering occurs

$$\mathbf{j}(t) = -\frac{e^2}{2\pi^2\hbar t}e^{-t/\tau_{\mathrm{D}}}\,\mathbf{E}_0\,. \tag{11.44}$$

on the large time scale $\tau_{\mathrm{D}} \equiv L^2/D_0$, the time it takes an electron to diffuse across the sample (for even larger times $t \gg \tau_{\mathrm{D}}$ quantum corrections beyond the first dominates the current).

Exercise 11.2. Show that, in dimensions one and three, the frequency dependence of the first quantum correction to the conductivity is

$$\frac{\delta\sigma(\omega)}{\sigma_0} = \begin{cases} -\frac{1+i}{2\sqrt{2}}\frac{1}{\sqrt{\omega\tau}} & d=1 \\[2mm] (1-i)\frac{3\sqrt{3}}{2\sqrt{2}}\frac{\sqrt{\omega\tau}}{(k_{\mathrm{F}}l)^2} & d=3\,. \end{cases} \tag{11.45}$$

In dimension d the quantum correction to the conductivity is thus of relative order $1/(k_{\mathrm{F}}l)^{d-1}$. In strictly one dimension the weak localization regime is thus absent; i.e. there is no regime where the first quantum correction is small compared with the Boltzmann result, we are always in the strong localization regime.

From the formulas, Eq. (6.57) and Eq. (11.40), we find that in a quasi-two-dimensional system, where the thickness of the film is much smaller than the length scale introduced by the frequency of the time-dependent external field, $L_\omega = \sqrt{D_0/\omega}$, the quantum correction to the conductance exhibits the singular frequency behavior

$$\delta\langle G_{\alpha\beta}(\omega)\rangle = -\frac{e^2}{2\pi^2\hbar}\,\delta_{\alpha\beta}\ln\frac{1}{\omega\tau}\,. \tag{11.46}$$

The quantum correction to the conductance is in the limit of a large two-dimensional system only finite because we consider a time-dependent external field, and the conductance increases with the frequency. This feature can be understood in terms of the coherent backscattering picture. In the presence of the time-dependent electric field the electron can at arbitrary times exchange a quantum of energy $\hbar\omega$ with the field, and the coherence between two otherwise coherent alternatives will be partially disrupted. The more ω increases, the more the coherence of the backscattering process is suppressed, and consequently the tendency to localization, as a result of which the conductivity increases.

The first quantum correction plays a role even at finite temperatures, and in Section 11.2 we show that from an experimental point of view there are important quantum corrections to the Boltzmann conductivity even at weak disorder. We have realized that if the time-reversal invariance for the electron dynamics can be broken, the coherence in the backscattering process is disrupted, and localization is suppressed. The interaction of an electron with its environment invariably breaks the coherence, and we discuss the effects of electron–phonon and electron–electron interaction in Section 11.3. A more distinct probe for influencing localization is to apply a magnetic field, which we discuss in Section 11.4.

We have realized that the precursor effect of localization, weak localization, is caused by coherent backscattering. The constructive interference between propagation along time-reversed loops increases the probability for a particle to return to its starting position. The phenomenon of localization can be understood qualitatively as follows. The main amplitude of the electronic wave function incipient on the first impurity in Figure 11.2 is not scattered into the loop depicted, but continues in its forward direction. However, this part of the wave also encounters coherent backscattering along another closed loop feeding constructively back into the original loop, and thereby increasing the probability of return. This process repeats at any impurity, and the random potential acts as a mirror, making it impossible for a particle to diffuse away from its starting point. This is the physics behind how the singularity in the Cooperon drives the Anderson metal–insulator transition.[22]

11.2   Weak localization

We start this section by discussing the weak-localization contribution to the conductivity in the position representation, before turning to discuss the effects of interactions on the weak-localization effect, the destruction of the phase coherence of the wave function due to electron–phonon and electron–electron interaction. Then anomalous magneto-resistance is considered; this is an important diagnostic tool in material science. Finally we discuss mesoscopic fluctuations.

The theory of weak localization dates back to the seminal work on the scaling theory of localization [72], and developed rapidly into a comprehensive understanding of the quantum corrections to the Boltzmann conductivity. Based on the insight provided by the diagrammatic technique, the first quantum correction, the weak-localization effect, was soon realized to be the result of a simple type of quantum mechanical interference (as already noted in Section 11.1.2), and the resulting physical insight eventually led to a quantitative understanding of mesoscopic phenomena in disordered conductors. In order to develop physical intuition of the phenomena, we shall use the quantum interference picture in parallel with the quantitative diagrammatic technique, to discuss the weak-localization phenomenon.

11.2.1   Quantum correction to conductivity

In Section 7.4 we derived the Boltzmann expression for the classical conductivity as the weak-disorder limiting case where the quantum mechanical wave nature of the motion of an electron is neglected. In terms of diagrams this corresponded to neglecting conductivity diagrams where impurity correlators cross, because such

---

[22]For a quantitative discussion of strong localization we refer the reader to chapter 9 of reference [1].

contributions are smaller by the factor λ_F/l, and thus constitute quantum corrections to the classical conductivity.

A special class of diagrams where impurity correlators crossed a maximal number of times was seen, in Section 11.1.2, in the time-reversal invariant situation, to exhibit singular behavior although the diagrams nominally are of order $\hbar/p_F l$.[23]

$$\tag{11.47}$$

We shall consider the explicitly time-dependent situation where the frequency ω of the external field is not equal to zero, in order to cut off the singular behavior. In this case (and others to be studied shortly) the first quantum correction to the conductivity in the parameter λ_F/l is a small correction to the Boltzmann conductivity (recall Eq. (11.39)), and we speak of the weak-localization effect.

In the discussion of interaction effects and magneto-resistance it will be convenient to use the spatial representation for the conductivity. The free-electron model and a delta-correlated random potential, Eq. (11.15), will be used for convenience.

In the position representation the impurity-averaged current density

$$j_\alpha(\mathbf{x}, \omega) \equiv \langle j_\alpha(\mathbf{x}, \omega) \rangle = \sum_\beta \int d\mathbf{x}' \, \langle \sigma_{\alpha\beta}(\mathbf{x}, \mathbf{x}', \omega) \rangle \, E_\beta(\mathbf{x}', \omega) \tag{11.48}$$

is, besides regular corrections of order $\mathcal{O}(\hbar/p_F l)$, specified by the conductivity tensor

$$\sigma_{\alpha\beta}(\mathbf{x} - \mathbf{x}', \omega) \equiv \langle \sigma_{\alpha\beta}(\mathbf{x}, \mathbf{x}', \omega) \rangle = \frac{1}{\pi} \left(\frac{e\hbar}{m} \right)^2 \int_{-\infty}^{\infty} dE \, \frac{f_0(E) - f_0(E + \hbar\omega)}{\omega}$$

$$\times \; \langle G^R(\mathbf{x}, \mathbf{x}'; E + \hbar\omega) \stackrel{\leftrightarrow}{\nabla}_{x_\alpha} \stackrel{\leftrightarrow}{\nabla}_{x'_\beta} G^A(\mathbf{x}', \mathbf{x}; E) \rangle \; . \tag{11.49}$$

The contribution to the conductivity from the maximally crossed diagrams is conveniently exhibited in twisted form where they become ladder-type diagrams.

---

[23]In addition to these maximally crossed diagrams, there are additional diagrams of the same order of magnitude (also coming from the regular terms). However, they give contributions to the conductivity which are insensitive to low magnetic fields and temperatures in comparison to the contribution from the maximally crossed diagrams.

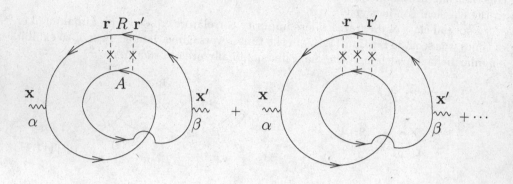

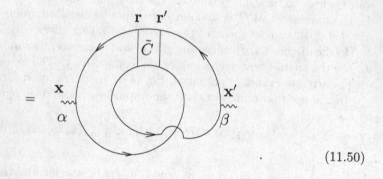

$$(11.50)$$

The sum of the maximally crossed diagrams, the Cooperon $\tilde{C}_\omega(\mathbf{r}, \mathbf{r}'; E)$, is in the position representation specified by the diagrams

The analytical expression for the quantum correction to the conductivity is therefore $(E_+ \equiv E + \hbar\omega)$.

$$\delta\sigma_{\alpha\beta}(\mathbf{x} - \mathbf{x}', \omega) = \frac{1}{\pi}\left(\frac{e\hbar}{m}\right)^2 \int d\mathbf{r} \int d\mathbf{r}' \int_{-\infty}^{\infty} dE \, \frac{f_0(E) - f_0(E + \hbar\omega)}{\omega} \, \tilde{C}_\omega(\mathbf{r}, \mathbf{r}'; E)$$

$$\times \, G^{\mathrm{R}}_{E_+}(\mathbf{x} - \mathbf{r}) G^{\mathrm{R}}_{E_+}(\mathbf{r}' - \mathbf{x}') \overset{\leftrightarrow}{\nabla}_{x_\alpha} \overset{\leftrightarrow}{\nabla}_{x'_\beta} \, G^{\mathrm{A}}_{E}(\mathbf{x}' - \mathbf{r}) G^{\mathrm{A}}_{E}(\mathbf{r}' - \mathbf{x}) \, . \qquad (11.52)$$

The impurity-averaged propagator decays exponentially as a function of its spatial variable with the scale set by the impurity mean free path. The spatial scale of variation of the sum of the maximally crossed diagrams is typically much larger. For the present case where we neglect effects of inelastic interactions, we recall from Eq. (11.23) that the spatial range of the Cooperon is $L_\omega = \sqrt{D_0/\omega}$, which for $\omega\tau \ll 1$ is much larger than the mean free path, since $D_0 = v_F l/d$ is the diffusion constant in d dimensions.[24] The impurity-averaged propagators attached to the maximally crossed diagrams will therefore require the starting and end points of $\tilde{C}_\omega(\mathbf{r}, \mathbf{r}', E)$ to be within the distance of a mean free path, in order for a non-vanishing contribution to the integral. On the scale of variation of the Cooperon this amounts to setting its arguments equal, and we can therefore substitute $\mathbf{r} \to \mathbf{x}, \mathbf{r}' \to \mathbf{x}$, and obtain

$$\delta\sigma_{\alpha\beta}(\mathbf{x} - \mathbf{x}', \omega) = \frac{1}{\pi}\left(\frac{eh}{m}\right)^2 \int_{-\infty}^{\infty} dE\, \frac{f_0(E_-) - f_0(E_+)}{\omega}\, \tilde{C}_\omega(\mathbf{x}, \mathbf{x}; E) \int d\mathbf{r} \int d\mathbf{r}'$$

$$\times G_{E_+}^{R}(\mathbf{x} - \mathbf{r}) G_{E_+}^{R}(\mathbf{r}' - \mathbf{x}')\, \overset{\leftrightarrow}{\nabla}_{x_\alpha} \overset{\leftrightarrow}{\nabla}_{x'_\beta}\, G_{E_-}^{A}(\mathbf{x}' - \mathbf{r}) G_{E_-}^{A}(\mathbf{r}' - \mathbf{x}) \quad (11.53)$$

The gate combination of the Fermi functions renders for the degenerate case, $\hbar\omega, kT \ll \epsilon_F$, the energy variable in the thermal layer around the Fermi surface, and we have for the first quantum correction to the conductivity of a degenerate electron gas

$$\delta\sigma_{\alpha\beta}(\mathbf{x} - \mathbf{x}', \omega) = \frac{\hbar}{\pi}\left(\frac{e}{m}\right)^2 \tilde{C}_\omega(\mathbf{x}, \mathbf{x}; \epsilon_F)\, \Phi_{\alpha,\beta}(\mathbf{x} - \mathbf{x}'), \quad (11.54)$$

where

$$\Phi_{\alpha,\beta}(\mathbf{x} - \mathbf{x}') \equiv \int d\mathbf{r} \int d\mathbf{r}'\, G_{\epsilon_F}^{R}(\mathbf{x} - \mathbf{r}) G_{\epsilon_F}^{R}(\mathbf{r}' - \mathbf{x}')\, \overset{\leftrightarrow}{\nabla}_{x_\alpha} \overset{\leftrightarrow}{\nabla}_{x'_\beta} G_{\epsilon_F}^{A}(\mathbf{x}' - \mathbf{r}) G_{\epsilon_F}^{A}(\mathbf{r}' - \mathbf{x}). \quad (11.55)$$

Clearly this function is local with the scale of the mean free path, and to lowest order in $\hbar/p_F l$ we have[25]

$$\Phi_{\alpha,\beta}(\mathbf{x} - \mathbf{x}') = -\frac{(2\pi N_0\tau)^2}{2\hbar^2}\, \frac{(x - x')_\alpha (x - x')_\beta}{|\mathbf{x} - \mathbf{x}'|^4}\, e^{-|\mathbf{x}-\mathbf{x}'|/l}\, \cos^2 k_F|\mathbf{x} - \mathbf{x}'|. \quad (11.56)$$

Since the function $\Phi_{\alpha,\beta}(\mathbf{x}-\mathbf{x}')$ decays on the scale of the mean free path, and appears in connection with the Cooperon, which is a smooth function on this scale, it acts effectively as a delta function

$$\Phi_{\alpha,\beta}(\mathbf{x} - \mathbf{x}') = -\frac{(2\pi N_0\tau)^2 l}{3\hbar^2}\, \delta_{\alpha\beta}\, \delta(\mathbf{x} - \mathbf{x}'). \quad (11.57)$$

We therefore obtain the fact that the first quantum correction, the weak-localization contribution, to the conductivity is local

$$\delta\sigma_{\alpha\beta}(\mathbf{x} - \mathbf{x}', \omega) = \delta\sigma(\mathbf{x}, \omega)\, \delta_{\alpha\beta}\, \delta(\mathbf{x} - \mathbf{x}') \quad (11.58)$$

---

[24] For samples of size larger than the mean free path, $L > l$, the diffusion process is effectively three-dimensional, so that one should use the value $d = 3$ in the expression for the diffusion constant. In strictly two-dimensional systems, such as for the electron gas in the inversion layer in a heterostructure at low temperatures, the value $d = 2$ should be used.

[25] For details we refer the reader to chapter 11 of reference [1].

and specified by[26]

$$\delta\sigma(\mathbf{x},\omega) = -\frac{2e^2 D_0 \tau}{\pi\hbar} \, C_\omega(\mathbf{x},\mathbf{x}) \, . \tag{11.59}$$

As we already noted in Section 11.1.2 the Cooperon is independent of the energy of the electron (here the Fermi energy since only electrons at the Fermi surface contribute to the conductivity) $C_\omega(\mathbf{x},\mathbf{x}') \equiv C_\omega(\mathbf{x},\mathbf{x}',\epsilon_F)$, and we have introduced $C_\omega(\mathbf{x},\mathbf{x}') \equiv u^{-2}\tilde{C}_\omega(\mathbf{x},\mathbf{x}')$.

The quantum correction to the conductance of a disordered degenerate electron gas is

$$\delta G_{\alpha\beta}(\omega) \equiv \langle \delta G_{\alpha\beta}(\omega)\rangle = L^{-2} \int d\mathbf{x} \int d\mathbf{x}' \, \langle \delta\sigma_{\alpha\beta}(\mathbf{x},\mathbf{x}',\omega)\rangle$$

$$= -\frac{2e^2 D_0 \tau}{\pi\hbar} L^{-2}\delta_{\alpha\beta} \int d\mathbf{x} \, C_\omega(\mathbf{x},\mathbf{x}) \, . \tag{11.60}$$

11.2.2 Cooperon equation

Many important results in the theory of weak localization can be obtained once the effect on the Cooperon of a time-dependent external field is obtained. Later we present the derivation of the Cooperon equation in the presence of a time-dependent electromagnetic field based on the quantum interference picture of the weak-localization effect. But first we provide the quantitative derivation of this result by employing the equation obeyed by the quasi-classical Green's function in Nambu or particle–hole space in the dirty, i.e. diffusive limit, Eq. (8.197). This will, in addition, extend our awareness of the information contained in the various components of the matrix Green's function in Nambu or particle–hole space.[27]

The goal is to generate the equation for the Cooperon by functional differentiation of the quasi-classical Green's function, and we therefore add a two-particle source to the Nambu space Hamiltonian, Ψ denoting the Nambu field, Eq. (8.32),

$$V(t_1,t_{1'}) = \int d\mathbf{x}_1 \, \Psi^\dagger(\mathbf{x}_1,t_1) \, V(\mathbf{x}_1,t_1,t_{1'}) \, \Psi(\mathbf{x}_1,t_{1'}) \, , \tag{11.61}$$

which therefore, according to Section 8.1.1, needs only off-diagonal Nambu matrix elements

$$V(\mathbf{x}_1,t_1,t_{1'}) = \begin{pmatrix} 0 & V_{12}(\mathbf{x}_1,t_1,t_{1'}) \\ V_{21}(\mathbf{x}_1,t_1,t_{1'}) & 0 \end{pmatrix} . \tag{11.62}$$

In linear response to the two-particle source we thus encounter the two-particle Green's function in the form of the particle–particle impurity ladder, and the Cooperon can be obtained by differentiation with respect to the source, which therefore is taken local in the space variable.

---

[26]We could also have evaluated the conductivity, Eq. (11.52), directly by Fourier-transforming the propagators, and recalling Eq. (11.27).

[27]This provides an alternative derivation to the ones in the literature. We follow the derivation in reference [9].

For the retarded component of Eq. (8.197) we have (we leave out the subscript indicating it is the s-wave, local in space part of the quasi-classical Green's function, g_s)

$$[g_0^{-1} + iV^\mathrm{R} - D_0\,\boldsymbol{\partial}\circ g^\mathrm{R}\circ\boldsymbol{\partial}\ \overset{\circ}{,}\ g^\mathrm{R}]_- = 0\,, \tag{11.63}$$

where the scalar potential enters in

$$g_0^{-1}(\mathbf{x}_1, t_1, t_{1'}) = (\tau_3\partial_{t_1} + ie\phi(\mathbf{x}_1, t_1))\delta(t_1 - t_{1'}) \tag{11.64}$$

and the vector potential through the diffusive term according to

$$\boldsymbol{\partial} = (\nabla_{\mathbf{x}_1} - ie\tau_3\mathbf{A}(\mathbf{x}_1, t_1))\,. \tag{11.65}$$

The equation of motion, which is homogeneous, is supplemented by the normalization condition, Eq. (8.182),

$$g^\mathrm{R}\circ g^\mathrm{R} = \delta(t_1 - t_{1'})\,. \tag{11.66}$$

The self-energy term associated with superconductivity has been expelled from Eq. (8.185) since for our case of interest the conductor is assumed in the normal state. Instead a source-term, V^R, has been introduced, a matrix in Nambu-space. Taking the functional derivative of the 12-component of g^R with respect to the Nambu component V_{12}^R is seen to generate the Cooperon

$$C(\mathbf{R}, \mathbf{R}', t_1, t_{1'}, t_2, t_{2'}) = \frac{1}{2i\tau}\frac{\delta g_{12}^\mathrm{R}(\mathbf{R}, t_1, t_{1'})}{\delta V_{12}^\mathrm{R}(\mathbf{R}', t_2, t_{2'})} \tag{11.67}$$

since the off-diagonal Nambu components of the source term add or subtract pairs of particles, and in the diffusive limit only ladder diagrams are considered. By construction, the functional derivative on the right in Eq. (11.67) is the ξ-integrated particle–particle ladder (including external legs) with the influence of the electromagnetic field fully included in the quasi-classical approximation.[28] The result of the functional derivative operation involved in Eq. (11.67) is depicted diagrammatically in Figure 11.3.

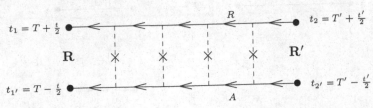

Figure 11.3 Cooperon obtained as derivative with respect to the two-particle source.

In order to obtain the equation satisfied by the functional derivative, the equation of motion is linearized with respect to the solution in the absence of the source term, g_0^R. We thus write

$$g^\mathrm{R} = g_0^\mathrm{R} + \delta g^\mathrm{R} \tag{11.68}$$

---

[28]This is usually no restriction since interest is in weak fields.

and use our knowledge that the normal state solution in the absence of the source term is

$$g_0^R = \tau_3 \delta(t_1 - t_{1'}) . \tag{11.69}$$

Inserting into Eq. (11.63) and linearizing the equation with respect to the source gives

$$[g_0^{-1} - D_0 \, \vec{\partial} \circ g_0^R \circ \vec{\partial} \, \overset{\circ}{,} \, \delta g^R]_- + i[V^R \, \overset{\circ}{,} \, g_0^R]_- - D_0[\vec{\partial} \circ \delta g^R \circ \vec{\partial} \, \overset{\circ}{,} \, g_0^R]_- = 0 . \tag{11.70}$$

Taking the 12-Nambu component gives

$$\left(\partial_{t_1} - \partial_{t_{1'}} + ie(\phi_{t_1} - \phi_{t_{1'}}) - D_0(\nabla_{\mathbf{x}} - ie(\mathbf{A}_{t_1} + \mathbf{A}_{t_{1'}}))^2\right)\delta g^R(\mathbf{x}, t_1, t_{1'})$$

$$= 2i V_{12}^R(\mathbf{x}, t_1, t_{1'}) , \tag{11.71}$$

where the spatial dependence $\mathbf{x}$ of the fields has been suppressed. Taking the functional derivative with respect to the 12-Nambu component of the source we get

$$\left(\partial_{t_1} - \partial_{t_{1'}} + ie(\phi_{t_1} - \phi_{t_{1'}}) - D_0(\nabla_{\mathbf{x}} - ie(\mathbf{A}_{t_1} + \mathbf{A}_{t_{1'}}))^2\right) \frac{\delta g_{12}^R(\mathbf{x}, t_1, t_{1'})}{\delta V_{12}^R(\mathbf{x}', t_2, t_{2'})}$$

$$= 2i \, \delta(\mathbf{x} - \mathbf{x}') \, \delta(t_1 - t_2) \, \delta(t_{1'} - t_{2'}) . \tag{11.72}$$

Because of the double time dependence of the external field, the functional derivative and the Cooperon have the time labeling depicted in Figure 11.3 and the following diagram

$$\tag{11.73}$$

Introducing new time variables

$$T = \frac{1}{2}(t_1 + t_{1'}) \quad , \quad T' = \frac{1}{2}(t_2 + t_{2'}) \quad , \quad t = t_1 - t_{1'} \quad , \quad t' = t_2 - t_{2'} , \tag{11.74}$$

we get

$$\left\{ 2\frac{\partial}{\partial t} + ie\phi_T(\mathbf{x},t) - D_0 \left(\nabla_{\mathbf{x}} - \frac{ie}{\hbar}\mathbf{A}_T(\mathbf{x},t) \right)^2 \right\} \frac{\delta g_{12}^{\mathrm{R}}(\mathbf{x},T,t)}{\delta V_{12}^{\mathrm{R}}(\mathbf{x}',T',t')}$$

$$= \frac{1}{\tau}\,\delta(\mathbf{x}-\mathbf{x}')\,\delta(t-t')\,\delta(T-T')\,, \tag{11.75}$$

where we have introduced the abbreviations

$$\phi_T(\mathbf{x},t) = \phi(\mathbf{x},T+t/2) - \phi(\mathbf{x},T-t/2) \tag{11.76}$$

and

$$\mathbf{A}_T(\mathbf{x},t) = \mathbf{A}(\mathbf{x},T+t/2) + \mathbf{A}(\mathbf{x},T-t/2)\,. \tag{11.77}$$

Accordingly for the Cooperon we get the equation

$$\left\{ 2\frac{\partial}{\partial t} + ie\phi_T(\mathbf{x},t) - D_0 \left(\nabla_{\mathbf{x}} - \frac{ie}{\hbar}\mathbf{A}_T(\mathbf{x},t) \right)^2 \right\} C_{t,t'}^{T,T'}(\mathbf{x},\mathbf{x}')$$

$$= \frac{1}{\tau}\,\delta(\mathbf{x}-\mathbf{x}')\,\delta(t-t')\,\delta(T-T')\,, \tag{11.78}$$

where we have introduced

$$C_{t,t'}^{T,T'}(\mathbf{x},\mathbf{x}') \equiv C(\mathbf{x},\mathbf{x}';t_1,t_{1'},t_2,t_{2'})\,. \tag{11.79}$$

Since there is no differentiation with respect to the variable T in Eq. (11.78), it is only a parameter in the Cooperon equation, and we have

$$C_{t,t'}^{T,T'}(\mathbf{x},\mathbf{x}) = C_{t,t'}^{T}(\mathbf{x},\mathbf{x}')\,\delta(T-T')\,, \tag{11.80}$$

where $C_{t,t'}^{T}(\mathbf{x},\mathbf{x}')$ satisfies the équation

$$\left\{ 2\frac{\partial}{\partial t} - D_0 \left(\nabla_{\mathbf{x}} - \frac{ie}{\hbar}\mathbf{A}_T(\mathbf{x},t) \right)^2 \right\} C_{t't'}^{T}(\mathbf{x},\mathbf{x}') = \frac{1}{\tau}\delta(\mathbf{x}-\mathbf{x}')\,\delta(t-t')\,. \tag{11.81}$$

Here we have left out the effect of a time-dependent scalar potential on the Cooperon since in the following we represent the electromagnetic field solely by the vector potential. We note that it can be restored by invoking the gauge co-variance property of the Cooperon.

We now derive the conductivity formula relevant for the case in question. We are here beyond linear response since we are taking into account to all orders how the Cooperon is influenced by the electromagnetic field. In the case of an external electromagnetic field represented by a vector potential influencing the Cooperon as well we consider the quantum correction to the kinetic propagator which is given by the contributions specified in the following diagram

$$\delta G^{K}(\mathbf{x}_1, t_1, \mathbf{x}_{1'}, t_{1'}) \quad = \quad \sum \qquad \qquad \qquad \qquad \qquad \qquad \qquad (11.82)$$

where the summation sign indicates the summation of all maximally crossed diagrams. For the quantum correction to the current we then have

$$\delta \mathbf{j}(\mathbf{x}, t) = \frac{e\hbar}{2im} \left(\frac{\partial}{\partial \mathbf{x}} - \frac{\partial}{\partial \mathbf{x}'} \right) \delta G^{K}(\mathbf{x}, t, \mathbf{x}', t) \bigg|_{\mathbf{x}'=\mathbf{x}} . \qquad (11.83)$$

The structure of the general maximally crossed diagram with n impurity correlators is

$$\delta G^{K} = \sum_{j=0}^{2n+1} (G^{R})^{j} G^{K} (G^{A})^{2n-j} . \qquad (11.84)$$

If the equilibrium kinetic propagator G_0^{K} occurs in the above diagram at a place different from the ones indicated by circles, the contribution vanishes to the order of accuracy. In that case, viz. we encounter the product of two retarded or two advanced propagators sharing the same momentum integration variable, and since the impurity correlator effectively decouples the momentum integrations, such terms are smaller by the factor $\hbar/\epsilon_F \tau$.

Displaying a maximally crossed kinetic propagator diagram on twisted form we have (we use the notation $1 \equiv (\mathbf{x}_1, t_1)$ etc.); the diagram in depicted in Figure 11.4.

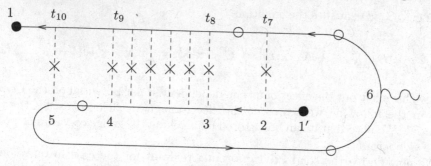

Figure 11.4 Twisted maximally crossed kinetic propagator diagram.

Because of the four different places where the kinetic propagator can occur we explicitly keep the four outermost impurity correlators, and obtain for the quantum correction to the kinetic propagator

$$\delta G^{\text{K}}(\mathbf{x}_1, t_1, \mathbf{x}_{1'}, t_{1'}) = \frac{eu^8}{2im} \int d\Gamma \, G^{\text{R}}(\mathbf{x}_1, t_1; \mathbf{x}_5, t_{10}) \, G^{\text{R}}(\mathbf{x}_5, t_{10}; \mathbf{x}_4, t_9) \, G^0(\mathbf{x}_3, t_8; \mathbf{x}_2, t_7)$$

$$\times \quad \mathbf{A}(\mathbf{x}_6, t_6) \cdot G^0(\mathbf{x}_2, t_7; \mathbf{x}_6, t_6) \, \overset{\leftrightarrow}{\nabla}_{\mathbf{x}_6} \, G^0(\mathbf{x}_6, t_6; \mathbf{x}_5, t_5)$$

$$\times \quad G^0(\mathbf{x}_5, t_5; \mathbf{x}_4, t_4) \, \frac{\delta g^{\text{R}}_{12}(\mathbf{x}_4, t_9, t_4)}{\delta V^{\text{R}}_{12}(\mathbf{x}_3, t_8, t_3)}$$

$$\times \quad G^{\text{A}}(\mathbf{x}_3, t_3; \mathbf{x}_2, t_2) \, G^{\text{A}}(\mathbf{x}_2, t_2; \mathbf{x}_{1'}, t_{1'}) \,, \tag{11.85}$$

where the propagators labeled by a zero as the superscript index indicate where the kinetic propagator can appear (i.e. we have a sum of four terms, and the kinetic propagator is always sandwiched in between retarded propagators to the left and advanced propagators to the right), and we have introduced the abbreviation

$$d\Gamma = dt_7 dt_8 dt_9 dt_{10} \prod_{i=1}^{6} d\mathbf{x}_i dt_i \,. \tag{11.86}$$

Since the propagators carry the large momentum p_{F}, we can take for the explicitly appearing linear response vector potential

$$\mathbf{A}(t) = \mathbf{A}_{\omega_1} e^{-i\omega_1 t} \,. \tag{11.87}$$

The eight exhibited propagators in Eq. (11.85) can be taken to be the equilibrium ones, and by Fourier transforming the propagators, and performing the integration over the momenta, we obtain for the quantum correction to the current density at frequency ω_2, $E_1^{\pm} = E_1 \pm \hbar\omega_1/2$,

$$\delta \mathbf{j}(\mathbf{x}, \omega_2) = \frac{4e^2 D_0 \tau}{i\pi} \mathbf{A}(\omega_1) \int_{-\infty}^{\infty} \frac{dE_1}{2\pi\hbar} \left(f_0(E_1^-) - f_0(E_1^+) \right) \int dt_1 dt_{1'} \, dt_2 dt_{2'} \, \delta(t_{2'} - t_{1'})$$

$$\times \quad e^{\frac{i}{\hbar}(E_1^- t_{1'} - E_1^+ t_2 - \hbar\omega_2 t_1)} \, C(\mathbf{x}, \mathbf{x}; t_1, t_{1'}, t_2, t_{2'}) \tag{11.88}$$

or equivalently

$$\delta \mathbf{j}(\mathbf{x}, \omega_2) = \frac{4e^2 D_0 \tau}{i\pi} \mathbf{A}(\omega_1) \int_{-\infty}^{\infty} \frac{dE_1}{2\pi\hbar} \left(f_0(E_1^-) - f_0(E_1^+) \right)$$

$$\times \quad \int_{-\infty}^{\infty} dt \int_{-\infty}^{\infty} dT \, C^T_{t,-t}(\mathbf{x}, \mathbf{x}) \, e^{iT(\omega_1 - \omega_2) + i\frac{t}{2}(\omega_1 + \omega_2)} \,. \tag{11.89}$$

For the quantum correction to the conductivity in the presence of a time-dependent electromagnetic field

$$\delta \mathbf{j}(\mathbf{x}, \omega_2) = \delta\sigma(\mathbf{x}, \omega_2, \omega_1)\, \mathbf{E}(\omega_1) \tag{11.90}$$

we therefore obtain[29]

$$\delta\sigma(\mathbf{x}, \omega_2, \omega_1) = -\frac{4e^2 D_0 \tau}{\pi\omega} \int\limits_{-\infty}^{\infty} \frac{dE_1}{2\pi\hbar}\, \left(f_0(E_1^-) - f_0(E_1^+)\right)$$

$$\times \int\limits_{-\infty}^{\infty} dt \int\limits_{-\infty}^{\infty} dT\; e^{iT(\omega_2-\omega_1)+\frac{i}{2}t(\omega_1+\omega_2)} C_{t,-t}^T(\mathbf{x}, \mathbf{x})\,. \tag{11.91}$$

In the degenerate case we have

$$\delta\sigma(\mathbf{x}, \omega_2, \omega_1) = -\frac{4e^2 D_0 \tau}{\pi\hbar} \int\limits_{-\infty}^{\infty} dt \int\limits_{-\infty}^{\infty} dT\; C_{t,-t}^T(\mathbf{x}, \mathbf{x})\, e^{iT(\omega_1-\omega_2)+i\frac{t}{2}(\omega_1+\omega_2)}\,. \tag{11.92}$$

In the event that the included effect of an electromagnetic field on the Cooperon is caused by a time-independent magnetic field, we recover the expression Eq. (11.59) for the quantum correction to the conductivity.

We shall exploit the derived formula when we consider the influence of electron–electron interaction on the quantum correction to the conductivity.

11.2.3 Quantum interference and the Cooperon

In this section, we shall elucidate in more detail than in Section 11.1.2 the physical process in real space described by the maximally crossed diagrams, and in addition consider the influence of external fields. The weak-localization effect can be understood in terms of a simple kind of quantum mechanical interference. By following the scattering sequences appearing in the diagrammatic representation of the Cooperon contribution to the conductivity, see Eq. (11.47), we realize that the quantum correction to the conductivity consists of products of the form "amplitude for scattering sequence of an electron off impurities in real space *times* the complex conjugate of the amplitude for the opposite scattering sequence." The quantum correction to the conductivity is thus the result of quantum mechanical interference between amplitudes for an electron traversing a loop in opposite directions. To lowest order in λ_F/l we need to include only the stationary, i.e. classical, paths determined by the electron bumping into impurities, as illustrated in Figure 11.2 where the trajectories involved in the weak-localization quantum interference process are depicted. The solid line, say, in Figure 11.2 corresponds to the propagation of the electron represented by

---

[29]For an electron gas in thermal equilibrium f_0 is the Fermi function, but in principle we could at this stage have any distribution not violating Pauli's exclusion principle. However, that would then necessitate a discussion of energy relaxation processes tending to drive the system toward the equilibrium distribution.

the retarded propagator in the conductivity diagram, and the dashed line to the propagation represented by the advanced propagator, the complex conjugate of the amplitude for scattering off impurities in the opposite sequence. The starting and end points refer to the points $\mathbf{x}$ and $\mathbf{x}'$ in Eq. (11.52), respectively.[30] According to the formula, Eq. (11.59), for the quantum correction to the conductivity, we need to consider only scattering sequences which start and end at the same point on the scale of the mean free path, as demanded by the impurity-averaged propagators attached to the maximally crossed diagrams in Eq. (11.52).

In the time-reversal invariant situation, the contribution to the return probability from the maximally crossed diagrams equals the contribution from the ladder diagrams, and the return probability including the weak-localization contribution is thus twice the classical result[31]

$$P_{\mathrm{cl+wl}}(\mathbf{x},t;\mathbf{x},t') = 2\,P_{\mathrm{cl}}(\mathbf{x},t;\mathbf{x},t') = 2\left(\frac{1}{4\pi D_0(t-t')}\right)^{d/2}, \qquad (11.93)$$

where the last expression is valid in the diffusive limit. To see how this comes about in the interference picture, let us consider the return probability in general. The quantity of interest is therefore the amplitude K for an electron to arrive at a given space point $\mathbf{x}$ at time $t/2$ when initially it started at the same space point at time $-t/2$. According to Feynman, this amplitude is given by the path integral expression

$$K(\mathbf{x},t/2;\mathbf{x},-t/2) = \int_{\mathbf{x}_{-t/2}=\mathbf{x}}^{\mathbf{x}_{t/2}=\mathbf{x}} \mathcal{D}\mathbf{x}_t\, e^{\frac{i}{\hbar}S[\mathbf{x}_t]} \equiv \sum_c A_c \qquad (11.94)$$

where the path integral includes all paths which start and end at the same point. For the return probability we have

$$P = |K|^2 = \left|\sum_c A_c\right|^2 = \sum_c |A_c|^2 + \sum_{c\neq c'} A_c A_{c'}^* \qquad (11.95)$$

where A_c is the amplitude for the path c. In the sum over paths we only need to include to order λ_F/l the stationary, i.e. classical, paths determined by the electron bumping into impurities. The sum of the absolute squares is then the classical contribution to the return probability, and the other terms are quantum interference terms. In the event that the particle only experiences the impurity potential, we have for the amplitude for the particle to traverse the path c,

$$A_c = e^{\frac{i}{\hbar}\int_{-\frac{t}{2}}^{\frac{t}{2}}d\bar{t}\,\{\frac{1}{2}m\dot{\mathbf{x}}_c^2(\bar{t}) - V(\mathbf{x}_c(\bar{t}))\}}. \qquad (11.96)$$

---

[30]The angle between initial and final velocities is exaggerated since we recall that in order for the Cooperon to give a large contribution the angle must be less than $1/k_F l$.

[31]The fact that impurity lines cross, does not *per se* make a diagram of order $1/k_F l$ relative to a non-crossed diagram. In case of the conductivity diagrams this is indeed the case for the maximally crossed diagrams because the circumstances needed for a large contribution set a constrain on the correlation of the initial and final velocity, $\mathbf{p}' \simeq -\mathbf{p} + \hbar\mathbf{Q}$ (recall also when estimating self-energy diagrams the importance of the incoming and outgoing momenta being equal, see reference [1]). However, in the quantity of interest here the *position* is fixed.

Owing to the impurity potential, the amplitude has a random phase. A first conjecture would be to expect that, upon impurity averaging, the interference terms in
general average to an insignificant small value, and we would be left with the classical contribution to the conductivity. However, there are certain interference terms
which are resilient to the impurity average. It is clear that impurity averaging can
not destroy the interference between time-reversed trajectories since we have for the
amplitude for traversing the time-reversed trajectory, $\mathbf{x}_{\bar{c}}(t) = \mathbf{x}_c(-t)$,

$$A_{\bar{c}} = e^{\frac{i}{\hbar} \int_{-\frac{t}{2}}^{\frac{t}{2}} d\bar{t} \{ \frac{1}{2} m \dot{\mathbf{x}}_{\bar{c}}^2(\bar{t}) - V(\mathbf{x}_{\bar{c}}(\bar{t})) \}} = e^{\frac{i}{\hbar} \int_{-\frac{t}{2}}^{\frac{t}{2}} d\bar{t} \{ \frac{1}{2} m[-\dot{\mathbf{x}}_c(-\bar{t})]^2 - V(\mathbf{x}_c(-\bar{t})) \}} = A_c. \quad (11.97)$$

In this time-reversal invariant situation the amplitudes for traversing a closed loop in
opposite directions are identical, $A_{\bar{c}} = A_c$, and the corresponding interference term
contribution to the return probability is independent of the disorder, $A_c A_{\bar{c}}^* = 1$!
The two amplitudes for the time-reversed electronic trajectories which return to the
starting point thus interfere constructively in case of time-reversal invariance. In
correspondence to this enhanced localization, there is a decrease in conductivity
which can be calculated according to Eq. (11.59).

The foregoing discussion based on the physical understanding of the weak localization effect will now be substantiated by deriving the equation satisfied by the
Cooperon. The Cooperon $C_\omega(\mathbf{x}, \mathbf{x}')$ is generated by the iterative equation

$$(11.98)$$

where we have introduced the diagrammatic notation

$$(11.99)$$

The Cooperon equation, Eq. (11.98), is most easily obtained by adding the term

$$
\begin{array}{c} \mathbf{x} \\ \mid \\ \ast \\ \mid \\ \mathbf{x}' \end{array} \quad = \quad u^2\,\delta(\mathbf{x}-\mathbf{x}') \tag{11.100}
$$

to the infinite sum of terms represented by the function $\tilde{C}$, Eq. (11.51). Alternatively, one can proceed as in Section 11.1.2, now exploiting the local character of the propagators. In any event, we have in the singular region $\tilde{C} \simeq u^2 C$.

The Cooperon equation in the spatial representation is

$$
C_\omega(\mathbf{x},\mathbf{x}') = \delta(\mathbf{x}-\mathbf{x}') + \int d\mathbf{x}''\,\tilde{J}^{\mathrm{C}}_\omega(\mathbf{x},\mathbf{x}'')\,C_\omega(\mathbf{x}'',\mathbf{x}')\,, \tag{11.101}
$$

where according to the Feynman rules the insertion is given by

$$
\tilde{J}^{\mathrm{C}}_\omega(\mathbf{x},\mathbf{x}') = u^2\,G^{\mathrm{R}}_{\epsilon_{\mathrm{F}}+\hbar\omega}(\mathbf{x},\mathbf{x}')\,G^{\mathrm{A}}_{\epsilon_{\mathrm{F}}}(\mathbf{x},\mathbf{x}')\,. \tag{11.102}
$$

The Cooperon is slowly varying on the scale of the mean free path, the spatial range of the function $\tilde{J}^{\mathrm{C}}_\omega(\mathbf{x},\mathbf{x}')$, and a low-order Taylor-expansion of the Cooperon on the right-hand side of Eq. (11.101) is therefore sufficient. Upon partial integration, the integral equation then becomes, for a second-order Taylor expansion, a differential equation for the Cooperon

$$
\left\{ -i\omega - D_0\nabla_{\mathbf{x}}^2 \right\} C_\omega(\mathbf{x},\mathbf{x}') = \frac{1}{\tau}\,\delta(\mathbf{x}-\mathbf{x}')\,. \tag{11.103}
$$

This equation is of course simply the position representation of the equation for the Cooperon already derived in the momentum representation, Eq. (11.23). Indeed we recover, that the Cooperon only varies on the large length scale $L_\omega = (D_0/\omega)^{1/2}$. The typical size of an interference loop is much larger than the mean free path, and we only need the large-scale behavior of the Boltzmannian paths of Figure 11.2, the smooth diffusive loops of Figure 11.5.

The fact that in the time-reversal invariant case we have obtained that the Cooperon satisfies the same diffusion-type equation as the Diffuson is not surprising. The Diffuson is determined by a similar integral equation as the Cooperon, however, with the important difference that one of the particle lines, say the advanced one, is reversed (recall Exercise 7.8 on page 197). The Diffuson will therefore be determined by the same integral equation as the Cooperon, except for $\tilde{J}^{\mathrm{C}}_\omega$ now being substituted by the diffusion insertion $\tilde{J}^{\mathrm{D}}_\omega$, given by

$$
\tilde{J}^{\mathrm{D}}_\omega(\mathbf{x},\mathbf{x}') = u^2\,G^{\mathrm{R}}_{\epsilon_{\mathrm{F}}+\hbar\omega}(\mathbf{x},\mathbf{x}')\,G^{\mathrm{A}}_{\epsilon_{\mathrm{F}}}(\mathbf{x}',\mathbf{x})\,. \tag{11.104}
$$

In a time-reversal invariant situation the two insertions are equal, $\tilde{J}^{\mathrm{C}}_\omega = \tilde{J}^{\mathrm{D}}_\omega$, and we recover that the Diffuson and the Cooperon satisfy the same equation (and we have hereby re-derived the result, Eq. (11.93)).

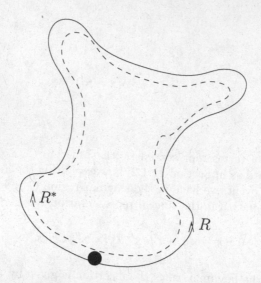

Figure 11.5 Diffusive loops.

In the time-reversal invariant situation the amplitudes for traversing a closed loop in opposite directions are identical, and in such a coherent situation one must trace the complete interference pattern of wave reflection in a random medium, and one encounters the phenomenon of localization discussed earlier. However, we also realize that the interference effect is sensitive to the breaking of time-reversal invariance. By breaking the coherence between the amplitudes for traversing time-reversed loops the tendency to localization of an electron can be suppressed.[32] In moderately disordered conductors we can therefore arrange for conditions so that the tendency to localization of the electronic wave function has only a weak though measurable influence on the conductivity. The first quantum correction then gives the dominating contribution in the parameter λ_F/l, and we speak of the so-called weak-localization regime. The destruction of phase coherence is the result of the interaction of the electron with its environment, such as electron–electron interaction, electron–phonon interaction, interaction with magnetic impurities, or interaction with an external magnetic field. From an experimental point of view the breaking of coherence between time-reversed trajectories by an external magnetic field is of special importance, and we start by discussing this case.

11.2.4   Quantum interference in a magnetic field

The influence of a magnetic field on the quantum interference process described by the Cooperon is readily established in view of the already presented formulas. In the weak magnetic field limit, $l^2 < l_B^2$, where $l_B = (\hbar/2eB)^{1/2}$ is the magnetic length, or equivalently $\omega_c\tau < \hbar/\epsilon_F\tau$, the bending of a classical trajectory with energy ϵ_F can be

---

[32]By disturbance, the coherence can be disrupted, and the tendency to localization can be suppressed, thereby decreasing the resistance. Normally, *disturbances* increase the resistance.

neglected on the scale of the mean free path. Classical magneto-resistance effects are then negligible, because they are of importance only when $\omega_c \tau \geq 1$. The amplitude for propagation along a straight-line classical path determined by the impurities is then changed only because of the presence of the magnetic field by the additional phase picked up along the straight line, the line integral along the path of the vector potential $\mathbf{A}$ describing the magnetic field. In the presence of such a weak static magnetic field the propagator is thus changed according to

$$G_E^{\mathrm{R}}(\mathbf{x}, \mathbf{x}') \rightarrow G_E^{\mathrm{R}}(\mathbf{x}, \mathbf{x}') \exp\left\{ \frac{ie}{\hbar} \int_{\mathbf{x}'}^{\mathbf{x}} d\bar{\mathbf{x}} \cdot \mathbf{A}(\bar{\mathbf{x}}) \right\} . \tag{11.105}$$

The resulting change in the Cooperon insertion is then

$$\tilde{J}_\omega^C(\mathbf{x}, \mathbf{x}') \rightarrow \tilde{J}_\omega^C(\mathbf{x}, \mathbf{x}') \exp\left\{ \frac{2ie}{\hbar} \int_{\mathbf{x}'}^{\mathbf{x}} d\bar{\mathbf{x}} \cdot \mathbf{A}(\bar{\mathbf{x}}) \right\}$$

$$= \tilde{J}_\omega^C(\mathbf{x}, \mathbf{x}') \exp\left\{ \frac{2ie}{\hbar} (\mathbf{x} - \mathbf{x}') \cdot \mathbf{A}(\mathbf{x}) \right\} . \tag{11.106}$$

The factor of 2 reflects the fact that in weak-localization interference terms between time-reversed trajectories, the additional phases due to the magnetic field add.

Repeating the Taylor-expansion leading to Eq. (11.103), we now obtain in the Cooperon equation additional terms due to the presence of the magnetic field

$$\left\{ -i\omega - D_0 \left(\nabla_{\mathbf{x}} - \frac{2ie}{\hbar} \mathbf{A}(\mathbf{x}) \right)^2 \right\} C_\omega(\mathbf{x}, \mathbf{x}') = \frac{1}{\tau} \delta(\mathbf{x} - \mathbf{x}') . \tag{11.107}$$

Introducing the Fourier transform

$$C_{t,t'}(\mathbf{x}, \mathbf{x}') = \int \frac{d\omega}{2\pi} e^{-i\omega(t-t')} C_\omega(\mathbf{x}, \mathbf{x}') \tag{11.108}$$

we obtain in the space-time representation the Cooperon equation[33]

$$\left\{ \frac{\partial}{\partial t} - D_0 \left(\nabla_{\mathbf{x}} - \frac{2ie}{\hbar} \mathbf{A}(\mathbf{x}) \right)^2 \right\} C_{t,t'}(\mathbf{x}, \mathbf{x}') = \frac{1}{\tau} \delta(\mathbf{x} - \mathbf{x}') \delta(t - t') . \tag{11.109}$$

We note that this equation is formally identical to the imaginary-time Green's function equation for a particle of mass $\hbar/2D_0$ and charge $2e$ moving in the magnetic field described by the vector potential $\mathbf{A}$. The solution of this equation can be expressed as the path integral

$$C_{t,t'}(\mathbf{x}, \mathbf{x}') = \frac{1}{\tau} \int_{\mathbf{x}_{t'}=\mathbf{x}'}^{\mathbf{x}_t=\mathbf{x}} \mathcal{D}\mathbf{x}_t \, e^{-\int_{t'}^{t} d\bar{t} \left(\frac{\dot{\mathbf{x}}_{\bar{t}}^2}{4D_0} + \frac{ie}{\hbar} \dot{\mathbf{x}}_{\bar{t}} \cdot \mathbf{A}(\mathbf{x}_{\bar{t}}) \right)} . \tag{11.110}$$

---

[33]This is of course just a special case of the general equation, Eq. (11.81), the case of a time-independent magnetic field

11.2.5 Quantum interference in a time-dependent field

Let us now obtain the equation satisfied by the Cooperon when the particle interacts with an environment as described by the Lagrangian L_1. The total Lagrangian is then $L = L_0 + L_1$, where[34]

$$L_0(\mathbf{x}, \dot{\mathbf{x}}) = \frac{1}{2} m \dot{\mathbf{x}}^2 - V(\mathbf{x}) \tag{11.111}$$

describes the particle in the impurity potential. We first present a derivation of the Cooperon equation based on the interference picture of the weak-localization effect, before presenting the diagrammatic derivation.[35]

The conditional probability density for an electron to arrive at position $\mathbf{x}$ at time t given it was at position $\mathbf{x}'$ at time t' is given by the absolute square of the propagator

$$P(\mathbf{x}, t; \mathbf{x}', t') = |K(\mathbf{x}, t; \mathbf{x}', t')|^2 . \tag{11.112}$$

In the quasi-classical limit, which is the one of interest, $\lambda_F \ll l$, we can, in the path integral expression for the propagator, replace the path integral by the sum over classical paths

$$K(\mathbf{x}, t; \mathbf{x}, t') = \int_{\mathbf{x}_{t'}=\mathbf{x}}^{\mathbf{x}_t=\mathbf{x}} \mathcal{D}\mathbf{x}_t \, e^{\frac{i}{\hbar} S[\mathbf{x}_t]} \simeq \sum_{\mathbf{x}_t^{cl}} A[\mathbf{x}_t^{cl}] \, e^{\frac{i}{\hbar} S[\mathbf{x}_t^{cl}]} \tag{11.113}$$

where the prefactor takes into account the Gaussian fluctuations around the classical path. We assume that we may neglect the influence of L_1 on the motion of the electrons, and the classical paths are determined by L_0, i.e. by the large kinetic energy and the strong impurity scattering. The paths in the summation are therefore solutions of the classical equation of motion

$$m \ddot{\mathbf{x}}_t^{cl} = -\nabla V(\mathbf{x}_t^{cl}) . \tag{11.114}$$

The quantum interference contribution to the return probability in time span t from the time-reversed loops is in the quasi-classical limit

$$P\left(\mathbf{x}, \frac{t}{2}; \mathbf{x}, -\frac{t}{2}\right) = \sum_{\mathbf{x}_t^{cl}} |A[\mathbf{x}_t^{cl}]|^2 \, e^{\frac{i}{\hbar}(S[\mathbf{x}_t^{cl}] - S[\mathbf{x}_{-t}^{cl}])} , \tag{11.115}$$

where $\mathbf{x}_{-t/2}^{cl} = \mathbf{x} = \mathbf{x}_{t/2}^{cl}$. We are interested in the return probability for an electron constrained to move on the Fermi surface, i.e. its energy is equal to the Fermi energy ϵ_F. For the weak-localization quantum interference contribution to the return probability we therefore obtain

$$C(t) = \frac{1}{N_0} \sum_{\mathbf{x}_t^{cl}} |A[\mathbf{x}_t^{cl}]|^2 \, e^{i\varphi[\mathbf{x}_t^{cl}]} \, \delta(\epsilon[\mathbf{x}_t^{cl}] - \epsilon_F) , \tag{11.116}$$

---

[34] A possible dynamics of the environment plays no role for the present discussion, and its Lagrangian is suppressed.

[35] We follow the presentation of reference [87].

where the sum is over classical trajectories of duration t that start and end at the same point, and

$$\epsilon[\mathbf{x}_t^{\rm cl}] = \frac{1}{2}m\,[\dot{\mathbf{x}}_t^{\rm cl}]^2 + V(\mathbf{x}_t^{\rm cl}) \tag{11.117}$$

is the energy of the electron on a classical trajectory. The normalization factor follows from the fact that the density of classical paths in the quasi-classical limit equals the density of states.[36] We have introduced the phase difference between a pair of time-reversed paths

$$\varphi[\mathbf{x}_t^{\rm cl}] = \frac{1}{\hbar}\left(S[\mathbf{x}_t^{\rm cl}] - S[\mathbf{x}_{-t}^{\rm cl}]\right)\,. \tag{11.118}$$

As noted previously in Section 11.2.4, a substantial cancellation occurs in the phase difference since L_0 is an even function of the velocity and the quenched disorder potential is independent of time. Hence, the phase difference is a small quantity given by

$$\varphi[\mathbf{x}_t^{\rm cl}] = \frac{1}{\hbar}\int_{-t/2}^{t/2} d\bar{t}\,\{L_1(\mathbf{x}_{\bar{t}}^{\rm cl},\dot{\mathbf{x}}_{\bar{t}}^{\rm cl},\bar{t}) - L_1(\mathbf{x}_{-\bar{t}}^{\rm cl},-\dot{\mathbf{x}}_{-\bar{t}}^{\rm cl},\bar{t})\}$$

$$= \frac{1}{\hbar}\int_{-t/2}^{t/2} d\bar{t}\,\{L_1(\mathbf{x}_{\bar{t}}^{\rm cl},\dot{\mathbf{x}}_{\bar{t}}^{\rm cl},\bar{t}) - L_1(\mathbf{x}_{\bar{t}}^{\rm cl},-\dot{\mathbf{x}}_{\bar{t}}^{\rm cl},-\bar{t})\}$$

$$\equiv \int_{-t/2}^{t/2} d\bar{t}\,\tilde{\varphi}(\mathbf{x}_t^{\rm cl})\,, \tag{11.119}$$

where in the last term in the second equality we have replaced the integration variable $\bar{t}$ by $-\bar{t}$. We recognize that L_1 though small, plays an important role here since it destroys the phase coherence between the time-reversed trajectories.

We must now average the quantum interference term with respect to the impurity potential. Since the dependence on the impurity potential in Eq. (11.116) is only implicit through its determination of the classical paths, averaging with respect to the random impurity potential is identical to averaging with respect to the probability functional for the classical paths in the random potential. In view of the expression appearing in Eq. (11.116), we thus encounter the probability of finding a classical path $\mathbf{x}_t$ of duration t which start and end at the same point, and for which the particle has the energy $\epsilon_{\rm F}$

$$P_t[\mathbf{x}_t] = \frac{1}{N_0}\left\langle \sum_{\mathbf{x}_t^{\rm cl}} |A[\mathbf{x}_t^{\rm cl}]|^2\,\delta(\epsilon[\mathbf{x}_t^{\rm cl}] - \epsilon_{\rm F})\,\delta[\mathbf{x}_t^{\rm cl} - \mathbf{x}_t]\right\rangle_{\rm imp} = \langle C(t)\rangle_{\rm imp}^{\varphi=0}\,. \tag{11.120}$$

---

[36]The Bohr–Sommerfeld quantization rule.

The second delta function, as indicated, in the functional sense, allows only the classical path in question to contribute to the path integral. The classical probability of return in time t of a particle with energy ϵ_F is given by

$$
P_R^{(cl)}(t) = \int\limits_{x_{-t/2}=x}^{x_{t/2}=x} \mathcal{D}\mathbf{x}_t \, P_t[\mathbf{x}_t] \, .
\tag{11.121}
$$

We therefore get, according to Eq. (11.116), for the impurity average of the weak-localization quantum interference term, the Cooperon,

$$
C_{\frac{t}{2},-\frac{t}{2}}(\mathbf{x},\mathbf{x}) = \langle C(t) \rangle_{\mathrm{imp}} = \int\limits_{x_{-t/2}=x}^{x_{t/2}=x} \mathcal{D}\mathbf{x}_t \, P_t[\mathbf{x}_t] \, e^{i\varphi[\mathbf{x}_t^{cl}]} \, .
\tag{11.122}
$$

In many situations of interest, an adequate expression for the probability density of classical paths in a random potential, $P_t[\mathbf{x}_t]$, is obtained by considering the classical paths as realizations of Brownian motion;[37] i.e. the classical motion is assumed a diffusion process, and the probability distribution of paths is given by Eq. (7.103). Performing the impurity average gives in the diffusive limit for the weak-localization interference term[38]

$$
C_{\frac{t}{2},\frac{-t}{2}}(\mathbf{x},\mathbf{x}) = \int\limits_{x_{-t/2}=x}^{x_{t/2}=x} \mathcal{D}\mathbf{x}_t \, e^{-\int\limits_{-t/2}^{t/2} dt \left(\frac{\dot{x}_t^2}{4D_0} - i\tilde{\varphi}(\mathbf{x}_t^{cl}) \right)} \, ,
\tag{11.123}
$$

where D_0 is the diffusion constant for a particle with energy ϵ_F, $D_0 = v_F^2 \tau/d$.

Let us now obtain the equation satisfied by the Cooperon in the presence of a time-dependent electromagnetic field. In that case we have for the interaction the Lagrangian

$$
L_1(\mathbf{x}_t, \dot{\mathbf{x}}_t, t) = e\dot{\mathbf{x}}_t \cdot \mathbf{A}(\mathbf{x}_t, t) - e\phi(\mathbf{x}_t, t) \, .
\tag{11.124}
$$

Since the coherence between time-reversed trajectories is partially upset, it is convenient to introduce arbitrary initial and final times, and we have for the phase difference between a pair of time-reversed paths

$$
\varphi[\mathbf{x}_t^{cl}] = \frac{1}{\hbar} \{ S[\mathbf{x}_t^{cl}] - S[\mathbf{x}_{t_i+t_f-t}^{cl}] \}
$$

$$
= \int\limits_{t_i}^{t_f} dt \, \left(L_1(\mathbf{x}_t^{cl}, \dot{\mathbf{x}}_t^{cl}, t) - L_1(\mathbf{x}_{t_i+t_f-t}^{cl}, \dot{\mathbf{x}}_{t_i+t_f-t}^{cl}, t) \right)
\tag{11.125}
$$

---

[37] An exception to this is discussed in Section 11.3.1.
[38] In case the classical motion in the random potential is adequately described as the diffusion process, we immediately recover the result Eq. (11.93) for the return probability.

as the contributions to the phase difference from L_0 cancels, and we are left with

$$\varphi[\mathbf{x}_t^{\mathrm{cl}}] \;=\; \frac{e}{\hbar} \int\limits_{t_i}^{t_f} dt \Big\{ \dot{\mathbf{x}}_{\mathrm{cl}}(t) \cdot \mathbf{A}(\mathbf{x}_{\mathrm{cl}}(t), t) \;+\; \phi(\mathbf{x}_{\mathrm{cl}}(t_i + t_f - t), t_i + t_f - t) \;-\; \phi(\mathbf{x}_{\mathrm{cl}}(t), t)$$

$$-\; \dot{\mathbf{x}}_{\mathrm{cl}}(t_i + t_f - t) \cdot \mathbf{A}(\mathbf{x}_{\mathrm{cl}}(t_i + t_f - t), t_i + t_f - t) \Big\} . \tag{11.126}$$

Introducing the shift in the time variable

$$t' \equiv t - T , \qquad T \equiv \frac{1}{2}(t_f + t_i) \tag{11.127}$$

we get

$$\varphi[\mathbf{x}_t^{\mathrm{cl}}] \;=\; \frac{e}{\hbar} \int\limits_{\frac{t_i - t_f}{2}}^{\frac{t_f - t_i}{2}} dt' \Big\{ \dot{\mathbf{x}}_{\mathrm{cl}}(t' + T) \cdot \mathbf{A}(\mathbf{x}_{\mathrm{cl}}(t' + T), t' + T)$$

$$-\; \dot{\mathbf{x}}_{\mathrm{cl}}(T - t') \cdot \mathbf{A}(\mathbf{x}_{\mathrm{cl}}(T - t'), T - t')$$

$$-\; \phi(\mathbf{x}_{\mathrm{cl}}(t' + T), t' + t) \;+\; \phi(\mathbf{x}_{\mathrm{cl}}(T - t'), T - t') \Big\} . \tag{11.128}$$

The electromagnetic field is assumed to have a negligible effect on determining the classical paths, and we can shift the time argument specifying the position on the path to be symmetric about the moment in time T, and thereby rewrite the phase difference, $t \equiv t_f - t_i$,

$$\varphi[\mathbf{x}_t^{\mathrm{cl}}] \;=\; \frac{e}{\hbar} \int\limits_{-\frac{t}{2}}^{\frac{t}{2}} d\bar{t} \Big\{ \dot{\mathbf{x}}_{\bar{t}}^{\mathrm{cl}} \cdot \mathbf{A}_T(\mathbf{x}_{\bar{t}}^{\mathrm{cl}}, \bar{t}) \;-\; \phi(\mathbf{x}_{\bar{t}}^{\mathrm{cl}}, \bar{t}) \Big\} , \tag{11.129}$$

where

$$\phi_T(\mathbf{x}, t) \;=\; \phi(\mathbf{x}, T + t) \;-\; \phi(\mathbf{x}, T - t) \tag{11.130}$$

and

$$\mathbf{A}_T(\mathbf{x}, t) \;=\; \mathbf{A}(\mathbf{x}, T + t) \;+\; \mathbf{A}(\mathbf{x}, T - t) . \tag{11.131}$$

An electric field can be represented solely by a scalar potential, and we immediately conclude that only if the field is different on time-reversed trajectories can it lead to destruction of phase coherence. In particular, an electric field constant in time does not affect the phase coherence, and thereby does not influence the weak-localization effect.

The differential equation corresponding to the path integral, Eq. (11.123), therefore gives for the Cooperon equation for the case of a time-dependent electromagnetic field

$$\left\{\frac{\partial}{\partial t} + \frac{e}{\hbar}\phi_T(\mathbf{x}_t, t) - D_0\left(\nabla_\mathbf{x} - \frac{ie}{\hbar}\mathbf{A}_T(\mathbf{x}, t)\right)^2\right\} C_{t,t'}^T(\mathbf{x}, \mathbf{x}') = \delta(\mathbf{x} - \mathbf{x}')\,\delta(t - t')\,.$$

$$(11.132)$$

When the sample is exposed to a time-independent magnetic field, we recover the static Cooperon equation, Eq. (11.107).

11.3 Phase breaking in weak localization

The phase coherence between the amplitudes for pairs of time-reversed trajectories is interrupted when the environment of the electron, besides the dominating random potential, is taken into account. At nonzero temperatures, energy exchange due to the interaction with the environment will partially upset the coherence between time-reversed paths involved in the weak-localization phenomenon. The constructive interference is then partially destroyed.

Quantitatively the effect on weak localization by inelastic interactions with energy transfers ΔE of the order of the temperature, $\Delta E \sim kT$, strongly inelastic processes, can be understood by the observation that the single-particle Green's function will be additionally damped owing to interactions. If in addition to disorder we have an interaction, say with phonons, the self-energy will in lowest order in the interaction be changed according to

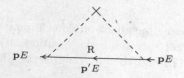

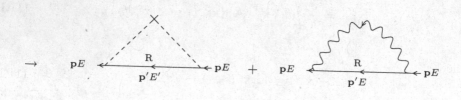

$$(11.133)$$

and we will get an additional contribution to the imaginary part of the self-energy

$$\Im m\Sigma^{\mathrm{R}} = -\frac{\hbar}{2\tau} - \frac{\hbar}{2\tau_{\mathrm{in}}}\,.$$

$$(11.134)$$

Upon redoing the calculation leading to Eq. (11.22) for the case in question, we obtain in the limit $\tau_{in} \gg \tau$

$$\zeta(\mathbf{Q}, \omega) = 1 - \frac{\tau}{\tau_{in}} + i\omega\tau + D_0\tau Q^2 . \tag{11.135}$$

This will in turn lead to the change in the Cooperon equation, $\omega \to \omega + i/\tau_{in}$, and we get the real space Cooperon equation[39]

$$\left\{ -i\omega - D_0\nabla_{\mathbf{x}}^2 + \frac{1}{\tau_{in}} \right\} C_\omega(\mathbf{x}, \mathbf{x}') = \frac{1}{\tau}\delta(\mathbf{x} - \mathbf{x}') . \tag{11.136}$$

The effect on weak localization of electron–electron interaction and electron–phonon interaction have been studied in detail experimentally [88, 89], and can phenomenologically be accounted for adequately by introducing a temperature-dependent phase-breaking rate $1/\tau_\varphi$ in the Cooperon equation, describing the temporal exponential decay $C(t) \to C(t) \exp\{-t/\tau_\varphi\}$ of phase coherence. In many cases the inelastic scattering rate, $1/\tau_{in}$, is identical to the phase-breaking rate, $1/\tau_\varphi$. This is for example the case for electron–phonon interaction, as we shortly demonstrate. However, one should keep in mind that the inelastic scattering rate is defined as the damping of an energy state for the case where all scattering processes are weighted equally, irrespective of the amount of energy transfer. In a clean metal the energy relaxation rate due to electron–phonon or electron–electron interaction is determined by energy transfers of the order of the temperature as a consequence of the exclusion principle (at temperatures below the Debye temperature).[40] In Section 11.5 we shall soon learn that in a three-dimensional sample the energy relaxation rate in a dirty metal is larger than in a clean metal owing to a strong enhancement of the electron–electron interaction with small energy transfer. When calculating the weak localization phase-breaking rate we must therefore pay special attention to the low-energy electron–electron interaction. In a thin film or in the two-dimensional case the energy relaxation rate even diverges in perturbation theory, owing to the abundance of collisions with small energy transfer. However, the physically measurable phase-breaking rate does of course not suffer such a divergence since the phase change caused by an inelastic collision is given by the energy transfer *times* the remaining time to elapse on the trajectory. Collisions with energy transfer of the order of (the phase-breaking rate) $\hbar\omega \sim \hbar/\tau_\varphi$ or less are therefore inefficient for destroying the phase coherence between the amplitudes for traversing typical time-reversed trajectories of duration the phase coherence time τ_φ.[41] In terms of diagrams this is reflected by the fact that interaction lines can connect the upper and lower particle lines in the Cooperon, whereas there are no such processes for the diagrammatic representation of the inelastic scattering rate, as discussed in Section 11.5. This distinction is of importance in the case of a thin metallic film, the quasi two-dimensional

---

[39]In the Cooperon, contributions from diagrams where besides impurity correlator lines interaction lines connecting the retarded and advanced particle line also appear should be included for consistency. However, for strongly inelastic processes these contributions are small.

[40]For details see, for example, chapter 10 of reference [1].

[41]A similar situation is the difference between the transport and momentum relaxation time. The transport relaxation time is the one appearing in the conductivity, reflecting that small angle scattering is ineffective in degrading the current.

case, where there is an abundance of scatterings with small energy transfer due to diffusion-enhanced electron–electron interaction.

In the time-reversal invariant situation, the Cooperon is equal to the classical probability that an electron at the Fermi level in time t returns to its starting point. If coherence is disrupted by interactions, the constructive interference is partially destroyed. This destruction of phase coherence results in the decay in time of coherence, described by the factor $\exp\{-t/\tau_\varphi\}$ in the expression for the Cooperon, the probability of not suffering a phase-breaking collision, described by the phase-breaking rate $1/\tau_\varphi$. In view of the quantum interference picture of the weak localization effect, we shall also refer to τ_φ as the wave function phase relaxation time.

A comprehensive understanding of the phase coherence length in weak localization, the length scale $L_\varphi \equiv \sqrt{D_0\tau_\varphi}$ over which the electron diffuses quantum mechanically coherently, has been established, and this has given valuable information about inelastic scattering processes. The phase coherence length L_φ is, at low temperatures, much larger than the impurity mean free path l, explaining the slow spatial variation of the Cooperon on the scale of the mean free path, which we have repeatedly exploited.

11.3.1 Electron–phonon interaction

In this section we calculate the phase-breaking rate due to electron–phonon interaction using the simple interference picture described in the previous section.[42] We start from the one-electron Lagrangian, which is given by

$$L(\mathbf{x}, \dot{\mathbf{x}}) = \frac{1}{2}m\dot{\mathbf{x}}^2 - V(\mathbf{x}) - e\phi(\mathbf{x}, t) , \qquad (11.137)$$

where V is the impurity potential, and the deformation potential is specified in terms of the lattice displacement field, Eq. (2.72),

$$e\phi(\mathbf{x}, t) = \frac{n}{2N_0} \nabla_{\mathbf{x}} \cdot \mathbf{u}(\mathbf{x}, t) . \qquad (11.138)$$

It is important to note that the impurities move in phase with the distorted lattice; hence the impurity potential has the form

$$V(\mathbf{x}) = \sum_i V_{\text{imp}}(\mathbf{x} - (\mathbf{R}_i + \mathbf{u}(\mathbf{x}, t)) , \qquad (11.139)$$

where $\mathbf{R}_i$ is the equilibrium position of the ith ion. The impurity scattering is thus only elastic in the frame of reference that locally moves along with the lattice. We therefore shift to this moving frame of reference by changing the electronic coordinate according to $\mathbf{x} \to \mathbf{x} + \mathbf{u}$. The impurity scattering then becomes static on account of generating additional terms of interaction. Expanding the Lagrangian Eq. (11.137) in terms of the displacement, and neglecting terms of relative order m/M, such as the term $m\dot{\mathbf{u}} \cdot \mathbf{v}/2$, the transformed Lagrangian can be written as $L = L_0 + L_1$, where

---

[42] We follow references [87] and [90].

L_0 is given in Eq. (11.111), and[43]

$$L_1(\mathbf{x}_t, \dot{\mathbf{x}}_t) = m\dot{\mathbf{x}}_t \cdot (\dot{\mathbf{x}}_t \cdot \nabla)\,\mathbf{u}(\mathbf{x}_t, t) - \frac{1}{3}\dot{\mathbf{x}}_t^2\,\nabla \cdot \mathbf{u}(\mathbf{x}_t, t)\,. \tag{11.140}$$

In the last line we have used the relation $n/2N_0 = mv_F^2/3$, and the fact that the magnitude of the velocity is conserved in elastic scattering. We therefore obtain for the phase difference[44]

$$\varphi[\mathbf{x}_t^{\mathrm{cl}}] = \frac{1}{\hbar}\int\limits_{-t/2}^{t/2} dt\,\{\nabla_\beta\,u_\alpha(\mathbf{x}_t^{\mathrm{cl}}, t) - \nabla_\beta\,u_\alpha(\mathbf{x}_t^{\mathrm{cl}}, -t)\}\left[\dot{x}_t^\alpha\,\dot{x}_t^\beta - \frac{1}{3}\,\delta_{\alpha\beta}\,\dot{\mathbf{x}}_t^2\right]\,, \tag{11.141}$$

where summation over repeated Cartesian indices is implied, and we have chosen the classical paths to satisfy the boundary condition, $\mathbf{x}_{-t/2}^{\mathrm{cl}} = \mathbf{0} = \mathbf{x}_{t/2}^{\mathrm{cl}}$.

We must now average the quantum interference term as given in Eq. (11.116) with respect to the lattice vibrations, and with respect to the random positions of the impurities. Since the Lagrangian for the lattice vibrations is a quadratic form in the displacement $\mathbf{u}$, and the phase difference $\varphi[\mathbf{x}_t^{\mathrm{cl}}]$ is linear in the displacement, the phonon average can be computed by Wick's theorem according to (see Exercise 4.108 on page 103)[45]

$$\langle e^{i\varphi[\mathbf{x}_t^{\mathrm{cl}}]}\rangle_{\mathrm{ph}} = e^{-\frac{1}{2}\langle\varphi[\mathbf{x}_t^{\mathrm{cl}}]^2\rangle_{\mathrm{ph}}}\,. \tag{11.142}$$

For the argument of the exponential we obtain ($\mathbf{v}_t \equiv \dot{\mathbf{x}}_t^{\mathrm{cl}}$)

$$\langle\varphi[\mathbf{x}_t^{\mathrm{cl}}]^2\rangle_{\mathrm{ph}} = \frac{m^2}{\hbar^2}\int\limits_{-t/2}^{t/2} dt_1\int\limits_{-t/2}^{t/2} dt_2\left[\sum_\pm (\pm)D_{\alpha\beta\gamma\delta}(\mathbf{x}_{t_1}^{\mathrm{cl}} - \mathbf{x}_{t_2}^{\mathrm{cl}}, t_1 \mp t_2)\right]$$

$$\times\left[v_{t_1}^\alpha\,v_{t_1}^\beta - \frac{1}{3}\,\delta_{\alpha\beta}\,\mathbf{v}_{t_1}^2\right]\left[v_{t_2}^\gamma\,v_{t_2}^\delta - \frac{1}{3}\,\delta_{\gamma\delta}\,\mathbf{v}_{t_2}^2\right]\,, \tag{11.143}$$

where the phonon correlator

$$D_{\alpha\beta\gamma\delta}(\mathbf{x}, t) = \langle\nabla_\beta\,u_\alpha(\mathbf{x}, t)\nabla_\delta\,u_\gamma(\mathbf{0}, 0)\rangle \tag{11.144}$$

is an even function of the time difference t.

Concerning the average with respect to impurity positions, we will resort to an approximation which, since the exponential function is a convex function, can be expressed as the inequality

$$\langle C(t)\rangle_{\mathrm{imp}} \geq \langle C(t)\rangle_{\mathrm{imp}}^{\varphi=0}\,e^{-\frac{1}{2}\langle\langle\varphi[\mathbf{x}_t^{\mathrm{cl}}]^2\rangle_{\mathrm{ph}}\rangle_{\mathrm{imp}}}\,, \tag{11.145}$$

---

[43]This result can also be obtained without introducing the moving frame of reference. By simply Taylor-expanding Eq. (11.139) and using Newton's equation we obtain a Lagrangian which differs from the one in Eq. (11.140) by only a total time derivative, and therefore generates the same dynamics.

[44]In neglecting the Jacobian of the nonlinear transformation to the moving frame, we neglect the influence of the lattice motion on the paths.

[45]We have suppressed the hat on $\mathbf{u}$ indicating that the displacement is an operator with respect to the lattice degrees of freedom (or we have envisaged treating the lattice vibrations in the path integral formulation).

where we have introduced the notation for the impurity average

$$\langle\langle(\varphi[\mathbf{x}_t^{cl}])^2\rangle_{ph}\rangle_{imp} = \frac{\int_{\mathbf{x}_{-t/2}=\mathbf{x}}^{\mathbf{x}_{t/2}=\mathbf{x}}\mathcal{D}\mathbf{x}_t\, P_t[\mathbf{x}_t]\,\langle(\varphi[\mathbf{x}_t^{cl}])^2\rangle_{ph}}{\int_{\mathbf{x}_{-t/2}=\mathbf{x}}^{\mathbf{x}_{t/2}=\mathbf{x}}\mathcal{D}\mathbf{x}_t\, P_t[\mathbf{x}_t]}. \tag{11.146}$$

The phase difference Eq. (11.141) depends on the local velocity of the electron, which is a meaningless quantity in Brownian motion.[46] It is therefore necessary when considering phase breaking due to electron–phonon interaction to consider the time-reversed paths involved in the weak-localization quantum interference process as realizations of Boltzmannian motion. At a given time, a Boltzmannian path is completely specified by its position and by the direction of its velocity as discussed in Section 7.4.1. We are dealing with the Markovian process described by the Boltzmann propagator $F(\mathbf{v},\mathbf{x},t;\mathbf{v}',\mathbf{x}',t')$, where we now use the velocity as variable instead of the momentum as used in Section 7.4.1. On account of the Markovian property, the four-point correlation function required in Eq. (11.146) (the start and (identical) end point and two intermediate points according to Eq. (11.143)) may be expressed as a product of three conditional probabilities of the type Eq. (7.70), and we obtain

$$\langle\langle\varphi[\mathbf{x}_t^{cl}]^2\rangle_{ph}\rangle_{imp} = \frac{4m^2}{\hbar^2}\int_{-t/2}^{t/2}dt_1\int_{-t/2}^{t/2}dt_2\int d\mathbf{x}_1\int d\mathbf{x}_2\int\frac{d\hat{\mathbf{v}}_1 d\hat{\mathbf{v}}_2}{(4\pi)^2}$$

$$\times\ \overline{F}(\mathbf{0},\frac{t}{2};\mathbf{x}_1,\mathbf{v}_1,t_1)F(\mathbf{x}_1,\mathbf{v}_1,t_1;\mathbf{x}_2,\mathbf{v}_2,t_2)\,\overline{F}(\mathbf{x}_2,\mathbf{v}_2,t_2;\mathbf{0},-\frac{t}{2})$$

$$\times\ \left[\sum_\pm(\pm)D_{\alpha\beta\gamma\delta}(\mathbf{x}_{t_1}^{cl}-\mathbf{x}_{t_2}^{cl},t_1\mp t_2)\right]\left[v_{t_1}^\alpha v_{t_1}^\beta-\frac{1}{3}\,\delta_{\alpha\beta}\,\mathbf{v}_{t_1}^2\right]\left[v_{t_2}^\gamma v_{t_2}^\delta-\frac{1}{3}\,\delta_{\gamma\delta}\,\mathbf{v}_{t_2}^2\right].$$

$$\tag{11.147}$$

We use the notation that an angular average of the Boltzmann propagator F with respect to one of its velocities is indicated by a bar. For example, we have for the return probability

$$\langle C(t)\rangle_{imp}^{(\varphi=0)} = \overline{\overline{F}}(\mathbf{x},t;\mathbf{x}',0) \equiv \int\frac{d\hat{\mathbf{v}}'}{4\pi}\,\overline{F}(\mathbf{x},t;\mathbf{v}',\mathbf{x}',t'). \tag{11.148}$$

The space-dependent quantities may be expressed by Fourier integrals according to Eq. (7.72). Since the Boltzmann propagator is retarded, $F(\mathbf{v},\mathbf{x},t;\mathbf{v}',\mathbf{x}',t')$ vanishes for t earlier than t', we can expand the upper t_1-integration to infinity and the

---

[46]The velocity entering in the Wiener measure, Eq. (7.103), is not the local velocity, but an average of the velocity on a Boltzmannian path; recall Exercise 7.6 on page 197.

lower t_2-integration to minus infinity. Only thermally excited phonons contribute to the destruction of phase coherence, and we conclude that $D_{\alpha\beta\gamma\delta}(\mathbf{x}_{t_1}^{cl} - \mathbf{x}_{t_2}^{cl}, t_1 \mp t_2)$ is essentially zero for $|t_1 \pm t_2| \geq \hbar/kT$. We can therefore extend the domain of integration to infinity with respect to $|t_1 \pm t_2|$ provided that $|t| \gg \hbar/kT$, and obtain in the convex approximation

$$\langle C(t) \rangle_{\mathrm{imp}} = \langle C(t) \rangle_{\mathrm{imp}}^{(\varphi=0)} \exp \left\{ - \frac{2m^2}{\hbar^2 \langle C(t) \rangle_{\mathrm{imp}}^{(\varphi=0)}} \int \frac{d\mathbf{k} d\mathbf{k}' d\omega d\omega'}{(2\pi)^8} \int \frac{d\hat{\mathbf{v}}_1 d\hat{\mathbf{v}}_2}{(4\pi)^2} \right.$$

$$\times \quad \overline{F}(\mathbf{v}_1; \mathbf{k}, \omega) F(\mathbf{v}_1, \mathbf{v}_2, \mathbf{k} + \mathbf{k}', \omega + \omega') D_{\alpha\beta\gamma\delta}(-\mathbf{k}', -\omega')$$

$$\times \quad \left[\overline{F}(\mathbf{v}_2, \mathbf{k}, \omega) e^{-i\omega t} - \overline{F}(\mathbf{v}_2, \mathbf{k}, \omega + 2\omega') e^{-i(\omega + \omega')t} \right]$$

$$\times \quad \left[v_{t_1}^{\alpha} v_{t_1}^{\beta} - \frac{1}{3} \delta_{\alpha\beta} \mathbf{v}_{t_1}^2 \right] \left[v_{t_2}^{\gamma} v_{t_2}^{\delta} - \frac{1}{3} \delta_{\gamma\delta} \mathbf{v}_{t_2}^2 \right] \right\}. \tag{11.149}$$

We expect that the argument of the exponential above increases linearly in t for large times. Since the classical return probability in three dimensions has the time dependence $\langle C(t) \rangle_{\mathrm{imp}}^{(\varphi=0)} \propto t^{-3/2}$ (recall the form of the diffusion propagator), the integral above should not decrease faster than $t^{-1/2}$. Such a slow decrease is obtained from the $(\mathbf{k}, \omega)$-integration only from the combination $\overline{F}(\mathbf{v}_1; \mathbf{k}, \omega) \overline{F}(\mathbf{v}_2; \mathbf{k}, \omega)$, which according to Eq. (7.76) features an infrared singular behavior $(-i\omega + D_0 k^2)^{-2}$ for small k and ω. In fact, it is just this combination that leads to a time-dependence proportional to $t^{-1/2}$ and, compared with that, all other contributions may be neglected. For the important region of integration we thus have $\omega \ll \omega'$, since ω' is determined by the phonon correlator, which gives the large contribution to the integral for the typical value $\hbar\omega' \simeq kT$. We are therefore allowed to approximate $F(\mathbf{v}_1, \mathbf{v}_2; \mathbf{k} + \mathbf{k}', \omega + \omega')$ by $F(\mathbf{v}_1, \mathbf{v}_2; \mathbf{k}', \omega')$. In addition, the same arguments show that the second term in the square bracket may be omitted. We thus obtain

$$\langle C(t) \rangle_{\mathrm{imp}} = \langle C(t) \rangle_{\mathrm{imp}}^{(\varphi=0)} e^{-t/\tau_\varphi}, \tag{11.150}$$

where the phase-breaking rate due to electron–phonon interaction is given by

$$\frac{1}{\tau_\varphi} = \frac{2m^2}{\hbar^2} \int \frac{d\mathbf{k}' d\omega'}{(2\pi)^4} \int \frac{d\hat{\mathbf{v}}_1 d\hat{\mathbf{v}}_2}{(4\pi)^2} F(\mathbf{v}_1, \mathbf{v}_2; \mathbf{k}', \omega') D^{\alpha\beta\gamma\delta}(\mathbf{k}', \omega') \left[v_1^{\alpha} v_1^{\beta} - \frac{1}{3} \delta_{\alpha\beta} \mathbf{v}_1^2 \right]$$

$$\times \quad \left[v_2^{\alpha} v_2^{\beta} - \frac{1}{3} \delta_{\alpha\beta} \mathbf{v}_2^2 \right]. \tag{11.151}$$

For simplicity we consider the Debye model where the lattice vibrations are specified by the density n_i and the mass M of the ions, and by the longitudinal c_l and the transverse c_t sound velocities.[47] We assume the phonons to have three-dimensional

---

[47] The jellium model does not allow inclusion of Umklapp processes in the electron–phonon scattering.

character. In case of longitudinal vibrations, we have the normal mode expansion of the displacement field

$$\mathbf{u}(\mathbf{r}, t) = \frac{i}{\sqrt{N}} \sum_{\mathbf{k}} \hat{\mathbf{k}} \, Q_{\mathbf{k}}(t) \, e^{i\mathbf{k}\cdot\mathbf{r}} , \qquad (11.152)$$

where N is the number of ions in the normalization volume. For the phonon average we have

$$\langle Q_{\mathbf{k}}(t) \, Q_{\mathbf{k}'}(t') \rangle = \delta_{\mathbf{k},-\mathbf{k}'} \frac{\hbar}{2M\omega_{\mathbf{k}}} H(\omega_{\mathbf{k}}) \, \cos\omega_{\mathbf{k}}(t - t') , \qquad (11.153)$$

where $\omega_{\mathbf{k}} = c_l k$, provided that k is less than the cut-off wave vector k_{D}, and we obtain for the Fourier transform of the longitudinal phonon correlator

$$D_{\mathrm{L}}^{\alpha\beta\gamma\delta}(\mathbf{k}, \omega)] = \frac{1}{2} k^\alpha k^\beta k^\gamma k^\delta \, H(\omega_{\mathbf{k}}) \left[\delta(\omega - \omega_{\mathbf{k}}) + \delta(\omega + \omega_{\mathbf{k}}) \right] . \qquad (11.154)$$

Strictly speaking, we encounter in the above derivation $H(\omega) = 2n(\omega) + 1$, where n is the Bose distribution function. However, the present single electron theory does not take into account that the fermionic exclusion principle forbids scattering of an electron into occupied states. Obedience of the Pauli exclusion principle is incorporated by the replacement[48]

$$\frac{1}{2} H(\omega) \; \rightarrow \; \coth \frac{\hbar\omega}{2kT} - \tanh \frac{\hbar\omega}{2kT} = \frac{2}{\sinh \frac{\hbar\omega}{kT}} . \qquad (11.155)$$

Upon inserting Eq. (11.154) in the expression Eq. (11.151) for the phase-breaking rate, we encounter the directional average of expressions of the type

$$\hat{k}_\alpha \hat{k}_\beta \left[v_\alpha v_\beta - \delta_{\alpha\beta} \frac{\mathbf{v}^2}{3} \right] = k^{-2} \left[(\mathbf{k} \cdot \mathbf{v})^2 - k^2 \frac{\mathbf{v}^2}{3} \right] . \qquad (11.156)$$

Altogether the angular averages appear in the combination

$$\Phi_{\mathrm{L}}(kl) = \frac{18}{\pi v_{\mathrm{F}}^3 k^3} \left\{ I(k, \omega) \left[\int \frac{d\hat{\mathbf{v}}}{4\pi} \frac{(\mathbf{k}\cdot\mathbf{v})^2 - k^2 \frac{\mathbf{v}^2}{3}}{-i\omega + i\mathbf{v}\cdot\mathbf{k} + 1\tau} \right]^2 + \int \frac{d\hat{\mathbf{v}}}{4\pi} \frac{[(\mathbf{k}\cdot\mathbf{v})^2 - k^2 \frac{\mathbf{v}^2}{3}]^2}{-i\omega + i\mathbf{v}\cdot\mathbf{k} + 1/\tau} \right\}$$

$$= \frac{2}{\pi} \left(\frac{kl \arctan kl}{kl - \arctan kl} - \frac{3}{kl} \right) , \qquad (11.157)$$

---

[48]The argument is identical to the similar feature for the inelastic scattering rate or imaginary part of the self-energy. In terms of diagrams, we recall that, in the above discussion, we have included only the effect of the kinetic or Keldysh component of the phonon propagator. Including the retarded and advanced components makes the electron experience its fermionic nature introducing the electron kinetic component which carries the tangent hyperbolic factor. As a consequence, a point also elaborated in reference [91], the zero-point fluctuations of the lattice can not disrupt the weak-localization phase coherence. A detailed discussion of the Pauli principle and the inelastic scattering rate is given in Section 11.5 in connection with the electron–electron interaction.

where the result in the last line is obtained since $\omega = c_l k \ll v_F k$. For the phase-breaking rate due to longitudinal phonons we thus obtain

$$\frac{1}{\tau_{\varphi,l}} = \frac{\pi \hbar^2}{6 m M c_l} \int_0^{k_D} dk\, k^2\, \Phi_L(kl)\, \frac{1}{\sinh \hbar c_l k / kT} . \tag{11.158}$$

We note the limiting behaviors

$$\frac{1}{\tau_{\varphi,l}} = \begin{cases} \frac{7\pi\zeta(3)}{12} \frac{(kT)^3}{\hbar n M c_l^4} & \hbar c_l/l \ll kT \ll \hbar c_l k_D \\[2ex] \frac{\pi^4}{30}\, l\, \frac{(kT)^4}{\hbar n M c_l^5} & kT \ll \hbar c_l/l . \end{cases} \tag{11.159}$$

The expression Eq. (11.157) for the function Φ_L demonstrates in a direct way the important compensation that takes place in the case of longitudinal phonons between the two mechanisms contained in L_1. First, the term $(\mathbf{k}\cdot\mathbf{v})^2$ corresponds to $m\mathbf{v}\cdot(\mathbf{v}\cdot\nabla)\mathbf{u}$ and represents the coupling of the electrons to the vibrating impurities. Second, the term $-k^2 \mathbf{v}^2/3$ is connected with $-m\mathbf{v}^2\nabla\cdot\mathbf{u}/3$, and originates from the interaction of the electrons with the lattice vibrations. Without this compensation, each of the mechanisms would appear to be enhanced in an impure metal, and would lead to an enhanced phase-breaking rate proportional to $(kT)^2/(nMc_l^3 l)$.

For the case of transverse vibrations, we note that $D_T^{\alpha\beta\gamma\delta}$ is of similar form as Eq. (11.154) where, however, $\hat{k}_\alpha \hat{k}_\gamma$ has to be replaced by $(\delta_{\alpha\gamma} - \hat{k}_\alpha \hat{k}_\gamma)$ and an additional factor of 2, which accounts for the multiplicity of transverse modes. We then obtain a phase-breaking rate due to interaction with transverse phonons, $\tau_{\varphi,t}$, which is similar to the expression in Eq. (11.158) with c_l and ϕ_L replaced by c_t and

$$\Phi_T(kl) = \frac{3}{\pi} \frac{2k^3 l^3 + 3kl - 3(k^2 l^2 + 1)\arctan kl}{k^4 l^4} \tag{11.160}$$

respectively. In particular, we obtain the limiting behaviors for the phase-breaking rate due to transverse phonons

$$\frac{1}{\tau_{\varphi,t}} = \begin{cases} \frac{\pi^2}{2} \frac{(kT)^2}{mMc_t^3 l} & \hbar c_t/l \ll kT \ll \hbar c_t k_D \\[2ex] \frac{\pi^4}{20}\, l\, \frac{(kT)^4}{\hbar^2 m M c_t^5} & kT \ll \hbar c_t/l . \end{cases} \tag{11.161}$$

We note that in the high-temperature region, $\hbar c_t/l \ll kT \ll \hbar c_t k_D$, the transverse contribution is negligible in comparison with the longitudinal one if $c_t \simeq c_l$. But the transverse rate dominates in the case where the transverse sound velocity is much smaller than the longitudinal one. Such a situation may quite well be realized in some amorphous metals; then, it is possible to observe a phase-breaking rate of the form $\tau_\varphi^{-1} \propto T^2/l$ at higher, but not too high, temperatures.[49] The predictions of the theory are in good agreement with magneto-resistance measurements and carefully conducted experiments of the temperature dependence of the resistance [92].

---

[49] A quadratic temperature dependence of the phase-breaking rate is often observed experimentally.

The physical meaning of the second term in Eq. (11.149) is as follows. It is appreciable only if the lattice deformation stays approximately constant during the time the electron spends on its path and leads, in this case, to a cancellation of the first term. Equivalently, electron–phonon interactions with small energy transfers do not lead to destruction of phase coherence. The effect of this term is thus effectively to introduce a lower cut-off in the integral of Eq. (11.158) at wave vector $k_0 = 1/c_l\tau_{\varphi,l}$. However, there are no realistic models of phonon spectra where this effect is of importance. We therefore have the relationships $\omega' \simeq kT/\hbar \gg \omega \simeq 1/\tau_\varphi$. It is therefore no surprise that the calculated phase-breaking rates are identical to the inelastic electron–phonon collision rates in a dirty metal [93]. When considering phase breaking due to electron–electron interaction, which we now turn to, the small energy transfer interactions are of importance.

11.3.2　Electron–electron interaction

In this section we consider the temperature dependence of the phase-breaking rate due to electron–electron interaction.[50] As already discussed at the beginning of this section, special attention to electron–electron interaction with small energy transfer must be exercised due to the diffusion enhancement. In diagrammatic terms we therefore need to take into account diagrams where the electron–electron interaction connects also the upper and lower particle lines in the Cooperon.

In Section 11.5 we shall show that the effective electron–electron interaction at low energies can be represented by a fluctuating field. Its correlation function in a dirty metal will be given by the expression in Eq. (11.269), which we henceforth employ. We can therefore obtain the effect on the Cooperon of the quasi-elastic electron–electron interaction by averaging the Cooperon with respect to a time-dependent electromagnetic field using the proper correlator. We therefore consider the equation for the Cooperon in the presence of an electromagnetic field, Eq. (11.81),

$$\left\{ 2\frac{\partial}{\partial t} - D_0 \left(\nabla_{\mathbf{x}} - \frac{ie}{\hbar} \mathbf{A}_T(\mathbf{x}, t) \right)^2 + \frac{1}{\tau^{\mathrm{e-e}}} \right\} C^T_{t,t'}(\mathbf{x}, \mathbf{x}') = \frac{1}{\tau} \delta(\mathbf{x} - \mathbf{x}')\,\delta(t - t')\,,$$

(11.162)

where we have chosen a gauge in which the scalar potential vanishes, and $1/\tau^{\mathrm{e-e}}$ is the energy relaxation rate due to high-energy electron–electron interaction processes, i.e. processes with energy transfers $\sim kT$.[51]

To account for the electron–electron interaction with small energy transfers, we must perform the Gaussian average of the Cooperon with respect to the fluctuating field. This is facilitated by writing the solution of the Cooperon equation as the path integral

$$C^T_{t,t'}(\mathbf{R}, \mathbf{R}') = \frac{1}{2\tau} \int_{\mathbf{x}_{t'}=\mathbf{x}'}^{\mathbf{x}_t=\mathbf{x}} \mathcal{D}\mathbf{x}_t\, e^{-S[\mathbf{x}_t]}\,,$$

(11.163)

---

[50]We follow reference [94].

[51]As will become clear in the following, the separation in high- and low-energy transfers takes place at energies of the order of the temperature. However, in the following we shall not need to specify the separation explicitly.

where the Euclidean action consists of two terms

$$S = S_0 + S_A, \qquad (11.164)$$

where

$$S_0[\mathbf{x}_t] = \int\limits_{t}^{t'} dt_1 \left(\frac{\dot{\mathbf{x}}_{t_1}^2}{4D_0} + \frac{1}{\tau^{e-e}} \right) \qquad (11.165)$$

and

$$S_A[\mathbf{x}_t] = \frac{ie}{\hbar} \int\limits_{t}^{t'} dt_1 \, \dot{\mathbf{x}}_{t_1} \cdot \mathbf{A}_T(\mathbf{x}_{t_1}, t_1) . \qquad (11.166)$$

In terms of diagrams, the Gaussian average corresponds to connecting the external field lines pairwise in all possible ways by the correlator of the field fluctuations, thereby producing the effect of the low-energy electron–electron interaction. Since the fluctuating vector potential appears linearly in the exponential Cooperon expression, the Gaussian average with respect to the fluctuating field is readily done

$$C_{t,t'}^T(\mathbf{R}, \mathbf{R}') = \frac{1}{2\tau} \int\limits_{\mathbf{r}_{t'}=\mathbf{R}'}^{\mathbf{r}_t=\mathbf{R}} \mathcal{D}\mathbf{r}_t \, e^{-(S_0[\mathbf{x}_t] + \langle S_A \rangle[\mathbf{x}_t])} \qquad (11.167)$$

where the averaged action $\langle S_A \rangle$ is expressed in terms of the correlator of the vector potential

$$\langle S_A \rangle[\mathbf{x}_t] = \frac{e^2}{2\hbar^2} \int\limits_{t'}^{t} dt_1 \int\limits_{t'}^{t} dt_2 \, \dot{x}_\mu(t_1) \, \dot{x}_\nu(t_2) \, \langle A_\mu^T(\mathbf{x}_{t_2}, t_1) A_\nu^T(\mathbf{x}_{t_2}, t_2) \rangle . \qquad (11.168)$$

If we recall the definition of $\mathbf{A}_T(\mathbf{x}_t, t)$, Eq. (11.77), we have

$$\langle A_\mu^T(\mathbf{x}_{t_2}, t_1) A_\nu^T(\mathbf{x}_{t_2}, t_2) \rangle = 2 \int \frac{d\mathbf{q}}{(2\pi)^d} \int \frac{d\omega}{2\pi} \, e^{i\mathbf{q}\cdot(\mathbf{x}_{t_1}-\mathbf{x}_{t_2})} \langle A_\mu A_\nu \rangle_{\mathbf{q}\omega}$$

$$\times \left(\cos\omega\frac{t_1+t_2}{2} + \cos\omega\frac{t_1-t_2}{2} \right), \qquad (11.169)$$

where we have introduced the notation

$$\langle A_\mu A_\nu \rangle_{\mathbf{q}\omega} \equiv \langle A_\mu(\mathbf{q}, \omega) A_\nu(-\mathbf{q}, -\omega) \rangle . \qquad (11.170)$$

The electric field fluctuations could equally well have been represented by a scalar potential

$$\langle A_\mu(\mathbf{q}, \omega) A_\nu(-\mathbf{q}, -\omega) \rangle = \frac{1}{\omega^2} \langle E_\mu(\mathbf{q}, \omega) E_\nu(-\mathbf{q}, -\omega) \rangle$$

$$= \frac{q_\mu q_\nu}{\omega^2} \langle \phi(\mathbf{q}, \omega) \phi(-\mathbf{q}, -\omega) \rangle . \qquad (11.171)$$

In Section 11.5 we show that the electron–electron interaction with small energy transfers, $\hbar\omega \ll kT$, is determined by the temperature, T, and the conductivity of the sample, σ_0, according to[52]

$$\langle A_\mu A_\nu \rangle_{\mathbf{q}\omega} = \frac{2kT}{\omega^2 \sigma_0} \frac{q_\mu q_\nu}{q^2} .$$ (11.172)

Upon partial integration we notice the identity (the boundary terms are seen to vanish as $\mathbf{x}_{-t} = \mathbf{x}_t$)

$$\int_{t'}^{t} dt_1 \int_{t'}^{t} dt_2 \, q_\mu \, q_\nu \, \dot{x}_\mu(t_1) \, \dot{x}_\nu(t_2) e^{i\mathbf{q}\cdot(\mathbf{x}_{t_1} - \mathbf{x}_{t_2})} \left(\cos\frac{\omega(t_1 + t_2)}{2} + \cos\frac{\omega(t_1 - t_2)}{2} \right)$$

$$= -\int_{t'}^{t} dt_1 \int_{t'}^{t} dt_2 \, e^{i\mathbf{q}\cdot(\mathbf{x}_{t_1} - \mathbf{x}_{t_2})} \frac{\omega^2}{4} \left(\cos\frac{\omega(t_1 + t_2)}{2} - \cos\frac{\omega(t_1 - t_2)}{2} \right)$$ (11.173)

and obtain

$$\langle S_\mathbf{A} \rangle [\mathbf{x}_t] = -\frac{e^2 kT}{2\sigma_0} \int' \frac{d\mathbf{q}}{(2\pi)^d} \int \frac{d\omega}{2\pi} \int_{t'}^{t} dt_1 \int_{t'}^{t} dt_2 \frac{e^{i\mathbf{q}\cdot(\mathbf{x}_{t_1} - \mathbf{x}_{t_2})}}{q^2} \left(\cos\frac{\omega(t_1 + t_2)}{2} - \cos\frac{\omega(t_1 - t_2)}{2} \right) .$$ (11.174)

Performing the integration over ω and t_2, the expression for the Cooperon becomes

$$C_{t,-t}^{T}(\mathbf{x}, \mathbf{x}') = \frac{1}{2\tau} \int_{\mathbf{x}_{-t}=\mathbf{x}'}^{\mathbf{x}_t=\mathbf{x}} \mathcal{D}\mathbf{x}_t \, e^{-\int_{-t}^{t} dt_1 \left\{ \frac{\dot{x}_{t_1}}{4D_0} + \frac{1}{\tau^{e-e}} + \frac{2e^2 kT}{\sigma_0} \int' \frac{d\mathbf{q}}{(2\pi)^2} \, q^{-2} (1 - \cos(\mathbf{q}\cdot(\mathbf{x}_{t_1} - \mathbf{x}_{-t_1}))) \right\}} .$$ (11.175)

The singular term is regularized by remembering that in Eq. (11.174) the ω-integration actually should have been terminated, in the present context, at the large frequency $kT/\hbar$. The factor $\exp\{i\mathbf{q}\cdot(\mathbf{x}_{t_1} - \mathbf{x}_{t_2})\}$ does therefore not reduce strictly to 1 for the first term in the parenthesis in Eq. (11.174) as $|\mathbf{x}_{t_1} - \mathbf{x}_{t_2}| \geq (D_0\hbar/kT)^{1/2}$, and this oscillating phase factor provides the convergence of the integral. We should therefore cut off the $\mathbf{q}$-integral at the wave vector satisfying $q = (kT/\hbar D_0)^{1/2} \equiv L_T^{-1}$, as indicated by the prime on the $\mathbf{q}$-integration in the two previous equations.

Introducing new variables

$$\mathbf{R}_t = \frac{\mathbf{x}_t + \mathbf{x}_{-t}}{\sqrt{2}} \quad , \quad \mathbf{r}_t = \frac{\mathbf{x}_t - \mathbf{x}_{-t}}{\sqrt{2}}$$ (11.176)

---

[52]Since the time label T now has disappeared, no confusion should arise in the following where T denotes the temperature. We recall Section 6.5, and note that the relation Eq. (11.172) is equivalent to the statement that the low-frequency electron–electron interaction in a disordered conductor is identical to the Nyquist noise in the electromagnetic field fluctuations.

the path integral separates in two parts[53]

$$C_{t,-t}(\mathbf{R},\mathbf{R}) \;=\; \frac{1}{2\sqrt{2}\tau} \int\limits_{-\infty}^{\infty} d\mathbf{R}_0 \int\limits_{\mathbf{R}_{t=0}=\mathbf{R}_0}^{\mathbf{R}_t=\sqrt{2}\mathbf{R}} \mathcal{D}\mathbf{R}_t \; e^{-\int_0^t dt' \frac{\dot{\mathbf{R}}_{t'}^2}{2D_0}}$$

$$\times \int\limits_{\mathbf{r}_0=0}^{\mathbf{r}_t=0} \mathcal{D}\mathbf{r}_t \; e^{-\int_0^t dt' \left\{ \frac{\dot{\mathbf{r}}_{t'}^2}{4D_0} + \frac{2}{\tau^{e-e}} + \frac{2e^2 kT}{\sigma_0} \int\frac{d\mathbf{q}}{(2\pi)^2} \, q^{-2}\left(1-\cos(\sqrt{2}\mathbf{q}\cdot\mathbf{r}_{t'})\right) \right\}} \qquad (11.177)$$

The path integral with respect to $\mathbf{R}_t$ gives the probability that a particle started at position $\mathbf{R}_0$ at time $t = 0$ by diffusion reaches the point $\sqrt{2}\mathbf{R}$ (recall Eq. (7.103)). Integrating this probability over all possible starting points is identical to integrating over all final points and by normalization gives unity. We are thus left with the expression for the Cooperon

$$C_{t,-t} \;=\; \frac{1}{2\sqrt{2}\tau} \int\limits_{\rho_0=0}^{\rho_t=0} \mathcal{D}\mathbf{r}_{\tilde{t}} \; e^{-\int_0^t d\tilde{t} \left(\frac{\dot{\mathbf{r}}_{\tilde{t}}^2}{4D_0} + V(\mathbf{r}_{\tilde{t}}) \right)} \,, \qquad (11.178)$$

where we have introduced the notation

$$V(\mathbf{r}) \;=\; \frac{2}{\tau^{e-e}} + \frac{2e^2 kT}{\sigma_0} \int\frac{d\mathbf{q}}{(2\pi)^d} \, q^{-2} \left(1 - \cos(\sqrt{2}\,\mathbf{q}\cdot\mathbf{r})\right) \,. \qquad (11.179)$$

As expected from translational invariance, the Cooperon is independent of position.

We have thus reduced the problem of calculating the quantum correction to the conductivity,

$$\delta\sigma(\omega) \;=\; -\frac{4e^2 D_0 \tau}{\pi\hbar} \int\limits_{-\infty}^{\infty} dt \; e^{i\omega t} \, C_{t,-t}(\mathbf{r},\mathbf{r}) \,, \qquad (11.180)$$

in the presence of electron–electron interaction, to solving for the Green's function the imaginary time Schrödinger problem

$$\{\partial_t - D_0 \triangle_{\mathbf{r}} + V(\mathbf{r})\} \, C_{t,t'}(\mathbf{r},\mathbf{r}') \;=\; \frac{1}{2\sqrt{2}\,\tau} \, \delta(\mathbf{r}-\mathbf{r}') \, \delta(t-t') \,. \qquad (11.181)$$

In the three-dimensional case the first term in the integrand of Eq. (11.179) gives rise to a temperature dependence of the form $T^{3/2}$. This is the same form as the one we shall find in Section 11.5 for the inelastic scattering rate due to electron–electron interaction in a dirty metal. This term can thus be joined with the first term of Eq. (11.179). We note that the description of the low-energy behavior thus joins up smoothly with the description of the high-energy behavior, as it should.

---

[53]This is immediately obtained by using the standard discretized representation of a path integral.

· We thus have for the potential in the three-dimensional case

$$V_3(\mathbf{r}) = \frac{2}{\tau^{e-e}} + \tilde{V}_3(\mathbf{r}) \tag{11.182}$$

where

$$\tilde{V}_3(\mathbf{r}) = \frac{-e^2 kT}{\sqrt{2}\pi\hbar^2\sigma_0} \begin{cases} \frac{1}{r} & r \gg L_T \\ \\ \frac{2\sqrt{2}}{\pi} L_T^{-1} & r \ll L_T. \end{cases} \tag{11.183}$$

Fourier-transforming Eq. (11.181) with respect to time and taking the static limit we obtain

$$\{-D_0 \triangle_{\mathbf{r}} + V_3(\mathbf{r})\} C_{\omega=0}(\mathbf{r}, \mathbf{r}') = \frac{1}{2\sqrt{2}\tau} \delta(\mathbf{r} - \mathbf{r}') \,. \tag{11.184}$$

Solving this equation to first order in the potential $\tilde{V}_3$ gives

$$C_1(\mathbf{0}, \mathbf{0}, \omega = 0) = -\frac{e^2 kT}{4\pi\hbar^2\tau D_0^2\sigma_0} \left\{ \frac{L_\epsilon}{\pi L_T} \left(e^{-2\sqrt{2}\frac{L_\epsilon}{L_T}} - 1\right) + Ei\left(-2\sqrt{2}\frac{L_\epsilon}{L_T}\right) \right\} \tag{11.185}$$

where Ei is the exponential integral[54] and

$$L_\epsilon = \sqrt{D_0 \tau^{e-e}} \,. \tag{11.186}$$

In accordance with the calculation of the inelastic lifetime in section 11.5 we have

$$\frac{L_T}{L_\epsilon} \sim \frac{(\frac{kT\tau}{\hbar})^{1/4}}{k_F l} \,. \tag{11.187}$$

We can therefore expand the expression in Eq. (11.185), and obtain for the quantum correction to the conductivity

$$\delta\sigma = \frac{e^2}{2\pi^2\hbar L_\epsilon} \left(1 + \frac{4\pi e^2 kT L_\epsilon}{\hbar^2 D_0\sigma_0} \ln\frac{L_T}{L_\epsilon}\right), \tag{11.188}$$

where the second term is the correction due to collisions with small energy transfer, proportional to $T^{1/4}\ln T$. In the two-dimensional case we obtain from Eq. (11.179) for the potential

$$V_2(\mathbf{r}) = \frac{2}{\tau^{e-e}} + \frac{e^2 kT}{\pi\hbar^2\sigma_0} \int\limits_0^{L_T^{-1}} dq \, \frac{1 - J_0(\sqrt{2}qr)}{q}, \tag{11.189}$$

where J_0 denotes the Bessel function. We observe the limiting behavior of the potential

$$V_2(\mathbf{r}) = \frac{2}{\tau^{e-e}} + \frac{e^2 kT}{\pi\hbar^2\sigma_0} \begin{cases} \frac{1}{4}\left(\frac{r}{L_T}\right)^2 & r \ll L_T \\ \\ \left(1 - J_0(\frac{\sqrt{2}r}{L_T})\right) \ln\frac{\sqrt{2}r}{L_T} - \mathcal{C} + \ln 2 & r \gg L_T, \end{cases} \tag{11.190}$$

---

[54] $Ei(x) = \int_{-\infty}^{x} dt \, \frac{e^t}{t}$ for $x < 0$.

where $\mathcal{C}$ is the Euler constant.

We then get the following equation for the Cooperon in the region of large values of r

$$\left\{ -D_0 \triangle_{\mathbf{r}} + \frac{2}{\tau^{\mathrm{e-e}}} + \frac{e^2 kT}{\pi \hbar^2 \sigma_0} \ln \frac{\sqrt{2} r}{L_T} \right\} C_{\omega=0}(\mathbf{r}, \mathbf{r}') = \frac{1}{2\sqrt{2}\tau} \delta(\mathbf{r} - \mathbf{r}') . \quad (11.191)$$

The electron typically diffuses coherently the distance $\sqrt{D_0 \tau^{\mathrm{e-e}}}$. According to Section 11.5, for the relaxation time in two dimensions for processes with large energy transfers, we have

$$\sqrt{D_0 \tau^{\mathrm{e-e}}} \sim \sqrt{\frac{D_0^2 N_2(0)\hbar^2}{kT}} \sim (k_{\mathrm{F}} l)^{1/2} L_T , \quad (11.192)$$

where $N_2(0)$ denotes the density of states at the Fermi energy in two dimensions. The electron thus diffuses coherently far into the region where the potential is logarithmic, and the slow change of the potential allows the substitution

$$\frac{2}{\tau^{\mathrm{e-e}}} + \frac{e^2 kT}{\pi \hbar^2 \sigma_0} \ln \frac{\sqrt{2} r}{L_T} \rightarrow \frac{e^2 kT}{\pi \hbar^2 \sigma_0} \ln \frac{\sqrt{2\tau^{\mathrm{e-e}} D_0}}{L_T} . \quad (11.193)$$

Inserting into Eq. (11.191), we can read off the phase-breaking rate due to electron–electron interaction in a dirty conductor in two dimensions[55]

$$\frac{1}{\tau_\varphi} = \frac{kT}{4\pi \hbar^2 D_0 N_2(0)} \ln 2\pi \hbar D_0 N_2(0) . \quad (11.194)$$

The phase-breaking rate due to diffusion-enhanced electron–electron interaction thus depends in two dimensions linearly on the temperature at low temperatures, $kT < \hbar/\tau$.

The above result for the phase-breaking rate can be understood as a consequence of the phase-breaking rate setting the lower energy cut-off, $\hbar/\tau_\varphi$, for the efficiency of inelastic scattering events in destroying phase coherence. To show this we note that the path integral expression for the Cooperon, Eq. (11.167), is the weighted average with respect to diffusive paths. Since this weight is convex, we have

$$C_t \geq C_t^{(0)} e^{-\langle\langle(\varphi[\mathbf{x}_t^{cl}])^2\rangle\rangle_{\mathrm{imp}}} , \quad (11.195)$$

where the second bracket signifies the average with respect to diffusive paths of the phase difference between the two interfering alternatives, Eq. (11.129),

$$\langle\langle(\varphi[\mathbf{x}_t^{cl}])^2\rangle_{\mathrm{ee}}\rangle_{\mathrm{imp}} = \frac{\int\limits_{\mathbf{x}_{-t/2}=\mathbf{x}}^{\mathbf{x}_{t/2}=\mathbf{x}} \mathcal{D}\mathbf{x}_t \, P_t[\mathbf{x}_t] \, \langle(\varphi[\mathbf{x}_t^{cl}])^2\rangle_{\mathrm{ee}}}{\int\limits_{\mathbf{x}_{-t/2}=\mathbf{x}}^{\mathbf{x}_{t/2}=\mathbf{x}} \mathcal{D}\mathbf{x}_t \, P_t[\mathbf{x}_t]} \quad (11.196)$$

---

[55] Many experiments are performed on thin metallic films. For such a quasi-two-dimensional case we can express the result for the phase breaking due to electron–electron interaction in a film of thickness a as $\frac{1}{\tau_\varphi} = \frac{e^2 kT}{2\pi a \sigma_0 \hbar^2} \ln \frac{\pi a \sigma_0 \hbar}{e^2}$.

and $C_t^{(0)}$ is the return probability in the absence of the fluctuating field, i.e. the denominator in the above equation. The first bracket signifies the Gaussian average over the fluctuating field, i.e. the low-energy electron–electron interaction,

$$\langle(\varphi[\mathbf{x}_t^{cl}])^2\rangle_{ee} = \frac{e^2}{\hbar^2}\int\limits_{-t/2}^{t/2}dt_1\int\limits_{-t/2}^{t/2}dt_2\,(\langle\phi(\mathbf{x}_{t_1}^{cl}-\mathbf{x}_{t_2}^{cl},t_1-t_2)\,\phi(\mathbf{0},0)\rangle_{ee}$$

$$-\;\langle\phi(\mathbf{x}_{t_1}^{cl}-\mathbf{x}_{t_2}^{cl},t_1+t_2)\,\phi(\mathbf{0},0)\rangle_{ee})\,, \tag{11.197}$$

where we now choose to let the scalar potential represent the fluctuating field. Fourier-transforming we encounter

$$\langle\langle\phi(\mathbf{x}_{t_1}^{cl}-\mathbf{x}_{t_2}^{cl},t_1-t_2)\phi(\mathbf{0},0)\rangle_{ee}\rangle_{imp} = 2\int\frac{d\mathbf{q}}{(2\pi)^d}\int\frac{d\omega}{2\pi}\langle e^{i\mathbf{q}\cdot(\mathbf{x}_{t_1}^{cl}-\mathbf{x}_{t_2}^{cl})}\rangle_{imp}\,\langle\phi\phi\rangle_{\mathbf{q}\omega}$$

$$\times\,(\cos\omega(t_1+t_2)-\cos\omega(t_1-t_2))\,,\tag{11.198}$$

where the correlator for the fluctuating potential is specified in Eq. (11.269). For a diffusion process we have, according to Eq. (7.104),[56]

$$\langle e^{i\mathbf{q}\cdot(\mathbf{x}_{t_1}^{cl}-\mathbf{x}_{t_2}^{cl})}\rangle_{imp} = e^{i\mathbf{q}\cdot\langle(\mathbf{x}_{t_1}^{cl}-\mathbf{x}_{t_2}^{cl})\rangle_{imp}} = e^{-D_0\,q^2|t_1-t_2|} \tag{11.199}$$

and we get

$$\langle\langle(\varphi[\mathbf{x}_t^{cl}])^2\rangle_{ee}\rangle_{imp} = \frac{2e^2kT}{\pi\sigma_0}\int\limits_{-t/2}^{t/2}dt_1\int\limits_{-t/2}^{t/2}dt_2\int\frac{d\mathbf{q}}{(2\pi)^d}\int\frac{d\omega}{2\pi}e^{-\frac{1}{2}D_0\,q^2|t_1-t_2|-i\omega(t_1-t_2)}\,,\tag{11.200}$$

where the ω-integration is limited to the region $1/\tau_\varphi\leq|\omega|\leq kT/\hbar$. The averaged phase difference is seen to increase linearly in time:

$$\frac{1}{2}\,\langle\langle(\varphi[\mathbf{x}_t^{cl}])^2\rangle_{ee}\rangle_{imp} = \frac{t}{\tau_\varphi} \tag{11.201}$$

at a rate in accordance with the previous result for the phase-breaking rate, Eq. (11.194).

The lack of effectiveness in destroying phase coherence by interactions with small energy transfers is reflected in the compensation at small frequencies between the two cosine terms appearing in the expression for the phase difference, Eq. (11.198). In the case of diffusion-enhanced electron–electron interaction this compensation is crucial as there is an abundance of scattering events with small energy transfer, whereas the compensation was immaterial for electron–phonon interaction where the typical energy transfer is determined by the temperature.

---

[56]The last equality is an approximation owing to the constraint, $\mathbf{x}_{-t/2}=\mathbf{x}_{t/2}$, however, for large times a very good one.

Whereas the phase-breaking rate for electron–phonon interaction is model dependent, i.e. material dependent, we note the interesting feature that the phase-breaking rate for diffusion-enhanced electron–electron interaction is universal. In two dimensions we can rewrite

$$\frac{1}{\tau_\varphi} = \frac{e^2 \sigma_0 kT}{2\pi\hbar^2} \ln \frac{k_{\rm F} l}{2} . \tag{11.202}$$

Phase-breaking rates in accordance with Eq. (11.194) have been extracted from numerous magneto-resistance measurements; see, for example, references [88] and [89]. We note that at sufficiently low temperatures the electron–electron interaction dominates the phase-breaking rate in comparison with the electron–phonon interaction.

11.4 Anomalous magneto-resistance

From an experimental point of view, the disruption of coherence between time-reversed trajectories by an externally controlled magnetic field is the tool by which to study the weak-localization effect. Magneto-resistance measurements in the weak-localization regime has considerably enhanced the available information regarding inelastic scattering times (and spin-flip and spin-orbit scattering times). The weak-localization effect thus plays an important diagnostic role in materials science.

The influence of a magnetic field on the Cooperon was established in Section 11.2.4, and we have the Cooperon equation

$$\left\{ -i\omega - D_0 \{\nabla_{\bf x} - \frac{2ie}{\hbar} {\bf A}({\bf x})\}^2 + 1/\tau_\varphi \right\} C_\omega({\bf x}, {\bf x}') = \frac{1}{\tau} \delta({\bf x} - {\bf x}') . \tag{11.203}$$

We can now safely study the d.c. conductivity, i.e. assume that the external electric field is static, so that its frequency is equal to zero, $\omega = 0$, as the Cooperon in an external magnetic field is no longer infrared divergent. The Cooperon is formally identical to the imaginary-time Schrödinger Green's function for a fictitious particle with mass equal to $\hbar/2D_0$ and charge $2e$ moving in a magnetic field (see Exercise C.1 on page 515). To solve the Cooperon equation for the magnetic field case, we can thus refer to the equivalent quantum mechanical problem of a particle in an external homogeneous magnetic field. Considering the case of a homogeneous magnetic field,[57] and choosing the z-direction along the magnetic field and representing the vector potential in the Landau gauge, ${\bf A} = B(-y, 0, 0)$, the corresponding Hamiltonian is

$$H = \frac{D_0}{\hbar} (\hat{p}_x + 2eB\hat{y})^2 + \frac{D_0}{\hbar} (\hat{p}_y^2 + \hat{p}_z^2) . \tag{11.204}$$

The problem separates

$$\psi(x, y) = e^{\frac{i}{\hbar} p_x x} e^{\frac{i}{\hbar} p_z z} \chi(y) , \tag{11.205}$$

where the function χ satisfies the equation

$$-\frac{\hbar D_0}{2} \frac{d^2 \chi(y)}{dy^2} + \frac{1}{2} \frac{\hbar}{2D_0} \tilde{\omega}_{\rm c}^2 \left(y - \frac{p_x}{2eB} \right)^2 \chi(y) = \tilde{E} \chi(y) \tag{11.206}$$

---

[57]The case of an inhomogeneous magnetic field is treated in reference [95].

the shifted harmonic oscillator problem where $\tilde{\omega}_c$ is the *cyclotron* frequency for the fictitious particle, $\tilde{\omega}_c \equiv 4D_0|e|B/\hbar$, so that the energy spectrum is $E = \tilde{E} + \hbar D_0 Q_z^2 = \hbar\tilde{\omega}_c(n+1/2) + \hbar D_0 Q_z^2$, $n = 0, 1, 2, ...$; $Q_z = 2\pi n_z/L_z$, $n_z = 0, \pm 1, \pm 2, ...$. In the *particle in a magnetic field* analogy, n is the orbital quantum number and p_x is the quantum number describing the position of the cyclotron orbit, and describes here the possible locations of closed loops. The Cooperon in the presence of a homogeneous magnetic field of strength B thus has the spectral representation

$$C_0(\mathbf{x}, \mathbf{x}') = \sum_{Q_z}' \sum_{n=0}^{n_{max}} \int \frac{dp_x}{2\pi\hbar} \frac{\psi_{n,p_x}(\mathbf{x})\,\psi_{n,p_x}^*(\mathbf{x}')}{4D_0|e|B\tau\hbar^{-1}(n+1/2) + D_0\,\tau Q_z^2 + \tau/\tau_\varphi} , \quad (11.207)$$

where the ψ_{n,p_x} are the Landau wave functions

$$\psi_{n,p_x}(\mathbf{x}) = \frac{1}{\sqrt{L_z}}\, e^{\frac{i}{\hbar}p_x x} e^{iQ_z z} \chi_n(y - p_x/2eB) \quad (11.208)$$

and $\chi_n(y)$ is the harmonic oscillator wave function. In accordance with the derivation of the Cooperon equation, we can describe variations only on length scales larger than the mean free path. The sum over the *orbital* quantum number n should therefore terminate when $D_0\tau|e|Bn_{max} \sim \hbar$, i.e. at values of the order of $n_{max} \simeq l_B^2/l^2$, where $l_B \equiv (\hbar/|e|B)^{1/2}$ is the magnetic length.

To calculate the Cooperon for equal spatial values, $C_0(\mathbf{x}, \mathbf{x})$, we actually do not need all the information contained in Eq. (11.207), since by normalization of the wave functions in the completeness relation we have

$$\int_{-\infty}^{\infty} \frac{dp_x}{2\pi\hbar} \chi_n^*\left(y - \frac{p_x}{2eB}\right) \chi_n\left(y - \frac{p_x}{2eB}\right) = -\frac{2eB}{2\pi\hbar} \int_{-\infty}^{\infty} dy\, |\chi_n(y)|^2 = -\frac{2eB}{2\pi\hbar}$$

$$(11.209)$$

and thereby

$$C_0(\mathbf{x}, \mathbf{x}) = -\frac{2eB}{2\pi\hbar} \sum_{Q_z}' \sum_{n=0}^{n_{max}} \frac{1}{4D_0|e|B\tau\hbar^{-1}(n+1/2) + D_0\,\tau Q_z^2 + \tau/\tau_\varphi} . \quad (11.210)$$

11.4.1 Magneto-resistance in thin films

We now consider the magneto-resistance of a film of thickness a, choosing the direction of the magnetic field perpendicular to the film.[58] Provided the thickness of the film is smaller than the phase coherence length, $a \ll L_\varphi$ (the thin film, or quasi-two-dimensional criterion), or the usually much weaker restriction that it is smaller than the magnetic length, $a \ll l_B$, only the smallest value of $Q_z = 2\pi n/L_z$, $n = 0, \pm 1, \pm 2, ...$ contributes to the sum. Since the smallest value is $Q_z = 0$, we obtain, according to Eq. (11.59), for the quantum correction to the conductivity

$$\delta\sigma(B) = \frac{e^3 B D_0 \tau}{\pi^2 \hbar^2 a} \sum_{n=0}^{n_{max}} \frac{1}{4D_0|e|B\tau\hbar^{-1}(n+1/2) + \tau/\tau_\varphi} . \quad (11.211)$$

---

[58]The strictly two-dimensional case can also be realized experimentally, for example by using the two-dimensional electron gas accumulating in the inversion layer in a MOSFET or heterostructure.

Employing the property of the di-gamma function ψ (see, for example, reference [96])

$$\psi(x+n) = \psi(x) + \sum_{n=0}^{n-1} \frac{1}{x+n} \qquad (11.212)$$

we get for the magneto-conductance

$$\delta G_{\alpha\beta}(B) = \frac{e^2}{4\pi^2\hbar} \tilde{f}_2(4D_0|e|B\hbar^{-1}\tau_\varphi)\,\delta_{\alpha\beta}\,, \qquad (11.213)$$

where

$$\tilde{f}_2(x) = \psi\left(\frac{1}{2} + \frac{1}{x}\right) + \psi\left(\frac{3}{2} + n_{\max} + \frac{1}{x}\right). \qquad (11.214)$$

The magneto-conductance of a thin film is now obtained by subtracting the zero field conductance. In the limit $B \to 0$, the sum can be estimated to become

$$\sum_{n=0}^{n_{\max}} \frac{1}{4D_0|e|B\tau\hbar^{-1}(n+1/2) + \tau/\tau_\varphi} \quad \to \quad \ln(n_{\max}4D_0|e|B\hbar^{-1}\tau_\varphi)\,. \qquad (11.215)$$

Using the property of the di-gamma function

$$\lim_{n\to\infty} \psi\left(\frac{3}{2} + n + \frac{1}{x}\right) \simeq \ln n \qquad (11.216)$$

we finally arrive at the low-field magneto-conductance of a thin film

$$\Delta G_{\alpha\beta}(B) \equiv \delta G_{\alpha\beta}(B) - \delta G_{\alpha\beta}(B \to 0) = \frac{e^2}{2\pi^2\hbar}\,f_2(B/B_\varphi)\,\delta_{\alpha\beta}\,, \qquad (11.217)$$

where

$$f_2(x) = \ln x + \psi\left(\frac{1}{2} + \frac{1}{x}\right) \qquad (11.218)$$

and $B_\varphi = \hbar/4D_0|e|\tau_\varphi$, the (temperature-dependent) characteristic scale of the magnetic field for the weak-localization effect, is determined by the inelastic scattering. This scale is indeed small compared with the scale for classical magneto-resistance effects $B_{cl} \sim m/|e|\tau$, as $B_\varphi \sim B_{cl}\hbar/\epsilon_F\tau_\varphi$.[59] The weak-localization magneto-conductance is seen to be sensitive to very small magnetic fields, namely when the magnetic length becomes comparable to the phase coherence length, $l_B \sim L_\varphi$, or equivalently, $\omega_c\tau \sim \hbar/\epsilon_F\tau_\varphi$. Since the impurity mean free time, τ, can be much smaller than the phase coherence time τ_φ, the above description can be valid over a wide magnetic field range where classical magneto-conductance effects are absent. Classical magneto-conductance effects are governed by the orbit bending scale,

---

[59]In terms of the mass of the electron we have for the mass of the fictitious particle $\hbar/2D_0 \sim m\hbar/\epsilon_F\tau$, and the low magnetic field sensitivity can be viewed as the result of the smallness of the fictitious mass in the problem.

$\omega_c\tau \sim 1$, whereas the weak-localization quantum effect sets in when a loop of typical area L_φ^2 encloses a flux quantum.[60] We note the limiting behavior of the function

$$f_2(x) = \begin{cases} \frac{x^2}{24} & \text{for} \quad x \ll 1 \\[2mm] \ln x & \text{for} \quad x \gg 1 \ . \end{cases} \tag{11.219}$$

The magneto-conductance is positive, and seen to have a quadratic upturn at low fields, and saturates beyond the characteristic field in a universal fashion, i.e. independent of sample parameters.[61] The magneto-resistance is therefore negative, $\Delta R = -\Delta G/G_{\rm cl}^2$, which is a distinct sign that the effect is not classical, since we are considering a macroscopic system.[62]

Weak localization magneto-conductance is also relevant for a three-dimensional sample, and cleared up a long-standing mystery in the field of magneto-transport in doped semiconductors. For details on the three-dimensional case we refer the reader to chapter 11 of reference [1].

The negative anomalous magneto-resistance can be understood qualitatively from the simple interference picture of the weak-localization effect. The presence of the magnetic field breaks the time-reversal invariance, and upsets the otherwise identical values of the phase factors in the amplitudes for traversing the time-reversed weak-localization loops. The quantum interference term for a loop c is the result of the presence of the magnetic field changed according to

$$A_c A_{\bar{c}}^* \ \to \ |A_c^{(B=0)}|^2 \exp\left\{ \frac{2ie}{\hbar} \oint_c d\bar{\mathbf{x}} \cdot \mathbf{A}(\bar{\mathbf{x}}) \right\} \ = \ |A_c^{(B=0)}|^2 \, e^{\frac{2ie}{\hbar}\Phi_c} \ , \tag{11.220}$$

where Φ_c is the flux enclosed by the loop c. The weak-localization interference term acquires a random phase depending on the loop size, and the strength of the magnetic field, decreasing the probability of return, and thereby increasing the conductivity. The negative contribution from each loop in the impurity field to the conductance is modulated in accordance with the phase shift prescription for amplitudes by the oscillatory factor, giving the expression

$$\langle G(B)\rangle - \langle G(O)\rangle \ = \ \frac{e^2}{2\pi^2\hbar} \langle \sum_c |A_c^{(B=0)}|^2 \left\{ 1 - \cos(2\pi\Phi_c/\Phi_0) \right\} e^{-t_c/\tau_\varphi} \rangle_{\rm imp} \ . \tag{11.221}$$

The summation is over all classical loops in the impurity field returning to within a distance of the mean free path to a given point, and t_c is the duration for traversing

---

[60]Beyond the low-field limit, $\omega_c\tau < \hbar/\epsilon_{\rm F}\tau$, the expression for the magneto-conductance can not be given in closed form, and its derivation is more involved, since we must account for the orbit bending due to the magnetic field, the Lorentz force [97]. When the impurity mean free time τ becomes comparable to the phase coherence time τ_φ, we are no longer in the diffusive regime, and a Boltzmannian description must be introduced [98].

[61]Experimental observations of the low field magneto-resistance of thin metallic films are in remarkable good agreement with the theory. The weak-localization effect is thus of importance for extracting information about inelastic scattering strengths, which is otherwise hard to come at. For reviews of the experimental results, see references [88] and [89].

[62]The classical magneto-resistance of a macroscopic sample calculated on the basis of the Boltzmann equation is positive.

the loop c, and Φ_0 is the flux quantum $\Phi_0 = 2\pi\hbar/2|e|$. The sum should be performed weighted with the probability for the realization of the loop in question, as expressed by the brackets. The weight of loops that are longer than the phase coherence length is suppressed, as their coherence are destroyed by inelastic scattering. In weak magnetic fields, only the longest loops are influenced by the phase shift due to the magnetic field. It is evident from Eq. (11.221) that the low field magneto-conductance is positive and quadratic in the field.[63] The continuing monotonic behavior as a function of the magnetic field until saturation is simply a geometric property of diffusion, viz. that small diffusive loops are prolific. Instead of verifying this statement, let us turn the argument around and use our physical understanding of the weak localization effect to learn about the distribution of the areas of diffusive loops in two dimensions. Rewriting Eq. (11.210) we have in two dimensions

$$C_0(\mathbf{x},\mathbf{x}) = \int_0^\infty dt\, e^{-t/\tau_\varphi} \frac{B}{\tau\Phi_0} \sum_{n=0}^{n_{max}} e^{-\frac{4\pi B D_0}{\Phi_0}(n+1/2)t} . \tag{11.222}$$

For times $t > \tau$ we can let the summation run over all natural numbers and we can sum the geometric series to obtain

$$C_0(\mathbf{x},\mathbf{x}) = \int_0^\infty dt\, \frac{e^{-t/\tau_\varphi}}{4\pi\tau D_0 t} \frac{2\pi B D_0 t}{\Phi_0} \frac{1}{\sinh \frac{2\pi B D_0 t}{\Phi_0}} . \tag{11.223}$$

The factors independent of the magnetic field are the return probability and the dephasing factor. Representing the factors depending on the field strength, which describes the influence of the magnetic field on the quantum interference process, by its cosine transform

$$\frac{2\pi B D_0 t}{\Phi_0} \frac{1}{\sinh \frac{2\pi B D_0 t}{\Phi_0}} = \int_{-\infty}^\infty dS \cos \frac{SB}{\Phi_0} f_t(S) \tag{11.224}$$

and inverting gives

$$f_t(S) = \frac{1}{4D_0 t} \frac{1}{\cosh^2\left(\frac{S}{4D_0 t}\right)} . \tag{11.225}$$

For the weak localization contribution to the conductance we can therefore write

$$\delta G(B) = -\frac{2e^2 D_0 \tau}{\pi\hbar} \int_0^\infty dt\, \frac{e^{-t/\tau_\varphi}}{4\pi\tau D_0 t} \int_0^\infty dS\, f_t(S) \cos\left(\frac{BS}{\Phi_0}\right) \tag{11.226}$$

and we note that $f_t(S)$ is normalized, and has the interpretation of the probability for a diffusive loop of duration t to enclose the area S.

For the average size of a diffusive loop of duration t we have

$$\langle S \rangle_t \equiv \int_0^\infty dS\, S f_t(S) = 4D_0 t \ln 2 , \tag{11.227}$$

---

[63]The minimum value of the magneto-resistance occurs *exactly* for zero magnetic field value, and the weak localization effect is thus one of the few effects that can be used as a reference for zero magnetic field.

i.e. the typical size of a diffusive loop of duration t is proportional to $D_0 t$.

For the fluctuations we have

$$\langle S^2 \rangle_t \equiv \int_{-\infty}^{\infty} dS \, S^2 f_t(S) = 8\pi^2 (D_0 t)^2 \tag{11.228}$$

and we can write

$$f_t(S) = \frac{\pi}{\sqrt{2 \langle S^2 \rangle_t}} \frac{1}{\cosh^2 \frac{\pi S}{\sqrt{2 \langle S^2 \rangle_t}}} . \tag{11.229}$$

The probability distribution for diffusive loops is thus a steadily decreasing function of the area.

The weak localization effect in cylinders and rings leads through the Aharonov–Bohm effect to an amazing manifestation of the quantum mechanical superposition principle at the macroscopic level. Furthermore, the weak localization effect can be *reversed* to weak anti-localization if the impurities, such as is the case in heavy compounds, give rise to spin–orbit scattering. Discussion of these effects can be found in chapter 11 of reference [1].

11.5   Coulomb interaction in a disordered conductor

The presence of impurities changes the effective electron–electron interaction. We shall study this effect in the weak disorder limit, $\epsilon_F \tau \gg \hbar$, which is the common situation in conductors such as metals and semiconductors. The change from ballistic to diffusive motion leads to diffusion enhancement of the electron–electron interaction. This leads to interesting observable effects such as the temperature dependence of the conductivity of a three-dimensional sample being proportional to the square root of the temperature [99], $\sigma \propto \sqrt{T}$, instead of the usually unnoticeable T^2-term due to Umklapp processes in a clean metal. For experimental evidence of the square root temperature dependence see references [100, 101].

Let us assume that the inverse screening length is much smaller than the Fermi wavelength; i.e. the range of the screened Coulomb potential, $\overline{V}$, is much larger than the spacing between the electrons. The exchange correction to the electron energy ϵ_λ due to electron–electron interaction is then much larger than the direct or Hartree term. We shall use the method of exact impurity eigenstates and, since diagonal elements dominate, $\Sigma_\lambda \equiv \Sigma_{\lambda\lambda}$, we have for the exchange self-energy

$$\Sigma_\lambda^{ex} = - \sum_{\lambda' \text{occ.}} \int d\mathbf{x} \int d\mathbf{x}' \, \overline{V}(\mathbf{x} - \mathbf{x}') \, \psi_\lambda^*(\mathbf{x}) \, \psi_{\lambda'}^*(\mathbf{x}') \, \psi_\lambda(\mathbf{x}') \, \psi_{\lambda'}(\mathbf{x}) , \tag{11.230}$$

where the summation is over all occupied states λ', i.e. all the states below the Fermi level since for the moment we assume zero temperature. We are interested in the mean energy shift averaged over all states with energy ξ (measured from the Fermi energy)

$$\Sigma^{ex}(\xi) = \frac{1}{N_0 V} \sum_\lambda \langle \delta(\xi - \xi_\lambda) \Sigma_\lambda^{ex} \rangle \tag{11.231}$$

for which we obtain the expression, say $\xi > 0$,

$$\Sigma^{\text{ex}}(\xi) = -\frac{1}{N_0 V} \int\limits_{-\infty}^{0} d\xi' \int d\mathbf{x} \int d\mathbf{x}' \, \overline{V}(\mathbf{x} - \mathbf{x}')$$

$$\times \left\langle \sideset{}{'}\sum_{\lambda,\lambda'} \delta(\xi - \xi_\lambda) \, \delta(\xi' - \xi_{\lambda'}) \, \psi_\lambda^*(\mathbf{x}) \, \psi_{\lambda'}^*(\mathbf{x}') \, \psi_\lambda(\mathbf{x}') \, \psi_{\lambda'}(\mathbf{x}) \right\rangle , \qquad (11.232)$$

where the prime on the summation sign indicates that the sum is only over states λ' occupied and states λ unoccupied. The impurity-averaged quantity is the product of two spectral weight functions in the exact impurity eigenstate representation, except for the restrictions on the summations. However, these are irrelevant as the main contribution comes from $\xi' \simeq \xi$. In the standard impurity averaging technique we encounter in the weak-disorder limit, $1/k_F l \ll 1$, the diffusion ladder, and we obtain

$$\Sigma^{\text{ex}}(\xi) = -\frac{1}{2\pi} \int\limits_{\xi/\hbar}^{\infty} d\omega \int \frac{d\mathbf{q}}{(2\pi)^d} \, \overline{V}(\mathbf{q}) \, \frac{D_0 q^2}{\omega^2 + (D_0 q^2)^2} . \qquad (11.233)$$

In the above model of a static interaction the average change in energy is purely real. The result obtained can be used to calculate the change in the density of states. To lowest order in the electron–electron interaction we have for the change in density of states due to the electron–electron interaction

$$\delta N(\xi) \equiv \langle N(\xi) \rangle - N_0(\xi) = -N_0(\xi) \frac{\partial \Sigma^{\text{ex}}(\xi)}{\partial \xi}$$

$$= \frac{N_0}{2\pi\hbar} \int \frac{d\mathbf{q}}{(2\pi)^d} \, \overline{V}(\mathbf{q}) \, \frac{D_0 q^2}{\left(\frac{\xi}{\hbar}\right)^2 + (D_0 q^2)^2} \qquad (11.234)$$

as the change in the density of states due to disorder is negligible in the weak-disorder limit.

Exercise 11.3. Verify that if $\overline{V}$ is a short-range potential, the change in the density of states near the Fermi surface due to electron–electron interaction is in the weak-disorder limit

$$\frac{\delta N_3(\xi)}{N_3(0)} = \frac{\overline{V}(\mathbf{q} = 0)}{4\sqrt{2}\pi^2} \frac{\sqrt{|\xi|}}{(\hbar D_0)^{3/2}} \qquad (11.235)$$

in three dimensions and, in two dimensions,

$$\frac{\delta N_2(\xi)}{N_2(0)} = \frac{\overline{V}(\mathbf{q} = 0)}{(2\pi)^2 \hbar D_0} \ln \frac{|\xi|\tau}{\hbar} . \qquad (11.236)$$

The singularity in the density of states is due to the spatial correlation of the exact impurity wave functions of almost equal energy, as described by the singular behavior of the spectral correlation function. The singularity in the density of states gives rise to the zero-bias anomaly, a dip in the conductivity of a tunnel junction at low voltages [102].

Quite generally the propagator in the energy representation satisfies, in the presence of disorder and electron–electron interaction, the equation

$$G^{\mathrm{R}}_{\lambda\lambda'}(E) = G^{(0)\mathrm{R}}_{\lambda\lambda'}(E) + \sum_{\lambda_1 \lambda'_1} G^{(0)\mathrm{R}}_{\lambda\lambda_1}(E)\, \Sigma^{\mathrm{R}}_{\lambda_1 \lambda'_1}(E)\, G^{\mathrm{R}}_{\lambda'_1 \lambda'}(E)\,, \qquad (11.237)$$

where the propagator in the absence of electron–electron interaction is diagonal, $G^{(0)\mathrm{R}}_{\lambda\lambda'}(E) = G^{(0)\mathrm{R}}_{\lambda}(E)\,\delta_{\lambda\lambda'}$, and specified in terms of the exact impurity eigenstates (here in the momentum representation)

$$G^{\mathrm{R(A)}}_0(\mathbf{p}, \mathbf{p}', E) = \sum_\lambda \frac{\psi_\lambda(\mathbf{p})\, \psi^*_\lambda(\mathbf{p}')}{E - \epsilon_\lambda \,(\overset{-}{\underset{+}{}})\, i0} \equiv \sum_\lambda \psi_\lambda(\mathbf{p})\, \psi^*_\lambda(\mathbf{p}')\, G^{(0)\mathrm{R(A)}}_\lambda(E)\,.$$
$$(11.238)$$

Since energy eigenstates are only spatially correlated if they have the same energy, only the diagonal terms, $\Sigma^{\mathrm{R}}_\lambda(E) \equiv \Sigma^{\mathrm{R}}_{\lambda\lambda}(E)$, contribute in Eq. (11.237), and we obtain the result that the propagator is approximately diagonal and specified by

$$G^{\mathrm{R}}_\lambda(E) = \frac{1}{E - \epsilon_\lambda - \Sigma^{\mathrm{R}}_\lambda(E)}\,. \qquad (11.239)$$

The imaginary part of the self-energy describes the decay of an exact impurity eigenstate due to electron–electron interaction. When calculating the inelastic decay rate, we should only count processes starting with the same energy, and on the average in the random potential we are therefore interested in the quantity

$$\Sigma^{\mathrm{R}}_{E'}(E) = \frac{1}{N_0 V} \sum_\lambda \left\langle \delta(E' - \xi_\lambda)\, \Sigma^{\mathrm{R}}_\lambda(E) \right\rangle. \qquad (11.240)$$

To lowest order in the electron–electron interaction we can set E equal to E' in Eq. (11.240) because their difference is the real part of the self-energy, and we get for the inelastic electron–electron collision rate

$$\frac{1}{\tau_{\mathrm{e-e}}(E, T)} = -2\,\Im m\, \Sigma^{\mathrm{R}}_E(E) = i\left(\Sigma^{\mathrm{R}}_E(E) - \Sigma^{\mathrm{A}}_E(E)\right)$$

$$= -\frac{1}{2\pi\hbar N_0 V} \sum_\lambda \left\langle \left(\Sigma^{\mathrm{R}}_\lambda(E) - \Sigma^{\mathrm{A}}_\lambda(E)\right)\left(G^{(0)\mathrm{R}}_\lambda(E) - G^{(0)\mathrm{A}}_\lambda(E)\right)\right\rangle,$$

$$(11.241)$$

where we have expressed the delta function in Eq. (11.240) in terms of the spectral function. We thus have to impurity average a product of a self-energy and a propagator, say the retarded self-energy and the advanced propagator, presently both expressed in the exact impurity eigenstate representation. In the weak-disorder limit, $k_F l \gg 1$, the contributions to the collision rate are therefore specified in terms of the Diffuson and the effective electron–electron interaction as depicted in Figure 11.6. For the case of the product of the retarded self-energy and the advanced propagator there are contributions from the two diagrams depicted in Figure 11.6. In the case of the retarded interaction, the Diffuson occurs only for the case where the kinetic Green's function appears right at the emission vertex since impurity correlators effectively decouple momentum integrations (recall the similar analysis in connection with Eq. (11.82)).

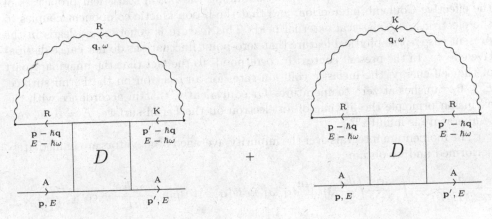

Figure 11.6 Lowest order interaction diagrams for the inelastic collision rate.

We then obtain for the inelastic collision rate or energy relaxation rate in terms of the Diffuson and the electron–electron interaction

$$\frac{1}{\tau_{e-e}(E,T)} = -\frac{1}{2\hbar V^2} \Im m \left(\int \frac{d\mathbf{q}}{(2\pi)^3} \int \frac{d\omega}{2\pi} \, D(\mathbf{q},\omega) (\overline{V}^R(\mathbf{q},\omega) - \overline{V}^A(\mathbf{q},\omega)) \, u^4 \right.$$

$$\times \sum_{\mathbf{pp'}} G^R(E - \hbar\omega, \mathbf{p} - \hbar\mathbf{q}) \, G^A(E,\mathbf{p'}) \, G^R(E - \hbar\omega, \mathbf{p'} - \hbar\mathbf{q}) \, G^A(E,\mathbf{p})$$

$$\times \left(\tanh \frac{E - \hbar\omega}{2kT} + \coth \frac{\hbar\omega}{2kT} \right) \right). \tag{11.242}$$

Here we have used that the effective Coulomb interaction has similar statistics properties as bosons, and in arriving at Eq. (11.242) we have in fact used the fluctuation–

dissipation relation that relates the kinetic component of the effective Coulomb interaction to the spectral component

$$\overline{V}^{K}(\mathbf{q},\omega) = \left(\overline{V}^{R}(\mathbf{q},\omega) - \overline{V}^{A}(\mathbf{q},\omega)\right) \coth \frac{\hbar\omega}{2kT} \tag{11.243}$$

accounting for the second term arising from the second diagram in Figure 11.6.[64]

At this point, we benefit in interpretation from an important feature of the developed real-time non-equilibrium diagram technique, viz. that for the choice of propagators we have made, the quantum statistics of fermions and bosons manifest itself in a distinct way in diagrams as noted in Section 5.4. In the first diagram in Figure 11.6, where the retarded interaction appears, it leads (according to the diagrammatic rules of Section 5.4) to the appearance of the quantum statistics of the fermions, accounting for the first term in Eq. (11.242). It is important that this term occurs in combination with the term containing the boson statistical properties of the effective Coulomb interaction, and that the boson kinetic component couples to the electrons as a classical external field. This feature is generic, and leads in the present case to the physical feature that zero-point fluctuations do not cause dissipative effects. In the present context it corresponds to the fact that the imaginary part of the self-energy, the inelastic collision rate, for an electron on the Fermi surface, $E = 0$, vanishes at zero temperature. Or equivalently, that in accordance with the exclusion principle the lifetime of an electron on the Fermi surface, $E = 0$, at zero temperature is infinite.[65]

The momentum integrals over the impurity-averaged propagators are immediately performed and we obtain

$$\frac{\hbar}{\tau_{e-e}(E,T)} = \int \frac{d\mathbf{q}}{(2\pi)^3} \int \frac{d\omega}{2\pi} \, \Im m \overline{V}^{R}(\mathbf{q},\omega) \, \Re e D(\mathbf{q},\omega) \left(\tanh \frac{E - \hbar\omega}{2kT} + \coth \frac{\hbar\omega}{2kT} \right) \tag{11.244}$$

from which we can calculate the collision rate.

The effective electron–electron interaction itself, specifying the electron self-energy, is also changed owing to the presence of impurities. It is thus the dynamically screened electron–electron interaction in the presence of impurities, as expressed by the dielectric function, $\epsilon(\mathbf{q},\omega)$,

$$\overline{V}^{R}(\mathbf{q},\omega) = \frac{V(\mathbf{q})}{\epsilon(\mathbf{q},\omega)}, \tag{11.245}$$

---

[64]In the calculation in Section 11.3.2 of the weak localization phase-breaking rate due to electron–electron interaction with small energy transfers, only the kinetic component of the interaction, $\overline{V}^{K}$, was included, but this is justified by the presence of its quantum statistics factor making it the dominant component in the low frequency regime. This is the reason for the success of the single-particle description used for the calculation, where the electron–electron interaction is represented by a Gaussian distributed classical stochastic potential since it has identical properties with respect to the dynamical indices as the kinetic component.

[65]Such spurious zero-point fluctuation effects are with frequency conjectured in the literature for various physical quantities. For an early rebuttal in the context of weak localization see reference [91].

which appears in Eq. (11.244), and not the bare Coulomb interaction, $V(\mathbf{q})$. The basic excitation of the bare Coulomb potential in an electron gas is the particle–hole excitation, and it will lead to screening of the interaction. It is sufficient to use the random phase approximation where additional interaction decorations by electron–electron interactions are negligible since the disorder effects are driven by the long ranged Diffuson.[66] Before averaging with respect to the random impurity potential we thus have the diagrammatic matrix representation of the effective electron–electron interaction

where the thick wiggly line represents the effective Coulomb interaction, i.e. in the triagonal representation the matrix

$$\overset{\sim}{}_{\mathbf{q}\omega} = \begin{pmatrix} \overline{V}^{R}(\mathbf{q},\omega) & \overline{V}^{K}(\mathbf{q},\omega) \\ 0 & \overline{V}^{A}(\mathbf{q},\omega) \end{pmatrix} \qquad (11.247)$$

and similarly for the thin line representing the bare Coulomb interaction for which we note $V^{K}(\mathbf{q}) = 0$. Analytically the Dyson equation for the matrix Coulomb propagator has the form

$$\overline{V}(\mathbf{q},\omega) = V(\mathbf{q}) + V(\mathbf{q})\,\Pi(\mathbf{q},\omega)\,\overline{V}(\mathbf{q},\omega)\,, \qquad (11.248)$$

where the polarization, Π, in the triagonal representation has the form

$$\Pi(\mathbf{q},\omega) = \begin{pmatrix} \Pi^{R}(\mathbf{q},\omega) & \Pi^{K}(\mathbf{q},\omega) \\ 0 & \Pi^{A}(\mathbf{q},\omega) \end{pmatrix} \qquad (11.249)$$

---

[66]The random phase approximation can also be stated as the linearized mean-field approximation as discussed for example in chapter 10 of reference [1].

and in the random phase approximation specified in terms of the dynamical indices according to

$$\Pi_{kk'} = -2i\tilde{\gamma}^k_{ii'}\, G_{i'j'}\, G_{ji}\, \gamma^{k'}_{j'j} \,. \tag{11.250}$$

Solving the Dyson equation for the effective interaction in the random phase approximation gives

$$= \frac{1}{\overset{\displaystyle \mathbf{p}_+ E_+}{\underset{\mathbf{p}_- E_-}{}}} \,. \tag{11.251}$$

According to our universal rules for boson–fermion coupling in the dynamical indices, Eq. (5.51) and Eq. (5.52), the retarded polarization bubble is given by

$$\Pi^{\mathrm{R}}(\mathbf{q},\omega) = -i\int \frac{d\mathbf{p}}{(2\pi)^3} \int\limits_{-\infty}^{\infty} \frac{dE}{2\pi} \Big(G^{\mathrm{R}}(p)\, G^{\mathrm{K}}(p-q) \,-\, G^{\mathrm{K}}(p)\, G^{\mathrm{A}}(p-q)\Big), \tag{11.252}$$

where $p = (E, \mathbf{p})$ and $q = (\omega, \mathbf{q})$. In the diagrammatic expansion of the effective electron–electron interaction, we must then impurity average the electron–hole or polarization bubble diagram. To lowest order in the disorder parameter $1/k_F l$, we should insert the impurity ladder into the bubble diagram; i.e. we encounter the diagrams of the type

$$. \tag{11.253}$$

The impurity-averaged bubble diagram is evaluated using the standard impurity Green's function technique, and we thus have in the diffusive limit, $ql, \omega\tau \ll 1$ (in the three-dimensional case), for the dielectric function, $ql, \omega\tau \ll 1$,

$$\epsilon(\mathbf{q},\omega) = 1 + \frac{e^2}{\epsilon_0\, q^2}\, \frac{2N_0 D_0\, q^2}{-i\omega + D_0\, q^2} = 1 + \frac{D_0\, \kappa_s^2}{-i\omega + D_0\, q^2}\,, \tag{11.254}$$

relating the bare Coulomb interaction to the effective interaction.[67] Inserting into Eq. (11.244), we can calculate the inelastic collision rate.

---

[67]The calculation is equivalent to the calculation of the density–density response function of a disorder conductor giving the expression

$$\chi(\mathbf{q},\omega) = \frac{2N_0 D_0\, q^2}{-i\omega + D_0\, q^2}\,.$$

This is understandable since we note that a fluctuation in the density of electrons creates an electric potential, which in turn is felt by an electron. Fluctuations in the density or current of the electrons give rise to fluctuations in an electromagnetic field inside the electron gas, as discussed quite generally in Section 6.5 in connection with the fluctuation–dissipation relations of linear response.

We could also calculate the inelastic collision rate or energy relaxation rate in the dirty limit by solving the Boltzmann equation with the two-particle interaction modified by the impurity scattering

$$\frac{\partial f(\epsilon)}{\partial t} = 2\pi \int_{-\infty}^{\infty} d\omega \int_{-\infty}^{\infty} \frac{d\epsilon'}{2\pi\hbar} \, P(\omega) \, R(\epsilon, \epsilon', \omega) \,, \tag{11.255}$$

where

$$R(\epsilon, \epsilon', \omega) = f(\epsilon) \, f(\epsilon' - \omega) \, (1 - f(\epsilon - \omega)) \, (1 - f(\epsilon'))$$

$$- \, f(\epsilon - \omega) \, f(\epsilon') \, (1 - f(\epsilon)) \, (1 - f(\epsilon' - \omega)) \tag{11.256}$$

and

$$P(\omega) = \frac{2N_0\tau^2}{\pi\hbar} \int \frac{d\mathbf{q}}{(2\pi)^3} \left(\frac{V(\mathbf{q})}{|\epsilon(\mathbf{q}, \omega)|} \frac{(D_0\,q)^2}{\omega^2 + (D_0\,q)^2} \right)^2 \tag{11.257}$$

is analogous to Eliashberg function, $\alpha^2 F$, for the electron–phonon case. We notice that we can rewrite

$$P(\omega) = \frac{\tau}{\pi\omega} \Im m \int \frac{d\mathbf{q}}{(2\pi)^3} \, V^{\mathrm{R}}(\mathbf{q}, \omega) \frac{\zeta(\mathbf{q}, \omega)}{1 - \zeta(\mathbf{q}, \omega)} \,, \tag{11.258}$$

where ζ is the insertion Eq. (11.20) (here the relevant case is the particle–hole channel, but the result is identical to that of the particle–particle channel) and given (in two and three dimensions) by

$$\zeta(\mathbf{q}, \omega) = \frac{i}{2ql} \ln \frac{ql + \omega\tau + i}{-ql + \omega\tau + i} \,, \qquad q \equiv |\mathbf{q}| \tag{11.259}$$

with the limiting behavior

$$\zeta(\mathbf{q}, \omega) = \begin{cases} \frac{\pi}{2ql} & ql > \omega\tau, ql > 1 \\ 1 + i\omega\tau - D_0\tau q^2 & ql, \omega\tau < 1 \,. \end{cases} \tag{11.260}$$

In the three-dimensional case we have, $\omega\tau < 1$,

$$P(\omega) = \frac{\omega^{-1/2}}{8\sqrt{2}\pi^2\hbar N_0 D_0^{3/2}} \,. \tag{11.261}$$

We therefore get for an electron on the Fermi surface in a dirty metal the electron–electron collision rate at temperatures $kT < \hbar/\tau$[68]

$$\frac{1}{\tau_{e-e}(T)} = \int_0^{\infty} d\omega \, P(\omega) \frac{2\omega}{\sinh \frac{\hbar\omega}{kT}} = c \, \frac{\tau^{1/2}}{k_{\mathrm{F}}l} \frac{(kT)^{3/2}}{\sqrt{\hbar\epsilon_{\mathrm{F}}\tau}} \,, \tag{11.262}$$

The dielectric function and the density and current response functions are thus all related

$$\epsilon(\mathbf{q}, \omega) = 1 + \frac{i\sigma(\mathbf{q}, \omega)}{\omega\,\epsilon_0} = 1 + \frac{e^2}{\epsilon_0\,q^2} \, \chi(\mathbf{q}, \omega) \,.$$

For a discussion we refer the reader to chapter 10 of reference [1].
[68] From the region of large ω and q we get the clean limit rate, Eq. (7.206), which dominates at temperatures $kT \gg \hbar/\tau$.

where c is a constant of order unity ($\zeta(3/2) \simeq 2.612$)

$$c = \frac{3\sqrt{3}\pi}{16}\,\zeta(3/2)(\sqrt{8}-1)\,. \tag{11.263}$$

For an electron in energy state ξ, $\xi < \hbar/\tau$, we get analogously in the dirty limit for the electron–electron collision rate at zero temperature[69]

$$\frac{1}{\tau_{\text{e--e}}(\xi)} = \frac{\sqrt{6}}{4}\,\frac{\tau^{1/2}}{\hbar^{3/2}(k_{\text{F}}l)^2}\,\xi^{3/2}\,. \tag{11.264}$$

The scattering rate due to electron–electron interaction is thus enhanced in a dirty metal compared with the clean case [103, 104, 105], diffusion enhanced electron–electron interaction.[70] Equivalently, the screening is weakened owing to the diffusive motion of the electrons. The interpretation of this enhancement can be given in terms of the previous phase space argument of Exercise 7.10 on page 214 for the relaxation time and the breaking of translational invariance due to the presence of disorder. The violation of momentum conservation in the virtual scattering processes due to impurities gives more phase space for final states. Alternatively, viewing the collisions in real space, owing to the motion being diffusive instead of ballistic the electrons spend more time close together where the interaction is strong, or, wave functions of equal energy in a random potential are spatially correlated thereby leading to an enhanced electron–electron interaction. The scattering process now includes quantum interference between the elastic and inelastic processes as signified by the collision rate $\hbar/\tau_{\text{e--e}}$ being dependent on $\hbar$.

We note that the expression for the energy relaxation rate in two dimensions diverges in the infrared for a dirty metal in the above lowest-order perturbative calculation. For the Coulomb potential for electrons constricted to movement in two dimensions the bare Coulomb potential is

$$V(\mathbf{q}) = \frac{2\pi e^2}{|\mathbf{q}|} \tag{11.265}$$

and for $\omega\tau < 1$

$$P_2(\omega) = \frac{1}{8\epsilon_{\text{F}}\tau}\,\frac{1}{\omega} \tag{11.266}$$

---

[69] At temperatures and energies $kT, \xi > \hbar/\tau$, the expressions for relaxation rates are those of the clean limit, recall Exercise 7.10 on page 214.

[70] In the case of electron–phonon interaction, local charge neutrality forces the electrons to follow adiabatically the thermal motion of the ions, and because of the coherent motion with the lattice of the fixed impurities, the interaction with the longitudinal phonons is in fact decreased owing to this compensation mechanism. The imaginary part of the electron self-energy will therefore be given by the results obtained in Section 11.3.1 for the phase-breaking rate. As shown there, the interaction with transverse phonons are either enhanced or diminished depending on the temperature regime. The influence of impurities will not be universal for the case of interaction with phonons as will be the case for the diffusion enhanced electron–electron interaction.

giving the divergent expression for the relaxation rate, $kT < \hbar/\tau$,[71]

$$\frac{1}{\tau_{e-e}(T)} = \frac{1}{2k_F l} \int_0^\infty d\omega \, \frac{1}{\sinh \frac{\hbar\omega}{kT}} \, . \tag{11.267}$$

However, this is not alarming since we do not expect the relaxation rate to be the relevant measurable quantity, as in this quantity scattering at all energies is weighted equally. We do not expect such divergences in physically measurable rates, and indeed the phase relaxation rate of the electronic wave function in a dirty two-dimensional metallic film does not diverge because of collisions with small energy transfer, as discussed in Section 11.3.2. There we made use of the expression for the effective electron–electron interaction at low energies and momenta in a dirty metal for which, according to Eq. (11.243), we have[72]

$$\bar{V}^K(\mathbf{q}, \omega) = \; = \frac{-4ie^2 kT}{\sigma_0 |\mathbf{q}|^2} \, . \tag{11.268}$$

The low frequency electron–electron interaction in a disordered conductor is thus identical to the Nyquist noise in the electromagnetic field fluctuations, the correlator we used in Section 11.3.2 (here represented by the scalar potential),

$$\langle \phi(\mathbf{q}, \omega) \phi(-\mathbf{q}, -\omega) \rangle = \frac{2kT}{\sigma_0 q^2} \, . \tag{11.269}$$

We observe the generality of the result of Section 6.5.

11.6   Mesoscopic fluctuations

In the following we shall show that when the size of a sample becomes comparable to the phase coherence length, $L \sim L_\varphi$, the individuality of the sample will be manifest in its transport properties. Such a sample is said to be mesoscopic. Characteristically the conductance will exhibit sample-specific, noise-like but reproduceable, aperiodic oscillations as a function of, say, magnetic field or chemical potential (i.e. density of electrons). The sample behavior is thus no longer characterized by its average characteristics, such as the average conductance, i.e. the average impurity concentration. The statistical assumption of phase-incoherent and therefore independent subsystems, allowing for such an average description, is no longer valid when the transport takes place quantum mechanically coherently throughout the whole sample. As a consequence, a mesoscopic sample does not possess the property of being self-averaging; i.e. the relative fluctuations in the conductance do not vanish in a

---

[71]We note that the relaxation rate due to processes with energy transfers of the order of the temperature is

$$\frac{1}{\tau_T} \sim \frac{kT}{mD_0} \, .$$

[72]The factor of $-2i$ between Eq. (11.268) and Eq. (11.269) simply reflects our choice of Feynman rules.

central limit fashion inversely proportional to the volume in the large-volume limit. To describe the fluctuations from the average value we need to study the higher moments of the conductance distribution such as the variance $\Delta G_{\alpha\beta,\gamma\delta}$. We shall first study the fluctuations in the conductance at zero temperature, and consider the variance

$$\Delta G_{\alpha\beta,\gamma\delta} \;=\; \langle (G_{\alpha\beta} - \langle G_{\alpha\beta}\rangle)(G_{\gamma\delta} - \langle G_{\gamma\delta}\rangle)\rangle \,. \tag{11.270}$$

For the conductance fluctuations we have the expression

$$\langle G_{\alpha\beta}\, G_{\gamma\delta}\rangle \;=\; (L^{-2})^2 \int d\mathbf{x}_2 \int d\mathbf{x}_2' \int d\mathbf{x}_1 \int d\mathbf{x}_1' \langle \sigma_{\alpha\beta}(\mathbf{x}_2,\mathbf{x}_2')\, \sigma_{\gamma\delta}(\mathbf{x}_1,\mathbf{x}_1')\rangle \,. \tag{11.271}$$

The diagrams for the variance of the conductance fluctuations can still be managed within the standard impurity diagram technique in the weak disorder limit, $\epsilon_F \tau \gg \hbar$, and a typical conductance fluctuation diagram is depicted in Figure 11.7 (here the box denotes the Diffuson).[73]

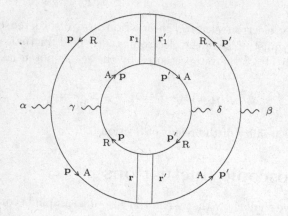

Figure 11.7 Conductance fluctuation diagram.

The construction of the conductance fluctuation diagrams follows from impurity averaging two conductivity diagrams. Draw two conductivity bubble diagrams, where the propagators include the impurity scattering. Treating the impurity scattering perturbatively, we get impurity vertices that we, upon impurity averaging as usual have to pair in all possible ways. Since we subtract the squared average conductance in forming the variance, ΔG, the diagrams for the variance consist only of diagrams where the two conductance loops are connected by impurity lines. As already noted in the discussion of weak localization, the dominant contributions to such loop-type diagrams are from the infrared and long-wavelength divergence of the Cooperon, and here additionally from the Diffuson.

---

[73]The diagram is in the position representation, and the momentum labels should presently be ignored, but will be explained shortly.

To calculate the contribution to the variance from the Diffuson diagram depicted in Figure 11.7, we write the corresponding expression down in the spatial representation in accordance with the usual Feynman rules for conductivity diagrams. Let us consider a hypercube of size L. If we assume that the sample size is bigger than the impurity mean free path, $L > l$, the spatial extension of the integration over the external, excitation and measuring, vertices can be extended to infinity, since the propagators have the spatial extension of the mean free path. We can therefore introduce the Fourier transform for the propagators since no reference to the finiteness of the system is necessary for such local quantities. Furthermore, since the spatial extension of the Diffuson is long range compared with the mean free path, we can set the spatial labels of the Diffusons equal to each other, i.e. $\mathbf{r}_1 = \mathbf{r}$ and $\mathbf{r}'_1 = \mathbf{r}'$. All the spatial integrations, except the ones determined by the Diffuson, can then be performed, leading to the momentum labels for the propagators as depicted in Figure 11.7 Let us study the fluctuations in the d.c. conductance, so that the frequency, ω, of the external field is zero. The energy labels have for visual clarity been deleted from Figure 11.7, since we only have elastic scattering and therefore one label, say ϵ, for the outer ring and one for the inner, ϵ'. According to the Feynman rules, we obtain for the Diffuson diagram the following analytical expression:

$$
\langle G_{\alpha\beta} G_{\gamma\delta} \rangle_{\mathrm{D}} = L^{-4} \left(\frac{e^2 \hbar^2 u^2}{4\pi m^2} \right)^2 \int_{-\infty}^{\infty} d\epsilon \, \frac{\partial f(\epsilon)}{\partial \epsilon} \int_{-\infty}^{\infty} d\epsilon' \, \frac{\partial f(\epsilon')}{\partial \epsilon'} \int \frac{d\mathbf{p}}{(2\pi\hbar)^3} \int \frac{d\mathbf{p}'}{(2\pi\hbar)^3}
$$

$$
\times \quad G_{\epsilon}^{\mathrm{R}}(\mathbf{p}') G_{\epsilon}^{\mathrm{A}}(\mathbf{p}') G_{\epsilon'}^{\mathrm{A}}(\mathbf{p}') G_{\epsilon'}^{\mathrm{R}}(\mathbf{p}') G_{\epsilon}^{\mathrm{R}}(\mathbf{p}) G_{\epsilon'}^{\mathrm{A}}(\mathbf{p}) G_{\epsilon}^{\mathrm{A}}(\mathbf{p}) G_{\epsilon'}^{\mathrm{R}}(\mathbf{p})
$$

$$
\times \quad p_\alpha \, p_\gamma \, p'_\delta \, p'_\beta \int d\mathbf{r} \int d\mathbf{r}' \, |D(\mathbf{r}, \mathbf{r}', \epsilon - \epsilon')|^2 . \tag{11.272}
$$

In order to obtain the above expression we have noted that

$$
D(\mathbf{r}, \mathbf{r}', \epsilon' - \epsilon) = [D(\mathbf{r}, \mathbf{r}', \epsilon - \epsilon')]^* , \tag{11.273}
$$

which follows from the relationship between the retarded and advanced propagators. At zero temperature, the Fermi functions set the energy variables in the propagators in the conductance loops to the Fermi energy, and the Diffuson frequency to zero. At zero temperature we therefore get for the considered Diffuson diagram the following analytical expression, $D(\mathbf{r}, \mathbf{r}') \equiv D(\mathbf{r}, \mathbf{r}', 0)$,

$$
\langle G_{\alpha\beta} G_{\gamma\delta} \rangle_{\mathrm{D}} = L^{-4} \left(\frac{e^2 \hbar^2 u^2}{4\pi m^2} \right)^2 \int \frac{d\mathbf{p}}{(2\pi\hbar)^3} \int \frac{d\mathbf{p}'}{(2\pi\hbar)^3} \, p_\alpha \, p_\gamma \, p'_\delta \, p'_\beta
$$

$$
\times \quad [G_{\epsilon_{\mathrm{F}}}^{\mathrm{R}}(\mathbf{p}) G_{\epsilon_{\mathrm{F}}}^{\mathrm{A}}(\mathbf{p}) G_{\epsilon_{\mathrm{F}}}^{\mathrm{R}}(\mathbf{p}') G_{\epsilon_{\mathrm{F}}}^{\mathrm{A}}(\mathbf{p}')]^2 \int d\mathbf{r} \int d\mathbf{r}' \, |D(\mathbf{r}, \mathbf{r}')|^2 . \tag{11.274}
$$

It is important to note that the same Diffuson appears twice. This is the leading singularity we need to keep track of. If we try to construct variance diagrams containing, say, three Diffusons, we will observe that they cannot carry the same wave vector, and will give a contribution smaller by the factor $\hbar/\epsilon_F \tau$. The momentum integrations at the current vertices can easily be performed by the residue method (recall Eq. (11.27))

$$j_{\alpha\gamma} = \int \frac{d\mathbf{p}}{(2\pi\hbar)^3} \, p_\alpha \, p_\gamma \, [G^R_{\epsilon_F}(\mathbf{p}) G^A_{\epsilon_F}(\mathbf{p})]^2 = \frac{4\pi}{3} \frac{p_F^2 N_0}{\hbar^3} \tau^3 \, \delta_{\alpha\gamma} \qquad (11.275)$$

and for the considered Diffuson diagram we obtain the expression

$$\langle G_{\alpha\beta} G_{\gamma\delta} \rangle_D = L^{-4} \left(\frac{e^2 D_0 \tau}{2\pi\hbar} \right)^2 \delta_{\alpha\gamma} \, \delta_{\delta\beta} \int d\mathbf{r} \int d\mathbf{r}' \, |D(\mathbf{r}, \mathbf{r}')|^2 \, . \qquad (11.276)$$

To calculate the Diffuson integrals we need to address the finite size of the sample and its attachment to the current leads, since the Diffuson has no inherent length scale cut-off. At the surface where the sample is attached to the leads, the Diffuson vanishes

$$D(\mathbf{r}, \mathbf{r}') = 0 \qquad \mathbf{r} \quad or \quad \mathbf{r}' \text{ on lead surfaces} \qquad (11.277)$$

in accordance with the assumption that once an electron reaches the lead it never returns to the disordered region phase coherently. On the other surfaces the current vanishes; i.e. the normal derivative of the Diffuson must vanish (recall Eq. (7.96) and Eq. (7.97))

$$\frac{\partial D(\mathbf{r}, \mathbf{r}')}{\partial \mathbf{n}} = 0 \qquad \mathbf{r} \quad or \quad \mathbf{r}' \text{ on non-lead surfaces with surface normal } \mathbf{n} \, .$$

$$(11.278)$$

We assume that the leads have the same size as the sample surface.[74] Therefore by solving the diffusion equation for the Diffuson, with the above mixed (Dirichlet–von Neumann) boundary condition, we obtain the expression

$$\int d\mathbf{r} \int d\mathbf{r}' \, |D(\mathbf{r}, \mathbf{r}')|^2 = \left(\sum_n \frac{1/\tau}{D_0 \, q_n^2} \right)^2 , \qquad (11.279)$$

where $n \equiv (n_x, n_y, n_z)$ is the eigenvalue index in the three-dimensional case

$$q_{n_\alpha} = \frac{\pi}{L} n_\alpha \qquad n_\alpha = n_x, n_y, n_z \qquad (11.280)$$

where

$$n_x = 1, 2, ..., \qquad n_{y,z} = 0, 1, 2, ... \qquad (11.281)$$

---

[74]This "thick lead" assumption is not of importance. Because of the relationship between the fluctuations in the density of states and the time scale for diffusing out of the sample, the result will be the same for any kind of lead attachment [106].

and we have assumed that the current leads are along the x-axis. Less than three dimensions corresponds to neglecting the n_y and n_z. We therefore obtain from the considered Diffuson diagram the contribution to the conductance fluctuations[75]

$$\langle G_{\alpha\beta}G_{\gamma\delta}\rangle_{\mathrm D} = \left(\frac{e^2}{4\pi\hbar}\right)^2 c_d\, \delta_{\alpha,\gamma}\, \delta_{\delta,\beta}\,, \qquad (11.282)$$

where the constant c_d depends on the sample dimension. The summation in Eq. (11.279) should, in accordance with the validity of the diffusion regime, be restricted to values satisfying $n_x^2 + n_y^2 + n_z^2 \leq N$, where N is of the order of $(L/l)^2$. However, the sum converges rapidly and the constants c_d are seen to be of order unity. The dimensionality criterion is essentially the same as in the theory of weak localization, as we shall show in the discussion below of the physical origin of the fluctuation effects. The important thing to notice is that the long-range nature of the Diffuson provides the L^4 factor that makes the variance, average of the squared conductance, independent of sample size (recall Eq. (6.57)). The diagram depicted in Figure 11.7 is only one of the two possible pairings of the current vertices, and we obtain an additional contribution from the diagram where, say, current vertices γ and δ are interchanged.

In addition to the contribution from the diagram in Figure 11.7 there is the other possible singular Diffuson contribution to the variance from the diagram depicted in Figure 11.8.

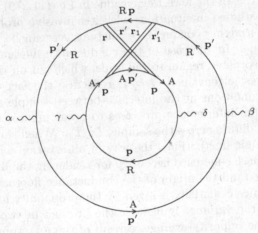

Figure 11.8 The other possible conductance fluctuation diagram.

---

[75]Because of these inherent mesoscopic fluctuations, we realize that the conductance discussed in the scaling theory of localization is the average conductance.

This diagram contributes the same amount as the one in Figure 11.7, but with a different pairing of the current vertices. We note that the diagram in Figure 11.8 allows for only one assignment of current vertices.[76]

Reversing the direction in one of the loops gives rise to similar diagrams, but now with the Cooperon appearing instead of the Diffuson. Because the boundary conditions on the Cooperon are the same as for the Diffuson, in the absence of a magnetic field, the Cooperon contributes an equal amount. For the total contribution to the variance of the conductance, we therefore have (allowing for the spin degree of freedom of the electron would quadruple the value) at zero temperature

$$\Delta G_{\alpha\beta,\gamma\delta} = \left(\frac{e^2}{2\pi\hbar}\right)^2 c_d \left(\delta_{\alpha\gamma}\,\delta_{\delta\beta} + \delta_{\alpha\delta}\,\delta_{\gamma\beta} + \delta_{\alpha,\beta}\,\delta_{\gamma,\delta}\right). \tag{11.283}$$

The variance of the conductance at zero temperature, and for the chosen geometry of a hypercube, is seen to be independent of size and dimension of the sample and degree of disorder, and the conductance fluctuations appear in the metallic regime described above to be universal.[77]

Since the average classical conductance is proportional to L^{d-2}, Ohm's law, we find that the relative variance, $\Delta G \langle G \rangle^{-2}$, is proportional to L^{4-2d}. This result should be contrasted with the behavior L^{-2d} of thermodynamic fluctuations, compared with which the quantum-interference-induced mesoscopic fluctuations are huge, reflecting the absence of self-averaging.

The dominating role of the lowest eigenvalue in Eq. (11.279) indicates that mesoscopic fluctuations, studied in situations with less-invasive probes than the current leads necessary for studying conductance fluctuations, can be enhanced compared to the universal value. In the case of the conductance fluctuations, the necessary connection of the disordered region to the leads, which cut off the singularity in the Diffuson by the lowest eigenvalue, $n_x = 1$, reflecting the fact that because of the physical boundary conditions at the interface between sample and leads, the maximal time for quantum interference processes to occur uninterrupted is the time it takes the electron to diffuse across the sample, L^2/D_0. When considering other ways of observing mesoscopic fluctuations, the way of observation will in turn introduce the destruction of phase coherence necessary for rendering the fluctuations finite.

In order to understand the origin of the conductance fluctuations, we note that, just as the conductance essentially is given by the probability for diffusing between points in a sample, the variance is likewise the product of two such probabilities. When we perform the impurity average, certain of the quantum interference terms will not be averaged away, since certain pairs of paths are coherent. This is similar to the case of coherence involved in the weak-localization effect, but in the present case of the variance of quite a different nature. For example, the quantum interference

---

[76]The contribution from the diagram in Figure 11.7 can, through the Einstein relation, be ascribed to fluctuations in the diffusion constant, whereas the diagram in Figure 11.8 gives the contribution from the fluctuations in the density of states, the two types of fluctuation being independent [107].

[77]However, for a non-cubic sample, the variance will be geometry dependent [108, 109].

process described by the diagram in Figure 11.7 is depicted in Figure 11.9, where the solid line corresponds to the outer conductance loop, and the dashed line corresponds to the inner conductance loop. The wavy portion of the lines corresponds to the long-range diffusion process.

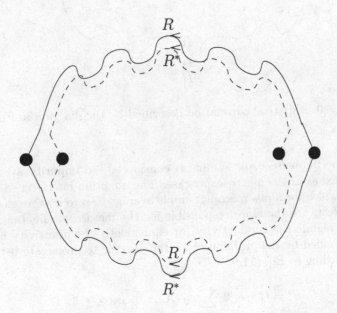

Figure 11.9 Statistical correlation described by the diagram in Figure 11.7.

When one takes the impurity average of the variance, the quantum interference terms can pair up for each diffusive path in the random potential, but now they correspond to amplitudes for propagation in different samples. The diagrams for the variance, therefore, do not describe any physical quantum interference process, since we are not describing a probability but a product of probabilities. The variance gives the statistical correlation between amplitudes in different samples. The interference process corresponding to the diagram in Figure 11.8 is likewise depicted in Figure 11.10.

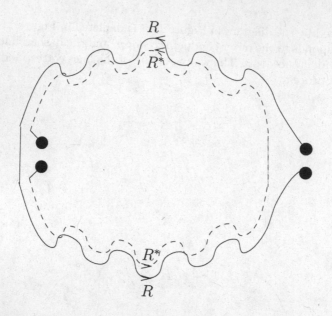

Figure 11.10 Statistical correlation described by the diagram in Figure 11.8.

When a specific mesoscopic sample is considered, no impurity average is effectively performed as in the macroscopic case. The quantum interference terms in the conductance, which for a macroscopic sample average to zero if we neglect the weak-localization effect, are therefore responsible for the mesoscopic fluctuations. In the weak-disorder regime the conductivity (or equivalently the diffusivity by Einstein's relation) is specified by the probability for the particle to propagate between points in space. According to Eq. (11.95)

$$P = P_{\rm cl} + 2 \sum_{c,c'} \sqrt{|A_c A_{c'}|} \, \cos(\phi_c - \phi_{c'}) \qquad (11.284)$$

as

$$A_c = |A_c| \, e^{i\phi_c} \ , \quad \phi_c = \frac{1}{\hbar} S[\mathbf{x}_c(t)] \ , \qquad (11.285)$$

where $|A_c|$ specifies the probability for the classical path c, and its phase is specified by the action. When the points in space in questions are farther apart than the mean free path, the ensemble average of the quantum interference term in the probability vanishes. The weak localization can be neglected because for random phases we have $\langle \cos(\phi_c - \phi_{c'}) \rangle_{\rm imp} = 0$. However, for the mean square of the probability, we encounter $\langle \cos^2(\phi_c - \phi_{c'}) \rangle_{\rm imp} = 1/2$, and obtain

$$\langle P^2 \rangle_{\rm imp} = \langle P \rangle_{\rm imp}^2 + 2 \sum_{c,c'} |A_c| |A_{c'}| \ . \qquad (11.286)$$

Because of quantum interference there is thus a difference between $\langle P^2 \rangle_{\rm imp}$ and

$\langle P \rangle^2_{\text{imp}}$ resulting in mesoscopic fluctuations. Since the effect is determined by the phases of paths, it is nonlocal.

The result in Eq. (11.283) is valid in the metallic regime, where the average conductance is larger than $e^2/\hbar$. To go beyond the metallic regime would necessitate introducing the quantum corrections to diffusion, the first of which is the weak-localization type, which diagrammatically corresponds to inserting Cooperons in between Diffusons. Such an analysis is necessary for a study of the fluctuations in the strongly disordered regime, as performed in reference [84].

The Diffuson and Cooperon in the conductance fluctuation diagrams do not describe diffusion and return probability, respectively, in a given sample, but quantum-statistical correlations between motion in different samples, i.e. different impurity configurations, as each conductance loop in the Figures 11.7 and 11.8 corresponds to different samples. In order to stress this important distinction, we shall in the following mark with a tilde the Diffusons and Cooperons appearing in fluctuation diagrams.

We now assess the effects of finite temperature on the conductance fluctuations. Besides the explicit temperature dependence due to the Fermi functions appearing in Eq. (11.272), the ladder diagrams will be modified by interaction effects. The presence of the Fermi functions corresponds to an energy average over the thermal layer near the Fermi surface, and through the energy dependence of the Diffuson and Cooperon introduces the temperature-dependent length scale $L_T = \sqrt{D_0\hbar/kT}$. Since the loops in the fluctuation diagrams correspond to different conductivity measurements, i.e. different samples, interaction lines (for example caused by electron–phonon or electron–electron interaction) are not allowed to connect the loops in a fluctuation diagram. The diffusion pole of the Diffuson appearing in a fluctuation diagram is therefore not immune to interaction effects. This was only the case when the Diffuson describes diffusion within a sample, since then the diffusion pole is a consequence of particle conservation and therefore unaffected by interaction effects. The consequence is that, just as in the case for the Cooperon, inelastic scattering will lead to a cut-off given by the phase-breaking rate $1/\tau_\varphi$. In short, the temperature effects will therefore ensure that up to the length scale of the order of the phase-coherence length, the conductance fluctuations are determined by the zero-temperature expression, and beyond this scale the conductance of such phase-incoherent volumes add as in the classical case.[78] A sample is therefore said to be mesoscopic when its size is in between the microscopic scale, set by the mean free path, and the macroscopic scale, set by the phase-coherence length, $l < L < L_\varphi$. A sample is therefore self-averaging only with respect to the impurity scattering for samples of size larger than the phase-coherence length.[79] A sample will therefore exhibit the weak-localization effect only when its size is much larger than the phase-coherence length but much smaller than the localization length $L_\varphi < L < \xi$.

An important way to reveal the conductance fluctuations experimentally is to measure the magneto-resistance of a mesoscopic sample. To study the fluctuation effects in magnetic fields, we must study the dependence of the variance on the

---

[78]For example for a wire we have $g(L) = g(L_\varphi)\, L/L_\varphi$.

[79]The conductance entering the scaling theory of localization is thus assumed averaged over phase-incoherent volumes.

magnetic fields $\Delta G_{\alpha\beta}(\mathbf{B}_+, \mathbf{B}_-)$, where $\mathbf{B}_+$ is the sum and $\mathbf{B}_-$ is the difference in the magnetic fields influencing the outer and inner loops. Since the conductance loops can correspond to samples placed in different field strengths, the diffusion pole appearing in a fluctuation diagram will not be immune to the presence of magnetic fields, as in the case when the Diffuson describes diffusion within a given sample, since particle conservation is, of course, unaffected by the presence of a magnetic field. According to the low-field prescription for inclusion of magnetic fields, Eq. (11.105), we get for the Diffuson

$$D_0 \left\{ (-i\nabla_{\mathbf{x}} - \frac{e}{\hbar}\mathbf{A}_-(\mathbf{x}))^2 + 1/\tau_\varphi \right\} \tilde{D}(\mathbf{x}, \mathbf{x}') \;=\; \frac{1}{\tau}\,\delta(\mathbf{x} - \mathbf{x}')\,, \qquad (11.287)$$

where $\mathbf{A}_-$ is the vector potential corresponding to the difference in magnetic fields, $\mathbf{B}_- = \nabla_{\mathbf{x}} \times \mathbf{A}_-$, and we have introduced the phase-breaking rate in view of the above consideration. In the case of the Diffuson, the magnetic field induced phases subtract, accounting for the appearance of the difference of the vector potentials $\mathbf{A}_-$. For the case of the Cooperon, the two phases add, and we obtain

$$D_0 \left\{ (-i\nabla_{\mathbf{x}} - \frac{e}{\hbar}\mathbf{A}_+(\mathbf{x}))^2 + 1/\tau_\varphi \right\} \tilde{C}(\mathbf{x}, \mathbf{x}') \;=\; \frac{1}{\tau}\,\delta(\mathbf{x} - \mathbf{x}')\,, \qquad (11.288)$$

where $\mathbf{A}_+$ is the vector potential corresponding to the sum of the fields, $\mathbf{B}_+ = \nabla \times \mathbf{A}_+$.

The *magneto-fingerprint* of a given sample, i.e. the dependence of its conductance on an external magnetic will show an erratic pattern with a given peak to valley ratio and a correlation field strength B_c. This, however, is not immediately the information we obtain by calculating the variance

$$\Delta G_{\alpha\beta,\gamma\delta}(\mathbf{B}_+, \mathbf{B}_-) \;=\; \langle [G_{\alpha\beta}(\mathbf{B}_1) - \langle G_{\alpha\beta}(\mathbf{B}_1)\rangle][G_{\gamma\delta}(\mathbf{B}_2) - \langle G_{\gamma\delta}(\mathbf{B}_2)\rangle]\rangle\,, \quad (11.289)$$

where $\mathbf{B}_1$ is the field in, say, the inner loop, $\mathbf{B}_1 = (\mathbf{B}_+ + \mathbf{B}_-)/2$, and $\mathbf{B}_2$ is the field in the outer loop, $\mathbf{B}_2 = (\mathbf{B}_+ - \mathbf{B}_-)/2$. In the variance, the magnetic fields are fixed in the two samples, and we are averaging over different impurity configurations, thus describing a situation in which the actual impurity configuration is changed, a hardly controllable endeavor from an experimental point of view. However, if the magneto-conductance of a given sample, $G(B)$, varies randomly with magnetic field, the two types of average – one with respect to magnetic field and the other with respect to impurity configuration – are equivalent, and the characteristics of the magneto-fingerprint can be extracted from the correlation function in Eq. (11.289). The physical reason for the validity of such an *ergodic* hypothesis [110, 111], that changing magnetic field is equivalent to changing impurity configuration, is that since the electronic motion in the sample is quantum mechanically coherent the wave function pattern is sensitive to the position of all the impurities in the sample, just as the presence of the magnetic field is felt throughout the sample by the electron.[80] The extreme sensitivity to impurity configuration is also witnessed by the fact that changing the position of one impurity by an atomic distance, $1/k_F$, is equivalent to shifting all the impurities by arbitrary amounts, i.e. to create a completely different sample [113, 114].

---

[80]The validity of the ergodic hypothesis has been substantiated in reference [112].

The ergodic hypothesis can be elucidated by the following consideration. In the mean square of the probability for propagating between two points in space we encounter the correlation function

$$\langle \Big(\cos(\phi_c(B_1) - \phi_{c'}(B_1)) \Big) \Big(\cos(\phi_c(B_2) - \phi_{c'}(B_2)) \Big) \rangle_{\mathrm{imp}} \tag{11.290}$$

where $(\phi_c(B) - \phi_{c'}(B))$ depends on the phases picked up due to the magnetic field, i.e. the flux through the area enclosed by the trajectories c and c'. When the magnetic field B_1 changes its value to B_2 (where the correlation function equals one half), the phase factor changes by 2π times the flux through the area enclosed by the trajectories c and c' in units of the flux quantum. This change, however, is equivalent to what happens when changing to a different impurity configuration for fixed magnetic field, i.e. the quantity we calculate.[81]

In order to calculate the variance in Eq. (11.289) we must solve Eq. (11.287) and Eq. (11.288) with the appropriate mixed boundary value conditions in the presence of magnetic fields, and insert the solutions into contributions like that in Eq. (11.276). However, determination of the characteristic correlations of the aperiodic magneto-conductance fluctuations can be done by inspection of Eq. (11.287) and Eq. (11.288). The correlation field B_c is determined by the sample-to-sample change in the magnetic field, i.e. $\mathbf{B}_-$. According to Eq. (11.287) and Eq. (11.288), this field is determined either by the sample size, through the gradient term, or the phase coherence length. When the phase-coherence length is longer than the sample size, the correlation field is therefore of order of the flux quantum divided by the sample area, $B_c \sim \phi_0/L^2$, where ϕ_0 is the normal flux quantum $\phi_0 = 2\pi\hbar/|e|$, since the typical diffusion loops, like those depicted in Figures 11.9 and 11.10, enclose an area of the order of the sample, L^2. We note that in magnetic fields exceeding $\max\{\phi_0/L^2, \phi_0/L_\varphi^2\}$, the Cooperon no longer contributes to the field dependence of the conductance fluctuations, because its dependence on magnetic field is suppressed according to the weak-localization analysis.[82]

We note that the weak-localization and mesoscopic fluctuation phenomena are a general feature of wave propagation in a random media, be the wave nature classical, such as sound and light,[83] or of quantum origin such as for the motion of electrons. The weak-localization effect was in fact originally envisaged for the multiple scattering of electromagnetic waves [81].[84] The coherent backscattering effect has been studied experimentally for light waves (for a review on classical wave propagation in random media, see reference [116]). For the wealth of interesting weak-localization and mesoscopic fluctuation effects, we refer the reader to reference [1], and to the references to review articles cited therein.

---

[81] Another way of revealing the mesoscopic fluctuations is to change the Fermi energy (i.e. the density of conduction electron as is feasible in an inversion layer). The typical energy scale E_c for these fluctuations is analogously determined by the typical time τ_{trav} it takes an electron to traverse the sample according to $E_c \sim \hbar/\tau_{\mathrm{trav}}$. In the diffusive regime we have $\tau_{\mathrm{trav}} \sim L^2/D_0$.

[82] For an account of the experimental discovery of conductance fluctuations, see reference [115].

[83] Here we refer to conditions described by Maxwell's equations.

[84] It is telling that it took the application of Feynman diagrams in the context of electronic motion in disordered conductors to understand the properties of classical waves in random media.

11.7 Summary

Quantum effects on transport coefficients have been studied in this chapter, especially the weak localization effect, which is the most important for practical diagnostics in material science as it is revealed at such small magnetic fields where the diffusion enhancement of the electron–electron interaction is unaffected and classical magneto-resistance effects absent. Though the weak localization effect is a quantum interference effect, the kinetics of the involved trajectories were the classical ones, be they Boltzmannian or Brownian, and we could therefore make ample use of the quasi-classical Green's function technique developed in Chapters 7 and 8. We calculated the phase breaking rates due to interactions, the phase relaxation of the wave function measured in magneto-resistance measurements, thereby opening the opportunity to probe the inelastic interactions experienced by electrons. We studied how the interactions are changed as a result of disorder. In the case of Coulomb interaction a universal diffusion enhancement or weakening of screening resulted, whereas for the case of electron–phonon interaction, the longitudinal interaction was weakened owing to the compensation mechanism of the vibrating impurities, whereas the interaction with transverse phonons could be enhanced or weakened depending on the temperature regime. Finally, we discussed the phenomena that sets in when the electronic motion is coherent in the sample and the signature of mesoscopic fluctuations are present in transport coefficients, such as the quantum fluctuations in the conductance, the universal conductance fluctuations.

12

Classical statistical dynamics

The methods of quantum field theory, originally designed to study quantum fluctuations, are also the tool for studying the thermal fluctuations of statistical physics, for example in connection with understanding critical phenomena. In fact, the methods and formalism of quantum fields are the universal language of fluctuations. In this chapter we shall capitalize on the universality of the methods of field theory as introduced in Chapters 9 and 10, and use them to study non-equilibrium phenomena in *classical* statistical physics where the fluctuations are those of a classical stochastic variable. We shall show that the developed non-equilibrium real-time formalism in the classical limit provides the theory of classical stochastic dynamics.

Newton's law, which governs the motion of the heavenly bodies, is not the law that seems to govern earthly ones. They sadly seem to lack inertia, get stuck and feebly ramble around according to Brownian dynamics as described by the Langevin equation. Their dynamics show transient effects, but if they are on short time scale too fast to observe, dissipative dynamics is typically specified by the equation $\mathbf{v} \propto \mathbf{F}$ where the proportionality constant could be called the friction coefficient. This is Aristotelian dynamics, average velocity proportional to force, believed to be correct before Galileo came along and did thorough experimentation. If a sponge is dropped from the tower of Pisa, it will almost instantly reach its saturation final velocity. If a heavier sponge is dropped simultaneously, it will fall faster reaching the ground first. If on the other hand an apple is dropped and when reaching the ground is given its opposite velocity it will according to Newton's equation spike back up to the position it was dropped from, before repeating its trip to the ground. If a sponge has its impact velocity at the ground reversed, it will fizzle immediately back to the ground. Unlike Newtonian mechanics, which is time reversal symmetric, sponge or dissipative dynamics chooses a direction of time.

We now turn to consider dissipative dynamics, in particular Langevin dynamics. In this chapter we will study systems with the additional feature of quenched disorder, in particular vortex dynamics in disordered superconductors. The field-theoretic formulation of the problem will allow the disorder average to be performed exactly. The functional methods will allow construction of a self-consistent theory for the effective action describing the influence of thermal fluctuations and quenched disorder on vortex motion. This will allow the determination of the vortex response to external forces, the vortex fluctuations, and the pinning of vortices due to quenched disorder, and allow to consider the dynamic melting of vortex lattices.

12.1   Field theory of stochastic dynamics

In this section we shall map the stochastic problem, formulated first in terms of a stochastic differential equation, onto a path integral formulation, and obtain the field theoretic formulation of classical statistical dynamics. We show that the resulting formalism is equivalent to that of a quantum field theory. In particular we shall consider quenched disorder and the resulting diagrammatics. The field theoretic formulation will allow us to perform the average over the quenched disorder exactly.

12.1.1   Langevin dynamics

A heavy particle interacting with a gas of light particles, say a pollen dust particle submerged in water, will viewed under a microscope execute erratic or Brownian motion. Or in general, when a particle interacting with a heat bath, i.e. weakly with a multitude of degrees of freedom in its environment (of high enough temperature so that quantum effects are absent), will exhibit dynamics governed by the Langevin equation

$$m\ddot{\mathbf{x}}_t = \mathbf{F}(\mathbf{x}_t, t) - \eta\dot{\mathbf{x}}_t + \boldsymbol{\xi}_t \,, \tag{12.1}$$

where m is the mass of the particle, $\mathbf{F}$ is a possible external force, η is the viscosity or the friction coefficient, and $\boldsymbol{\xi}_t$ is the fluctuating force describing the thermal agitation of the particle due to the interaction with the environment, the thermal noise, or some other relevant source of noise. For a system interacting with a classical environment assumed in thermal equilibrium at a temperature T, the fluctuating force is a Gaussian stochastic process described by the correlation function for its Cartesian components[1]

$$\langle \xi_t^{(\alpha)} \xi_{t'}^{(\beta)} \rangle = 2\eta k_{\mathrm{B}} T \, \delta(t - t') \, \delta_{\alpha\beta} \tag{12.2}$$

relating friction and fluctuations according to the fluctuation–dissipation theorem, as proper for linear response.

Being the dissipative dynamics for a system coupled to a heat bath, Langevin dynamics is relevant for describing a vast range of phenomena, and of course not just that of a particle as considered above. For example, randomly stirred fluids in which case the relevant equation would be the Navier–Stokes equation with proper noise term [117]. The field theoretic formulation of the following section runs identical for all such cases. Also, the coordinate above need not literally be that of a particle, but could for example describe the position of a vortex in a type-II superconductor, as discussed in Section 12.2. However, we shall in the following keep referring to the degree of freedom as that of a *particle*.

---

[1]The quantum case and the classical limit are discussed in Appendix A.

12.1.2 Fluctuating linear oscillator

For a given realization of the fluctuating force, $\boldsymbol{\xi}_t$, there is a solution to the Langevin equation, Eq. (12.1), specifying the realization of the corresponding motion of the particle $\mathbf{x}_t$. In other words, $\mathbf{x}_t$ is a functional of $\boldsymbol{\xi}_t$, $\mathbf{x}_t = \mathbf{x}_t[\boldsymbol{\xi}_{\bar{t}}]$, and vice versa $\boldsymbol{\xi}_t = \boldsymbol{\xi}_t[\mathbf{x}_{\bar{t}}]$. The properties of the fluctuating force is described by its probability distribution, $P_\xi[\boldsymbol{\xi}_t]$, which is assumed to be Gaussian

$$P_\xi[\boldsymbol{\xi}_t] = \int \mathcal{D}\boldsymbol{\xi}_t \, e^{-\frac{1}{2}\int dt_1 \int dt_2 \, \boldsymbol{\xi}_{t_1} K^{-1}_{t_1,t_2} \boldsymbol{\xi}_{t_2}} \, , \qquad (12.3)$$

where $K^{-1}_{t_1,t_2}$ is the inverse of the correlator of the stochastic force

$$K_{t,t'} = \langle \boldsymbol{\xi}_t \boldsymbol{\xi}_{t'} \rangle \qquad (12.4)$$

and we have used dyadic notation to express the matrix structure of the force correlations in Cartesian space. This structure is, however, irrelevant as the quantity is diagonal.

Using the one-to-one map between the fluctuating force and the particle path, $\mathbf{x}_t \longleftrightarrow \boldsymbol{\xi}_t$, the probability of given paths, $\mathbf{x}_t$s, equals that of the corresponding forces, $\boldsymbol{\xi}_t$s,

$$P_x[\mathbf{x}_t] \, \mathcal{D}\mathbf{x_t} = \mathbf{P}_\xi[\boldsymbol{\xi}_t] \, \mathcal{D}\boldsymbol{\xi_t} \, . \qquad (12.5)$$

In general, this does not allow us to state proportionality between the two probability distributions, since the volume change in the transformation from $\mathcal{D}\boldsymbol{\xi}_t$ to $\mathcal{D}\mathbf{x}_t$ must be taken into account, the change in measure described by the Jacobian. Only if the Langevin equation, Eq. (12.1), is linear is this a trivial matter, restricting the force in Eq. (12.1) to that of a harmonic oscillator, $\mathbf{F}(\mathbf{x}_t, t) = -m\omega_0^2 \mathbf{x}_t$ (and a possible external space-independent force, $\mathbf{F}(t)$, which we suppress in the following), i.e. the equation of motion is that of a harmonic oscillator in the presence of a fluctuating force

$$-D_{\mathrm{R}}^{-1}\mathbf{x} = \boldsymbol{\xi} \qquad (12.6)$$

where we have introduced the retarded Green's function for the damped harmonic oscillator

$$D_{\mathrm{R_{d.h.o.}}}^{-1}(t,t') = -(m\partial_t^2 + \eta\partial_t + m\omega_0^2)\,\delta(t-t') \qquad (12.7)$$

and suppressed the time variable and used matrix multiplication notation in the time variable.[2] In the considered linear case the Jacobian is a constant and

$$P_x[x_t] \propto P_\xi[\boldsymbol{\xi}_t] = P_\xi[m\ddot{\mathbf{x}}_t + \eta\dot{\mathbf{x}}_t + m\omega_0^2\mathbf{x}_t] \, , \qquad (12.8)$$

where the last equality is obtained by using the equation of motion, Eq. (12.6). Using the fact that the fluctuations are Gaussian, gives for the probability distribution of paths,

$$P_x[\mathbf{x}_t] \propto e^{-\frac{1}{2}\int dt_1 dt_2 \, \boldsymbol{\xi}_{t_1}[\mathbf{x}_{\bar{t}_1}] \, K^{-1}_{t_1,t_2} \, \boldsymbol{\xi}_{t_2}[\mathbf{x}_{\bar{t}_2}]}$$

$$= e^{-\frac{1}{2}\int dt_1 dt_2 (m\ddot{\mathbf{x}}_{t_1}+\eta\dot{\mathbf{x}}_{t_1}+m\omega_0^2\mathbf{x}_{t_1}) K^{-1}_{t_1,t_2} (m\ddot{\mathbf{x}}_{t_2}+\eta\dot{\mathbf{x}}_{t_2}+m\omega_0^2\mathbf{x}_{t_2})} \qquad (12.9)$$

---

[2] It could of course be considered a matrix in Cartesian coordinates, but since it would be diagonal it is superfluous.

for the case of a harmonic oscillator coupled to a heat bath.

Completing the square in the following Gaussian path integral

$$P[\mathbf{x}_t] = \mathcal{N}^{-1} \int \mathcal{D}\tilde{\mathbf{x}}_t \, e^{-\frac{1}{2} \int_{-\infty}^{\infty} d\tilde{t} \int_{-\infty}^{\infty} d\tilde{t}' \, \tilde{\mathbf{x}}_{\tilde{t}} K_{\tilde{t},\tilde{t}'} \tilde{\mathbf{x}}_{\tilde{t}'} \, -i \int_{-\infty}^{\infty} d\tilde{t} \, \tilde{\mathbf{x}}_{\tilde{t}} \cdot (m\ddot{\mathbf{x}}_{\tilde{t}} + \eta \dot{\mathbf{x}}_{\tilde{t}} + m\omega_0^2 \mathbf{x}_{\tilde{t}})} \quad (12.10)$$

gives the previous expression, and the path integral representation for the probability distribution of paths has been obtained. The proportionality factor is fixed by normalization of the probability distribution.

We can also arrive at the expression for the probability distribution of paths in the following way. For a given realization of the fluctuating force, $\boldsymbol{\xi}_t$, the probability distribution for the particle path corresponds with certainty to the one fulfilling the equation of motion as expressed by the delta functional

$$P[\mathbf{x}_t] = \mathcal{N}^{-1} \delta[m\ddot{\mathbf{x}}_t + \eta \dot{\mathbf{x}}_t + m\omega_0^2 \mathbf{x}_t - \boldsymbol{\xi}_t] , \quad (12.11)$$

where $\mathcal{N}^{-1}$ is the constant resulting from the Jacobian. Introducing the functional integral representation of the delta functional we get

$$P[\mathbf{x}] = \mathcal{N}^{-1} \int \mathcal{D}\tilde{\mathbf{x}} \, e^{-i \int_{-\infty}^{\infty} dt \, \tilde{\mathbf{x}} \cdot (m\ddot{\mathbf{x}} + \eta \dot{\mathbf{x}} + m\omega_0^2 \mathbf{x} - \boldsymbol{\xi})} . \quad (12.12)$$

The average over the thermal noise, being Gaussian, can now be performed and we obtain

$$P_x[\mathbf{x}_t] = \langle P[\mathbf{x}] \rangle_\xi = \mathcal{N}^{-1} \int \mathcal{D}\tilde{\mathbf{x}} \, e^{-i \int_{-\infty}^{\infty} d\tilde{t} \int_{-\infty}^{\infty} dt \, \tilde{\mathbf{x}} \cdot (m\ddot{\mathbf{x}} + \eta \dot{\mathbf{x}} + m\omega_0^2 \mathbf{x} - \tilde{\mathbf{x}} i \langle \boldsymbol{\xi}\boldsymbol{\xi} \rangle \tilde{\mathbf{x}})} . \quad (12.13)$$

Realizing that the correlation function, Eq. (12.2), is the high-temperature classical limit of the inverse of the correlation function for a harmonic quantum oscillator coupled to a heat bath (see Appendix A), i.e. the kinetic component of the real-time matrix Green's function, we introduce the notation

$$-i D_K^{-1}(t,t') = K^{-1}(t,t') = \langle \xi_t \, \xi_{t'} \rangle = 2\eta k_B T \, \delta(t-t') \quad (12.14)$$

where reference to the irrelevant Cartesian coordinates is left out, i.e. K^{-1} now denotes the scalar part in Eq. (12.2).

In addition the advanced inverse Green's function is introduced

$$D_A^{-1}(t,t') = D_R^{-1}(t',t) \quad (12.15)$$

and both functions are diagonal matrices in Cartesian space and will therefore be treated as scalars. We can then rewrite for the path probability distribution (reintroducing an external force $\mathbf{F}(\mathbf{x},t)$)

$$P_x[\mathbf{x}_t] = \int \mathcal{D}\tilde{\mathbf{x}} \, e^{iS_0[\tilde{\mathbf{x}},\mathbf{x}] + i\tilde{\mathbf{x}} \cdot \mathbf{F} + i\mathbf{x} \cdot \mathbf{j}} , \quad (12.16)$$

where

$$S_0[\tilde{\mathbf{x}}, \mathbf{x}] = \frac{1}{2}(\tilde{\mathbf{x}} D_R^{-1} \mathbf{x} + \mathbf{x} D_A^{-1} \tilde{\mathbf{x}} + \tilde{\mathbf{x}} D_K^{-1} \tilde{\mathbf{x}}) \quad (12.17)$$

and we have absorbed the normalization factor in the path integral notation, it is fixed by the normalization of the probability distribution

$$\int \mathcal{D}\mathbf{x}_t \, P_x[\mathbf{x}_t] = 1 \,. \tag{12.18}$$

We have in addition to the physical external force $\mathbf{F}(t)$ introduced a source, $\mathbf{J}(t)$, and have the generating functional

$$Z[\mathbf{F}, \mathbf{J}] = \int \mathcal{D}\mathbf{x} \int \mathcal{D}\tilde{\mathbf{x}} \, e^{iS_0[\tilde{\mathbf{x}},\mathbf{x}]+i\tilde{\mathbf{x}}\cdot\mathbf{F}+i\mathbf{x}\cdot\mathbf{J}} \tag{12.19}$$

with the normalization

$$Z[\mathbf{F}, \mathbf{J} = 0] = 1 \,. \tag{12.20}$$

The source is introduced in order to generate the correlation functions of interest, for example

$$\langle \mathbf{x}_t \rangle_\xi = \int \mathcal{D}\mathbf{x} \, \mathbf{x}_t \, P_x[\mathbf{x}_t] = -i \frac{\delta Z[J]}{\delta \mathbf{J}(t)} \bigg|_{\mathbf{J}=0} = -\int_{-\infty}^{\infty} d\bar{t} \, D^{\mathrm{R}}(t,\bar{t}) \, \mathbf{F}(\bar{t}) \,, \tag{12.21}$$

where the last equality follows from the equation of motion, and the retarded propagator D^{R} is thus the linear response function (for the considered linear oscillator, the linear response is the exact response).

Because of the normalization condition, Eq. (12.20), all correlation functions of the auxiliary field $\tilde{\mathbf{x}}$ (generated by differentiation with respect to the physical external force $\mathbf{F}$), vanish when the source $\mathbf{J}$ vanishes.

We now realize that the above theory is equivalent to the celebrated Martin–Siggia–Rose formulation of classical statistical dynamics [118], here in its path integral formulation [119, 120, 121], albeit for the moment only for the case of a damped harmonic oscillator.[3] This restriction was of course self-inflicted and the formalism has numerous general applications, such as to critical dynamics for example studying critical relaxation [122].

We note that whereas equilibrium quantum statistical physics is described by Euclidean field theory (recall Section 1.1 and see Exercise A.1 on page 506 in Appendix A), non-equilibrium classical stochastic phenomena are described by a field theory formally equivalent to real-time quantum field theory.

We hasten to consider a nontrivial situation, viz. that of the presence of quenched disorder. The corresponding field theory will be of the most complicated form, in diagrammatic terms it will have vertices of arbitrarily high connectivity.

---

[3]In other words, the Martin–Siggia–Rose formalism is simply the classical limit of the real-time technique for non-equilibrium states, where the doubling of the degrees of freedom necessary to describe non-equilibrium situations is provided by the dynamics of the system. We note that the presented field theoretic formulation of the Langevin dynamics is the classical limit of Schwinger's closed time path formulation of quantum statistical mechanics of a particle coupled linearly to, for the considered type of damping, an Ohmic environment. Equivalently, it is the classical limit of the Feynman–Vernon path integral formulation of a particle coupled linearly to a heat bath, as discussed in Appendix A.

12.1.3   Quenched disorder

We now return to the in general nonlinear classical stochastic problem specified by the Langevin equation

$$m\ddot{\mathbf{x}}_t + \eta\dot{\mathbf{x}}_t = -\nabla V(\mathbf{x}_t) + \mathbf{F}_t + \boldsymbol{\xi}_t \,, \tag{12.22}$$

where V eventually will be taken to describe quenched disorder. Owing to the presence of the nonlinear term $V(\mathbf{x}_t)$, the argument for the Jacobian being a constant is less trivial. However, by using forward discretization,[4] one obtains the result that, owing to the presence of a finite mass term, the Jacobian can be chosen as a constant and the analysis of the previous section can be taken over giving[5]

$$P[\mathbf{x}] = \mathcal{N}^{-1} \int \mathcal{D}\tilde{\mathbf{x}}\, e^{-i\int_{-\infty}^{\infty} dt\, \tilde{\mathbf{x}} \cdot (m\ddot{\mathbf{x}} + \eta\dot{\mathbf{x}} + \nabla V - \mathbf{F} - \boldsymbol{\xi})} \,. \tag{12.23}$$

The averages over the thermal noise and the disorder can now be performed and we obtain the following expression for the path probability density

$$P_x[\mathbf{x}_t] = \langle\langle P[\mathbf{x}]\rangle\rangle = \mathcal{N}^{-1} \int \mathcal{D}\tilde{\mathbf{x}}\, e^{iS[\tilde{\mathbf{x}},\mathbf{x}] + i\tilde{\mathbf{x}}\cdot\mathbf{F}} \,, \tag{12.24}$$

where the action, $S = S_0 + S_V$, is a sum of a part owing to the quenched disorder and the quadratic part

$$S_0[\tilde{\mathbf{x}}, \mathbf{x}] = \frac{1}{2}(\tilde{\mathbf{x}}D_R^{-1}\mathbf{x} + \mathbf{x}D_A^{-1}\tilde{\mathbf{x}} + \tilde{\mathbf{x}}i\langle\boldsymbol{\xi}\boldsymbol{\xi}\rangle\tilde{\mathbf{x}}) \,, \tag{12.25}$$

where we have introduced the inverse propagator for the problem in the absence of the disorder

$$D_R^{-1}(t, t') = -(m\partial_t^2 + \eta\partial_t)\delta(t - t') \,, \tag{12.26}$$

i.e. the retarded *free* propagator satisfies

$$-(m\partial_t^2 + \eta\partial_t)D_{tt'}^R = \delta(t - t') \tag{12.27}$$

with the boundary condition

$$D_{tt'}^R = 0, \ t < t'. \tag{12.28}$$

The corresponding inverse advanced Green's function

$$D_A^{-1}(t, t') = D_R^{-1}(t', t) \tag{12.29}$$

has been introduced, and we shall also use the notation introduced in Eq. (12.14).

---

[4]Stochastic differential equations should be approached with care, since different discretizations can lead to different types of calculus.

[5]Often in Langevin dynamics the over-damped case is the relevant one, i.e. in the present case corresponding to the absence of the mass term, $m = 0$. In such cases it can be convenient to throw in a mass term at intermediate calculations as a regularizer. In Section 12.1.6 we show that for the over-damped case the Jacobian leads in diagrammatic terms to the absence of tadpole diagrams.

The quenched disorder is assumed described by a Gaussian distributed stochastic potential with zero mean, $\langle V(\mathbf{x}) \rangle = 0$, and thus characterized by its correlation function

$$\nu(\mathbf{x} - \mathbf{x}') = \langle V(\mathbf{x})V(\mathbf{x}') \rangle \,, \tag{12.30}$$

where now the brackets denote averaging with respect to the quenched disorder. The interaction part is then

$$S_V[\tilde{\mathbf{x}}, \mathbf{x}] = -\frac{i}{2} \int_{-\infty}^{\infty} dt \int_{-\infty}^{\infty} dt' \, \tilde{x}_t^\alpha \, \frac{\partial^2}{\partial x^\alpha \partial x^\beta} \nu(\mathbf{x}) \Big|_{\mathbf{x}=\mathbf{x}_t - \mathbf{x}_{t'}} \tilde{x}_{t'}^\beta \,. \tag{12.31}$$

The above model thus describes a classical object subject to a viscous medium and a random potential. In the case of two spatial dimensions it could be a particle on a rough surface experiencing Ohmic dissipation, a case relevant to tribology. However, the theory is applicable to any system where quenched disorder is of importance, say such as when studying critical dynamics of spin-glasses.

The generating functional for the theory is thus

$$Z[\mathbf{F}, \mathbf{J}] = \int \mathcal{D}\mathbf{x} \int \mathcal{D}\tilde{\mathbf{x}} \, e^{iS[\tilde{\mathbf{x}}, \mathbf{x}] + i\tilde{\mathbf{x}} \cdot \mathbf{F} + i\mathbf{x} \cdot \mathbf{J}} \tag{12.32}$$

where the action is $S = S_0 + S_V$ with S_0 and S_V given by Eq. (12.25) and Eq. (12.31), respectively. The normalization

$$Z[\mathbf{F}, \mathbf{J} = 0] = 1 \,. \tag{12.33}$$

allows us to avoid the replica trick for performing the average over the quenched disorder [123].

The generating functional generates the correlation functions of the theory

$$\langle \mathbf{x}_{t_1} \cdots \mathbf{x}_{t_n} \rangle_\xi = \int \mathcal{D}\mathbf{x} \, \mathbf{x}_{t_1} \cdots \mathbf{x}_{t_n} \, P_x[\mathbf{x}_t] = (-i)^n \frac{\delta^n Z[J]}{\delta \mathbf{J}(t_1) \cdots \delta \mathbf{J}(t_n)} \Big|_{\mathbf{J}=0} \,. \tag{12.34}$$

The retarded full Green's function, $G_{\alpha\alpha'}^{\mathrm{R}}$, is seen to be the linear response function to the physical force $F_{\alpha'}$, i.e. to linear order in the external force we have

$$\langle x_\alpha(t) \rangle = \int_{-\infty}^{\infty} dt' \, G_{\alpha\alpha'}^{\mathrm{R}}(t, t') \, F_{\alpha'}(t') \tag{12.35}$$

and $G_{\alpha\alpha'}^{\mathrm{K}}$ is the correlation function, both matrices in Cartesian indices as indicated.

12.1.4 Dynamical index notation

It is useful to introduce compact matrix notation by introducing the dynamical index notation. We collect the path and auxiliary field into the vector field

$$\phi = \begin{pmatrix} \tilde{\mathbf{x}} \\ \mathbf{x} \end{pmatrix} = \begin{pmatrix} \phi_1 \\ \phi_2 \end{pmatrix} \tag{12.36}$$

as well as the forces

$$f \equiv \begin{pmatrix} \mathbf{F} \\ \mathbf{J} \end{pmatrix} \equiv \begin{pmatrix} \mathbf{f}_1 \\ \mathbf{f}_2 \end{pmatrix}.$$ (12.37)

This corresponds to introducing the real-time dynamical index notation we used to describe the non-equilibrium states of a quantum field theory. Here they appear as the Schwinger–Keldysh indices in the classical limit of quantum mechanics.[6] In this notation the quadratic part of the action becomes

$$S_0[\phi] = \frac{1}{2}\phi D^{-1}\phi,$$ (12.38)

where

$$D^{-1} = \begin{pmatrix} D_K^{-1} & D_R^{-1} \\ D_A^{-1} & 0 \end{pmatrix} = \begin{pmatrix} i\langle \xi_t \xi_{t'} \rangle & D_R^{-1} \\ D_A^{-1} & 0 \end{pmatrix}$$ (12.39)

is the free inverse matrix propagator

$$D^{-1}D = \delta(t-t')\underline{\underline{1}}$$ (12.40)

and

$$D(t,t') = \begin{pmatrix} 0 & D^A(t,t') \\ D^R(t,t') & D^K(t,t') \end{pmatrix}.$$ (12.41)

Exercise 12.1. Show by Fourier transformation of Eq. (12.27) that

$$D_\omega^R = \frac{1}{(\omega + i0)(m\omega + i\eta)}$$ (12.42)

and thereby that the solution of Eq. (12.27) is

$$D_{tt'}^R = -\frac{1}{\eta}\theta(t-t')\left(1 - e^{-\eta(t-t')/m}\right).$$ (12.43)

The generator for the free theory is

$$Z_0[f] = \int \mathcal{D}\phi \, e^{iS_0[\phi]+i\phi f} = \sqrt{\det iD}\, e^{-\frac{1}{2}fDf},$$ (12.44)

where the matrix D is specified in Eq. (12.41). The diagonal component, the kinetic component, is given by the equation

$$D^K = -D^R i\langle \boldsymbol{\xi}\boldsymbol{\xi}\rangle D^A$$ (12.45)

and its Fourier transform is therefore

$$D_\omega^K = -2i\eta k_B T D_\omega^R D_\omega^A.$$ (12.46)

The free correlation function of the particle positions

$$\langle \mathbf{x}_t \mathbf{x}_{t'}\rangle = (-i)^2 \frac{\delta^2 Z_0[f]}{\delta \mathbf{f}_2(t)\,\delta \mathbf{f}_2(t')}\bigg|_{\mathbf{f}_2=0} = iD_{tt'}^K + \langle \mathbf{x}_t\rangle\langle \mathbf{x}_{t'}\rangle$$ (12.47)

---

[6]See also the Feynman–Vernon theory, discussed in Appendix A.

has connected and disconnected parts.

The generating functional in the presence of disorder becomes

$$Z[f] = \int \mathcal{D}\phi \, e^{iS[\phi]+i\phi f} , \tag{12.48}$$

where $S = S_0 + S_V$, and the action due to the quenched disorder is

$$S_V[\phi] = -\frac{i}{2} \int_{-\infty}^{\infty} dt \int_{-\infty}^{\infty} dt' \, \phi_1^\alpha(t) \frac{\partial^2}{\partial x^\alpha \partial x^\beta} \nu(\mathbf{x}) \Big|_{\mathbf{x}=\mathbf{x}_t-\mathbf{x}_{t'}} \phi_1^\beta(t') \tag{12.49}$$

and the normalization condition becomes

$$Z[\mathbf{f}_1, \mathbf{f}_2 = 0] = 1 . \tag{12.50}$$

The generator generates the correlation functions, for example the two-point Green's function

$$\langle \phi_t \, \phi_{t'} \rangle = -\frac{\delta^2 Z}{\delta f_t \, \delta f_{t'}} \Big|_{\mathbf{f}_2=0} . \tag{12.51}$$

The generator of connected Green's functions, $iW[f] = \ln Z[f]$, for example generates the average field

$$\langle \phi_t \rangle = -i \frac{1}{Z[f]} \frac{\delta Z[f]}{\delta f_t} \Big|_{\mathbf{f}_2=0} = -i \frac{\delta Z[f]}{\delta f_t} \Big|_{\mathbf{f}_2=0} . \tag{12.52}$$

12.1.5 Quenched disorder and diagrammatics

Let us investigate the structure of the diagrammatic perturbation expansion resulting from the quenched disorder, i.e. the vertices originating from the quenched disorder. The perturbative expansion of the generating functional in terms of the disorder correlator is

$$Z[f] = \int \mathcal{D}\phi \, e^{i\phi D^{-1}\phi + if\phi} \left(1 + iS_V[\phi] + \frac{1}{2!}(iS_V[\phi])^2 + \cdots \right) . \tag{12.53}$$

The vertices in a diagrammatic depiction of the perturbation expansion are determined by S_V, Eq. (12.31) and can be expressed as

$$S_V[\phi] = \frac{i}{2} \int dt \int dt' \int \frac{d\mathbf{k}}{(2\pi)^2} \, \nu(\mathbf{k})\mathbf{k} \cdot \tilde{\mathbf{x}}_t \, e^{i\mathbf{k}(\mathbf{x}_t-\mathbf{x}_{t'})} \mathbf{k} \cdot \tilde{\mathbf{x}}_{t'} . \tag{12.54}$$

The vertices of the theory thus have one auxiliary field, $\tilde{\mathbf{x}}$, attached and an arbitrary number of fields $\mathbf{x}$ attached, and are depicted as a circle with the time in question marked inside and a dash-dotted line to describe the attachment of an impurity correlator

$$\tag{12.55}$$

As any vertex contains attachment for the impurity correlator, vertices occur in pairs

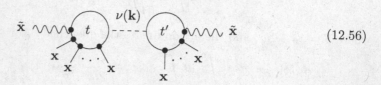

$$(12.56)$$

resulting in vertices of second order in the auxiliary field $\tilde{\mathbf{x}}$ but of arbitrary order in position of the particle, $\mathbf{x}$. The diagrammatic representation of the perturbation expansion in terms of the disorder is thus specified by this basic vertex, and the propagators of the theory are in this classical limit of the real-time technique, the propagators D^{R}, D^{A} and D^{K}. Diagrams representing terms in the perturbation expansion of the generating functional consist of the vertices described above and connected to one another or to sources by lines representing retarded, advanced and kinetic Green's functions. An example of a typical such vacuum diagram of the theory, containing two impurity correlators, is displayed in Figure 12.1.

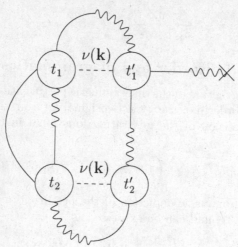

Figure 12.1 Example of a vacuum diagram. The solid line represents the correlation function or kinetic component, G^{K}, of the matrix Green's function. The retarded Green's function, G^{R}, is depicted as a wiggly line ending up in a straight line, and vice versa for the advanced Green's function G^{A}. A dashed line attached to circles represents the impurity correlator. The cross in the figure represents the external force $\mathbf{F}$.

As an application of the above Langevin dynamics in a random potential, we shall study the dynamics of a vortex lattice. But before we discuss the phenomenology of vortex dynamics, we consider the relation of the theory with a mass term to the over-damped case.

12.1.6   Over-damped dynamics and the Jacobian

We have noted in Section 12.1.3 that the presence of the mass terms can be used as a regularizer leaving the Jacobian for the transformation between paths and stochastic force an irrelevant constant. However, many situations of interest are concerned with over-damped dynamics and we shall therefore here deal with that situation explicitly. We show in this section that the neglect of the mass term in the equation of motion gives a Jacobian, which in diagrammatic terms leads to the cancellation of the tadpole diagrams.

In the over-damped case the inverse retarded Green's function, Eq. (12.26), becomes

$$D_R^{-1}(t, t') = -\eta \, \partial_t \delta(t - t') \tag{12.57}$$

corresponding to setting the mass of the particle equal to zero. The Jacobian, J, is for the considered situation the determinant

$$J = \det \left(\frac{\delta \boldsymbol{\xi}_t}{\delta \mathbf{x}_{t'}} \right) \tag{12.58}$$

which by use of the equation of motion can be rewritten

$$J = -(D^R)_{tt'}^{-1} + \frac{\delta \nabla V(\mathbf{x}_t)}{\delta \mathbf{x}_{t'}} = \eta \, \partial_t \delta(t - t') + \frac{\delta \nabla V(\mathbf{x}_t)}{\delta \mathbf{x}_{t'}} \tag{12.59}$$

or equivalently

$$J = \det \left(\eta \partial_t \delta(t - t') \delta^{\alpha\beta} + \frac{\partial^2 V(\mathbf{x}_t)}{\partial x_t^\alpha \partial x_t^\beta} \delta(t - t') \right)$$

$$= \det \left(\eta \partial_t \delta(t - t') \delta^{\alpha\beta} \right)$$

$$\times \det \left(\delta(t'' - t') \delta^{\alpha\beta} + \eta^{-1} \int_{\tilde{t}} \partial_t^{-1}(t'', \tilde{t}) \frac{\partial^2 V(\mathbf{x}_t)}{\partial x^\alpha \partial x^\beta} \delta(\tilde{t} - t') \right), \tag{12.60}$$

where the inverse time differential operator is

$$\partial_t^{-1}(t_1, t_2) = \theta(t_1 - t_2). \tag{12.61}$$

Using the trace-log formula, $\ln \det M = \text{Tr} \ln M$, the Jacobian then becomes

$$J = \det \left(\eta \partial_t \delta(t - t') \delta^{\alpha\beta} \right)$$

$$\times \exp \left(\text{Tr} \ln \left(\delta(t'' - t') \delta^{\alpha\beta} + \eta^{-1} \partial_t^{-1}(t'', \tilde{t}) \frac{\partial^2 V(\mathbf{x}_t)}{\partial x^\alpha \partial x^\beta} \delta(\tilde{t} - t') \right) \right)$$

$$= \det \left(\eta \partial_t \delta(t - t') \delta^{\alpha\beta} \right)$$

$$\times \exp \left(-\sum_{n=1}^\infty \frac{1}{n} \text{Tr} (-\eta^{-1} \partial_t^{-1}(t'', \tilde{t}) \frac{\partial^2 V(\mathbf{x}_t)}{\partial x \partial x} \delta(t - t'))^n \right). \tag{12.62}$$

The Jacobian adds a term to the action, and the diagrams generated by the Jacobian are seen to be exactly the tadpole diagrams generated by the original action except for an overall minus sign, and the Jacobian can thus be neglected if we simultaneously omit all tadpole diagrams. This is equivalent to choosing the step function in Eq. (12.62) to be defined according to the prescription

$$\theta(t) = \begin{cases} 0 & t \leq 0 \\ 1 & t > 1 \end{cases} \tag{12.63}$$

since then the first term of the Taylor expansion of the logarithm will be

$$\mathrm{Tr}(\partial^{-1}(t - t'')V''(\mathbf{x}_{t''})\delta(t'' - t')) = \int dt\, \theta(0)V''(\mathbf{x}_t) = 0 \,. \tag{12.64}$$

The higher-order terms in the Taylor expansion are similarly shown to be zero. The result we obtain for the Jacobian for this particular choice of the step function is therefore independent of the disorder potential V

$$J = \det\left(\eta\partial_t(t - t'')\,\delta(t'' - t')\right) = \text{const} \,. \tag{12.65}$$

The derivation of the self-consistent equations can therefore be carried out in the same way as for the case of a nonzero mass when we have chosen this particular definition of the step or Heaviside function. The only difference is that the following form of the free retarded propagator is used:

$$D^{\mathrm{R}}(t, t') = -\frac{1}{\eta}\theta(t - t') \,. \tag{12.66}$$

The equations obtained by setting the mass equal to zero in the previous equations are then exactly the same as the ones obtained for the over-damped case.

12.2 Magnetic properties of type-II superconductors

The advent of high-temperature superconductors has led to a renewed interest in vortex dynamics since high-temperature superconductors have large values of the Ginzburg–Landau parameter and the magnetic field versus temperature (B–T) phase diagram is dominated by the vortex phase.[7] In this section we consider the phenomenology of type-II superconductors, in particular the forces on vortices and their dynamics. Since vortex dynamics in the flux flow regime is Langevin dynamics with quenched disorder, they provide a realization of the model discussed in the previous sections.

12.2.1 Abrikosov vortex state

The essential feature of the magnetic properties of a type-II superconductors is the existence of the Abrikosov flux-line phase [124]. At low magnetic field strengths,

---

[7]The Ginzburg–Landau parameter, $\kappa = \lambda/\xi$, is the ratio between the penetration depth and the superconducting coherence length. The magnetic field penetration depth was first introduced in the phenomenological London equations, $\mu_0 \mathbf{j}_s = \mathbf{E}/\lambda^2$ and $\mu_0 \nabla \times \mathbf{j}_s = -\mathbf{B}/\lambda^2$, the latter the important relation between the magnetic field and a supercurrent describing the Meissner effect of flux expulsion as obtained employing the Maxwell equation to get $\mathbf{B} + \lambda^2 \nabla \times \nabla \times \mathbf{B} = \mathbf{0}$.

just as for a type-I superconductor, a type-II superconductor exhibits the Meissner effect, magnetic flux expulsion. A counter supercurrent on a sample's surface makes a superconductor exhibit perfect diamagnetism, giving it a magnetic moment (which can provide magnetic levitation). Above a critical magnetic field, H_{c1}, the superconducting properties of a type-II superconductor weakens, say for example its magnetic moment on increase of magnetic field, and the superconductor has entered the Shubnikov phase (1937). In this state, magnetic flux will penetrate a type-II superconductor in the form of magnetic flux lines, each carrying a magnetic flux quantum, $\phi_0 = h/(2e)$, with associated vortices of supercurrents. This phase is the Abrikosov lattice flux-line phase, and persists up to an upper critical field, H_{c2}, where superconductivity breaks down, and the superconductor enters the normal state. The supercurrents circling the vortex cores, where the order parameter is depressed and vanishing at the center, screen the magnetic field throughout the bulk of the material. The coupling of magnetic field and current results in a repulsive interaction between vortices which for an isotropic superconductor leads to a stable lattice for the regular triangular array, the Abrikosov flux lattice.

The energetics of two vortices are governed by the magnetic field energy and the kinetic energy of the supercurrent, and as governed by the London equation give a repulsive force, assuming the same sign of vorticity, on each vortex of strength

$$F = \phi_0 \, j_s \,, \tag{12.67}$$

where j_s is the supercurrent density associated with one vortex at the position of the other vortex. In the presence of a transport current, $\mathbf{j}$, through the superconductor the vortices will therefore per unit length be subject to a Lorentz force of magnitude

$$F_L = \phi_0 \, j \,, \tag{12.68}$$

where j is the transport current density, and the direction of the force is specified by $\mathbf{j} \times \mathbf{B}$. Even a small transport current will give rise to motion of the vortex lattice perpendicular to the current in a pure type-II superconductor in the Abrikosov–Shubnikov phase. This motion causes dissipative processes due to the normal currents in the core, which phenomenologically can be described, at low velocities, by a friction force (per unit length) opposing the motion of a vortex with velocity $\mathbf{v}$

$$\mathbf{F}_f = -\eta \, \mathbf{v} \,. \tag{12.69}$$

The friction coefficient is given by[8]

$$\eta = \frac{\phi_0^2}{2\pi a^2 \rho_n} \tag{12.70}$$

where ρ_n is the normal resistance of the metal, and a is the size of the normal core (approximately equal to the superconducting coherence length).

---

[8]For a phenomenological justification of the friction term we refer to the Bardeen–Stephen model [125], or analysis based on the time dependent Ginzburg–Landau equation [126, 127, 128]. As proclaimed, we describe only the phenomenology of the relevant forces, no derivation based on the microscopic theory will be done, instead we refer the reader in general to reference [129].

In addition, there can also be a Hall force

$$\mathbf{F}_H = \alpha \, \mathbf{v} \times \hat{\mathbf{n}} \tag{12.71}$$

acting on the vortex [130].

In a real superconductor there are always imperfections, referred to as impurities, causing the vortices to have energetically preferred positions. The pinning force is caused by defects such as twinning or grain boundaries, or dislocation lines. These can *pin* a vortex, which would otherwise move in the presence of a transport current.[9] At low enough temperatures and below a critical value of the transport supercurrent, the vortex lattice is pinned and the current carrying state dissipationless. At larger currents or higher temperatures, the motion of the vortices occur by thermal excitation of (bundles of) vortices hopping between pinning centers, the state of flux *creep*. In the regime where the pinning force, $\mathbf{F}_p$, is weak compared with the driving force, the motion of the vortex lattice is steady, characterized by a velocity, $\mathbf{v}$, the superconductor is in the dreaded flux flow regime. The moving magnetic field structure associated with the vortices, leads by induction to the presence of an electric field, $\mathbf{E} = -\mathbf{v} \times \mathbf{B}$. The electric field has, as a result of the friction force, a component parallel to the current, and the work, $\mathbf{E} \cdot \mathbf{j}$, performed by the electric field is dissipated by the friction force. The resistance is of the order of the normal state resistance, and the dissipation will drive the superconductor to its normal state.

There is also interaction between the vortices as discussed previously. We shall be interested in the case where the deformation of the Abrikosov lattice is weak, leading to a harmonic interaction between the vortices described by continuum elasticity theory.

12.2.2   Vortex lattice dynamics

We now turn to the case of interest, the dynamics of the Abrikosov vortex lattice in the flux flow regime. The formalism is identical to the previously considered case of one particle, except the occurrence of the whole lattice of vortices with the additional feature of their interaction.

We consider a two-dimensional (2D) description of the vortices, since we have in mind a thin superconducting film, or a three-dimensional (3D) layered superconductor with uncorrelated disorder between the layers. We shall be interested in the influence of quenched disorder on the vortex dynamics in the flux flow regime. The description of the vortex dynamics is, according to the previous section, described by the Langevin equation of the form

$$m\ddot{\mathbf{u}}_{\mathbf{R}t} + \eta\dot{\mathbf{u}}_{\mathbf{R}t} + \sum_{\mathbf{R}'} \Phi_{\mathbf{R}\mathbf{R}'} \mathbf{u}_{\mathbf{R}'t} = \mathbf{F} + \alpha \dot{\mathbf{u}}_{\mathbf{R}t} \times \hat{\mathbf{z}} - \nabla V(\mathbf{R} + \mathbf{u}_{\mathbf{R}t}) + \boldsymbol{\xi}_{\mathbf{R}t} \, , \tag{12.72}$$

where $\mathbf{u}_{\mathbf{R}t}$ is the two-dimensional displacement, normal to $\hat{\mathbf{z}}$, at time t of the vortex (or bundle of vortices), which initially has equilibrium position $\mathbf{R}$, η is the friction

---

[9]The existence of the Abrikosov vortex state and the pinning of vortices is, from the point of applications using superconducting coils as magnets, the most important property. They can produce magnetic fields in the excess of tens of Tesla. Usual copper coils can not produce the stable field produced by the supercurrent, not to mention its mess of water-cooling.

coefficient, and m is a possible mass of the vortex (both per unit length). The mass of a vortex is small and will eventually be set to zero. The interaction between the vortices is treated in the harmonic approximation and described by the dynamic matrix $\Phi_{\mathbf{RR}'}$ whose relevant elasticity moduli is discussed in Section 12.6. The force (per unit length) on the right-hand side of Eq. (12.72) consists of the Lorentz force, $\mathbf{F} = \phi_0 \mathbf{j} \times \hat{\mathbf{z}}$, due to the transport current density $\mathbf{j}$, which we eventually assume constant, and the second term on the right-hand side is a possible Hall force, characterized by the parameter α, and V is the pinning potential due to the quenched disorder. The pinning is described by a Gaussian distributed stochastic potential with zero mean, $\langle V(\mathbf{x}) \rangle = 0$, and thus characterized by its correlation function

$$\nu(\mathbf{x} - \mathbf{x}') = \langle V(\mathbf{x}) V(\mathbf{x}') \rangle . \tag{12.73}$$

The thermal noise, $\boldsymbol{\xi}$, is the white noise stochastic process with zero mean and correlation function specified according to the fluctuation–dissipation theorem (where the brackets now denote averaging with respect to the thermal noise)

$$\langle \xi^{\alpha}_{\mathbf{R}t} \xi^{\alpha'}_{\mathbf{R}'t'} \rangle = 2\eta T \delta(t - t') \, \delta_{\mathbf{RR}'} \, \delta_{\alpha\alpha'} \tag{12.74}$$

and, since the forces are per unit length, the *temperature* T has the dimension of energy per unit length.

Upon averaging with respect to the thermal noise and the quenched disorder, the average restoring force of the lattice vanishes

$$-\sum_{\mathbf{R}'} \Phi_{\mathbf{RR}'} \langle\!\langle \mathbf{u}_{\mathbf{R}'t} \rangle\!\rangle = \mathbf{0} \tag{12.75}$$

since the average displacement is the same for all vortices, and a rigid translation of the vortex lattice does not change its elastic energy, leaving the dynamic matrix with the symmetry property

$$\sum_{\mathbf{R}'} \Phi_{\mathbf{RR}'} = 0 . \tag{12.76}$$

Owing to dissipation, the vortex lattice reaches a steady state velocity $\mathbf{v} = \langle\!\langle \dot{\mathbf{u}}_{\mathbf{R}t} \rangle\!\rangle$, corresponding to the average force on any vortex vanishes

$$\mathbf{F} + \mathbf{F}_{\mathrm{f}} + \mathbf{F}_{\mathrm{H}} + \mathbf{F}_{\mathrm{p}} = \mathbf{0} , \tag{12.77}$$

i.e. there will be a balance between the Lorentz force, $\mathbf{F}$, the average friction force, $\mathbf{F}_{\mathrm{f}} = -\eta \mathbf{v}$, the average Hall force, $\mathbf{F}_{\mathrm{H}} = \alpha \mathbf{v} \times \hat{\mathbf{z}}$, and the pinning force

$$\mathbf{F}_{\mathrm{p}} = -\langle\!\langle \nabla V(\mathbf{R} + \mathbf{u}_{\mathbf{R}t}) \rangle\!\rangle . \tag{12.78}$$

The pinning force is determined by the relative positions of the vortices with respect to the pinning centers and is invariant with respect to the change of the sign of α. The average velocity, $\mathbf{v}$, is the only vector characterizing the vortex motion which is invariant with respect to the change of the sign of α, and the pinning force is therefore antiparallel to the velocity. Thus, the pinning yields a renormalization of the friction coefficient

$$-\eta \mathbf{v} + \mathbf{F}_{\mathrm{p}} = -\eta_{\mathrm{eff}} \, \mathbf{v} . \tag{12.79}$$

The effective friction coefficient depends on the average velocity of the lattice, the disorder, the temperature, the interaction between the vortices, the Hall force, and a possible mass of the vortex. In the absence of disorder, the effective friction coefficient reduces to the bare friction coefficient η.

The pinning problem has no simple analytical solution. One way of attacking the problem is a perturbation calculation in powers of the disorder potential. A second-order perturbation calculation works well for high velocities, as we show in Section 12.5.1.[10] At low enough velocities the higher-order contributions in the disorder become important. We shall employ the self-consistent effective action method of Cornwall *et al.* [53] to sum up an infinite subset of the contributions in V. Such self-consistent methods are uncontrolled but many times they yield surprisingly good results. In order to apply the field theoretic methods of Cornwall *et al.* we need to reformulate the stochastic problem in terms of a generating functional, which is achieved by the field theoretical formulation of classical statistical dynamics.

In the following the influence of pinning on vortex dynamics in type-II superconductors is investigated. The vortex dynamics is described by the Langevin equation, and we shall employ a field-theoretic formulation of the pinning problem which allows the average over the quenched disorder to be performed exactly. By using the diagrammatic functional method for this classical statistical dynamic field theory, we can, from the effective action discussed in the previous chapter, obtain an expression for the pinning force in terms of the Green's function describing the motion of the vortices.

12.3   Field theory of pinning

The average vortex motion is conveniently described by reformulating the stochastic problem in terms of the field theory of classical statistical dynamics introduced in Section 12.1. The probability functional for a realization $\{u_{Rt}\}_R$ of the motion of the vortex lattice is expressed as a functional integral over a set of auxiliary variables $\{\tilde{u}_{Rt}\}_R$, and we are led to consider the generating functional[11]

$$\mathcal{Z}[\mathbf{F},\mathbf{J}] = \int \prod_R \mathcal{D}u_{Rt} \int \prod_R \mathcal{D}\tilde{u}_{Rt}\, \mathcal{J}e^{iS[u,\tilde{u}]}, \qquad (12.80)$$

where in the action

$$S[u,\tilde{u}] = \tilde{u}(D_R^{-1}u + \mathbf{F} - \nabla V + \xi) + \mathbf{J}u \qquad (12.81)$$

the inverse free retarded Green's function is specified by

$$-D_R^{-1}u_{Rt} = m\ddot{u}_{Rt} + \eta\dot{u}_{Rt} + \sum_{R'}\Phi_{RR'}u_{R't} + \alpha\hat{z}\times\dot{u}_{Rt}, \qquad (12.82)$$

---

[10]Vortex pinning in the flux flow regime was originally considered treating the disorder in lowest order perturbation theory [131, 132], and later by applying field theoretical methods [133, 134].
[11]In the following we essentially follow reference [134].

i.e.

$$D_R^{-1}(\mathbf{R}, t; \mathbf{R}', t') = -\Phi_{\mathbf{RR}'}\,\delta(t - t') - \left[(m\partial_t^2 + \eta\partial_t)1 - i\alpha\sigma^y\partial_t\right]\delta_{\mathbf{R},\mathbf{R}'}\,\delta(t - t')\,,$$

$$(12.83)$$

where matrix notation is used for its Cartesian components, i.e. 1 and σ^y denote the unit matrix (occasionally suppressed for convenience) and the Pauli matrix in Cartesian space, respectively. The Fourier transform of the inverse free retarded Green's function is therefore the two by two matrix in Cartesian space given by the expression

$$D_R^{-1}(\mathbf{q}, \omega) = \begin{pmatrix} m\omega^2 + i\eta\omega & -i\alpha\omega \\ i\alpha\omega & m\omega^2 + i\eta\omega \end{pmatrix} - \Phi_{\mathbf{q}}\,.$$

$$(12.84)$$

In Eq. (12.81) we have introduced matrix notation in order to suppress the integrations over time and summations over vortex positions and Cartesian indices. Thus, for example, $\tilde{\mathbf{u}}D_R^{-1}\mathbf{u}$ denotes the expression

$$\tilde{\mathbf{u}}D_R^{-1}\mathbf{u} = \sum_{\substack{\mathbf{R}\mathbf{R}' \\ \alpha,\alpha'=x,y}} \int_{-\infty}^{\infty}dt \int_{-\infty}^{\infty}dt'\,\tilde{u}_\alpha(\mathbf{R}, t)\,D_R^{-1\alpha\alpha'}(\mathbf{R}, t; \mathbf{R}', t')\,u_{\alpha'}(\mathbf{R}', t')\,.$$

$$(12.85)$$

The Jacobian, $\mathcal{J} = |\delta\xi_{\mathbf{R}t}/\delta\tilde{u}_{\mathbf{R}'t'}|$, guaranteeing the normalization of the generating functional

$$Z[\mathbf{F}, \mathbf{J} = 0] = 1$$

$$(12.86)$$

is given by

$$\mathcal{J} \propto \exp\left[-\sum_{\mathbf{R}\alpha\alpha'}\int_{-\infty}^{\infty}dt\,D_{\mathbf{R}t;\mathbf{R}t}^{R\alpha\alpha'}\frac{\partial^2 V(\mathbf{R} + \mathbf{u}_{\mathbf{R}t})}{\partial x_{\alpha'}\partial x_\alpha}\right]\,,$$

$$(12.87)$$

where the proportionality constant is the determinant of the inverse free retarded Green's function, $|(D_R^{-1})_{\mathbf{R}t,\mathbf{R}'t'}^{\alpha\alpha'}|$. As discussed in Section 12.1.6, in the case of a nonzero mass, $m \neq 0$, the Jacobian is an irrelevant constant; and in the case of zero mass, dropping the Jacobian from the integrand is equivalent to defining the retarded free Green's function to vanish at equal times, $D_{tt}^R = 0$, which in turn leads to the full retarded Green's function satisfying the same initial condition. In terms of diagrams, the contribution from the Jacobian exactly cancels the tadpole diagrams as discussed in Section 12.1.6.

The average with respect to both the thermal noise and the disorder is immediately performed, and we obtain the averaged functional, dropping the irrelevant Jacobian,

$$Z[f] = \langle\!\langle \mathcal{Z} \rangle\!\rangle = \int \mathcal{D}\phi\,e^{iS[\phi]+if\phi}\,.$$

$$(12.88)$$

We have employed the compact notation for the fields

$$\phi_{\mathbf{R}t} = (\tilde{\mathbf{u}}_{\mathbf{R}t}, \mathbf{u}_{\mathbf{R}t}) = (\phi_1(\mathbf{R}, t), \phi_2(\mathbf{R}, t))$$

$$(12.89)$$

and for the external force and an introduced source, $\mathbf{J}(\mathbf{R}, t)$,

$$f(\mathbf{R}, t) = (\mathbf{F}(\mathbf{R}, t), \mathbf{J}(\mathbf{R}, t)) . \tag{12.90}$$

The action obtained upon averaging, which we also denote by S, consists of two terms

$$S[\phi] = S_0[\phi] + S_V[\phi] . \tag{12.91}$$

The first term is quadratic in the field

$$S_0[\phi] = \frac{1}{2}\phi D^{-1}\phi , \tag{12.92}$$

where the matrix notation now in addition includes the dynamical indices, i.e. $\phi D^{-1}\phi$ denotes the expression

$$\phi D^{-1}\phi = i \sum_{\substack{\mathbf{RR'} \\ \alpha\alpha'ij}} \int_{-\infty}^{\infty} dt \int_{-\infty}^{\infty} dt' \phi_i^\alpha(\mathbf{R}, t) \, D_{ij}^{-1\,\alpha\alpha'}(\mathbf{R}, t; \mathbf{R'}, t') \, \phi_j^{\alpha'}(\mathbf{R'}, t') . \tag{12.93}$$

The inverse free matrix Green's function in dynamical index space

$$D^{-1} = \begin{pmatrix} D_{11}^{-1} & D_{12}^{-1} \\ D_{21}^{-1} & D_{22}^{-1} \end{pmatrix} = \begin{pmatrix} 2i\eta T\delta(t-t')\,\delta_{\alpha\alpha'}\,\delta_{\mathbf{RR'}} & D_{\mathrm{R}}^{-1}(\mathbf{R}, t; \mathbf{R'}, t') \\ D_{\mathrm{A}}^{-1}(\mathbf{R}, t; \mathbf{R'}, t') & 0 \end{pmatrix} \tag{12.94}$$

is a symmetric matrix in all indices and variables, since the inverse free advanced Green's function is obtained by interchanging Cartesian indices as well as position and time variables

$$D_{\mathrm{A}}^{-1\alpha'\alpha}(\mathbf{R'}, t'; \mathbf{R}, t) = D_{\mathrm{R}}^{-1\alpha\alpha'}(\mathbf{R}, t; \mathbf{R'}, t') . \tag{12.95}$$

The interaction term originating from the disorder is

$$S_V[\phi] = -\frac{i}{2} \sum_{\substack{\mathbf{RR'} \\ \alpha\alpha'}} \int_{-\infty}^{\infty} dt \int_{-\infty}^{\infty} dt' \, \tilde{u}_{\mathbf{R}t}^\alpha \frac{\partial^2 \nu(\mathbf{u}_{\mathbf{R}t} - \mathbf{u}_{\mathbf{R'}t'})}{\partial u_{\mathbf{R}t}^\alpha \partial u_{\mathbf{R}t}^{\alpha'}} \tilde{u}_{\mathbf{R'}t'}^{\alpha'} . \tag{12.96}$$

The source term introduced in Eq. (12.80)

$$\mathbf{Ju} = \sum_{\mathbf{R}} \int_{-\infty}^{\infty} dt \, \mathbf{J}(\mathbf{R}, t) \cdot \mathbf{u}(\mathbf{R}, t) , \tag{12.97}$$

where the source, $\mathbf{J}(\mathbf{R}, t)$, couples to the vortex positions, $\mathbf{u}(\mathbf{R}, t)$, is added to the action in order to generate the vortex correlation functions. For example, we have for the average position

$$\langle\langle \mathbf{u}_{\mathbf{R}t} \rangle\rangle = -i \left. \frac{\delta Z}{\delta \mathbf{J}_{\mathbf{R}t}} \right|_{\mathbf{J}=0} \tag{12.98}$$

and the two-point unconnected Green's function

$$\langle\!\langle \mathbf{u}_{\mathbf{R}t}\, \mathbf{u}_{\mathbf{R}'t'}\rangle\!\rangle = -\left.\frac{\delta^2 Z}{\delta \mathbf{J}_{\mathbf{R}t}\, \delta \mathbf{J}_{\mathbf{R}'t'}}\right|_{\mathbf{J}=0}. \tag{12.99}$$

Here and in the following we use dyadic notation, i.e. $\mathbf{u}_{\mathbf{R}t}\, \mathbf{u}_{\mathbf{R}'t'}$ is the Cartesian matrix with the components $u_\alpha(\mathbf{R},t)\, u_{\alpha'}(\mathbf{R}',t')$.

12.3.1 Effective action

In order to obtain self-consistent equations involving the two-point Green's function in a two-particle irreducible fashion, we add a two-particle source term K to the action in the generating functional (recall Section 10.5.1)

$$Z[f,K] = \int \mathcal{D}\phi\, \exp\left(iS[\phi] + if\phi + \frac{i}{2}\phi K\phi\right). \tag{12.100}$$

The generator of connected Green's functions

$$iW[f,K] = \ln Z[f,K] \tag{12.101}$$

has accordingly derivatives

$$\frac{\delta W}{\delta f_i^\alpha(\mathbf{R},t)} = \overline{\phi}_i^\alpha(\mathbf{R},t) \tag{12.102}$$

and

$$\frac{\delta W}{\delta K_{ii'}^{\alpha\alpha'}(\mathbf{R},t;\mathbf{R}',t')} = \frac{1}{2}\overline{\phi}_i^\alpha(\mathbf{R},t)\, \overline{\phi}_{i'}^{\alpha'}(\mathbf{R}',t') + \frac{i}{2} G_{ii'}^{\alpha\alpha'}(\mathbf{R},t;\mathbf{R}',t'), \tag{12.103}$$

where $\overline{\phi}$ is the average field, with respect to the action $S[\phi] + f\phi + \phi K\phi/2$,

$$\overline{\phi}_i^\alpha(\mathbf{R},t) = \int \mathcal{D}\phi\, \phi_i^\alpha(\mathbf{R},t) \exp\left(iS[\phi] + if\phi + \frac{i}{2}\phi K\phi\right) \tag{12.104}$$

and G is the full connected two-point matrix Green's function

$$G_{ij} = -\frac{\delta^2 W}{\delta f_i\, \delta f_j} = -i\begin{pmatrix} \langle\!\langle \delta\tilde{u}_{\mathbf{R}t}^\alpha\, \delta\tilde{u}_{\mathbf{R}'t'}^{\alpha'}\rangle\!\rangle & \langle\!\langle \delta\tilde{u}_{\mathbf{R}t}^\alpha\, \delta u_{\mathbf{R}'t'}^{\alpha'}\rangle\!\rangle \\ \langle\!\langle \delta u_{\mathbf{R}t}^\alpha\, \delta\tilde{u}_{\mathbf{R}'t'}^{\alpha'}\rangle\!\rangle & \langle\!\langle \delta u_{\mathbf{R}t}^\alpha\, \delta u_{\mathbf{R}'t'}^{\alpha'}\rangle\!\rangle \end{pmatrix}, \tag{12.105}$$

where

$$\delta\mathbf{u}_{\mathbf{R}t} = \mathbf{u}_{\mathbf{R}t} - \langle\!\langle \mathbf{u}_{\mathbf{R}t}\rangle\!\rangle , \qquad \delta\tilde{\mathbf{u}}_{\mathbf{R}t} = \tilde{\mathbf{u}}_{\mathbf{R}t} - \langle\!\langle \tilde{\mathbf{u}}_{\mathbf{R}t}\rangle\!\rangle. \tag{12.106}$$

In the physical problem of interest, the sources K and J vanish, $K = 0$ and $\mathbf{J} = 0$, and the full matrix Green's function has, owing to the normalization of the generating functional

$$Z[\mathbf{F},\mathbf{J}=0,K=0] = 1, \tag{12.107}$$

the structure in the dynamical index space

$$
G_{ij} = -i \begin{pmatrix} 0 & \langle\langle \tilde{u}^{\alpha}_{\mathbf{R}t} u^{\alpha'}_{\mathbf{R}'t'} \rangle\rangle \\ \langle\langle u^{\alpha}_{\mathbf{R}t} \tilde{u}^{\alpha'}_{\mathbf{R}'t'} \rangle\rangle & \langle\langle \delta u^{\alpha}_{\mathbf{R}t} \delta u^{\alpha'}_{\mathbf{R}'t'} \rangle\rangle \end{pmatrix}
$$

$$
= \begin{pmatrix} 0 & G^{A}_{\alpha\alpha'}(\mathbf{R},t;\mathbf{R}',t') \\ G^{R}_{\alpha\alpha'}(\mathbf{R},t;\mathbf{R}',t') & G^{K}_{\alpha\alpha'}(\mathbf{R},t;\mathbf{R}',t') \end{pmatrix} , \tag{12.108}
$$

where we observe that the connected and unconnected retarded (or advanced) Green's functions are equal. Similarly, in the absence of sources the expectation value of the auxiliary field vanishes, and the average field is therefore given by

$$
\bar{\phi}_{\mathbf{R}t} = (\langle\langle \tilde{\mathbf{u}}_{\mathbf{R}t} \rangle\rangle, \langle\langle \mathbf{u}_{\mathbf{R}t} \rangle\rangle) = (\mathbf{0}, \mathbf{v}t) , \tag{12.109}
$$

where $\mathbf{v}$ is the average velocity of the vortex lattice.

The retarded Green's function $G^{R}_{\alpha\alpha'}$ yields the linear response to the force $F_{\alpha'}$, i.e. to linear order in the external force we have

$$
\langle\langle u_{\alpha}(\mathbf{R},t) \rangle\rangle = \sum_{\mathbf{R}'} \int_{-\infty}^{\infty} dt' \, G^{R}_{\alpha\alpha'}(\mathbf{R},t;\mathbf{R}',t') \, F_{\alpha'}(\mathbf{R}',t') , \tag{12.110}
$$

and $G^{K}_{\alpha\alpha'}$ is the correlation function, both matrices in Cartesian indices as indicated. The matrix Green's function in dynamical index space, Eq. (12.108), has only two independent components, since the advanced Green's function is given by

$$
G^{A}_{\alpha\alpha'}(\mathbf{R},t;\mathbf{R}',t') = G^{R}_{\alpha'\alpha}(\mathbf{R}',t';\mathbf{R},t) . \tag{12.111}
$$

Pursuing an equation for the pinning force, we introduce the effective action, Γ, the generator of two-particle irreducible vertex functions, i.e. the double Legendre transform of the generator of connected Green's functions, W (recall Section 10.5.1),

$$
\Gamma[\bar{\phi}, G] = W[f, K] - f\bar{\phi} - \frac{1}{2}\bar{\phi}K\bar{\phi} - \frac{i}{2}\mathrm{Tr}GK , \tag{12.112}
$$

where Tr denotes the trace over all variables and indices, i.e. $\mathrm{Tr}GK$ denotes the expression

$$
\mathrm{Tr}GK = \sum_{\substack{\mathbf{R},\mathbf{R}' \\ \alpha,\alpha'=x,y \\ i,i'=1,2}} \int_{-\infty}^{\infty}\int_{-\infty}^{\infty} dt\, dt' \, G^{\alpha\alpha'}_{ii'}(\mathbf{R},t;\mathbf{R}',t') K^{\alpha'\alpha}_{i'i}(\mathbf{R}',t';\mathbf{R},t) . \tag{12.113}
$$

The effective action satisfies the equations

$$
\frac{\delta\Gamma}{\delta\bar{\phi}} = -f - K\bar{\phi} \tag{12.114}
$$

and

$$\frac{\delta \Gamma}{\delta G} = -\frac{i}{2} K \,. \tag{12.115}$$

The effective action was shown in Section 10.5.1 to have the form

$$\begin{aligned}
\Gamma[\bar{\phi}, G] &= S[\bar{\phi}] + \frac{i}{2} \mathrm{Tr} D_S^{-1} G - \frac{i}{2} \mathrm{Tr} \ln D^{-1} G - \frac{i}{2} \mathrm{Tr} 1 \\
&\quad - i \ln \langle e^{i S_{\mathrm{int}}[\bar{\phi}, \psi]} \rangle_G^{\mathrm{2PI}} \,,
\end{aligned} \tag{12.116}$$

where the quantity D_S^{-1} is the second derivative of the action at the average field

$$D_S^{-1}[\bar{\phi}](t, t') = \frac{\delta^2 S[\bar{\phi}]}{\delta \bar{\phi}_t \, \delta \bar{\phi}_{t'}} \tag{12.117}$$

and $S_{\mathrm{int}}[\bar{\phi}, \psi]$ is the part of the action $S[\bar{\phi} + \psi]$ that is higher than second order in ψ in an expansion around the average field. The superscript "2PI" on the last term indicates that only the two-particle irreducible vacuum diagrams should be included in the interaction part of the effective action, the last term in Eq. (12.116), and the subscript that propagator lines represent G, i.e. the brackets with subscript G denote the average

$$\langle e^{i S_{\mathrm{int}}[\bar{\phi}, \psi]} \rangle_G = (\det iG)^{-1/2} \int \mathcal{D}\psi \, e^{\frac{i}{2} \psi G^{-1} \psi} \, e^{i S_{\mathrm{int}}[\bar{\phi}, \psi]} \,. \tag{12.118}$$

The first dynamical index component of Eq. (12.114) together with the equation for the average motion Eq. (12.77) provide an expression for the pinning force, Eq. (12.78), in term of the dynamical matrix propagator of the theory. The general expression is still intractable, and in the next section we shall introduce the main approximation.

12.4 Self-consistent theory of vortex dynamics

Because of the disorder, the equation of motion describing the vortex dynamics has no simple analytical solution. The employed field theoretical formulation of the pinning problem will therefore be used in combination with a self-consistent approximation for the effective action for studying vortex motion in type-II superconductors. Since we have constructed the two-particle irreducible effective action, we expect that its lowest-order approximation contains the main influence of the quenched disorder on the vortex dynamics. The validity of the self-consistent theory is ascertained by comparing with numerical simulations of the Langevin equation. The effective action method will be used to study the dynamics of single vortices and vortex lattices, and yields results for the pinning force, fluctuations in position and velocity, etc. The dependence of the pinning force on vortex velocity, temperature and disorder strength is calculated for independent vortices as well as for a vortex lattice, and both analytical and numerical results for the pinning of vortices in the flux flow regime are obtained. Finally, the influence of pinning on the dynamic melting of a vortex lattice is studied in Section 12.7.

12.4.1   Hartree approximation

In order to obtain a closed expression for the self-energy in terms of the two-point Green's function, we expand the exponential and keep only the lowest-order term

$$-i\ln\langle e^{iS_{\text{int}}[\bar{\phi},\psi]}\rangle_G^{\text{2PI}} \simeq -i\ln\langle 1+iS_{\text{int}}[\bar{\phi},\psi]\rangle_G^{\text{2PI}} \simeq \langle S_{\text{int}}[\bar{\phi},\psi]\rangle_G , \qquad (12.119)$$

i.e. we consider the Hartree approximation, which in diagrammatic terms corresponds to neglecting diagrams where different impurity correlators are connected by Green's functions.

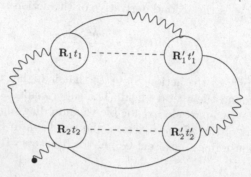

Figure 12.2 Typical vacuum diagram not included in the Hartree approximation for the effective action. The solid line represents the correlation function or kinetic component, G^{K}, of the matrix Green's function. The retarded Green's function, G^{R}, is depicted as a wiggly line ending up in a straight line, and vice versa for the advanced Green's function G^{A}. The curly line ending up on the dot represents the first kinetic component of the average field. A dashed line attached to circles represents the impurity correlator and the additional dependence on the second component of the average field as explicitly specified in Eq. (12.120).

A typical vacuum diagram not included in the Hartree approximation for the effective action is shown in Figure 12.2, and represents the expression

$$\left(\frac{i}{2}\right)^2\left(\frac{1}{4!}\right)^2\int\frac{d\mathbf{k}_1}{(2\pi)^2}\frac{d\mathbf{k}_2}{(2\pi)^2}\,\mathbf{k}_2\cdot\bar{\phi}_1(\mathbf{R}_2,t_2)$$

$$\times(\mathbf{k}_2 G^{\text{R}}(\mathbf{R}_2,t_2;\mathbf{R}_1,t_1)\mathbf{k}_1)(\mathbf{k}_1 G^{\text{R}}(\mathbf{R}_1,t_1;\mathbf{R}_1',t_1')\mathbf{k}_1)$$

$$\times(\mathbf{k}_1 G^{\text{R}}(\mathbf{R}_1',t_1';\mathbf{R}_2',t_2')\mathbf{k}_2)(\mathbf{k}_2 G^{\text{K}}(\mathbf{R}_2,t_2;\mathbf{R}_2',t_2')\mathbf{k}_2)$$

$$\times\nu(\mathbf{k}_1)e^{i\mathbf{k}_1\cdot(\mathbf{R}_1-\mathbf{R}_1'+\mathbf{v}(t_1-t_1'))}\nu(\mathbf{k}_2)e^{i\mathbf{k}_2\cdot(\mathbf{R}_2-\mathbf{R}_2'+\mathbf{v}(t_2-t_2'))} , \qquad (12.120)$$

where integrations over time and summations over vortex positions are implied, and we have introduced the notation

$$\mathbf{k}G^{\text{R}}(\mathbf{R},t;\mathbf{R}',t')\mathbf{k}' = \sum_{\alpha\alpha'} k_\alpha G_{\alpha\alpha'}^{\text{R}}(\mathbf{R},t;\mathbf{R}',t')\,k_{\alpha'}' \qquad (12.121)$$

for Cartesian scalars.

In the Hartree approximation, Eq. (12.119), we drop the superscript "2PI" since the action $S_{\text{int}}[\bar{\phi}, \psi]$ only generates two-particle-irreducible vacuum diagrams, due to the appearance of only one impurity correlator. The Hartree approximation can be expressed as a Gaussian fluctuation corrected saddle-point approximation [135].

The effective action can in the Hartree approximation be rewritten on the form

$$\Gamma[\bar{\phi}, G] = S_0[\bar{\phi}] - \frac{i}{2}\text{Tr}\ln D^{-1}G + \frac{i}{2}\text{Tr}D^{-1}G - \frac{i}{2}\text{Tr}1 + \langle S_V[\bar{\phi} + \psi]\rangle_G \quad (12.122)$$

since

$$\langle S_{\text{int}}[\bar{\phi}, \psi]\rangle_G = \langle S_V[\bar{\phi} + \psi]\rangle_G - S_V[\bar{\phi}] - \frac{i}{2}\text{Tr}\int_{-\infty}^{\infty}dt\int_{-\infty}^{\infty}dt'\,\frac{\delta^2 S_V[\bar{\phi}]}{\delta\bar{\phi}_t\,\delta\bar{\phi}_{t'}}\,G_{t't}\,, \quad (12.123)$$

where the trace in the time variable has been written explicitly for clarity.

In the physical situation of interest the two-particle source, K, vanishes, and since Γ is two-particle-irreducible, Eq. (12.115) therefore becomes the Dyson equation

$$G^{-1} = D^{-1} - \Sigma\,, \quad (12.124)$$

where the self-energy in the Hartree approximation is the matrix in dynamical index space

$$\Sigma_{ij} = \begin{pmatrix} \Sigma^K & \Sigma^R \\ \Sigma^A & 0 \end{pmatrix} = 2i\,\frac{\delta\langle S_V[\bar{\phi} + \psi]\rangle_G}{\delta G_{ij}}\bigg|_{K=0,\,\mathbf{J}=0}. \quad (12.125)$$

The Dyson equation, Eq. (12.124), and the self-energy expression, Eq. (12.125), and the equation relating the effective action to the external force, Eq. (12.114), constitute a set of self-consistent equations for the Green's functions, the self-energies, and the average field, in this non-equilibrium theory the latter specifies the velocity of the vortex lattice.

The matrix self-energy in dynamical index space has only two independent components since

$$\Sigma^A_{\alpha\alpha'}(\mathbf{R}, t; \mathbf{R}', t') = \Sigma^R_{\alpha'\alpha}(\mathbf{R}', t'; \mathbf{R}, t)\,, \quad (12.126)$$

a simple consequence of Eq. (12.111) and the Dyson equation. From Eq. (12.125) we obtain for a vortex lattice having a unit cell of area a^2 and consisting of N vortices, the self-energy components (each a matrix in Cartesian space)

$$\Sigma^K(\mathbf{R}, t; \mathbf{R}', t') = -\frac{i}{Na^2}\sum_{\mathbf{k}}\nu(\mathbf{k})\,\mathbf{k}\mathbf{k}\,e^{-\bar{\varphi}(\mathbf{R},t;\mathbf{R}',t';\mathbf{k};\mathbf{v})} \quad (12.127)$$

and

$$\Sigma^R(\mathbf{R}, t; \mathbf{R}', t') = \sigma^R(\mathbf{R}, t; \mathbf{R}', t') - \delta_{\mathbf{R}\mathbf{R}'}\delta(t - t')\sum_{\tilde{\mathbf{R}}}\int_{-\infty}^{\infty}d\tilde{t}\,\sigma^R(\mathbf{R}, t; \tilde{\mathbf{R}}, \tilde{t})\,, \quad (12.128)$$

where

$$\sigma^{\mathrm{R}}(\mathbf{R}, t; \mathbf{R}', t') = \frac{1}{Na^2} \sum_{\mathbf{k}} \nu(\mathbf{k}) \, \mathbf{k}\mathbf{k} \, (\mathbf{k}G^{\mathrm{R}}(\mathbf{R}, t; \mathbf{R}', t')\mathbf{k}) = e^{-\tilde{\varphi}(\mathbf{R}, t; \mathbf{R}', t'; \mathbf{k}; \mathbf{v})} \; .$$

$$(12.129)$$

We use dyadic notation, i.e. $\mathbf{k}\mathbf{k}$ denotes the matrix with the Cartesian components $k_\alpha k_{\alpha'}$. The influence of thermal and disorder-induced fluctuations are described by the fluctuation or damping exponent

$$\varphi_{\mathbf{k}}(\mathbf{R}, t; \mathbf{R}', t') = i\mathbf{k}(G^{\mathrm{K}}(\mathbf{R}, t; \mathbf{R}, t) - G^{\mathrm{K}}(\mathbf{R}, t; \mathbf{R}', t'))\mathbf{k} \qquad (12.130)$$

contained in

$$\tilde{\varphi}(\mathbf{R}, t; \mathbf{R}', t'; \mathbf{k}; \mathbf{v}) = -i\mathbf{k} \cdot (\mathbf{R} - \mathbf{R}' + \mathbf{v}(t - t')) + \varphi_{\mathbf{k}}(\mathbf{R}, t; \mathbf{R}', t') \; . \quad (12.131)$$

The pinning force on a vortex, Eq. (12.78), is determined by the averaged equation of motion, Eq. (12.77), and the first dynamical index component of Eq. (12.114), which in the Hartree approximation yields

$$-\sum_{\mathbf{R}'} \sum_{\alpha'} \int_{-\infty}^{\infty} dt' \; D_R^{-1\alpha\alpha'}(\mathbf{R}, t; \mathbf{R}', t') \, v_{\alpha'} \, t' = F_{\mathbf{R}}^\alpha + \left. \frac{\delta \langle S_V[\overline{\phi} + \psi] \rangle_G}{\delta \overline{\phi}_1^\alpha(\mathbf{R}, t)} \right|_{\overline{\phi}_{\mathbf{R}t} = (0, \mathbf{v}t)} \quad (12.132)$$

resulting in the expression for the pinning force

$$\mathbf{F_p} = i \sum_{\mathbf{R}'} \int_{-\infty}^{\infty} dt' \int \frac{d\mathbf{k}}{(2\pi)^2} \, \mathbf{k} \, \nu(\mathbf{k}) (\mathbf{k}G_{\mathbf{R}t\mathbf{R}'t'}^R \mathbf{k}) e^{-\tilde{\varphi}(\mathbf{R}, t; \mathbf{R}', t'; \mathbf{k}; \mathbf{v})} \; . \quad (12.133)$$

The self-consistent theory in the Hartree approximation is still intractable to analytical treatment, except in the limiting cases considered in the following, but it is manageable numerically.[12] In the following we shall study numerically the vortex dynamics in the Hartree approximation. The results obtained from the self-consistent theory will then be compared with analytical results obtained in perturbation theory, and with simulations of the vortex dynamics.

12.5   Single vortex

In order to study the essential features of the model and the self-consistent method, we first consider the case of a single vortex, since this example will allow the important test of comparing the results of the self-consistent theory with simulations. The case of non-interacting vortices is appropriate for low magnetic fields, where the vortices are so widely separated that the interaction between them can be neglected. The dynamics of a single vortex is described by the Langevin equation

$$m\ddot{\mathbf{x}}_t + \eta\dot{\mathbf{x}}_t = -\nabla V(\mathbf{x}_t) + \mathbf{F}_t + \boldsymbol{\xi}_t \; , \qquad (12.134)$$

---

[12]In the rest of this chapter we follow reference [134].

where $\mathbf{x}_t$ is the vortex position at time t. We defer the discussion of the Hall force to Section 12.5.5.

When presenting analytical and numerical results obtained from the self-consistent theory, we shall always choose the vortex mass (per unit length) to be small, in fact so small, $m \ll \eta^2 r_{\mathrm{p}}^3/\sqrt{\nu_0}$, that the case of zero mass only deviates slightly from the presented results, i.e. at most a few percent.

In the analytical and numerical calculations, the correlator of the pinning potential shall be taken as the Gaussian function with range r_{p} and strength ν_0

$$\nu(\mathbf{x} - \mathbf{x}') = \frac{\nu_0}{2\pi r_{\mathrm{p}}^2} e^{-(\mathbf{x}-\mathbf{x}')^2/2r_{\mathrm{p}}^2} \quad , \quad \nu(\mathbf{k}) = \nu_0 e^{-r_{\mathrm{p}}^2 k^2}. \tag{12.135}$$

12.5.1 Perturbation theory

At high velocities, the pinning force can be obtained from lowest-order perturbation theory in the disorder, since the pinning force then is small compared with the friction force, and makes, according to Eq. (12.77), only a small contribution to the total force on the vortex. We first consider the case of zero temperature, where we obtain the following set of equations by collecting terms of equal powers in the pinning potential

$$-\int_{-\infty}^{\infty} dt' \, D_{\mathrm{R}}^{-1}(t,t') \, \mathbf{x}_{t'}^{(0)} = \mathbf{F}_t \tag{12.136}$$

and

$$-\int_{-\infty}^{\infty} dt' \, D_{\mathrm{R}}^{-1}(t,t') \, \mathbf{x}_{t'}^{(1)} = -\nabla V(\mathbf{x}_t^{(0)}) \tag{12.137}$$

and

$$-\int_{-\infty}^{\infty} dt' \, D_{\mathrm{R}}^{-1}(t,t') \, \mathbf{x}_{t'}^{(2)} = -\nabla \left(\mathbf{x}_t^{(1)} \cdot \nabla V(\mathbf{x}_t^{(0)}) \right). \tag{12.138}$$

Assuming that the external force is independent of time, the average vortex velocity will be constant in time, and in the absence of disorder the average vortex position is

$$\langle\!\langle \mathbf{x}_t^{(0)} \rangle\!\rangle = \mathbf{v}t = \frac{\mathbf{F}t}{\eta}, \tag{12.139}$$

i.e. the friction force balances the external force, $\eta\mathbf{v} = \mathbf{F}$. The first-order contribution to the vortex position vanishes upon averaging with respect to the pinning potential, and the second-order contribution to the average vortex velocity becomes, according to Eqs. (12.136) to (12.138),

$$\langle\!\langle \dot{\mathbf{x}}_t^{(2)} \rangle\!\rangle = -\frac{i}{\eta} \int_{-\infty}^{\infty} dt' \, D_{tt'}^{\mathrm{R}} \int \frac{d\mathbf{k}}{(2\pi)^2} \, \mathbf{k} \, k^2 \nu_0 \, e^{-k^2 r_{\mathrm{p}}^2 + i\mathbf{k}\cdot\mathbf{v}(t-t')}$$

$$= \frac{\nu_0}{4\pi r_{\mathrm{p}}^5 \eta} \int_0^{\infty} dt \, D_{t0}^{\mathrm{R}} \left(\frac{vt}{r_{\mathrm{p}}} - \left(\frac{vt}{2r_{\mathrm{p}}} \right)^2 \right) e^{-\left(\frac{vt}{2r_{\mathrm{p}}} \right)^2}. \tag{12.140}$$

The second-order contribution is immediately calculated, and for example for the case of a vanishing mass, $m \ll \eta^2 r_{\mathrm{p}}^3 / \sqrt{\nu_0}$, we obtain

$$\langle\!\langle \dot{\mathbf{x}}_t^{(2)} \rangle\!\rangle = -\frac{\nu_0}{4\pi r_{\mathrm{p}}^4 \eta^2 v^2} \, \mathbf{v} \,. \tag{12.141}$$

The pinning force is then, according to Eq. (12.77), to lowest order in the disorder strength, ν_0, given by

$$\mathbf{F}_{\mathrm{p}} = -\frac{\nu_0}{4\pi r_{\mathrm{p}}^4 \eta v^2} \, \mathbf{v} \,, \tag{12.142}$$

i.e. the magnitude of the pinning force is inversely proportional to the magnitude of the velocity. The perturbation result is therefore valid for large velocities, $v \gg \sqrt{\nu_0}/\eta r_{\mathrm{p}}^2$, i.e. when the friction force is much larger than the average force, $\sqrt{\nu_0}/r_{\mathrm{p}}^2$, owing to the disorder.

12.5.2   Self-consistent theory

The self-energy equations for a single vortex reduces in the Hartree approximation to

$$\Sigma^{\mathrm{R}}(t,t') = \int \frac{d\mathbf{k}}{(2\pi)^2} \left[\sigma_{\mathbf{k}}^{\mathrm{R}}(t,t') - \delta(t-t') \int_{-\infty}^{\infty} d\bar{t} \, \sigma_{\mathbf{k}}^{\mathrm{R}}(t,\bar{t}) \right] \tag{12.143}$$

and

$$\sigma_{\mathbf{k}}^{\mathrm{R}}(t,t') = \nu(\mathbf{k}) \, \mathbf{k}\mathbf{k} \, (\mathbf{k} G^{\mathrm{R}}(t,t')\mathbf{k}) \, e^{i\mathbf{k}\cdot\mathbf{v}(t-t') \, - \, \varphi_{\mathbf{k}}(t,t')} \tag{12.144}$$

and

$$\Sigma^{\mathrm{K}}(t,t') = -i \int \frac{d\mathbf{k}}{(2\pi)^2} \, \nu(\mathbf{k}) \, \mathbf{k}\mathbf{k} \, e^{i\mathbf{k}\cdot\mathbf{v}(t-t') \, - \, \varphi_{\mathbf{k}}(t,t')} \tag{12.145}$$

with the fluctuation exponent

$$\varphi_{\mathbf{k}}(t,t') = i\mathbf{k}\Big(G^{\mathrm{K}}(t,t) - G^{\mathrm{K}}(t,t') \Big)\mathbf{k} \,. \tag{12.146}$$

Writing out the components of the matrix Dyson equation in the dynamical indices, Eq. (12.124), we obtain the Cartesian matrix Green's functions

$$G^{\mathrm{K}}(\omega) = G^{\mathrm{R}}(\omega)\Big(\Sigma^{\mathrm{K}}(\omega) - 2i\eta T \mathbf{1}\Big) G^{\mathrm{A}}(\omega) \tag{12.147}$$

and

$$G^{\mathrm{R}}(\omega) = \frac{\hat{\mathbf{v}}\hat{\mathbf{v}}}{m\omega^2 + i\eta\omega - \Sigma_{\|}^{\mathrm{R}}(\omega)} + \frac{1 - \hat{\mathbf{v}}\hat{\mathbf{v}}}{m\omega^2 + i\eta\omega - \Sigma_{\perp}^{\mathrm{R}}(\omega)} \,, \tag{12.148}$$

where the subscripts $\|$ and $\perp$ denote longitudinal and transverse components of the retarded self-energy with respect to the direction of the velocity

$$\Sigma_{\|}^{\mathrm{R}}(\omega) = \sum_{\alpha,\alpha'} \hat{v}_\alpha \, \Sigma_{\alpha\alpha'}^{\mathrm{R}}(\omega) \, \hat{v}_{\alpha'} \tag{12.149}$$

and

$$\Sigma_{\perp}^{R}(\omega) = \sum_{\alpha,\alpha'} \Sigma_{\alpha\alpha'}^{R}(\omega) (\delta_{\alpha\alpha'} - \hat{v}_{\alpha} \hat{v}_{\alpha'}) . \tag{12.150}$$

The advanced Green's function is obtained from the retarded by complex conjugation and interchange of Cartesian indices

$$G_{\alpha\alpha'}^{A}(\omega) = [G_{\alpha'\alpha}^{R}(\omega)]^{*} . \tag{12.151}$$

The expression for the pinning force, Eq. (12.133), reduces for a single vortex to

$$\mathbf{F}_{\mathrm{p}} = i \int_{-\infty}^{\infty} dt' \int \frac{d\mathbf{k}}{(2\pi)^2} \mathbf{k}\, \nu(\mathbf{k}) \, (\mathbf{k}\, G_{tt'}^{R}\, \mathbf{k})\, e^{i\mathbf{k}\cdot\mathbf{v}(t-t') - \varphi_{\mathbf{k}}(t,t')} . \tag{12.152}$$

The previous discussion of the high-velocity regime, where lowest-order perturbation theory in the disorder is valid, can be generalized to nonzero temperature. At high velocities, $v \gg \sqrt{\nu_0}/\eta r_{\mathrm{p}}^2$, the self-energies are, according to Eqs. (12.143)–(12.145), inversely proportional to the velocity, and they can accordingly be neglected in the calculation of the pinning force. We can therefore in this limit insert the free retarded Green's functions in the self-consistent expression for the pinning force, Eq. (12.152), thereby obtaining an expression valid to lowest order in the disorder strength, ν_0,

$$\mathbf{F}_{\mathrm{p}} = -\frac{i}{\eta} \int \frac{d\mathbf{k}}{(2\pi)^2} \mathbf{k}\, k^2\, \nu_0\, e^{-r_{\mathrm{p}}^2 k^2} \int_{0}^{\infty} dt\, e^{i\mathbf{k}\cdot\mathbf{v}t - k^2 T t/\eta} , \tag{12.153}$$

where again we only display the result for vanishing mass, $m \ll \eta^2 r_{\mathrm{p}}^3/\sqrt{\nu_0}$. The integration over time can then be performed, and we obtain the result that the pinning force for large velocities, $v \gg T/(r_{\mathrm{p}}\eta)$, is given by the perturbation theory expression, Eq. (12.142).

It is also possible to obtain an analytical expression for the pinning force at moderate velocities, provided the temperature is high enough. At high temperatures, $T \gg \sqrt{\nu_0}/r_{\mathrm{p}}$, the kinetic component of the self-energy is inversely proportional to the temperature, $\Sigma^{K}(\omega = v/r_{\mathrm{p}}) \sim \nu_0 \eta/(r_{\mathrm{p}}^2 T)$, and its contribution to the fluctuation exponent is much smaller than the contribution from the thermal fluctuations. Similarly, at temperatures $T \gg \sqrt{\nu_0}/(\eta r_{\mathrm{p}}^3 v)$, the retarded self-energy is of order $\Sigma^{R}(\omega = v/r_{\mathrm{p}}) \sim \nu_0/(r_{\mathrm{p}}^4 T)$. At moderate velocities, $v \lesssim \sqrt{\nu_0}/(\eta r_{\mathrm{p}}^2)$, the free retarded Green's function can therefore be inserted in the expression for the pinning force, and we can expand the exponential $\exp\{i\mathbf{k}\cdot\mathbf{v}t\}$, and keep only the lowest-order term in the velocity, since the inequality $v \ll T/(\eta r_{\mathrm{p}})$ is satisfied, and obtain the result that the pinning force is proportional to the velocity and inversely proportional to the square of the temperature

$$\mathbf{F}_{\mathrm{p}} = -\frac{\nu_0 \eta}{8\pi r_{\mathrm{p}}^2 T^2} \mathbf{v} . \tag{12.154}$$

Thus, when the thermal energy exceeds the average disorder barrier height, $\sqrt{\nu_0}/r_{\mathrm{p}}$, the pinning force is very small compared with the friction force, and pinning just leads to a slight renormalization of the bare friction coefficient. In this high-temperature limit, which can be realized in high-temperature superconductors, we observe that the self-consistent theory, at not too high velocities, yields a pinning force that has a linear velocity dependence, in contrast to the case of low temperatures where we obtain from the self-consistent theory, as apparent from for example Figure 12.3, the fact that the velocity dependence of the pinning force is sub-linear.

12.5.3   Simulations

In order to ascertain the validity of the self-consistent theory beyond the high-velocity regime, where perturbation theory is valid, we perform numerical simulations of the Langevin equation, Eq. (12.134). The pinning force is obtained from Eq. (12.77), once the simulation result for the average velocity as a function of the external force is determined. We simulate the two-dimensional motion of a vortex in a region of linear size $L = 20r_p$, and use periodic boundary conditions. The disorder is generated on a grid consisting of 1024×1024 points.

The disorder correlator is diagonal in the wave vectors, since averaged quantities are translationally invariant,

$$\langle V(\mathbf{k})V(\mathbf{k}')\rangle = \nu(\mathbf{k})L^2\delta_{\mathbf{k}+\mathbf{k}'=0} \tag{12.155}$$

and the real and imaginary parts of the disorder potential can be generated independently according to

$$\Re e\, V(\mathbf{k}) = \frac{\sqrt{\nu_0}L}{\sqrt{2}}e^{-r_p^2 k^2/2}\sigma \quad , \quad \Im m\, V(\mathbf{k}) = \frac{\sqrt{\nu_0}L}{\sqrt{2}}e^{-r_p^2 k^2/2}\delta\, , \tag{12.156}$$

where σ and δ are normally distributed stochastic variables with zero mean and unit standard deviation. The gradient of the disorder potential at the grid points is obtained by employing the finite difference scheme. The potential gradient at the vortex position is then obtained by interpolation of the values of the potential at the four nearest grid points.

The simulations show that the vortex follows a fairly narrow channel through the potential landscape. In the absence of the Hall force, the vortex will traverse only a very limited region of the generated potential owing to the imposed periodic boundary condition. To make better use of the generated potential, we therefore randomize the vortex position at equidistant moments in time, and run the simulation for a short time without measuring the velocity, in order for the velocity to relax, before again starting to measure the velocity. In this way the number of generated potentials can be kept at a minimum of twenty.

12.5.4   Numerical results

For any given average velocity of the lattice, the coupled equations of Green's functions and self-energies may be solved numerically by iteration. We start the iteration procedure by first calculating the Green's functions for vanishing self-energies, corresponding to the absence of disorder, and the self-energies are then calculated from Eqs. (12.143)–(12.145). The procedure is then iterated until convergence is reached. The pinning force on a single vortex can then be evaluated numerically from Eq. (12.152).

In the numerical calculations we shall always assume that the correlator of the pinning potential is the Gaussian function, Eq. (12.135), with range r_p and strength ν_0. In order to simplify the numerical calculation, the self-consistent equations for the self-energies and the Green's functions, Eq. (12.147) and Eq. (12.148), are brought on dimensionless form by introducing the following units for length, time and mass, r_p, $\eta r_p^3/\nu_0^{1/2}$, $\eta^2 r_p^4/\nu_0^{1/2}$.

We have solved the set of self-consistent equations numerically by iteration. In Figure 12.3, the pinning force as a function of velocity is shown for different values of the temperature.

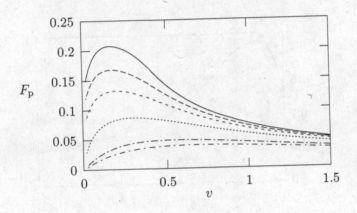

Figure 12.3 Pinning force (in units of $\nu_0^{1/2} r_p^{-2}$) on a single vortex as a function of velocity (in units of $\eta^{-1} r_p^{-2} \nu_0^{1/2}$) obtained from the self-consistent theory. The curves correspond to the different temperatures $T = 0.005, 0.05, 0.1, 0.2, 0.4, 0.5$ (in units of $\nu_0^{1/2}/r_p$), where the uppermost curve corresponds to $T = 0.005$, and $m = 0.1\eta^2 r_p^3 \nu_0^{-1/2}$.

We find that the pinning force has a non-monotonic dependence as a function of velocity, and that the peak in the pinning force decreases rapidly with increasing temperature, and develops into a plateau once the thermal energy is of the order of the average barrier height. At the highest temperature, the velocity dependence of the pinning force is seen in Figure 12.3 to approach the linear regime at low velocities in accordance with the analytical result obtained in the high temperature limit, Eq. (12.154). At high velocities, the pinning force is independent of the temperature as apparent from Figure 12.3.

In fact, the pinning force is inversely proportional to the velocity at high velocities in agreement with the perturbation theory result, Eq. (12.142), as apparent from Figure 12.4, where a comparison is made between the pinning force obtained from lowest-order perturbation theory and the numerically evaluated self-consistent result. The two results agree as expected in the large velocity regime, whereas the perturbation theory result has an unphysical divergence at low velocities due to the neglect of fluctuations, and a consequent absence of damping by the fluctuation exponent in Eq. (12.152).

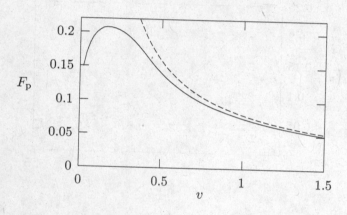

Figure 12.4 Pinning force (in units of $\nu_0^{1/2} r_p^{-2}$) on a single vortex as a function of velocity (in units of $\eta^{-1} r_p^{-2} \nu_0^{1/2}$). The solid line represents the result obtained from the self-consistent theory, while the dashed line represents the result of lowest-order perturbation theory in the disorder ($T = 0.005 \nu_0^{1/2} r_p^{-1}$ and $m = 0.1 \eta^2 r_p^3 \nu_0^{-1/2}$).

In order to check the validity of the self-consistent theory beyond lowest-order perturbation theory, we have performed numerical simulations. In Figure 12.5, a comparison between the self-consistent theory and a numerical simulation of the pinning force as a function of velocity is presented. The agreement between the self-consistent theory and the simulation is good, except around the maximum value of the pinning force, where the simulation is found to yield a higher pinning force in comparison to the self-consistent theory. In this region the relative velocity fluctuations are large, and in fact the self-consistent theory predicts that the relative velocity fluctuations are diverging at zero velocity even at zero temperature, as we discuss shortly. The self-consistent equations and their numerical solution, as well as the simulations, can therefore be expected to be less accurate at low velocities.

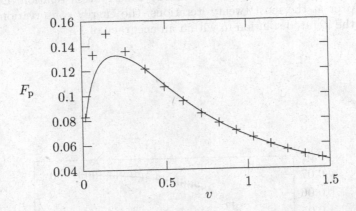

Figure 12.5 Comparison of the pinning force (in units of $\nu_0^{1/2} r_p^{-2}$) on a single vortex as a function of velocity (in units of $\eta^{-1} r_p^{-2} \nu_0^{1/2}$) obtained from the self-consistent theory, solid line, and the numerical simulation, plus signs ($T = 0.1\nu_0^{1/2} r_p^{-1}$ and $m = 0.1\eta^2 r_p^3 \nu_0^{-1/2}$).

The convergence of the iterative procedure can be monitored by checking that energy conservation is fulfilled. The energy conservation relation is obtained by multiplying the Langevin equation by the velocity of the vortex and averaging over the thermal noise and the quenched disorder

$$m\langle\!\langle \dot{\mathbf{x}}_t \cdot \ddot{\mathbf{x}}_t \rangle\!\rangle + \eta\langle\!\langle \dot{\mathbf{x}}_t^2 \rangle\!\rangle = -\langle\!\langle \dot{\mathbf{x}}_t \cdot \nabla V(\mathbf{x}_t) \rangle\!\rangle + \mathbf{F} \cdot \mathbf{v} + \langle\!\langle \dot{\mathbf{x}}_t \cdot \boldsymbol{\xi}_t \rangle\!\rangle . \tag{12.157}$$

The first term is proportional to $\partial_t \langle\!\langle \dot{\mathbf{x}}_t^2 \rangle\!\rangle$, and vanishes since averaged quantities are independent of time, as the external force is assumed to be independent of time. The first term on the right-hand side, the term originating from the disorder, vanishes for the same reason, since it can be rewritten as $-\partial_t \langle\!\langle V(\mathbf{x}_t) \rangle\!\rangle$. The energy conservation relation therefore becomes, $\mathbf{v} = \langle\!\langle \dot{\mathbf{x}}_t \rangle\!\rangle$,

$$\eta\langle\!\langle (\dot{\mathbf{x}}_t - \mathbf{v})^2 \rangle\!\rangle - \langle\!\langle \dot{\mathbf{x}}_t \cdot \boldsymbol{\xi}_t \rangle\!\rangle = -\mathbf{v} \cdot \mathbf{F}_p \tag{12.158}$$

or, in terms of the Green's functions,

$$-i\eta\partial_t^2\mathrm{tr}G_{tt'}^K\big|_{t'=t} + 2\eta T\partial_t\mathrm{tr}G_{tt'}^R\big|_{t'=t} = -\mathbf{v}\cdot\mathbf{F}_{\mathrm{p}}.\qquad(12.159)$$

where tr denotes the trace with respect to the Cartesian indices. The energy conservation relation simply states that, on average, the work performed by the external and thermal noise forces is dissipated owing to friction.

In order to ascertain the convergence of the iteration process, employed when solving the self-consistent equations, we test how accurately the iterated solution satisfies the energy conservation relation. In Figure 12.6 the velocity dependence of the left- and right-hand sides of the energy conservation relation, Eq. (12.159), is shown. After at the most twenty iterations, the energy conservation relation is satisfied by the iterated solution to within an accuracy of 1%.

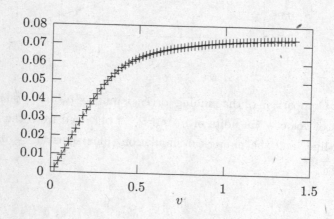

Figure 12.6 The values (in units of $\nu_0\eta^{-1}r_{\mathrm{p}}^{-4}$) of the expressions on the two sides of the energy conservation relation, Eq. (12.159), are shown as a function of the velocity (in units of $\eta^{-1}r_{\mathrm{p}}^{-2}\nu_0^{1/2}$). The dashed line and the plus symbols correspond to the left- and right-hand side, respectively ($T = 0.05\nu_0^{1/2}r_{\mathrm{p}}^{-1}$ and $m = 0.1\eta^2r_{\mathrm{p}}^3\nu_0^{-1/2}$). The energy conservation relation is fulfilled to within an accuracy of 1%.

In Section 12.7 we shall consider dynamic melting of the vortex lattice, and it is therefore of interest to check the validity of the fluctuations predicted by the self-consistent theory against direct simulations of the Langevin equation. In order to check the accuracy of the velocity fluctuations calculated within the self-consistent theory, we have performed simulations of the velocity fluctuations. In Figure 12.7, the velocity fluctuations obtained from the self-consistent theory are compared with simulations.

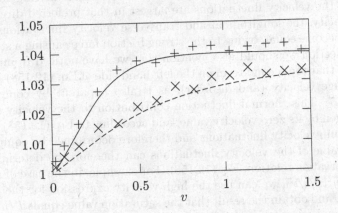

Figure 12.7 Longitudinal and transverse velocity fluctuations (in units of $\eta^{-2}r_{\rm p}^{-4}\nu_0$) as a function of the average velocity (in units of $\eta^{-1}r_{\rm p}^{-2}\nu_0^{1/2}$). The solid and dashed lines represent the results for the longitudinal (parallel to the external force), $\langle\!\langle(\dot{x}_t - v)^2\rangle\!\rangle$, and transverse, $\langle\!\langle\dot{y}_t^2\rangle\!\rangle$, velocity fluctuations obtained from the self-consistent theory, respectively. The plus signs and crosses represent the simulation results for the longitudinal and transverse velocity fluctuations, respectively ($T = 0.1\nu_0^{1/2}r_{\rm p}^{-1}$ and $m = 0.1\eta^2 r_{\rm p}^3\nu_0^{-1/2}$). At low average velocities the fluctuations approach their thermal value, T/m, which for the parameters and units in question equals 1. At intermediate average velocities the longitudinal velocity fluctuations are larger than the transverse, owing to the jerky motion of the particle along the preferred direction of the external force, before reaching the same value at high average velocities where the effect of the disorder simply acts as an additional contribution to the temperature.

The agreement between the self-consistent theory and the numerical simulations is seen to be good, indicating that fluctuations calculated from the self-consistent theory are quantitatively correct. At low average velocities the velocity fluctuations approach their thermal value T/m. The relative velocity fluctuations diverge at zero velocity even at zero temperature. This can be inferred from the energy conservation relation, Eq. (12.158), and the sub-linear velocity dependence of the pinning force at low velocities, as for example is apparent from Figure 12.3. At intermediate average velocities, the velocity fluctuations in the direction parallel to the average velocity (chosen along the $\hat{\mathbf{x}}$-axis), the longitudinal velocity fluctuations, $\langle\langle(\dot{x}_t-v)^2\rangle\rangle$, are found to be larger than the fluctuations perpendicular to the average velocity, the transverse velocity fluctuations, $\langle\langle\dot{y}_t^2\rangle\rangle$. The reason behind this is that at not too high velocities, where the force due to the disorder is strong compared with the friction force, the motion of the particle is jerky since the particle slowly makes it to the disorder potential tops, and subsequently is accelerated by the disorder potential. Since the average motion of the particle is caused by the external driving force, the jerky motion and the velocity fluctuations áre largest in that preferred direction. At high average velocity, the longitudinal and transverse velocity fluctuations saturate and are seen to become equal, owing to the strong friction force causing a steadier motion. In this connection we should also mention that we have noticed from our numerical calculations that the second term on the left-hand side of Eq. (12.158) is independent of the average velocity (and disorder), as is also apparent by comparing Figures 12.6 and 12.7. This thermal fluctuation contribution to the velocity fluctuations is therefore given by its zero velocity value, and according to Eq. (12.158) is specified by the equilibrium velocity fluctuations and therefore determined by equipartition. The saturation value of the velocity fluctuations can therefore be determined from the energy conservation relation, Eq. (12.158). For example, in the case of a small vortex mass, $m \ll \eta^2 r_{\mathrm{p}}^3/\sqrt{\nu_0}$, we can use the high velocity expression for the pinning force, Eq. (12.142), and obtain the result that the saturation value equals $T/m+\nu_0/8\pi r_{\mathrm{p}}^4\eta^2$, a result in good agreement with Figure 12.7. At high average velocity, the velocity fluctuations saturate, and the effect of the disorder simply acts as an additional contribution to the temperature.

12.5.5 Hall force

In this section the effect of a Hall force is considered, and the previous analysis of the dynamics of a single vortex is extended to include the Hall force

$$m\ddot{\mathbf{x}}_t + \eta\dot{\mathbf{x}}_t = \alpha\dot{\mathbf{x}}_t \times \hat{\mathbf{z}} - \nabla V(\mathbf{x}_t) + \mathbf{F}_t + \boldsymbol{\xi}_t. \qquad (12.160)$$

We shall use the self-consistent theory to calculate the pinning force, the velocity fluctuations, and the Hall angle

$$\theta = \arctan\frac{F_{\mathrm{H}}}{\hat{\mathbf{v}}\cdot\mathbf{F}} = \arctan\frac{\alpha}{\eta_{\mathrm{eff}}}, \qquad (12.161)$$

which can be expressed in terms of the effective friction coefficient.

Analytical results

The inverse of the free retarded Green's function acquires, according to Eq. (12.160), off-diagonal elements

$$D_{\mathrm{R}}^{-1}(\omega) = \begin{pmatrix} m\omega^2 + i\eta\omega & -i\alpha\omega \\ i\alpha\omega & m\omega^2 + i\eta\omega \end{pmatrix} \qquad (12.162)$$

and the free retarded Green's function is given by

$$D_\omega^R = \frac{1}{(\omega+i0)\,((m\omega+i\eta)^2-\alpha^2)} \begin{pmatrix} m\omega+i\eta & i\alpha \\ -i\alpha & m\omega+i\eta \end{pmatrix}. \qquad (12.163)$$

In the high-velocity regime, $v \gg \sqrt{\nu_0}/(\eta r_p^2)$, where lowest-order perturbation theory in the disorder is valid, we can neglect the self-energies in the self-consistent expression for the pinning force, Eq. (12.152), i.e. we can insert the free retarded Green's function and neglect the fluctuation exponent. Since the free retarded Green's function is antisymmetric in the Cartesian indices, only the diagonal elements make a contribution to the pinning force. The diagonal elements of the free retarded Green's function are identical, $D_{t0}^{Rxx} = D_{t0}^{Ryy}$, and given by

$$D_{t0}^{Rxx} = \theta(t)\frac{-\eta}{\eta^2+\alpha^2}\left(1 + \left(\frac{\alpha}{\eta}\sin\frac{\alpha t}{m} - \cos\frac{\alpha t}{m}\right)e^{-\eta t/m}\right) \qquad (12.164)$$

and we obtain for the pinning force, for vanishing mass, $m \ll \eta^2 r_p^3/\sqrt{\nu_0}$,

$$\mathbf{F}_p = -\frac{\eta\nu_0}{4\pi(\eta^2+\alpha^2)r_p^4 v^2}\mathbf{v}. \qquad (12.165)$$

We observe that the pinning force is suppressed by the Hall force in the high-velocity limit, $v \gg \sqrt{\nu_0}(\eta^2+\alpha^2)^{-1/2}r_p^{-2}$, and the high-velocity regime therefore sets in at a lower value in the presence of the Hall force.

At high temperatures, $T \gg \sqrt{\nu_0}/r_p$, and moderate velocities, $v < \eta\sqrt{\nu_0}/((\eta^2 + \alpha^2)r_p^2)$, the Hall force has the opposite effect, i.e. it increases the pinning force, as a calculation similar to the one leading to Eq. (12.154) shows that the pinning force is ($m \ll \eta^2 r_p^3/\sqrt{\nu_0}$):

$$\mathbf{F}_p = -\frac{\nu_0(\eta^2+\alpha^2)}{8\pi\eta T^2 r_p^2}\mathbf{v}. \qquad (12.166)$$

We have found by solving the self-consistent equations numerically at high temperature, $T = 10\sqrt{\nu_0}/r_p$, that the pinning force is linear at low velocities and increases with increasing Hall force. The deviation from the linear behavior in the presence of the Hall force starts at a lower velocity value in accordance with the high-velocity regime starting at a lower value in the presence of the Hall force.

Numerical results

For any given average velocity of the vortex, the pinning force can be calculated from the self-consistent theory. We have numerically calculated the pinning force for various strengths of the Hall force. In Figure 12.8, the resulting pinning force as a function of the velocity is shown for different strengths of the Hall force for a temperature lower than the average barrier height, $T < \sqrt{\nu_0}/r_p$. The Hall force is seen to reduce the pinning force in this temperature regime except, of course, at low velocities.

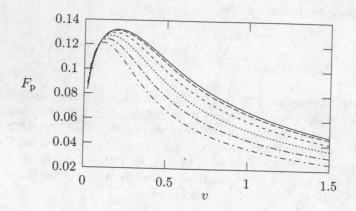

Figure 12.8 Pinning force (in units of $\nu_0^{1/2} r_{\mathrm{p}}^{-2}$) on a single vortex as a function of velocity (in units of $\eta^{-1} r_{\mathrm{p}}^{-2} \nu_0^{1/2}$) obtained from the self-consistent theory for various strengths of the Hall force. The different curves correspond to $\alpha/\eta = 0, 0.2, 0.4, 0.6, 0.8, 1$, where the uppermost curve corresponds to $\alpha = 0$ ($m = 0.1 \eta^2 r_{\mathrm{p}}^3 \nu_0^{-1/2}$ and $T = 0.1 \nu_0^{1/2} r_{\mathrm{p}}^{-1}$).

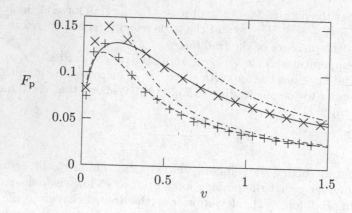

Figure 12.9 Pinning force (in units of $10^{-4} \nu_0^{1/2} r_{\mathrm{p}}^{-2}$) on a single vortex as a function of velocity. Comparison of the simulation results and the results of the self-consistent and lowest order perturbation theory, Eq. (12.165), for the case of no Hall force, $\alpha = 0$, and a moderately strong Hall force, $\alpha = \eta$ ($m = 0.1 \eta^2 r_{\mathrm{p}}^3 \nu_0^{-1/2}$ and $T = 0.1 \nu_0^{1/2} r_{\mathrm{p}}^{-1}$). The solid line represents the self-consistent result and the crosses the simulation result, while the upper dash-dotted line represents the perturbation theory result, all for the case $\alpha = 0$. The dashed line and the plus symbols represent the self-consistent and simulation results, while the lower dash-dotted line represents the perturbation theory result, all for the case $\alpha = \eta$.

In Figure 12.9 we compare the pinning force obtained from the self-consistent theory with the result of perturbation theory valid at high velocities, Eq. (12.165), and simulations. According to Figure 12.9, the reduction of the pinning force due to the Hall force predicted by the self-consistent and the perturbation theory is in good agreement at high velocities. The pinning force obtained from the self-consistent theory and the simulations are also in good agreement in the presence of a Hall force, even at lower velocities. In fact in much better agreement than in the absence of the Hall force, in accordance with the fact that the Hall force suppresses the velocity fluctuations, as we demonstrate shortly.

The Hall angle calculated from the self-consistent theory approaches from below the disorder-independent value $\arctan(\alpha/\eta)$ at high velocities, as shown in Figure 12.10.

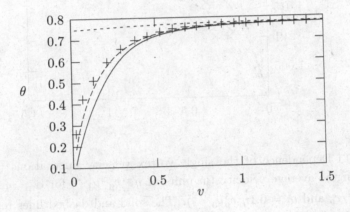

Figure 12.10 Hall angle as a function of velocity for a single vortex. The curves represent the self-consistent results for the three temperatures $T = 0, 0.1, 1$ (in units of $\nu_0^{1/2} r_p^{-1}$), where the uppermost curve corresponds to the highest temperature. The plus symbols represent the simulation results for the temperature $T = 0.1\nu_0^{1/2} r_p^{-1}$. The parameter α/η is one and $m = 0.1\eta^2 r_p^3 \nu_0^{1/2}$.

In Figure 12.10, the Hall angle obtained from the self-consistent theory is also compared with simulations, and the agreement is seen to be good. As apparent from Figure 12.10, an increase in the temperature increases the Hall angle at low velocities, because the effective friction coefficient decreases with increasing temperature, and this feature vanishes at high velocities. From Figure 12.10 we can also infer the following behavior of the Hall angle at zero velocity: at low temperatures it is zero, since the dependence of the pinning force at low velocities is sub-linear. At a certain temperature, the Hall angle at zero velocity jumps to a finite value, since the pinning force then depends linearly on the velocity, and saturates at high temperatures at the disorder independent value.

We have also determined the influence of the Hall force on the velocity fluctuations as shown in Figure 12.11.

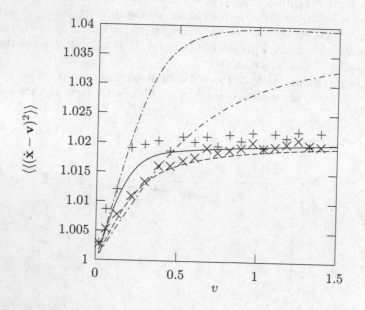

Figure 12.11 Dependence of the single vortex velocity fluctuations (in units of $\eta^{-2}r_{\mathrm{p}}^{-4}\nu_0$) on the average velocity (in units of $\eta^{-1}r_{\mathrm{p}}^{-2}\nu_0^{1/2}$) for $\alpha = \eta$ and $\alpha = 0$ ($T = 0.1\nu_0^{1/2}/r_{\mathrm{p}}$ and $m = 0.1\eta^2 r_{\mathrm{p}}^3\nu_0^{-1/2}$). The solid and dashed lines represent the longitudinal and transverse velocity fluctuations, respectively, calculated by using the self-consistent theory for the case $\alpha = \eta$, and the plus symbols and crosses represent the corresponding simulation results. The two dash-dotted lines represent the longitudinal and transverse velocity fluctuations, respectively, calculated by using the self-consistent theory in the absence of the Hall force, $\alpha = 0$, which were compared with simulations in Figure 12.7.

We observe that the Hall force at low velocities slightly increases the transverse velocity fluctuations, and decreases the longitudinal fluctuations, whereas the longitudinal and transverse velocity fluctuations are strongly suppressed by the Hall force at higher velocities, in particular the longitudinal fluctuations. The suppression of the velocity fluctuations is caused by the blurring by the Hall force of the preferred direction of motion due to the external force, resulting in a less jerky motion. At high average velocity, the longitudinal and transverse velocity fluctuations saturate and become equal because of the strong friction. As previously discussed in the absence of the Hall force, the saturation value can be determined from the energy conservation relation (which take the same form, Eq. (12.159), as in the absence of the Hall force, since the Hall force does not perform any work) and the high-velocity expression for the pinning force, Eq. (12.165), since our numerical results show that the second term on the left-hand side of Eq. (12.158) is independent of the Hall force and velocity (and disorder). This observation tells us that the suppression of the velocity fluctuations caused by the Hall force, according to the energy conservation relation, Eq. (12.158), is in correspondence with the suppression of the pinning force. We note from Figure 12.11 that the high-velocity regime sets in at lower velocities than in the absence of the Hall force. In Figure 12.11, the velocity fluctuations calculated by using the self-consistent theory are also compared with simulations, and the agreement is seen to be good.

We have ascertained the convergence of the numerical iteration process by testing that the obtained solutions satisfy the energy conservation relation. We find that the energy conservation relation is fulfilled within an accuracy of 2%, except at the lowest velocities.

12.6   Vortex lattice

After having gained confidence in the Hartree approximation studying the case of a single vortex, we consider in this section the influence of pinning on a vortex lattice in the flux flow regime, where the lattice moves with a constant average velocity, $\langle\langle \dot{\mathbf{u}}_{\mathbf{R}t} \rangle\rangle = \mathbf{v}$, since the external force is assumed independent of time. We consider a triangular Abrikosov vortex lattice, and treat the interaction between the vortices in the harmonic approximation. The free retarded Green's function of the vortex lattice

$$D_{\mathbf{q}\omega}^{\mathrm{R}} = \sum_b \frac{\mathbf{e}_b(\mathbf{q})\,\mathbf{e}_b(\mathbf{q})}{m\omega^2 + i\eta\omega - K_b(\mathbf{q})} \tag{12.167}$$

is obtained by diagonalizing the dynamic matrix, and inverting the inverse free retarded Green's function specified by Eq. (12.84) (for the moment we neglect the Hall force). The sum in Eq. (12.167) is over the two modes, $b = 1, 2$, corresponding to eigenvectors $\mathbf{e}_b(\mathbf{q})$ and eigenvalues $K_b(\mathbf{q})$, respectively. The eigenvalues and eigenvectors of the dynamic matrix are periodic with respect to translations by reciprocal lattice vectors.

Since the lattice distortions of interest are of small wave length compared to the lattice constant, the dynamic matrix of the vortex lattice is specified by the continuum theory of elastic media, i.e. through the compression modulus c_{11} and the shear modulus c_{66}, and in accordance with reference [136],

$$\Phi_{\mathbf{q}} = \frac{\phi_0}{B} \begin{pmatrix} c_{11}q_x^2 + c_{66}q_y^2 & (c_{11} - c_{66})q_x q_y \\ (c_{11} - c_{66})q_x q_y & c_{66}q_x^2 + c_{11}q_y^2 \end{pmatrix}, \tag{12.168}$$

where $\mathbf{q}$ belongs to the first Brillouin zone, and B is the magnitude of the external magnetic field, and ϕ_0/B is therefore equal to the area, a^2, of the unit cell of the vortex lattice. In the continuum limit we obtain a longitudinal branch, $\mathbf{e}_l(\mathbf{q}) \cdot \hat{\mathbf{q}} = 1$, with corresponding eigenvalues $K_l(\mathbf{q}) = c_{11}a^2q^2$, and a transverse branch, $\mathbf{e}_t(\mathbf{q}) \cdot \hat{\mathbf{q}} = 0$, with corresponding eigenvalues $K_t(\mathbf{q}) = c_{66}a^2q^2$.

12.6.1   High-velocity limit

At high velocities, $v \gg \sqrt{\nu_0}/(\eta r_p^2)$, where lowest-order perturbation theory in the disorder is valid, we can neglect the self-energies in the self-consistent expression for the pinning force, Eq. (12.133), i.e. we can insert the free retarded Green's function for the lattice and, assuming $v \gg T/(\eta r_p)$, neglect the fluctuation exponent, and obtain for the pinning force

$$\mathbf{F}_p = -\int \frac{d\mathbf{k}}{(2\pi)^2} \, \mathbf{k} \, \nu(\mathbf{k}) \sum_{b=l,t} \frac{\eta \mathbf{k} \cdot \mathbf{v} \, (\mathbf{k} \cdot \mathbf{e}_b(\mathbf{k}))^2}{(\eta \mathbf{k} \cdot \mathbf{v})^2 + (K_b(\mathbf{k}))^2} \, . \tag{12.169}$$

The maximum values, attained at the boundaries of the Brillouin zones, of the transverse and longitudinal eigenvalues are specified by the compression and shear moduli, $K_t \sim c_{66}$ and $K_l \sim c_{11}$. The compression modulus is much greater than the shear modulus, $c_{11} \gg c_{66}$, in thin films and high-temperature superconductors (see for example reference [137]). The order of magnitude of the first term in the denominator of Eq. (12.169) is $\eta v^2 r_p^{-2}$, since the range of the impurity correlator is r_p, and at intermediate velocities, $c_{66}r_p/\eta \ll v \ll c_{11}r_p/\eta$, only the transverse mode therefore contributes to the pinning force, and we obtain

$$\mathbf{F}_p = -\int \frac{d\mathbf{k}}{(2\pi)^2} \, \mathbf{k} \, \nu(\mathbf{k}) \, \frac{(\mathbf{k} \cdot \mathbf{e}_t(\mathbf{k}))^2}{\eta \mathbf{k} \cdot \mathbf{v}} \, . \tag{12.170}$$

The eigenvalues $\mathbf{e}_t(\mathbf{k})$ are periodic in the reciprocal lattice and, assuming short-range disorder, $r_p \ll a$, the rest of the integrand is slowly varying, and we obtain for the pinning force

$$\mathbf{F}_p = -\frac{1}{2} \int \frac{d\mathbf{k}}{(2\pi)^2} \mathbf{k} \frac{\nu(\mathbf{k})k^2}{\eta \mathbf{k} \cdot \mathbf{v}} = -\frac{\nu_0}{8\pi r_p^4 \eta v^2} \mathbf{v} \, . \tag{12.171}$$

At very high velocities, $v \gg c_{11}r_p/\eta$, the eigenvalues of the dynamic matrix in Eq. (12.169) can be neglected compared with the velocity-dependent term in the denominator, and the longitudinal and transverse parts of the free retarded Green's function give equal contributions to the pinning force, and we obtain

$$\mathbf{F}_p = -\frac{\nu_0}{4\pi r_p^4 \eta v^2} \mathbf{v} \, . \tag{12.172}$$

This result is identical to the expression for the pinning force on a single vortex, Eq. (12.142), in the high velocity regime, $v \gg \sqrt{v_0}/(\eta r_p^2)$, since the influence of the elastic interaction is negligible.

12.6.2 Numerical results

In this section we consider the pinning force on the vortex lattice obtained from the self-consistent theory. For any given average velocity of the lattice, the coupled equations of Green's functions and self-energies, Eq. (12.124) and Eq. (12.125), may be solved numerically by iteration. In order to simplify the numerical calculation, the self-consistent equations are brought on dimensionless form by introducing the following units for length, time, and mass, a, $\eta a^3/v_0^{1/2}$, and $\eta^2 a^4/v_0^{1/2}$. Starting by neglecting the self-energies, we obtain numerically the response and correlation functions. From Eq. (12.133) we can then determine the pinning force as a function of the velocity. We have calculated the velocity dependence of the pinning force for vortex lattices of sizes 4×4, 8×8, and 16×16 using the self-consistent theory, and the results are shown in Figure 12.12.

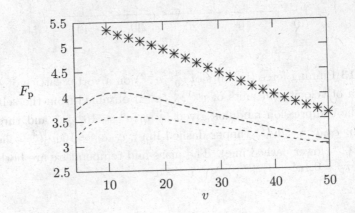

Figure 12.12 Pinning force (in units of $v_0^{1/2} a^{-2}$) as a function of velocity (in units of $\eta^{-1} v_0^{1/2} a^{-2}$) obtained from the self-consistent theory for three different lattice sizes. The stars correspond to a 4×4 lattice, and the two curves correspond to 8×8 and 16×16 lattices, respectively. The mass and temperature are chosen to be zero, and the elastic constants are given by $c_{66} a^3 = 100 v_0^{1/2}$ and $c_{11} a^3 = 10^4 v_0^{1/2}$, and the range of the disorder correlator is chosen to be $r_p = 0.1a$.

The difference between the results obtained for the 8×8 and the 16×16 lattice is small, and we conclude that the pinning force is fairly insensitive to the size of the lattice.

In Figure 12.13 we compare the pinning force as a function of the velocity for lattices of different stiffnesses, and we find that the pinning force decreases with increasing stiffness of the lattice.

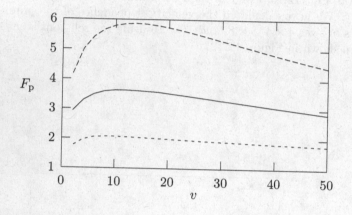

Figure 12.13 Pinning force (in units of $\nu_0^{1/2} a^{-2}$) on a vortex lattice of size 16×16 as a function of velocity (in units of $\eta^{-1} \nu_0^{1/2} a^{-2}$) obtained from the self-consistent theory for the compression modulus given by $c_{11} a^3 = 10^4 \nu_0^{1/2}$ and three different shear moduli: $c_{66} a^3 = 50 \nu_0^{1/2}$ (upper dashed line), $c_{66} a^3 = 100 \nu_0^{1/2}$ (solid line) and $c_{66} a^3 = 200 \nu_0^{1/2}$ (lower dashed line). The mass and temperature are both chosen to be zero, and $r_p = 0.1a$.

Generally, the interaction between the vortices lowers the pinning force, since the neighboring vortices in a moving lattice drag a vortex over the potential barriers. This can be inferred from the self-consistent theory by comparing the pinning forces depicted in Figures 12.3 and 12.12, and in perturbation theory by noting the extra term originating from the elastic interaction in the denominator of the expression for the pinning force, Eq. (12.169).

When the temperature is increased, the pinning force decreases, except at very high velocity, as apparent from Figure 12.14. This feature is common to the single vortex case, and simply reflects that thermal noise helps a vortex over the potential barriers.

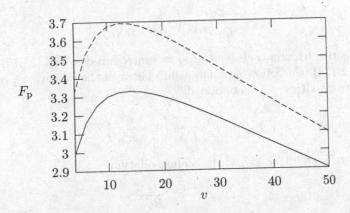

Figure 12.14 Pinning force (in units of $\nu_0^{1/2} a^{-2}$) on a vortex lattice of size 16×16 as a function of velocity (in units of $\eta^{-1} \nu_0^{1/2} a^{-2}$) obtained from the self-consistent theory for two different temperatures. The elastic constants are given by $c_{66} a^3 = 100 \nu_0^{1/2}$ and $c_{11} a^3 = 10^4 \nu_0^{1/2}$, and $r_p = 0.1a$ and $m = 1.0 \cdot 10^{-4} \eta^2 a^3 \nu_0^{-1/2}$. The dashed line corresponds to $T = 0$, and the solid line to $T = 0.5 \nu_0^{1/2} a^{-1}$.

The convergence of the iterative procedure is monitored by checking that energy conservation is fulfilled. The energy conservation relation for a vortex lattice is obtained as in Section 12.5.5, and since the term originating from the harmonic interaction between the vortices disappears owing to the symmetry property of the dynamic matrix, Eq. (12.76), we obtain for a vortex lattice the energy conservation relation

$$\eta \, \partial_t \, \mathrm{tr} \left(-i\partial_t G^{\mathrm{K}}(\mathbf{R}, t; \mathbf{R}, t') + 2T \, G^{\mathrm{R}}(\mathbf{R}, t; \mathbf{R}, t') \right) |_{t'=t} = -\mathbf{v} \cdot \mathbf{F}_{\mathrm{p}} \, . \quad (12.173)$$

The convergence of the iteration procedure, employed when solving the self-consistent equations, has been checked by numerically calculating the terms in Eq. (12.173). We find that the right- and left-hand sides of the energy conservation relation differ by no more than a few percent after twenty iterations.

12.6.3   Hall force

We now consider the influence of a Hall force on the dynamics of a vortex lattice. The motion of the vortex lattice, with its associated magnetic field, induces an average electric field. The relationship between the average vortex velocity and the induced electric field, $\mathbf{E} = \mathbf{v} \times \mathbf{B}$, and the expression for the Lorentz force, yields for the resistivity tensor of a superconducting film

$$\rho = \frac{\phi_0 B}{\eta_{\text{eff}}^2 + \alpha^2} \left(\begin{array}{cc} \eta_{\text{eff}} & \alpha \\ -\alpha & \eta_{\text{eff}} \end{array} \right) , \tag{12.174}$$

where the effective friction coefficient, η_{eff}, was introduced in Eq. (12.79).[13] According-ing to Eq. (12.174), the following relationship between the transverse, ρ_{xy}, and the longitudinal resistivities, ρ_{xx}, is obtained

$$\rho_{xy} = \rho_{xx}^2 \frac{\alpha}{B\phi_0} \left(1 + \frac{\alpha^2}{\eta_{\text{eff}}^2} \right). \tag{12.175}$$

If the Hall force is small, $\alpha \ll \eta_{\text{eff}}$, the scaling relation

$$\rho_{xy} = \rho_{xx}^2 \frac{\alpha}{B\phi_0} \tag{12.176}$$

is seen to be obeyed. This scaling law is valid for all velocities of the vortex, provided the Hall force is small compared with the friction force, $\alpha \ll \eta$. We note that the scaling law is also valid at small vortex velocities for arbitrary values of the Hall force, if the effective friction coefficient diverges at small velocities. This occurs if the pinning force decreases slower than linearly in the vortex velocity. This is indeed the case, according to the self-consistent theory, at temperatures lower than the average barrier height, $T \ll \sqrt{\nu_0}/r_p$, as indicated by the low velocity behavior of the pinning force in Figure 12.15. This behavior of the pinning force is also obtained for non-interacting vortices as apparent from Figure 12.8.

In Figure 12.15 is shown the pinning force obtained from the self-consistent theory as a function of velocity for the case of zero temperature. As expected there is no influence of the Hall force on the pinning force at low velocities, but we find a suppression at intermediate velocities, and at very high velocities, $v \gg c_{11}a/\eta$, we recover the high velocity limit of the single vortex result, i.e. Eq. (12.165). By comparison of Figures 12.8 and 12.15, we find that the Hall force has a much weaker influence at intermediate velocities on the pinning of an interacting vortex lattice than on a system of non-interacting vortices. Furthermore, the influence of the Hall force on the pinning force is more pronounced for a stiff lattice than a soft lattice, as seen from the inset in Figure 12.15.

---

[13]The effective friction coefficient was determined to lowest order in the disorder in reference [138].

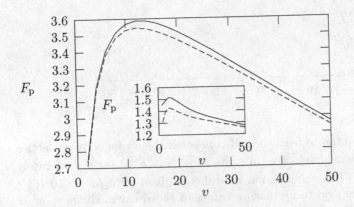

Figure 12.15 Pinning force (in units of $\nu_0^{1/2}a^{-2}$) on a vortex lattice of size 16×16 as a function of velocity (in units of $\eta^{-1}\nu_0^{1/2}a^{-2}$) obtained from the self-consistent theory. The solid and dashed lines correspond to $\alpha = 0$ and $\alpha = \eta$, respectively. The temperature and mass are both chosen to be zero, and $r_{\mathrm{p}} = 0.1a$. The elastic constants are given by $c_{11}a^3 = 10^4\nu_0^{1/2}$ and $c_{66}a^3 = 100\nu_0^{1/2}$. Inset: pinning force as a function of velocity for $\alpha = 0$ and $\alpha = \eta$, respectively. Here $c_{66}a^3 = 300\nu_0^{1/2}$ and the other parameters are unchanged.

In Figure 12.16 the dependence of the Hall angle on the velocity is presented for various stiffnesses of the vortex lattice; the stiffest lattice has the greatest Hall angle. Since the pinning force is reduced by the interaction between the vortices, the Hall angle for a lattice is larger than for an independent vortex, except at high velocities where they saturate at the same value. A similar behavior of the Hall angle at zero velocity, as observed for a single vortex in Section 12.5.5, pertains to a vortex lattice.

12.7   Dynamic melting

In this section we consider the influence of quenched disorder on the dynamic melting of a vortex lattice. This non-equilibrium phase transition has been studied experimentally [139, 140, 141, 142, 143, 144], as well as through numerical simulation and a phenomenological theory and perturbation theory [145, 146, 147]. The notion of dynamic melting refers to the melting of a moving vortex lattice where, in addition to the thermal fluctuations, fluctuations in vortex positions are induced by the disorder. A temperature-dependent critical velocity distinguishes a transition between a phase

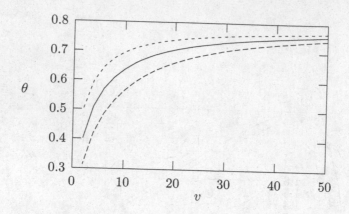

Figure 12.16 Hall angle obtained from the self-consistent theory for a vortex lattice of size 16×16 as a function of velocity (in units of $\eta^{-1}\nu_0^{1/2}a^{-1}$) for a moderately strong Hall force, $\alpha = \eta$. The compression modulus is given by $c_{11}a^3 = 10^4\nu_0^{1/2}$, and the three curves correspond to decreasing values of the shear modulus $c_{66}a^3 = 200\nu_0^{1/2}$, $100\nu_0^{1/2}$, $50\nu_0^{1/2}$. The mass and temperature are both chosen to be zero, and $r_{\rm p} = 0.1a$.

where the vortices form a moving lattice, the solid phase, and a vortex liquid phase.

Before solving the self-consistent equations by numerical iteration in order to obtain the phase diagram, we consider the heuristic argument for determining the phase diagram for dynamic melting of a vortex lattice presented in reference [145]. There, the disorder induced fluctuations were estimated by considering the correlation function

$$\kappa_{\alpha\alpha'}(\mathbf{x}, t) = \langle\!\langle f_\alpha^{(\rm p)}(\mathbf{x}, t)\, f_{\alpha'}^{(\rm p)}(\mathbf{0}, 0)\rangle\!\rangle \tag{12.177}$$

of the pinning force density

$$\mathbf{f}^{(\rm p)}(\mathbf{x}, t) = -\sum_{\mathbf{R}} \delta(\mathbf{x} - \mathbf{R} - \mathbf{u}_{\mathbf{R}t})\,\nabla V(\mathbf{x} - \mathbf{v}t)\,. \tag{12.178}$$

Neglecting the interdependence of the fluctuations of the vortex positions and the fluctuations in the disorder potential, the pinning force correlation function factorizes

$$\kappa_{\alpha\alpha'}(\mathbf{x}, t) \simeq \sum_{\mathbf{R}\mathbf{R}'} \langle\!\langle\delta(\mathbf{x} - \mathbf{R} - \mathbf{u}_{\mathbf{R}t})\,\delta(\mathbf{R}' - \mathbf{u}_{\mathbf{R}'0})\rangle\!\rangle \nabla_\alpha\nabla_{\alpha'} \langle\!\langle V(\mathbf{x} - \mathbf{v}t)V(\mathbf{0})\rangle\!\rangle\,. \tag{12.179}$$

Introducing the Fourier transform (A is the area of the film)

$$C_{\mathbf{R}\mathbf{R}'}(\mathbf{q}, t) = A^{-1}\langle\!\langle e^{-i\mathbf{q}\cdot(\mathbf{R}+\mathbf{u}_{\mathbf{R}t}-\mathbf{R}'-\mathbf{u}_{\mathbf{R}'0})}\rangle\!\rangle \tag{12.180}$$

of the vortex density-density correlation function

$$C_{\mathbf{RR'}}(\mathbf{x},t) = \langle\!\langle \delta(\mathbf{x}-\mathbf{R}-\mathbf{u}_{\mathbf{R}t})\,\delta(\mathbf{R'}-\mathbf{u}_{\mathbf{R'}0})\rangle\!\rangle \qquad (12.181)$$

and employing the translational invariance yields

$$\kappa_{\alpha\alpha'}(\mathbf{x},t) = -n_{\mathrm{v}} \sum_{\mathbf{RR'}} \langle\!\langle \delta(\mathbf{x}-\mathbf{R}-\mathbf{u}_{\mathbf{R}t}-\mathbf{R'}-\mathbf{u}_{\mathbf{R'}0})\rangle\!\rangle \nabla_\alpha \nabla_{\alpha'}\,\nu(\mathbf{x}-\mathbf{v}t)\,, \qquad (12.182)$$

where n_{v} is the density of vortices. In the fluidlike phase the motion of different vortices is *incoherent* and the off-diagonal terms, $\mathbf{R} \neq \mathbf{R'}$, can be neglected yielding

$$\kappa_{\alpha\alpha'}(\mathbf{x},t) = -n_{\mathrm{v}}\,\delta(\mathbf{x})\,\nabla_\alpha \nabla_{\alpha'}\,\nu(\mathbf{v}t)\,. \qquad (12.183)$$

In analogy with the noise correlator, the effect of disorder-induced fluctuations is then represented by a *shaking temperature*

$$T_{\mathrm{sh}} = \frac{1}{4\eta n_{\mathrm{v}}} \sum_\alpha \int d\mathbf{x}\int_{-\infty}^{\infty} dt\,\kappa_{\alpha\alpha}(\mathbf{x},t)\frac{1}{4\sqrt{2\pi}}\frac{\nu_0}{\eta v r_{\mathrm{p}}^3} = \frac{1}{4\sqrt{2\pi}}\frac{\nu_0}{F r_{\mathrm{p}}^3}\,, \qquad (12.184)$$

where in the last equality it is assumed that the pinning force is small compared with the friction force, i.e. $\eta v \simeq F$. An effective temperature is then obtained by adding the *shaking temperature* to the temperature, $T_{\mathrm{eff}} = T + T_{\mathrm{sh}}$, and according to Eq. (12.184) the effective temperature decreases with increasing external force, i.e. with increasing average velocity of the vortices. As the external force is increased the fluid thus freezes into a lattice. The value of the external force for which the moving lattice melts, the transition force F_{t}, is in this *shaking* theory defined as the value for which the effective temperature equals the melting temperature, T_{m}, in the absence of disorder

$$T_{\mathrm{eff}}(F=F_{\mathrm{t}}) = T_{\mathrm{m}} \qquad (12.185)$$

and has therefore in the shaking theory the temperature dependence

$$F_{\mathrm{t}}(T) = \frac{\nu_0}{4\sqrt{2\pi}r_{\mathrm{p}}^3(T_{\mathrm{m}}-T)} \qquad (12.186)$$

for temperatures below the melting temperature of the ideal lattice. We note that the transition force for strong enough disorder exceeds the critical force for which the lattice is pinned $F_{\mathrm{t}} > F_{\mathrm{c}} \sim \nu_0^{1/2}/r_{\mathrm{p}}^2$.

We now describe the calculation within the self-consistent theory of the physical quantities of interest for dynamic melting. The conventional way of determining a melting transition is to use the Lindemann criterion, which states that the lattice melts when the displacement fluctuations reach a critical value $\langle \mathbf{u}^2 \rangle = c_{\mathrm{L}}^2 a^2$, where c_{L} is the Lindemann parameter, which is typically in the interval ranging from 0.1 to 0.2, and a^2 is the area of the unit cell of the vortex lattice. In two dimensions the position fluctuations of a vortex diverge even for a clean system, and the Lindemann criterion implies that a two-dimensional vortex lattice is always unstable against thermal fluctuations. However, a quasi-long-range translational order persists up to a certain melting temperature [146]. As a criterion for the loss of long-range translational order a modified Lindemann criterion involving the relative vortex fluctuations

$$\langle (\mathbf{u}(\mathbf{R} + \mathbf{a}_0, t) - \mathbf{u}(\mathbf{R}, t))^2 \rangle = 2 c_{\mathrm{L}}^2 a^2 , \tag{12.187}$$

where $\mathbf{a}_0$ is a primitive lattice vector, has successfully been employed [146], and its validity verified within a variational treatment [148]. The relative displacement fluctuations of the vortices are specified in terms of the correlation function according to

$$\langle\langle (\mathbf{u}(\mathbf{R} + \mathbf{a}_0, t) - \mathbf{u}(\mathbf{R}, t))^2 \rangle\rangle = 2 i \mathrm{tr} \left(G^{\mathrm{K}}(\mathbf{0}, 0) - G^{\mathrm{K}}(\mathbf{a}_0, 0) \right) , \tag{12.188}$$

where the translation invariance of the Green's functions has been exploited. The correlation function is determined by the Dyson equation, Eq. (12.147), where the influence of the quenched disorder appears explicitly through Σ^{K} and implicitly through Σ^{R} and Σ^{A} in the retarded and advanced response functions. Furthermore, the self-energies depend self-consistently on the response and correlation functions. We have calculated numerically the Green's functions and self-energies and thereby the vortex fluctuations for a vortex lattice of size 8×8, and evaluated the pinning force from Eq. (12.133).

We determine the phase diagram for dynamic melting of the vortex lattice by calculating the relative displacement fluctuations for a set of velocities, and interpolate to find the transition velocity, v_{t}, i.e. the value of the velocity at which the fluctuations fulfill the modified Lindemann criterion (the determination of the Lindemann parameter is discussed shortly). An example of such a set of velocities is presented in the lower inset in Figure 12.17, where the relative displacement fluctuations as a function of velocity are shown. The magnitude of the transition force is determined by the averaged equation of motion

$$F_{\mathrm{t}} = \eta v_{\mathrm{t}} + F_{\mathrm{p}}(v_{\mathrm{t}}) \tag{12.189}$$

and is then obtained by using the numerically calculated pinning force. Repeating the calculation of the transition force for various temperatures determines the melting curve, i.e. the temperature dependence of the transition force, $F_{\mathrm{t}}(T)$, separating two phases in the FT-plane: a high-velocity phase where the vortices form a moving solid when the external force exceeds the transition force, $F > F_{\mathrm{t}}(T)$, and a liquid phase for forces less than the transition force.

In order to be able to compare the results of the self-consistent theory with the simulation results, we use the same parameters as input to the self-consistent theory as used in the literature [145]. There, the melting temperature in the absence of disorder is given by $T_{\mathrm{m}} = 0.007$ (the unit of energy per unit length is taken to be $2(\phi_0/4\pi\lambda)^2$) as obtained by simulations of clean systems [149], and assumed equal to the Kosterlitz–Thouless temperature [150, 151]

$$T_{\mathrm{KT}} = \frac{c_{66} a^2}{4\pi} . \tag{12.190}$$

The shear modulus is therefore determined to have the value $c_{66} = 0.088$ (as a is taken as the unit of length). The range of the vortex interaction, λ, was approximately equal to the lattice spacing, a_0, giving for the compression modulus [130]

$$c_{11} = \frac{16\pi\lambda^2 c_{66}}{a_0^2} \simeq 50\, c_{66} \simeq 4.4 \ . \tag{12.191}$$

The range and strength of the disorder correlator in the simulations are in the chosen units, $r_p = 0.2$ and $\nu_0 = 1.42 \cdot 10^{-5}$, and since the simulations are done for an over-damped system, the vortex mass in the self-consistent theory should be set to zero.

As described above, our numerical results for the relative displacement fluctuations can be used to obtain the dynamic phase diagram once the Lindemann parameter is determined. In order to do so we calculate *melting* curves by using the self-consistent theory for a set of different values of the Lindemann parameter. We find that these curves have the same shape, close to the melting temperature, as the melting curve obtained from the shaking theory, Eq. (12.186),

$$T = C_1 - \frac{C_2}{F_t} \ . \tag{12.192}$$

The curve which intersects at the melting temperature $T_m = 0.007$, the one depicted in the upper inset in Figure 12.17, i.e. the one for which C_1 is closest to the value 0.007, is then chosen, determining the Lindemann parameter to be given by the value $c_L = 0.124$.

Having determined the Lindemann parameter, we can determine the melting curve, and the corresponding phase diagram obtained from the self-consistent theory is shown in Figure 12.17. The simulation results of reference [145] are also presented, as well as the melting curve obtained from the shaking theory. We note the agreement of the simulation with the self-consistent theory, as well as with the shaking theory, although the simulation data are not in the large-velocity regime and the shaking argument is therefore not *a priori* valid.

In view of the good agreement between the self-consistent theory, the shaking theory and the simulation, and the fact that we have only one fitting parameter at our disposal, the melting temperature in the absence of disorder, it is of interest to recall that while the melting curve obtained from the shaking theory was based on an argument only valid in the liquid phase, i.e. freezing of the vortex liquid was considered, the melting curve we obtained from the self-consistent theory is calculated in the solid phase, i.e. we consider melting of the moving lattice. Furthermore, the melting of the vortex lattice was indicated in the simulation by an abrupt increase in the structural disorder [145], yet another melting criterion, and the agreement of the self-consistent theory with the simulation data are therefore further validating the use of the modified Lindemann criterion.

As is apparent from the upper inset in Figure 12.17, the critical exponent obtained from the self-consistent theory, 1.0, is in excellent agreement with the prediction of the shaking theory, where the critical exponent equals one. Furthermore, we find that the self-consistent theory yields the value $1.65 \cdot 10^{-4}$ for the magnitude of the slope C_2, which is in good agreement with the value, $\nu_0/(4\sqrt{2\pi}r_p^3) = 1.77 \cdot 10^{-4}$, predicted by the shaking theory, represented by the lower dashed line. That the values are so close testifies to the appropriateness of characterizing the disorder-induced fluctuations effectively by a temperature.

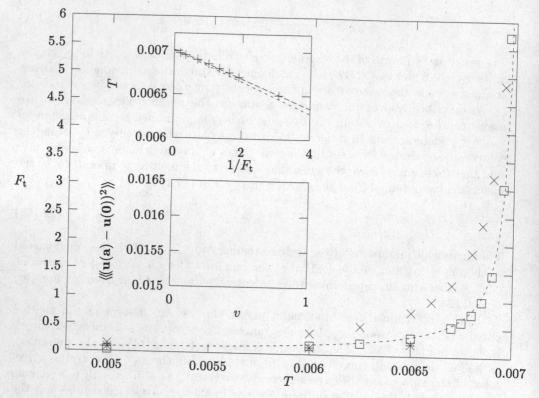

Figure 12.17 Phase diagram for the dynamic melting transition. The melting curve separates the two phases – for values of the external force larger than the transition force the moving vortices form a solid, and for smaller values they form a liquid. The dots in the boxes represent points on the melting curve obtained from the self-consistent theory using a vortex lattice of size 8×8, while the three stars represent the simulation results of reference 6. The crosses represent the lowest-order perturbation theory results. The dashed line is the curve $F_t(T) = 1.77 \cdot 10^{-4}/(0.007 - T)$, the melting curve predicted by the shaking theory. Upper inset: relationship between temperature and the inverse transition force obtained from the self-consistent theory, close to the melting temperature, for the particular value of the Lindemann parameter $c_L = 0.124$, for which the curve intersects the vertical axis at $T_m = 0.00701$. The set of points calculated from the self-consistent theory (plus signs) coincides with a straight line in excellent agreement with the prediction for the critical exponent by the shaking theory being 1. Lower inset: relative displacement fluctuations as a function of velocity. The dots to the left are calculated by using the self-consistent theory and the dots to the right are calculated by using lowest-order perturbation theory (for the temperature $T = 0.0065$).

It is of interest to compare the melting curves obtained from the self-consistent theory and perturbation theory. Expanding the kinetic component of the Dyson equation, Eq. (12.124), to lowest order in the disorder we obtain

$$G_{\mathbf{q}\omega}^{K(1)} = D_{\mathbf{q}\omega}^{R} \left(\Sigma_{\mathbf{q}\omega}^{K(1)} - 2i\eta T \right) D_{\mathbf{q}\omega}^{A}$$

$$- 2i\eta k_{B} T D_{\mathbf{q}\omega}^{R} \left(\Sigma_{\mathbf{q}\omega}^{R(1)} D_{\mathbf{q}\omega}^{R} + D_{\mathbf{q}\omega}^{A} \Sigma_{\mathbf{q}\omega}^{A(1)} \right) D_{\mathbf{q}\omega}^{A} , \qquad (12.193)$$

where $\Sigma^{R(1)}$, $\Sigma^{A(1)}$ and $\Sigma^{K(1)}$ are the lowest-order approximations of the self-energies, i.e. calculated to first order in ν_0. The relative vortex displacement fluctuations, Eq. (12.188), can then be obtained in perturbation theory from Eq. (12.193). In Figure 12.17 is shown the melting curve predicted by perturbation theory, i.e. where for the transition velocity interpolation we use the relative vortex fluctuations obtained from perturbation theory, an example of which is shown in the lower inset. As is to be expected, the perturbation theory result is in good agreement with the self-consistent theory, and the shaking theory, at high velocities. However, we observe from Figure 12.17 that the melting curve obtained from lowest-order perturbation theory deviates markedly at intermediate velocities from the prediction of the non-perturbative self-consistent theory, and thereby from the shaking theory, which is known to account well for the measured melting curve [142].

The shaking theory is seen to be in remarkable good agreement with the self-consistent theory for the parameter values considered above. We have investigated whether this feature persists for stronger disorder. As apparent from Figure 12.18, there is a more pronounced difference between the shaking theory and the self-consistent theory at stronger disorder. Whereas the deviation between the self-consistent and shaking theory for the previous parameter values typically is 5%, in the case of a five-fold stronger disorder, $\nu_0 = 7.1 \cdot 10^{-5}$, it is more than 15%.

We have studied the influence of pinning on vortex dynamics in the flux flow regime. A self-consistent theory for the vortex correlation and response functions was constructed, allowing a non-perturbative treatment of the disorder using the powerful functional methods of quantum field theory presented in Chapter 10. The validity of the self-consistent theory was established by comparison with numerical simulations of the Langevin equation.

The self-consistent theory was first applied to a single vortex, appropriate for low magnetic fields where the vortices are so widely separated that the interaction between them can be neglected. The result for the pinning force was compared with lowest-order perturbation theory and good agreement was found at high velocities, whereas perturbation theory failed to capture the non-monotonic behavior at low velocities, a feature captured by the self-consistent theory. The influence of the Hall force on the pinning force on a single vortex was then considered using the self-consistent theory. The Hall force was observed to suppress the pinning force, an effect also confirmed by our simulations. The suppression of the pinning force was shown at high velocities to be in agreement with the analytical result obtained from lowest-order perturbation theory. The suppression of the pinning force was caused by the Hall force through its reduction of the response function, while the effect of fluctuations through the fluctuation exponent at not too high temperatures could be

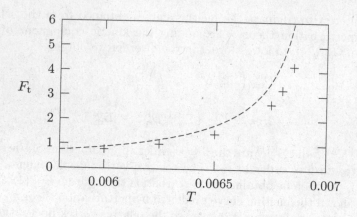

Figure 12.18 Phase diagram for the dynamic melting transition for the disorder strength $\nu_0 = 7.1 \cdot 10^{-5}$. The plus signs represent points on the melting curve obtained from the self-consistent theory for a vortex lattice of size 8×8, while the dashed curve is the curve $F_t(T) = 8.85 \cdot 10^{-4}/(0.007 - T)$, the melting curve predicted by the shaking theory.

neglected. The situation at high temperatures was the opposite, since in that case the thermal fluctuations were of importance, and the Hall force then increased the pinning force because it suppressed the fluctuation exponent.

We also studied a vortex lattice treating the interaction between the vortices in the harmonic approximation. The pinning force on the vortex lattice was found to be reduced by the interaction. The pinning force as a function of velocity displayed a plateau at intermediate velocities, before eventually approaching at very high velocities the pinning force on a single vortex. Analytical results for the pinning force were obtained in different velocity regimes depending on the magnitude of the compression modulus of the vortex lattice. Furthermore, we included the Hall force and showed that its influence on the pinning force was much weaker on a vortex lattice than on a single vortex.

We developed a self-consistent theory of the dynamic melting transition of a vortex lattice, enabling us to determine numerically the melting curve directly from the dynamics of the vortices. The presented self-consistent theory corroborated the phase diagram obtained from the phenomenological shaking theory far better than lowest-order perturbation theory. The melting curve obtained from the self-consistent theory was found to be in good quantitative agreement with simulations and experimental data.

12.8   Summary

In this chapter we have considered the theory of classical statistical dynamics treating systems coupled to a heat bath and classical stochastic forces. In particular we studied Langevin dynamics and quenched disorder, and applied the method to study the dynamics of the Abrikosov flux line lattice. As to be expected, the formalism of classical statistical dynamics is the classical limit of the general formalism of non-equilibrium states, Schwinger's closed time path formulation of quantum statistical mechanics, the general technique to treat non-equilibrium states we have developed and applied in this book. The language of quantum field theory is thus the tool to study fluctuations whatever their nature might be.

Appendix A

Path integrals

Quantum dynamics was stated in Chapter 1 in terms of operator calculus, viz. through the Schrödinger equation or equivalently via the Hamiltonian as in the evolution operator. Alternatively, quantum dynamics can be expressed in terms of path integrals which directly exposes the basic principle of quantum mechanics, the superposition principle[1]. To acquaint ourselves with path integrals we show here for the case of a single particle the equivalence of the two formulations by deriving the path integral formulation from the operator expression for Dirac's transformation function of Eq. (1.16), $\langle \mathbf{x}, t | \mathbf{x}', t' \rangle = \langle \mathbf{x} | \hat{U}(t, t') | \mathbf{x}' \rangle = G(\mathbf{x}, t; \mathbf{x}', t') \equiv K(\mathbf{x}, t; \mathbf{x}', t')$. Propagating in small steps by inserting complete sets at intermediate times we have for the propagator

$$\langle \mathbf{x}, t | \mathbf{x}', t' \rangle = \int d\mathbf{x}_1 \int d\mathbf{x}_2 \ldots \int d\mathbf{x}_N \, \langle \mathbf{x}, t | \mathbf{x}_N, t_N \rangle \langle \mathbf{x}_N, t_N | \mathbf{x}_{N-1}, t_{N-1} \rangle$$

$$\times \quad \langle \mathbf{x}_{N-1}, t_{N-1} | \mathbf{x}_{N-2}, t_{N-2} \rangle \cdots \langle \mathbf{x}_1, t_1 | \mathbf{x}', t' \rangle . \qquad (A.1)$$

We are consequently interested in the transformation function for infinitesimal times, and from Eq. (1.16) we obtain

$$\langle \mathbf{x}_n, t_n | \mathbf{x}_{n-1}, t_{n-1} \rangle = \langle \mathbf{x}_n | e^{-\frac{i}{\hbar} \Delta t \hat{H}(t_n)} | \mathbf{x}_{n-1} \rangle$$

$$= \delta(\mathbf{x}_n - \mathbf{x}_{n-1}) + \frac{\Delta t}{i\hbar} \langle \mathbf{x}_n | \hat{H}(t_n) | \mathbf{x}_{n-1} \rangle + \mathcal{O}(\Delta t^2) , \quad (A.2)$$

where $\Delta t = t_n - t_{n-1} = (t - t')/(N+1))$, as we have inserted N intermediate resolutions of the identity.

In the following we shall consider a particle of mass m in a potential V for which we have the Hamiltonian

$$\hat{H}(t) = \frac{\hat{\mathbf{p}}^2}{2m} + V(\hat{\mathbf{x}}, t) , \qquad (A.3)$$

i.e. $\hat{H} = H(\hat{\mathbf{p}}, \hat{\mathbf{x}}, t)$, where H by correspondence is Hamilton's function.

---

[1]For a detailed exposition of how the superposition principle for alternative paths leads to the Schrödinger equation, we refer the reader to chapter 1 of reference [1].

Inserting a complete set of momentum states, we get

$$\langle \mathbf{x}_n | H(\hat{\mathbf{x}}, \hat{\mathbf{p}}, t_n) | \mathbf{x}_{n-1} \rangle = \langle \mathbf{x}_n | H(\mathbf{x}_n, \hat{\mathbf{p}}, t_n) | \mathbf{x}_{n-1} \rangle$$

$$= \int \frac{d\mathbf{p}_n}{(2\pi\hbar)^d} \, e^{\frac{i}{\hbar} \mathbf{p}_n \cdot (\mathbf{x}_n - \mathbf{x}_{n-1})} \, H(\mathbf{x}_n, \mathbf{p}_n, t_n) \,, \quad \text{(A.4)}$$

where we encounter Hamilton's function on phase space

$$H(\mathbf{x}_n, \mathbf{p}_n, t_n) = \frac{\mathbf{p}_n^2}{2m} + V(\mathbf{x}_n, t_n) \,. \quad \text{(A.5)}$$

Inserting into Eq. (A.2), we get

$$\langle \mathbf{x}_n, t_n | \mathbf{x}_{n-1}, t_{n-1} \rangle = \int \frac{d\mathbf{p}_n}{(2\pi\hbar)^d} \, e^{\frac{i}{\hbar} \mathbf{p}_n \cdot (\mathbf{x}_n - \mathbf{x}_{n-1})} \left(1 + \frac{\Delta t}{i\hbar} H(\mathbf{x}_n, \mathbf{p}_n, t_n) + \mathcal{O}(\Delta t^2) \right)$$

$$= \int \frac{d\mathbf{p}_n}{(2\pi\hbar)^d} \, e^{\frac{i}{\hbar} [\mathbf{p}_n \cdot (\mathbf{x}_n - \mathbf{x}_{n-1}) - \Delta t H(\mathbf{x}_n, \mathbf{p}_n, t_n)]} + \mathcal{O}(\Delta t^2) \,. \quad \text{(A.6)}$$

Inserting additional internal times, we approach the limit $\Delta t \to 0$, or equivalently $N \to \infty$, obtaining for the transformation function

$$\langle \mathbf{x}, t | \mathbf{x}', t' \rangle = \lim_{N \to \infty} \int \prod_{n=1}^{N} d\mathbf{x}_n \prod_{n=1}^{N+1} \frac{d\mathbf{p}_n}{(2\pi\hbar)^d} \, e^{\frac{i}{\hbar} [\mathbf{p}_n \cdot (\mathbf{x}_n - \mathbf{x}_{n-1}) - \Delta t H(\mathbf{x}_n, \mathbf{p}_n, t_n)]}$$

$$\equiv \int \frac{\mathcal{D}\mathbf{x}_{\bar{t}} \, \mathcal{D}\mathbf{p}_{\bar{t}}}{(2\pi\hbar)^d} \, e^{\frac{i}{\hbar} \int_{t'}^{t} d\bar{t} \, [\mathbf{p}_{\bar{t}} \cdot \dot{\mathbf{x}}_{\bar{t}} - H(\mathbf{x}_{\bar{t}}, \mathbf{p}_{\bar{t}}, \bar{t})]} \,, \quad \text{(A.7)}$$

where $\mathbf{x}_0 \equiv \mathbf{x}'$, and $\mathbf{x}_{N+1} \equiv \mathbf{x}$. In the last equation we have just written the limit of the sum as a path integral, and the integration measure has been identified by the explicit limiting procedure.

The Hamilton function is quadratic in the momentum variable, and we have Gaussian integrals which can be performed

$$\int_{-\infty}^{\infty} \frac{d\mathbf{p}_n}{(2\pi\hbar)^d} \, e^{\frac{i}{\hbar} \Delta t (\mathbf{p}_n \cdot \frac{\mathbf{x}_n - \mathbf{x}_{n-1}}{\Delta t} - \frac{\mathbf{p}_n^2}{2m})} = \left(\frac{m}{2\pi i \hbar \Delta t} \right)^{d/2} e^{\frac{i}{2\hbar} m \left(\frac{\mathbf{x}_n - \mathbf{x}_{n-1}}{\Delta t} \right)^2 \Delta t} \quad \text{(A.8)}$$

and we thus get for the propagator

$$K(\mathbf{x}, t; \mathbf{x}', t') = \lim_{N \to \infty} \frac{1}{\left(\frac{m}{2\pi i \hbar \Delta t} \right)^{-d/2}} \int \prod_{n=1}^{N} \frac{d\mathbf{x}_n}{\left(\frac{m}{2\pi i \hbar \Delta t} \right)^{-d/2}} e^{\frac{i}{\hbar} \Delta t \sum_{n=1}^{N+1} \left[\frac{m (\mathbf{x}_n - \mathbf{x}_{n-1})^2}{2\Delta t} - V(\mathbf{x}_n, t_n) \right]}$$

$$\equiv \int_{\mathbf{x}_{t'} = \mathbf{x}'}^{\mathbf{x}_t = \mathbf{x}} \mathcal{D}\mathbf{x}_{\bar{t}} \, e^{\frac{i}{\hbar} \int_{t'}^{t} d\bar{t} \, L(\mathbf{x}_{\bar{t}}, \dot{\mathbf{x}}_{\bar{t}}, \bar{t})} \,, \quad \text{(A.9)}$$

where L in the continuum limit is seen to be Lagrange's function

$$L(\mathbf{x}_t, \dot{\mathbf{x}}_t, t) = \frac{1}{2}m\dot{\mathbf{x}}_t^2 - V(\mathbf{x}_t, t) = \dot{\mathbf{x}}_t \cdot \mathbf{p}_t - H(\mathbf{x}_t, \mathbf{p}_t, t) \tag{A.10}$$

related to Hamilton's function through a Legendre transformation. The integration measure has here been obtained for the case where we take the piecewise linear approximation for a path.[2]

Instead of formulating quantum dynamics in terms of operator calculus we have thus exhibited it in a way revealing the underlying superposition principle, viz. according to Feynman's principle: each possible alternative path contributes a pure phase factor to the propagator, $\exp\{iS/\hbar\}$, where

$$S[\mathbf{x}_t] = \int_{t'}^{t} d\bar{t}\, L(\mathbf{x}_{\bar{t}}, \dot{\mathbf{x}}_{\bar{t}}, \bar{t}) \tag{A.11}$$

is the classical action expression for the path, $\mathbf{x}_t$, in question.[3]

The classical path is determined by stationarity of the action

$$\left.\frac{\delta S}{\delta \mathbf{x}_t}\right|_{\mathbf{x}_t = \mathbf{x}_t^{cl}} = \mathbf{0} \tag{A.12}$$

the principle of least action,[4] or explicitly through the Euler-Lagrange equations

$$\frac{d}{dt}\left(\frac{\partial L}{\partial \dot{\mathbf{x}}}\right) - \frac{\partial L}{\partial \mathbf{x}} = 0 \tag{A.13}$$

the classical equation of motion.

Formulating quantum mechanics of a single particle as the zero-dimensional limit of quantum field theory amounts to focussing on the correlation functions of, for example, the position operator in the Heisenberg picture, say the time-ordered correlation function

$$G_{\mathrm{H}}(t, t') \equiv \langle T(\hat{\mathbf{x}}_{\mathrm{H}}(t)\, \hat{\mathbf{x}}_{\mathrm{H}}(t')) \rangle \tag{A.14}$$

where the bracket refers to averaging with respect to some state of the particle, pure or mixed, say for the ground state

$$G_{\mathrm{H}}(t, t') = \langle \psi_0 | T(\hat{\mathbf{x}}_{\mathrm{H}}(t)\, \hat{\mathbf{x}}_{\mathrm{H}}(t')) | \psi_0 \rangle . \tag{A.15}$$

---

[2]Other measures can be used, such as expanding the paths on a complete set of functions, so that the sum over all paths becomes the integral over all the expansion coefficients.

[3]In classical mechanics only the classical paths between two space-time points in question are of physical relevance; however, stating the quantum law of motion involves all paths. The way in which the various alternative paths contribute to the expression for the propagator was conceived by Dirac [152], who realized that the conditional amplitude for an infinitesimal time step is related to Lagrange's function, L, according to

$$\langle \mathbf{x}, t + \Delta t | \mathbf{x}', t \rangle \propto e^{\frac{i}{\hbar}\Delta t\, L(\mathbf{x}, (\mathbf{x} - \mathbf{x}')/\Delta t)}$$

however, with L expressed in terms of the coordinates at times t and $t + \Delta t$. This gem of Dirac's was turned into brilliance by Feynman.

[4]Or principle of stationary action, but typically the extremum is a minimum.

Noting, by inserting complete sets of eigenstates for the Heisenberg operators, $\hat{x}_H(t)\,|\mathbf{x},t\rangle = \mathbf{x}\,|\mathbf{x},t\rangle$, we have, for $t_i < t, t' < t_f$,

$$\langle \mathbf{x}_f, t_f | T(\hat{\mathbf{x}}_H(t)\,\hat{\mathbf{x}}_H(t')) | \mathbf{x}_i, t_i \rangle \;=\; \int_{\mathbf{x}_{t_i}=\mathbf{x}_i}^{\mathbf{x}_{t_f}=\mathbf{x}_f} \mathcal{D}\mathbf{x}_{\bar{t}}\; \mathbf{x}_t\, \mathbf{x}_{t'}\; e^{\frac{i}{\hbar}\int_{t_i}^{t_f} d\bar{t}\, L(\mathbf{x}_{\bar{t}}, \dot{\mathbf{x}}_{\bar{t}}, \bar{t})}\,, \tag{A.16}$$

where on the right-hand side the order of the real position variables $\mathbf{x}_t$ and $\mathbf{x}_{t'}$ is immaterial since the path integral automatically gives the time-ordered correlation function due to its built-in time-slicing defining procedure (recall Eq. (A.9)). We therefore have

$$G_H(t,t') \;=\; \int d\mathbf{x}_f \int d\mathbf{x}_i\; \psi_0^*(\mathbf{x}_f)\,\psi_0(\mathbf{x}_i) \int_{\mathbf{x}_{t_i}=\mathbf{x}_i}^{\mathbf{x}_{t_f}=\mathbf{x}_f} \mathcal{D}\mathbf{x}_{\bar{t}}\; \mathbf{x}_t\, \mathbf{x}_{t'}\; e^{\frac{i}{\hbar}\int_{t_i}^{t_f} d\bar{t}\, L(\mathbf{x}_{\bar{t}}, \dot{\mathbf{x}}_{\bar{t}}, \bar{t})} \tag{A.17}$$

and equivalently for any number of time ordered Heisenberg operators, thereby representing any time-ordered correlation function on path integral form.

Exercise A.1. Derive for a particle in a potential the path integral expression for the imaginary-time propagator (consider the one-dimensional case for simplicity for a start)

$$\mathcal{G}(x, x', \hbar/kT) \;\equiv\; G(x, -i\hbar/kT; x', 0) \;=\; \langle x | e^{-\hat{H}/kT} | x' \rangle \;=\; \int_{x(0)=x'}^{x(\hbar/kT)=x} \mathcal{D}x_\tau\; e^{-S_\mathcal{E}[x_\tau]/\hbar}$$

$$\tag{A.18}$$

where the Euclidean action

$$S_\mathcal{E}[x_\tau] = \int_0^{\hbar/kT} d\tau\; L_\mathcal{E}(x_\tau, \dot{x}_\tau) \tag{A.19}$$

is specified in terms of the Euclidean Lagrange function

$$L_\mathcal{E}(x_\tau, \dot{x}_\tau) = \frac{1}{2}m\dot{x}_\tau^2 + V(x_\tau)\,, \tag{A.20}$$

where the potential energy is *added* to the kinetic energy.

Interpreting τ as a length, we note that the Euclidean Lagrange function $L_\mathcal{E}$ equals the potential energy of a string of *length* $L \equiv \hbar/kT$ and tension m, placed in the external potential V, and we have established that the imaginary-time propagator is specified in terms of the classical partition function for the string.

In general, only for the case of a quadratic Lagrange function, i.e. for homogeneous external fields, can the path integral for the propagator be performed, or rather simply circumvented by shifting the variable of integration to that of the deviation from the classical path, $\mathbf{x}(t) = \mathbf{x}_{cl}(t) + \delta\mathbf{x}_t$, and recalling that the action is stationary for the classical path, leading to

$$K(\mathbf{x}, t; \mathbf{x}', t') = A(t, t')\, e^{\frac{i}{\hbar}S_{cl}(\mathbf{x},t;\mathbf{x}',t')}\,, \tag{A.21}$$

i.e. specified in terms of the action for the classical path and a prefactor, the contribution from the Gaussian fluctuations around the classical path, which can be determined from the initial condition for the propagator Eq. (1.15).

Exercise A.2. Obtain the expression for the propagator, $T \equiv t - t'$,

$$
\begin{aligned}
K(x,t,x',t') &= \sqrt{\frac{m\omega}{2\pi i \hbar \sin \omega T}} \exp \Bigg\{ \frac{im\omega}{2\hbar \sin \omega T} \Big((x^2 + x'^2) \cos \omega T - 2xx' \\
&\quad + \frac{2x}{m\omega} \int_t^{t'} d\bar{t} f(\bar{t}) \sin(\omega(\bar{t} - t')) + \frac{2x'}{m\omega} \int_{t'}^t d\bar{t} f(\bar{t}) \sin(\omega(t - \bar{t})) \\
&\quad - \frac{2}{m^2 \omega^2} \int_{t'}^t dt_2 \int_{t_2}^{t'} dt_1 f(t_2) f(t_1) \sin(\omega(t - t_2)) \sin(\omega(t_1 - t')) \Big) \Bigg\}
\end{aligned}
$$

(A.22)

for a forced harmonic oscillator

$$
L(x_t, \dot{x}_t, t) = \frac{1}{2} m \dot{x}_t^2 - \frac{1}{2} m \omega^2 x_t^2 + f(t) x_t
\tag{A.23}
$$

by evaluating the classical action.

Consider a particle coupled weakly to N other degrees of freedom, i.e., linearly to a set of N harmonic oscillators collectively labeled $\mathbf{R} = (R_1, R_2, ..., R_N)$. The total Lagrange function, $L = L_\mathrm{S} + L_\mathrm{I} + L_\mathrm{E}$, is then

$$
L_\mathrm{S} = \frac{1}{2} m \dot{x}^2 - V(x,t) \quad , \quad L_\mathrm{E} = \frac{1}{2} \sum_{\alpha=1}^N \left(m_\alpha \dot{R}_\alpha^2 - m \omega_\alpha^2 R_\alpha^2 \right) ,
\tag{A.24}
$$

where the particle in addition is coupled to an applied external potential, $V(x,t)$, and the linear interaction with the environment oscillators is specified by

$$
L_\mathrm{I} = -x \sum_{\alpha=1}^N \lambda_\alpha R_\alpha .
\tag{A.25}
$$

At some past moment in time, t', the density matrix is assumed separable, $\rho(x, \mathbf{R}, x', \mathbf{R}', t') = \rho_\mathrm{S}(x, x') \rho_\mathrm{E}(\mathbf{R}, \mathbf{R}')$, i.e. prior to that initial time the particle did not interact with the environment of oscillators, the system and the environment are uncorrelated. The equation, Eq. (3.13), for the density matrix specifies, by tracing out the oscillator degrees of freedom, the density matrix for the particle at time t in terms of its density matrix at the initial time according to

$$
\rho(x_\mathrm{f}, x_\mathrm{f}', t) = \int dx_\mathrm{i} \int dx_\mathrm{i}' \, J(x_\mathrm{f}, x_\mathrm{f}', t; x_\mathrm{i}, x_\mathrm{i}', t') \, \rho(x_\mathrm{i}, x_\mathrm{i}')
\tag{A.26}
$$

and the propagator of the particle density matrix is

$$J(x_f, x_f', t; x_i, x_i', t') = \int_{x_{t'}^{(1)}=x_i'}^{x_t^{(1)}=x_f'} \mathcal{D}x_{\bar{t}}^{(1)} \int_{x_{t'}^{(2)}=x_i}^{x_t^{(2)}=x_f} \mathcal{D}x_{\bar{t}}^{(2)} \, e^{\frac{i}{\hbar}(S(x_{\bar{t}}^{(2)})-S(x_{\bar{t}}^{(1)}))} \, \mathcal{F}[x_{\bar{t}}^{(1)}, x_{\bar{t}}^{(2)}] \quad \text{(A.27)}$$

where S is the action for the particle in the absence of the environment, and the so-called influence functional $\mathcal{F}$ is

$$\mathcal{F}[x_t^{(1)}, x_t^{(2)}] = \int d\mathbf{R}_f \int d\mathbf{R}_i \int d\mathbf{R}_i' \, \rho_E(\mathbf{R}_i, \mathbf{R}_i') \int_{\mathbf{R}(t')=\mathbf{R}_i}^{\mathbf{R}(t)=\mathbf{R}_f} \mathcal{D}\mathbf{R}(t) \int_{\mathbf{Q}(t')=\mathbf{R}_i'}^{\mathbf{Q}(t)=\mathbf{R}_f} \mathcal{D}\mathbf{Q}(t)$$

$$\times \quad \exp\left\{ \frac{i}{\hbar} \left(S_I[x_t^{(2)}, \mathbf{R}(t)] + S_E[\mathbf{R}(t)] - S_I[x_t^{(1)}, \mathbf{Q}(t)] - S_E[\mathbf{Q}(t)] \right) \right\},$$

$$\text{(A.28)}$$

where S_E is the action for the isolated environment oscillators, and S_I is the action due to the interaction, analogous to the external force term in Eq. (A.23) upon the substitution $f \to \lambda x$ for each of the couplings to the oscillators.

Assuming that the initial state of the oscillators is the thermal equilibrium state, $\rho_E(\mathbf{R}, \mathbf{R}') = \prod_\alpha \rho_T(R_\alpha, R_\alpha)$, the equilibrium density matrix is immediately obtained from Eq. (1.21) and the result of Exercise A.2, in the absence of the force, as it is obtained upon the substitution $t - t' \to -i\hbar/kT$ in Eq. (A.22), here T denotes the temperature (or equivalently in view of Exercise A.1, the imaginary time variable being interpreted as variable on the appendix part of the contour depicted in Figure 4.4)

$$\rho_T(R_\alpha, R_\alpha') = \exp\left\{ -\frac{m_\alpha \omega_\alpha}{2\hbar \sinh \frac{\hbar\omega_\alpha}{kT}} \left((R_\alpha^2 + R_\alpha'^2) \cosh \frac{\hbar\omega_\alpha}{kT} - 2R_\alpha R_\alpha' \right) \right\}$$

$$\times \quad \left(\frac{m_\alpha \omega_\alpha}{2\pi\hbar \sinh \frac{\hbar\omega_\alpha}{kT}} \right)^{1/2}. \quad \text{(A.29)}$$

The path integrals with respect to the oscillators are immediately obtained using the result of Exercise A.2, and the remaining three ordinary integrals in Eq. (A.28) are Gaussian and can be performed giving for the influence functional

$$\mathcal{F}[x_{\bar{t}}^{(1)}, x_{\bar{t}}^{(2)}] = \exp\left\{ \frac{i}{\hbar} \int_{t'}^t dt_2 \int_{t'}^{t_2} dt_1 [x_{t_2}^{(2)} - x_{t_2}^{(1)}] \, D(t_2 - t_1) \, [x_{t_1}^{(2)} + x_{t_1}^{(1)}] \right.$$

$$\left. - \frac{1}{\hbar} \int_{t'}^t dt_2 \int_{t'}^{t_2} dt_1 [x_{t_2}^{(2)} - x_{t_2}^{(1)}] \, D^K(t_2 - t_1) \, [x_{t_1}^{(2)} - x_{t_1}^{(1)}] \right\}, \quad \text{(A.30)}$$

where

$$D^{\mathrm{K}}(t - t') \;=\; \sum_\alpha \frac{\lambda_\alpha^2}{2m_\alpha \omega_\alpha} \coth\left(\frac{\hbar\omega_\alpha}{2kT}\right) \cos(\omega_\alpha(t - t'))$$

$$= \; \frac{1}{\hbar} \sum_\alpha \lambda_\alpha^2 \, \langle\{\hat{R}_\alpha(t), \hat{R}_\alpha(t')\}\rangle \tag{A.31}$$

and

$$D(t - t') \;=\; \sum_\alpha \frac{\lambda_\alpha^2}{2m_\alpha \omega_\alpha} \sin(\omega_\alpha(t - t')) \;=\; \frac{1}{\hbar} \sum_\alpha \lambda_\alpha^2 \, \langle[\hat{R}_\alpha(t), \hat{R}_\alpha(t')]\rangle \tag{A.32}$$

specifies the non-Markovian dynamics of the oscillator through a systematic dissipative or friction term and the kinetic Green's function, Eq. (A.31), describing the fluctuation effects of the environment, the two physically distinctly different terms being related by the fluctuation–dissipation relation.[5] The influence functional is also immediately obtained by observing that the expression in Eq. (A.28) can be put on contour form by letting the time variable reside on the contour depicted in Figure 4.4, noting the force is vanishing on the appendix part of the contour. We then obtain the exponent of the form as in Eq. (9.38), and by combining the retarded and advanced terms the form in Eq. (A.30). This observation accounts for the identification in terms of the operator expressions for the thermal equilibrium oscillator Green's functions in Eq. (A.31) and Eq. (A.32).[6]

Introducing a continuum of oscillators and the coupling in such a way that the spectral weight function of the oscillators

$$J(\omega) \;=\; \pi \sum_\alpha \frac{\lambda_\alpha^2}{2m_\alpha \omega_\alpha} \delta(\omega - \omega_\alpha) \;=\; i\left(D^{\mathrm{R}}(\omega) - D^{\mathrm{A}}(\omega)\right) \;=\; D(\omega) \tag{A.33}$$

becomes the linear or Ohmic spectrum

$$J(\omega) \;=\; \eta\,\omega\,\theta(\omega_{\mathrm{c}} - \omega) \tag{A.34}$$

the friction term becomes local as $D(t) = -\eta\,\dot{\delta}(t)$ in the limit of a large cut-off frequency, i.e. for times much larger than ω_{c}^{-1}. We then obtain for the propagator of the density matrix for the particle

$$J(x_{\mathrm{f}}, x_{\mathrm{f}}', t; x_{\mathrm{i}}, x_{\mathrm{i}}', t') \;=\; \int \mathcal{D}x(t) \int \mathcal{D}y(t)\, \exp\left\{\frac{i}{\hbar}(S_1 + S_2)\right\}, \tag{A.35}$$

---

[5] Instead of brute force, the result follows straightforwardly from the expresssion for the generating functional for a harmonic oscillator, Eq. (4.108), and handling the linear coupling according to Eq. (9.41) and Eq. (9.27).

[6] Essential for the structure in the expression in Eq. (A.30) is only that the coupling to the environment oscillators is linear. The non-equilibrium, i.e. driven, spin-boson problem, representing for example a monitored qubit coupled to a decohering dissipative environment, is discussed in reference [153].

where $x_t = (x_t^{(2)} + x_t^{(1)})/2$ and $y_t = x_t^{(2)} - x_t^{(1)}$, and

$$S_2 = i \int_{t'}^{t} dt_2 \int_{t'}^{t} dt_1 \, y_{t_2} \, D^K(t_2 - t_1) \, y_{t_1} \tag{A.36}$$

and (up to a boundary term which vanishes for initial and final states satisfying $y(t) = 0 = y(t')$, which will be assumed in the following)

$$S_1 = - \int_{t'}^{t} d\bar{t} \, y_{\bar{t}} \Big(m\ddot{x}_{\bar{t}} + \eta \dot{x}_{\bar{t}} + V_R(x_{\bar{t}} + y_{\bar{t}}/2) - V_R(x_{\bar{t}} - y_{\bar{t}}/2) \Big) \tag{A.37}$$

where the Ohmic spectrum guarantees a friction force proportional to the velocity. For the chosen type of coupling, the potential is the result of the interaction renormalized by a harmonic contribution, $V_R(x) = V(x) - \omega_c \eta x^2/\pi$. We have arrived at the Feynman–Vernon path integral theory of dissipative quantum dynamics for the case of an Ohmic environment [154, 155, 156, 157].

If the external potential is at most harmonic, the path integral with respect to y_t is Gaussian and can be performed giving an expression for the path probability analogous to Eq. (12.9). We therefore obtain that the dissipative dynamics of the quantum oscillator is a Gaussian stochastic process described by the Langevin equation, Eq. (12.1), however the noise is not just the classical thermal one of Eq. (12.2), but includes the quantum noise due to the environment as the stochastic force is described by the correlation function

$$D^K(t) = \int_{-\omega_c}^{\omega_c} \frac{d\omega}{2\pi} e^{-i\omega t} D^K(\omega) \quad , \quad D^K(\omega) = \eta \omega \coth \frac{\hbar \omega}{2kT} . \tag{A.38}$$

The time scale of the correlations in the environment, t_c, the measure of the non-Markovian character of the dynamics, is set by the temperature according to

$$\frac{\hbar}{2\eta kT} \int_{-\infty}^{\infty} dt \, t^2 D^K(t) = -t_c^2 \quad , \quad t_c = \frac{\hbar}{2kT} \quad , \quad \frac{\hbar}{2\eta kT} \int_{-\infty}^{\infty} dt \, D^K(t) = 1 . \tag{A.39}$$

We note that, owing to quantum effects, the noise is not white but blue

$$D^K(t) = -2\eta \left(\frac{kT}{\hbar}\right)^2 \frac{1}{\sinh^2 \frac{\pi kT|t|}{\hbar}} \quad , \quad \omega_c|t| \gg 1 . \tag{A.40}$$

We note that the damping term S_2 in Eq. (A.36) limits the excursions of $y(t)$. In the high temperature limit, $kT \gg \hbar\omega_c$, quantum excursions y_t of the particle are suppressed, and the integration with respect to $y(t)$ is Gaussian and the remaining integrand in the path integral in Eq. (A.35) is the probability distribution for a given realization of a classical path, Eq. (12.9). The corresponding Markovian stochastic process in the Wigner coordinate, x_t, is described by the Langevin equation, Eq. (12.1), and we recover the theory of classical stochastic dynamics discussed in Section 12.1. At high enough temperatures, all quantum interference effects of the particle are suppressed by the thermal fluctuations, and the classical dissipative dynamics of the particle emerges. We note how potentials, alien to classical dynamics but essential in quantum dynamics, disappear as the classical limit emerges, delivering only the effect of the corresponding classical force, $-V'(x_t)$.

Appendix B

Path integrals and symmetries

A virtue of the path integral formulation is that symmetries of the action easily lead to exact relations between various Green's functions, the Ward identities.

An infinitesimal symmetry transformation

$$\phi_1 \rightarrow \phi_1 + \epsilon F_1[\phi] \tag{B.1}$$

is one that leaves the action invariant, i.e.

$$\delta S = \epsilon \int d1 \, \frac{\delta S[\phi]}{\delta \phi_1} \, F_1[\phi] = 0. \tag{B.2}$$

If the infinitesimal symmetry transformation is not global, i.e. ϵ is not a constant infinitesimal, but an infinitesimal function of space and time, $\epsilon(t, \mathbf{x})$, the variation of the action under the transformation

$$\phi(t, \mathbf{x}) \rightarrow \phi(t, \mathbf{x}) + \epsilon(t, \mathbf{x}) \tag{B.3}$$

will in general not vanish, but takes the form, $x = (t, \mathbf{x}) = x_\mu$

$$\delta S = -\sum_{\mu=0}^{3} \int dx \, j_\mu(x) \, \frac{\partial \epsilon(x)}{\partial x_\mu} \tag{B.4}$$

in order to vanish for the global case considered above. If the field $\phi(t, \mathbf{x})$ satisfies the classical equation of motion

$$\frac{\delta S[\varphi]}{\delta \varphi(t, \mathbf{x})} = 0 \tag{B.5}$$

the action is stationary with respect to arbitrary variations, and assuming $\epsilon(t, \mathbf{x})$ to vanish for large arguments, a partial integration leads to the continuity equation

$$\sum_{\mu=0}^{3} \frac{\partial j_\mu(x)}{\partial x_\mu} = 0 \tag{B.6}$$

and the existence of the conserved quantity, the constant of motion

$$Q = \int d\mathbf{x}\, j_0(t, \mathbf{x}) .$$ (B.7)

A symmetry of the action thus implies a conservation law, Noether's theorem.

Returning to the global transformation, Eq. (B.1), the measure in the path integral representation of the generating functional

$$Z[J] = \int \mathcal{D}\phi\, e^{i[\phi] + i\phi\, J}$$ (B.8)

changes with the Jacobian according to

$$\mathcal{D}\phi \to \mathcal{D}\phi\, \mathrm{Det}\left(\delta_{12} - \epsilon \frac{\delta F_1[\phi]}{\delta \phi_2}\right) = \mathcal{D}\phi\left(1 - \epsilon \frac{\delta F_1[\phi]}{\delta \phi_1}\right) + \mathcal{O}(\epsilon^2)$$ (B.9)

and since the generating functional is invariant with respect to the transformation Eq. (B.1), we obtain

$$Z[J] = \int \mathcal{D}\phi\left(1 - \epsilon \frac{\delta F_1[\phi]}{\delta \phi_1}\right) e^{i[\phi] + i\phi\, J}\left(1 + \left(i\frac{\delta S[\phi]}{\delta \phi_1} + iJ\right)\epsilon F_1[\phi]\right) + \mathcal{O}(\epsilon^2)$$

and thereby

(B.10)

$$\int \mathcal{D}\phi\, e^{i[\phi] + i\phi\, J}\left(\left(\frac{\delta S[\phi]}{\delta \phi_1} + J_1\right) F_1[\phi] + i\frac{\delta F_1[\phi]}{\delta \phi_1}\right) = 0$$ (B.11)

or equivalently

$$\left(\left(\frac{\delta S\left[\frac{\delta}{i\delta J}\right]}{\delta \phi_1} + J_1\right) F_1\left[\frac{\delta}{i\delta J}\right] + i\frac{\delta F_1\left[\frac{\delta}{i\delta J}\right]}{\delta \phi_1}\right) Z[J] = 0 .$$ (B.12)

In the event that the transformation, Eq. (B.1), is a translation, i.e. just a field independent constant, $F_1[\phi] = f_1$, Eq. (B.12) simply becomes the Dyson–Schwinger equation, Eq. (9.32), (recall also Eq. (10.42)).

The real advantage of the path integral formulation presents itself if the transformation, $F_1[\phi]$, is a symmetry of the action

$$\frac{\delta S[\phi]}{\delta \phi_1} F_1[\phi] = 0$$ (B.13)

which leaves also the measure $\mathcal{D}\phi$ invariant, in which case Eq. (B.12) becomes the Ward identity

$$J_1 F_1\left[\frac{\delta}{i\delta J}\right] Z[J] = 0$$ (B.14)

relating various Green's functions, for example the vertex function and the one-particle Green's functions.

Appendix C

Retarded and advanced Green's functions

In this appendix we shall consider the properties of the retarded and advanced Green's functions for the case of a single particle. When it comes to calculations Green's functions are convenient, and even more so when many-body systems and their interactions are considered as studied in the main text.

The retarded Green's function or propagator for a single particle is defined as (the choice of phase factor is for convenience of perturbation expansions)

$$G^{\mathrm{R}}(\mathbf{x},t;\mathbf{x}',t') \equiv \begin{cases} -iG(\mathbf{x},t;\mathbf{x}',t') & \text{for } t \geq t' \\ 0 & \text{for } t < t', \end{cases} \tag{C.1}$$

where the propagator for a single particle already was considered in Appendix A, $G(\mathbf{x},t;\mathbf{x}',t') = \langle \mathbf{x},t|\mathbf{x}',t' \rangle = \langle \mathbf{x}|\hat{U}(t,t')|\mathbf{x}' \rangle$. The retarded propagator for a particle whose dynamics is specified by the Hamiltonian H, satisfies the equation

$$\left\{ i\hbar\frac{\partial}{\partial t} - H \right\} G^{\mathrm{R}}(\mathbf{x},t;\mathbf{x}',t') = \hbar\,\delta(\mathbf{x}-\mathbf{x}')\,\delta(t-t') \tag{C.2}$$

which in conjunction with the condition

$$G^{\mathrm{R}}(\mathbf{x},t;\mathbf{x},t') = 0 \qquad \text{for} \quad t < t' \tag{C.3}$$

specifies the retarded propagator. The source term on the right-hand side of Eq. (C.2) represents the discontinuity in the retarded propagator at time $t = t'$, and is recognized by integrating the left-hand side of Eq. (C.2) over an infinitesimal time interval around t', and using the initial condition[1]

$$G^{\mathrm{R}}(\mathbf{x},t'+0;\mathbf{x}',t') = -i\,\delta(\mathbf{x}-\mathbf{x}'). \tag{C.4}$$

---

[1] The retarded propagator also has the following interpretation: prior to time t' the particle is absent, and at time $t = t'$ the particle is created at point $\mathbf{x}'$, and is subsequently propagated according to the Schrödinger equation. In contrast to the relativistic quantum theory, this point of view of propagation is not mandatory in non-relativistic quantum mechanics where the quantum numbers describing the particle species are conserved.

Or one recalls that the derivative of the step function is the delta function. The retarded Green's function is thus the fundamental solution of the Schrödinger equation and rightfully the mathematical function introduced by Green. The inverse operator to a differential equation is expressed as an integral operator with the Green's function as the kernel. In the context of many-body theory we have used the label Green's in the less specific sense, just referring to correlation functions.

The retarded Green's function propagates the wave function forwards in time, as we have for $t > t'$ for the wave function at time t

$$\psi(\mathbf{x}, t) = i \int d\mathbf{x}' \, G^{\mathrm{R}}(\mathbf{x}, t; \mathbf{x}', t') \, \psi(\mathbf{x}', t') \tag{C.5}$$

in terms of the wave function at the earlier time t', and has the physical meaning of a probability amplitude for propagating between the two space-time points in question.

According to Eq. (C.1), the retarded propagator is given by

$$G^{\mathrm{R}}(\mathbf{x}, t; \mathbf{x}', t') = -i\theta(t - t') \, \langle \mathbf{x} | \hat{U}(t, t') | \mathbf{x}' \rangle . \tag{C.6}$$

By direct differentiation with respect to time it also immediately follows that the retarded propagator satisfies Eq. (C.2).

We note, according to Appendix A, the path integral expression for the retarded propagator

$$G^{\mathrm{R}}(\mathbf{x}, t; \mathbf{x}', t') = -i\theta(t - t') G(\mathbf{x}, t; \mathbf{x}', t')$$

$$= -i\,\theta(t - t') \int_{\mathbf{x}_{t'}=\mathbf{x}'}^{\mathbf{x}_t=\mathbf{x}} \mathcal{D}\mathbf{x}_{\bar{t}} \, e^{\frac{i}{\hbar} \int_{t'}^{t} d\bar{t}\, L(\mathbf{x}_{\bar{t}}, \dot{\mathbf{x}}_{\bar{t}})} . \tag{C.7}$$

We shall also need the advanced propagator

$$G^{\mathrm{A}}(\mathbf{x}, t; \mathbf{x}', t') \equiv \begin{cases} 0 & \text{for } t > t' \\ iG(\mathbf{x}, t; \mathbf{x}', t') & \text{for } t \leq t' , \end{cases} \tag{C.8}$$

which propagates the wave function backwards in time, as we have for $t < t'$ for the wave function at time t

$$\psi(\mathbf{x}, t) = -i \int d\mathbf{x}' \, G^{\mathrm{A}}(\mathbf{x}, t; \mathbf{x}', t') \, \psi(\mathbf{x}', t') \tag{C.9}$$

in terms of the wave function at the later time t'.

The retarded and advanced propagators are related according to

$$G^{\mathrm{A}}(\mathbf{x}, t; \mathbf{x}', t') = [G^{\mathrm{R}}(\mathbf{x}', t'; \mathbf{x}, t)]^* . \tag{C.10}$$

The advanced propagator is also a solution of Eq. (C.2), but zero in the opposite time region as compared to the retarded propagator.

We note that, in the position representation, we have

$$G(\mathbf{x}, t; \mathbf{x}', t') = \langle \mathbf{x} | \hat{U}(t, t') | \mathbf{x}' \rangle = i[G^{\mathrm{R}}(\mathbf{x}, t; \mathbf{x}', t') - G^{\mathrm{A}}(\mathbf{x}, t; \mathbf{x}', t')]$$

$$\equiv A(\mathbf{x}, t; \mathbf{x}', t') , \tag{C.11}$$

where we now have introduced the notation A for the Green's function G, and also refer to it as the spectral function.

Introducing the retarded and advanced Green's operators

$$\hat{G}^{\mathrm{R}}(t,t') \equiv -i\theta(t-t')\,\hat{U}(t,t') \ , \quad \hat{G}^{\mathrm{A}}(t,t') \equiv i\theta(t'-t)\,\hat{U}(t,t') \qquad (\mathrm{C}.12)$$

we have for the evolution operator

$$\hat{U}(t,t') = i(\hat{G}^{\mathrm{R}}(t,t') - \hat{G}^{\mathrm{A}}(t,t')) \ \equiv \ \hat{G}(t,t') \ \stackrel{\Delta}{=} \ A(t,t') \qquad (\mathrm{C}.13)$$

and the unitarity of the evolution operator is reflected in the hermitian relationship of the Green's operators

$$\hat{G}^{\mathrm{A}}(t,t') = [\hat{G}^{\mathrm{R}}(t',t)]^{\dagger} \ . \qquad (\mathrm{C}.14)$$

The retarded and advanced Green's operators are characterized as solutions to the same differential equation

$$\left(i\hbar\frac{\partial}{\partial t} - \hat{H}\right) \hat{G}^{\mathrm{R(A)}}(t,t') \ = \ \hbar\,\delta(t-t')\,\hat{I} \qquad (\mathrm{C}.15)$$

but are zero for different time relationship.

The various representations of the Green's operators are obtained by taking matrix elements. For example, in the momentum representation we have for the retarded propagator the matrix representation

$$G^{\mathrm{R}}(\mathbf{p},t;\mathbf{p}',t') \ = \ -i\theta(t-t')\langle\mathbf{p},t|\mathbf{p}',t'\rangle \ = \ \langle\mathbf{p}|\hat{G}^{\mathrm{R}}(t,t')|\mathbf{p}'\rangle \ . \qquad (\mathrm{C}.16)$$

Exercise C.1. Defining in general the imaginary-time propagator

$$\mathcal{G}(\mathbf{x},\tau;\mathbf{x}',\tau') \ \equiv \ \theta(\tau-\tau')\langle\mathbf{x}|e^{-\frac{\hat{H}(\tau-\tau')}{\hbar}}|\mathbf{x}'\rangle \qquad (\mathrm{C}.17)$$

show that for the Hamiltonian for a particle in a magnetic field described by the vector potential $\mathbf{A}(\hat{\mathbf{x}})$

$$\hat{H} \ = \ \frac{1}{2m}\Big(\hat{\mathbf{p}} - e\mathbf{A}(\hat{\mathbf{x}})\Big)^{2} \qquad (\mathrm{C}.18)$$

the imaginary-time propagator satisfies the equation

$$\left(\hbar\frac{\partial}{\partial\tau} + \frac{1}{2m}\left(\frac{\hbar}{i}\nabla_{\mathbf{x}} - e\mathbf{A}(\mathbf{x})\right)^{2}\right) \mathcal{G}(\mathbf{x},\tau;\mathbf{x}',\tau') \ \equiv \ \hbar\,\delta(\mathbf{x}-\mathbf{x}')\,\delta(\tau-\tau') \qquad (\mathrm{C}.19)$$

and write down the path integral representation of the solution.

The free particle propagator in the momentum representation

$$G_{0}^{\mathrm{R}}(\mathbf{p},t;\mathbf{p}',t') \ = \ -i\theta(t-t')\langle\mathbf{p}|e^{-\frac{i}{\hbar}\frac{\hat{p}^{2}}{2m}(t-t')}|\mathbf{p}'\rangle \qquad (\mathrm{C}.20)$$

is given by

$$G_0^R(\mathbf{p}, t; \mathbf{p}', t') = G_0^R(\mathbf{p}, t, t')\langle \mathbf{p}|\mathbf{p}'\rangle = G_0^R(\mathbf{p}, t - t')\begin{cases} \delta(\mathbf{p} - \mathbf{p}') \\ \delta_{\mathbf{p},\mathbf{p}'} \end{cases}, \qquad (C.21)$$

where the Kronecker or delta function (depending on whether the particle is confined to a box or not) reflects the spatial translation invariance of free propagation. The compatibility of the energy and momentum of a free particle, $[\hat{H}_0, \hat{p}] = 0$, is reflected in the definite temporal oscillations of the propagator

$$G_0^R(\mathbf{p}, t, t') = -i\theta(t - t')\, e^{-\frac{i}{\hbar}\epsilon_{\mathbf{p}}(t - t')} \qquad (C.22)$$

determined by the energy of the state in question

$$\epsilon_{\mathbf{p}} = \frac{\mathbf{p}^2}{2m} \qquad (C.23)$$

the dispersion relation for a free non-relativistic particle of mass m.

Fourier transforming, i.e. inserting a complete set of momentum states, we obtain for the free particle propagator in the spatial representation

$$G_0^R(\mathbf{x}, t; \mathbf{x}', t') = -i\theta(t - t')\langle \mathbf{x}|e^{-\frac{i}{\hbar}\hat{H}_0(t - t')}|\mathbf{x}'\rangle$$

$$= -i\theta(t - t')\left(\frac{m}{2\pi\hbar i(t - t')}\right)^{d/2} e^{\frac{im}{2\hbar}\frac{(\mathbf{x} - \mathbf{x}')^2}{t - t'}}. \qquad (C.24)$$

Exercise C.2. Show that the free retarded propagator in the momentum representation satisfies the equation

$$\left\{i\hbar\frac{\partial}{\partial t} - \epsilon_{\mathbf{p}}\right\} G_0^R(\mathbf{p}, t; \mathbf{p}', t') = \hbar\,\delta(\mathbf{p} - \mathbf{p}')\,\delta(t - t'). \qquad (C.25)$$

Appendix D

Analytic properties of Green's functions

In the following we shall in particular consider the analytical properties of the Green's functions for a single particle. However, by introducing the Green's operators, results are taken over to the general case of a many-body system.

For an isolated system, where the Hamiltonian is time independent, we can for any complex number E with a positive imaginary part, transform the retarded Green's operator, Eq. (C.12), according to

$$\hat{G}_E^{\mathrm{R}} = \frac{1}{\hbar} \int_{-\infty}^{\infty} d(t - t') \, e^{\frac{i}{\hbar}E(t-t')} \, \hat{G}^{\mathrm{R}}(t - t') . \tag{D.1}$$

The Fourier transform is obtained as the analytic continuation from the upper half plane, $\Im m E > 0$. According to Eq. (C.15) we have, for $\Im m E > 0$, the equation

$$\left(E - \hat{H}\right) \hat{G}_E^{\mathrm{R}} = \hat{I} . \tag{D.2}$$

Analogously we obtain that the advanced Green's operator is the solution of the same equation

$$\left(E - \hat{H}\right) \hat{G}_E = \hat{I} \tag{D.3}$$

for values of the energy variable E in the lower half plane, $\Im m E < 0$, and by analytical continuation to the real axis

$$\hat{G}_E^{\mathrm{A}} \equiv \frac{1}{\hbar} \int_{-\infty}^{\infty} dt \, e^{\frac{i}{\hbar}Et} \, \hat{G}^{\mathrm{A}}(t) . \tag{D.4}$$

We note the Fourier inversion formulas

$$\hat{G}^{\mathrm{R(A)}}(t) = \frac{1}{2\pi} \int_{-\infty \, (\pm) \, i0}^{\infty \, (\pm) \, i0} dE \, e^{-\frac{i}{\hbar}Et} \, \hat{G}_E^{\mathrm{R(A)}} \tag{D.5}$$

and the hermitian property, Eq. (C.14), leads to the relationship

$$\hat{G}_E^A = [\hat{G}_{E^*}^R]^\dagger \ . \tag{D.6}$$

We introduce the Green's operator

$$\hat{G}_E \equiv \begin{cases} \hat{G}_E^R & \text{for } \Im m E > 0 \\[2ex] \hat{G}_E^A & \text{for } \Im m E < 0 \end{cases} \tag{D.7}$$

for which we have the spectral representation

$$\hat{G}_E = \frac{1}{E - \hat{H}} = \sum_\lambda \frac{|\epsilon_\lambda\rangle\langle\epsilon_\lambda|}{E - \epsilon_\lambda} \tag{D.8}$$

in terms of the eigenstates, $|\epsilon_\lambda\rangle$, of the Hamiltonian

$$\hat{H} |\epsilon_\lambda\rangle. = \epsilon_\lambda |\epsilon_\lambda\rangle \ . \tag{D.9}$$

The analytical properties of the retarded and advanced Green's operators leads, by an application of Cauchy's theorem, to the spectral representations

$$\hat{G}_E^{R(A)} = \int_{-\infty}^\infty \frac{dE'}{2\pi} \frac{\hat{A}_{E'}}{E - E' \stackrel{(\pm)}{} i0} \tag{D.10}$$

where we have introduced the spectral operator, the discontinuity of the Green's operator across the real axis

$$\hat{A}_E \equiv i(\hat{G}_E^R - \hat{G}_E^A) = i(\hat{G}_{E+i0} - \hat{G}_{E-i0})$$

$$= 2\pi \delta(E - \hat{H}) = 2\pi \sum_\lambda |\epsilon_\lambda\rangle\langle\epsilon_\lambda| \delta(E - \epsilon_\lambda) \ . \tag{D.11}$$

Equivalently, we have the relationship between real and imaginary parts of, say, position representation matrix elements

$$\Re e\, G^R(\mathbf{x}, \mathbf{x}', E) = \mathcal{P} \int_{-\infty}^\infty \frac{dE'}{\pi} \frac{\Im m\, G^R(\mathbf{x}, \mathbf{x}', E')}{E' - E} \tag{D.12}$$

and

$$\Im m\, G^R(\mathbf{x}, \mathbf{x}', E) = -\mathcal{P} \int_{-\infty}^\infty \frac{dE'}{\pi} \frac{\Re e\, G^R(\mathbf{x}, \mathbf{x}', E')}{E' - E} \ . \tag{D.13}$$

The Kramers–Kronig relations due to the retarded propagator is analytic in the upper half-plane.

The perturbation expansion of the propagator in a static potential is seen to be equivalent to the operator expansion for the Green's operator

$$\hat{G}_E = \frac{1}{E - \hat{H}} = \frac{1}{E - \hat{H}_0 + \hat{V}} = \frac{1}{(E - \hat{H}_0)(1 - (E - \hat{H}_0)^{-1}\hat{V})}$$

$$= \frac{1}{1 - (E - \hat{H}_0)^{-1}\hat{V}} \frac{1}{E - \hat{H}_0}$$

$$= \left(1 + (E - \hat{H}_0)^{-1}\hat{V} + (E - \hat{H}_0)^{-1}\hat{V}(E - \hat{H}_0)^{-1}\hat{V} + \dots\right) \frac{1}{E - \hat{H}_0}$$

$$= \hat{G}_0(E) + \hat{G}_0(E)\hat{V}\hat{G}_0(E) + \hat{G}_0(E)\hat{V}\hat{G}_0(E)\hat{V}\hat{G}_0(E) + \dots, \quad \text{(D.14)}$$

where

$$\hat{G}_0(E) = \frac{1}{E - \hat{H}_0} \quad \text{(D.15)}$$

is the free Green's operator.

The momentum representation of the retarded (advanced) propagator or Green's function in the energy variable can be expressed as the matrix element

$$G^{\mathrm{R(A)}}(\mathbf{p}, \mathbf{p}', E) = \langle \mathbf{p} | \hat{G}_E^{\mathrm{R(A)}} | \mathbf{p}' \rangle \quad \text{(D.16)}$$

of the retarded (advanced) Green's operator

$$\hat{G}_E^{\mathrm{R(A)}} = \frac{1}{E - \hat{H} \, (\overset{+}{\underset{-}{}}) \, i0} \equiv (E - \hat{H} \, (\overset{+}{\underset{-}{}}) \, i0)^{-1} \quad \text{(D.17)}$$

the analytical continuation from the various half-planes of the Green's operator. Other representations are obtained similarly, for example,

$$G^{\mathrm{R(A)}}(\mathbf{x}, \mathbf{x}', E) = \langle \mathbf{x} | \hat{G}_E^{\mathrm{R(A)}} | \mathbf{x}' \rangle . \quad \text{(D.18)}$$

The hermitian property Eq. (D.6) gives the relationship

$$[G^{\mathrm{R}}(\mathbf{x}, \mathbf{x}', E)]^* = G^{\mathrm{A}}(\mathbf{x}', \mathbf{x}, E^*) \quad \text{(D.19)}$$

and similarly in other representations.

Employing the resolution of the identity in terms of the eigenstates of $\hat{H}$

$$\hat{I} = \sum_\lambda |\epsilon_\lambda\rangle\langle\epsilon_\lambda| \quad \text{(D.20)}$$

we get the spectral representation in, for example, the position representation

$$G^{\mathrm{R(A)}}(\mathbf{x}, \mathbf{x}', E) = \sum_\lambda \frac{\psi_\lambda(\mathbf{x})\psi_\lambda^*(\mathbf{x}')}{E - \epsilon_\lambda \, (\overset{+}{\underset{-}{}}) \, i0} . \quad \text{(D.21)}$$

The Green's functions thus have singularities at the energy eigenvalues (the energy spectrum), constituting a branch cut for the continuum part of the spectrum, and simple poles for the discrete part, the latter corresponding to states which are normalizable (possible bound states of the system).

Along a branch cut the spectral function measures the discontinuity in the Green's operator

$$
\begin{aligned}
A(\mathbf{x}, \mathbf{x}', E) &\equiv \langle \mathbf{x}|i(\hat{G}_{E+i0} - \hat{G}_{E-i0})|\mathbf{x}'\rangle \\
&= i\left(G^{\mathrm{R}}(\mathbf{x}, \mathbf{x}', E) - G^{\mathrm{A}}(\mathbf{x}, \mathbf{x}', E)\right) \\
&= -2\,\Im m G^{\mathrm{R}}(\mathbf{x}, \mathbf{x}', E) \\
&= 2\pi \sum_{\lambda} \psi_{\lambda}(\mathbf{x})\psi_{\lambda}^{*}(\mathbf{x}')\,\delta(E - \epsilon_{\lambda})\,.
\end{aligned}
\tag{D.22}
$$

From the expression $(\hat{P}(\mathbf{x}) = |\mathbf{x}\rangle\langle\mathbf{x}|)$

$$
A(\mathbf{x}, \mathbf{x}, E) = 2\pi\,Tr(\hat{P}(\mathbf{x})\delta(E - \hat{H})) = 2\pi \sum_{\lambda} |\langle\mathbf{x}|\epsilon_{\lambda}\rangle|^{2}\delta(E - \epsilon_{\lambda})
\tag{D.23}
$$

we note that the diagonal elements of the spectral function, $A(\mathbf{x}, \mathbf{x}, E)$, is the local density of states per unit volume: the unnormalized probability per unit energy for the event to find the particle at position $\mathbf{x}$ with energy E (or vice versa, the probability density for the particle in energy state E to be found at position $\mathbf{x}$). Employing the resolution of the identity we have

$$
\int d\mathbf{x}\, A(\mathbf{x}, \mathbf{x}, E) = 2\pi \sum_{\lambda} \delta(E - \epsilon_{\lambda}) \equiv 2\pi\mathcal{N}(E)\,,
\tag{D.24}
$$

where $\mathcal{N}(E)$ is seen to be the number of energy levels per unit energy, and Eq. (D.24) is thus the statement that the relative probability of finding the particle somewhere in space with energy E is proportional to the number of states available at that energy.

We also note the completeness relation

$$
\int_{\sigma} \frac{dE}{2\pi}\, A(\mathbf{x}, \mathbf{x}', E) = \delta(\mathbf{x} - \mathbf{x}')
\tag{D.25}
$$

where the integration (and summation over discrete part) is over the energy spectrum.

The position and momentum representation matrix elements of any operator are related by Fourier transformation. For the spectral operator we have (assuming the system enclosed in a box of volume V)

$$
\begin{aligned}
A(\mathbf{x}, \mathbf{x}', E) &= \sum_{\mathbf{p}\mathbf{p}'} \langle\mathbf{x}|\mathbf{p}\rangle A(\mathbf{p}, \mathbf{p}', E)\langle\mathbf{p}'|\mathbf{x}'\rangle \\
&= \frac{1}{V} \sum_{\mathbf{p}\mathbf{p}'} e^{\frac{i}{\hbar}\mathbf{p}\cdot\mathbf{x} - \frac{i}{\hbar}\mathbf{p}'\cdot\mathbf{x}'} A(\mathbf{p}, \mathbf{p}', E)\,,
\end{aligned}
\tag{D.26}
$$

and inversely we have

$$
A(\mathbf{p}, \mathbf{p}', E) = \langle\mathbf{p}|\hat{A}_{E}|\mathbf{p}'\rangle = N^{-1} \int d\mathbf{x} \int d\mathbf{x}'\, e^{-\frac{i}{\hbar}\mathbf{p}\cdot\mathbf{x} + \frac{i}{\hbar}\mathbf{p}'\cdot\mathbf{x}'} A(\mathbf{x}, \mathbf{x}', E)\,,
\tag{D.27}
$$

where the normalization depends on whether the particle is confined or not, $N = V, (2\pi\hbar)^d$.

For the diagonal momentum components of the spectral function we have ($\hat{P}(\mathbf{p}) = |\mathbf{p}\rangle\langle\mathbf{p}|$)

$$A(\mathbf{p}, \mathbf{p}, E) = 2\pi \, Tr(\hat{P}(\mathbf{p})\, \delta(E - \hat{H})) = 2\pi \sum_\lambda |\langle\mathbf{p}|\epsilon_\lambda\rangle|^2 \delta(E - \epsilon_\lambda) \qquad (D.28)$$

describing the unnormalized probability for a particle with momentum $\mathbf{p}$ to have energy E (or vice versa). Analogously to the position representation we obtain

$$\sum_{\mathbf{p}} A(\mathbf{p}, \mathbf{p}, E) = 2\pi \mathcal{N}(E) . \qquad (D.29)$$

We have the momentum normalization condition

$$\int_\sigma \frac{dE}{2\pi} A(\mathbf{p}, \mathbf{p}', E) = \begin{cases} \delta(\mathbf{p} - \mathbf{p}') \\ \\ \delta_{\mathbf{p},\mathbf{p}'} \end{cases} \qquad (D.30)$$

depending on whether the particle is confined or not.

Let us finally discuss the analytical properties of the free propagator. Fourier transforming the free retarded propagator, Eq. (C.22), we get (in three spatial dimensions for the pre-exponential factor to be correct), $\Im mE > 0$,

$$G_0^R(\mathbf{x}, \mathbf{x}', E) = \frac{-m}{2\pi\hbar^2} \frac{e^{\frac{i}{\hbar}p_E|\mathbf{x}-\mathbf{x}'|}}{|\mathbf{x} - \mathbf{x}'|} \quad , \qquad p_E = \sqrt{2mE} \qquad (D.31)$$

the solution of the spatial representation of the operator equation, Eq. (D.3),

$$\left(E - \frac{\hbar^2}{2m}\Delta_{\mathbf{x}}\right) G_0(\mathbf{x}, \mathbf{x}', E) = \delta(\mathbf{x} - \mathbf{x}') , \qquad (D.32)$$

which is analytic in the upper half plane.

The square root function, $\sqrt{E}$, has a half line branch cut, which according to the spectral representation, Eq. (D.21), must be chosen along the positive real axis, the energy spectrum of a free particle, as we choose the lowest energy eigenvalue to have the value zero. In order for the Green's function to remain bounded for infinite separation of its spatial arguments, $|\mathbf{x} - \mathbf{x}'| \to \infty$, we must make the following choice of argument function

$$\sqrt{E} \equiv \begin{cases} \sqrt{E} & \text{for } \Re eE > 0 \\ \\ i\sqrt{|E|} & \text{for } \Re eE < 0 \end{cases} . \qquad (D.33)$$

rendering the free spectral function of the form

$$A_0(\mathbf{x}, \mathbf{x}', E) = \frac{m}{\pi\hbar^2} \frac{\sin(\frac{1}{\hbar}p_E|\mathbf{x} - \mathbf{x}'|)}{|\mathbf{x} - \mathbf{x}'|} \theta(E) \qquad (D.34)$$

and we can read off the free particle density of states, the number of energy levels per unit energy per unit volume,[1]

$$
N_0(E) \equiv \frac{1}{2\pi} A_0(\mathbf{x}, \mathbf{x}; E) = \theta(E)
\begin{cases}
\sqrt{\frac{m}{2\pi^2 \hbar^2 E}} & d = 1 \\[2ex]
\frac{m}{2\pi \hbar^2} & d = 2 \\[2ex]
\frac{m\sqrt{2mE}}{2\pi^2 \hbar^3} & d = 3
\end{cases}
, \qquad (D.35)
$$

where for completeness we have also listed the one- and two-dimensional cases.

The spectral function for a free particle in the momentum representation follows, for example, from Eq. (D.28)

$$
A_0(\mathbf{p}, E) \equiv A_0(\mathbf{p}, \mathbf{p}, E) = 2\pi \, \delta(E - \epsilon_{\mathbf{p}}) , \qquad (D.36)
$$

and describes the result that a free particle with momentum $\mathbf{p}$ with certainty has energy $E = \epsilon_{\mathbf{p}}$, or vice versa.

---

[1]This result is of course directly obtained by simple counting of the momentum states in a given energy range, because for a free particle constrained to the volume L^d, there is one momentum state per momentum volume $(2\pi\hbar/L)^d$. However, the above argument makes no reference to a finite volume.

References

[1] J. Rammer, *Quantum Transport Theory*, Frontiers in Physics Vol. 99 (Reading: Perseus Books, 1998; paperback edition 2004).

[2] R. P. Feynman in *Elementary Particles and the Laws of Physics*. The 1986 Dirac Memorial Lectures (Cambridge: Cambridge University Press, 1987).

[3] J. Rammer, *Non-Equilibrium Superconductivity*, Master's thesis, University of Copenhagen, Denmark (1981) (in Danish).

[4] M. Gell-Mann and F. Low, *Phys. Rev.*, **84** (1951), 350.

[5] J. Schwinger, *J. Math. Phys.*, **2** (1961), 407.

[6] V. Korenman, *Ann. Phys. (Paris)*, **39** (1966), 72.

[7] K.-C. Chou, Z.-B. Su, B.-L. Hao and L. Yu, *Phys. Rep.*, **118** (1985), 1.

[8] J. Rammer and H. Smith, *Rev. Mod. Phys.*, **58** (1986), 323.

[9] J. Rammer, *Applications of quantum field theoretical methods in transport theory of metals*, Ph. D. thesis, University of Copenhagen, Denmark (1985). (Published in part).

[10] L. V. Keldysh, *Zh. Eksp. Teor. Fiz.*, **47** (1964), 1515 [*Sov. Phys. JETP*, **20** (1965), 1018].

[11] H. Umezawa, H. Matsumoto and Y. Takahashi, *Thermo Field Dynamics and Condensed States* (North-Holland, 1982).

[12] A. I. Larkin and Yu. N. Ovchinnikov, *Zh. Eksp. Teor. Fiz.*, **68** (1975), 1915 [*Sov. Phys. JETP*, **41** (1975), 960].

[13] J. Rammer, *Rev. Mod. Phys.*, **63** (1991), 781.

[14] L. P. Kadanoff and G. Baym, *Quantum Statistical Mechanics* (New York: Benjamin, 1962).

[15] A. A. Abrikosov, L. P. Gorkov and I. E. Dzyaloshinski, *Methods of Quantum Field Theory in Statistical Physics* 2nd edition (New York: Pergamon, 1965).

[16] T. Matsubara, *Prog. Theor. Phys.*, **14** (1955), 351.

[17] E. S. Fradkin, *Zh. Eksp. Teor. Fiz.*, **36** (1959), 1286 [*Sov. Phys. JETP*, **9** (1959), 912].

[18] P. C. Martin and J. Schwinger, *Phys. Rev.*, **115** (1959), 1342.

[19] G. M. Eliashberg, *Zh. Eksp. Teor. Fiz.*, **61** (1971), 1254 [*Sov. Phys. JETP*, **34** (1972), 668].

[20] D. C. Langreth, in *Linear and Nonlinear Electron Transport in Solids*, NATO Advanced Study Institute Series B, Vol. 17, J. T. Devreese and E. van Doren, eds. (New York/London: Plenum, 1976).

[21] S. W. Lovesey, *Theory of Neutron Scattering from Condensed Matter* (Oxford: Clarendon Press, 1984).

[22] L. W. Boltzmann, Ber. Wien. Akad., **66** (1872), 275, and *Vorlesungen über Gastheorie* (Leipzig: Barth, 1896). English translation: *Lectures on Gas Theory* (Berkeley: University of California Press, 1964).

[23] R. E. Prange and L. P. Kadanoff, *Phys. Rev.*, **134** (1964), A566.

[24] G. Eilenberger, *Z. Phys.*, **214** (1968), 195.

[25] A. B. Migdal, *Zh. Eksp. Teor. Fiz.*, **34** (1958), 1438 [*Sov. Phys. JETP*, **7** (1958), 966].

[26] T. Holstein, *Ann. Phys. (Paris)*, **29** (1964), 410.

[27] D. C. Langreth, *Phys. Rev.*, **148** (1966), 707.

[28] B. L. Altshuler, *Zh. Eksp. Teor. Fiz.*, **75** (1978), 1330 [*Sov. Phys. JETP*, **48** (1978), 670].

[29] J. L. Opsal, B. J. Thaler and J. Bass, *Phys. Rev. Lett.*, **36** (1976), 1211.

[30] R. Fletcher, *Phys. Rev.*, B **14** (1976), 4329.

[31] J. Misguich, G. Pelletier, and P. Schuck (eds.) *Statistical Description of Transport in Plasma, Astro-, and Nuclear Physics* (Nova Science Publishers, 1993).

[32] J. Bardeen, *Encyclopedia of Physics*, S. Flügge, ed., vol. XV (Berlin: Springer-Verlag, 1956), p. 274.

[33] C. N. Yang, *Rev. Mod. Phys.*, **34** (1962), 694.

[34] J. R. Schrieffer, *Superconductivity*, Frontiers in Physics Vol. 20 (Addison-Wesley, 1964). (Third printing, revised, 1983.)

[35] P. Nozires and D. Pines, *Theory of Quantum Liquids, Volume I: Normal Fermi Liquids* (W. A. Benjamin, 1966).

[36] J. W. Serene and D. Rainer, *Phys. Rep.*, **101** (1983), 221.

[37] G. M. Eliashberg, *Zh. Eksp. Teor. Fiz.*, **38** (1960), 966 [*Sov. Phys. JETP*, **11** (1960), 696].

[38] U. Eckern, *J. Low Temp. Phys.*, **62** (1986), 525.

[39] A. A. Abrikosov and L. P. Gor'kov, *Zh. Eksp. Teor. Fiz.*, **39** (1960), 1781 [*Sov. Phys. JETP*, **12** (1961), 1243].

[40] A. L. Shelankov, *J. Low Temp. Phys.*, **60** (1985), 29.

[41] K. S. Usadel, *Phys. Rev. Lett.*, **25** (1970), 507.

[42] K. E. Gray, ed., *Nonequilibrium Superconductivity, Phonons, and Kapitza Boundaries* (New York: Plenum Press, 1981).

[43] A. Schmid (1981), *Kinetic equations for dirty superconductors.* (Included in reference [42].)

[44] I. Schuller and K. E. Gray, *Solid State Commun.*, **23** (1977), 337.

[45] A. Schmid and G. Schön, *Phys. Rev. Lett.*, **43** (1979), 793.

[46] A. Schmid and G. Schön, *J. Low Temp. Phys.*, **20** (1975), 207.

[47] J. Clarke (1981), *Charge imbalance.* (Included in reference [42].)

[48] D. N. Langenberg and A. I. Larkin, eds., *Nonequilibrium Superconductivity* (Elsevier Science Publishers B.V., 1986).

[49] N. B. Kopnin, *Theory of Nonequilibrium Superconductivity* (Oxford: Oxford University Press, 2001).

[50] J. Schwinger, *Proc. Nat. Acad. Sci.*, **37** (1951), 452.

[51] K. B. Efetov, *Adv. Phys.*, **32** (1983), 53.

[52] R. Jackiw, *Phys. Rev. D*, **9** (1974), 1686.

[53] J. M. Cornwall, R. Jackiw and E. Tomboulis, *Phys. Rev. D*, **10** (1974), 2428.

[54] E. Lundh and J. Rammer, *Phys. Rev. A*, **66** (2002), 033607.

[55] N. N. Bogoliubov, *J. Phys. (Moscow)*, **11** (1947), 23.

[56] E. P. Gross, *Nuovo Cimento*, **20** (1961), 454; *J. Math. Phys.*, **4** (1963), 195.

[57] L. P. Pitaevskii, *Zh. Eksp. Teor. Fiz.*, **40** (1961), 646 [*Sov. Phys. JETP*, **13** (1961), 451].

[58] S. T. Beliaev, *Zh. Eksp. Teor. Fiz.*, **34** (1958), 417 [*Sov. Phys. JETP*, **7** (1958), 289].

[59] S. T. Beliaev, *Zh. Eksp. Teor. Fiz.*, **34** (1958), 433 [*Sov. Phys. JETP*, **7** (1958), 299].

[60] V. N. Popov and L. D. Faddeev, *Zh. Eksp. Teor. Fiz.*, **47** (1964), 1315 [*Sov. Phys. JETP*, **20** (1965), 890].

[61] V. N. Popov, *Zh. Eksp. Teor. Fiz.*, **47** (1964), 1759 [*Sov. Phys. JETP*, **20** (1965), 1185].

[62] V. N. Popov, *Functional Integrals and Collective Excitations* (New York: Cambridge University Press, 1987).

[63] M. H. Anderson, J. R. Ensher, M. R. Matthews, C. E. Wieman, and E. A. Cornell, *Science*, **269** (1995), 198.

[64] O. Penrose and L. Onsager, *Phys. Rev.*, **104** (1956), 576.

[65] F. Dalfovo and S. Stringari, *Phys. Rev. A*, **53** (1996), 2477.

[66] N. P. Proukakis, S. A. Morgan, S. Choi, and K. Burnett, *Phys. Rev. A*, **58** (1998), 2435.

[67] G. Baym and C. J. Pethick, *Phys. Rev. Lett.*, **76** (1996), 6.

[68] F. Dalfovo, S. Giorgini, L. P. Pitaevskii, and S. Stringari, *Rev. Mod. Phys.*, **71** (1999), 463.

[69] S. Stenholm, *Phys. Rev. A*, **57** (1998), 2942.

[70] S. Inouye, M. R. Andrews, J. Stenger, H.-J. Miesner, D. M. Stamper-Kurn, and W. Ketterle, *Nature*, **392** (1998), 151.

[71] P. W. Anderson, *Phys. Rev.*, **102** (1958), 1008.

[72] E. Abrahams, P. W. Anderson, D. C. Licciardello and T. V. Ramakrishnan, *Phys. Rev. Lett.*, **42** (1979), 673. Reprinted in P. W. Anderson, *Basic Notions of Condensed Matter Physics* (Benjamin-Cummings, 1984).

[73] F. Wegner, *Z. Phys. B*, **25** (1976), 327.

[74] D. J. Thouless in *Ill Condensed Matter*, R. Balian, R. Maynard, and G. Toulouse, eds., Les Houches, Session **XXXI** (North-Holland, 1987).

[75] N. F. Mott and W. D. Twose, *Adv. Phys.*, **10** (1961), 107.

[76] R. Landauer, *Philos. Mag.*, **21** (1970), 863.

[77] V. L. Berezinskii, *Zh. Eksp. Teor. Fiz.*, **65** (1973), 1251 [*Sov. Phys. JETP*, **38** (1974), 620].

[78] A. A. Abrikosov and I. A. Ryzhkin, *Adv. Phys.*, **27** (1978), 147.

[79] N. F. Mott, in *Electronics and structural properties of amorphous semiconductors*, P. G. Le Comber and J. Mort, eds. (London: Academic Press, 1973).

[80] J.S. Langer and T. Neal, *Phys. Rev. Lett.*, **16** (1966), 984.

[81] K. M. Watson, *J. Math. Phys.*, **10** (1969), 688.

[82] A. I. Larkin and D. E. Khmel'nitskii, *Usp. Fyz. Nauk.*, **136** (1982), 536 [*Sov. Phys. Usp.*, **25** (1982), 185].

[83] D. E. Khmelnitskii, *Physica* B, **126** (1984), 235.

[84] B. L. Al'tshuler, V. E. Kravtsov and I. V. Lerner in *Mesoscopic Phenomena in Solids*, B. L. Al'tshuler, P. A. Lee, and R. A. Webb, eds. (North-Holland: Elsevier Science Publishers B.V., 1991).

[85] D. Belitz and T. R. Kirkpatrick, *Rev. Mod. Phys.*, **66** (1994), 261.

[86] L. P. Gor'kov, A. I. Larkin and D. E. Khmel'nitskii, *Pis'ma Zh. Eksp. Teor. Fiz.*, **30** (1979), 1251 [*Sov. Phys. JETP Lett.*, **30** (1979), 228].

[87] J. Rammer and A. Schmid, *Destruction of phase coherence by electron-phonon interaction in disordered conductors.* Contributed paper to the *International Conference on Localization, Interaction and Transport Phenomena*, Braunschweig, Abstracts p. 155 (1984.) (or NORDITA preprint-85/39 (1985)).

[88] G. Bergmann, *Phys. Rep.*, **107** (1984), 1.

[89] B. L. Al'tshuler, A. G. Aronov, M. E. Gershenson and Yu. V. Sharvin, *Sov. Sci. Rev. A. Phys.*, **9** (1987), 223 (I. M. Khalatnikov, ed.).

[90] J. Rammer and A. Schmid, *Phys. Rev.* B, **34** (1986), 1352.

[91] J. Rammer, A. L. Shelankov and A. Schmid, *Phys. Rev. Lett.*, **60** (1988), 1985 (C).

[92] K. S. Il'in, N. G. Ptitsina, A.V. Sergeev *et al.*, *Phys. Rev.* B, **57** (1998), 15623.

[93] A. Schmid, *Z. Phys.*, **259** (1973), 421.

[94] B. L. Al'tshuler, A. G. Aronov and D. E. Khmel'nitskii, *J. Phys.* C, **15** (1982), 7367.

[95] J. Rammer and A. L. Shelankov, *Phys. Rev.* B, **36** (1987), 3135.

[96] I. S. Gradstheyn and I. M. Ryzhik, *Table of Integrals, Series and Products* (New York: Academic Press, 1980).

[97] A. Kawabate, *J. Phys. Soc. Japan*, **53** (1984), 3540.

[98] H.-P. Wittmann and A. Schmid, *J. Low Temp. Phys.*, **69** (1987), 131.

[99] B. L. Al'tshuler and A. G. Aronov, *Zh. Eksp. Teor. Fiz.*, **77** (1979), 2028 [*Sov. Phys. JETP*, **50** (1979), 968].

[100] I. L. Bronevoi, *Zh. Eksp. Teor. Fiz.*, **79** (1980), 1936. [*Sov. Phys. JETP*, **52** (1980), 977]

[101] I. L. Bronevoi, *Zh. Eksp. Teor. Fiz.*, **83** (1982), 338 [*Sov. Phys. JETP*, **56** (1982), 185].

[102] B. L. Al'tshuler and A. G. Aronov in *Electron–Electron Interactions in Disordered Systems*, A. L. Efros and M. Pollak, eds. (North-Holland: Elsevier Science Publishers B.V., 1985).

[103] A. Schmid, *Z. Physik*, **271** (1974), 251.

[104] B. L. Al'tshuler and A. G. Aronov, *Zh. Eksp. Teor. Fiz.*, **75** (1978), 1610 [*Sov. Phys. JETP*, **48** (1978), 812].

[105] B. L. Al'tshuler and A. G. Aronov, *Pis'ma Zh. Eksp. Teor. Fiz.*, **30** (1979), 514 [*Sov. Phys. JETP Lett.*, **30** (1979), 482].

[106] R. A. Serota, S. Feng, C. Kane and P. A. Lee, *Phys. Rev. B*, **36** (1987), 5031.

[107] B. L. Al'tshuler and B. I. Shklovskii, *Zh. Eksp. Teor. Fiz.*, **91** (1986), 220 [*Sov. Phys. JETP*, **64** (1986), 127].

[108] B. L. Al'tshuler and D. E. Khmel'nitskii, *Pis'ma Zh. Eksp. Teor. Fiz.*, **42** (1985), 291 [*Sov. Phys. JETP Lett.*, **42** (1985), 359].

[109] P. A. Lee, A. D. Stone and H. Fukuyama, *Phys. Rev. B*, **35** (1987), 1039.

[110] B. L. Al'tshuler, *Pis'ma Zh. Eksp. Teor. Fiz.*, **41** (1985), 530 [*Sov. Phys. JETP Lett.*, **41** (1985), 648].

[111] P. A. Lee and A. D. Stone, *Phys. Rev. Lett.*, **55** (1985), 1622.

[112] B. L. Al'tshuler, V. E. Kravtsov and I. V. Lerner, *Pis'ma Zh. Eksp. Teor. Fiz.*, **43** (1986), 342 [*Sov. Phys. JETP Lett.*, **43** (1986), 441].

[113] B. L. Al'tshuler and B. Z. Spivak, *Pis'ma Zh. Eksp. Teor. Fiz.*, **42** (1985), 363 [*Sov. Phys. JETP Lett.*, **42** (1985), 447].

[114] S. Feng, P. A. Lee and A. D. Stone, *Phys. Rev. Lett.*, **56** (1986), 1960; **56** (1986), 2772 (E).

[115] S. Washburn and A. Webb, *Adv. Phys.*, **35** (1986), 375.

[116] P. Shen (ed.), *Scattering and Localization of Classical Waves in Random Media* (World Scientific, 1989).

[117] D. Forster, D. R. Nelson and M. J. Stephen, *Phys. Rev. A*, **16** (1977), 732.

[118] P. C. Martin, E. D. Siggia and H. A. Rose, *Phys. Rev. A*, **8** (1973), 423.

[119] H. K. Janssen, *Z. Phys. B*, **23** (1976), 377.

[120] R. Bausch, H. K. Janssen and H. Wagner, *Z. Phys. B*, **24** (1976), 113.

[121] C. De Dominicis, *Phys. Rev. B*, **18** (1978), 4913.

[122] P. C. Hohenberg and B. I. Halperin, *Rev. Mod. Phys.*, **49** (1977), 435.

[123] S. F. Edwards and P. W. Anderson, *J. Phys. (Paris)* F, **5** (1975), 965.

[124] A. A. Abrikosov, *Zh. Eksp. Teor. Fiz.*, **32** (1957), 1422 [*Sov. Phys. JETP*, **5** (1957), 1174].

[125] J. Bardeen and M. J. Stephen, *Phys. Rev.*, **140** (1965), A1197.

[126] A. Schmid, *Phys. Kondensierten Materie*, **5** (1966), 302.

[127] C. Caroli and K. Maki, *Phys. Rev.*, **159** (1967), 306.

[128] C. R. Hu and R. S. Thompson, *Phys. Rev.* B, **6** (1972), 110.

[129] M. Tinkham, *Introduction to Superconductivity*, 2nd edition (McGraw-Hill, 1996).

[130] G. Blatter, M. V. Feigel'man, V. B. Geshkenbein, A. I. Larkin and V. M. Vinokur, *Rev. Mod. Phys.*, **66** (1994), 1125.

[131] A. Schmid and W. Hauger, *J. Low Temp. Phys.*, **11** (1973), 667.

[132] A. L. Larkin and Yu. N. Ovchinnikov, *Zh. Eksp. Teor. Fiz.*, **65** (1973), 1704 [*Sov. Phys. JETP*, **38** (1974), 854].

[133] J. Müllers and A. Schmid, *Ann. Physik*, **4** (1995), 757.

[134] S. Grundberg and J. Rammer, *Phys. Rev.* B, **61** (2000), 699.

[135] C. R. Werner and U. Eckern, *Ann. Physik*, **6** (1997), 595.

[136] E. H. Brandt, *Rep. Prog. Phys.*, **58** (1995), 1465.

[137] E. H. Brandt, *Int. J. Mod. Phys.* B, **5** (1999), 751.

[138] Wu Liu, T. W. Clinton and C. J. Lobb, *Phys. Rev.* B, **52** (1995), 7482.

[139] S. Bhattacharya and M. J. Higgins, *Phys. Rev. Lett.*, **70** (1993), 2617.

[140] W. K. Kwok, J. A. Fendrich, C. J. van der Beek and G. W. Crabtree, *Phys. Rev. Lett.*, **73** (1994), 2614.

[141] U. Yaron, P. L. Gammel, D. A. Huse *et al.*, *Nature*, **376** (1995), 753.

[142] M. C. Hellerqvist, D. Ephron, W. R. White, M. R. Beasley and A. Kapitulnik, *Phys. Rev. Lett.*, **76** (1996), 4022.

[143] M. J. Higgins and S. Bhattacharya, *Physica* C, **257** (1996), 232.

[144] M. Marchevsky, J. Aarts, P. H. Kes and M. V. Indenbom, *Phys. Rev. Lett.*, **78** (1997), 531.

[145] A. E. Koshelev and V. M. Vinokur, *Phys. Rev. Lett.*, **73** (1994), 3580.

[146] S. Scheidl and V. M. Vinokur, *Phys. Rev.* B, **57** (1998), 13800.

[147] S. Scheidl and V. M. Vinokur, *Phys. Rev.* E, **57** (1998), 2574.

[148] J. Kierfeld, T. Nattermann and T. Hwa, *Phys. Rev.* B, **55** (1997), 626.

[149] J. M. Caillol, D. Levesque, J. J. Weis and J. P. Hansen, *J. Stat. Phys.*, **28** (1982), 325.

[150] J. M. Kosterlitz and D. J. Thouless, *J. Phys.* C, **6** (1973), 1181.

[151] D. S. Fisher, *Phys. Rev.* B, **22** (1980), 1190.

[152] P. A. M. Dirac, *Physik. Zeits. Sowjetunion*, **3** (1933), 64. Reprinted in J. Schwinger, ed., *Selected Papers on Quantum Electrodynamics* (New York: Dover Publications, 1958).

[153] P. Ao and J. Rammer, *Phys. Rev.* B, **43** (1991), 5397.

[154] R. P. Feynman and F. L. Vernon, *Annals of Physics*, **24** (1963), 118.

[155] R. P. Feynman and A. R. Hibbs, *Quantum Mechanics and Path Integrals* (New York: McGraw-Hill, 1965).

[156] A. Schmid, *J. Low Temp. Phys.*, **49** (1982), 609.

[157] A. O. Caldeira and A. J. Leggett, *Physica*, **121A** (1983), 587.

Index

Abrikosov flux lattice, 461
Abrikosov–Shubnikov phase, 461
absorption vertex, 129
action, 306, 321, 325
    free, 317
adiabatic switching, 83
adjoint operator, 8
analytical continuation, 142
analytical properties of the free propagator, 521
annihilation operator
    bose, 23
    fermion, 16
anti-time-ordering, 56
antisymmetric subspace, 9
appendix contour, 91
Aristotelian dynamics, 449
auxiliary field, 453
average field, 297

BCS-energy gap, 220
BCS-pairing, 219
BCS-state, 39
BCS-theory, 217, 219
Bogoliubov equations, 364
Bogoliubov–Valatin transformation, 221
Boltzmann conductivity, 377
Boltzmann equation, 188
Boltzmann factor, 323
Boltzmann propagator, 193, 412
Boltzmannian paths, 401
Boltzmannian motion, 192
    path of, 192
bose field, 24
Bose function, 98
Bose gas, 351
Bose–Einstein condensate, 51, 77

Bose–Einstein condensation, 51, 301, 351, 353
Bose–Einstein distribution, 73, 98
bosons, 6
Brownian motion, 194, 450

canonical ensemble, 49, 70
    grand, 50, 70
canonical formulation, 281
central limit theorem, 329
charge imbalance, 250
charge-density wave, 231
chemical potential, 50
classical electrodynamics, 28
classical equation of motion, 310, 511
classical field, 297
closed contour, 85
closed time path, 84
coherence length, 231, 368
coherent backscattering, 384, 387
collision integral
    electron–phonon, 208
    electron-electron, 214
    electron-impurity, 188
collision rate, 435
commutator, 3
condensate density, 356
condensate wave function, 353, 363
condensation energy, 221
conductance, 159
conductance fluctuations, 442
conductance tensor, 159
conductivity, 191
    minimum metallic, 377
conductivity diagram, 161
conductivity diagrams, 377
conductivity tensor, 158

continuity equation, 13, 14, 191, 195
contour ordered Green's function
    inverse free, 105
contour ordering, 85, 88
contour variable, 87
Cooperon, 379, 381, 390
Cooperon equation, 403
creation operator
    bose, 23
    fermion, 14
critical phenomena, 258, 290
critical velocity, 493
current correlation function, 169
current density, 13, 40, 63, 67
current response, 155
current response function, 155
current vertex, 157
cyclotron frequency, 215

d'Alembertian, 27
Debye cut-off, 59
Debye model, 26
deformation potential, 46, 410
delta functional, 314
density, 67
    probability, 13
density matrix, 22, 48, 56
density operator, 38
    current, 40
density response, 153
density response function, 153
density–density response, 434
diamagnetic current, 40
dielectric function, 434
diffusion approximation, 194
diffusion constant, 194, 198
diffusion equation, 196
diffusion propagator, 194, 196
Diffuson, 197, 381, 382, 401, 431
diluteness parameter, 351
dirty superconductor, 242
displacement field, 26, 27, 410
dissipation, 169
distribution function, 187
distribution functions for superconductors,
    245

Drude theory, 190
dual space, 9
dyadic notation, 467
dynamic melting, 493
Dyson equation, 116
    equilibrium, 138
    left-right subtracted, 179
    matrix, 135
    non-equilibrium, 303
Dyson equations
    non-equilibrium, 116
Dyson's formula, 84
Dyson–Beliaev equation, 357
Dyson–Schwinger equation, 279

effective action, 296, 299, 323
    two-particle irreducible, 343
Eilenberger equations, 232
Einstein relation, 244
elastic medium, 26
electric field fluctuations, 172
electron–electron interaction
    diffusion enhanced, 436
electron–hole excitations, 112
electron–phonon interaction, 45, 46, 200,
    410
electron–photon Hamiltonian, 48
electron–photon interaction, 48
emission vertex, 129
energy, 56
energy gap, 220, 222
energy relaxation rate, 431, 435, 436
Euclidean action, 197, 506
evolution operator, 3, 55
exact impurity eigenstates, 430
exclusion principle, 16, 18, 21

Fermi energy, 22
Fermi field, 21
Fermi function, 98
Fermi gas, 22
Fermi momentum, 37
Fermi sea, 22, 111
Fermi surface, 22
Fermi wavelength, 231, 232, 377
Fermi's Golden Rule, 168
Fermi–Dirac distribution, 73

fermion–boson interaction, 45, 107, 125
fermions, 6
Feynman rules, 113
Feynman-Vernon theory, 510
field, 297
fluctuation–dissipation relation, 180
fluctuation–dissipation theorem, 70, 72,
        171, 450
flux flow regime, 462
flux quantum, 427
Fock space, 15
free energy, 50, 290
free particle density of states, 522
free particle propagator, 515
functional derivative, 272
functional determinant, 317
functional integral, 314
        restricted, 330
functional integration, 313
fundamental dynamic equation, 279

gauge invariance, 14
gauge transformation, 231
Gaussian integral, 315
Gell-Mann–Low theorem, 83
generating functional, 69, 103, 270, 281,
        317, 320
generator of 1PI-vertices, 299
generator of connected amplitudes, 284,
        290
generator of connected Green's functions,
        322
generator of cumulants, 290
Ginzburg–Landau regime, 247
Gorkov equations, 224
gradient approximation, 184
Grassmann field, 319
Grassmann algebra, 282
Grassmann variable, 282, 319
        function of a, 319
Green's function, 4, 62
        advanced, 67
        anomalous, 224, 357
        anti-time-ordered, 65
        closed time path, 87
        contour ordered, 87, 90

full, 118
grand canonical, 72
Greater, 64
imaginary time, 140
inverse, 66
kinetic, 67, 72
Lesser, 63
matrix, 122
one-particle, 22
phonon, 66
quasi-classical, 199
retarded, 66, 513
single-particle, 62
symmetric matrix, 280
time-ordered, 65, 81
trajectory, 240
two-particle, 64
Green's function technique
        standard diagrammatic impurity, 377
Green's operator
        advanced, 517
        retarded, 517
Green's operators, 515, 518

Hamilton's function, 61, 504
Hamiltonian, 3, 36
        grand canonical, 72
Hartree approximation, 470
Hartree diagrams, 110
Heisenberg picture, 57
hermitian conjugation, 318
hermitian operator, 9
high-temperature superconductors, 476

identical particles, 6
imaginary-time, 140
imaginary-time formalism, 140
imaginary-time propagator, 506, 515
impurity correlator, 106
impurity scattering, 188
inelastic differential cross section, 175
inelastic scattering rate, 409
influence functional, 508
interaction picture, 82
inverse propagator, 261
irreducible vertex functions
        one-particle, 298

Jacobian, 316, 459
jellium model, 45
Johnson noise, 172
Joule heating, 169

Kadanoff–Baym equations, 148
kinematic momentum, 40
kinetic energy operator, 36
kinetic equation
    classical, 188
kinetic propagator, 395
Kramers–Kronig relations, 75, 158, 182, 518
Kubo–Martin–Schwinger boundary conditions, 140

Lagrange density, 321
Lagrange's function, 505
Landau criterion, 191, 209
Landau gauge, 423
Landau–Boltzmann equation, 209
Langevin equation, 450
Legendre transform, 300
Legendre transformation, 505
Lindemann criterion, 495
linear response, 151
local density of states, 520
localization, 388
localization length, 375, 385
London equations, 460
longitudinal phonons, 45, 71
loop expansion, 331

magnetic impurities, 224, 233
magnetic length, 402, 424
magnetic moments, 37
magneto-fingerprint, 446
magneto-resistance, 424
    anomalous, 426
Markov process, 192, 193
master equation, 168
matrix element, 3
matrix Green's function, 122
    inverse free, 125
Matsubara frequency, 141
maximally crossed diagrams, 378
mean field theories, 350

mean free path, 193
Meissner effect, 460
melting curve, 496
mesoscopic fluctuations, 442
mesoscopic sample, 437
metal–insulator transition, 373, 376
Migdal's theorem, 200, 218
mixed coordinates, 68, 181
mixture, 48, 56
momentum operator, 35
monomial, 282
multi-particle space, 15
multiplication in parallel, 147
multiplication in series, 146
multiplication rule, 259

N-state amplitude, 258, 261
N-state diagram, 258
Nambu field, 225
Nambu space, 225
Nernst–Ettingshausen effect, 215
neutron scattering, 175
Noether's theorem, 512
normal state, 301
normal-ordered, 43
normalization condition, 235, 242
number operator, 19, 24
    total, 38
Nyquist noise, 173, 418, 437

occupation number representation, 29
Ohm's law, 172
Ohmic environment, 453, 510
one-body operator, 34
one-parameter scaling theory, 374
one-particle irreducible, 113, 114, 296
    vertices, 296
operator
    antisymmetrization, 7
    symmetrization, 7
optical mass, 207
order parameter, 219, 223, 224, 235, 246

pair correlation function, 23
pair creation, 112
pair-breaking parameter, 251
paramagnetic current, 40

particle–hole space, 225
particle–hole symmetry, 201
partition function, 4, 49, 290
path integral, 504
Pauli's exclusion principle, 7, 22
Pauli's master equation, 190
Peierls instability, 231
phase coherence length, 410
phase relaxation
    rate, 437
    time, 410
phase-breaking rate, 409, 410, 413, 415,
    421
phonon, 26, 28, 46
phonon field, 66
phonon field operator, 46
phonon Hamiltonian, 26
photon, 47
pinning, 462
pinning force, 463, 469, 472, 475
pinning potential, 463
plane wave, 2
Poisson bracket, 185, 213
polarization, 433
polarization vector, 47
Popov approximation, 362
principle of least action, 505
probability amplitude, 258
propagator, 4
    advanced, 514
    bare, 255
    free, 255
    retarded, 513

quanta, 28
quantum electrodynamics, 28
quantum fields, 21
quantum kinetic equation, 179, 202, 233,
    236
quantum of action, 3
quantum phase transition, 377
quantum statistics, 262
quasi-particles, 46
quenched disorder, 93

radiation gauge, 29
ray, 2

real rules, 134
real-time contour, 93
renormalization, 205, 207, 257, 306
    optical mass, 207
resistance, 159, 169
resistance tensor, 159
response function
    current-current, 156
    density-density, 153
road diagram, 286

s-wave scattering length, 358
scalar product, 2
scaling function, 375
scaling theory of localization, 374, 445
scattering experiment, 174
Schrödinger picture, 56
Schrödinger equation, 3, 13, 54
screening, 383, 433
second quantization, 27, 61
self-averaging, 437, 445
self-consistent approximations, 350
self-energy, 114, 117, 119, 302
    impurity, 188
    matrix, 135
skeleton diagrams, 118
skeleton diagrams
    two-particle irreducible, 118
Slater determinant, 10
Sommerfeld model, 44
sound velocity, 25
spectral function, 68, 515, 520
spectral operator, 518
spectral representation, 182, 518
spectral weight function, 68, 75, 139, 180,
    185, 509
spin, 37
spin-flip scattering, 233
spontaneously broken symmetry, 301
state label, 255
stationary state, 153
statistical operator, 48
    thermal equilibrium, 49
strong localization, 375, 387
structure factor
    dynamic, 177

superconductivity, 217
    strong coupling, 224
superfluid velocity, 232
superposition principle, 110, 259, 260, 322, 428, 505
symmetric representation, 128
symmetric subspace, 9

t-matrix, 359
tadpole diagram, 266
tadpole diagrams, 110, 460
thermal equilibrium, 49, 165
thermo-field approach, 122
thermodynamic potential, 51
thermodynamics, 4
third rank vertex, 126
three-line vertex, 113
time-dependent Ginzburg–Landau equation, 248
time-dependent Gross–Pitaevskii equation, 363
time-ordering, 55
trace, 49
transformation function, 503
transient phenomena, 92
transition force, 495
transition operator, 34
transport relaxation time, 164, 191
tree diagrams, 299, 331
triagonal representation, 128
triagonal representation, 129
two-body interaction, 42
two-body potential, 42
two-particle interaction, 13, 64, 130
two-particle irreducible vertices, 341
two-particle source, 339

unitary operator, 3
unlinked diagrams, 260
Usadel equation, 244

vacuum diagrams, 266, 270, 280, 293, 326
vacuum state, 15, 37, 38, 43, 55
vertex, 257
    two-particle irreducible, 342
vertices
    bare, 256

von Neumann equation, 56

Ward identity, 512
wave function, 2
weak localization, 387
weak-localization, 391
    effect, 378, 389, 445
    regime, 402
Wick's theorem, 95, 96, 324, 326, 328
Wiener measure, 197
Wigner coordinates, 68, 181
Wigner function, 187

zero-bias anomaly, 430